Seafood Science

Advances in Chemistry, Technology and Applications

T0314347

Seafood Science
Advances in Chemistry, Technology and Applications

Editor

Se-Kwon Kim
Department of Marine-bio Convergence Science
Specialized Graduate School of
Convergence Science and Technology
Pukyong National University
Pusan, Republic of Korea

CRC Press
Taylor & Francis Group
Boca Raton London New York

CRC Press is an imprint of the
Taylor & Francis Group, an **informa** business
A SCIENCE PUBLISHERS BOOK

CRC Press
Taylor & Francis Group
6000 Broken Sound Parkway NW, Suite 300
Boca Raton, FL 33487-2742

First issued in paperback 2020

© 2015 by Taylor & Francis Group, LLC
CRC Press is an imprint of Taylor & Francis Group, an Informa business

ISBN-13: 978-1-4665-9582-8 (hbk)
ISBN-13: 978-0-367-73995-9 (pbk)

Library of Congress Cataloging-in-Publication Data

Seafood science : advances in chemistry, technology, and applications / editor, Se-Kwon Kim.
  pages cm
Includes bibliographical references and index.
ISBN 978-1-4665-9582-8 (hardback)
  1. Seafood. 2. Food--Composition. 3. Fishery processing. 4. Fisheries. I. Kim, Se-Kwon.

TX385.S435 2014
664'.94--dc23   2014018083

Visit the Taylor & Francis Web site at
http://www.taylorandfrancis.com

and the CRC Press Web site at
http://www.crcpress.com

Preface

Seafood is simply "food from the sea". It comes in various forms and is used as food for human beings. Seafood predominantly includes fish, crustaceans, mollusks and seaweeds. It is one of the important sources of protein and nutrients for human health.

This book contains a total 21 chapters. Chapter 1 provides a general introduction to the topics covered in this book and also the seafood production statistics around the globe. Chapter 2, 4 and 7 give information related to seafood processing methods such as fermentation, cooking, and MALDI-TOF for the seafood production and safety assessment. Chapter 5, 6, 8 and 10 cover fish protein and fish oil. Chapter 11 deals about chitosan as bio based nanocomposites in seafood industry and aquaculture. Chapter 14, 15 and 16 deal with seaweed production and their applications. Chapter 17 and 18 cover the biological application of seafood in food industry. Chapter 19 and 21 describe the health benefit value and discuss health risk of seafood.

Overall, this book provides details about seafood (fish, crustaceans, mollusks and seaweed) processing methods, biological applications, health benefits and health risk. These contributions will form essential reading for seafood scientists and can serve as instructional course for students.

I am grateful to all the authors who have provided the state-of-art contributions in the field of seafood. Their relentless effort was the result of scientific attitude, drawn from the past history in this field.

Prof. Se-Kwon Kim
Busan, South Korea

Contents

Preface   v

1. **Introduction to Seafood Science**   1
   Se-Kwon Kim and *Jayachandran Venkatesan*

2. **Fermentation of Seaweeds and its Applications**   14
   Motoharu Uchida

3. **Recent Advantages of Seafood Cooking Methods based**   47
   on Nutritional Quality and Health Benefits
   Abdul Bakrudeen Ali Ahmed, Teoh Lydia and *Rosna Mat Taha*

4. **Oil Tannage for Chamois Leather**   80
   Eser Eke Bayramoğlu and *Seher Erkal*

5. **Fish Protein Coating to Enhance the Shelf Life of Fishery**   90
   Products
   V. Venugopal Menon

6. **Recovery of Fish Protein using pH Shift Processing**   117
   Yeung Joon Choi and *Sang-Keun Jin*

7. **Usage of MALDI-TOF Mass Spectrometry in Sea Food**   132
   Safety Assessment
   Karola Böhme, Marcos Quintela-Baluja,
   Inmaculada C. Fernández-No, Jorge Barros-Velázquez,
   Jose M. Gallardo, Benito Cañas and *Pilar Calo-Mata*

8. **Production and Application of Microbial Transglutaminase**   170
   to Improve Gelling Capabilities of Some Indonesian
   Minced Fish
   Ekowati Chasanah and *Yusro Nuri Fawzya*

9. **Lactic Acid Bacteria in Seafood Products: Current Trends**   182
   and Future Perspectives
   Panchanathan Manivasagan, Jayachandran Venkatesan and
   Se-Kwon Kim

10. **Feeding Trial of Red Sea Bream with Dioxin Reduced**   202
    Fish Oil
    T. Honryo

11. **Chitosan as Bio-based Nanocomposite in Seafood Industry**   211
    and Aquaculture
    Alireza Alishahi, Jade Proulx and *Mohammed Aider*

12. **Recent Developments in Quality Evaluation, Optimization**   232
    and Traceability System in Shrimp Supply Chain
    Imran Ahmad, Chawalit Jeenanunta and *Athapol Noomhorm*

13. **Anti-aging & Immunoenhancing Properties of Marine**   261
    Bioactive Compounds
    Ranithri Abeynayake and *Eresha Mendis*

14. **Arsenic in Seaweed: Presence, Bioavailability and**   276
    Speciation
    Cristina García Sartal, María Carmen Barciela Alonso and
    Pilar Bermejo Barrera

15. **Application of Bacterial Fermentation in Edible Brown**   352
    Algae
    Sung-Hwan Eom and *Young-Mog Kim*

16. **Production, Handling and Processing of Seaweeds**   359
    in Indonesia
    Hari Eko Irianto and *Syamdidi*

17. **Food Applications of By-Products From the Sea**   376
    C. Senaka Ranadheera and *Janak K. Vidanarachchi*

18. **Mining Products from Shrimp Processing Waste and**   397
    Their Biological Activities
    Asep Awaludin Prihanto, Rahmi Nurdiani and *Muhamad Firdaus*

19. **Selenium-Health Benefit Values as Seafood Safety Criteria**   433
    Nicholas V.C. Ralston, Alexander Azenkeng, Carla R. Ralston,
    J. Lloyd Blackwell III and *Laura J. Raymond*

20. **Role of Bacteria in Seafood Products**   458
    Françoise Leroi

21. **Health Risks Associated with Seafood**   483
    Samanta S. Khora

Index   571

About the Editor   575

Color Plate Section   577

1

Introduction to Seafood Science

Se-Kwon Kim[1,2,]* and *Jayachandran Venkatesan*[1]

1 Introduction

Seafood is defined as "food from the sea"; it occurs in various forms and is used as food for human beings. Seafood predominantly includes fish, shellfish (crustaceans, molluscs, echinoderms) and seaweed. Seafood is a rich source of nutrients and has been a part of traditional Asian cuisine since ancient times (Fig. 1). The main seafood consumed daily by humans is fish. Harvesting, processing and consuming are the important things to be considered while using seafood (Table 1).

The total yearly worldwide production of finfish, shellfish and plants now exceeds 75 million metric tons (http://www.globalaginvesting. com/news/blogdetail?contentid=1439). Globally, aquaculture is majorly concentrated in the Asia Pacific region. According to the FAO's most recent statistics, the Asia Pacific region accounts for 89.1% of global aquaculture production with China alone contributing 62.3%. The type of species produced is highly variable, based on the region. China leads in the production of carp; Thailand, India, Indonesia, Vietnam, and China lead with shrimp and prawns, while Norway and Chile lead in salmon production (Table 1).

From the global viewpoint, South Korea is the 9th largest importer of fish and seafood in the world, importing over CAD $3.2 billion in 2010, an increase of 7.48% from 2009. South Korea's main fish and seafood imports in 2010 included frozen fish and bones, frozen shrimp and prawns, frozen

---
[1] Department of Marine-bio Convergence Science, Pukyong National University, Busan 608-737, Republic of Korea.
[2] Marine Bioprocess Research Center, Pukyong National University, Busan 608-737, Republic of Korea.
* Corresponding author

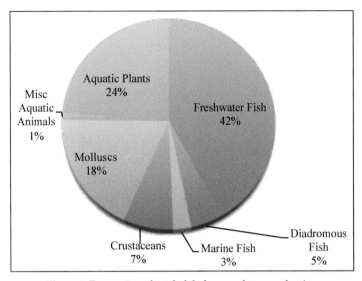

Figure 1. Proportion of total global aquaculture production.

Color image of this figure appears in the color plate section at the end of the book.

Table 1. Annual fisheries landings (kg per capita) and effort (Watts per capita) for fleets of continents in the 1950s and the 2000s.

	Landings		Effort	
Fleets	1950s	2000s	1950s	2000s
Europe	16.2	19.8	2.3	6.4
Asia	6.1	9.8	0.4	2.4
Africa	5.6	6.1	0.6	1.3
South America	8.2	44.1	0.7	1.8
Oceania	7.8	39.2	1.4	31
North America	15.5	16.2	1	3.9

dried and salted octopus, frozen fish meat, live fish and molluscs (Fig. 2). The main suppliers of seafood are China, Russia, Vietnam, Japan and the United States.

Annual global aquaculture production has increased 3 times within the past 15 years, and by 2015, aquaculture is predicted to account for 39% of total global seafood production by weight. Lack of adequate nutrition is a leading contributor to the global burden of disease; increased food production through aquaculture is a seemingly welcome sign (Garrett et al. 1997; Sapkota et al. 2008).

Capture fisheries and aquaculture supplied the world with about 148 million tonnes of fish in 2010, of which about 128 million tonnes were utilized as food for people (Table 2).

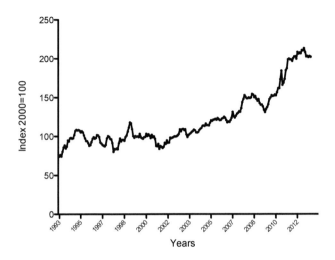

Figure 2. Export (Harmonised system): Fish, crustaceans, molluscs and other aquatic invertebrates (IDO3).
Source: U.S. Department of Labour: Bureau of Labour statistics

Table 2. The total fish production in 2004 and projections for 2010 and later simulated target years. All figures are in million tonnes. FAO—Food and Agricultural Organization; SOFIA—State of World Fisheries and Aquaculture.

	2000	2004	2010	2015	2020	2030
Information source	FAO	FAO	SOFIA	FAO	SOFIA	SOFIA
Marine capture	86.8	85.8	86		86	87
Inland capture	8.8	9.2	6		6	6
Total capture	95.6	95	93	105	93	93
Aquaculture	35.5	45.5	53	74	70	83
Total production	131.1	140.5	146	179	163	176
Food fish production	96.9	105.6	120		138	150
Non-food use	34.2	34.8	26		26	26

2 Fish

Over 32,000 fish species have been described, making them the most diverse group of vertebrates. However, humans commonly eat only small numbers of fishes. Fish is generally consumed in raw, fried, grilled and boiled form. Fish is the main source of protein. In addition, polyunsaturated fatty acids (PUFA) are an important constituent in the fish. Several studies have demonstrated that PUFA content significantly decreases in the fried form of fish foods (Candela et al. 1998). Fish oils also consist of omega-3 fatty acids: eicosapentaenoic acid (EPA) and docosachexenoic acid (DHA). Seafood is a high source of omega 3 long-chain polyunsaturated fatty acids, which are involved in the prevention of cardiovascular diseases.

Fish are aquatic vertebrates, and sub grouped as marine pelagic, marine demersal, diadromous and fresh water. The pelagic fish lives near the surface water column of the sea. The seafood groups can be divided into larger predator fish (shark, tuna, marlin, swordfish, mackerel, salmon) and smaller forage fish (herring, sardines, sprats, anchovies, menhaden). Demersal fish live at the bottom of the sea. Disdromous fish are fish which migrate between the sea and fresh water. Fresh water fish live in rivers, lakes and ponds. The important freshwater seafood species are carp, tilapia, catfish and trout.

According to the Food and Agriculture Organization (FAO), the world harvest of fish in 2005 consisted of 93.2 million tonnes (Fig. 3). The number of individual fish caught in the wild has been estimated at 0.97–2.7 trillion per year (not counting fish farms or marine invertebrates).

Fish and fishery products represent a very valuable source of protein and essential micronutrients for balanced nutrition and good health. In 2009,

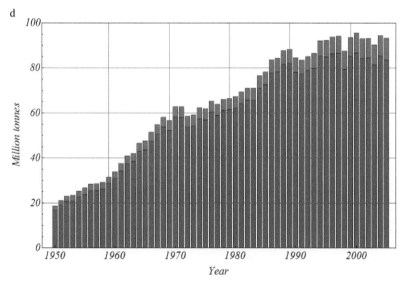

Figure 3. Marine (blue) and inland water wild fish catches 1950–2005.

Color image of this figure appears in the color plate section at the end of the book.

fish accounted for 16.6 percent of the world population's intake of animal protein and 6.5 percent of all protein consumed. Globally, fish provides about 3.0 billion people with almost 20 percent of their intake of animal protein, and 4.3 billion people with about 15 percent of such protein.

In 2010, the top ten producing countries accounted for 87.6 percent by quantity and 81.9 percent by value of the world's farmed food fish. Asia accounted for 89 percent of world aquaculture production by volume in 2010, and this was dominated by the contribution of China, which accounted for more than 60 percent of global aquaculture production volume. Other major producers in Asia are India, Vietnam, Indonesia, Bangladesh, Thailand, Myanmar, the Philippines and Japan. In Asia, the share of freshwater aquaculture has been gradually increasing, up to 65.6 percent in 2010 from around 60 percent in the 1990s. In terms of volume, Asian aquaculture is dominated by finfishes (64.6 percent), followed by molluscs (24.2 percent), crustaceans (9.7 percent) and miscellaneous species (1.5 percent).

2.1 Fish Protein

Protein is composed of amino acids that are necessary for the proper growth and function of the human body. While the human body has the capacity to produce amino acids required for protein production, a set of essential amino acids needed are to be obtained from meat or vegetables. Fish is a food of excellent nutritional value, providing high quality protein and a wide variety of vitamins and minerals that includes vitamin A and D, phosphorous, magnesium, selenium and iodine in marine fish. Researchers suggest that intake of small quantities of fish can have a significant positive impact in improving the quality of dietary protein by complementing the essential amino acids. In the current book, the detailed description of fish protein is discussed in Chapters 5 and 6. According to the World Resources Institute, a typical American gets 4.1% of her or his total protein supply from fish (Fig. 4).

The Maldives ranks first, with 54.8% of their protein coming from fish. Korea ranks in 9th position, getting 18.2% of the protein from fish. The word average is 6.5% (http://rankingamerica.wordpress.com/2009/04/09/the-us-ranks-92nd-in-fish-protein/). Transglutaminase enzyme has been helpful in modifying and meeting various needs such as improving flavor, texture, appearance and function of the modified food, which have been explained in Chapter 8.

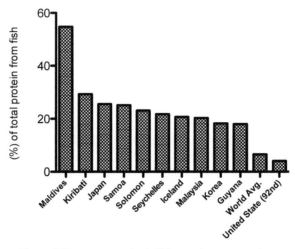

Figure 4. Top ten countries in fish protein consumption.

3 Molluscs

Molluscs are invertebrates with soft bodies. The body of a mollusc is generally composed of the shell and the fleshy, living part. The fleshy part of a mollusc can be further divided into the foot and the visceral mass. The foot is a distinctive molluscan feature, adapted in a variety of ways for locomotion. The important subgroups of molluscs are bivalves, gastropods and cephalopods. Bivalves are commonly called shells. The main shells among seafood are oysters, scallops, mussels and cockles. Gastropods are also known as sea snails. The important seafood groups among gastropods are abalone, cockle, conch, cuttlefish, loco, mussel, limpets, whelks and periwinkles. Cephalopods are not protected by shells. They have been widely used as seafood by Korean, Japanese and Chinese people since ancient times. The important seafoods are octopus, squid and cuttlefish. While being used as food, octopus must be boiled properly to rid it of slime, smell, and residual ink. Episodes of poisoning occasionally happen to the consumers of molluscs, the main hazard being represented by bivalve molluscs (Ciminiello and Fattorusso 2006; Orr et al. 2005; von Salvini-Plawen 1980).

PescaGalicia show that 96.8% of marine aquaculture production in Galicia is bivalve molluscs with a harvest of 215,681 tonnes in 2010 and a value of EUR 106.6 million at first sale. The average price in 2010 was EUR 0.41/kg, an increase of 3.69% compared with 2009, when mussels were trading on an average at EUR 0.39/kg (http://webcache.googleusercontent. com/search?q=cache:kUlKxUyS82QJ:www.globefish.org/bivalves-august-011.html+molluscs+production+statistics&cd=7&hl=en&ct=clnk).

3.1 Processing Molluscs

All molluscs are processed prior to use. Processing may vary from boxing bivalves for the live market to further processing, including freezing and canning. Processing of molluscs is done for several reasons: to convert the raw materials to a more desirable form, to preserve the products, to maintain quality, to completely utilize the raw product, and to assure safety. Molluscs such as clams and oysters are shucked to make them usable in market form. Molluscan shellfish are smoked, frozen, or canned to prolong shelf life and stabilize the quality. Temperature control is the most critical factor for providing a good product; however, few harvesting boats have the facility of refrigerated storage. In general, molluscs contain much less fat than oily fish and approximately the same quantity of fat as lean fish (Bemrah et al. 2009; Sirot et al. 2008).

4 Crustaceans

Crustaceans are invertebrates with segmented bodies protected by hard shells, usually made of chitin. The subgroups of crustaceans are shrimps, crabs, lobsters and krill. Shrimp and prawns are small with spiny rostrums. They play an important role in food industries. Because consumers are increasingly conscious of the relationship between good diet and health, the consumption of marine-based foods has been growing continuously. Consumers identify seafood as nutritious, complete foods, that are an excellent source of high-quality proteins and valuable lipids with high amounts of polyunsaturated fatty acids (PUFA).

Crustacean aquaculture has a high added value: while it accounts for 6.3% of world production in terms of quantity, it corresponds to 20% in terms of value. Twenty-six percent of the output from this type of aquaculture is from fresh water. This activity started in the 1980s and following slow development during 1990s, is now increasing by more than 8% per year (Fig. 5).

4.1 Shrimp

Dramatic increase in commercial shrimp farming has been recorded in the decades following the 1970's, particularly due to the market demands of Europe, United States and Japan. Currently, approximately 80% of farmed shrimp is cultured in Asia, including China and Thailand. Shrimp are marketed and sold in frozen condition. Shrimp is one of the main sources for omega-3-fatty acids and has low levels of mercury. As compared with other seafoods, shrimp is high in calcium, iodine and protein and low in

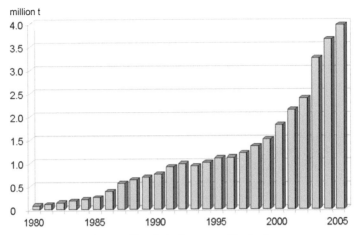

Figure 5. Trends in Crustacean Production.

food energy. Shrimp based meal is also a significant source of cholesterol (e Silva et al. 1996; Shang et al. 1998; Smith and Guentzel 2010).

Aquaculture production is weak due to adverse impacts of disease and environmental conditions. Disease outbreaks in recent years have affected farmed Atlantic salmon in Chile, oysters in Europe, and marine shrimp farming in several countries in Asia, South America and Africa, resulting in partial or sometimes total loss of production. In 2010, aquaculture in China suffered production losses of 1.7 million tonnes caused by natural disasters, diseases and pollution. Disease outbreaks virtually wiped out marine shrimp farming production in Mozambique in 2011.

4.2 Crab

Crabs are decapod crustaceans that live in seawaters, freshwaters, or on land. Crabs make up 20% of all marine crustaceans caught, farmed, and consumed worldwide. The common species is *Portunus tribuberculatus,* which accounts for one fifth of the total. Other commercially important taxa include *Portunus pelagicus,* several species in the genus *Chionoecetes,* the blue crab (*Callinectes sapidus*), *Charybdis* spp., *Cancer pagurus,* the Dungeness crab (*Metacarcinus magister*) and *Scylla serrata,* each of which yields more than 20,000 tonnes annually. Crabs are prepared and consumed as a dish in several different ways all over the world. Some species are eaten whole, including the shell, such as soft shell crab; with other species, just the claws or legs are eaten. The latter is particularly common for larger crabs, such as the snow crab. The biggest importers are Japan, France, Spain, Hong Kong, the US, Canada and Portugal.

Sea water king crab is a superfamily of crab-like decapod crustaceans, chiefly found in cold seas. Because of their large size and the taste of their meat, many species are widely caught and sold as food, the most common being the red king crab, *Paralithodes camtschaticus*. King crabs are generally thought to be derived from hermit crab-like ancestors, which may explain the asymmetry still found in the adult forms. King crabs are the most widely quoted example of carcinisation among the Decapoda. The evidence for this explanation comes from the asymmetry of the king crab's abdomen, which is thought to reflect the asymmetry of hermit crabs, which must fit into a spiral shell (Dean et al. 2009).

5 Aquatic Plants

Aquatic plants are referred to as seaweed and algae species. Asian people use seaweed and algae as food. There are several methods that are available for the processing of the seaweed and algae. Seaweed can be divided into three groups, brown algae (Phaeophyta), red algae (Rhodophyta) and green algae (Chlorophyta). Brown algae mainly consist of alginate and fucoidan as major fiber components. Red algae contains glactan, and green algae are made up of cellulose and hemicellulose. Polysaccharides are the major constituents of seaweed. Average proximate contents on a dry basis are 15.3% protein, 57.0% carbohydrate, 1.7% lipid, and 25.6% ash. Seaweed is directly taken by Asian people as food, whereas in the western countries, peoples extract the important components from the seaweed and use them as food hydrocolloids; the products include agar, carrageenan and alginates. Seaweed has been used in various dishes including appetizers, casseroles, muffins, pilafs and soups in Korea, China and Japan due to its high mineral content.

For some thousands of years seaweed has been highly valued and widely consumed as a direct human food by oriental communities. In the West, no such tastes or traditions have been acquired. The Greeks and Romans had little regard for seaweed and the use of algae as a specific food item was, in general, discontinued in Europe and North America in the course of the last century. In Europe, human consumption of fresh or lightly treated seaweed appears to have withered away as standards of living have risen. Young stipes of *Laminaria saccharina* used to be sold in street markets in Scotland as 'tangle' and dulse (*Rhodymenia* spp.) was a significant food item in Ireland for many centuries.

The use of seaweed as food has been traced back to the fourth century in Japan and the sixth century in China. Today these two countries and the Republic of Korea are the largest consumers of seaweed as food. China is the largest producer of edible seaweeds, harvesting about 5 million wet tonnes. The greater part of this is for *kombu* produced from hundreds of hectares of the brown seaweed, *Laminaria japonica,* that is grown on suspended ropes in the ocean. The Republic of Korea grows about 800,000 wet tonnes of three different species, and about 50 percent of this is for *wakame,* produced from a different brown seaweed, *Undaria pinnatifida,* grown in a similar fashion to *Laminaria* in China. Japanese production is around 600,000 wet tonnes and 75 percent of this is for *nori,* the thin dark seaweed wrapped around a rice ball in sushi (Brownlee et al. 2012).

Aquatic algae production (by volume) increased at average annual rates of 9.5 percent in the 1990s and 7.4 percent in the 2000s—comparable with rates for farmed aquatic animals—with production increasing from 3.8 million tonnes in 1990 to 19 million tonnes in 2010. Cultivation has over shadowed production of algae collected from the wild, which accounted for only 4.5 percent of total algae production in 2010. Following downward adjustments by FAO of the estimated value of several major species from a few major producers with incomplete reported data, the estimated total value of farmed algae worldwide has been reduced for a number of years in the time series. The total value of farmed aquatic algae in 2010 is estimated at US$5.7 billion, while that for 2008 is now re-estimated at US$4.4 billion. A few species dominate algae culture, with 98.9 percent of world production in 2010 coming from Japanese kelp (*Saccharina/Laminaria japonica)* (mainly in the coastal waters of China), *Eucheuma* seaweeds (a mixture of *Kappaphycus alvarezii,* formerly known as *Eucheuma cottonii,* and *Eucheuma* spp.), *Gracilaria* spp., nori/laver (*Porphyra* spp.), wakame (*Undaria pinnatifida*) and unidentified marine macroalgae species (3.1 million tonnes, mostly from China).

6 Other Aquatic Animals

Sea mammals form a diverse group of 128 species that rely on the ocean for their existence. Whale meat is still harvested from legal, non-commercial hunts (Freeman 1993; Jefferies 2009). Dolphins are traditionally considered as food (Massuti et al. 1998; Ofori-Danson et al. 2003). Sea turtles have long been valued as food in many parts of the world. Echinoderms are headless invertebrates, found on the seafloor in all oceans and at all depths. Echinoderms used for seafood include sea cucumbers, sea urchins, and occasionally starfish. Jellyfish are soft and gelatinous, with a body shaped like an umbrella or bell, which pulsates for locomotion.

7 Health Risks

Mercury, selenium and lead from seafood consumption cause a variety of adverse effects on human health. A study investigated the impact of mercury concentration in seafood consumption by grouping the supply regions (Atlantic Ocean, Pacific Ocean, and foreign shores) (Grandjean et al. 1992; Sunderland 2007). Seafood allergies are most commonly seen in humans, mainly those living in regions where they are often eaten. Researchers have estimated that 75% of individuals who are allergic to one type of crustacean (shrimp, lobster, crawfish or crab) are also allergic to other types. Symptoms range from itching and swelling of the mouth and throat to life threatening reactions. They occur when the seafood is ingested, but can also occur when raw seafood is handled and even after inhaling steam while crustaceans such as shrimp are being cooked. Cooking does not appear to destroy the allergens in crustaceans and molluscs. Shellfish allergens are usually found in the flesh and are part of the muscle protein system, whilst in foods such as shrimps, allergens have also been found in the shells. Potential health risks of seafood consumption have addressed by several researchers (Guo et al. 2007; Han et al. 1998; Martí-Cid et al. 2007; Sidhu 2003; Storelli 2008).

8 Future Prospects in Seafood

The latter half of the 20th century witnessed widespread expansion of capture fisheries supply, and correspondingly positive social and economic impacts associated with the global availability of high-quality aquatic foods. Based on FAO statistics for 1950–2006, the first overview of marine fisheries resources by country confirmed that, globally, the maximum average level of bottom fish and small pelagic fish production had been reached within the final decade. Several analyses are troubling from a resource exploitation perspective and suggest a global system that is overstressed, reducing in biodiversity and in imminent danger of collapse. There is also a strong societal argument for maximizing beneficial use of natural resources, and the clear need for food, which would justify the fullest possible level of harvesting consistent with the ability for these harvests to be sustained.

9 Conclusion

Seafood, which includes fish, crustaceans and seaweed, has historically been consumed around the globe. It is an important source of protein and nutrients for human health. However, health risks of seafood metal toxin have been addressed extensively.

Acknowledgements

This work was supported by a grant from Marine Bioprocess Research Center of the Marine Bio 21 Center funded by the Ministry of Land, Transport and Maritime, Republic of Korea.

References

Bemrah, N., V. Sirot, J.-C. Leblanc and J.-L. Volatier. 2009. Fish and seafood consumption and omega 3 intake in French coastal populations: CALIPSO survey. Public Health Nutrition. 12(05): 599–608.

Brownlee, I., A. Fairclough, A. Hall and J. Paxman. 2012. The potential health benefits of seaweed and seaweed extract.

Candela, M., I. Astiasaran and J. Bello. 1998. Deep-fat frying modifies high-fat fish lipid fraction. Journal of Agricultural and Food Chemistry. 46(7): 2793–2796.

Ciminiello, P. and E. Fattorusso. 2006. Bivalve molluscs as vectors of marine biotoxins involved in seafood poisoning. *In*: G. Cimino and M. Gavagnin (eds.). Molluscs. 43: 53–82: Springer Berlin Heidelberg.

Dean, P.N., T. Ahyong Shane, C. Tin-Yam, A. Crandall Keith, C. Dworschak Peter, L. Felder Darryl, M. Feldmann Rodney, H. Fransen Charles, Y. Goulding Laura and L. Rafael. 2009. A classification of living and fossil genera of decapod crustaceans.

e Silva, E.D.O., C.E. Seidman, J.J. Tian, L.C. Hudgins, F.M. Sacks and J.L. Breslow. 1996. Effects of shrimp consumption on plasma lipoproteins. The American Journal of Clinical Nutrition. 64(5): 712–717.

Freeman, M.M. 1993. The International Whaling Commission, Small-type Whaling, and Coming to Terms with Subsistence. Human Organization. 52(3): 243–251.

Garrett, E.S., C.L. dos Santos and M.L. Jahncke. 1997. Public, animal, and environmental health implications of aquaculture. Emerging Infectious Diseases. 3(4): 453.

Grandjean, P., P. Weihe, P.J. Jørgensen, T. Clarkson, E. Cernichiari and T. Viderø. 1992. Impact of Maternal Seafood Diet on Fetal Exposure to Mercury, Selenium, and Lead. Archives of Environmental Health: An International Journal. 47(3): 185–195.

Guo, J.Y., E.Y. Zeng, F.C. Wu, X.Z. Meng, B.X. Mai and X.J. Luo. 2007. Organochlorine pesticides in seafood products from southern China and health risk assessment. Environmental Toxicology and Chemistry. 26(6): 1109–1115.

Han, B.-C., W. Jeng, R. Chen, G. Fang, T. Hung and R. Tseng. 1998. Estimation of target hazard quotients and potential health risks for metals by consumption of seafood in Taiwan. Archives of Environmental Contamination and Toxicology. 35(4): 711–720.

Jefferies, C.S. 2009. Strange Bedfellows or Reluctant Allies: Assessing whether environmental non-governmental organizations (ENGOs) should serve as official monitors of whaling for the International Whaling Commission (IWC). Windsor Rev. Legal & Soc. Issues. 26: 75.

Martí-Cid, R., A. Bocio, J.M. Llobet and J.L. Domingo. 2007. Intake of chemical contaminants through fish and seafood consumption by children of Catalonia, Spain: health risks. Food and Chemical Toxicology. 45(10): 1968–1974.

Massuti, E., S. Deudero, P. Sanchez and B. Morales-Nin. 1998. Diet and feeding of dolphin (Coryphaena hippurus) in western Mediterranean waters. Bulletin of Marine Science. 63(2): 329–341.

Ofori-Danson, P., K. Van Waerebeek and S. Debrah. 2003. A survey for the conservation of dolphins in Ghanaian coastal waters. Journal of the Ghana Science Association. 5(2).

Orr, J.C., V.J. Fabry, O. Aumont, L. Bopp, S.C. Doney, R.A. Feely, A. Gnanadesikan, N. Gruber, A. Ishida and F. Joos. 2005. Anthropogenic ocean acidification over the twenty-first century and its impact on calcifying organisms. Nature. 437(7059): 681–686.

Sapkota, A., A.R. Sapkota, M. Kucharski, J. Burke, S. McKenzie, P. Walker and R. Lawrence. 2008. Aquaculture practices and potential human health risks: current knowledge and future priorities. Environment International. 34(8): 1215–1226.

Shang, Y.C., P. Leung and B.H. Ling. 1998. Comparative economics of shrimp farming in Asia. Aquaculture. 164(1): 183–200.

Sidhu, K.S. 2003. Health benefits and potential risks related to consumption of fish or fish oil. Regulatory Toxicology and Pharmacology: RTP. 38(3): 336.

Sirot, V., M. Oseredczuk, N. Bemrah-Aouachria, J.-L. Volatier and J.-C. Leblanc. 2008. Lipid and fatty acid composition of fish and seafood consumed in France: CALIPSO study. Journal of Food Composition and Analysis. 21(1): 8–16.

Smith, K.L. and J.L. Guentzel. 2010. Mercury concentrations and omega-3 fatty acids in fish and shrimp: Preferential consumption for maximum health benefits. Marine Pollution Bulletin. 60(9): 1615–1618.

Storelli, M. 2008. Potential human health risks from metals (Hg, Cd, and Pb) and polychlorinated biphenyls (PCBs) via seafood consumption: estimation of target hazard quotients (THQs) and toxic equivalents (TEQs). Food and Chemical Toxicology. 46(8): 2782–2788.

Sunderland, E.M. 2007. Mercury exposure from domestic and imported estuarine and marine fish in the US seafood market. Environmental Health Perspectives. 115(2): 235.

von Salvini-Plawen, L. 1980. A reconsideration of systematics in the Mollusca (phylogeny and higher classification). Malacologia: International Journal of Malacology. 19(2).

2

Fermentation of Seaweeds and its Applications

Motoharu Uchida

1 Introduction

There have been many kinds of fermentation technology and products since ancient times. For example, fermented food items from soybean are common in the East Asian countries of China, Korea and Japan, while those from fish are common in Southeast Asian countries (Ishige 1993). Despite the long history of fermentation technology, fermented food items produced from seaweeds are yet to be developed (Fig. 1, Uchida and Miyoshi 2012). Many studies were conducted on methane production via fermentation from seaweeds during the 1970s and 1980s (Aquaculture Associates 1982; Chynoweth et al. 1993). However, methane production via fermentation (methane fermentation) is a technology for supplying energy, not for foods and food production.

Seaweeds, or macroalgae (macrophytes), can be divided into three groups: brown algae (Phaeophyta), red algae (Rhodophyta), and green algae (Chlorophyta). Seagrass (Magnoliophyta), biologically differentiated from seaweeds, will be discussed as an item of seaweeds for convenience in the present chapter. Carbohydrates are the major component of seaweeds (ca. 50–70% on dry weight basis), containing quantitative polysaccharides to construct frond tissue (Japanese Ocean Industries Association 1984). But, these major components of seaweed polysaccharides are empirically known as unfavorable substrates for fermentation. This may be one of the reasons

National Research Institute of Fisheries and Environment of Inland Sea, Maruishi, Ohno, Saeki Hiroshima, Japan.

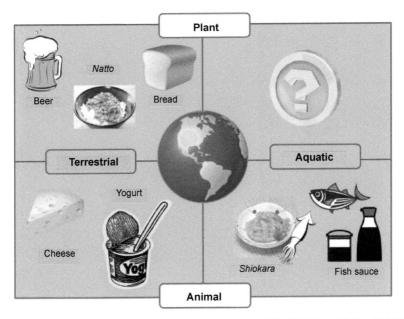

Figure 1. Categories of fermented foods based on raw materials (Uchida and Miyoshi 2012). Fermented foods prepared from aquatic plants (algae) are yet to be developed.

Color image of this figure appears in the color plate section at the end of the book.

why seaweed fermentation technology has yet to be developed. But it was recently reported that seaweed can be used as a substrate for lactic acid and ethanol fermentation, provided that the frond tissue is saccharified by using cellulase enzymes (Uchida and Murata 2002; Uchida and Murata 2004). This finding opens the possibility of obtaining foods and related items from seaweed fermentation.

This chapter reviews studies on the lactic acid fermentation of seaweeds. It also discusses other kinds of algal fermentation that are now being developed, such as ethanol fermentation (Uchida 2009; Uchida 2011; Uchida and Miyoshi 2012). With all these advances in seaweed fermentation, we suggest that in the near future, there is great potential for creating a new industry based on algal fermentation technology.

2 Chemical Composition of Seaweeds

It is important to know the proximate components of seaweeds before conducting fermentation. Sugars, especially, are key substances from the viewpoint of a substrate for fermentation. Table 1 shows the moisture and proximate contents of seaweeds (Miyoshi et al. 2013). The average value

Table 1 Moisture and proximate contents of seaweeds (Miyoshi et al. 2013).

Groups	Moisture (%)	Proximate contents (on dry basis %)				Fiber (%)	Non-fiber (%)
		Protein	Lipid	Carbohydrate	Ash		
Seaweeds and seagrasses (Average, n=104)	87.5	15.3	1.7	57.1	25.6	43.5	13.6
Brown seaweeds (Average, n=30)	85.5	13.7	2.3	56.9	27.1	45.7	9.8
Red seaweeds (Average, n=23)	88.2	18.4	2.6	54.0	24.6	44.3	9.4
Green seaweeds (Average, n=33)	89.2	15.7	0.8	57.2	26.2	40.3	17.1
Seagrasses (Average, n=17)	86.5	13.4	1.1	60.3	23.9	44.7	15.4
Microalgae (n=3)	78.5	48.2	7.9	22.8	21.1	NT	15.4

Carbohydrate=100-(protein+lipid+ash) , Fiber; measured by the modified Prosky method, Non-fiber; calculated as carbohydrate-fiber of moisture is 87.5% (wt/wt, n = 104) and the values are similar among the different seaweed groups. Average proximate contents on a dry weight basis are 15.3% protein, 57.0% carbohydrate, 1.7% lipid, and 25.6% ash, respectively (n = 104). Seaweeds are rich in carbohydrates, while microalgae are rich in protein (48.2%, n = 3) and lipids (7.9%). A major part of the seaweed carbohydrate is fiber (43.5% of total weight, measured by the modified Prosky method (The Council for Science and Technology, Ministry of Education, Culture, Sports, Science, and Technology, Japan 2004). Brown algae contain alginate and fucoidan as major fiber components. Red algae contain galactan (e.g., agar, carrageenan, etc.) as a major fiber component. Green algae and seagrasses contain cellulose and hemicellulose as major fiber components. These seaweed fibers are not suitable substrates for fermentation. This may be the reason why seaweed fermentation has not yet been developed to date. On the other hand, seaweeds also contain storage sugars such as mannitol, laminaran (brown algae), floridoside, trehalose (red algae), starch (green and other group seaweeds), and sucrose (seagrasses). These storage sugars can be readily utilized as a substrate for fermentation. However, the quantity of storage sugars shows large differences depending on the sample lot (Fukushi 1988).

3 Saccharification of Seaweeds

Saccharification is an important process before conducting seaweed fermentation. A major part of seaweed carbohydrate is polysaccharides and this needs to be degraded to sugars before fermentation. Treatment of fronds with cellulase is reported to be effective for saccharification for many kinds of seaweeds (Uchida and Murata 2004). Agarase, mannase, alginate lyase, and hemicellulase are also expected to be effective for saccharification of seaweeds; however, only a limited number of enzymes are commercially available. The use of protease is effective in producing soluble proteins including peptides and free amino acids in the case of protein-rich (23.2% protein on a dry weight basis) laver (Fig. 2, in preparation; Laver (nori) is a food product of red algae, *Porphyra yezoensis*), but also effective for saccharification. Quantities of glucose are produced and exuded to the supernatant of fermentation cultures, especially in the case of protein-poor (14.9% protein) laver, after the treatment with protease (Fig. 2b, in preparation). Details of the process of how glucose is produced from the laver by the protease treatment remain to be clarified. This enzymatic treatment was conducted without sterilization, and a decrease of soluble protein and glucose was observed, along with the incubation time, probably, due to microbial growth, suggesting that the enzymatic treatment time should preferably be less than 12 h at 50°C. The saccharification process is commonly conducted for 5–8 hours at 55–58°C for the manufacture of

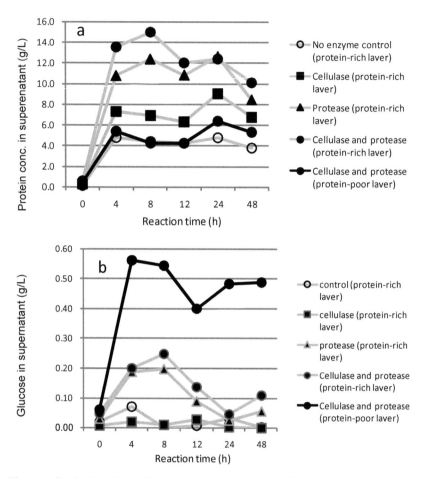

Figure 2. Production of soluble protein and glucose from laver after enzyme treatment (unpublished). Three grams of laver were mixed with or without enzymes, 0.1 g cellulase and/or 0.14 g protease, and total weight was adjusted to 50 g with distilled water. Protein-rich (23.2% protein) and -poor (14.9% protein) laver was used. The mixtures were reacted for 48 h at 50°C with moderate agitation. Protein (a) and glucose contents (b) in the supernatant were measured using commercial kits (Pierce BCA Protein Assay Kit and F kit, respectively). Data is shown as a mean of triplicate experiments.

Color image of this figure appears in the color plate section at the end of the book.

Japanese *sake.* This is important information to determine the optimum conditions for accomplishing seaweed saccharification while avoiding the growth of contaminant microorganisms.

4 Lactic Acid Fermentation of Seaweeds

Beginning of the study

A fermented material of *Ulva* spp. (Chlorophyta) was first obtained on 9 October 1998, after treating the fronds with enzymes containing cellulase activity and leaving the material for 17 months at 2°C (Uchida 2002; Uchida and Murata 2004). The fermented material obtained had a sweet odor, suggesting some kind of fermentation was occurring with the culture. Then, this culture was transferred with cellulase to a new *Ulva* culture at an interval of several weeks, with the induction of fermentation being demonstrated repeatedly. Microbial analysis based on rRNA gene nucleotide sequences identified the predominant microorganisms in the culture as *Lactobacillus brevis* (lactic acid bacteria, LAB), *Debaryomyces hansenii* var. *hansenii* (typical marine yeast), and *Candida zeylanoides* related specimens (yeast) (Fig. 3). Therefore, the primarily observed 'fermentation' was regarded as a mixture of lactic acid fermentation and ethanol fermentation. Inoculation of three kinds of separately cultured microorganisms induced fermentation on various kinds of seaweed (Table 2, Uchida and Murata 2004). This is the first report on an intentional induction of lactic acid fermentation of seaweed.

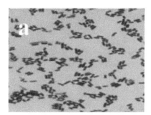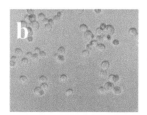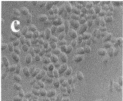

Figure 3. Microorganisms initially isolated from an algal fermented culture as a starter culture (Uchida and Murata 2004). A bacterium (a: *Lactobacillus brevis*) and two kinds of yeast (b: *Debaryomyces hansenii*, and c: *Candida zeylanoides*-related yeast).

Color image of this figure appears in the color plate section at the end of the book.

Optimum conditions for obtaining successful fermentation

The addition of cellulase (>0.5%) promoted the predominance of LAB and yeast, and successfully induced fermentation (Table 3a, Uchida and Murata 2002). The addition of the starter microorganisms (initial concentration, ca. 10^{6-7} CFU/ml) had the effect of enabling a reduction in the quantity of cellulase use (>0.25% cellulase addition was enough for successful fermentation, Table 3b, Uchida and Murata 2002). The addition of salt (5%) also promoted the predominance of LAB (Uchida et al. 2007). Based on these fermentation experiments, the optimum composition of seaweed cultures

Table 2 Fermentation results of various kinds of seaweed with the addition of cellulase and the microbial starter culture (Uchida and Murata 2004)

Seaweeds	Group	pH		Gas	Production (g/100mL) of:	
		Initial	After 7Ds		Lactic acid	Ethanol
Chondracanthus teedii	Rhodophyta	5.7	4.5	+	+ (0.16)	+ (0.18)
Chondracanthus tenellus		5.7	3.9	+	+ (0.25)	+ (0.18)
Gelidium linoides		6.5	5.3	+	+ (0.18)	+ (0.12)
Gracilaria incurvata		6.2	4.0	−	+ (0.25)	+ (0.12)
Gracilaria vermiculophylla		6.0	4.0	+	+ (0.31)	+ (0.23)
Hypnea charoides		6.4	6.1	+	+ (0.22)	+ (0.16)
Prionitis angusta		6.3	3.8	+	+ (0.25)	+ (0.17)
Prionitis divaricata		6.4	4.7	+	+ (0.25)	+ (0.41)
Pterocladia capillacea		6.4	5.6	±	+ (0.12)	+ (0.08)
Dilophus okamurae	Phaeophyta	6.7	6.1	−	+ (0.02)	+ (0.04)
Eisenia bicyclis		5.1	4.2	−	+ (0.02)	+ (0.03)
Hizikia fusiformis		5.4	5.0	+	+ (0.01)	+ (0.24)
Ishige okamurae		4.8	4.9	−	+ (0.01)	+ (0.10)
Laminaria japonica		5.4	3.3	+	+ (0.16)	+ (0.15)
Padina arborescens		6.2	5.8	±	− (<0.01)	+ (0.08)
Sargassum ringgoldianum		5.1	5.0	−	+ (0.01)	+ (0.04)
Undaria pinnatifida (Whole No. 1)		5.7	3.6	+	+ (0.23)	+ (0.38)
Undaria pinnatifida (Whole No. 2)		5.8	3.9	+	+ (0.18)	+ (0.07)
Undaria pinnatifida (Stem)		5.9	3.5	+	+ (0.25)	+ (0.12)
Ulva sp. No. 1	Chlorophyta	5.6	3.3	+	+ (0.76)	+ (0.16)
Ulva spp. No. 2		5.8	4.7	+	+ (0.45)	+ (0.41)
Zostera marina	Vascular plant	6.0	3.3	−	+ (1.14)	+ (0.26)

0.5 g of seaweed (dried) and 0.1g of cellulase R-10 were suspended with 9 ml of autoclaved 3.5% NaCl solution. 0.05 ml each of the cultured cell suspensions (OD660nm=1) of the strains *L. casei* B5201, *D. hansenii* Y5201 and *Candida* sp. Y5206 were added as a starter. The culture tubes were incubated for 7 days at 20°C rotating at 5 rpm with the caps closed. Data is the mean of the duplicate tests.

Table 3a Microbial flora of the *Undaria* culture water samples prepared with different concentrations of cellulase without the addition of the microbial mixture (Uchida and Murata 2002).

No. of Microbes (CFU/ml)	Cellulase % (wt/vol) of culture solution						
	0	0.01	0.1	0.25	0.5	1.0	3.0
Heterotrophic	$2.2\pm1.7\times10^7$	$4.6\pm2.4\times10^7$	$2.3\pm0.9\times10^7$	$1.8\pm0.4\times10^7$	3.5×10^6, 1.1×10^6	$<10^5$	$2.5\pm0.4\times10^7$
Marine heterotrophic	$2.4\pm1.6\times10^7$	$7.0\pm5.1\times10^7$	$1.8\pm1.4\times10^7$	$1.6\pm0.4\times10^7$	3.6×10^6, $<10^5$	$<10^5$	$<10^5$
(/Heterotrophic %)	(129 ± 26)	(129 ± 43)	(64.8 ± 38.4)	(109 ± 23)	(103)		
Lactic acid bacteria	$<10^6$	$<10^6$	$<10^6$	$<10^6$	1.3×10^5, 7.0×10^5	$<10^5$	$2.5\pm0.5\times10^7$
(/Heterotrophic %)					(3.7), (63.6)		(99.4 ± 4.2)
Yeast	$<10^2$	$<10^2$	$<10^2$	$<10^2$	1.9×10^4, 4.2×10^4	$<10^2$	$<10^2$

The numbers of microbes in the cultures after 5 days of incubation are shown. Results from the replicate bottles were different and are shown individually for the experimental case of 0.5% cellulase addition. For other data, means±SEM are given. Heterotrophic, marine heterotrophic, lactic acid bacteria, and yeast were counted on Standard Method Agar, Marine Agar 2216, PlateCount Agar with BCP containing cycloheximide at 50 mg/l, and Sabouraud Agar containing NaCl at 5% and antibacterial agents at 50mg/l level, respectively.

Table 3b Microbial flora of the *Undaria* culture water samples prepared at different concentrations of cellulase with the addition of the microbial mixture (Uchida and Murata 2002).

No. of Microbes (CFU/ml)	Cellulase% (wt/vol) of culture solution					
	0	0.25	0.5	1.0	3.0	5.0
Heterotrophic	8.7×10^6, 1.4×10^7	$2.9\pm0.4\times10^8$	$2.8\pm0.3\times10^8$	$2.6\pm0.3\times10^8$	$2.3\pm0.1\times10^8$	$1.8\pm0.1\times10^8$
Marine heterotrophic	1.7×10^6, 1.3×10^8	$1.2\pm0.4\times10^7$	2.0×10^7	$9.2\pm1.9\times10^6$	$8.3\pm1.3\times10^6$	$8.7\pm2.4\times10^6$
(/Heterotrophic %)	(19.5), (92.9)	(4.2 ± 0.8)	(8)	(3.7 ± 1.1)	(3.6 ± 0.6)	(4.9 ± 1.6)
Lactic acid bacteria	7.0×10^6, 3.9×10^7	$3.1\pm0.2\times10^8$	$2.4\pm0.3\times10^8$	$2.8\pm0.2\times10^8$	$2.5\pm0.6\times10^8$	$2.0\pm0.1\times10^8$
(/Heterotrophic %)	(80.5), (27.9)	(108 ± 8)	(85.5 ± 1.5)	(108 ± 5)	(106 ± 19)	(111 ± 1)
Yeast	2.4×10^6, 3.5×10^6	$9.8\pm3.3\times10^6$	$1.2\pm0.4\times10^7$	$2.2\pm0.5\times10^7$	$1.1\pm0.2\times10^7$	$3.7\pm2.7\times10^6$

The number of microbes were counted in the same method as described in table 3a. Results from the replicate bottles were different and are shown individually for the experimental case of 0% cellulase addition. Data of marine heterotrophic microorganisms at 0.5% cellulase concentration is a single results out of replicate bottles. For other data, means±SEM are given.

for successful fermentation is seaweed (at 5–10% wt/wt), cellulase (0.1–1%), NaCl (2.5–5%, when non-halophilic LAB is used as starter), and LAB starter (10^{4-6} CFU/ml). A wide range of incubation temperature (5–35°C) can be used for fermentation. Use of higher temperatures (e.g., 30–35°C) results in reducing the incubation time for fermentation, but empirically has a higher risk of the growth of spoiling microorganisms such as *Bacillus* group bacteria (unpublished observations).

Microorganisms used for lactic acid fermentation

In order to determine the suitable combinations of LAB and yeast strains for use as a starter culture for the lactic acid fermentation of seaweed, fermentation culture was prepared with different combinations of LAB strains (i.e., *L. brevis* FERM BP-7301, *Lactobacillus acidophilus* IAM10074, *Lactobacillus plantarum* IAM12477^T) and yeast strains (i.e., the two isolates above, plus *Saccharomyces cerevisiae*) (Uchida et al. 2004a). The inoculation of LAB with or without yeast strains yielded successful induction of fermentation, while inoculation of the yeast strains alone yielded unsatisfactory results, along with some contaminant bacteria growth. It therefore follows that the inoculation of yeast is not necessary when intending to induce the lactic acid fermentation of seaweeds.

The species specific primer sets were developed to identify the LAB strains by using PCR techniques, prior to the examination of suitable LAB strains for use as a starter culture for seaweed fermentation (Uchida et al. 2004b). Fourteen LAB strains including 11 species were tested under culture conditions prepared with or without salt (Fig. 4, Uchida et al. 2007). A commercial product of *Undaria pinnatifida* (Phaeophyta) powder was used as a substrate for fermentation without sterilizing it. The starter suitability of the LAB strains was assessed from their predominance after 11 days of culture at 20°C. The predominance was assessed by PCR using the developed species specific primer sets. Among the tested strains, *L. brevis*, *L. plantarum*, *Lactobacillus casei*, and *Lactobacillus rhamnosus* showed a high (>90%) predominance in their cultures, while the control cultures prepared without the inoculation of LAB showed no detectable LAB growth and then spoiled. In those spoiled cultures, *Bacillus* strains such as *Bacillus cereus* related and *B. fusiformis* related species were observed to dominate (Uchida et al. 2007). The *Undaria* powder prior to fermentation contained culturable microorganisms at 1.4–3.1 x 10^2 CFU/g, but the *Bacillus cereus* related strain was not a major component, suggesting a concern about the selective growth of the *Bacillus cereus* related strain during spoiled fermentation. This study did not test whether the *Bacillus cereus* related strains isolated from seaweed possessed the genetic capability to produce toxics.

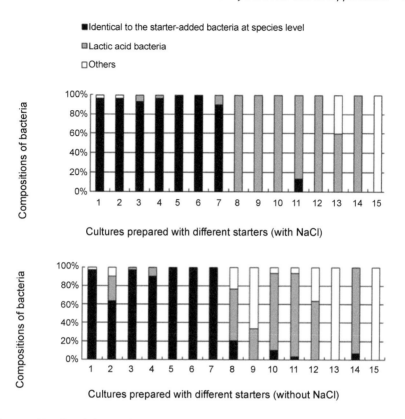

Figure 4. Results of the test for examining suitable starter culture of lactic acid bacteria for seaweed fermentation (Uchida et al. 2007). To prepare cultures with NaCl, 2.0 g of the commercial product of *Undaria* powder (Hamamidori, Riken Shokuhin) was mixed with 40 ml of autoclaved 3.5 % (wt/vol) NaCl solution, 40 mg of cellulase (12S, Yakult Honsha Co., Ltd.) and 0.4 ml of bacterial cell suspension. The bacterial cell suspension was prepared for the 14 LAB strains: No. 1; *Lactobacillus brevis* FRA 000033, 2; *Lact. brevis* IAM 12005, 3; *Lact. plantarum* ATCC 14917T, 4; *Lact. plantarum* IAM 12477T, 5; *Lact. casei* IFO 15883T, 6; *Lact. casei* FRA 000035, 7; *Lact. rhamnosus* IAM 1118T, 8; *Lact. zeae* IAM 12473T, 9; *Lact. acidophilus* IFO 13951T, 10; *Lact. kefir* NRIC 1693T, 11; *Lact. fermentum* ATCC 14931T, 12; *Lact. delbrueckii* subsp. *bulgaricus* ATCC 11842T, 13; *Streptococcus thermophilus* NCFB 2392, and 14; *Leuconostoc mesenteroides* IAM 13004T. These strains were pre-cultured with MRS medium (Merck Co.), collected by centrifuge (8,000 g x 20 min.), washed with autoclaved 0.85% NaCl solution, re-suspended to make a concentration of O.D.660 nm = 1.0 (containing 7.3×10^7–1.1×10^9 CFU/ml), and then used. To prepare culture without NaCl, autoclaved distilled water was used instead of 3.5% NaCl solution. Cultures without the inoculation of lactic acid bacteria were prepared as being without starter culture controls (No. 15). After incubating for 11 days at 20°C, the microbial composition was investigated. Ten colonies, each formed on the SMA plates prepared for viable counting, were chosen at random from the triplicated trials (Total n = 30), and then transferred to the BCP plates with the % proportion of yellow-colored colonies shown as average ± SE of lactic acid bacteria. *Lact. brevis,* (Trial Nos. 1, 2), *Lact. plantarum* (2, 4), *Lact. casei,* (5, 6), and *Lact. rhamnosus* (7) showed marked ability to be dominant in the *Undaria* cultures.

Color image of this figure appears in the color plate section at the end of the book.

For lactic acid fermentation in highly salted conditions

Common fermented sauces are highly salted and can be preserved for a long time without putrefaction. For example, typical soybean and fish sauces are salted at 12–16% (Subdivision on Resources, The Council for Science and Technology, Ministry of Education, Culture, Sports, Science and Technology, Japan 2002) and 15–31% wt/vol (Itoh et al. 1993; Ren et al. 1993; Uchida et al. 2005), respectively. In these fermentation products, it is known that halophilic lactic acid bacteria (HLAB) such as *Tetragenococcus* species are dominant and supposed to play an important role for a successful fermentation process. Three species are known to date as the *Tetragenococcus* group bacteria. *Tetragenococcus halophilus* is the most common specimen among the group (e.g., Marcello et al. 1985; Uchida et al. 2005). *Tetragenococcus muriaticus* was isolated from fish sauce (Satomi et al. 1997). *Tetragenococcus osmophilus* was isolated from a juice containing sugar at a high concentration (Juste et al. 2012). But HLAB that can grow in highly salted cultures containing aquatic plants such as seaweed and microalgae are not known. The author conducted two kinds of preliminary studies. Firstly, laver-cultures were prepared without autoclaving with salt concentrations at 10% and 15% wt/wt, and left for two months. However, growth of HLAB was not observed in any of the salt enhanced cultures. Secondly, the authors obtained four strains of HLAB (*Tetragenococcus halophilus* NBRC5888[T], NBRC2015, NBRC20245, and *Tetragenococcus muriaticus* NBRC10006) from a culture collection and attempted to conduct fermentation with the highly salted (15%) laver cultures using these microorganisms as a starter culture. However, no growth of HLAB was observed after two months' incubation. After these unpublished observations, the author attempted to isolate HLAB from environmental samples. Table 4 (Uchida et al. 2014) shows the results of screening tests for HALB isolation. Twenty five samples were taken from environments including six sands and soils from a salt farm, ten seaweeds, six fish fermentation products, one pelagic water sample taken from 321 m depth, one summer snow sample taken from 2450 m high mountains, one microalgae flock sample from the surface of a pond located at the 2450 m high mountains. The isolation culture was prepared with laver and salted at 15% wt/wt and autoclaved for 20 min. at 100°C. This heat treatment was incomplete for sterilizing spore-forming bacteria, and microbial growth (at 10^5 CFU/l level) was observed for three of four control cultures prepared without inoculation of environmental samples after 9 months' incubation at 23°C. But the microorganisms grown in these control cultures were not acid producing bacteria, i.e., not HLAB. The cultures inoculated with the three authentic strains of *T. halophilus* did not show any growth of HLAB. Growth of LAB (i.e., tentatively defined as acid producing on the 8% NaCl-added Plate Count Agar with BCP plates and catalase negative bacteria)

Table 4 Results of screening test for isolation of halophilic lactic acid bacteria able to grow in highly salted laver cultures (Uchida et al. 2014).

Tube No.	Sampling date	Locations	Sources	Microbial counts (CFU/g)							
				After 1 month				After 9 months			
				Total	Acid producing bacteria	Lactic acid bacteria	Identification based on 16S rDNA[*1]	Total	Acid producing bacteria	Lactic acid bacteria	Identification based on 16S rDNA
1	20110712	Ishikawa	Sand (From salt farm)	$<10^5$				7.2×10^5	$<10^3$	$<10^3$	
2	20110712	Ishikawa	Soil (From a bucket surface, salt farm)	$<10^5$				ca. 2×10^6	$<10^3$	$<10^3$	
3	20110712	Ishikawa	Soil (From the boiling house inside, salt farm)	$<10^5$				4×10^3	$<10^3$	$<10^3$	
4	20110712	Ishikawa	Soil (From the boiling house outside, salt farm)	$<10^5$				1.6×10^5	$<10^3$	$<10^3$	
5	20110712	Ishikawa	Soil (From a wood gear surface, salt farm)	$<10^5$				4.4×10^4	$<10^3$	$<10^3$	
6	20110712	Ishikawa	Sand (From salt farm)	$<10^5$				ca. 2×10^6	$<10^3$	$<10^3$	
7	20110712	Ishikawa	Seaweed (*Sargassum thunbergii*)	$<10^5$				ca. 2×10^6	$<10^3$	$<10^3$	
8	20110712	Ishikawa	Seaweed (*Sargassum* sp.)	$<10^5$				1.2×10^5	$<10^3$	$<10^3$	
9	20110712	Ishikawa	Fish sauce (Sardine, Lot H2106)	$<10^5$				$<10^3$			
10	20110712	Ishikawa	Fish sauce (Sardine, Lot H2207)	4.0×10^5	4.0×10^5	4.0×10^5	*T. muriaticus*	1.0×10^3	$<10^3$	$<10^3$	
11	20110712	Ishikawa	Fermented fish with rice bran (Mackerel, spoiled)	$<10^5$				1.6×10^6	1.0×10^6	$<10^4$	
12	20110712	Ishikawa	Fish sauce (Squid visceral)	$<10^5$				ca. 2×10^6	$<10^3$		
13	20110712	Ishikawa	Fermented fish with rice bran (Horse mackerel)	$<10^5$				3.4×10^7	3.4×10^7	3.0×10^7	*T. halophilus*
14	20110712	Ishikawa	Fish sauce (Mackerel and horse mackerel)	$<10^5$				3.2×10^7	3.2×10^7	3.2×10^7	*T. halophilus*
15	20110713	Toyama	Pelagic water (From 321m depth)	$<10^5$				1.7×10^8	1.7×10^8	1.7×10^8	*T. halophilus*
16	20110714	Toyama	Snow (From 2450m heigh mountains)	$<10^5$				1.9×10^8	1.9×10^8	1.9×10^8	*T. halophilus*
17	20110714	Toyama	Microalgal flocs (From the pond surface at 2450m heig	$<10^5$				2.5×10^8	2.5×10^8	2.4×10^8	*T. halophilus*
18	20110716	Hiroshima	Seaweed (*Gracilaria* sp., slim)	$<10^5$				2.2×10^5	2.2×10^5	2.2×10^5	*T. halophilus*
19	20110716	Hiroshima	Seaweed (*Gracilaria incurvata*)	$<10^5$				$<10^3$			
20	20110716	Hiroshima	Seaweed (*Gracilaria* sp., thick)	$<10^5$				$<10^3$			
21	20110716	Hiroshima	Seaweed (*Solieria pacifica*)	$<10^5$				$<10^3$			
22	20110716	Hiroshima	Seaweed (*Chondrus giganteus*)	$<10^5$				$<10^3$			
23	20110716	Hiroshima	Seaweed (*Martensia denticulata*)	$<10^5$				$<10^3$			

Table 4 contd....

Table 4 contd.

Tube No.	Sampling date	Locations	Sources	After 1 month				After 9 months			
				Total	Acid producing bacteria	Lactic acid bacteria	Identification based on 16S rDNA[*1]	Total	Acid producing bacteria	Lactic acid bacteria	Identification based on 16S rDNA
24	20110716	Hiroshima	Seaweed (*Codium fragile*)	$<10^5$				$<10^3$			
25	20110716	Hiroshima	Seaweed (*Ecklonia kurome*)	$<10^5$				$<10^3$			
26	20110706	Control	(No sample inoculation)	$<10^5$				1.5×10^5	$<10^3$		
27	20110706	Control	(No sample inoculation)	$<10^5$				2.6×10^5	$<10^3$		
28	20110706	Control	(No sample inoculation)	$<10^5$				1.2×10^5	$<10^3$		
29	20110706	Control	(No sample inoculation)	$<10^5$				$<10^3$			
30	20110803	Inoculated with *T. halophilus* NBRC5888[T]		$<10^5$				5.0×10^4	$<10^3$		
31	20110803	Inoculated with *T. halophilus* NBRC2015		$<10^5$				ca. 2×10^6	$<10^3$		
32	20110803	Inoculated with *T. halophilus* NBRC20245		$<10^5$				4.2×10^5	$<10^3$		

Microbial counts (CFU/g)

Cultures used for isolation: Fifty gram laver was mixed with 3.5g protease, 7.5g NaCl, and 371.5g distilled water, autoclaved for 20 min. at 100°C, and dispensed 5ml each to sterile screw-capped tubes. Environmental samples, ca. 0.5g, were inoculated and incubated at 23°C. Micorbial number was counted with 8% NaCl-added Plate Count Agar with BCP "Nissui" plates (8%BCP). Micorooganisms forming yellow colored colony were regarded as acid producing bacteria. Yellow colored and catalase negative colonies were regarded as lactic acid bacteria. For identifying the dominant bacteria at species level, five colonies each were isolated at random and determined for the partial (ca. 800 base pairs) nucleic acid sequence of 16S rRNA gene. All the 30 isolates obtained from the nine months cultures were allotted to *T. halophilus*.

was observed after 9 months for six cultures prepared with the inoculation of two fish fermentation products, one pelagic water, one snow, one microalgal flock, one red algae. Five colonies each were isolated at random from the six cultures and all of the isolates (i.e., 30 isolates) were allotted to *T. halophilus* based on phylogenic study of 16S rRNA gene (Uchida et al. 2014). Two representative strains each obtained from the pelagic water and snow cultures (strains DSW01-02 and SNW01-02, respectively) were inoculated at three different inoculation sizes to newly prepared laver cultures salted at 15% concentration and autoclaved for 20 min at 121°C. Significant growth was observed for the cases of the four isolates, i.e., strains DSW01-02 and SNW01-02, while a decrease in the number of cells was observed for the case of the three authentic *T. halophilus* strains (Uchida et al. 2014). The *T. halophilus* strains isolated in this study are expected to be suitable in seaweed fermentation such as laver sauce manufacturing.

Optimum conditions for the fermentative production of algal single cell products

It was observed that cellulase activity can easily fragment algal frond tissues, and that a large number of fragments of single cell detritus (SCD: algal detrital products originating from one cell unit) were produced in the case of *U. pinnatifida* (Fig. 5, Uchida et al. 2004a). The optimum conditions for fermentation were studied from two perspectives: the production efficiency of SCD, and the dominance of lactic acid bacteria (LAB).

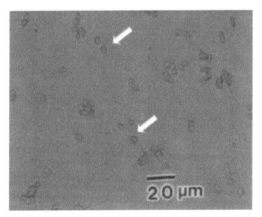

Figure 5. Microscopic observation of single cell detritus (SCD) prepared from *Undaria pinnatifida* (arrows, photo after 8 days of incubation) (Uchida et al. 2004a).

Color image of this figure appears in the color plate section at the end of the book.

Table 5   Results of growth test for isolated and reference strains of T. ha bph i&us in highly salted laver cultures (Uchida et al. 2014).

Tube No.	Specimen	Strains	Isolation sources	Laver culture (ml)	Inoculum (ml)	Colony counts CFU/ml	
						Initial	After 27days
1 Blank (No addition)				10	0	0	$<10^2$
2				10	0	0	$<10^2$
3				10	0	0	$<10^2$
4	T. halophilus	NBRC5888^T	(Type strain)	10	0.001	1.4×10^3	$<10^2$
5				10	0.01	1.4×10^4	$<10^2$
6				10	0.1	1.4×10^5	2.0×10^3
7	T. halophilus	NBRC2015	Thai fish sauce	10	0.001	1.5×10^4	1.5×10^2
8				10	0.01	1.5×10^5	6×10^2
9				10	0.1	1.5×10^6	$<10^2$
10	T. halophilus	NBRC20245	Soy sauce (M orom i)	10	0.001	6.0×10^2	$<10^2$
11				10	0.01	6.0×10^3	1.2×10^3
12				10	0.1	6.0×10^4	2.8×10^4
13	T. halophilus	DSW01	Laver incubated with pelagic water	10	0.001	2.5×10^2	1.7×10^4
14				10	0.01	2.5×10^3	7.7×10^5
15				10	0.1	2.5×10^4	1.2×10^7
16	T. halophilus	DSW02	Laver incubated with pelagic water	10	0.001	6.5×10^2	2.3×10^3
17				10	0.01	6.5×10^3	5.5×10^4
18				10	0.1	6.5×10^4	7.2×10^5
19	T. halophilus	SNW01	Laver incubated with mountain snow	10	0.001	3.5×10^3	5.6×10^6
20				10	0.01	3.5×10^4	1.8×10^7
21				10	0.1	3.5×10^5	3.6×10^7
22	T. halophilus	SNW02	Laver incubated with mountain snow	10	0.001	6.0×10^3	8.7×10^6
23				10	0.01	6.0×10^4	5.9×10^6
24				10	0.1	6.0×10^5	3.7×10^7

Each cultured cell suspension (conc. at OD_{660nm}=1.0) was added to sterilized (autoclaved for 20 min at 121°C) laver cultures at 0.001ml, 0.01ml, and 0.1ml, and incubated at 23°C. Initial number of the inolulated cells were calculated from the results of colony counting for each cell suspension. Colony counting was conducted with 8%NaCl-GAM plates.

As for the salt concentration, cultures have a high viscosity and difficulty in absorbing moisture homogeneously, resulting in the formation of aggregations of algal particles at low salt range (Table 6, Uchida and Murata 2002). It is therefore relatively difficult to prepare a homogeneous algal suspension at a salt concentration of less than 2.5% (wt/vol). Furthermore, contaminant bacteria often grow and spoil cultures that are prepared without salt. On the other hand, *Lactobacillus* group bacteria are not halotolerant and their growth will be significantly restricted under a salt concentration exceeding 5%. Furthermore, cellulase activity will be markedly restricted under a salt concentration of 5% or more (unpublished). The total volume of the solid part of *Undaria pinnatifida* frond suspension cultures decreased to 100%, 98%, 90%, 71%, 60%, and 39% after being left for six days when the salt concentrations of the cultures were 0%, 1%, 2.5%, 3.5%, 5%, and 10%, respectively. Based on these observations, the suitable salt concentration in the culture water used for SCD preparation was in the range of 2.5 to 3.5%.

As for the incubation temperature, a temperature range of 5–50°C can be used for SCD production. The maximum production rate of SCD is achieved at 20°C, and remains almost constant in the range of 20–50°C

Table 6 Effect of NaCl concentration of the culture medium for the productivity of the Undaria-SCD diet (Uchida and Murata 2002).

Characteristics of the Undaria-suspension	NaCl% (wt/vol) of culture solution					
	0	1	2.5	3.5	5	10
Aggregation	++	++	+	-	-	-
Viscosity	++	++	+	-	-	-
Suspension stability (%)	100	98	90	71	60	39

180ml of NaCl solutions prepared at the above concentrations were mixed with 10.0g Undaria-materials, resulting in 200ml total volume (Final concentration of NaCl were 0, 0.9, 2.25, 3.15, 4.5, and 9.0% wt/vol, respectively). Suspension stability is an index to represent the dehydration and shrinking of the frond tissues, which is expressed as the per cent proportion of solid per total contents on volume basis after being left for 6 days.

(Fig. 6a, Uchida and Murata 2002). The final number of SCD products is almost the same in the range of 5–50°C, provided that incubation lasts more than two weeks.

As for incubation time, a period of six to 14 days is sufficient to produce the maximum number of SCD products. LAB will achieve growth above the 10^8 CFU/ml level within five days, and lactic acid production and pH value will become almost constant after six to 14 days (Uchida and Murata 2002).

As for the cellulase concentration, the addition of more than 1% was necessary to avoid spoiling and obtain a fermented culture if the LAB starter was not used (Table 3a). Commercial products of cellulase are

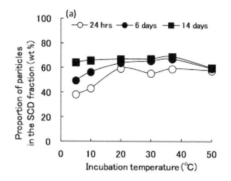

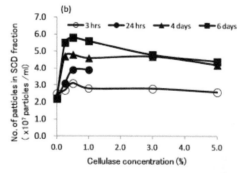

Figure 6. Relation between single cell detritus (SCD) production and incubation temperature (a) or cellulase concentration (b) in *Undaria* fermentation culture (Uchida and Murata 2002). Ten grams of dried particles of *U. pinnatifida* (Wakamidori, <74 μm, Riken Co.) were dispensed into each 500-ml bottle containing 180 ml of autoclaved distilled water with different concentrations of cellulase (R-10, Yakult Honsha Co. Ltd). The culture bottles were incubated at 20°C. The number of detrital particles of *Undaria* was counted using a Coulter Multisizer (Coulter Electronics Ltd.). The algal particles in fractions of 5.8–11.5 μm in diameter were tentatively regarded as SCD products. The weight distribution was calculated from the distribution of detrital particles, based on the hypothesis that the detrital particles were of a spherical form with a specific gravity of 1.0. All cultures were prepared in duplicate and the average values of data are shown.

expensive and saving the use thereof is important in terms of cost. The use of cellulase at 0.1% concentration was observed as being effective enough for the fermentation and production of SCD from *U. pinnatifida* fronds in a 10-L-scale culture (Uchida et al. 2004a). Under the optimum conditions of a culture containing 5% *U. pinnatifida* (on a dry basis), 3.5% NaCl, and 0.5% cellulase with additional LAB starter, SCD was produced at a rate of 5.8 $\times 10^7$ cells/ml after six days of incubation at 20°C (Fig. 6b). The produced SCD was stable without microbial spoiling and could be preserved for more than 18 months at room temperature although a slight decrease of the particle number was observed along with the storage time (Fig. 7, Uchida

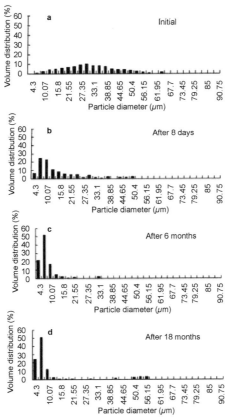

Figure 7. Time course change of volume-size distribution of *U. pinnatifida* particles in the fermented culture (a-d) (Uchida et al. 2004a). One kg of commercially available dried particles of *U. pinnatifida* (particle size <74 mm) was dispensed into bottles containing 9 l of autoclaved 3.3% wt/vol NaCl solution and 10 g of cellulase (R-10), and then mixed well. The culture was inoculated with a microbial mixture composed of *L. brevis* strain B5201 (FERM BP-7301, 5.0 $\times 10^{10}$ CFU), *Debaryomyces hansenii* var. *hansenii* strain Y5201 (FERM BP-7302, 5.5 $\times 10^7$ CFU), and *Candida* sp. strain Y5206 (FERM BP-7303, 1.2 $\times 10^8$ CFU). The culture bottle was incubated and left for 18 months at 20°C.

et al. 2004a). It is reported that mass preparation of SCD is possible also from *Ulva* spp. by using macerozyme besides cellulase (Uchida 2005). Mass preparation of the SCD products is possible usually from thin and fragile fronds, and commercial development of effective enzymes is necessary to obtain the SCD products from a wider range of seaweed species.

Lactic acid fermentation of microalgae

High potentiality of microalgae as a raw material for fermentation is well noted (Harun et al. 2010; Jones and Mayfield 2012). Lactic acid fermentation was demonstrated on microalgae (i.e., *Chlorella* sp., *Tetracelmis* sp., *Pavlova lutheri*, *Chaetoceros* sp., *Nannochloropsis* sp.) as well as seaweeds in a preliminary study (Uchida 2007). The pretreatment of microalgae (as well as seaweeds) with cellulase was effective in inducing fermentation. Lactic acid was produced in the range of 1.5–5.4 g/l in the case of microalgae, while 3.5 g/l was produced in the case of *U. pinnatifida* when fermentation cultures were prepared with the composition of 2 g *Undaria* powder, 0.2 g cellulase in 40 ml 3.5% NaCl solution. The use of macerozyme and lactase in addition to cellulase increased the production of lactic acid from 5.4 g/l to 9.6 g/l (*Chlorella* sp.), and from 1.5 g/l to 4.3 g/l (*Tetracelmis* sp.). However, microscopic observation showed quite limited cell wall decomposition in all cases. Suitable enzyme products must therefore be developed to obtain products with the cell wall eliminated or otherwise well-dissolved products.

Housing units for conducting algal fermentation

For conducting algal fermentation, plastic tanks or buckets with lids can be used for culture. Plastic bags such as polypropylene bags are cheaper and more convenient for dispersing the aggregate of seaweed and preparing a homogeneous frond suspension. Two 200-L-scale algal fermentation tanks named the 'marine silos' were manufactured and set up along the waterside of Lake Hamana (Shirasu-cho, Hamamatsu, Shizuoka, Japan) in 2003 (Uchida 2010). These marine silos are equipped with a temperature control and mixing facility, and 200-L-scale fermentation was demonstrated in treatment using *Ulva* spp. fronds collected from Lake Hamana.

5 Application of Fermented Products in Foods and Food-related Industry

Laver sauce

Many kinds of fermented foods have been developed to date. Most fermented foods are made from such terrestrial (agricultural and stockbreeding) materials as rice, soybean, barley, vegetables, and milk. Some products are produced from aquatic biomaterials, but no products have yet been produced from aquatic plant materials (i.e., algae) (Fig. 1). The development of the lactic acid fermentation methods described above open the possibility of producing fermented foods from algae. Seaweed sauce is one possible product to be developed. A major difficulty in developing seaweed sauce having a commercial value is the shortage of amino acid compounds contained in the supernatant of fermented products prepared from seaweed. Typical *Koikuchi* soy sauce contains more than 1.2 g/100 g of total nitrogen (Ren et al. 1993). In contrast, the supernatant of fermented products prepared from seaweed, such as *U. pinnatifida* and *Ulva* spp., only contains 0.1–0.3 g/100 g of total nitrogen (unpublished), which can be explained by the low protein content of these seaweeds: Fresh seaweed has a high moisture content (contains 87.5% on an average, Table 1). And on an average, seaweeds contain 15.3% of protein on a dry weight basis (Table 1), while soy bean contains less moisture and ca. 40% of protein on a dry weight basis (Subdivision on Resources, The Council for Science and Technology, Ministry of Education, Culture, Sports, Science, and Technology, Japan 2002). But *Susabinori* seaweed (*Porphyra yezoensis*, Rhodophyta) or laver contains an exceptionally high quantity of protein exceeding 50% (Subdivision on Resources, The Council for Science and Technology, Ministry of Education, Culture, Sports, Science, and Technology, Japan 2002). The supernatant of control cultures prepared from laver initially contained 0.3 g/100 g of total nitrogen, but contained 1.0 g/100 g of total nitrogen after protease treatment with the fronds (Table 7, unpublished). By adding surplus 50% quantity of laver to the culture after protease treatment, the total nitrogen value increased to 1.3 g/100 g. The total nitrogen value further increased to 1.7 g/100 ml by changing the housing unit from a plastic tank to a plastic bag, which made it easier to mix the laver contents by hand during the fermentation and promoted the decomposition of laver. The total nitrogen value of 1.3–1.7 g/100 g is comparable to soy sauce products, and their taste was dense and regarded as a level of products having commercial value. Free amino acid compositions contained in the supernatant of a laver sauce obtained with protease and *Nidan-jikomi* treatment are shown in Table 8 (unpublished). The total amino acid concentration of the laver sauce was 4536 mg/100 g, while values of soy sauces and fish sauces were 1569–6569 and 3054–5180 mg/100 g, respectively, supporting the idea that the quality of the laver sauce is the same level as conventional commercial products of soy sauce and fish sauce. The glutamic acid content of the laver sauce and soy sauce (*Koikuchi*) was 580 and 1238–1733 mg/100 g, respectively, suggesting that *Umami*-taste is still stronger in the soy sauce (*Koikuchi*) products. Tsuchiya et al. (2007) reported gamma–aminobutyric acid (GABA)

Table 7 Increase of total nitrogen contents in laver sauce after manufacturing treatments (unpublished) and reference products.

Treatments	Experimental products (Laver sauce)				Reference products	
	No. 1 (Control)	No. 2	No. 3	No. 4	Soy sauce (Koikuchi)	Fish sauce (Japanese products)
Protease treatment	-	o	o	o	-	-
Nidan-jikomi (surplus 50% addition of laver)	-	-	o	o	-	-
Hausing units used for fermentation	Plastic tank	Plastic tank	Plastic tank	Plastic bag	-	-
Total nitrogen (%)	0.3	1.0	1.3	1.7	1.5	1.3 – 2.1

Laver sauces No. 1-No. 3 were prepared in triplicate with the following recipe: No. 1; 6.5 g laver, 1.5 g NaCl, 5 mg lactic acid bacteria starter (*L. casei*, 5x10^6 CFU) and adusted the total weight to 48g with distilled water. No. 2; The same as No. 1 except that 280 mg protease P (Amano Enzyme Inc.) was added instead of the relevant quantity of water, No. 3; The same as No. 2 except that further 3.25 g laver was added on Day 5 of incubation. The incubation was conducted for 21 days at 23°C   No. 4 was prepared with the recipe of 1.3 kg laver, 240 g NaCl, 56 g protease P, 0.8 g *L. casei* starter, and 6.8 kg distilled water using a 20-L-volume plastic bag. Another 1.2 kg laver and 1.24 kg NaCl were added with 10mg alkali-hemicellulase (Wako) at Day 5 of incubation. No. 4 was incubated for 70 days at 23°C. The supernatant of laver sauces was collected and analyzed for total nitrogen content by Kjeldahl method. Data of the reference products were cited from Ren et al. (1993).

Table 8 Comparison of free amino acid composition in laver sauce, soy sauces, and fish sauces (unpublished).

| | Laver sauce | Amino acids (mg/100g) | | | | | |
| | | Soy sauce | | | Fish sauce | | |
		Koikuchi	Usukuchi	Shiro	Squid	Sardine	Anchovy
Isoleucine	260	306-368	191	58	242	327	96
Leucine	520	454-565	248	100	244	408	107
Lysine	380	371-431	276	115	581	676	451
Methionine	140	94-114	82	24	246	307	132
Cystine	16	0-5	6	9	35	48	23
Phenylalanine	210	122-214	148	83	170	224	31
Tyrosine	38	66-81	76	64	252	160	77
Threonine	290	252-293	160	53	227	243	243
Tryptophan	47	Not tested	Not tested	Not tested	Not tested	Not tested	Not tested
Valine	390	373-427	205	70	372	425	239
Arginine	28	193-239	260	72	52	53	16
Histidine	87	100-132	129	25	145	435	306
Alanine	650	415-649	169	80	680	432	275
Aspartic acid	350	335-420	215	50	211	268	248
Glutamic acid	580	1238-1733	420	515	820	411	367
Glycine	200	203-258	98	36	307	210	143
Proline	130	346-430	201	136	451	174	106
Serine	220	361-417	192	79	145	84	194
Total	4536	5728-6569	3076	1569	5180	4885	3054

The laver sauce was prepared with the recipe of 1.8kg laver, 84g protease P (Amano Enzyme Inc.), mannanase RBG (Yakult Honsha Co. Ltd.), 360g NaCl, 10.13 kg distilled water, and 5ml starter solution (*T. halophilus* NBRC2015). Surplus 900 g laver was added at 24h of incubation (i.e. *Nidan-jikomi* treatment) and another 1kg of NaCl was added on Day 8 to the culture, then, further incubated for 22 days at 23°C. The supernatant was collected by centrifuge (12000rpm, 20min) and filtration (GF/C) and measured for the free amino acid composition. Data of soy sauces and fish sauces were cited from Ren et al. (1993).

production from discolored laver by lactic acid fermentation, which has a physiological function of suppressing hypertension.

Marine silage as fisheries diet

Marine silage (MS) is a new dietary item prepared by decomposing seaweed to a cellular unit (Uchida 1997; Uchida et al. 1997a,b) and by performing lactic acid fermentation (Fig. 8, Uchida 2007; Uchida and Miyoshi 2008). A 10-L-scale preparation and the long-term preservation of MS was demonstrated using *U. pinnatifida* as a substrate (Fig. 7, Uchida et al. 2004a). The mass preparation of MS is readily accomplishable and it was shown that the prepared MS can be preserved for a long time (more than 18 months). In addition, two separate feeding trials were conducted to demonstrate the dietary value of MS for the Japanese pearl oyster (*Pinctada fucata martensii*) (Uchida et al. 2004a). In the first trial, pearl oysters grew significantly when fed MS at a rate of 3×10^4 cells/ml per day. The shell growth and survival rates (69 ± 11 µm/day; mean $\pm$ SE and 100%, respectively, $n = 30$) of the pearl oysters were higher than those of the unfed control oysters (-26 ± 8 µm/day and 100%, $n = 30$), while oysters fed *Chaetoceros calcitrans* (known as one of the most preferred microalgal diets for bivalves and used as a positive

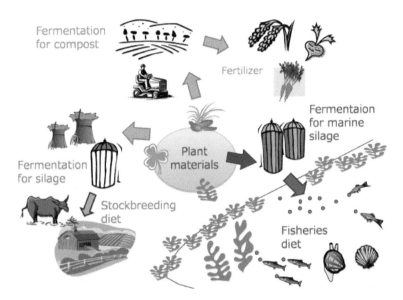

Figure 8. Use of fermentation technologies on plant materials for foods and diet production (Uchida and Miyoshi 2010). Fermentation is rarely or never used in the fisheries industry, and Uchida proposed a conceptual diet of marine silage.

Color image of this figure appears in the color plate section at the end of the book.

control diet) grew 205 ± 12 μm/day ($n = 29$) and 96.7% survived. Proximate analysis of MS suggested a shortage of such nutritional elements as lipid content. The second rearing trial demonstrated remarkable shell growth when the pearl oysters were fed MS supplemented with a small quantity of *C. calcitrans*, as shown in Table 9. The dietary effect of MS prepared from *U. pinnatifida* was also demonstrated on rotifers (*Brachionus* sp., a common hatchery diet for fish rearing) (Technical Research Association for New Food Creation 2005). But MS prepared from *Ulva* spp. showed a negative effect on the growth of short necked clam (*Ruditapes philippinarum*) (unpublished). Small particles of *Ulva* prepared without fermentation also showed a negative effect on short necked clam growth, suggesting that *Ulva* fronds contain growth inhibitory compounds, and not a compound artificially produced through fermentation treatment.

The diet supplemented with MS prepared from *Ecklonia* sp. (Phaeophyta) at 10% wt/wt were fed to red sea bream challenged by an iridovirus. The survival rate of the fish fed a diet containing MS was found to be significantly ($P<0.05$) higher than that of the control group (Fig. 9, Technical research association for new food creation 2005).

Table 9 Results of feeding trials of MS and MSNF with young pearl oyster, *Pinctada fucata martensii* (Uchida et al. 2004a).

Treatments	Feeding conditions (/ml/day)	Growth rate (μmday⁻¹)	Survival (%)
Unfed	No feed	-10 ± 14 [de]	53.3
Chaetoceros (C)	*C. calcitrans* 3×10^4 cells	168 ± 33 [a]	66.7
1/10C	*C. calcitrans* 3×10^3 cells	-7 ± 10 [cde]	66.7
MS	SCD prepared from *U. pinnatifida* (=MS) 2×10^4 particles	23 ± 13 [bd]	53.3
MS+1/10C	MS 2×10^4 particles+*C. calcitrans* 3×10^3 cells	125 ± 13 [a]	80.0
x2.5 MS	MS 5×10^4 particles	47 ± 17 [bc]	66.7
MSNF	SCD prepared from *U. pinnatifida* without fermentation (MSNF) 2×10^4 particles	18 ± 16 [be]	66.7
MSNF+1/10C	MSNF 2×10^4 particles+*C. calcitrans* 3×10^3 cells	71 ± 14 [b]	93.3
x2.5 MSNF	MSNF 5×10^4 particles	23 ± 18 [bd]	53.7

Results are based on single trial and shown as the average±SEM. Treatments with common superscripts are not significantly different (Duncan's multiple arrangement test, $P<0.05$)

Diet for stockbreeding

Seaweeds have been utilized as a stockbreeding diet (Ohno 2004). One reason to use seaweed is to supply breeding animals with valuable minerals such as iodine (Ohno 2004). Usually, seaweed is added to the diet in the form of grain or powder, and no fermented product of seaweed has yet been used. Fermented products prepared from *Porphyra yezoensis* and *Undaria pinnatifida* were used to supplement diets of both broiler and egg-laying chickens, and the dietary effects were examined. The addition of *P. yezoensis*

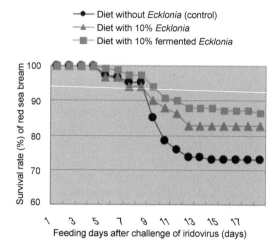

Figure 9. Results of the feeding test for red sea bream (Technical research association for new food creation 2005). The fish were challenged by iridovirus and then fed a control diet, a diet containing *Ecklonia* sp. at 10% wt/wt, and a diet containing fermented *Ecklonia* sp. at 10% wt/wt.

Color image of this figure appears in the color plate section at the end of the book.

did not show any significant difference (unpublished). In contrast, the addition of *U. pinnatifida* showed some significant effects, such as growth promotion, increasing the moisture-retention ability of meat, and the lipid metabolizing activity. But all of these preferable effects were not observed every time, and more repeated testing is needed. Although the usefulness of seaweed as a stockbreeding diet is well accepted, whether the treatment by fermentation of seaweed is superior remains open for future study.

Fertilizer

Many kinds of seaweed have been utilized as fertilizer, such as *Enteromorpha* spp., *Ulva* spp. (Chlorophyta), *Gracilaria chilensis* (Rhodophyta), *Ascophyllum nodosum*, *Ecklonia maxima*, *Laminaria schinzii*, *Durvillaea potatorum*, and *Sargassum* spp. (Phaeophyta) (Zemke-White and Ohno 1999). Most algal fertilizer products are produced from seaweeds through a simple drying or extraction treatment (Ohno 2004). Products called compost are produced by a kind of fermentation treatment. For example, compost was prepared from *Ulva rigida*, collected in the Venice Lagoon, after three weeks of fermentation (Cuomo et al. 1995). Such treatment using fermentation is expected to promote the uptake of nutrients through the degradation of algal components, and prove useful in avoiding excessive heat emission in soils and subsequent plant damage. We have conducted some preliminary studies on the fertilizing effects of seaweed products prepared by lactic acid fermentation. For example, the supernatant of fermented products prepared from *Ulva* spp. was given to a Japanese green tea farm. The farmer commented that the spines of the green tea leaves became sharper and more developed. However, this observation was not based on statistical analysis. Although the usefulness of seaweed as a fertilizer is well accepted, whether the treatment by fermentation of seaweed is superior remains open for future study.

6 Ethanol Fermentation

Current activity

Technology for the ethanol fermentation of seaweeds began development in the 2000s with the aim of producing a biofuel (Uchida 2009; Roesijadi et al. 2010; Lee et al. 2011; Notoya 2011; Cormier 2012). Korean research groups are focusing on the utilization of red algae as a feedstock (Wie et al. 2009; Yoon et al. 2010). Red algae contain galactan as a major component (25–35%, Ohno 2004). The basic structure of galactan is a polymer of D-galactose and 3,6-L-anhydrogalactose. Therefore, the quantity of D-galactose can be easily obtained by acid or enzymatic treatment of red algae, and the D-galactose can be metabolized to ethanol at high efficiency (51%) as well as to D-glucose. Brown algae contain alginate (3–38%, JOIA 1984) as a major component, but microorganisms rarely convert the alginate to ethanol. Takeda et al. (2011) demonstrated that alginate can be metabolized to ethanol using genetically engineered bacteria (*Sphingomonas* sp.). Wargacki et al. (2012) developed another type of genetically engineered bacteria from *Escherichia coli* and demonstrated the direct production of ethanol from a brown algae frond. In those studies, ethanol concentration in the fermentation culture was observed at 13 g/l (= 16.5 ml/l, Takeda et al. 2011) and 44 ml/l (Wargacki et al. 2012). Yanagisawa et al. (2011) reported the production of ethanol at 55 ml/l from agar weed.

Ethanol from seagrass seeds

The author is focusing on the utilization of seagrass seeds as a raw material for ethanol fermentation. As mentioned at the beginning of this chapter, average moisture content is 87.5%, and carbohydrate content is 57.1% on a dry weight basis in the whole body of seaweeds (Table 1). On the hypothesis that seaweed carbohydrates can be converted to ethanol (as well as glucose) at the efficiency of 51%, the expected ethanol production from seaweeds is 36.4 g/kg fresh fronds (given by (1–0.875) x 0.571 x 0.51 x 1000), or 46.2 ml/l (when 1 kg fronds have 1 liter volume). This value is the highest estimate and other studies estimate 80 g (Goh and Lee 2010) to 120 g ethanol/kg dry seaweeds (Morchio and Cáceres 2009), which equates to 10–15 g ethanol/kg fresh fronds. In contrast, seeds of *Zostera marina* (seagrass) contain lower moisture (62% on a wet weight basis), higher carbohydrate (83.5% on a dry weight basis, Table 10), or 48.3% starch (on a dry weight basis) (Uchida et al. submitted). This high sugar content is comparable to cereals (82.5–91.2% carbohydrate, Table 10). The seagrass seeds were saccharified in a culture condition containing 33% dry matter at start with glucoamylase treatment for 96 h at 50°C, and sugar juice containing glucose at 103.4 g/l level was obtained (Table 11, in preparation). The supernatant of the newly prepared saccharified juice (contained 92 g/l of glucose) was fermented with yeast *Saccharomyces cerevisiae* for 15 days (i.e., monographic double-fermentation), and production of ethanol was observed at 61.7–65 g/l level. Furthermore, both the supernatant and solid part of the saccharified

Table 10   Proximate contents among seagrass seeds, cereals, and seaweeds (unpublished data).

Potential feedstocks	Proximate contents (% on a dry basis)			
	Protein	Carbohydrate	Lipid	Ash
Zostera marina (Seed, seagrass)	10.1	83.5	1.3	5.1
Rice (Polished)[*]	8.0	87.3	3.2	1.4
Rice (Unpolished)[*]	7.2	91.2	1.1	0.5
Wheat flour[*]	12.1	82.5	3.5	1.8
Corn (Maize kernels)[*]	10.1	82.6	5.8	15.2
Zostera marina (Whole, seagrass)	15.1	57.4	1.4	26.1
Water hyacinth (Freshwater plant)	9.6	76.7	0.7	13.0
Ulva sp. (Green seaweed)	17.0	58.3	0.3	24.5
Gracilaria incurvata (Red seaweed)	18.9	56.6	0.3	24.1
Undaria pinnatifida (Brown seaweed)	18.9	50.3	1.4	29.3

Data of potential feedstocks with asterisks are cited from the literature (Subdivision on Resources, The Council for Science and Technology, Ministry of Education, Culture, Sports, Science, and Technology, Japan 2002).

Table 11   Production of glucose from *Zostera marina* seeds during the saccharification treatment (unpublished data).

Enzymes	Glucose conc. in the culture supernatant (g/l)											
	Initial		After 3h		After 6h		After 9h		After 24h		After 96h	
	Mean	SD	Mean	SD	Mean	SD	Mean	SD	Mean	SD	Mean	SD
Control (No enzyme)	1.3	0.3	4.3	0.4	5.8	0.6	7.0	0.5	3.9	2.0	0.8	0.8
Cellulase	1.3	0.3	6.2	0.1	9.5	1.2	12.3	0.6	7.7	1.6	4.5	5.4
Glucoamylase	8.0	0.5	34.8	0.1	58.0	3.2	69.0	6.0	88.0	2.2	103.4	10.6

Seagrass seeds (milled and dried) 2.0g, distilled water 4g, and enzyme 0.06g were dispensed to screw-capped glass tubes 16mm-diameter) and incubated at 50°C with 120rpm shake. Cellulase 12S (Yakult Pharmaceutical CO., Ltd) and Gluczyme AF6 (Amano Enzyme Inc.) was tested for saccharification. Data is based on duplicated test. Glucose was measured by F-kit.

juice (containing 87.3 g/l of glucose) was fermented with yeast for 15 days (i.e., parallel double-fermentation), and production of ethanol was observed at 116.3–119.3 g/l, which equates to 147–151 ml/l (Table 12, in preparation). This data is based on duplicated tanks-fermentation, and the highest value of ethanol concentration was marked at 130.4 g/l (i.e., 165 ml/l). This ethanol concentration is much higher than the above previous records (16.5–44 ml/l) obtained from seaweed feedstock using genetically engineered microbes, and comparable to that of Japanese *Sake* products. It is noteworthy that seagrass seeds are a possible cereal resource harvested from aquatic squares, and development of novel alcoholic beverage products may be possible from them in future.

7 Future Prospects of Seaweed Fermentation

Study on seaweed fermentation has just started and has a large potential in the future. For example, development of the ethanol fermentation of seaweeds will raise expectations for the acetic fermentation of seaweeds, because acetic acid is easily produced from ethanol by acetic acid bacteria (Adachi et al. 2003; Younesi et al. 2005). Saccharification process is the primary step of fermentation. Seaweeds contain unique polysaccharides such as alginate and galactan, which are rarely contained in terrestrial plants (JOIA 1984; Uchida 2011). Therefore, the development of enzyme products suitable for seaweed saccharification is important for the further development of seaweed fermentation technology. Development of the *koji* fermentation system (i.e., culture mold utilizing the mold's enzymes for saccharification, Rokas 2009) and the parallel double-fermentation system (for conducting fermentation in parallel with the saccharification process) could prove to be key technologies for the further development of seaweed fermentation technology (Ichishima 1997; Uchida 2011). Fermentation technology for microalgae also remains open for future study. Microalgae are richer in proteins and lipids than seaweeds (Moura Jr. et al. 2007; Mayur et al. 2011), and have a high potential for obtaining economically valuable products in terms of foods, diets, fertilizers, bio-plastics, and the energy industry.

Fermentation can be conducted without supplying electric power and therefore can be considered an eco-friendly processing system. In particular, many Asian countries have traditional food cultures based on fermentation using agricultural and aquatic (fish) products. The author believes that there is significant potential for an algal fermentation industry being expanded in the future (Uchida 2010).

Table 12 Results of ethanol fermentation tests of Z. *marina* seeds (in preparation).

Yeast strains	Fermentation methods	Initial		After five days				After 15 days			
		Glucose (g/l)	Galactose (g/l)	EtOH (g/l)		Glucose (g/l)		EtOH (g/l)		Glucose (g/l)	
		Mean	Mean	Mean	SD	Mean	SD	Mean	SD	Mean	SD
Control (No yeast addition)	–	92.0	0.0	0.1	0.0	93.1	1.8	0.4	0.1	92.5	0.7
S. *cerevisiae* NBRC10217[T]	Monographic double-fermentation	92.0	0.0	39.4	3.3	25.7	1.9	61.7	1.9	2.1	0.6
S. *cerevisiae* K7	Monographic double-fermentation	92.0	0.0	57.6	1.7	0.0	0.0	65.0	0.7	0.0	0.0
S. *cerevisiae* NBRC10217[T]	Parallel double-fermentation	87.3	NT	12.5	1.1	96.5	2.0	116.3	7.9	4.7	5.0
S. *cerevisiae* K7	Parallel double-fermentation	87.3	NT	44.8	1.4	18.0	3.7	119.3	14.4	0.3	0.0

Saccharified solutions obtained from Z. *marina* seeds after glucoamylase treatment (24h, 50°C) were used as substrates for parallel double-fermentation. Only the supernatant of the saccharified solution was used for monographic double fermentation. Two kinds of yeast strains (*S. cerevisiae* NBRC10217[T] and K7) were added and cultured for five days at 37°C, and further 10 days at 23°C. The data is based on duplicated test tanks. NT: not measured. The EtOH concentration of 61.7, 65.0, 116.3, and 119.3 g/l are relevant to 78.1, 82.3, 147.2, and 151.0 ml/l, respectively.

Acknowledgements

This study was partially funded by the Pioneer *tokubetsu kenkyu* program of the Agriculture, Forestry, and Fisheries Research Council, and by the *Suisan biomass no sigen-ka gijutsu kaihatsu jigyo* program (Development of conversion technology for utilization of fisheries biomass) of the Fisheries Agency. Part of the study on lactic acid bacteria was conducted in cooperation with Yakult Honsha Co., Ltd. and part of the study on fisheries diet was conducted in cooperation with Nippon Suisan Kaisha Ltd. The marine silos were developed in cooperation with Fuyo Ocean Development & Engineering Co., Ltd., EBARA JITSUGYO Co., Ltd., and Mr. Yoshimura (Hamanako Aosa Riyo Kyogikai). The study on the dietary effects for broiler chicken was conducted in cooperation with the Okayama Prefectural Research Institute for Daily and Stockbreeding. The author is grateful to Dr. Miyoshi for the substantial contribution in the laver sauce experiments.

References

Adachi, O., D. Moonmangmee, H. Toyama, M. Yamada, E. Shinagawa and K. Matsushita. 2003. New developments in oxidative fermentation. Appl. Microbiol. Biotechnol. 60(6): 643–53.

Aquaculture Associates. 1982. Energy from marine biomass. GRI Contract No. 5081-310-0458.

Chynoweth, D.P., C.E. Turick, J.M. Owens and M.W. Peck. 1993. Biochemical methane potential of biomass and waste feed stocks. Biomass and Bioenergy. 5: 95–111.

Cormier, Z. 2012. Biofuel from beneath the waves. Nature. doi:10.1038/nature.2012.9860.

Cuomo, V., A. Perretti, I. Palomba, A. Verde and A. Cuomo. 1995. Utilization of *Ulva rigida* biomass in the Venice Lagoon (Italy): biotransformation in compost. J. Appl. Phycol. 7: 479–485.

Fukushi, A. 1988. Seasonal variations of compositions in different parts of cultivated Japanese kelp (*Laminaria japonica*). Scientific reports of Hokkaido Fisheries Experimental Station. 31: 55–61.

Goh, C.S. and K.T. Lee. 2010. A visionary and conceptual macroalgae-based third-generation bioethanol (TGB) biorefinery in Sabah, Malaysia as an underlay for renewable and sustainable development. Renewable and Sustainable Energy Reviews. 14: 842–848.

Harun, R., M.K. Danquah and G.M. Forde. 2010. Microalgal biomass as a fermentation feedstock for bioethanol production. J. Chem. Technol. Biotechnol. 85: 199–203.

Ichishima, E. 1997. *Hakko Shokuhin e no shotai*. Shokabo, Tokyo: 156 pp [In Japanese].

Ishige, N. 1993. Cultural aspects of fermented fish products in Asia. pp. 13–32. *In*: Lee, Cherl-Ho, K.H. Steinkraus and P.J.A. Reilly (eds.). Fish Fermentation Technology. United Nations University Press, Tokyo.

Itoh, H., H. Tachi and S. Kikuchi. 1993. Fish fermentation technology in Japan. pp. 177–186. *In*: Lee, Cherl-Ho, K.H. Steinkraus and P.J.A. Reilly (eds.). Fish Fermentation Technology. United Nations University Press, Tokyo.

Japan Ocean Industries Association (JOIA). 1984. Investigation into fuel production from marine biomass, the second report: "total system" (FY *Showa 58 nendo, Kaiyo biomass niyoru nenryo-yu seisan ni kansuru chosa, seika houkokusho, dai-2-bu;* "total system") [In Japanese].

Jones, C.S. and S.P. Mayfield. 2012. Algae biofuels: versatility for the future of bioenergy. Curr. Opin. Biotechnol. 23: 346–351.

Justé, A., S. Van Trappen, C. Verreth, I. Cleenwerck, P. Devos, B. Lievens and K.A. Willems. 2012. Characterization of *Tetragenococcus* strains from sugar thick juice reveals a novel species, *Tetragenococcus osmophilus* sp. nov., and divides *Tetragenococcus halophilus* into two subspecies, *T. halophilus* subsp. *halophilus* subsp. nov. and *T. halophilus* subsp. *flandriensis* subsp. nov. Int. J. Syst. Evol. Microbiol. 62: 129–137.

Lee, S.Y. Oh, D. Kim, D. Kwon, C. Lee and J. Lee. 2011. Converting carbohydrates extracted from marine algae into ethanol using various ethanolic *Escherichia coli* strains. Appl. Biochem. Biotechnol. 164: 878–888.

Marcello, V., A.P. de Ruiz Holgado, J.J. Sanchez, R.E. Trucco and G. Oliver. 1985. Isolation and characterization of *Pediococcus halophilus* from salted anchovies (*Engraulis anchoita*). Appl. Environ. Microbiol. 49(3): 664–666.

Mayur, M., P.R.S. Chutia, B.K. Konwara and R. Kataki. 2011. Microalgae *Chlorella* as a potential bio-energy feedstock. Applied Energy. 88: 3307–3312.

Miyoshi, T., M. Uchida, M. Kaneniwa and G. Yoshida. 2013. Collection and component analysis of aquatic plants with a cope for fermentative utilization. J. Fish. Technol. 6: 109–124.

Morchio, R. and C. Cáceres. 2009. Macroalgae Current State in Latin America. International Workshop on Sustainable Bioenergy from Algae. Berlin, Germany.

Moura, A.M., Jr., E.B. Neto, M.L. Koening and E.E. Leça. 2007. Chemical Composition of Three Microalgae Species for Possible Use in Mariculture. Brazilian Archives of Biology and Technology. 50: 461–467.

Notoya, M. 2011. Seaweed biofuel. CMC Publishing, Tokyo: 204 pp [In Japanese].

Ohno, M. 2004. *Yuyo-kaiso-shi*, Uchidarokakuho, Toyko, Japan: 575 pp [In Japanese].

Ren H., T. Hayashi, H. Endo and E. Watanabe. 1993. Characteristics of Chinese and Korean soy and fish sauces on the basis of their free amino acid composition. Nippon Suisan Gakkaishi. 59: 1929–1935.

Roesijadi, G., S.B. Jones, L.J. Snowden-Swan and Y. Zhu. 2010. Macroalgae as a biomass feedstock: A preliminary analysis, prepared for the U.S. Department of Energy under contract DE-AC05-76RL01830 by Pacific Northwest National Laboratory.

Rokas, A. 2009. The effect of domestication on the fungal proteome. Trends in Genetics. TIG. 25(2): 60–63.

Satomi, M., B. Kimura, M. Mizoi, T. Sato and T. Fujii. 1997. *Tetragenococcus muriaticus* sp. nov., a new moderately halophilic lactic acid bacterium isolated from fermented squid liver sauce. Int. J. Syst. Bacteriol. 47: 832–836.

Subdivision on Resources, The Council for Science and Technology, Ministry of Education, Culture, Sports, Science, and Technology, Japan. 2002. Standard tables of food composition in Japan-Fifth Revised. Kagawa Education Institute of Nutrition, Tokyo: 464 pp. [In Japanese].

Takeda, H., F. Yoneyama, S. Kawai, W. Hashimoto and K. Murata. 2011. Bioethanol production from marine biomass alginate by metabolically engineered bacteria. Energy Environ. Sci. 4: 2575–2581.

Technical Research Association for New Food Creation (New Food Creation *Gijutsu Kenkyu Kumiai*). 2005. Development of a new next-generation fermentation technology (*Jisedai-gata hakko gijutsu no kaihatsu*). In the annual report of a research project funded by the Ministry of Agriculture, Forestry and Fisheries (FY *Heisei 16 nendo kenkyu seika hokokusho*). Ministry of Agriculture, Forestry and Fisheries, Tokyo, Japan. pp. 81–91.

The Council for Science and Technology, Ministry of Education, Culture, Sports, Science, and Technology, Japan. 2004. Analytical manual for Standard Tables of Food Composition in Japan—Reference Materials for the Subdivision on Resources—Revised and Enlarged Edition. Kenpakusha, Tokyo: 176 pp.

Tsuchiya, K., S. Matsuda, G. Hirakawa, O. Shimada, R. Horio, T. Fujii, A. Ishida and M. Iwahara. 2007. GABA production from discolored laver by lactic acid fermentation and physiological function of fermented laver. Food Preservation Science. 33: 121–125.

Uchida, M. 1997. Microbial conversion of macroalgae into a detrital hatchery diet. JARQ. 33: 295–301.

Uchida, M., K. Nakata and M. Maeda. 1997a. Introduction of detrital food webs into an aquaculture system by supplying single cell algal detritus produced from *Laminaria japonica* as a hatchery diet for *Artemia* nauplii. Aquaculture. 154: 125–137.

Uchida, M., K. Nakata and M. Maeda. 1997b. Conversion of *Ulva* fronds to a hatchery diet for *Artemia* nauplii utilizing the degrading and attaching abilities of *Pseudoalteromonas espejiana*. J. Appl. Phycol. 9: 541–549.

Uchida, M. 2002. Fermentation of seaweeds. J. Jap. Lactic Acid Bac. 13: 92–113.

Uchida, M. and M. Murata. 2002. Fermentative preparation of single cell detritus from seaweed, *Undaria pinnatifida*, suitable as a replacement hatchery diet for unicellular algae. Aquaculture. 207: 345–357.

Uchida, M. and M. Murata. 2004. Isolation of a lactic acid bacterium and yeast consortium from a fermented material of *Ulva* spp. (Chlorophyta). Journal of Applied Microbiology. 97: 1297–1310.

Uchida, M., [The late]K. Numaguchi and M. Murata. 2004a. Mass preparation of marine silage from *Undaria pinnatifida* and its dietary effect for young pearl oyster. Fish. Sci. 70: 456–462.

Uchida, M., H. Amakasu, Y. Satoh and M. Murata. 2004b. Combinations of lactic acid bacteria and yeast suitable for preparation of marine silage. Fish. Sci. 70: 507–517.

Uchida, M. 2005. Studies on lactic acid fermentation of seaweed. Bull. Fish. Res. Agen. 14: 21–85.

Uchida, M., J. Ou, B. Chen, C. Yuan, X. Zhang, S. Chen, Y. Funatsu, K. Kawasaki, M. Satomi and Y. Fukuda. 2005. Effects of soy sauce *koji* and lactic acid bacteria on the fermentation of fish sauce from freshwater silver carp *Hypophthalmichthys molitrix*. Fish. Sci. 71: 422–430.

Uchida, M. 2007. Preparation of marine silage and its potential for industrial use. Proceedings of the Thirty-fourth U.S.-Japan Aquaculture Panel Symposium. pp. 51–56.

Uchida, M., M. Murata and F. Ishikawa. 2007. Lactic acid bacteria effective for regulating the growth of contaminant bacteria during the fermentation of *Undaria pinnatifida* (*Phaeophyta*). Fish. Sci. 73: 694–704.

Uchida, M. and T. Miyoshi. 2008. Development of a new dietary material from unutilized algal resources using fermentation skills. Bull. Fish. Res. Agen. 31: 25–29.

Uchida, M. 2009. Recent topics in studies on biofuel production from aquatic biomass in Japan. Nippon Suisan Gakkaishi. J. Jpn. Soc. Fish. Sci. 75: 1106–1108 [In Japanese].

Uchida, M. 2010. *Nihon hatsu kaiso-hakko-san'gyo no sohshutsu*. pp. 442–456. *In*: Tanaka, M.S. Kawai, N. Taniguchi and T. Sakata (eds.). *Suisan no 21 Seiki*. Kyoto University Press, Kyoto [In Japanese].

Uchida, M. 2011. Analysis and collection of data on algal fiber and sugar contents for fermentative utilization of seaweed biomass. SEN'I GAKKAISHI. 67: 181–186.

Uchida, M. and T. Miyoshi. 2012. Algal fermentation—The seed for a new fermentation industry of foods and related products. JARQ. 47: 53–63.

Uchida, M., T. Miyoshi, G. Yoshida, K. Niwa, M. Mori and H. Wakabayashi. 2014. Isolation and characterization of halophilic lactic acid bacteria acting as a starter culture for sauce fermentation of the red alga Nori (*Porphyra yezoensis*). J. Appl. Microbiol. doi:10.1111/jam. 12466.

Wargacki, A.J., E. Leonard, M.N. Win, D.D. Regitsky, C.N.S. Santos, P.B. Kim, S.R. Cooper, R.M. Raisner, A. Herman, A.B. Sivitz, A. Lakshmanaswamy, Y. Kashiyama, D. Baker and

Y. Yoshikuni. 2012. An engineered microbial platform for direct biofuel production from brown macroalgae. Science. 35: 308–313.

Wie, S.G., H.J. Kim, S.A. Mahadevan, D.J. Yang and H.J. Bae. 2009. The potential value of the seaweed Ceylon moss (*Gelidium amansii*) as an alternative bioenergy resource. Bioresour. Technol. 100: 6658–6660.

Yanagisawa, M., K. Nakamura, O. Ariga and K. Nakasaki. 2011. Production of high concentrations of bioethanol from seaweeds that contain easily hydrolysable polysaccharides. Process Biochemistry. 46: 2111–2116.

Yoon, J.J., Y.J. Kim, S.H. Kim, H.J. Ryu, J.Y. Choi, G.S. Kim and M.K. Shin. 2010. Production of polysaccharides and corresponding sugars from red seaweed. Adv. Mater. Res. 93–94: 463–466.

Younesi, H., G. Najafpour and A.R. Mohameda. 2005. Ethanol and acetate production from synthesis gas via fermentation processes using anaerobic bacterium, *Clostridium ljungdahlii*. Biochemical Engineering Journal. 27: 110–119.

Zemke-White, L.W. and M. Ohno. 1999. World seaweed utilization: An-end-of-century summary. J. Appl. Phycology. 11: 369–376.

3

Recent Advantages of Seafood Cooking Methods based on Nutritional Quality and Health Benefits

Abdul Bakrudeen Ali Ahmed, * *Teoh Lydia* and
Rosna Mat Taha

1 Introduction

In Asian countries, seafood is directly used for several culinary purposes, whereas in the west, it is exclusively used for the extraction of important food hydrocolloids including agar, carrageenan, and alginates. Availability almost throughout the year and relatively easy collection potential make macroalgae an inexpensive food source. With the advancement of biological and marine sciences, identification and large scale culturing of edible microalgae (blue-green algae) have also become a reality, and later they have been introduced into different food applications.

Seafood is a rich source of nutrients included in Asians' traditional cuisine and is being extensively explored for its other merits as a food. Apart from its proven nutritional properties, bioactive molecules found in seafoods have attracted the interest of health conscious societies, as seafood is regarded as a remarkable marine medicinal food.

Institute of Biological Sciences, Faculty of Science, University of Malaya, 50603, Kuala Lumpur, Malaysia.
* Corresponding author: dr.bakrudeenaliahmed@yahoo.co.in

Our ecosystem is unique in its biodiversity. The most complex are the underwater ecosystems with numerous creatures and species, many of which are not yet discovered. We are unaware of many biological properties that may be buried deep in the oceans. Kelps are one such species which remain deep rooted in the marine environment. There are many varieties of kelps in underwater ecosystems that vary in size, shape and colour. They are a major keystone species in the ecosystem; without them, many organisms would die. They also possess abundant nutrients and minerals which make them highly bioactive. Anticoagulants, antibiotics, antiparasites, antihypertensives, reducers of blood cholesterol, dilatory drugs and insecticides are made with the help of such properties.

Sea kelp that grows on the bottom of the sea is especially rich in nutrients that are good for the human body. Beta carotene, fiber, chlorophyll, protein, enzymes, and amino acids are found in the ancient fruit of the sea. It contains a complex mix of sodium, potassium, iron, calcium, magnesium, phosphorus, and other minerals. The nutrients in sea kelp make for a perfect dietary recommendation. As a matter of fact, our body resembles in composition the water found on deep levels of the sea, and we share with deep water, approximately 56 components that also circulate through our bodies. Thus, in sea kelp, the chemical composition resembles that of human plasma and consuming it contributes to regulate our internal balance (www.articleclick.com).

Ninety percent of the world's living biomass is found in the oceans, with marine species comprising approximately half of the total global biodiversity (Kim and Wijesekara 2010; Swing 2003). Therefore, the wide diversity of marine organisms is being recognized as a rich source of functional materials, including polyunsaturated fatty acids (PUFA), polysaccharides, essential minerals and vitamins, enzymes, and bioactive peptides (Shahidi 2008; Shahidi and Alasalvar 2011; Shahidi and Janak Kamil et al. 2011).

Seafood consumption has been steadily increasing in the United States, and in 2007, Americans consumed 7.4 kg of fish and shellfish per person. This trend is particularly relevant in light of a growing human population and the indication in recent forecasts is that some of the current fisheries may collapse by mid-century if they are not managed more sustainably (http://www.noaanews.noaa.gov). Processing of raw fish into food products generates large quantities of by product. It has been estimated that the value of addition of human food developed from the by product will increase significantly in the future (Kristinsson and Rasco 2000; Gildberg 2002).

Natural products have great economic and ecological importance, and many natural products are yet to be discovered. The marine environment is a rich source for production of natural bioactive metabolites which are used in various clinical trials (Skropeta 2008). Over 60% of natural products can be considered as drugs in the pharmaceutical industry (Newman et al. 2003). Many novel compounds (drugs) have been isolated from the sea and screened for biological studies including their use as anti-obesity, anti-diabetes, anti-hypertension, anti-microbial, anti-fertility, anti-tumoral, anti-arthritic, hemolytic drugs and as an anti-inflammatory substance. With increasing health consciousness among consumers and the rapid progress of physiologically functional foods, the profile of medicinal products containing chitin oligosaccharides with biological activities seems to be greatly promising in worldwide (Clydesdale 2004). It could be present in food and nutritional. It is emerging as a great potential earner for the food industry (Tucker et al. 2006). Currently, the functional food market has significance in the earning of U.S. $100 billion/year (Aluko 2006).

Biotechnology is the application of science and technology to living organisms, as well as parts, products and models, and alters living or non-living materials for the production of knowledge, goods and human services. Marine biotechnology is an emerging discipline based on the use of marine natural resources. The ocean covers about 70% of the earth's surface, providing a diverse living environment for invertebrates (Lalli and Parsons 1993). The marine environment is a rich source of both biological and chemical diversity. This diversity has been the source of unique chemical compounds with the potential for industrial development as pharmaceuticals, cosmetics, nutritional supplements, molecular probes, fine chemicals and agrochemicals (Carte 1996). Seafood uses around the world include human foods, fertilizers and the extraction of valuable products such as industrial gums and chemicals. Moreover, recent research has pointed to new opportunities, particularly in the field of medicine, associated with bioactive properties of molecules extracted from seafoods. Seaweed may belong to one of several groups of multicellular algae; the red algae, green algae and brown algae. As these three groups are not thought to have a common multicellular ancestor, the seaweeds are a polyphyletic group. In addition, some tuft-forming blue-green algae (cyanobacteria) are sometimes considered as seaweeds. Seaweed is a colloquial term and lacks a formal definition (Madhusudan et al. 2011).

Seafoods are one of the potential contributors to the production of cosmetic compounds. The total worldwide seafood production is estimated to be 8.5 million metric tons. Of this, 88.65% (i.e., 7.5 million metric tons) is produced by cultivation in an area of 200×10^3 ha, while the remaining 0.96 million metric tons are exploited from the natural seafood (seaweed) beds world over (FAO 2003). The seafood (seaweed) industry uses 7.5–8 million metric tons of wet seafood annually either from the wild or from cultivated crops. The estimated value of a wide variety products derived from seafood (seaweeds) is US$ 5–6 billion (FAO 2004). In earlier days, seafoods (seaweed) were used for food preparation and later they were used to yield industrial medicinal, pharmaceutical and cosmetic products (Dhargalkar and Verlecar 2009).

Break even prices faced by prawn farmers are typically between $12/kg and $15/kg for whole prawns (Dasgupta and Templeton 2003). Selling prawns at food festivals is a unique way of direct marketing the product. Festivals provide the opportunity for people to sample new and different food products. Literature on marketing of freshwater prawns in the United States has been relatively sparse, perhaps as a consequence of the small size of the industry. The consumer acceptance of prawns in a Mississippi restaurant was examined. In that study, 77% of consumers rated prawn similar to marine shrimp. 88% of consumers indicated that they would be repeat customers for prawn if it was available in restaurants (Dillard et al. 1986). In contrast to the paucity of marketing studies on prawn in the United Stated, the literature on the use of conjoint analysis in aquaculture marketing has been relatively rich. One of the earlier conjoint analysis studies was by Anderson (1988), who evaluated buyer perceptions of several attributes of salmon (species, fresh vs. frozen, product form, etc.) among whole sellers, retailers and restaurants in the U.S. Northeast. Consumers considered taste and freshness to be the most important prawn attributes. Data from Kentucky in which cooked prawns (originally fresh and frozen) and comparably sized cooked pink marine shrimp (*Litopenaeus duorarum*) were rated by consumers in a blind taste test was analysed. The main conclusions were that consumers were able to distinguish cooked prawns from cooked marine shrimp and they rated prawns' appearance higher than the appearance of marine shrimp (Dasgupta and Templeton 2003).

1.1 Sea cucumber

Sea cucumbers are echinoderms from the class *Holothuroidea* (Bordbar et al. 2011). Sea cucumbers are marine invertebrates related to starfish and sea urchins. There are some 1,250 known species, and many are indeed soft-bodied animals and tube-shaped like a cucumber or worm (Masre et al. 2012). Sea cucumbers are habitually found in the benthic areas and deep seas across the world. They have high commercial value coupled with increasing global production and trade. Sea cucumbers, informally named as 'gamat', have long been used for food and folk medicine in the communities of Asia and Middle East (Bordbar et al. 2011). Many sea cucumbers are deposit feeders. The tube feet around the mouth are modified and called oral tentacles. The tentacles are coated with sticky mucus used to pick up organic matter from the bottom or scoop sediment into the mouth. Some sea cucumbers hide and extend their tentacles to get food directly from the water. They eat mud or sandy sediments to get at the attached organic material (Castro and Michael 2007; Karleskint et al. 2010). The sea cucumber reacts to predators by secreting toxic substances, by producing sticky discharge in the form of toxic filaments through the anus to discourage potential predators (Castro and Michael 2007). Therapeutic properties and medicinal benefits of sea cucumbers can be linked to the presence of a wide array of bioactives (Bordbar et al. 2011). Sea cucumbers and their products have long been a source of traditional medicine due to their various nutritional and medicinal values (Masre et al. 2012).

1.2 Oyster

Oysters are bivalve molluscs (Castro and Michael 2007). They are distantly related to clam and mussels and belong to the Ostreidae family. Oysters are filter feeders (Karleskint et al. 2010). They have no heads and no radula. The body is laterally compressed and enclosed in a shell with two valves that cover it; these are usually attached dorsally at a hinge by ligaments (Castro and Michael 2007; Karleskint et al. 2010). The two valves are closed by adductor muscles. When the muscles relax, the elasticity of the ligaments and the weight of the valves cause them to open. The foot is located laterally. The mantle forms separate inhalant and exhalant openings. Gills with cilia on the surface move the water through the oyster body and the opening directs the water flow into and out of the mantle cavity (Karleskint et al. 2010). Oysters have global distribution and they have had much higher annual production than any other freshwater or marine water for many years (Zhang et al. 2012). Oysters form extensive beds composed by providing a habitat for many other species (Karleskint et al. 2010). Oysters live attached to the hard-surfaced areas. There are many types of oysters; one of them is the pearl oyster. Pearl oysters are the most commercially valuable for producing pearls. Oysters have been eaten by the lovers of food for thousands of years (Castro and Micheal 2007).

1.3 Shrimp

Shrimps are the largest group of decapods of the order crustaceans (Castro and Michael 2007; Karleskint et al. 2010). Shrimp can be found on the bottom of the sea floor, among sediments, and on animal (Castro and Michael 2007). The body is segmented and bilaterally symmetrical, having jointed appendages, such as legs and mouth parts (Castro and Michael 2007). Shrimps have three main body regions: the head, thorax and abdomen (Karleskint et al. 2010). The carapace is well developed and encloses the section of the body known as the cephalothorax. The rest of the body is called the abdomen (Castro and Michael 2007). Each body segment usually contains a pair of appendages (Karleskint et al. 2010). The first five pairs of walking legs are modified to form chelipeds, pincers that are used for capturing prey and defence (Karleskint et al. 2010). They also have three pairs of maxillipeds, which are closer to the mouth, turned forward and specialized to sort out food and pushed it towards the mouth. Shrimps have laterally compressed bodies with distinct and elongated abdomens. Shrimps are specialists in feeding on bits of detritus on the bottom so they are scavengers. For shrimp, the food passes to a stomach that typically has chitinous teeth or ridges for grinding, and bristles for sifting. Decapods have two chambered stomachs, which are connected to digestive glands that secrete digestive enzymes and absorb nutrients. Digestion is essentially extracellular. The intestine ends in an anus. Shrimp have an exoskeleton which is composed of chitin and secreted by the underlying layer of tissue. Crustaceans have two pairs of antennae, which are involved in sensing (Castro and Michael 2007).

1.4 Clam

A clam is a bivalve mollusc (Castro and Michael 2007; Karleskint et al. 2010). Clams have shells divided into jointed halves called valves. There is no radula and no head. Clams' bodies are laterally compressed and the two halves of the shells that cover it are usually attached dorsally at a hinge. Clams have two major parts, the head foot region and visceral mass. The mollusc's soft parts are covered by a protective tissue called mantle. From the visceral mass, the mantle extends and hangs down on each side of the body, forming a space between the mantle and the body, known as the mantle cavity. On the surface of the animal's gills, the cilia move water through the animal's body and the water movement in and out the mantle cavity is directed by the opening (Karleskint et al. 2010). Clams use their shovel shaped foot to burrow in sand or mud. After a clam is buried, water is drawn in and out of the mantle cavity through siphons formed by the fusion of the edge of the mantle. By this, clams can feed and obtain oxygen while buried in sediment (Castro and Michael 2007). The entering water carries food. The food particles are mainly plankton, filtered from the water by the gills (Karleskint et al. 2010). Clams do not have a single brain but several sets of ganglia, clusters of nerve cells located in different parts of the body (Castro and Michael 2007).

1.5 Crab

Crabs belongs to the phylum Arthropoda, class Crustacean. Crabs are decapods, the largest group of crustacean. The crab is related to lobster, and shrimp (Castro and Michael 2007). The arthropods have a hard protective exterior skeleton called the exoskeleton which is composed of protein and tough polysaccharide called chitin (Karleskint et al. 2010). The crab's body is segmented and bilaterally symmetrical (Castro and Michael 2007). Crabs have a pair of mandibles on the head. The crustacean's body is divided into three main parts; the head, thorax and abdomen. The heads and thorax are fused to form a cephalothorax (Castro and Michael 2007&2). The crab's abdomen is small and tucked under the compact and typically broad cephalothorax. The abdomen is visible as a flat, V-shaped plate in males and expanded to u-shaped for carrying eggs in females. Crabs walk sideways when in a hurry. The largest and diverse group of decapods is crabs. Most are scavengers or predators. Some crabs have specialized diets such as seaweeds or organic materials found in mud or coral mucus. Many crabs live along rocky shores or sandy beaches exposed to air much of the time. Some crabs are land crabs. They spend most of their lives on land and return to the sea only to release their eggs (Castro and Michael 2007).

1.6 Scallop

Molluscs are one of the largest groups of animal. The term 'mollusca' refers to the animals' soft bodies, which in most cases are covered by a shell made of calcium carbonate (Karleskint et al. 2010). Scallop is an edible bivalve mollusc of the family Pectinidae and order ostreoida (Hart and Chute 2004). The molluscan body is divided into two major parts, the head-foot and the visceral mass. The head-foot parts contain the head with its mouth and sensory organs and the foot, which is the animal's locomotion organ. The mollusc's soft parts are covered by a protective tissue called the mantle. The mantle extends from the visceral mass and hangs down on each side of the body, forming a space between the mantle and the body known as the mantle cavity. Scallops do not have head and radula (Karleskint et al. 2010). The body is laterally compressed and enclosed in a shell with two parts. The expanded and folded gills are used to obtain oxygen and to filter and sort small food particles from the water (Karleskint et al. 2010; Castro and Michael 2007). Scallops are animals for human consumption as food (Karleskint et al. 2010). Some scallops lived unattached and can swim for short distances by rapidly ejecting water from the mantle cavity when the valves are clapped (Castro and Michael 2007).

2 Cooking Methods

Seafood is usually cooked in order to improve the flavour of the product. This kind of food can be subjected to different cooking processes such as frying, boiling, baking or grilling. These thermal treatments could produce an alteration in their total arsenic content or in their chemical forms, resulting in more toxic species that could pose a risk to the consumer's health. Different studies have been carried out to estimate possible changes in arsenic speciation after cooking. Cooking is normally applied to fish proteins such surimi to induce heat-gelation, resulting in the desired texture, flavour and other sensory attributes. More importantly though, cooking results in microbial reduction and hence, acceptable microbial safety. Since ISP results in a 5-log microbial reduction, the cooking step may not be necessary. This may be beneficial in terms of energy and equipment savings as well as simplified processing. Therefore, this study also aimed at comparing physicochemical properties of gels made with ISP-recovered protein isolates using cold-gelation and heat-gelation (Lopez-Lopez et al. 2011).

Fish is commonly consumed as fried, grilled, boiled or steamed. Generally, most information about PUFA content is available for raw fish. Thus, the consumer has little knowledge about the nutritive values of cooked fish. Several studies have mentioned that unsaturated fatty acids are more susceptible to oxidation than their saturated analogues and PUFA content in aquatic species has been shown to decrease during frying (Candella et al. 1996; Candella et al. 1998). However, it was found that in some kinds of fish, the PUFA levels remained unchanged after boiling, roasting and frying (Montano et al. 2001; Gladyshev et al. 2006). The effect of frying (Candella et al. 1996; Candella et al. 1998; Gladyshev et al. 2006; Gladyshev et al. 2007), grilling (Garcia-Arias et al. 2003), boiling (Gladyshev et al. 2006; Gladyshev et al. 2007) and oven cooking (Candella et al. 1998; Turkkan et al. 2008) on the nutritive value of different fish species have been previously studied. Dietary habits may include different food components and cooking processes. Fish is usually eaten after various cooking treatments such as boiling, grilling, baking or frying, but sometimes it is eaten raw (e.g., in sushi meals). It is known that cooking may affect the amount and speciation of chemical pollutants in foods. The investigation detailed below assesses the effect of cooking methods and food components on the bioaccessibility of Hg in three fish species using an *in vitro* digestion model. The first objective was to verify if different cooking methods altered Hg concentrations (on a dry weight basis) in fish tissues, prior to digestion. We predicted that cooking would alter water content and Hg levels on wet weight basis. The second objective was to assess Hg bioaccessibility after simulated digestion of raw, boiled and fried fish mussels (Burger et al. 2003; Ersoy and Ozeren 2009; Ersoy et al. 2006; Gokoglu et al. 2004; He et al. 2010; Perello et al. 2008).

Sea cucumber

Although there are many cultured and harvestable sea cucumber species, around 20 species are reported with relatively high economic and food value. Sea cucumbers, usually processed into a dried product known as "bêche-de-mer", are valued as an important seafood, particularly in Asian countries (Wen et al. 2010). Malaysians consume boiled skin extracts as a tonic in the treatment of asthma, hypertension, rheumatism and wound, cuts and burns. A variety of sea cucumber derived food and products are available in South Pacific and Asian markets, including China, Japan, Malaysia and Indonesia (Mehmet et al. 2011; Fredalina et al. 1999).

Oyster

Oyster can be eaten on the half shell, raw, smoked, boiled, baked, fried, roasted, stewed, canned, pickled, steam, broiled or used in a variety of drinks (http://www.torontosocialgroup.com/events/13834750/?eventId=13834750&action=detail). Raw oysters should always be served chilled on a bed of ice. Thinly-sliced, buttered pumpernickel or crisp, thin crackers complete the raw oyster eating experience. Relaxing the muscles to shuck oysters is easier if they are tossed in the freezer for about 10 to 15 minutes. If live oysters are to be used in a cooked dish, steam (a few seconds will do it) or microwave (about 30–60 seconds on high, depending on the oven wattage) them just until the shells open. Then cut them from the shells and proceed. Oysters are salty by nature, so most recipes using oysters will not need to be salted. Choose freshly shucked oysters for broiling, smoking or baking on the half-shell. As with many foods, size and age make a difference; smaller and younger oysters will most likely be tender. Most importantly, cook oysters gently to avoid turning them into a rubbery, chewy waste of good shellfish. When the edges begin to curl, its means the oysters have had enough heat (http://homecooking.about.com/od/seafood/a/oystertips.htm).

Shrimp

Shrimp can be cooked in variety of ways. They can be boiled, steamed, grilled, sautéed, baked or deep-fried. They can also be cooked with or without the shell, with the vein or deveined. Shrimp should always be cooked quickly in order to preserve their sweet, delicate flavour. They are very quick to cook and the flavour can easily be ruined by overcooking. Most shrimp cook in as little as 3 minutes. When they are pink, they are done (http://whatscookingamerica.net/ShrimpTips.htm). For pan frying, the shrimp must be rinsed and patted dry with paper towels. Heat a frying pan and add oil. Add shrimp, making sure they are not crowded in the pan, and fry by turning occasionally for 4 to 8 minutes, depending on size. Shrimp are done when they are opaque in the center. For deep frying, pour oil into a wok or deep fryer; the shrimp cooker should be less than half full of oil. Heat oil, dip peeled shrimp in batter, drain and slip into hot oil. Cook shrimp until brown for 2 to 3 minutes. Shrimp can also be cooked by simmering. For simmering, pour enough cooking liquid (water or broth, and herbs and spices) in the pan to cover shrimp. Bring to a boil, add shrimp and reduce heat. Simmer for 3 to 6 minutes until shrimp are opaque in the center, depending on size and whether or not they have been peeled. Shrimp can be grilled. String shrimp on a skewer, then place them above prepared hot coals or fire. Cook for 3 to 4 minutes, until opaque and moist on the inside. For broiling, place aluminium foil on a baking pan and spread shrimp on top. Broil for 2 minutes on each side (http://truestarhealth.com/Notes/1932001.html#).

Clam

Although there are many ways to cook clams, steaming, grilling and baking are the best ways to prepare them. After the clams are cooked, serve them on their own or add them to other dishes, such as soup or pasta. When buying, one must select only those that are still closed, unbroken and undamaged. Store clams in a refrigerator until they are ready to use. To prepare them for cooking, soak the clams in fresh water for 20 minutes. This eliminates much of the salt and the sand in the shells. One of the easiest ways to cook clams is steaming. Typically, steamer clams are the best cooked this way, but one can also steam littlenecks or cherrystones. For a delicious appetizer, grill some clams on the barbecue. They are extremely easy to prepare and grill very quickly. Clams must be placed in a single layer on a hot grill, so that none are overlapping. When clams are open wide, they are done. You can serve clams with sauce, cocktail sauce or traditional garlic butter. Baked clams are also delicious appetizers. Before baking the clams, clean the shells thoroughly. Bake clams for 10 to 12 minutes and they are ready to eat (http://familyfood.hiddenvalley.com/easy-ways-cook-clams-1230.html).

Crab

Crabs can be fried, steamed, boiled, grilled, sauteed or broiled. Live crabs can be steamed or boiled and eaten straightaway or used in sauce or salads. Fried crab cakes, made with picked meat, bread crumbs, butter and seasoning, are a traditional favourite. Soft shells are best sautéed, broiled or grilled at high heat, so the shells become crisp. They are often fried and served in sandwiches (http://www.seafoodsource.com/seafoodhandbook. aspx?id=10737418965). Crab meat, chunked, flaked or shredded, can be served hot or cold. For hot menu items, gentle heating is all that is required. Add to soups and stews during the last 5 minutes of cooking. Legs are often served in the shell with drawn butter. To steam, throw legs in a covered pot with an inch or so of water, bring to a boil and steam just until heated through, about 5 minutes (http://www.seafoodsource.com/seafoodhandbook.aspx?id=10737418968).

Scallop

The most important thing to remember when cooking scallops is to never overcook them, as this eliminates the tender texture that makes them delicious. The small and sweet shellfish are extremely versatile as they can be fried, seared, grilled and boiled (http://www.mademan.com/mm/how-cook-scallops.html). As soon as the scallop is opaque most of the way through, it should be removed from the source of heat and allowed to finish cooking on a serving plate for minute or two before being served immediately, hot off the grill, stove or broiler. The following suggestions are guidelines for cooking tender, flavourful scallops. To broil them, place scallops coated with olive oil on a broiling pan and broil on high, turning them once, until they are just opaque throughout. For baking, scallops should be baked at between 375°C to 400°C, with or without butter and breading, until scallops are firm and opaque. To grill scallops, always oil both the grill and the scallops before grilling. Place the scallops on water-soaked wooden grilling skewers and cook over direct heat, flipping to cook all sides, until scallops are just opaque throughout. Pan seared scallops are prepared by using a well-oiled skillet. Cook scallops on medium-high by turning and flipping frequently, until they are opaque and firm (http://www.life123.com/food/main-courses/scallops-recipes/quick-and-easy-tips-for-cooking-scallops.shtml).

3 Nutritional Value

Seafood is a famous delicacy in some parts of Asia and also a well-known source of important food hydrocolloids, such as agar, alginates and carrageenan. In addition to the food value of seafood, several health benefits have also been reported to be present in this valuable food source. It is presumed that the unique features of the marine environment, where the seafoods are grown, are mainly responsible for most of its properties. Among the functional effects of the seafood, nutritional and health related benefits have been widely studied. Compared to the terrestrial plant and animal based foods, seafood is rich in some health promoting molecules and materials such as dietary fiber, fatty acids, essential amino acids and vitamins A, B, C and E. In this chapter, the nutritive value of seafood and the functional effects of its soluble fiber are discussed with special reference to the digestive health promotion of humans. Dietary fiber, a group of non-starch carbohydrates basically of plant origin found in various vegetables, fruits, grains, nuts and root crops, is an essential part of a healthy diet (Taylor et al. 2011).

Digestive enzymes are supported to the nutritional improvement of the human body. During its passage through the gastrointestinal tract, as a result of its involvement in some important functions, dietary fiber indirectly supports human nutrition. These functions include reduction of incidences of colorectal cancers, suppression of bowel inflammation and related abdominal disorders, facilitation of bowel movement and growth promotion of health promoting gut micro flora. Proteins are an essential component of the diet needed for the survival of animals and humans (Friedman 1996) and under have major resources of carbon around percentage. Proteins may be divided into fibrous and globular proteins according to the shape of their tertiary structure (Zimmermann 2003). Protein could also be grouped into seven categories with reference to their functions; enzyme catalysis, defence, transport, support, motion, regulation and storage proteins (Losos et al. 2008). Proteins play a very important role in all processes in the human body. These processes are facilitated by enzymes, which are actually specialized proteins increasing the reaction rate without any change in themselves. Deficiency of enzymes causes a slower reaction. Enzymes are mainly globular proteins. This protein molecule is where the tertiary structure has given the molecules a general rounded ball shape. The other type of proteins has long, thin structures and are found in tissues like muscles and hair (Marshall 2005). One variety of globular proteins has a unique ability to transport substances across cell membranes. Others use their shape to recognize foreign microbes and cancer cells. Proteins also have regulatory roles within the cell of turning on and shutting off genes during genesis. Moreover, proteins also receive information and act as cell-surface receptors. However, fibrous proteins play a structural role. These fibers include collagen in skin, keratin in hair, fibrin in blood clots, ligaments, tendons and bonds (Fontcuberta et al. 2011).

Proteins, unlike fatty acids and carbohydrates, are not stored in the body, but are deaminated, followed by the oxidation of their carbon skeletons through the pathways of glucose or fat metabolism, i.e., they are stored in the form of glycogen or fat. This depends on the specific amino acid and the energy balance at the time. Nitrogen waste is excreted in urine as either urea or ammonia. Considering the body need of proteins to sustain the essential physiological functions, a diet low in protein could be deficient in many important vitamins and minerals in comparison to a protein rich diet. During fasting and starvation, muscles provides a source of amino acids. The balance between protein synthesis and resolution is closely regulated (Becker and Smith 2006). Protein energy malnutrition plays a fundamental role in developing countries due to poverty, minimal medical care, endemic infections and poor sanitary conditions. This manifests particularly in children, who are offered little food or food of inadequate quality for their needs. Studies in Nigeria found that protein energy malnutrition has been the second most frequent cause of death of children under 5 years of age (Nnakwe 1996). Dietary intake is a major route of exposure to perfluorinated compounds (PFCs). Although, fish and seafood contribute significantly to total dietary exposure to these compounds, there is uncertainty with respect to the effect of cooking on PFC concentrations these foods. Species were relatively unimportant to retail buyers. They were primarily concerned with quality and price of salmon. In contrast, expensive restaurants considered quality, freshness, flesh colour and region of origin to be of paramount importance over price (Becker and Smith 2006).

All resources collected were critically reviewed based on a strict range of criteria in an endeavour to minimize bias. Resources were also reviewed for accuracy, bias and obvious commercial interest. A wide range of sources were accessed to ensure that a thorough assessment was conducted. A range of readability formulas, assessment tools and guidelines were used to assess each resource. The identification process realized 120 current resources associated with the health benefits of regular consumption of seafood as part of a healthy diet that could be used by GPs and AHPs. A healthy diet is important to help maintain good health and this includes eating fish. Including foods rich in omega-3 fatty acids as part of a healthy diet may help reduce inflammation. Foods that are high in omega-3s include fish and seafood. A healthy diet is important to help maintain good health and this includes eating fish. Oily fish such as sardines and salmon have a greater amount of omega-3 fats. Try to eat them at least once a week. Fish oil supplements are high in omega-3 fats. Foods with good fat are part of a healthy diet. This includes oily fish like salmon, mackerel, sardines and tuna. Eat one to two meals of fish (preferably oily fish) a week (Gerber et al. 2012).

The nutritional value of sea cucumber data (Fig. 1–4); oyster data (Fig. 5–9); shrimp data (Fig. 10–19); clams data (Fig. 20–25); crab data (Fig. 26–28); scallop data (Fig. 29–34).

Sea cucumber

Total Weight of Sea Cucumber: 100 g

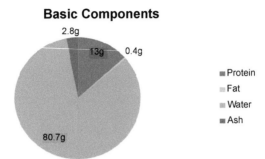

Figure 1. Shows the basic components in 100 g of Sea cucumber.
Color image of this figure appears in the color plate section at the end of the book.

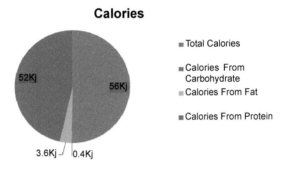

Figure 2. Shows the amount of calories in 100 g of sea cucumber.
Color image of this figure appears in the color plate section at the end of the book.

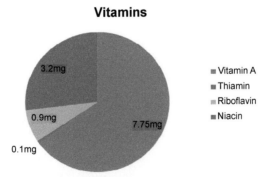

Figure 3. Shows the contents of vitamins in 100 g of sea cucumber.
Color image of this figure appears in the color plate section at the end of the book.

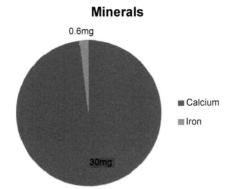

Figure 4. Shows the mineral content in 100 g of sea cucumber.
Color image of this figure appears in the color plate section at the end of the book.

<u>Oyster</u>
Amount of Oyster (raw): 1 cup
Total Weight of Oyster (raw): 248 g

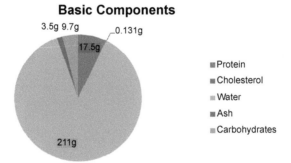

Figure 5. Shows the basic components in 248 g of oyster.
Color image of this figure appears in the color plate section at the end of the book.

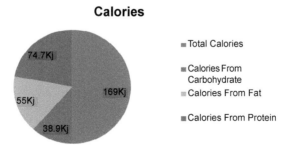

Figure 6. Shows the amount of calories in 248 g of oyster.
Color image of this figure appears in the color plate section at the end of the book.

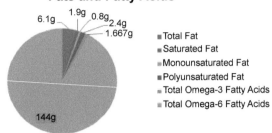

Figure 7. Shows the contents of fats and fatty acids in 248 g of oyster.
Color image of this figure appears in the color plate section at the end of the book.

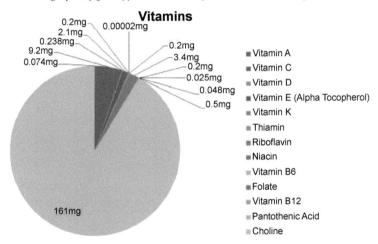

Figure 8. Shows the contents of vitamins in 248 g of oyster.
Color image of this figure appears in the color plate section at the end of the book.

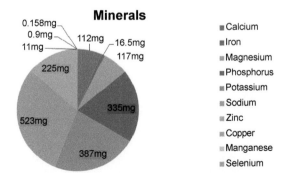

Figure 9. Shows the contents of minerals in 248 g of oyster.
Color image of this figure appears in the color plate section at the end of the book.

Shrimp
Amount of Shrimp: 4.00 oz.
Total Weight of Shrimp: 113.40 g

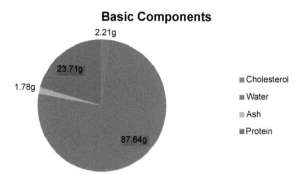

Figure 10. Shows the contents of basic components in 113.40 g of shrimp.

Color image of this figure appears in the color plate section at the end of the book.

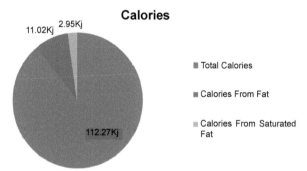

Figure 11. Shows the contents of calories in 113.40 g of shrimp.

Color image of this figure appears in the color plate section at the end of the book.

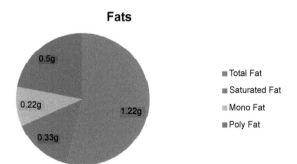

Figure 12. Shows the contents of fats in 113.40 g of shrimp.

Color image of this figure appears in the color plate section at the end of the book.

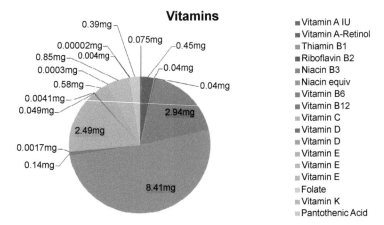

Figure 13. Shows the contents of vitamins in 113.40 g of shrimp.
Color image of this figure appears in the color plate section at the end of the book.

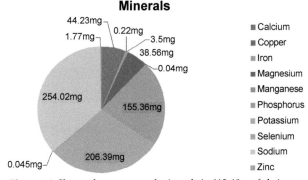

Figure 14. Shows the contents of minerals in 113.40 g of shrimp.
Color image of this figure appears in the color plate section at the end of the book.

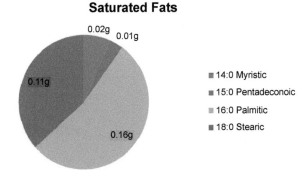

Figure 15. Shows the contents of saturated fats in 113.40 g of shrimp.
Color image of this figure appears in the color plate section at the end of the book.

Mono Fats

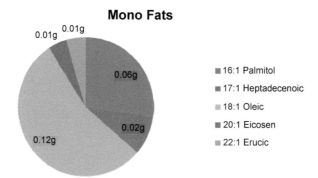

Figure 16. Shows the contents of mono fats in 113.40 g of shrimp.
Color image of this figure appears in the color plate section at the end of the book.

Poly Fats

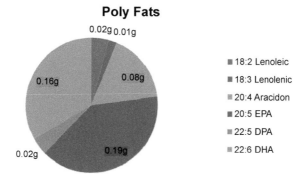

Figure 17. Shows the contents of poly fats in 113.40 g of shrimp.
Color image of this figure appears in the color plate section at the end of the book.

Other Fats

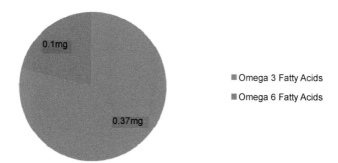

Figure 18. Shows the contents of other fats in 113.40 g of shrimp.
Color image of this figure appears in the color plate section at the end of the book.

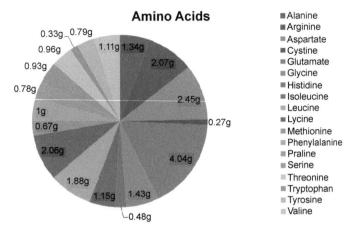

Figure 19. Shows the contents of amino acids in 113.40 g of shrimp.
Color image of this figure appears in the color plate section at the end of the book.

Clams
Amount of Clams: 227 g (1 cup)

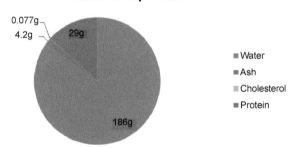

Figure 20. Shows the contents of basic components in 227 g of clams.
Color image of this figure appears in the color plate section at the end of the book.

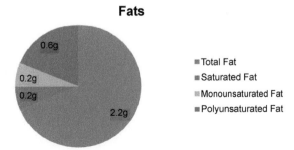

Figure 21. Shows the contents of fats in 227 g of clams.
Color image of this figure appears in the color plate section at the end of the book.

Calories

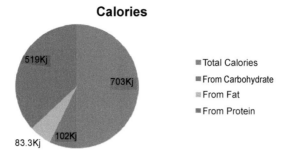

Figure 22. Shows the contents of calories in 227 g of clams.
Color image of this figure appears in the color plate section at the end of the book.

Fatty Acids

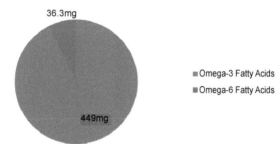

Figure 23. Shows the contents of fatty acids in 227 g of clams.
Color image of this figure appears in the color plate section at the end of the book.

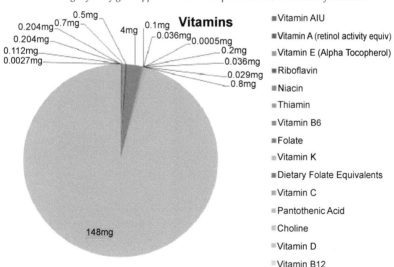

Figure 24. Shows the contents of vitamins in 227 g of clams.
Color image of this figure appears in the color plate section at the end of the book.

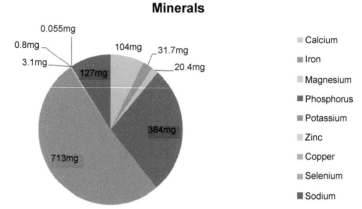

Figure 25. Shows the contents of minerals in 227 g of clams.
Color image of this figure appears in the color plate section at the end of the book.

Crab
Amount of Crab: 85 g

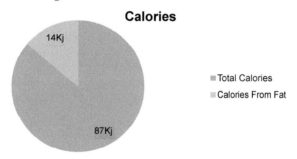

Figure 26. Shows the contents of calories in 85 g of clams.
Color image of this figure appears in the color plate section at the end of the book.

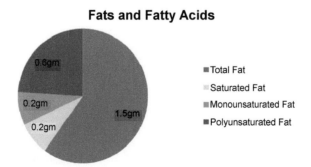

Figure 27. Shows the contents of fats and fatty acids in 85 g of clams.
Color image of this figure appears in the color plate section at the end of the book.

Protein and Minerals

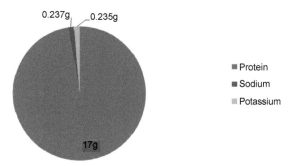

Figure 28. Shows the contents of proteins and minerals in 85 g of clams.
Color image of this figure appears in the color plate section at the end of the book.

Scallop
Amount of Scallop: 4.00 oz-wt
Total Weight of Scallop: 113.40g

Basic Components

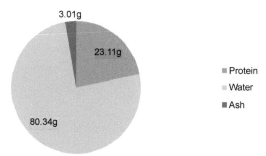

Figure 29. Shows the contents of basic components in 113.40 g of scallop.
Color image of this figure appears in the color plate section at the end of the book.

Calories

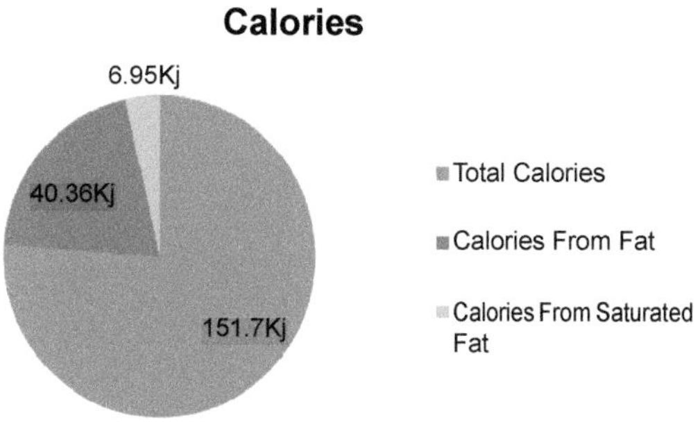

Figure 30. Shows the contents of calories in 113.40 g of scallop.
Color image of this figure appears in the color plate section at the end of the book.

Fats

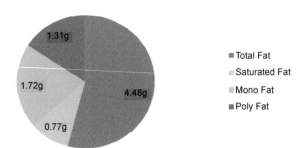

Figure 31. Shows the contents of fats in 113.40 g of scallop.
Color image of this figure appears in the color plate section at the end of the book.

Vitamins

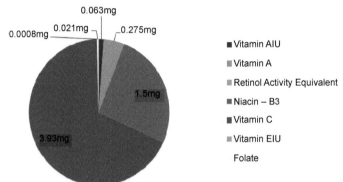

Figure 32. Shows the contents of vitamins in 113.40 g of scallop.
Color image of this figure appears in the color plate section at the end of the book.

Fats

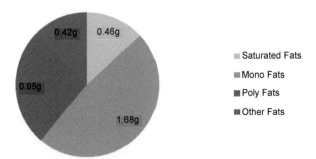

Figure 33. Shows the contents of fats in 113.40 g of scallop.
Color image of this figure appears in the color plate section at the end of the book.

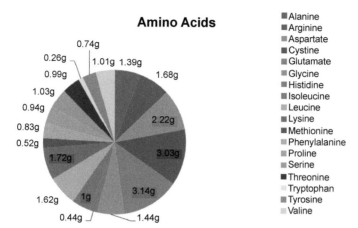

Figure 34. Shows the contents of amino acids in 113.40 g of scallop.
Color image of this figure appears in the color plate section at the end of the book.

1.4 Health benefits

Sea cucumbers have a high nutrition value. They have been recognized in folk medicine systems of Asian countries and as having an impressive range of medicinal health functions for nourishing the body, tonifying the kidneys, moistening dryness of the intestines, treatment of stomach ulcers, asthma, anti-hypertension, rheumatism and wound healing. Scientists believe that the sea cucumber extraction can reduce the growth of cancer cells. Sea cucumber has emerged as a potential anti angiogenic and anti-tumour agent. Some studies reveal the potential of anti-carcinogenic properties in sea cucumber derived as bioactive against certain cancers. Some studies suggest the sea cucumber possesses potent anti-inflammatory properties. It is also able to improve sexual performance in humans. The sea cucumber also used for its anticoagulant and antithrombotic effects. The glucosamine and chondroitin compound in sea cucumber is helpful for treating arthritis. The extraction of sea cucumber has proven itself as a potential antimicrobial agent. Sea cucumber has been explored as a potential source of valuable antioxidants among marine organisms (Bordbar et al. 2011).

Sea Cucumber

- Sea cucumbers are known for their anti-aging properties.
- The Chinese consider a sea cucumber to be a health panacea and hence, use it for treating kidney disorders, constipation and reproductive problems, including impotency.

- Sea cucumbers contain glucosamine and chondroitin that are helpful for treating arthritis.
- Sea cucumber is the most popular traditional remedy used by Malay women for treating injuries and wounds caused by episiotomy.
- Ointments prepared from sea cucumbers are used for treating back and joint pains by the Peninsular Malaysian communities.
- Sea cucumbers are considered as a general health tonic and benefit people with chronic joint pains, tendonitis, sprains, ligament stress and arthritis.
- They are commonly used for treating debility among elders, frequent urination and problems in intestinal tract lining.
- Sea cucumbers are an instant energy booster.
- Sea cucumbers assist in nourishing the brain and heart, as they contain fatty acids such as EPA (eicosapentaenoic acid) and DHA (docosahexaenoic acid), similar to fish oil.
- Sea cucumber extracts are combined with well known herbs to produce ointments effective for health and beauty that can be used on the face, hands, feet, joints, muscles, scalp, hair, mouth, gum and sensitive areas.
- Topical application of sea cucumber ointment acts as sun protector, night cream and moisturizer for dry and cracking skin. The ointment helps in restoring a pleasant, soft and supple feeling to the skin.
- Scientists from various countries like China, Japan, Russia and United States are working on the extracts of sea cucumbers. According to them, sea cucumbers might prove to be useful in stopping the growth of cancer cells.

Oysters

- A rich natural source of zinc, oysters assist in maintaining a strong immune system, healing wounds, maintaining the senses of taste and smell and inhibiting the abnormal clotting that contributes to cardiovascular disease.
- Oysters contain proteins that are high in tyrosine, an amino acid that is used by the brain to help in regulating mood and adapting to stress.
- Including oysters in your diet helps in maintaining collagen levels in the skin. This enables the skin to retain its elasticity and firmness and delays the onset of wrinkles.
- The high levels of calcium, iron and vitamin A in oysters assist in the healthy being of the bones, blood and eyesight, respectively.

- Oysters are low in fat, cholesterol and calories and can thus be enjoyed by everyone, if consumed in moderation.
- Since oysters are rich in amino acids, they are known as aphrodisiacs, triggering increased levels of sex hormones, testosterone and oestrogen.

Shrimp

- Shrimp is a very rich source of the trace mineral, selenium. Selenium helps in neutralizing the effects of free radicals, which are associated with cancer and other degenerative diseases.
- Shrimps are also an excellent source of low-fat and low-cal protein. Around 23.7 grams of protein is gained from a four-ounce serving of shrimp. This amount of shrimp contains 112 calories and less than a gram of fat.
- A good quantity of Vitamin D is found in shrimp; this is essential for strong teeth and bones, and also regulates the absorption of calcium and phosphorus, by the body.
- Being a good source of Vitamin B12, shrimp consumption facilitates proper brain function and the formation and maturation of blood cells in the body.
- Shrimp is loaded with Omega-3 fatty acids, which are essential to guard against the risk of cardiovascular problems, as they reduce the level of cholesterol in blood. The Omega-3 fatty acids present in shrimp also help to avoid blood clotting, prevent the development of rheumatoid arthritis, slow down the development of cancerous tumors and serve to prevent Alzheimer's disease as well.
- Shrimp has anti-inflammatory properties, which aid in reducing gum swelling.

Clams

- Clams make up for a heart-healthy choice, as they are low in fat and rich in omega-3 fatty acids.
- They serve as a good source of phosphorus for the body, the nutrient required for proper formation of bones and teeth. In addition, they also help the body utilize vitamins.
- The protein level of clams is much higher than that of red meat, with very few calories. It helps in building lean body mass in athletes.
- They are a rich source of iron for the body. In fact, the iron content in clams is much higher than beef. They are a boon for individuals suffering from iron deficiency.

- The high amount of potassium present in clams assists the body in maintaining blood pressure and regulating heart function, along with other body processes.
- They are rich in vitamin A, which is required by the body to maintain healthy skin and also promotes vision, growth and bone development.
- Clams have a fair amount of cholesterol, which is required to maintain healthy blood cholesterol in the body.

Crab

- Due to being very low in calories and fat, crabs are healthy for the heart. Despite their low calorie content, crabs make for a filling main course dish. A four ounce serving of crab meat contains only 98 calories and under two grams of fat.
- Crabs are a very rich source of lean protein, which makes them a perfect high protein alternative for athletes, sportsperson and body builders. They are also a good option for diabetics, as they are free of carbohydrates.
- Crabs are a good source of Omega-3 fatty acids, which are necessary for heart health and help to lower triglycerides and blood pressure, thereby reducing the risk of heart disease. The Omega-3 fatty acids contained in crabs are also believed to boost the immune system.
- Eating a seafood rich diet can expose the body to high amounts of mercury. However, mid-Atlantic blue crab contains a very low amount of mercury, which makes it a safer seafood choice.
- Crabs are a good source of chromium, which helps in the metabolism of sugar. It helps the body to maintain normal blood glucose levels. Chromium basically works with insulin to regulate the sugar levels of the body.
- Crabs contain selenium, a trace mineral that helps reduce oxidative damage to the cells and tissues. A diet containing crab provides enough selenium to the body to keep these systems functioning.
- Crab meat is a rich source of vitamins and minerals, especially of Vitamin B12, one of the vitamins that are critical for healthy nerve function. It is also a good source of minerals, such as zinc and copper.

Lobsters

- Since lobster contains very low amount of carbohydrates, it forms a perfect ingredient in no- or low-carbohydrate diet.

- Lobster is beneficial for both heart and brain, since it contains a high constitution of Omega-3 fatty acids.
- High in selenium, lobster aids the immune system and thyroid glands and helps in preventing heart disease.
- It contains copper which helps in avoiding bone and tissue diseases.
- The high amount of phosphorus present in lobster helps in the proper functioning of the kidneys and reduces arthritis pain.
- It is extremely beneficial in increasing brain activity, boosting the immune system and maintaining a healthy reproductive system.
- Lobster helps in avoiding various health disorders, like osteoporosis, osteoarthritis, rheumatoid arthritis, cardiovascular disease, chronic conditions involving bone, connective tissue, heart and blood vessels and colon cancer.

Scallops

- Being a very good source of vitamin B12, scallops promote cardiovascular health. The human body needs vitamin B12 to convert homocysteine (a chemical that can be hazardous to blood vessel walls) into other benign chemicals.
- Scallops are a rich source of omega-3 fatty acids, magnesium and potassium. These nutrients are also beneficial for the cardiovascular system. Studies have shown that omega-3 fatty acids increase heart rate variability (HRV) in just about 3 weeks.
- Consumption of scallops also protects the body against fatal abnormal heart rhythms. Any fish rich in omega-3 improves the electrical properties of heart cells, as per a study conducted in Greece.
- Omega-3 polyunsaturated fatty acids present in scallops also help in controlling high blood pressure. These essential fatty acids keep the blood pressure levels low and thereby prevent health problems related to high blood pressure.
- Another study showed that consuming scallops 2–3 times a month may provide protection against ischemic stroke. Ischemic stroke is caused by the lack of blood supply to the brain, for instance, clotting. For those who consume scallops 5 or more times a week, the stroke chances are reduced by 31%.
- A daily serve of scallops provides significant protection against the risks of coronary heart disease. Nutrients present in this fish reduce platelet aggregation and decrease the production of pro-inflammatory compounds called leukotrienes.

- Scallops are a potent source of selenium, which helps neutralize the hazardous effects of free radicals. Excessive accumulation of free radicals (which are by-products of metabolism) works to weaken the immune system, thereby making it prone to diseases.
- Scallops are also known to provide some relief from hypertension. The magnesium contained in scallops is associated with relaxation of blood vessels, which keeps hypertension under check.

Conclusion

For centuries, Mankind has relied on food as the main source of well-being and health. The major factors determining life expectancy, is an improved nutrition and better healthcare, which has increased all over the world (WHO 2011). However, the modern lifestyle and current feeding behaviours have fostered the development of illness due to physical inactivity, overweight and obesity, and other diet related factors, and tobacco and alcohol-related risks. The evolution of the human diets over the past years has adversely affected health, and chronic diseases cause substantial disability and death. Chronic diseases, including obesity, diabetes mellitus, cardiovascular disease (CVD), hypertension and stroke, and some types of cancer, have become the target in a fight where seaweeds may play a major role.

Acknowledgement

The authors would like to thank University of Malaya Research Grant BK015-2014.

References

Aluko, R. 2006. Functional Foods and Nutraceuticals. IFIS publishing. http://www.ifs.org/fsc/ixid 144335.

Anderson, J.T. 1988. A review of size dependent survival during pre-recruit stages of fishes in relation to recruitment. J. Northw. Atl. Fish. Sci. 8: 55–66.

Becker, G.W. and K. Smith. 2006. Basic metabolism III: Protein. Surgery. 24: 115–120.

Bordbar, S., Farooq Anwar and Nazamid Saari. 2011. High-Value Components and Bioactives from Sea Cucumbers for Functional Foods—A Review. Marine Drugs. 9: 1761–1805.

Burger, J., C. Dixon, C.S. Boring and M. Gochfeld. 2003. Effect of deep-frying fish on risk from mercury. J. Toxicol. Environ. Health part A. 66: 817–828.

Candella, M., I. Astiasaran and J. Bello. 1996. Effects of frying and warmholding on fatty acids and cholesterol of sole (*Solea solea*). Codfish (*Gadus morrhua*) and hake (*Merluccius merluccius*). Food Chem. 58(3): 227–231.

Candella, M., I. Astiasaran and J. Bello. 1998. Deep-fat frying modifies high-fat fish fraction. J. Agric. Food. Chem. 46: 2793–2796.

Carte, B.K. 1996. Biomedical potential of marine natural products. Bio. Sci. 46: 271–286.

Castro, P. and E.H. Michael. 2007. Marine Biology (6th ed.). New York; Mc Graw Hill.

Clydesdale, F.M. 2004. Functional Foods: Opportunities and Challenged. Food Technol. 58: 34.

Dasgupta, S. and S. Templeton. 2003. Comparing Kentucky grown freshwater prawn with marine shrimp: results of a taste test. Transactions of the Kentucky Academy of Science. 64(2): 128–134.

Dhargalkar, V.K. and X.N. Verlecar. 2009. Southern ocean seaweeds: Resource for exploration in food and drugs. Aquaculture. 287: 229–242.

Dillard, J.G., M.J. Fuller and D.W. Whitten. 1986. Consumer acceptance of freshwater shrimp in Mississippi restaurant Agricultural Economics Research Report 170. Department of Agricultural Economics, Mississippi Agricultural and Forestry Experiment Station, Mississippi State University, Starkville, Mississippi. USA.

Ersoy, B. and A. Ozeren. 2009. The effect of cooking methods on mineral and vitamin contents of African catfish. Food Chem. 115: 419–422.

Ersoy, B., Yanar, A. Kucukgulmez and M. Celik. 2006. Effects of four cooking methods on the heavy metal concentrations of sea bass fillets (*Dicentrarchus labrax* linne, 1785). Food Chem. 99(4): 748–751.

FAO. 2003. Year book fishery statistics. 96: 112–114, 393–397.

FAO. 2004. The state of world fisheries and aquaculture. Part 3. Highlights of special FAO studies. Publ. FAO, UN. 1–36.

Fontcuberta, M., J. Calderon, J.R. Villalbi, F. Centrich, S. Portana, A. Espelt, J. Duran and M. Nebot. 2011. Total and inorganic arsenic in marketed food and associated health risks for the catalan (Spain) population. Journal of Agricultural and Food Chemistry. 59(18): 10013–10022.

Fredalina, B.D., B.H. Ridzwan, A.A. Zainal Abidin, M.A. Kaswandi, H. Zaiton, I. Zali, P. Kittakoop and A.M. Jais. 1999. Fatty acidcomposition in local sea cucumber, *Stichopus Chloronotus* for wound healing. Gen. Pharmacol. 33: 337–340.

Friedman, M. 1996. Nutritional value of proteins from different food sources. A review. J. Agric. Food Chem. 44: 6–29.

Garcia-Arias, M.T., E. Alvarez Pontes, M.C. Garcia-linares, M.C. Garcia-Fernandez and F.J. Sanchez-Muniz. 2003. Grilling of Sardines fillets. Effects of frozen and thawed modality on their protein quality. Lebensm. Wiss. Technol. 36: 763–769.

Gerber, L.R., R. Karimi and T.P. Fitzgerald. 2012. Sustaining seafood for public health. Frontiers in Ecology and the Environment. 10(9): 487–493.

Gildberg, A. 2002. Enhancing returns from greater utilization. *In*: H.A. Bremner (ed.). Safety and Quality Issues in Fish Processing. Woodhead Publishing Limited: Cambridge.

Gladyshev, M.I., N.N. Sushchik, G.A. Gubanenko, S.M. Demirchieva and G.S. Kalachova. 2007. Effect of boiling and frying on the content of essential polyunsaturated fatty acids in muscle tissue of four fish species. Food Chem. 101: 1694–1700.

Gladyshev, M.I., N.N. Sushchik, G.A. Gubanenko, S.M. Demirchieva and G.S. Kalachova. 2006. Effect of way of cooking on content of essential polyunsaturated fatty acids in muscle tissue of humpback salmon (*Oncorhynchus gorbuscha*). Food Chem. 96: 446–451.

Gokoglu, N., P. Yerlikaya and E. Cengiz. 2004. Effects of cooking methods on the proximate composition and mineral contents of rainbow trout (*Oncorhynchus mykiss*). Food Chem. 84: 19–22.

Hart, D.R. and A.S. Chute. 2004. Essential fish habitat source document: sea scallop, Placopecten magellanicus, life history and habitat characteristics (2nd ed.). NOAA technical memorandum NMFS NE-189. U.S. Department of Commerce, Woods Hole, Massachusetts. 21 pp.

He, M., C.Y. Ke and W.X. Wang. 2010. Effects of cooking and subcellular distribution on the bioaccessibility of trace elements in two marine fish species. J. Agric. Food Chem. 58: 3517–3523.

Karleskint, G., R. Turner and J.W. Small. 2010. Introduction to Marine Biology (3rd ed.). Canada; Brooks/Cole.

Kim, S. and I. Wijesekara. 2010. Development and biological activities of marine-derived bioactive peptides: A review. J. Funct. Foods. 2: 1–9.

Kristinsson, H.G. and B.A. Rasco. 2000. Biochemical and functional properties of Atlantic salmon (*Salmo salar*) muscle proteins hydrolyzed with various alkaline proteases. J. Agric. Food Chem. 48: 657–666.

Lalli, C.M. and T.R. Parsons. 1993. Biological Oceanography. 1–10. Pergamon Press, Oxford, New York.

Lopez-Lopez, I., S. Cofrades, V. Caneque, M.T. Diaz, O. Lopez and F. Jimenez-Colmenero. 2011. Effect of cooking on the chemical composition of low-salt, low-fat, Wakame/olive oil added beef patties with special reference to fatty acid content. Meat Science. 89(1): 27–34.

Losos, J.B., K.A. Mason and S.R. Singer. 2008. Proteins. pp. 43–53. *In*: P.E. Reidy and A.l. Winch (eds.). Biology. McGraw Hill Companies, Inc., New York.

Madhusudan, C., S. Manoj, K. Rahul and C.M. Rishi. 2011. Seaweeds: A diet with nutritional, medicinal and industrial value. Research Journal of Medicinal Plant. 5(2): 153–157.

Marshall, K. 2005. Role of proteins in the body. pp. 15–20. *In*: J. Challem and T. Dukrin (eds.). Basic Health Publications User's Guide to Protein and Amino Acids. Basic Health Publications, Inc., Laguna Beach.

Masre, Siti Fathiah, W. George, K.N.S. Yip, Sirajudeen and Farid Che Ghazali. 2012. Quantitative analysis of sulphated glycosaminoglycans content of Malaysian sea cucumber *Stichopushermanni* and *Stichopusvastus*. Natural Product Research: Formerly Natural Product Letter. 26(7): 684–689.

Mehmet, A., S. Huseyin, T. Bekir, E. Yilmaz and K. Sevim. 2011. Proximate composition and fatty acid profile of three different fresh and dried commercial sea cucumber from Turkey. Int. J. Food Sci. Technol. 46: 500–508.

Montano, N., G. Gavino and V.C. Gavino. 2001. Polyunsaturated fatty acids contents of some traditional fish and shrimp paste condiments of the Philippines. Food. Chem. 75: 155–158.

Newman, D.J., G.M. Cragg and K.M. Snader. 2003. Natural Products as Sources of New Drugs Over the Period 1981–2002. J. Nat. Prod. 66: 1022.

Nnakwe, N. 1996. The effect and cause of protein-energy malnutrition in Nigerian children. Nutr. Res. 15: 785–794.

Perello, G., R. Marti-Cid, J.M. Llobet and J.L. Domingo. 2008. Effects of various cooking processes on the concentrations of arsenic, cadmium, mercury and lead in foods. J. Agric. Food Chem. 56: 11262–11269.

Shahidi, F. 2008. Bioactives from Marine Resources. Washington: ACS Publications, pp. 24–34.

Shahidi, F. and C. Alasalvar. 2011. Chapter 36 Marine oils and other marine nutraceuticals. pp. 444–454. *In*: Handbook of seafood Quality, Safety and Health Applications. New York: Wiley Online Library.

Shahidi, F. and Y. Janak Kamil. 2011. Enzymes from fish and aquatic invertebrates and their application in the food industry. Trend Food Sci. Technol. 12: 435–464.

Skropeta, D. 2008. Deep-sea Natural Products. Nat. Prod. Rep. 25: 989–1216.

Swing, J. 2003. What future for the oceans? Foreign Aff. 82: 139–152.

Taylor, J., A. McManus and C. Nicholson. 2011. A critical review of nutrition resources for general practitioners focusing on healthy diet including seafood. Australasian Medical Journal. 4(12): 694–699.

Tucker, M., S.R. Waley and J.S. Sharp. 2006. Consumer Perception of Food Related Risks. Int. J. FoodSci. Technol. 41: 35.

Turkkan, A.U., S. Cakli and B. Kilinc. 2008. Effects of cooking methods on the proximate composition and fatty acid composition of Sea bass (*Dicentrarchus labrax*, Linnaeus, 1758). Food Bioprod. Process. 86(3): 163–166.

Wen, J., C. Hu and S. Fan. 2010. Chemical composition and nutritional quality of sea cucumber. J. Sci. Food Agric. 90: 2469–2474.

WHO, World Health Orgonization. 2011. World Health Statictics 2011, ISBN: 9789241564199.

Zhang, G., X. Fang, X. Guo, L. Li, R. Luo, F. Xu, P. Yang, L. Zhang, X. Wang, H. Qi, Z. Xiong, H. Que, Y. Xie, P.W. Holland, J. Paps,Y. Zhu, F. Wu, Y. Chen, J. Wang, C. Peng, J. Meng, L. Yang, J. Liu, B. Wen, N. Zhang, Z. Huang, Q. Zhu, Y. Feng, A. Mount, D. Hedgecock, Z. Xu, Y. Liu, T. Domazet-Lošo, Y. Du, X. Sun, S. Zhang, B. Liu, P. Cheng, X. Jiang, J. Li, D. Fan, W. Wang, W. Fu, T. Wang, B. Wang, J. Zhang, Z. Peng, Y. Li, N. Li, J. Wang, M. Chen, Y. He, F. Tan, X. Song, Q. Zheng, R. Huang, H. Yang, X. Du, L. Chen, M. Yang, P.M. Gaffney, S. Wang, L. Luo, Z. She, Y. Ming, W. Huang, S. Zhang, B. Huang, Y. Zhang, T. Qu, P. Ni, G. Miao, J. Wang, Q. Wang, C.E. Steinberg, H. Wang, N. Li, L. Qian, G. Zhang, Y. Li, H. Yang, X. Liu, J. Wang, Y. Yin and J. Wang. 2012. The oyster genome reveals stress adaptation and complexity of shell formation. Nature. 490(7418): 49–54.

Zimmermann, K.H. 2003. Proteins. pp. 5–22. *In*: M. Ismail (ed.). An Introduction to Protein Informatics. Springer-Verlag, New York, LLC.

4

Oil Tannage for Chamois Leather

Eser Eke Bayramoğlu[1], * and *Seher Erkal[2]*

1 Introduction

The chamois are tanned by traditional methods using unsaturated oil such as cod liver oil to give a soft and natural feel to the product. Typically chamois leather is used to dry off surfaces after washing; this is due to the absorbency of the leather (Bayramoglu 2006). In addition, grime particles are drawn away from the surface being cleaned. The particles are held within the hollow fibre of the leather, eliminating abrasion. Other uses of oil tanned leathers are for glove-making and for filtering water from petrol, etc. Chamois leather has a characteristic yellow colour and because of this reason it is not necessary to dye it. It is an example of a leathering type of process because, although it resists microbial attack, the shrinkage temperature is not raised significantly above the value of raw pelt. In essence, the process involves filling wet pelt with unsaturated oil, then polymerising the oil *in situ* by oxidation (Covington 2009).

Chamois leather production is a little different from the other types of leather. The tanning medium is raw cod liver oil which is obtained by boiling cod livers in water so that the oil is released and can be collected by skimming from the surface (Shaphause 1975).

[1] Ege University, Faculty of Engineering, Department of Leather Engineering, 35100 Bornova-İZMİR–TÜRKİYE.
[2] Distance Learning Center, Okan University, Tuzla Campus, Akfirat-Tuzla/Istanbul, Turkey.
  Email: seher.eke@gmail.com
* Corresponding author: eser.eke@ege.edu.tr, eserekebay@gmail.com

Oxidation of the oil is essential for the tanning action. Some of the researchers reported that its oxidation process takes a relatively long time, i.e., 9 to 14 days. Sodium bicarbonate type of oxidizing agents may be used to shorten this period; however it further increases the cost (Suparno et al. 2011). It was reported that both temperature and airflow rate were statistically significant upon the oxidative stability of the cod liver oil (Garcia-Morena et al. 2013). In this context, steam and heat-adjustable drums for optimized oxidation during chamois production are currently used, and oil tannage-oxidation acts during chamois production can now be made in just a few days in these drums. In the oxidation of the oil, peroxides are formed, and further, smaller quantities of acrolein, aliphatic ketones and other pungent volatile organic compounds. These thermal decomposition products combine with the skin, imparting to it, the yellow colour characteristic of chamois leather (Gustavson 1956). The exact chemical nature of the tannage cannot be precisely defined but it is known that during the 'oxidation' of the oil it undergoes chemical changes, some of which liberate aldehydes (often referred to as acrolein) and the leather has many similarities to formaldehyde and glutaraldehyde tanned leather (Shaphause 1975). However, acrolein alone does not make acceptable chamois leather (Shaphause 1985).

2 Nature of Fatty Acids and Chemical Distributions

It is found that the various marine oils are not all equally well suited for chamois tannage. Cod oil is the most convenient oil for chamois production according to this author (O'Flaherty et al. 1958).

Fish oils and marine-animal oils are generally characterized by a rather large group of saturated fatty acids, which are commonly associated with mixed triglycerides (Gruger 1967). In marine oils, fatty acids with more unsaturated and longer chains (containing 20, 22 and 24 carbon atoms) are found in larger quantities. The degree of unsaturation can reach six double bonds per fatty acid residue (O'Flaherty et al. 1958).

In comparison to body oils, liver oils and oils from particular parts of fish and marine animals can often contain large amounts of fatty acids associated with phospholipids, glyceryl ethers (alkoxydiglycerides), and wax esters, depending on the source of oils and lipids (Lovern 1962).

Nature of Fatty Acids

The fatty acids derived from fish oils are of three principal types:

1. saturated
2. monosaturated
3. polyunsaturated

The Formula,

$$CH_3 (CH_2)_x (CH=CHCH_2)_n (CH_2)_y COOH$$

Where n = 0 to 6, illustrates the type of fatty acid structures common to fish oils.

The saturated fatty acids have carbon chain lengths that generally range from C_{12} (lauric acid) and to C_{24} (lignoceric acid). Also, traces of C_8 and C_{10} acids may be found in some fish oils. A C_5 acid (isovaleric), however, occurs in jaw-bone oils of dolphin and porpoise (Gruger 1967).

Fatty acids of triglycerides, lecithins and phosphatidyl ethanol-amines (cephalins) from livers of cod and lobster and from muscles of cod and scallop were analysed by Brockerhoff et al. 1963. They demonstrated that the polyunsaturated fatty acids were preferentially located in the β position of both the triglycerides and lecithins. The phosphatidyl ethanolamines were also found to be highly unsaturated (Gruger 1967).

Brockerhoff and Hoyle (1963) analyzed the fatty acid distribution in the α and β positions of triglycerides from fish body and liver oils and found that the plyenoic acids were preferentially distributed in the β position (Gruger 1967).

The fatty acid composition of some body oils from fish–liver is listed in Table 1 (Gruger 1967).

Mixed glycerides of the highly unsaturated fatty acids can undergo complex oxidation reactions, leading to products which have many industrial applications. Depending on the mixture of fatty acids which make up the glycerides, lipids may be classed as non-drying, semi-drying or drying in character, i.e., by the ease with which they may be oxidized. Thus the drying oils, when spread on a surface and exposed to air, slowly form dry, tough and durable films (Reed 1966).

Type of drumming and the heating of the oiled pelts in order to induce autoxidation also impact the end result because too little heating (leading to insufficient tanning) is just as disadvantageous as too much heating. (Excess heating causes severe damage to the leather fibres.) (O'Flaherty et al. 1958).

Marine oils also contain, as do all other natural triglyceride fat and oils, certain unsaponifiable constituents. The unsaponifiable material in the fish oil in general is indeed small, but it must not be overlooked, since it acts partly as an oxidative accelerator and partly as an inhibitor of oxidation (O'Flaherty et al. 1958).

In Table 2 the fatty acid composition of several of the fish oils which are especially suitable for oil tanning is presented. The figures in parentheses in the second row of numbers give an estimate of the number of double bonds (O'Flaherty et al. 1958).

Table 1. Fatty acid composition of oils from fish livers (Gruger 1967).

| Name of Species | | Fatty Acid Percentage Composition | | | | | | | | | | | | | | | | |
Common	Scientific	14:0	15:0	16:0	16:1	17:0	18:0	18:1	18:2	18:3	18:4	20:1	20:4	20:5	22:1	22:5	22:6	Reference
Cod	*Gadus callarias*	3	+	12	5	1	3	24	1	1	1	9	1	8	5	1	19	Klenk 1962
Cod	*Gadus callarias*(?)	2.8	0.3	11.6	8.6	0.3	2.7	25.2	2.5	0.7	2.2	13.1	NK	9.3	6.3	1.0	8.6	DeWitt 1963
Cod	NK	2.9	NK	14.6	6.2	NK	3.5	39.0	1.7	0.3	+	9.1	2.1	2.6	4.6	1.5	9.7	Ito 1963B
Cod,Atlantic	*Gadus morhua*	2.8	0.4	10.7	6.9	1.2	3.7	23.9	1.5	0.9	2.6	8.8	1.0	8.0	5.3	1.3	14.3	Gruger 1964
Cod,Atlantic	*Gadus morhua*	3.5	0.5	10.4	12.2	0.1	1.2	19.6	0.8	0.2	0.7	14.6	1.7	5.0	13.3	2.0	10.5	Ackman 1964A
Cod(eggs)	*Gadus morhua*	0.6	0.1	23.4	6.1	+	1.3	23.4	0.2	0.2	0.1	1.0	2.5	16.5	0.2	0.8	22.1	Ackman 1964C

[1] Not known or not mentioned in reference is designated as NK. Trace amounts designated by a+ sign
[2] Combined 18:3 & 20:1 acids
[3] Combined 20:4 & 22:1 acids

Table 2. (O'Flaherty et al. 1958).

	C_{14}	C_{16}	C_{18}	C_{20}	C_{22}
Newfoundland Cod Oil					
% Saturated fatty acids	5.8	8.4	0.6		
% Unsaturated fatty acids	0.2	20.0 (2.3)	29.1 (2.8)	25.4 (6.0)	9.6 (6.9)
Seal Oils					
% Saturated fatty acids	3.7	10.5	6.2		
% Unsaturated fatty acids	1.6 (2.0)	15.5 (2.2)	30.8 (2.7)	16.5 (5.7)	18.1 (10.6)
Herring Oil					
% Saturated fatty acids	7.3	13.0	Traces		
% Unsaturated fatty acids		4.9 (2.7)	20.7 (4.2)	30.1 (4.6)	23.2 (4.3)

3 Oxidation and Polymerization Mechanisms of Unsaturated Lipids

Reeds reported that the reactions involved in oil-tannage are complex and are still not known with any great certainty. The mechanism appears to involve reaction with molecular oxygen to form hydro-peroxides:

$$- CH_2 - CH = CH - \xrightarrow{O_2} - \underset{\underset{OOH}{|}}{CH} - CH = CH - \text{ a hydro-peroxide}$$

In the actual chamois process where the skins, containing both oil and water, are stoved at 50–60°C, further reactions occur. Thus aldehydes may be formed:

$$- \underset{\underset{OOH}{|}}{CH} - CH = CH - \rightarrow \underset{\underset{O}{||}}{CH} - CH = CH$$

These may lead to some cross-linking (tanning) of the collagen macromolecules. Highly viscous products, probably containing epoxide groupings

$$- CH - CH -$$
$$\diagdown \diagup$$
$$O$$

are also obtained by polymerization and these contribute to the character of chamois leather by coating the collagen fibrils with hydrophobic materials (Reed 1966).

Generally chamois tannages use small amounts of aldehydes (e.g., ½–1% on pressed weight) as a pretannage before normal cod liver oil tannage (Shaphause 1975). Nowadays formaldehydes are generally not used as tanning agents but there are some chemicals on the market which are glutaraldehyde based.

4 Iodine Values

For tanning operation of fish oils, iodine values are also important.

As a rule, which is valid to a certain extent, it can be said that fish oil should have an average iodine value of 120 to 160 and should not have too low an acid value (O'Flaherty et al. 1958).

The iodine value of a fat or wax is a measure of the degree of unsaturation, i.e., the number of double bonds in the compounds present. Most of these double bonds will be in the fatty acids (and fatty alcohols in the case of waxes), but unsaturation in any other compounds present will also be returned by the iodine value. Iodine value indicates the degree of unsaturation but not the type (White 1999).

For animal body fats, e.g., beef tallow, they are usually within the range 35–70; for vegetable oils, 85–200, while marine oils (sperm and fish oils) form a compact group of highly unsaturated glycerides, with iodine values 140–210 (Reed 1966).

Klenow has shown that the iodine value combined with the acid value makes possible a sufficient indication of the tanning behaviour of the oil. This view is energetically contradicted by Stather (O'Flaherty et al. 1958).

According to Sagoschen and Czepelak a 10 to 18 percent free fatty acid content is most favorable for chamois tanning. Chambard cites that even 20 to 25 percent of free fatty acids is desirable (O'Flaherty et al. 1958).

Content of Iodine Values, Vitamins A and D and Other Analytical Data For Some European Marine Oils are listed in Table 3.

There are three methods of determining iodine value in use, those of Wijs, Hanus and Hubl. The first two are the most common, and are official methods, with Wijs generally preferred; they differ principally in the type of iodine halide employed (White 1999).

Also valuable is the saponification value (number of fatty acids in ester combinations) in determining changes in the fish oils in autoxidation (O'Flaherty et al. 1958).

As the result of the changes of the characteristic indexes, increase of acid and saponification values and decrease of the iodine value and the discrepancy, it can be recognized that during autoxidation the highly unsaturated fatty acids, especially, are oxidized or decomposed. The rise of the acid and saponification values can be attributed to a considerable increase in fatty acids, which arise partially from saponification of glycerine esters and partly through oxidation. This explains the increase of the saponification value (O'Flaherty et al. 1958).

Table 3. Content of Iodine Values, Vitamins A and D and Other Analytical Data For Some European Marine Oils (Sand 1967).

Oil	Iodine Values	Sapon. Value	Unsap. Per Cent	I.U. per Gm.	
				Vitamin A	Vitamin D
Cod-liver	160–173	182–187	0.6–1.3	800–1500	60–120
Saith-liver	160–175	182–188	0.6–1.3	800–2500	80–200
Ling-liver	135–160	180–185	1.0–2.2	2000–5000	100–200
Haddock-liver	160–180	184–190	0.6–1.5	800–1200	50–100
Dogfish-liver	110–145	150–165	8–13	1000–2000	10–30
Porbeagle-liver	170–190	178–185	1–3	1000–1500	50–100
Halibut-liver	150–165	165–175	Ab. 10	20,000–200,000	1000–2000

5 Major Issues to Consider in Chamois Processing and Chamois Products

For chamois leather production, sheep skins are generally used. Chamois leather was originally the tanned skins of the goat like antelope called 'chamois' by the French. Basically, however, all types of light and loose-textured skins can be worked into chamois leather, especially the pelts of sheep, goats, reindeer, dogs, rabbits, etc. This process can also be used for fur tanning (O'Flaherty et al. 1958). Turkish, French or English originated sheep skins are preferable.

Beamhouse processes are nearly the same as garment leather production but these leather skins must be more thin and soft. The skins must be loosened up during liming much more intensively than for other tannages, to obtain the desired softness and stretch. Obtaining the required fineness and tenderness on the leather is one of the most important processes required. The grain layer is the removed section of the skiver by splitting or shaving, the flesh side being used for chamois and the outer piece for other types of leather, generally for garment leather.

The following actions are currently used at a modern leather tannery:

Sheep skin is preferred as the material at this tannery. Skin with excessive fat and veins is not appropriate for chamois production. Goat skin is not a proper product for chamois production either. High-quality chamois can be produced from British, French-origin skins or local Turkish sheep skins. Thus, British, French or Turkish pickled skins are preferably purchased and used.

Chamois production has certain key points. One of them is splitting the sheep skin, and the flesh layer of the split is ideal for chamois production. Sub-processes are conducted as normal, and the skins are held through protection-aimed pickle. In this process, the leathers' pH is brought to 1 with 10–12 Baume salt density. Such skins may be stored. The skins in protection pickle go through splitting process. However the skins need to be bloated for slitting. First the skins are washed in the mixers with plenty of water and detergent. The same process can also be realized in the drum. Non-ionic tensit is used at that moment. 30 kg tensit is given to 2000–2500 pieces of skins. The skins are washed for 30 minutes to one and a half hour and then the wash is filtered. After that, the skin continues being washed just with water until Baume equals to 0. After that process, the skins are run with HCl for half an hour. Around 15 kg HCl is given to 2000–2500 skins. At that moment, pH is at the 3–3.5 level. Then another material (SF 55) is added. This material is used for preventing the slickness of the leather. It provides the skin with a non-skid surface. 30–35 kg of this material (SF 55) is used for 2000–2500 skins. Not using this material causes a problem at the splitting machine as the skins become very slippery. So, the skins are made ready for the splitting process. The skins are swollen for approximately one finger thickness and have an anti-skidding structure. The skin is divided into two with the splitting processes. Here, the key point is to keep the layer with hair roots under the grain, and prevent its separation. The ratio of splitting in this respect is adjusted visually. The skins are classified as thin, medium and thick, and splitting is adjusted on the basis of practical experience. Here, the important point is to keep the hardness-giving portions of the skin within the grain. Depending on the skin's status, it can be adjusted with 60% grain side, 40% flesh side or 50% grain side and 50% flesh side. The skin's grain section is called grain flesher and it is separated for apparel leather production. The bottom section, or the 'sub-slit' is called flesher and chamois is made of that portion.

After splitting, the flesh section of the skins is separated and they are pickled again. Salt is adjusted to pH 1 at 10–12 baume density. Those skins are taken from the drum onto pallets, and rested for nearly one week until their water is drained. The skins in that form look like a thin handkerchief. The resting period is minimum one week and can be extended by up to one month. Then the flesher is weighed.

100% water, 3–4 baume
0.75% Relugan GT-50 (glutardialdehyde BASF)- 1 hour
2% Sodium formate- 30 min
1% Sodium formate-30 min
1.5–2% Sodium carbonate- 1 hour pH=6–6.5
Washes- 4–5 washes → Baume equals to 0, there will be no salt left in the flesher.

The drum is drained. Fleshers are squeezed in wringers with special mats to drain their water. The wringer has mat rollers on both sides (Minimising the water content to allow oil penetration, Covington 2009).

The fleshers are weighed.

The fleshers are taken to the oil tannage drum. The drum has a special rack system. The posts in other drums are not provided in this drum. A 3.5x3.5 drum size is enough for 2500 fleshers.

Here, another key point to note is the 10–20% difference in fleshers' weight before and after squeezing. Flesher weight to squeezed weight must change by a minimum of 10% and a maximum of 20%. The squeeze must be adjusted very carefully; otherwise, the fleshers will be too dry. If the weight ratio of fleshers is very low, then water is added to them.

After the fleshers are taken to special drums, 28% oil per squeezed weight must be given. The drum is automatic and it has a special system which works during its run. During oxidation, the water in the fleshers is removed and oil is infused instead. It attracts the water like a fan and discharges it. As the drum revolves, the water automatically vaporizes and discharges. Steam is continuously supplied to the drum. The temperature in the drum is 52°C. When the drum's internal temperature is 52°C, the system automatically shuts down vapour. So the heating of the skin is ensured and the system keeps itself at the same temperature. The drum continuously revolves for 2 days until the temperature in it reaches 52°C. After 2 days, when the discharge temperature reaches 52°C, the drum is revolved for another 6 hours. Automatic steam is cut and cooling is started. Another feature of that drum is the special cooling system in it. There are special cooling towers in the drum. Owing to this cooling system, the fleshers are cooled in 1 hour in winter and in 3–4 hours in summer. As a result of this process, leathers in dark yellow colour arise. The excessive oil of the skin must be removed. The skins in this condition are put into a Bowe machine and washed twice. 300 lt 5 min per chlore ethylene is used. The drying action is automatically carried out on the Bowe machine. Thanks to that, the excessive oil on the skin is removed, the skin dries and turns into light yellow colour. Then, the mechanical processes begin. The leather is slightly moisturized without being soaked. This process is critical because the chamois leather will have spots if it gets wet. The water is sprayed in grains in order to avoid any stains on the chamois, and the process is realized in the tanning drum. Then, the leather is taken for the staking process and then buffed. During buffing, the split surface (also the flesh surface if necessary) is removed with 220 grade paper. After dusting, the leathers become ready for sale.

Acknowledgements

The authoresses would like to express their gratitude to Serap Purçukyaş Tosun, Işıl Öztuvgan and Ali Yorgancioglu.

References

Bayramoğlu, E.E. 2006. Balık Yağının Deri Sanayinde Kullanımı (Use of Fish Oil in Leather Industry). AquaCulture and Fisheries. 1,1, 48.

Brockerhoff, H., R.G. Ackman and R.J. Hoyle. 1963. Specific distribution of fatty acids in marine lipids. Arc. Biochem. Biophys. 100: 9–12.

Brockerhoff, H. and R.J. Hoyle. 1963. On the structure of the depot fats of marine fish and mammals. Arch. Biochem. Biophys. 102: 452–455.

Covington, A.D. 2009. Tanning Chemistry, The Science of Leather. RSC Publishing. 315 p.

Garcia-Morena, P.J., R.P. Galvez, A. Guadix and E.M. Guadix. 2013. Influence of the parameters of the Rancimat test on the determination of the oxidative stability index of cod liver oil. LWT - Food Science and Technology. 51(1): 303–308.

Gruger, E.H. 1967. Chapter 1, Fatty Acid Composition. M.E. Stansby (ed.). Fish Oils, The Avi Publishing Company. 3–6, 15.

Gustavson, K.H. 1956. The Chemistry of Tanning Processes, Academic Press Inc., 295, New York.

Lovern, J.A. 1962. The lipids of fish and changes occuring in them during processing and storage. In Fish in Nutrition. E. Heen and R. Kreuzer (eds.). Fishing News Ltd. London.

O'Flaherty, F., W.T. Roddy and R.M. Lollar. 1958. The Chemistry and Technology of Leather, Volume II- Types of Tannages, Krieger Publishing Company.

Reed, R. 1966. Science for Students of Leather Technology. Pergamon Press. Great Britain. 117-118, 278p.

Sand, G. 1967. Fish Oil Industry in Europa. M.E. Stansby (ed.). Fish Oils. The Avi Publishing Company. 413.

Sharphause, J.H. 1985. Journal of Leather Tec. Chem. 69(2): 29.

Sharphause, J.H. 1975. Leather Technician's Handbook, London. 150–155.

Suparno, O., E. Gumbira-Sa'id, I.A. Kartika, M. Mubarek and S. Mubarek. 2011. An Innovative New Application of Oxidizing Agents to Accelerate Chamois Leather Tanning. Part 1: The Effects of Oxidizing Agents on Chamois Leather Quality. Jalca. 10: 360–367.

White, T. 1999. Oils, Fats, Waxes and Fatliquors. Leather Technologists Pocket Book, Chapter 6.

5

Fish Protein Coating to Enhance the Shelf Life of Fishery Products

V. Venugopal Menon

1 Introduction

Fishery products are highly perishable, because of their susceptibility to contamination by spoilage causing microorganisms. Fish lipids are highly sensitive to oxidation and associated flavor changes and to denaturation of proteins. Fishery products are also prone to contamination by diverse pathogenic microorganisms, threatening consumer safety (Huss 1995; Venugopal 1990). Several technologies are available for preservation of fishery products, which include chilling, freezing and individual quick freezing, cook-chilling, coating and battering, radiation processing, modified atmosphere packaging and high pressure processes, among others (Venugopal 2006). Conventional processes like chilling and freezing have limitations. Whereas chilling offers only limited extension in shelf life, prolonged frozen storage leads to dehydration, lipid oxidation, protein denaturation and accumulation of copious quantities of drip, which affect shelf life and consumer acceptability of the products. Water glazing of frozen items before storage is being commercially practiced to minimize the problems to some extent (Huss 1995; Ashie et al. 1996).

Formerly of Food Technology Division, Bhabha Atomic Research Centre, Mumbai 400085 India.
Present address: B-602 Skyline Villa, Opp. IIT Main Gate, Powai, Mumbai 400076 India.
   Email: vvenugopalmenon@gmail.com

2 Edible Films and Coating

In recent years, there has been an interest in using eco- and consumer friendly processes complimentary to conventional techniques to enhance safety and storage life with a view to better distribution of food products in the 'as-is' conditions. Use of edible films is an up-coming novel technology. These films are variedly designated as biodegradable, biocompatible, environmentally friendly, renewable, biopolymer, edible or green. Like synthetic films, these films help preserve food quality by preventing contamination by spoilage-causing and pathogenic microorganisms, controlling moisture transfer, reducing rancidity through minimizing oxygen uptake and loss of volatile aroma compounds. In developing such films, control of moisture migration from composite food products or between food and its environment remains a major challenge. Water vapor permeability (WVP), therefore, is the most important functional property of a food film, whether synthetic or edible. WVP comprises sorption, diffusion and adsorption of water molecules and is largely governed by the interactions between the polymer and water. Oxygen permeability of films has a large influence on lipid oxidation and associated sensory changes in the food. Films having poor oxygen permeability retard lipid oxidation. Poor gas permeability also prevents loss of odor bearing compounds from the food matrix. The effectiveness of edible films and coatings in the protection of food depends primarily on the thickness and permeability characteristics of the film. Care must be taken to prevent unattractive surface appearance of coatings, exogenous impact of the coating materials including color, aroma, taste, and texture, which can influence consumer acceptance of the coated products. Once a satisfactory film is applied, it remains on the product during storage and will be disintegrated or dissolved during cooking or the mastication process (Falguera et al. 2011; Dangaran et al. 2009; Cisneros-Zevallos and Krochta 2003).

Raw materials for environmentally friendly edible coatings may be derived from agro-wastes, dairy or microbial sources. These materials include macromolecules such as polysaccharides (starch, starch derivatives, cellulose, pectin, alginate), proteins (gelatin, casein, wheat gluten, zein, whey, soy protein), or lipids (beeswax, acetylated monoglycerides, fatty alcohols, fatty acids). Along with them, polyols such as glycerol or polyethylene glycol are used as plasticizers, while weak acids like acetic or lactic acid are used to regulate the pH. The advantage of the above mentioned macromolecules are that they can be recovered from the processing discards of ever-growing agriculture wastes and possess a high degree of biodegradability, a property by which they are amenable to degradation to carbon dioxide within a reasonable period of time (Venugopal 2011; Siracusa et al. 2008; Dutta et al. 2006; Tharanathan 2003). Edible films have been examined for various food items including fishery products. For instance, a sodium alginate coating could maintain the quality of rainbow trout fillets during chilled storage over a period of 20 days when the fillet samples were coated with an aqueous solution of 3% (w/v) sodium alginate (Hamzeh and Rezaei 2012; Phadke et al. 2011 a,b). Similarly, coating in aqueous solution of chitosan (1.5%, w/v) along with 1% acetic acid and 0.5% tea polyphenol significantly reduced growth of microorganisms and spoilage of chilled stored freshwater fish (Aider 2010; Zhang et al. 2009).

Several proteins have been examined as coating for diverse food products including muscle foods (Krochta et al. 1994). Traditionally, soy films have been employed in the Orient to wrap and shape ground meat or vegetables. Similarly, gelatin has been used as sausage casing. The storage life of meat has been extended with a surface coating of collagen and gelatin. The coating provides a water and oxygen barrier and could reduce purge, color deterioration, aroma deterioration, and spoilage (Antoniewski et al. 2007).

3 Edible Films from Myofibrillar Proteins

Before discussing the uses of myofibrillar proteins as coating material, general characteristics of these proteins will be briefly mentioned. Myofibrillar proteins constitute about 55% to 60% of the total muscle tissue proteins, or about 10% of the weight of skeletal muscle. These proteins are sparingly soluble in water and are highly sensitive to heat denaturation. They are soluble in salt solutions and their solubility in intermediate or high ionic strength solutions is used in their extraction from muscle tissues (Huss 1995; Ashie et al. 1996; Asghar et al. 1985). Myosin, the major structural protein of vertebrate muscle, including fish muscle, has a molecular weight of about 500 kDa. It is composed of two heavy (~200 kDa each) and four light (~20 kDa each) chains. The N-terminal halves of each heavy chain fold into two globular heads, and the C-terminal ends fold into distinct α-helices that wrap around one another to form a long (160 nm), fibrous, coiled-coil tail. Each globular head is composed of ~850 amino acid residues contributed by one of the heavy chains and two of the light chains. The myosin head is highly asymmetric, the secondary structure being dominated by many long α-helices. The globular heads contain the nucleotide-binding sites (for adenosine-5'-triphosphate hydrolysis) and regions required for binding actin to form the actomyosin complex (present in post-mortem muscle), whereas the long rod-like portions of myosin form the backbone of the thick filament of the muscle structure. More than half of the amino acids that constitute myosin are hydrophilic. Most of these residues are exposed to the molecular surface, which may be in contact with water. The high degree of flexibility of the myosin molecule arises through the rotation of the amino acid side chains and the torsional motions of the polypeptide backbone, which give rise to numerous conformations both at the head-tail junction and at locations within the tail end. The internal flexibility provides the protein with a dynamic foundation for biological functions in foods, ranging from enzyme catalysis to gelation, texture and oil emulsification activity (Muller 1992; Niwa 1992; Kristjansson and Kinsella 1991; Otting et al. 1991; Ludescher 1990).

Application of myofibrillar proteins as coating for muscle foods, which are rich in these proteins, is novel and product-friendly. Protein films based on myofibrillar proteins have to be cast by either solvent or thermoplastic processes. The "solvent process" is based on the casting of a film-forming solution. The thickness of the film is dependent on viscosity of the solution, which is influenced by protein concentration and pH. Standard conditions for casting the film involve use of protein solution at concentration of 2.0% (w/v), 100 g film forming solution, pH 3.0 and a casting temperature of 25°C. The functional properties of the prepared film are slightly better than those of other protein-based films, with its tensile strength being close to that of low density polyethylene films (Aymard et al. 1995). Study of myofibrillar protein films revealed that elongation at break and relaxation coefficient were not dependent on film thickness, while force at break was directly proportional to film thickness. The water vapor transport of the bio-packaging did not behave as an ideal material according to Fick's and Henry's laws (Cuq et al. 1996). The "thermoplastic process" makes use of glass transition temperature of myofibrillar proteins. The glass transition usually occurs between 215° and 250°C for the dry protein; addition of water or hydrophilic plasticizers (sucrose and sorbitol) induced large decreases in the glass transition temperature. Hydration and temperature effects on functional properties have been found to be correlated to the glass transition temperature (Gontard et al. 1998). Large scale production of myofibrillar proteins films, however, poses problems. Making such films by casting or thermo-molding methods, as mentioned above, are labor intensive and time consuming. Furthermore, such films are difficult to heat seal. In addition, extended storage of the films may result in their dehydration, making them brittle. Therefore, commercial applications of such films are rather difficult.

A novel approach to using myofibrillar proteins as film is to provide a direct wet coating of gel of myofibrillar proteins on muscle foods. The coating can provide almost all the benefits of a protein film. Such an approach has been suggested to enhance shelf stability and hence marketability of high value fishery products such as fillets, steaks, shell-fish, etc. by giving the products coatings of structural proteins from underutilized, low cost fish meat. The coating solution could be made from the washed meat of the low cost fish or meat trimmings obtained from fish filleting operations (Venugopal 1998). The process involves initial conversion of the proteins into a gel by acidification using weak acids, followed by the development of free-flowing myofibrillar protein dispersion as coating material. These aspects are described below.

4 Gelation of Fish Myofibrillar Proteins

4.1 Gelation at neutral pH conditions

The ability to form gel is an important functional property of macromolecules, including proteins. A protein gel is an intermediate form between solid and liquid phases in which strands of protein chains are cross-linked to form a continuous three-dimensional network. In the native fish muscle, several low molecular weight compounds and enzymes are adhered to myosin and actomyosin, which hinder their interactions with water, leading to poor solubility of these structural proteins under physiological conditions. Washing of fish meat removes the soluble components such as pigments, enzymes and lipids adhering to the proteins, liberating polar sites for interactions. The washed proteins have near neutral pH. Removal of soluble components alters the electrostatic balance, resulting in unfolding the myosin molecules, bringing the buried non-polar side chains into contact with water. The 'cages' of water molecules that surround the non-polar side chains are stabilized by hydrogen bonding. The surface hydration shells provide the flexible matrix that enables the polypeptide chain to respond efficiently to environmental changes (such as alterations in pH and temperature), and also helps to stabilize the hydrated proteins. Such stabilization involves a multitude of non-covalent interactions including hydrophobic interactions, van der Waals' forces, hydrogen bonds, ionic interactions and also formation of covalent disulfide bonds (Totosaus et al. 2002; Stone and Stanley 1992; Sharp and Offer 1992; Niwa 1992). Development of restructured foods from fish meat involves initial preparation of surimi, which is thoroughly washed concentrate of fish muscle structural (myofibrillar) proteins. Gelation of the proteins, which is optimal at pH 6 and at 60° to 70°C, essentially involves three steps via dissociation of myofibril structure by protein solubilization in the presence of salt, partial unfolding of myosin structure caused by heating and irreversible aggregation of unfolded myosin to form a three-dimensional structure (Lanier and Lee 1992; Hultin 1985). During the aggregation process water, oil and flavoring compounds may be entrapped in the gel matrix and these influence the yield, texture, and cohesion of the final product. The rigidity of the gel could be suitably adjusted by incorporating ingredients such as salt, starch, polyphosphate or proteins from other sources in the matrix (Venugopal 2006; Totosaus et al. 2002). Considerable research has been pursued to understand the gel formation by myofibrillar proteins because of its role in the development of texturized foods. Current interest in this area rallies around types of muscles (beef, pork, poultry, fish, and rabbit), interactions of actin and myosin with each other and with fat, gelatin, starch, hydrocolloids, soy protein, whey, and non-protein additives such as phosphates and acidifiers, and the influences of pH, ionic strength, rates of heating, protein oxidation, as well as the use of transglutaminase and the influence of high hydrostatic pressure. The observations that solubilization of these proteins is not always essential for gel formation, and that good gels can be formed in the absence of salt, are exciting possibilities for development of texturized meat products having low salt and fat (Sun and Holley 2011).

4.2 Gelation under mild acidic conditions

Classic surimi gelation occurs as a result of heat-induced unfolding of proteins at neutral pH conditions, as mentioned above. Theoretically it should also be possible to make the protein unfold under controlled, non-thermal conditions such as weak (mild) acidic pH (Fretheim et al. 1985). Gelation of fish muscle structural proteins at acidic pH has not been studied in detail. However, the two main advantages of this type of gelation are: (i) there is no need to add salt to dissolve the myosin as in the conventional gelation process, and (ii) because of the extreme pH of the gel, the product has comparatively higher microbial stability. Weak organic acids such as acetic and lactic, which have low ionization potential, are ideal to slowly lower the initial pH, facilitating unfolding of the myofibrillar proteins. Inorganic acids cannot be used since they bring about drastic fall of pH, causing precipitation of the proteins. The protein strands that unfold as a result of mild acidification can aggregate among themselves to form the gel. The aggregation can be accelerated by subjecting the acidified proteins to controlled heat treatment. This also offers the possibility for development of salt-free or low sodium restructured products (Venugopal et al. 1994; Riebroy et al. 2005).

Similar to surimi gel, for weak-acid induced gelation too, the fish mince needs to be initially washed thoroughly in plenty of potable water. Generally, the eviscerated fish is subjected to mechanical deboning and the meat is subjected to the washing process, which may be suitably modified, depending upon the fish employed. For fatty fish such as mackerel and herring, it is advisable to use dilute aqueous solution (such as 0.5% w/v) of sodium bicarbonate in the second washing to help remove adhering lipids. Cold water treatment, together with the shift in pH brought about by the bicarbonate, significantly aids the removal of lipids. Fish mince having an unappealing dark color, such as that of capelin, may be decolorized by washing with dilute aqueous solution (1–2% w/v) of sodium chloride and sodium bicarbonate, which facilitates removal of the pigments, before the bicarbonate wash (Venugopal et al. 1995). In the case of light colored, low fat fish meat such as that of threadfin bream or Bombay duck, washing with fresh cold water is repeated twice, keeping the holding time of 1 hr for each wash (Kakatkar et al. 2004a; Venugopal et al. 1995). In case mechanical deboning is difficult. as in the case of the Indian dog shark (*Scoliodon laticaudus*), the fish is skinned and cut into pieces (4–5 g, average size). The pieces are washed by overnight soaking in excess of cold water, followed by draining (Venugopal et al. 1994). The washing procedure employed for different fish species is given in Table 1. Repeated washing results in significant reduction in the contents of low-molecular-weight proteins and increase in concentration of myosin and actomyosin, thereby enhancing the ability of myofibrillar proteins to undergo gelation. Studies on dynamic viscoelastic behavior of washed and unwashed meat has also revealed a structure build-up reaction that was more pronounced in the washed meat. It has also been reported that washed fish meat also possesses a higher emulsion capacity, depending upon the number of washing cycles, which is relevant to product development (Mathew et al. 2002).

In order to make gel, the washed meat is homogenized in equal amounts (w/w) of cold (5–8°C) water, which is then acidified by drop-wise addition of small amounts of glacial acetic acid to bring the pH to about 3.5. During acidification, the homogenate is gently stirred. Usually, concentration of the acid at 0.5% of the slurry is sufficient. Incorporation of the acid induces gelation of the proteins, which could be detected by suitable rheological methods. In most cases, except in the case of washed shark meat, acidification by weak acids resulted in decrease of apparent viscosity of the homogenate. Weak acid induced gelation has been examined with the washed meat of a number of fish species, including Atlantic cod, Bombay duck, freshwater fish, rohu, shark, threadfin bream, capelin, Atlantic mackerel and Atlantic

Table 1. Procedure for washing fish meat in water and acetic acid-induced gelation of washed meat.

Fish	Washing and gelation procedure	References
Atlantic herring	Three step washing in chilled (<10°C) water in mince to water ratio of 1:3. Second wash in water containing 0.5% (w/v) each of bicarbonate and NaCl to remove bound lipids and pigments. Washed meat homogenized in equal amount of chilled water. Mixed with acetic acid to lower pH to 3.5.	Shahidi and Venugopal 1994
Atlantic mackerel	Three step washing in chilled (<10°C) water at mince to water ratio of 1:3. Second wash in 0.5% (w/v) bicarbonate to remove bound lipids.	Venugopal and Shahidi 1994
Capelin	Three step washing in chilled (<10°C) water in mince to water ratio of 1:3. Draining after each wash. Washed meat homogenized in equal amount of chilled water.	Venugopal et al. 1994
Indian mackerel	Three step washing in chilled (<10°C) water in mince to water ratio of 1:3. Second wash in 0.5% (w/v) bicarbonate to remove bound lipids.	Venugopal et al. 1998
Shark	Meat pieces (4–5 g) each soaked in equal amounts of chilled water, kept overnight at 0–2°C, drained and washed twice with equal amounts of fresh chilled water.	Venugopal et al. 1994; 2002
Threadfin bream	Three step washing in chilled (<10°C) water in meat to water ratio of 1:3. Second wash in 0.5% (w/v) bicarbonate to remove bound lipids.	Chawla et al. 1996
Bombay duck	Fillet pieces (4–5 g each) soaked in equal amounts of chilled water kept overnight at 0–2°C, drained and washed with equal amounts of fresh chilled water twice.	Kakatkar et al. 2004a

herring (Kakatkar et al. 2004a; Panchavarnam et al. 2003; Venugopal 1997; Chawla et al. 1996; Venugopal et al. 1995; Venugopal et al. 1994; Venugopal and Shahidi 1994; Shahidi and Venugopal 1994). The rheological properties of the gel will be discussed later.

4.3 Gelation under alkaline conditions

Though not studied in detail, gelation of fish proteins under alkaline conditions has also been observed. Washed fish meat treated with ammonium hydroxide has been found to form gel. Solubility of rockfish whole muscle and actomyosin was minimal at pH 5 and gradually increased as the pH was shifted to alkaline pH. Alkali-treated muscle proteins readily aggregated upon heating, showing dynamic rheological patterns, which were different as compared with those of whole muscle and washed mince. Disulfide linkages occurred at a greater extent in gel prepared by alkaline solubilization, resulting in higher breaking force and deformation. Like acid-induced gelation, alkaline solubilization was also associated with degradation of myosin heavy chain, with the formation of a protein band of about 120 kDa (Yongsawatdigul and Park 2004). Further, the gelation of the proteins under alkaline conditions could also be favored by electrostatic interactions of the proteins, as discussed above. Gel of this type has been used to prepare edible coatings (Aymard et al. 1995).

5 Application of Weak Acid-induced Fish Myofibrillar Protein Gel as Direct Edible Coatings

5.1 Development of free-flowing myofibrillar protein dispersion from the gel

The process to make a free flowing aqueous solution of myofibrillar proteins makes use of the gel forming properties of the proteins under acidic or alkaline conditions. The advantage of gelation at extreme pH values is that gels of fish proteins formed at this pH, in contrast to those made at neutral pH values, are less viscous. Acidification results in rapid fall in apparent viscosity of the water homogenates of most washed fish meat. The fall in viscosity was further accelerated by mild heating (up to 50°C) of the acidified slurries, resulting in a free-flowing protein dispersion (see Rheological aspects). The acid-treated heated washed meat dispersion could be suitably diluted with water to get free-flowing dispersions having protein concentrations in the range of 1–3% (w/v). Such protein dispersions have been developed using many fish species including Bombay duck, freshwater fish rohu, threadfin bream, Atlantic mackerel and Atlantic herring (Venugopal 1997). Figure 1 shows the process for making protein dispersion. In the development of such fish meat dispersions, an exception has been noted in the case of the Atlantic fish, capelin. The slurry of washed fish mince in water forms a gel when subjected to mild heating, without the need of acidification by acetic acid (Venugopal et al. 1995). The free-flowing aqueous gel dispersion having required protein concentration (usually, 1–3% w/v) can be directly applied on the required fish product by employing processes such as dipping, spraying, enrobing, etc. as will be pointed out later.

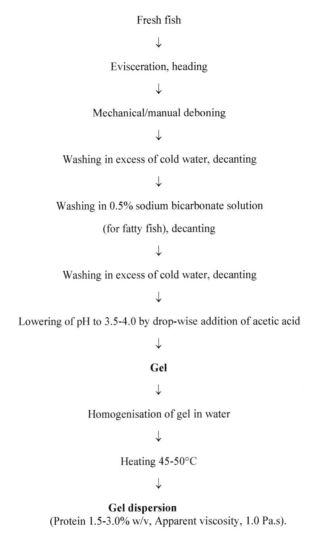

Fresh fish

↓

Evisceration, heading

↓

Mechanical/manual deboning

↓

Washing in excess of cold water, decanting

↓

Washing in 0.5% sodium bicarbonate solution

(for fatty fish), decanting

↓

Washing in excess of cold water, decanting

↓

Lowering of pH to 3.5-4.0 by drop-wise addition of acetic acid

↓

Gel

↓

Homogenisation of gel in water

↓

Heating 45-50°C

↓

Gel dispersion
(Protein 1.5-3.0% w/v, Apparent viscosity, 1.0 Pa.s).

Figure 1. General procedure for preparation of acetic acid induced gel and thermostable gel dispersion of fish meat in water. Procedures for washing different fish are also given in Table 1.

6 Characteristics of Weak Acid-induced Gel of Fish Meat and Its Aqueous Dispersion

6.1 Affinity for water

Most functional properties of proteins in foods are related to their interactions with water. Water is bound to the proteins in one or more basic forms, namely, structural (bound) water, which is unavailable for chemical reactions, hydrophobic monolayer water, non-freezable water, capillary water, hydrodynamic water and hydration water. Of these, hydration water makes proteins maintain their structural integrity so as to render them functionally active (Kinsella 1982; Asghar et al. 1985). Washing of fish meat results in an enhanced affinity of the proteins to water, resulting in increase in weight of washed fish meat, despite the removal of the sarcoplasmic fraction of the muscle (Wu et al. 1992). Mild acidification results in further imbibing of water in the gel network. This is true even in the case of fish such as Bombay duck, the fresh muscle of which has moisture content as high as 90%. Gelation of the washed Bombay duck meat in the presence of acetic acid also resulted in significant hydration of the proteins, which swelled, absorbing water up to 11 g per g protein during the gelation process. The water that was imbibed during gelation was strongly bound to the proteins, and could not be separated even by high speed centrifugation (Kakatkar et al. 2004a; Venugopal 1997).

6.2 Rheological properties

One of the important properties of fish protein gels, including the conventional surimi gel, is their viscoelastic character, which make the gels behave as elastic solids, exhibiting viscous flow. These properties have direct relations with the texture of the products developed from the gels. Measurement of viscoelasticity involves identification of responses of the gels to both, large and small stresses. The response to large stress is usually determined by 'texture profile analysis', using texture meters such as the Instron Universal testing machine (Bourne 2002). A major advance in this field is the dynamic measurement of small deformations in the gel under either constant or sinusoidal oscillating stress. The controlled stress approach, where the measurement is based on displacement (rotational speed) in response to an applied torque (stress), provides subtle changes in the gel, indicative of its viscoelasticity. Dynamic rheological tests based on controlled stress are widely used to study heat-induced gelation of surimi (Oakenfull et al. 1997; Hamann 1994; Lanier and Lee 1992; Holcomb 1991). The viscoelastic properties of acid induced shark gel showed higher storage modulus G' (a measure of elasticity or energy stored), than loss modulus, G" (a measure of viscosity, or energy lost). The G' increased with decrease in moisture content, suggesting higher rigidity of the gels at lower moisture contents. Gelation process was also monitored by measuring changes in the stress-strain phase angle during oscillatory testing, where the loss tangent (referred to as tan δ = G"/G'), gave information on sol-gel transition. Relatively low values of tan δ indicated the elastic nature of the gel (Kok et al. 2009; Venugopal et al. 2002). Heating of acid-induced surimi gel up to

43°C did not affect storage modulus, suggesting that this type of gelation did not involve heat-induced setting unlike the salt-induced one. Further, salts such as KCl and NaCl weakened the firmness of the acid-induced gel (Lian et al. 2002).

Flow behavior properties of protein gel dispersions such as viscosity and consistency are important in optimization of processes such as heating, pumping, mixing and concentration. These properties are governed by the composition of proteins as well as their molecular shape, size and charge, and also by factors such as temperature, concentration, pH and ionic strength (Tung 1978; Rao 1977). Rheological properties of protein gel dispersions of washed meat of fish including Bombay duck, capelin, Atlantic mackerel, Atlantic herring, Indian mackerel, and threadfin bream have been studied (Venugopal 1997). Acidification of water homogenates of washed minces of fish species (except shark) drastically reduces their apparent viscosities. Mild heating of the acidified slurry further accelerated fall in viscosity. In the case of capelin, apparent viscosity of the homogenate was significantly reduced by heating to 35°C or by overnight incubation at ambient temperature, even in the absence of acidification (Venugopal et al. 1995). All protein dispersions exhibited non-Newtonian flow behavior. The influence of shear rate on the apparent viscosity suggested the psuedoplastic nature of the dispersion (Venugopal et al. 1995). Table 2 shows the influence of acetic acid and heating on apparent viscosity of washed mince homogenates of different fish species.

Table 2. Influence of acetic acid on the apparent viscosity of washed fish homogenates in water.

Fish species	Protein content of water homogenate (%)	Apparent viscosity, before acetic acid addition	Apparent viscosity after addition of acetic acid to lower pH to 3.5	Apparent viscosity after addition of acetic acid (pH 3.5) followed by heating to 50°C
Atlantic herring	1.6	3.4	1.4	0.07
Atlantic mackerel	2.4	5.9	0.3	0.06
Capelin	1.9	3.0	A	0.04 (no acetic acid required)
Indian mackerel	1.5	4.0	0.5	0.05
Shark	2.3	12.0	600	>1200*
Threadfin bream	2.3	6.2	1.2	0.04

The viscosity values are denoted as apparent viscosity values, and expressed as Pa.s. *Viscosity of shark was measured using Branbender viscoamylograph and expressed as Brabender units. 'A' indicates samples precipitated in presence of acetic acid. Source: Adapted from Venugopal, V Functionality and potential applications of thermostable water dispersions of fish meat. Trends Food Sci. Technol. 14, 39, 1997, with permission from Elssevier.

6.3 Solubility of the proteins in the dispersion and their thermal stability

The proteins in the dispersion are highly soluble in water, which could not be precipitated from the gel or gel dispersion by centrifugation even at 12,100 X g for 30 min. It was interesting to note that the proteins remained in solution even when the dispersion was heated at 100°C for several minutes. The proteins in the dispersions were not precipitated by a combination of heating (100°C, 15 min) and centrifugation (up to 135,000 X g) either. It may be pointed out that such high solubility was not as a result of the hydrolytic breakdown of proteins into low molecular weight peptides, since the proteins could be precipitated out by precipitating agents such as trichloroacetic acid, suggesting that the proteins in the dispersions retained their high molecular weights. Table 3 shows the solubility of proteins in thermostable dispersions of mince from different fish species.

Table 3. Solubility of proteins in thermostable dispersions of mince from different fish species.

Fish species	Protein content of washed fish mince homogenate in water (%)	Protein content of gel dispersion (%)	Solubility (%) of proteins in dispersions after heating to 100°C for 10 min
Atlantic mackerel	2.40	2.35	98.0
Atlantic herring	1.30	1.28	99.0
Indian mackerel	1.50	1.30	86.0
Capelin	2.50	2.10	84.0
Shark	2.94	2.16	73.0

Source: Venugopal 1997, with permission from Elsevier

6.4 Influence of salts on acid-induced gelation and stability of the proteins in the dispersion

The presence of salts such as NaCl and $CaCl_2$ had drastic adverse effects on acid-induced gelation and stability of the proteins. Addition of traces of ionic salts to the washed meat prior to acidification totally prevented gelation. If the salts were incorporated in the prepared dispersion, it resulted in the collapse of the gel network and precipitation of the proteins. This property was characteristic of only ionic salts, while non-ionic salts did not adversely affect the protein stability. Therefore, there is potential to incorporate non-ionic additives including antimicrobials, antioxidants, texturizers, etc., before the acidification and heating, to enhance the functional properties of the dispersion and its film. Similar to ionic salts, almost all proteins in the dispersions were precipitated when they were heated after increasing their pH from 3.5 to 7.0 (Venugopal et al. 1997).

6.5 Chemical aspects of stability of proteins in the dispersion

The reasons for the extreme stability of the proteins in the dispersion can be attributed to involvement of electrostatic interactions among proteins (Venugopal 1997). Mild acidification can have a profound influence on the content of invisible amino acids in the myosin molecule. Formation of carboxylic groups due to acidification induces a net positive charge, which enhances electrostatic repulsions in the protein and leads to gelation and associated changes in viscosity of the gel as well as that of the dispersion. During the gelation process, water or solvent molecules become entrapped within the matrix, enhancing hydration of the protein molecules. The hydration water also protects the proteins against denaturation. In addition, hydrophobic interactions of head and tail portions of the molecules as well as formation of disulfide linkages in the gel could also favor stability of the proteins. These changes are strong enough to protect the proteins in the dispersion against precipitation even under extreme conditions of temperature and centrifugal force. Weak acid induced gelation is also associated with decrease in sulfydryl contents and the formation of disulfide bonds. This has been verified with respect to gels from threadfin bream, shark and Atlantic herring. Decrease in disulfide groups in the proteins of the acid-induced gel is comparable with that in conventional surimi gelation. Along with this, a decrease in myosin heavy chain (MHC) content is also indicated (Chawla et al. 1996; Shahidi and Venugopal 1994). The degradation of MHC in threadfin bream was indicated by concomitant appearance of a protein band of about 186 kDa. Solubility of the gel in various solvents containing sodium dodecyl sulfate, urea and β-mercapto-ethanol also suggested structural changes in myosin during gelation (Chawla et al. 1996). The thermostable dispersions of fish meat can provide information on the stability characteristics of proteins and their structure-function relationships (Doi 1993; Nosoh and Sekiguchi 1991). The ability of ionic salts on the dispersion could be explained by the fact these salts affect the balance of the electrostatic forces causing aggregation and precipitation of the proteins (Kristjansson and Kinsella 1991). However, more studies are needed to explain the properties of fish muscle proteins in acidified water dispersions. Table 4 shows a comparison of conventional surimi gelation with weak-acid induced gelation of fish structural proteins.

Table 4. Comparison of conventional surimi gelation with weak-acid induced gelation of fish structural proteins.

Characteristics	Conventional surimi gelation	Weak acid-induced gelation
Gelation pH	Neutral or slightly alkaline	Acidic pH (3.5–4.0)
Agents for gelation	Salt and mild heat for solubilization of the proteins in water	Weak organic acids such as acetic or lactic acid
Chemical changes	Formation of covalent (disulfide) and non-covalent linkages, degradation of myosin heavy chain, decrease in α-helix content and increase in hydrophobicity of myosin	Formation of covalent (disulfide) and non-covalent linkages, degradation of myosin heavy chain, decrease in α-helix content and increase in hydrophobicity of myosin?
Water holding capacity	Excellent	Excellent
Microbial stability of gel	Poor due to neutral or alkaline pH	Good due to acidic pH
Rheological characteristics	Viscoelastic nature	Viscoelastic nature
Influence of heat on rheological properties	No significant change	Rapid fall in viscosity of gel dispersion making it free flowing, exhibiting pseudoplastic behavior of water solubility of proteins in dispersion not affected by temperature
Influence of ionic compounds on the gel	Gel characteristics *not* affected	Adversely affect gel characteristics leading to precipitation of proteins
Applications	Restructured imitation products	Edible coating for fishery products, protein powders Restructured products Hygienic sauce by fermentation

7 Uses of Weak Acid-induced Fish Meat Gel Dispersion as Coating Material for Fishery Products

The thermostable, low-viscous fish meat dispersion prepared from weak acid-induced gel can find a number of applications, the most important use being as coating for preservation of fishery products during chilled or frozen storage. It has the potential to replace conventional water glazing of frozen fishery products. The dispersion coating provides thin layers of fish structural proteins on the surface of the food. The thickness of the coating can be controlled by using an appropriate concentration of protein in the dispersion. Ideally, it should have an optimum protein content of about 1.5–3.0% and an apparent viscosity of about 1.0 Pa.s. The protein used as coating is natural (from fish muscle itself) and hence can be eaten as part of the coated food product (Venugopal 1997). The presence of traces of acetic acid in the dispersion provides antimicrobial activity to the coating, enhancing microbial safety of the treated products during chilled storage (Lin and Chuang 2001). The dispersion has the potential to replace well studied coatings of other proteins. It can also be used for other animal products such as red meat and poultry products, which are also rich in myofibrillar proteins (Ustunol 2009; Cutter 2006; Gennadios et al. 1997). Some of the potential applications of the dispersions to improve the quality of fishery products are discussed below.

7.1 Shelf life enhancement of chilled fish

Acid-induced gel dispersion from the freshwater fish, rohu (*Labeo rohita*) meat has been successfully used to enhance the shelf life of fish fillets during chilled storage. The dispersion for coating was prepared from the washed rohu meat, according to the procedure described in Fig. 1. The dispersion had a pH, 3.5, protein content, 3%, and apparent viscosity of 1.0 Pa.s. Fresh rohu fillets were dipped in the chilled (0–3°C) dispersion for 1 hr. The treated steaks were packaged in 200 gauge polyethylene pouches and stored under ice. Chemical and sensory analyses showed that the dispersion-coated steaks gave a shelf life of 32 days as compared with 20 days for non-coated counterparts stored under the same conditions. The shelf life of the dispersion-coated steaks could be further enhanced by exposing them to gamma radiation at 1 kGy immediately after coating, which reduced the load of spoilage causing microorganisms in the fish. Although the acidic nature of the dispersion could cause some bleaching of fish pigments in the fillets during prolonged chilled storage, this could be prevented by incorporating antioxidants such as butyl hydroxyanisole or ascorbic acid at 0.5% (w/v) in the dispersion (Panchavarnam et al. 2003). Comparable results were obtained with steaks of seer fish also. Coating of seer (*Scomberomorus guttatus*) fish steaks with the dispersion prepared from the same fish enhanced the chilled storage life of the fish steaks from 15 to 20 days (Kakatkar et al. 2004b).

7.2 Frozen fishery products

The major quality losses during extended frozen storage of various fishery items, including whole or eviscerated fish, fillets and steaks, shellfish products such as peeled shrimp, lobster tails and cephalopods such as cuttlefish and squid items (both block frozen or individually quick frozen) are due to dehydration, freeze-burn (dehydration of the surfaces), hence there is weight loss, lipid oxidation and discoloration. Quality deterioration of fishery products during frozen storage is depicted in Table 5. Slow dehydration may take place in cold storage even under good operating conditions. Dehydration accelerates oxidation of fat and alteration of the proteins, causing adverse texture and appearance of the product. Frozen fish that have suffered severe drying in cold storage have a white, toughened, dry and wrinkled appearance on the surface, referred to as freezer burn. The skin of the thawed fish may have a similarly dry, wrinkled look, and if drying has been exceptionally severe, the flesh beneath can become spongy. Lipid oxidation reactions are particularly significant during prolonged frozen storage of fish fillets and blocks of meat mince obtained from fatty fish species (Connell 1995; Torry 1958). Glazing is an effective and economic means of protecting frozen fish and shellfish during storage and transport. Glazing is done by spraying cold water over the product or by dipping the products in chilled water, which freezes them in a few seconds. The treatment gives a protective layer of ice over the frozen product, which retards moisture loss and prevents oxidative rancidity. The skin of ice so formed will evaporate slowly while the fish is in cold storage. Regular inspection of the fish is necessary to ensure that the glaze is renewed when required (Torry 1958). Whole fish such as herring can be block frozen and stored successfully by placing the fresh fish in a polyethylene lined paper bag in a vertical plate freezer and topping up the bag with water to fill the voids before freezing (Santos and Regenstein 1990). Water glazing enhanced the shelf life of Indian mackerel, a popular fatty fish in the tropics, from 3 to 4 months to 6 months at –20°C (Joseph 2003). Commercially, shrimp is

Table 5. Quality deterioration of some fish during frozen storage.

Fish	Thawed state characteristics
Herring, whole	Skin and meat show rusting
Mackerel, whole	Surface dehydrated, skin and meat showing rusting, Rancid odor in skin and meat
Chinese pomfret, whole	Dehydration at lower part of belly and fin. Head and belly parts yellowish, discolored, sponge-like meat
White pomfret, whole	Skin and meat show rusting, sponge-like meat
Jew fish, whole	Slight rancid odor in skin
Lemon sole, whole	Surface dehydrated, spoiled and rancid odor, sponge like meat
Haddock fillet	No odor, sponge-like meat, cracks on surface
Cod fillet	No odor, sponge-like meat
Flounder fillet	No odor

Source: Adapted from Santos-Yap 1996

subjected to traditional block freezing or more popular individually quick freezing (IQF). After glazing the frozen products with water (ice cold water chlorinated at 5 ppm level), the material is stored ideally at –40°C. If the storage temperature is higher at –18°C, the storage life of IQF frozen shrimp is about a few months. Vacuum packing along with 8% water glazing can extend the shelf life up to 10 months at this temperature. Glazing may increase the product temperature to as high as 4° to 8°C in the case of small, cooked and peeled shrimp. Shrimp takes up water to the extent of 5 to 10% of its weight, or sometimes more, providing a coating of ice of 0.5 to 2 mm thickness (Londahl 1997). In a recent study, glaze uptake, effects of initial frozen shrimp temperature, glazing time and different glaze percentages on physical and chemical changes of frozen shrimp during storage were investigated. Shrimp were frozen in a spiral freezing machine (–35°C/15 min) and transferred to the air blast freezer until the core temperature reached –18°, –25° or –30°C. The product was submitted to the glazing process and stored at –18°C for 180 days. Glazing resulted in 15–20% uptake of water by the product (Goncalves et al. 2009). The beneficial effects of glazing could be enhanced by incorporation of additives such as disodium acidphosphate, sodium carbonate and calcium lactate, alginate, chitosan, antioxidants (such as ascorbic and citric acids), glutamic acid, and monosodium glutamate, corn syrup solids, chitosan, etc. in the water used for glazing (Aider 2010; Zhang et al. 2009; Santos and Regenstein 1990).

Fish protein dispersions can be an excellent replacement for water glaze to protect fishery products during the frozen storage. The advantage is that the main ingredient in the dispersion is fish protein itself and hence it is natural and safe. The use of fish dispersion to protect the mince of Indian mackerel during frozen storage has been reported. The mince, being that of a medium fatty fish, is prone to dehydration and associated lipid oxidation during the frozen storage. Frozen mackerel mince blocks without glazing lost 35% of its initial weight when stored at –18°C for 80 days. Conventional water glazing could not completely prevent dehydration and freeze-burn loss during frozen storage of the fatty fish products. Edible coatings from mackerel structural proteins could be advantageously used as coating for blocks of mackerel mince to prevent weight loss and rancidity development during prolonged frozen storage. Glazing of the mince blocks with the dispersion prepared from Indian mackerel structural protein dispersion (prepared according to Fig. 1) resulted in only 17% loss in weight, compared with 36% weight loss in water glazed samples, when stored up to 80 days under frozen conditions, as shown in Fig. 2. Furthermore, about 50% reduction in lipid oxidation was also observed in the dispersion-glazed mackerel mince blocks (Kakatkar et al. 2004b). In another study, effectiveness of protein dispersion coatings prepared from seer fish (*Scomberomorus guttatus*) against drip loss in frozen stored seer fillets has been reported. The

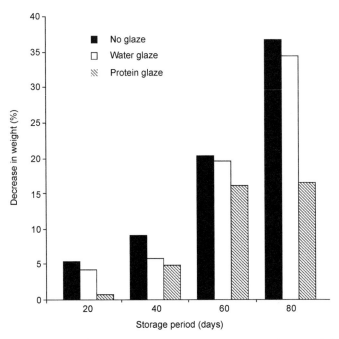

Figure 2. Influence of protein dispersion coating on drip loss based on weight loss in frozen mackerel meat mince during its frozen storage.

fish fillets were frozen in an air blast freezer (–35°C) and stored at –20°C (± 2°C) for six months. The fish fillets were coated using water glazing (WG) or seer fish protein dispersion coating (SPDC) and compared with the samples without coating (WC). It was observed that seer protein dispersion coated samples were significantly superior to WC and WG fillets and had minimum drip loss. Instead of seer fish protein dispersion, dispersion from any other fish such as croaker, a low cost fish, can also be used for coating the seer fish fillets (Phadke et al. 2011a,b). These results suggest that fish protein coating has significant advantages in controlling quality loss of frozen fishery products.

7.3 Potentials for modification of functional properties of the dispersion coating

As pointed out earlier, the protein dispersion coatings give protection to treated products against water loss, oxygen permeability, microbial invasion, etc. There is the possibility of enhancing these properties by the incorporation of additional preservatives in the dispersion before coating. For instance, the water barrier capability and tensile strength (TS) of gelatin/olive oil composite films were modified through emulsification. As the lipid droplets in the films decreased, the film exhibited an enhanced hydrophobic character and became an excellent barrier to ultraviolet light (Gomes-Estaca et al. 2009). Incorporation of natural additives including antimicrobials and antioxidants in the films offers another novel consumer-friendly technique to enhance stability and functional properties of the coating (Aider 2010; Zhang et al. 2009; Ma et al. 2012; Gomee-Estaca et al. 2006). Various bioactive compounds, and also nutraceuticals such as carotenoids, have been isolated from different sources, particularly of marine origin. Bioactive peptides having antimicrobial, antioxidant, antifungal, antiviral, and antitumor activities have been identified from several marine organisms (Kim and Wijesekaraa 2010; Venugopal 2006, 2009). Recent studies reported that trypsin hydrolyzates of fish proteins showed significant antioxidant activities in terms of free radical scavenging potential against polyunsaturated fatty acids (PUFA) peroxidation (Naquash and Nazeer 2010). Such peptides, and also chitosan, essential oils, etc. can be incorporated to enhance the preservation capacity, as also antioxidant capacity of the films. Active packaging materials containing antimicrobial peptides have shown effectiveness in inhibiting pathogenic microorganisms, thus resulting in improved food safety (Espitia et al. 2012; Gildberg 2004). Antimicrobial activity of composite edible films based on fish gelatin and chitosan, incorporated with clove essential oil, has been found useful in the preservation of raw sliced salmon (Gomez-Estaca et al. 2009). Ideally, for incorporation of peptides in the film, a portion of the dispersion could be initially treated with an enzyme such as pepsin to generate peptides. The hydrolyzed dispersion can be mixed with the coating material to derive the benefits. The dispersion can also be the carrier of other nutraceuticals to enhance the nutritive value of the coated products.

8 Techniques of Application of the Coating Solution

Edible coatings can be applied by several methods such as dipping, spraying, panning, fluidized bed, etc. The performance of these techniques depends principally on the physical properties of the coating (viscosity, density, and surface tension) and characteristics of the foods to be coated. Dipping the food portions in appropriate solutions of coating material is probably the most convenient technique. It offers a uniform film around a complex and rough surface (Zhao 2012). Giving a coat of alginate is a conventional process to enhance the storage life of vegetables. The polysaccharide gels in the presence of calcium ions, and the gelled film collapses and adheres to the vegetable tissue during storage. This technique offers as its main advantages uniform coating, thickness control, and the possibility of multilayer applications. Spray coating is a popular technique for applying food coatings. Aqueous bovine gelatin at 20% concentration of the protein, when spray-coated onto beef tenderloins, pork loins, salmon fillets, and chicken breasts, reduced color deterioration when the products were vacuum packed and stored at 4°C (Antoniewski et al. 2007). New spray systems such as electrosprays and micro-sprays are now available for industrial level applications of edible coatings (Andrade et al. 2012). A panning process consists of depositing the product to be coated into a large, rotating bowl, referred to as the 'pan'. The coating solution is then ladled or sprayed into the rotating pan, and the product is tumbled within the pan to evenly distribute the coating solution over the surface of the food material. Forced air, either ambient or at elevated temperature, is applied to dry the coating (Dangaran et al. 2009). Over the past several decades, fluidized bed coating has been a focus of research for a variety of key applications in the chemical, pharmaceutical, and food industries. It was originally used to apply a very thin film layer onto dry particles of very low density and/or small size. Nowadays, its application is extended to a wide variety of food products, such as functional ingredients and additives, including processing aids (leavening agents and enzymes), preservatives (acids and salts), fortifiers (vitamins and minerals), flavors (natural and synthetic), and spices. New trends in edible coatings focus on highly functional micro or nanostructured, multilayered composite coatings and techniques like atomic layer deposition that deposits ultra-thin films (Andrade et al. 2012). The protein dispersion discussed in this chapter is noted for its high thermostability and low viscosity. It is, therefore, amenable to any of the techniques for coating diverse muscle foods, particularly fishery products such as whole fish, fillets, steaks, shellfish items, etc. for prolonged chilled or frozen storage.

9 Other Applications of Weak Acid-induced Fish Protein Gel

Apart from its use as edible coating, a number of products can be made based on acid induced gelation of fish proteins. As mentioned earlier, acetic acid-induced gelation of shark meat results in the formation of high viscous mass of the fish proteins. Making use of this property, a restructured food product has been developed. The process involves acidification of washed, deskinned and deboned shark meat homogenate in water by traces of acetic acid to lower its pH to 3.5. Gelation of proteins was indicated by the formation of a thick mass of the meat, its hardness being dependent upon the moisture content of the washed fish homogenate. The gel was spread in stainless-steel trays, steamed for 15 min, cooled, cut into steaks and aerobically packaged in polyethylene pouches. The product was stable against microbial growth when stored for 2 months at 10°C or 1 week at ambient temperature. Longer storage at ambient temperature resulted in yellowing of the steamed gel. The 10°C stored steaks were de-acidified and salted by dipping in an aqueous solution of 50 g/L each of $NaHCO_3$ and NaCl for 15 minutes before shallow frying in vegetable oil. The product was acceptable in terms of texture and flavor by a panel of experienced tasters (Venugopal et al. 2002). The gel could be also used as fish paneer (similar to dairy paneer, precipitated milk proteins, popular in India) or as an agent to enhance shelf stability and texture when incorporated in other fish meat products (Smruti and Venugopal 2004). Acid-induced gel dispersion of fish meat has also been used to prepare hygienic sauce by fermentation with lactic acid bacteria (Sree Rekha et al. 2002). The proteins in the dispersions, because of their high thermostabily are also amenable to spray drying to prepare functional protein powders (Venugopal et al. 1997; Venugopal et al. 1996). Protein powder can also be made by making use of the high instability of proteins in the dispersion to salts. The proteins can be precipitated by the addition of salts followed by mild heating, and can be subsequently dehydrated. Alternately, enhancing the pH, followed by heating of the dispersion can be employed to precipitate the proteins.

Table 6. Potential applications of acid-induced fish myofibrillar protein gels and gel dispersions.

Products	Benefits
Coating for chilled fish items including fresh fish, steaks and fillets	Reduction of bacterial spoilage, control of pathogenic microorganisms, control of lipid oxidation and hence extended shelf life
Coating for chilled shellfish items such as peeled and deveined shrimp, lobster tail, processed cuttlefish, squid, such as rings, etc.	Reduction of lipid oxidation, bacterial spoilage and extended shelf life
Frozen fish items including fresh fish, steaks and fillets—block frozen or individually quick frozen	Reduction of drip loss, dehydration and freeze-burn, control of rancidity, enhanced shelf life
Frozen shellfish items such as peeled and deveined shrimp, lobster tail, processed cuttlefish, squid such as rings, fillets—block frozen or individually quick frozen	Reduction of drip loss, dehydration, control of rancidity, enhanced shelf life
Various chilled and frozen fishery products including finfish, shellfish and cephalopod products	Potentials for incorporation of antimicrobials and antioxidants in the dispersion for enhanced microbial control, retention of flavour and hence extended shelf life
Restructured and other ready-to-eat fishery products such as sausages, cutlets, etc.	Extended shelf life due to antimicrobial properties of acid-induced gel
Various fishery products	Coating with protein dispersion can give better presentation in retail packages
Battered and fried fishery products	Dispersion coating may reduce oil uptake and hence better nutritive value for the products

Extraction and recovery of fish muscle proteins by pH-shift has been successfully employed in the isolation of proteins from Atlantic croaker (Kristinsson and Liang 2006). Table 6 summarizes various applications of water dispersions of fish muscle proteins.

In conclusion, a process to make novel, stable, gel dispersion from washed fish muscle structural proteins is pointed out. Proteins in the dispersion are highly stable to heat. The low viscous gel dispersion can be used as edible coating for high value fishery products during chilled or frozen storage. The coating can prevent microbial spoilage during chilled storage of fresh fish. The dispersion is superior to conventional water glaze in its ability to suppress dehydration and rancidity development during prolonged frozen storage of fishery products, such as fish fillets, steaks and shellfish, whether bulk or individually quick frozen. The protective effect of coating can be further enhanced by incorporation in the dispersion of antimicrobials or antioxidants such as bioactive peptides, preferably from fishery sources themselves. Potentials also exist to enhance nutritional quality by using the coatings as carriers of nutraceuticals.

References

Aider, M. 2010. Chitosan application for active bio-based films production and potential in the food industry: Review LWT Food Sci. Technol. 43: 837–842.

Andrade, R., D., O. Skurtys and F.A. Osorio. 2012. Atomizing spray systems for application of edible coatings. Comp. Rev. Food Sci. Food Safety. 11: 323–337.

Antoniewski, M.N., S.A. Barringer, C.L. Knipe and H.N. Zerby. 2007. Effect of a gelatin coating on the shelf life of fresh meat. J. Food Sci. 72: E382–E387.

Asghar, A., K. Samejima, T. Yashui and R.L. Henrickson. 1982. Functionality of muscle proteins in gelation mechanisms of structured meat products. Crit. Rev. Food Sci. Nutr. 22: 27–106.

Ashie, I.N., J.P. Smith and B.K. Simpson. 1986. Spoilage and shelf life extension of fresh fish and shellfish. Crit. Rev. Food Sci. Nutr. 36: 87–99.

Aymard, C., J.L. Cuq and S. Guilbert. 1995. Edible packaging films based on fish myofibrillar proteins: Formulations and functional properties. J. Food Sci. 60: 1369–1373.

Bourne, M.C. 2002. Food Texture and Viscosity: Concept and Measurement, Academic Press, NY, 416 p.

Chawla, S.P., V. Venugopal and P.M. Nair. 1996. Gelation of proteins from washed muscle of threadfin bream (*nemipterus japonicus*) under mild acidic conditions. J. Food Sci. 61: 362–367.

Cisneros-Zevallos, L. and J.M. Krochta. 2003. Dependence of coating thickness on viscosity of coating solution applied to fruits and vegetables by dipping method. J. Food Sci. 68: 503–510.

Connell, J.J. 1995. Control of Fish Quality Fishing News Books: Oxford England.

Cutter, C.N. 2006. Opportunities for bio-based packaging technologies to improve the quality and safety of fresh and further processed muscle foods. Meat Sci. 74: 131–134.

Cuq, B., N. Gontard, J.L. Cuq and S. Guilbert. 1996. Functional properties of myofibrillar protein-based bio packaging as affected by film thickness. J. Food Sci. 61: 580–584.

Dangaran, K.P., M. Tomasula and P. Qi. 2009. Structure and function of protein-based edible films and coatings. pp. 5–56. *In:* M.E. Embuscado and K.C. Huber (eds.). Edible Films And Coatings For Food Applications. Springer, New York.

Doi, E. 1993. Gels and gelling of globular proteins. Trends Food Sci. Technol. 4: 1–5.

Dutta, A., U. Raychaudhuri and R. Chakraborty. 2006. Biopolymers for food packaging. Ind. Food Ind. 25: 32–36.

Espitia, P.J., N. de Fátima Ferreira Soares, J.S. dos Reis Coimbra, N.J. de Andrade, R. Souza Cruz and E.A. Alves Medeiros. 2012. Bioactive peptides: synthesis properties and applications in the packaging and preservation of food. Comp. Rev. Food Sci. Food Safety. 11: 187–204.

Falguera, V., J.P. Quintero, A. Jiménez, J.A. Muñoz and A. Ibarz. 2011. Edible films and coatings: structures active functions and trends in their use. Trends Food Sci. Technol. 22: 292–303.

Fretheim, K., B. Egelandsdal, O. Harbitz and K. Samejima. 1985. Slow lowering of pH induces gel formation of myosin. Food Chem. 18: 169–178.

Gennadios, A., M.A. Hanna and L.B. Kurth. 1997. Application of edible coatings on meats poultry and seafoods: A review. Lebensm. Wiss. U. Technol. 30: 337–341.

Gildberg, A. 2004. Enzymes and Bioactive Peptides from Fish Waste Related to Fish Silage Fish Feed and Fish Sauce Productio. J. Aquatic Food Products Technol. 13: 3–11.

Gómez-Estaca, J. A. A. López de Lacey, M.C. Gómez-Guillén, M.E. López-Caballero and P. Montero. 2009. Antimicrobial activity of composite edible films based on fish gelatin and chitosan incorporated with clove essential oil. J. Aquatic Food Product Technol. 18: 4–52.

Goncalves, A.A., C. Santiago and G. Gindri, Jr. 2009. The effect of glaze uptake on storage quality of frozen shrimp. J. Food Eng. 90: 285–290.

Gontard, N., J.-L. Cuq and S. Guilbert. 1998. Packaging films based on myofibrillar proteins: Fabrication properties and applications. Nahrung. 42: 260–263.

Hamann, D. 1994. Rheological studies of fish proteins. pp. 225–231. *In:* K. Nishinaki and E. Doi (eds.). Food Hydrocolloids: Structure Properties and Functions, Plenum Press, New York.

Hamzeh, A. and M. Rezaei. 2012. The effects of sodium alginate on quality of rainbow trout (Oncorhynchus mykiss) fillets stored at 4 ± 2°C. J. Aquatic Food Prod. Technol. 21: 14–21.

Holcomb, D.N. 1991. Rheology. *In:* Y.H. Hui (ed.). Encyclopedia of Food Science and Technology, Vol. 4, John Wiley & Sons, New York.

Hultin, H.O. 1985. Characteristics of muscle tissue. pp. 729–738. *In:* O.R. Fennema (ed.). Food Chemistry. Marcel Dekker, New York.

Huss, H.H. 1995. Quality and Quality Changes in Fresh Fish, FAO Fisheries Technical.

Joseph, J. 2003. Freezing methods for fish products. pp. 59–64. *In:* Product Development and Seafood Safety, Central Institute of Fisheries Technology: Cochin India.

Kakatkar, A., A.K. Sharma and V. Venugopal. 2004a. Hydration of Bombay duck muscleproteins during weak acid-induced gelation and characteristics of the gel dispersion. Food Chem. 83 99.

Kakatkar, A., S.V. Sherekar and V. Venugopal. 2004b. Fish protein dispersion as a coating to prevent quality loss in processed fishery products. Fishery Technol. (India). 41: 29–33.

Kim, S.-K. and I. Wijesekaraa. 2010. Development and biological activities of marine-derived bioactive peptides: A review. J. Functional Foods. 2: 1–9.

Kinsella, J.E. 1982. Relationship between structure and functional properties of food proteins. pp. 51–63. *In:* P.F. Fox and J.J. Condon (eds.). Food Proteins, Applied Science Publ., New York.

Kok, T.N., J.D. Park, T.M. Lin and J.W. Park. 2007. Multidisciplinary approaches for early determination of gelation properties of fish proteins. J. Aquatic Food Product Technol. 16: 5–18.

Kristinsson, H.G. and Y. Liang. 2006. Effect of pH-shift processing and surimi processing on Atlantic croaker (*Micropogonias undulates*) muscle proteins. J. Food Sci. 71: C304–C312.

Kristjansson, M.N. and J.E. Kinsella. 1991. Protein and enzyme stability: Structural thermodynamic and experimental aspects. Adv. Food Nutr. Res. 35: 237–244.

Krochta, J.M., E.A. Baldwin and M.O. Nisperos-carrieo. 1994. Edible Coatings and Films To Improve Food Quality. Technomic Publ. Lancaster, Pa.

Lanier, T.C. and C.M. Lee. 1992. Surimi Technology, Marcel Dekker New York.

Lian, P.Z., C.M. Lee and K.H. Chung. 2002. Textural and physical properties of acid-induced and potassium-substituted low-sodium surimi gels. J. Food Sci. 67: 109–114.

Lin, K.W. and C.H. Chuang. 2001. Effectiveness of dipping phosphate lactate and acetic acid solutions on the quality and shelf life of pork loin chops. J. Food Sci. 66: 595–600.

Londahl, G. 1997. Technological aspects of freezing and glazing shrimp Infofish Int. 3: 49–52.

Ludescher, R.D. 1990. Molecular dynamics of food proteins: Experimental techniques and observations. Trends Food Sci. Technol. 1: 145–149.

Ma, W., C.-H. Tang, S.-W. Yin, X.-D. Yang, Q. Wang, F. Liu and Z.-H. Wei. 2012. Characterization of gelatin-based edible films incorporated with olive oil. Food Res. Int. 49: 572–579.

Mathew, S., M. Karthikeyan and B.A. Shamasundar. 2002. Effect of water washing of shark (*Scoliodon laticaudus*) meat on the properties of proteins with special reference to gelation. Nahrung. 46: 78–82.

Naquash, S.Y. and R.A. Nazeer. 2010. Antioxidant activity of hydrolysates and peptide fractions of *Nemipterus japonicus* and *Exocoetus volitans* muscle. J. Aquatic Food Product Technol. 180–192.

Niwa, E. 1992. Chemistry of surimi gelation. pp. 389–402. *In*: T.C. Lanier and C.M. Lee (eds.). Surimi Technology, Marcel Dekker, New York.

Nosoh, Y. and T. Sekiguchi. 1991. Stable Proteins. pp. 103–112. *In*: Protein Stability and Stabilization through Protein Engineering Ellis Horwood, New York.

Oakenfull, D., J. Pearce and W. Burley. 1997. Gelation. pp. 111–142. *In*: S. Damodaran and A. Paraf (eds.). Food Proteins and their Application, Marcel Dekker, New York.

Otting, G., E. Liepinsh and K. Wuthrich. 1992. Protein hydration in aqueous solution. Science. 974–977.

Panchavarnam, S., A. Kakatkar and V. Venugopal. 2003. Preparation and use of freshwater fish rohu (Labeo rihita) protein dispersion in shelf life extension of the fish steaks. Lebensm. Wiss. U. Technol. 36: 433.

Phadke, G.G., A.U. Pagarkar, K. Sehgal and K.N. Mohanta. 2011a. Application of edible and biodegradable coatings in enhancing seafood quality and storage life: A review Ecology Environment and Conservation. 17: 619–623.

Phadke, G.G., A.U. Pagarkar and V.R. Joshi. 2011b. Effectiveness of glazing and protein dispersion coatings on drip loss in frozen stored Seer fish (Scomberomorus guttatus) fillets. Ecology Environment and Conservation. 17: 607–610.

Rao, M.A. 1977. Rheology of liquid foods—a review J. Text. Stud. 8: 135–145.

Riebroy, S., S. Benjakul, W. Visessanguan, U. Erikson and T. Rustad. 2009. Acid-induced gelation of natural actomyosin from Atlantic cod and burbot (*Lota lota*). Food Hydrocoll. 23: 26–39.

Santos, E. and J.M. Regenstein. 1990. Effects of vacuum packaging glazing and erythorbic acid on the shelf life of frozen white hake and mackerel. J. Food Sci. 55: 64–69.

Santos-Yap, En E.M. 1996. Fish and seafood, in Freezing Effects on Food Quality, Jeremiah, L. E. (ed.) Marcel Dekker, New York.

Shahidi, F and V. Venugopal. 1994. Solubilisation and thermostability of water dispersions of muscle structural proteins of Atlantic herring. J. Agric. Food Chem. 42: 1440–1444.

Sharp, A. and G. Offer. 1992. The mechanism of formation of gels from myosin molecules. J. Sci. Food Agri. 58: 63.

Siracusa, V., P. Rocculi, S. Romani and M.D. Rosa. 2008. Biodegradable polymers for food packaging: a review. Trends Food Sci. Technol. 19: 634–643.

Smruti, K. and V. Venugopal. 2004. Shelf life enhancement of hardhead catfish (Aris felis) patties, making use of acetic acid induced gelation of the fish proteins. Fishery Technol. (India). 41: 121.

Sree Rekha, P.S., M.D. Alur and V. Venugopal. 2002. A process for convenient production of hygienicfish sauce by lactic acid fermentation of shark meat gel. Fishery Technol. (India). 39: 124.

Stone, D.W. and A.P. Stanley. 1992. Gelation of fish muscle proteins. Food Res. Int. 25: 381–388.

Sun, X.D. and R.A. Holley. 2011. Factors influencing gel formation by myofibrillar proteins in muscle foods. Comp. Rev. Food Sci. and Food Safety. 10: 33–51.

Tharanathan, R.N. 2003. Biodegradable films and composite coatings: Past present and future. Trends Food sci. Technol. 14: 71–78.

Torry Advisory Note #28, http://www.fao.org/wairdocs/tan/x5907e/x5907e01.htm#Factors limiting storage life

Totosaus, A., J.G. Montejano, J.A. Salazar and I. Guerrero. 2002. A review of physical and chemical protein-gel induction. Int. J. Food Sci. Technol. 37: 589–601.

Tung, M.A. 1978. Rheology of protein dispersions. J. Texture Stud. 9: 3–12.

Ustunol, Z. 2009. Edible films and coatings for meat and poultry. pp. 248–265. *In*: M.E. Embuscado and K.C. Huber (eds.). Edible Films and Coatings for Food Applications. Springer, New York.

Venugopal, V., A. Kakatkar, D.R. Bongirwar, M. Karthikeyan, S. Mathew and B.A. Shamasundar. 2002. Gelation of shark meat under mild acidic conditions: physicochemical and rheological characterization of the gel. J. Food Sci. 67: 2681–2686.

Venugopal, V. 2006. Quick freezing and individually quick frozen products. In Seafood Processing Adding Value Through Quick Freezing Retortable Pouch Packaging And Cook-Chilling, CRC press, Boca Raton Florida.

Venugopal, V. 2009. Marine Products For Healthcare: Functional And Bioactive Nutraceuticals From The Ocean, CRC Press, Boca Raton Florida.

Venugopal, V. 2011. Marine Polysaccharides: Food Applications, CRC Press, Boca Raton Florida.

Venugopal, V. 1998. Underutilized fish meat as a source of edible films and coatings for the muscle food industry. Outlook on Agriculure. 27: 57–59.

Venugopal, V. 1997. Functionality and potential applications of thermostable water dispersions of fish meat. Trends Food Sci. Technol. 14: 39–51.

Venugopal, V. 1990. Extracellular proteases of contaminant bacteria in fish spoilage: J. Food Prot. 53: 341–350.

Venugopal, V., S.P. Chawla and P.M. Nair. 1996. Preparation of spray dried protein. Journal of Muscle Foods. 7: 501–509.

Venugopal, V. and F. Shahidi. 1994. Thermostable water dispersions of myofibrillar proteins from Atlantic mackerel. J. Food Sci. 59: 265–268.

Venugopal, V., S.N. Doke and P. Thomas. 1997. Thermostable water dispersions of shark meat and its application to prepare protein powder. J. Aquatic Food Product Technol. 6: 53–57.

Venugopal, V., A. Martin and T.R. Patel. 1995. Rheological and solubility characteristics of washed capelin (Mallotus villosus) mince in water. J. Food Biochem. 19: 175.

Venugopal, V., S.N. Doke and P.M. Nair. 1994. Gelation of shark myofibrillar proteins by weak organic acids. Food Chem. 50: 185–190.

Wu, M.C., M.T. Atallah and H.O. Hultin. 1992. The proteins of washed minced fish muscle have significant solubility in water. J. Food Biochem. 15: 209–215.

Yongsawatdigul, J. and J.W. Park. 2004. Effects of alkali and acid solubilization on gelation characteristics of rockfish muscle proteins. J. Food Sci. 69: 499–505.

Zhang, L., L. Yongkang, H. Sumei and H. Shen. 2009. Effects of chitosan coatings enriched with different antioxidants on preservation of grass carp (ctenopharyngodon idellus) during cold storage. J. Aquatic Food Product Technol. 21: 508–518.
Zhao, Y. 2012. Application of commercial coatings. pp. 319–332. *In*: E.A. Baldwin, R. Hagenmaier and J. Bai (eds.). Edible Coatings and Films to Improve Food Quality, CRC Press, Boca Raton Florida.

6

Recovery of Fish Protein using pH Shift Processing

Yeung Joon Choi[1,]* and *Sang-Keun Jin*[2]

1 Introduction

The demand for fish protein in the world rapidly increases, while the production of marine fish is almost constant during a decade. This demand has led to overfishing of many of the more traditional species, particularly white fish such as Alaska pollock, Pacific whiting, blue whiting and threadfin bream, etc., and has required governmental control to prevent excessive fishing and zero emissions approach to sustainable fisheries science. Also, bycatch is considered one of the most significant issues affecting a fishery manager's ability to optimize yield (Patrick and Benaka 2013). Dark-muscled fish species currently make up about 40% of the total fish catch worldwide. There is great interest in utilizing the large quantities of current available, low value, fatty, pelagic fish for human food (Park 2005). However, cost-effective processing to improve the quality of the seafood resources is required.

The pH shifting process using isoelectric solubilization/precipitation is a method that results in high amounts of recovered proteins by exposure to acidic or alkaline pH conditions that cause separation of protein from the insoluble fractions of the fish (i.e., bones, skin, scales, membrane lipids, etc.). Using the pH shift process on fish results in high protein recovery from

[1] Department of Seafood Science and Technology/Institute of Marine Industry, Gyeongsang National University, Tongyeong, Republic of Korea.

[2] Department of Animal Resources Technology, Gyeongnam National University of Science and Technology, Jinju, Republic of Korea.

* Corresponding author: yjchoi@gnu.ac.kr

muscle, and provides protein gel comparable to the ones obtained from the conventional surimi process. Also, this process greatly reduces water pollution caused by soluble protein from muscle, and has the advantage of protein recovery from byproducts such as frame and bone (Tahergorabi et al. 2012; Kristinsson et al. 2005; Perez-Mateos et al. 2004; Kim et al. 2003; Choi et al. 2002).

This chapter focuses on an overview of the pH shifting process, including the technological problems, a yield of recovered protein, gelling properties, and new insights on the basis of recent research.

2 pH Shifting Process

The pH shifting process has been developed to overcome some of the problems that are caused by the nature of the pelagic species (Hultin and Kelleher 1999; 2000). Most of protein is solubilized at extreme acidic and alkaline pH, and precipitated at its isoelectric point. For example, the minced muscle of small croaker and jack mackerel was highly solubilized at pH 2.0–2.5 and pH 10.5–11.0, and precipitated at pH 4.5–5.5 (Fig. 1) (Park et al. 2003a). The solubility of the minced Pacific whiting and herring muscle was the highest in pH 2.0 and pH 11, and in pH 2.7 and 10.8, but the lowest in pH 5.0–5.5 (Choi and Park 2002; Undeland et al. 2002). These results indicate that the solubility of muscle protein on pH depends on fish species due to variation in muscle composition. The increase of solubility in acidic and alkaline pH is caused by the increase of ionic interaction of charged ion and H-bonding with water.

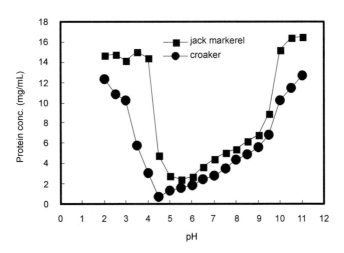

Figure 1. Effect of pH on solubility of minced white croaker and Jack mackerel muscle (from Park et al. 2003).

The concept of the pH shifting process is simple (Fig. 2). The muscle protein was mixed with 5 to 10 volumes of water, and adjusted to pH 10.5 or higher with alkali solution (NaOH solution), or to pH 3.0 or lower with acid solution (HCl solution). It was necessary to choose the pH at which the consistency of the solution decreases to a value that allows the removal of undesirable materials. If it is desired to remove cellular membranes, it is generally necessary to go to a consistency of about 50 mPa.sec or lower (Park 2005). The mixture is centrifuged. The middle layer, after discarding the top layer (fat) and bottom layer (connective tissue, membrane lipids), is adjusted to pH 5.0–5.5 using NaOH solution. The easiest way to precipitate proteins is by adjusting the pH to a value near the isoelectric point. Most salt soluble protein becomes insoluble under this condition. The muscle proteins are precipitated and collected by a process such as centrifugation. Finally, the sediment is adjusted to about pH 7.0 using NaOH solution (Choi and Park 2002; Kim et al. 2003). Recovered proteins from fish muscle contain water soluble protein such as a sarcoplasmic proteins and myoglobin, which are mostly washed away during a conventional surimi processing. Therefore, the color of a recovered protein is more yellow compared to the one of conventional surimi. However, non proteinous nitrogen, including free amino acid, oligopeptides, nucleotide and organic bases, remains in the supernatant fraction after centrifugation and may subsequently be removed.

Minced fish muscle
       Solubilize in acidic and alkaline pH
       Centrifuge (at least 4,000 x g)
Middle layer (solubilized protein) Top layer (fat) Sediment (membrane lipid, skin, bone, etc.)
       Adjust to pH of near isoelectric point of protein
       Centrifuge (at least 3,000 x g)
Sediment (isolated protein) Supernatant (soluble fraction)
       Adjust to about pH 7.0
Recovered protein

Figure 2. Scheme of pH shifting process for producing surimi from fish muscle (from Jang et al. 2006).

3 Properties of the Recovered Protein from Fish Muscle

Yield of the recovered protein by pH shift process reaches the average of 25%. While yields of jack mackerel (*Trachurus japonicus*), croceine croaker (*Pseudosciaena crocea*) and croaker (*Pennahia argentata*) are 31%, 30% and 33% in alkaline pH shift conditions respectively, and are higher than yields in acidic pH shift condition. However, yields of mi-iuy croaker (*Miichthys mmiiuy*) and mackerel in acidic pH shift condition are higher than those in alkaline pH shift condition. Most fish species in the pH shift processing show a high yield up to 3–15% compared to a conventional surimi processing (Fig. 3). The results indicate that yield in pH shift processing depends on fish species. The increase of yield results from a recovery of sarcoplasmic protein, and the change of yield from different species is proposed by muscle protein composition, particularly sarcoplasmic protein content.

The yield from conventional surimi processing was 21.4% for Pacific whiting (Change-Lee et al. 1990) and the range of 21–25% for some fish species (Toyoda et al. 1992; Park et al. 1997).

While breaking force values of the thermal gel from Jack mackerel, mackerel and blackspotted croaker are high in the surimi conventional process, those of crocein croaker and croaker are high in the surimi alkaline process (Fig. 4). However, the deformation values of the thermal gel from

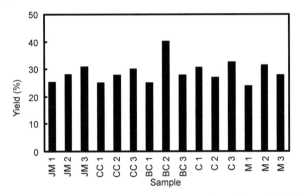

Figure 3. Yield of the recovered protein from conventional, acidic and alkaline processing. The muscle proteins of Jack mackerel (JM), crocein croaker (CC), blackspotted croaker (BC), croaker (C) and mackerel (M) were recovered by conventional (1), acidic (2) and alkaline (3) processing (from Park et al. 2003).

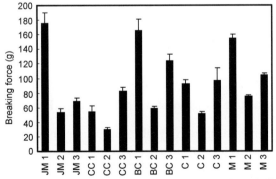

Figure 4. Breaking force of the thermal gel from conventional, acidic and alkaline processing. The muscle proteins of Jack mackerel (JM), crocein croaker (CC), blackspotted croaker (BC) croaker (C) and mackerel (M) were recovered by conventional (1), acidic (2) and alkaline (3) processing (from Park et al. 2003).

the recovered protein by pH shift process are low compared to those of the thermal gel from the conventional processing in the analysed fish species. These results indicate that muscle protein, particularly myosin, is sensitive to the acidic pH condition, and more heavily denatured.

A large molecular band is observed right below the myosin heavy chain (HMC) from fish proteins treated at pH 2.5 (Fig. 5) (Park et al. 2003; Kim et al. 2003). However, it was not clear whether the appearance of the small bands was the result of proteolysis by high proteases or acid hydrolysis. The proteases from fish muscle hydrolyze the myosin heavy chain, resulting in fragments and/or smaller peptides (Jiang et al. 1997; Ohkubo et al. 2005). A myofibril-bound serine proteinase in the skeletal muscle is distributed in silver carp. Optimal profiles of pH and temperature of the protease are 8.5 and 55°C, respectively. Hydrolysis of myofibrillar proteins such as myosin heavy chain, actin and tropomyosin by purified protease, occurred especially at around 55°C. This protease plays a significant role in the Modori phenomenon (Cao et al. 2005). Cathepsin activity shows considerable difference, depending on processing. The specific activity of cathepsin B and L from recovery protein after three-cycle washing is lower than the activity of those from acid-aided (pH shift process) recovery protein. Cathepsin H is completely removed after three-cycle washing and with acid-aided processing (Choi and Park 2002). This study indicates that acid-aided solubilization might have enhanced the cathepsin activity, particularly for Pacific whiting. Cathepsin L has been estimated to have 10 times greater activity per protease molecule against myosin than cathepsin

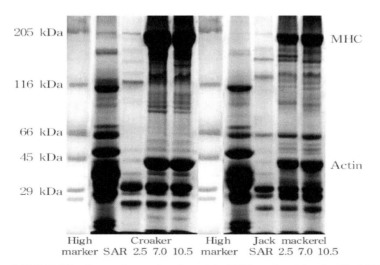

Figure 5. SDS-PAGE patterns of sarcoplasmic protein (SAR), and protein of acidic (pH 2.5), conventional (pH 7.0) and alkaline (pH 10.5) process from croaker and Jack mackerel (from Park et al. 2003).

B. The enzyme has several forms (pI 5.8 to 6.1) and a wide range of pH (3.0 to 6.5) for activity (Choi et al. 2005). Cathepsin L also contributed to the modori phenomena in kamaboko processed from walleye Pollock surimi (Hu et al. 2007). The numerous molecules appearing for both acidic and alkaline treated fish proteins can have been contributed by the retained sarcoplasmic proteins and myofibrillar proteins degraded by either acid or alkali. While the conventional surimi process avoids any denaturation during processing, pH shift process induces fish proteins to be denatured through strong acid or alkaline treatments.

4 Effect of Sarcoplasmic Protein and Ionic Strength on the Recovery Protein Gel

Fish muscle protein is composed of 20 to 30% sarcoplasmic, 66 to 77% myofibrillar, and 3 to 5% stromal proteins (Suzuki 1981). The content of sarcoplasmic protein in fish muscle varies with fish species, but is generally higher in pelagic fish, such as sardine and mackerel. The functions of sarcoplasmic proteins in regards to the formation of myofibrillar protein gels, though, are not clearly in agreement as to whether they work positively or negatively. According to one study, the heat-coagulable sarcoplasmic proteins adhere to the myofibrillar proteins when fish muscle is heated (Shimizu and Nishioka 1974). This phenomenon impedes the gel formation of fish paste and is considered to be one of the reasons why it is difficult to make a strong elastic gel from pelagic fish (Okada 1964; Hashimoto et al. 1985). However, there is a contrasting report that sarcoplasmic proteins positively contribute to the gel formation of myofibrillar protein (Morioka and Shimizu 1990). Heat coagulability of sarcoplasmic protein is found to be highly correlated with the puncture force of sarcoplasmic protein gels. The high gel strength and puncture force of sarcoplasmic protein gels is mainly thought to be due to a large amount of heat coagulable proteins (Morioka and Shimizu 1993). The addition of sarcoplasmic protein from rockfish muscle under low and high pH appears to delay the thermal denaturation of myosin and actin, and positively contributed to the gelation of myofibrillar proteins (Kim et al. 2005). Gelation of the recovery protein from alkali-treated process is estimated by the increase of β-sheet structure, S-S bond by oxidation of surface sulfhydryl group in heating, and polymerization of myosin heavy chain (Jung et al. 2004). Sarcoplasmic proteins are unfolded after extremely acidic and alkaline extraction, exposing tryptophan and aliphatic residues, the α-helical structure is converted to β-sheet following acidic extraction, whereas alkaline treatment did not disturb the α-helical structure of sarcoplasmic proteins. In addition to this, disulfide formation, hydrogen bonding via tyrosine residues, and hydrophobic interactions occurred under extreme pH extraction. Acidic extraction induced denaturation and aggregation of sarcoplasmic proteins to a greater extent than did alkaline treatment. Sarcoplasmic proteins from an alkaline pH shift process readily aggregated to form a gel at 45.10°, whereas higher thermal denaturation temperatures (>80°C) and gel points (about 78°C) are observed in acid-treated sarcoplasmic proteins (Tadpitchayankoon et al. 2010). The addition of sarcoplasmic protein to recovery protein increases the breaking and deformation values in force in the range of 1% to 9% (Fig. 6). The thermal gel strength is increased with the addition of sarcoplasmic protein in the range of 10% to 20% (Morioka et al. 1992).

Breaking force, deformation value and whiteness of the alkali-treated recovery protein gel are influenced by an ionic strength. Breaking force is significantly decreased with increase of NaCl ($p<0.05$) (Fig. 7). The rate of a decreased breaking force is high in jack mackerel compared to that of croaker. Breaking forces of jack mackerel and croaker are decreased up to 56% and 38% by the addition of 3% NaCl. However, deformation value is not significantly changed with the addition of NaCl ($p<0.05$). Whiteness of gel is increased with increase of NaCl.

In conventional surimi, the higher the salt concentration (up to 5–7%), the higher the gel strength. Based on the processing nature of surimi seafood such as kamaboko and crabstick, the level of salt for commercial practices is between 1.2 and 2% (Park and John Lin 2005). The gel strength of the pacific whiting surimi is the highest in 2.5% NaCl concentration (Chung et al. 1993). Myofibrillar proteins are salt-soluble proteins. During the heating of salted surimi paste, the proteins unfold, exposing the reactive surfaces of

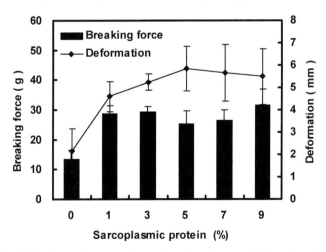

Figure 6. The effect of sarcoplasmic protein on breaking force and deformation of thermal gel by alkaline pH shift process (from Park et al. 2003).

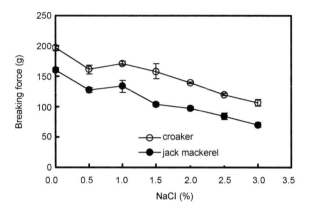

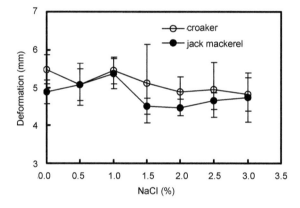

Figure 7. The effect of NaCl concentration on breaking force (top) and deformation (bottom) of thermal gel from jack mackerel and croaker by alkaline pH shift process (from Park et al. 2003).

neighboring protein molecules, which then interact to form intermolecular bonds. When sufficient bonding occurs, a three-dimensional network is formed, resulting in a visco-elastic gel.

5 Effect of Alkaline pH Shifting Process on Lysinoalanine (LAL)

Alkaline processing is used in the production of protein concentrates and isolates, in texturized soy proteins as an ingredient (Riaz 2004), in the production of proteins with specific functional properties such as foaming (Aymard et al. 1978), and in the pH shifting process to recover muscle protein from fish (Choi and Choi 2003; Kim et al. 2003). Processing of food proteins, especially alkaline pH, includes cross-link formation (Damodaran 1996). Heating of protein in alkaline pH conditions results in the abstraction of the proton from α-carbon atoms, and the formation of a carbanion. The carbanion derivatives lead to the formation of a highly reactive dehydroalanine residue (DHA). LAL is the major cross-link commonly found in alkali treated proteins because of the abundance of readily accessible lysyl residues. The formation of cross-linked amino acids such as LAL has raised a concern about the possible toxic effects of alkaline processing (Wood-Rethwill and Warthesen 1980). LAL has been shown to cause pathological changes in rat kidneys (Woodward and Short 1973). Only small amounts of LAL are formed under normal conditions used for processing of several foods. Thus, toxicity of LAL in alkaline treated foods is not believed to be a major concern. However, reduction in digestibility, loss of bioavailability of lysine, and racemization of amino acids are undesirable (Damodaran 1996). LAL was not found in alkali-treated fish products at pH 10 heated for 60 min at 90°C; however, measurable quantities of LAL were found in similarly heated fish products at pH 12 and 13 (Miller et al. 1983).

LAL is not detected in both fish muscle and recovered protein from alkaline pH shifting process (Table 1) (Kim et al. 2007). The contents of threonine, serine and arginine were similar to both proteins. Accompanying the formation of LAL was a 51% decrease in threonine and serine, 40% in arginine, 25% in lysine, and 28% in histidine (Miller et al. 1983). The carbanion derivative of cysteine, cystine, and phosphoserine undergoes β-elimination reaction, leading to the formation of a highly reactive dehydroalanine residue (DHA). DHA formation can also occur via a one-step mechanism without formation of the carbanion. The highly reactive DHA residues react with nucleophilic groups, such as the ε-amino group of a lysyl residue, the thiol group of cysteine residue, the δ-amino group

Table 1. Major amino acid composition as to LAL in alkali-treated croaker protein (g/100g-sample) (from Kim et al. 2007).

Amino acid	Recovered protein	Storage for 22 h			Storage for 66 h		
		A	B	C	D	E	F
Thr	5.41	4.80	5.10	4.39	5.14	4.84	4.45
Ser	4.64	4.72	4.82	4.15	4.80	4.57	4.23
Cys	1.50	0.70	0.82	0.55	0.74	0.64	0.62
His	0.89	0.79	0.80	0.66	0.86	0.77	0.67
Lys	10.64	8.80	9.45	8.03	9.31	8.86	8.10
Arg	6.22	5.97	6.41	5.38	6.40	6.09	5.40
LAL	-	-	-	0.20	-	-	0.23

1) A and D was solubilized at pH 11.0 and precipitated at pH 5.5
2) B and E was solubilized at pH 11, precipitated at pH 5.5, followed by heating at 90°C for 1 h
3) C and F was solubilized at pH 11, heated at 90°C for 1 h, and then precipitated at pH 5.5

of ornithine, or a histidyl residue, resulting in formation of lysinoalanine, lanthionine, ornithoalanine, and histidylalanine cross-links in proteins (Damodaran 1996).

LAL is not formed in recovered protein after storage during 1 h, 3 h, 5 h and 9 h at pH 11.0 (Kim et al. 2007). Furthermore, LAL is not detected in recovered protein cooked at 90°C after treating in pH 11 for 3 h. The results indicate that formation of LAL depends on thermal and alkaline conditions, and protein sources. LAL formation of 0.1 N NaOH-treated casein and soy protein increase linearly with increasing temperature (Friedman et al. 1981; 1984). Miller et al. (1983) report that fish muscle treated at pH 10.0 does not form LAL during 60 min at 90°C. The formation of LAL depends on pH, alkali-treated time, heating temperature and protein sources (Friedman et al. 1981). The results indicate that recovered protein from pH shifting process is very stable against the formation of LAL.

The thermal processing steps on LAL formation are investigated in soluble protein at pH 11.0 and recovered protein after heating at 90°C for 1 h. LAL is detected only in soluble protein at pH 11.0, followed heating at 90°C. The amount of LAL increased from 201 mg/100 g-sample for 24 h storage to 229 mg/100g-sample for 66 h storage (Kim et al. 2007). The results indicate that LAL is formed in alkaline conditions, followed by heating. The recovered protein and its heat-induced gel from alkaline pH shifting process are very stable in the formation of LAL.

6 Frozen Stability of Recovery Protein by pH Shift Process

Freezing and frozen storage have been widely used to slow down microbial and enzymatic deterioration of fish. Structural changes occurring during frozen storage of surimi lead to protein denaturation and subsequent loss of gelling capacity. A novel process using pH shift to recover fish protein has been intensively studied. However, little information of its stability has been revealed. When the recovered protein from croaker and Jack mackerel, by pH shift process with 4% sucrose and 5% sorbitol as cryoprotectants, are stored at –20°C, TBA values of residual lipid in the recovered protein from Jack mackerel increase up to 60 days, and then decrease. However, those of croaker do not show any significant difference compared to fresh recovered protein for 20 days. While the breaking force, deformation and whiteness of the thermal gels from croaker do not change up to 120 days, the recovered protein from Jack mackerel do not form the thermal gel after 60 days. Frozen storage of the recovered protein is limited to 90 days for croaker, and to 60 days for jack mackerel while considering the gelling ability and textural properties (Hur et al. 2006). The highest gel texture is found for samples frozen at pH 5.5 and 7.0 with cryoprotectants and without freeze/thaw. A slightly reduced stability of alkali-treated protein kept at pH 5.5 than at pH 7 is noticed; whether kept at pH 5.5 or 7.0, cryoprotectants are required to maintain frozen stability for longer shelf life (Thawornchinsombut and Park 2006).

7 New Approach of Recovery Protein

There is improved safety with the pH shift process. It allows selective, pH-induced water solubility of muscle proteins with removal of materials not intended for human consumption, including bones, skin, scale, etc. Muscle proteins from fish, particularly pelagic fish, have been recovered at pilot scale using batch mode (Jang et al. 2006). The pH shift processing results in mild, non-thermal pasteurization (Hur et al. 2006; Lansdowne et al. 2009a,b). Toxins such as polychlorinated biphenyls and dioxan are significantly reduced and cholesterol levels are also reduced (Park 2005; Marmon et al. 2009). Heavy metals such as cadmium might be reduced in preparation of protein hydrolysate from squid liver (Lee et al. 2012). The pH shift process offers several advantages over mechanical processing and may be a useful technology to recover functional and nutritious proteins from whole gutted fish or fish processing byproducts for subsequent development of food products (Tahergorabi et al. 2012). Most recently, the recovery proteins by pH shift process from fish and meat byproducts have been applied to functional and formulated foods.

For acceptable surimi gel at 78% moisture of recovery protein from alkali-treated fish muscle, while optimum ratio of ingredient is 89.4–90.0% for croaker recovery protein, 5.9–6.3% for potato starch and 5.0–5.4% for bovine plasma protein (Park et al. 2003b), optimum ratio of ingredients is 89.5–90.0% for Jack mackerel recovery protein, 4.6–6.0% for corn starch and 4.3–5.4% for bovine plasma protein (Choi and Choi 2003). Functionalities of drum dried fish muscle protein from pH shift process have been investigated by determining solubility, emulsion activity, rehydration, fat adsorption capacity, viscosity and color (Choi et al. 2006). Solubility of powder is higher at pH 7.0 than at pH 5.5, and not dependent on the ionic strength. Dried protein shows relatively low emulsion activity. Emulsion stability, foam capacity and stability are not observed. Viscosity is in the range of 50,200 to 39,000 cP. Water and fat adsorption are 2.63–2.89 g water/g and 2.13–2.17 g oil/g, respectively, and not dependent on particle size and pH. Drum dried powder of recovery protein has a potential application as an ingredient of meat patty products. Imitation crab sticks are made from the formulation of Alaska pollock, threadfin bream and recovered protein from spent laying hens' breast by pH shift process and the quality of the gel products are evaluated (Jin et al. 2009; Hur et al. 2011). Cape hake protein powder is prepared by alkaline pH shift process and freeze drying. It contains 90% protein and high levels of minerals. The structure of emulsions prepared with hake protein powder show increasing complexity as the protein level increases (Pires et al. 2012). Heat-set gels made of recovered protein are fortified with w-3 polyunsturated fatty acids (PUFAs)–rich oil (Tahergorabi et al. 2013). They demonstrate that w-3 PUFAs fortification of recovered protein has potential application in the development of functional foods.

Conclusion

A new procedure has been developed to isolate muscle proteins from raw muscle tissue by a process of shifting pH values. Using the pH shift process on fish results in high protein recovery from muscle, and provides comparable protein gel compared to the ones obtained from the conventional surimi process. A new procedure for recovery of fish protein is especially suited to low-value raw materials that are difficult to process by the standard surimi process of simply washing minced muscle tissue. Also, it greatly reduces water pollution caused by soluble protein from muscle, and has the advantage of protein recovery from pelagic fish and fish byproducts such as frame and bone. Typically, strong acids and bases are used in the process of solubilization and precipitation. These agents are substituted for mild agents to improve safety and consumer acceptability. The pH shift process results in mild, non-thermal pasteurization. Toxins such as polychlorinated biphenyls, dioxin and heavy metals are significantly reduced. The pH shift process offers several advantages over mechanical processing and may be a useful technology to recover functional and nutritious proteins from whole gutted fish or fish processing byproducts for subsequent development of food products. The recovery proteins by pH shift process from fish and meat byproducts have been applied as materials for nutraceutical and formulated foods.

References

Aymard, C., J.L. Cuq and J.C. Cheftel. 1978. Formation of lysinoalanine and lanthionine in various food proteins, heated at neutral or alkaline pH. Food Chem. 3(1): 1–5.

Cao, M.-J., L.-L. Wu, K. Hara, L. Weng and W.-J. Su. 2005. Purification and characterization of a myofibril-bound serine proteinase from the skeletal muscle of silver carp. J. Food Biochem. 29(5): 533–546.

Change-Lee, M.V., I.E. Lampila and D.L. Crawford. 1990. Yield and composition of surimi from Pacific whiting (Merluccius productus) and the effect of various protein additives on gel strength. J. Food Sci. 55: 83–86.

Choi, J.-D. and Y.J. Choi. 2003. Optimum formulation of starch and non-muscle protein for alkali surimi gel from Jack mackerel. J. Korean Soc. Food Sci. Nutr. 32: 1032–1038.

Choi, Y.J., I.-S. Kang and T.C. Lanier. 2005. Proteolytic enzymes and control in surimi. pp. 227–277. In: J.W. Park (ed.). Surimi and Surimi Seafood. Taylor & Francis Group, N.Y.

Choi, G., Y. Hong, K.W. Lee and Y.J. Choi. 2006. Food functionalities of dried fish protein powder. J. Korean Soc. Food Sci. Nutr. 35: 1394–1398.

Choi, Y.J. and J.W. Park. 2002. Acid-aided protein recovery from enzyme-rich Pacific whiting. J. Food Sci. 67: 2962–2969.

Chung, Y.C., I. Richardson and M.T. Morrissey. 1993. Effect of pH and NaCl on gel strength of Pacific whiting surimi. J. Aquatic Food Product Technol. 2: 19–35.

Damodaran, S. 1996. Amino acids, peptides, and proteins. pp. 405–409. *In*: D.R. Fennema (3th ed.). Food Chemistry, Marcel Dekker Inc., N.Y.

Friedman, M., C.E. Levin and A.T. Noma. 1984. Factors governing lysinoalanine formation in soy proteins. J. Food Sci. 49: 1282–1288.

Friedman, M., J.C. Zahnley and P.M. Masters. 1981. Relationship between *in vitro* digestibility of casein and its content of lysionalanine and D-amino acids. J. Food Sci. 46: 127–131, 134.

Hashimoto, A., N. Katoh, H. Nozaki and K. Arai. 1985. Inhibiting effect of various factors in muscle of Pacific mackerel on gel forming ability. Nippon Suisan Gakkaishi. 51: 425–432.

Hu, Y., K. Morioka and Y. Itoh. 2007. Hydrolysis of surimi paste from walleye Pollock (theragra chalcogramma) by cysteine proteinase cathepsin L and effect of the proteinase inhibitor (E-64) on gelation. Food Chem. 104: 702–708.

Hultin, H.O. and S.D. Kelleher. 1999. Process for isolating a protein composition from a muscle source and protein composition. U.S. Patent No. 6005073 A. December 21.

Hultin, H.O. and S.D. Kelleher. 2000. High efficiency alkaline protein extraction. U.S. Patent No. 6,136,959. October 24.

Hur, S.I., H.-S. Lim, J.H. Kim and Y.J. Choi. 2006. Frozen stability of proteins recovered from fish muscle by alkaline processing. J. Korean Soc. Food Sci. Nutr. 35: 903–907.

Hur, S.J., B.D. Choi, Y.J. Choi, B.G. Kim and S.K. Jin. 2011. Quality characteristics of imitation crab sticks made from Alaska pollock and spent laying hen meat. LWT-Food Sci. Technol. 44: 1482–1489.

Jang, Y.-B., G.-B. Kim, K.W. Lee and Y.J. Choi. 2006. Alkaline pilot processing for recovery of fish muscle protein and properties of recovered protein. J. Korean Soc. Food Sci. Nutr. 35: 1045–1050.

Jiang, S.T., B. Lee, C. Tsao and J. Lee. 1997. Mackerel cathepsins B and L effects on thermal degradation of surimi. J. Food Sci. 62: 310–315.

Jin, S.K., I.S. Kim, Y.J. Choi, B.G. Kim and S.J. Hur. 2009. The development of imitation crab sticks containing chicken breast surimi. LWT-Food Sci. Technol. 42: 150–156.

Jung, C.H., J.-S. Kim, S.-K. Jin, I.S. Kim, K.-J. Jung and Y.J. Choi. 2004. Gelation properties and industrial application of functional protein from fish muscle-1. Effect of pH on chemical bonds during thermal denaturation. J. Korean Soc. Food Sci. Nutr. 33: 1668–1675.

Kim, G.-B., K.W. Lee, S.L. Hur and Y.J. Choi. 2007. Lysinoalanine in protein recovered from frozen Belanger's croaker, Johnius grypotus, using alkaline processing. J. Kor. Fish Soc. 40: 337–342.

Kim, Y.S., J.W. Park and Y.J. Choi. 2003. New approaches for the effective recovery of fish proteins and their physicochemical characteristics. Fisheries Sci. 69: 1231–1239.

Kim, Y.S., J. Yonsawatdigul, J.W. Park and S. Thawonchinsombut. 2005. Characteristics of sarcoplasmic proteins and their interaction with myofibrillar proteins. J. Food Biochem. 29: 517–532.

Kristinsson, H.G., A.E. Theodore, N. Demir and B. Ingadottir. 2005. A comparative study between acid- and alkali-aided processing and surimi processing for the recovery of proteins from channel catfish muscle. J. Food Sci. 70: C298–C306.

Lansdowne, L., S. Beamer, J. Jaczynski and K.E. Matak. 2009a. Survival of *Escherichia coli* after isoelectric solubilization/precipitation of fish. J. Food Protection. 72: 1398–1403.

Lansdowne, L., S. Beamer, J. Jaczynski and K.E. Matak. 2009b. Survival of Listeria innocua after isoelectric solubilization and precipitation of fish protein. J. Food Sci. 74: M201–M205.

Lee, S.-S., S.-H. Park, J.-D. Park, K. Konno and Y.J. Choi. 2012. Functionalities of squid liver hydrolysate. J. Korean Soc. Food Sci. Nutr. 41: 1677–1685.

Marmon, S.K., P. Liljelind and I. Undeland. 2009. Removal of lipids, dioxins and polychlorinated biphenyls during production of protein isolates from Baltic herring (Clupea harengus) using pH-shift process. J. Agric. Food Chem. 57: 7819–7825.

Miller, R., J. Spinelli and J.K. Babbitt. 1983. Effect of heat and alkali on lysinoalanine formation in fish muscle. J. Food Sci. 48: 296–297.

Morioka, K. and Y. Shimizu. 1990. Contribution of sarcoplasmic proteins to gel formation of fish meat. Nippon Suisan Gakkaishi. 56: 929–933.

Morioka, K., K. Kurashima and Y. Shimizu. 1992. Heat gelling properties of fish sarcoplasmic protein. Nippon Suisan Gakkaishi. 58: 767–772.

Morioka, K. and Y. Shimizu. 1993. Relationship between the heat gelling properties and composition of fish sarcoplasmic proteins. Nippon Suisan Gakkaishi. 59: 1631.

Ohkubo, M., K. Osatomi, K. Hara, T. Ishihara and F. Aranishi. 2005. A novel type of myofibril-bound serine protease from white croaker (Argyrosomus argentatus). Comp. Biochem. Physiol. B. 141: 231–236.

Okada, M. 1964. Effect of washing on the jelly forming ability of fish meat. Nippon Suisan Gakkaishi. 30: 255–261.

Park, J.W. 2005. Surimi and surimi seafood, Taylor & Francis, Boca Raton, FL.

Park, J.W. and T.M. John Lin. 2005. Surimi: Manufacturing and evaluation. pp. 33–105. In: J.W. Park (ed.). Surimi and Surimi Seafood, Taylor & Francis, Boca Raton, FL.

Park, J.W., T.M. Lin and J. Yongsawatdigul. 1997. New developments in manufacturing of surimi and surimi seafood. Food Reviews International. 13: 577–610.

Park, J.D., C.-H. Jung, J.-S. Kim, D.-M. Cho, M.S. Cho and Y.J. Choi. 2003a. Surimi processing using acid and alkali solubilization of fish muscle protein. J. Korean Soc. Food Sci. Nutr. 32: 400–405.

Park, J.D., J.-S. Kim, Y.-J. Cho, J.-D. Choi and Y.J. Choi. 2003b. Optimum formulation of starch and non-muscle protein for alkali surimi gel from frozen white croaker. J. Korean Soc. Food Sci. Nutr. 32: 1026–1031.

Park, J.D., S.-S. Yoon, C.H. Jung, M.S. Cho and Y.J. Choi. 2003. Effect of sarcoplasmic protein and NaCl on heating gel from fish muscle surimi prepared by acid and alkaline processing. J. Korean Soc. Food Sci. Nutr. 32: 567–573.

Patrick, W.S. and L.R. Benaka. 2013. Estimating the economic impacts of bycatch in U.S. commercial fisheries. Marine Policy. 38: 470–475.

Perez-Mateos, M., P.M. Amato and T.C. Lanier. 2004. Gelling properties of Atlantic croaker surimi processed by acid or alkaline solubilization. J. Food Sci. 69: FCT328–FCT333.

Pires, C., S. Costa, A.P. Batista, M.C. Nunes, A. Raymundo and I. Batista. 2012. Properties of protein powder prepared from cape hake by-products. J. Food Engineering. 108: 268–275.

Riaz, M.N. 2004. Texturized soy protein as an ingredient. pp. 517–558. In: R.Y. Yada (ed.). Proteins in Food Processing, Woodhead Publishing Limited, Cambridge, England.

Shimizu, Y. and E. Nishioka. 1974. Interactions between horse mackerel actomyosin and sarcoplasmic proteins during heat coagulation. Nippon Suisan Gakkaishi. 40: 231–234.

Suzuki, T. 1981. Characteristics of fish meat and fish protein. pp. 1–38. In: T. Suzuki (ed.). Fish and Krill Protein. Applied Science Publisher, London.

Tadpitchayangkoon, P., J.W. Park, S.G. Mayer and J. Yongsawatdigul. 2010. Structural changes and dynamic rheological properties of sarcoplasmic proteins subjected to pH-shift method. J. Agric. Food Chem. 58: 4241–4249.

Tahergorabi, R., S.K. Beamer, E. Matak and J. Jaczynski. 2012. Functional food products made from fish protein isolate recovered with isoelectric solubilization/precipitation. LWT-Food Sci. and Technol. 48: 89–95.

Tahergorabi, R., S.K. Beamer, E. Matak and J. Jaczynski. 2013. Chemical properties of w-3 fortified gels made of protein isolate recovered with isoelectric solubilization/precipitation from whole fish. Food Chem. http://dx.doi.org/10.1016/j.foodchem.2013.01.077.

Thawornchinsombut, S. and J.W. Park. 2006. Frozen stability of fish protein isolate under various storage conditions. J. Food Sci. 71: C227–C232.

Toyoda, K., I. Kimura, T. Fujita, S.F. Noguchi and C.M. Lee. 1992. The surimi manufacturing process. pp. 79–112. *In*: T.C. Lanier and C.M. Lee (eds.). Surimi Technology, Marcel Dekker Inc., New York.

Undeland, I., S.D. Kelleher and H.O. Hultin. 2002. Recovery of functional proteins from herring (Clupea harengus) light muscle by an acid or alkaline solubilization process. J. Agric. Food Chem. 50: 7371–7379.

Woodard, J.C. and D.D. Short. 1973. Toxicity of alkali treated soy proteins in rats. J. Nutr. 103: 569–574.

Wood-Rethwill, J.C. and J.J. Warthesen. 1980. Lysinoalaine determination in proteins using high-pressure liquid chromatography. J. Food Sci. 45: 1637–1640.

7

Usage of MALDI-TOF Mass Spectrometry in Sea Food Safety Assessment

Karola Böhme,[1] Marcos Quintela-Baluja,[1] Inmaculada C. Fernández-No,[1] Jorge Barros-Velázquez,[1] Jose M. Gallardo,[2] Benito Cañas[3] and Pilar Calo-Mata[1],*

1 Seafood Spoilage

Seafood is considered an important part of our diet, representing 15% of the total protein ingested all over the world and exhibiting beneficial effects on the consumer health, due to its excellent nutritional properties. Nevertheless, seafood also harbors health risks for the consumers, being the most frequent source of food poisoning that causes diseases with varying degrees of severity, ranging from mild indisposition to chronic or life threatening illness (Gram and Huss 1996; Huss et al. 2000; Iwamoto et al. 2010). The most prevalent intoxication caused by the consumption of seafood is the scombroid fish poisoning that is induced by high levels of histamine, produced by mesophilic bacteria in inappropriately chilled fish

[1] Department of Analytical Chemistry, Nutrition and Food Science, School of Veterinary Sciences, University of Santiago de Compostela, Lugo, Spain.
[2] Department of Food Technology, Institute for Marine Research (IIM-CSIC), Vigo, Spain.
[3] Department of Analytical Chemistry, University Complutense of Madrid, Madrid, Spain.
* Corresponding author

(Lehane and Olley 2000; Flick et al. 2001; Kim et al. 2003). Besides seafood-borne diseases, the industry of fishing and aquaculture is confronted with high economic losses caused by the rapid alteration of products of marine origin (Huss 1995; Rodriguez et al. 2003). Seafood spoilage is caused by microorganisms, enzymes and chemical action, bacteria being the major cause of spoilage of most aquatic food products, due to its ability to form volatile substances that cause off-odors (Gram 1992; Gram and Dalgaard 2002; Fonnesbech Vogel et al. 2005).

The bacterial load in a product of marine origin is closely related to the conditions of the aquatic environment and the applied conservation method which influences the growth and survival of bacteria. In this sense, although pathogenic and spoilage bacterial species can already be present in the indigenous microbiota, most contamination was found to occur during the processing stage and before packaging (Huss et al. 2000). Thus, some important foodborne pathogens, such as *Staphylococcus aureus*, *Bacillus cereus* and *Listeria monocytogenes* are ubiquitously present in the food-processing chain and difficult to eliminate. In contrast, the indigenous microflora is mainly constituted by psychrotrophic, Gram-negative bacteria, such as *Aeromonas* spp., *Pseudomonas* spp. and *Shewanella* spp. that include important food spoilers, causing the alteration of the fresh products even at low temperatures (Gram and Dalgaard 2002). In seafood products coming from warm, coastal waters, the consumption of raw products and shellfish implies an elevated health risk due to the presence of toxin-producing *Vibrio* spp. (Lhafi and Kühne 2007; Eja et al. 2008). The growth of the normal psychrotrophic microbiota is inhibited by light preservation and/or the storage under anaerobic conditions (vacuum or CO_2 packaging), favoring the growth of *Photobacterium phosphoreum* and Gram-positive bacteria, such as lactic acid bacteria (LAB) (Gram and Huss 1996; Paludan-Müller et al. 1998). Contamination of the aquatic environment during processing by humans or animals can lead to the presence of *Enterobacteriaceae*, including human pathogens and histamine forming species (Ward 2001). In order to reduce the number of viable pathogens, seafood products are often subjected to heat treatment after packaging or before consumption. Remarkably, such heat treatments do not destroy potential toxins and spores of spore-forming bacteria, such as *Bacillus* spp. and *Clostridium* spp.

2 Bacterial Detection and Identification

In order to control and minimize the microbiological risk of pathogenic bacteria in seafood products, it is necessary to detect pathogenic and spoilage bacteria in a fast and accurate manner. Once the nature of the microbiota present in the seafood product is known, measures can be taken to eliminate or reduce the probable microbial risk and to predict and increase the stability during storage.

The study of post-mortem microbial spoilage of seafood traditionally relied, on one hand, on the analysis of spoilage metabolites such as trimethylamine, hydrogen sulfide, biogenic amines, volatile bases, etc. and on the other hand, on the isolation and identification of the spoilage microbiota (Fonnesbech Vogel et al. 2005).

For the main pathogenic bacteria, which are critical in marine products, limits and standardized methods for detection and enumeration were established (ICMSF 1986; Huss 1994). However, most of these methods are based on counting techniques after a cultivation step and are often laborious, requiring much time until the results are obtained. In addition, the existing methods for the detection of pathogenic bacteria in food allow the identification of only a few bacterial species or groups that are of interest in the safety and quality of marine products.

That is why, in recent times, the need for sensitive and rapid methods for early detection and bacterial identification has been highlighted (Barros-Velázquez et al. 2002; Lee et al. 2008). In the last decades, the progress of microbiological identification turned to more rapid and sensitive methods, including antibody-based assays and DNA-based methods, together with important advances in bioinformatic tools. Thus, some methodologies such as ELISA or PCR have already become classics. PCR, coupled to sequencing tools, has provided a big amount of information that has been deposited in public databases and is freely available. Recently, the development of rapid and high sensitive techniques, such as real-time PCR (RTi-PCR), DNA microarrays and biosensors, provoked the replacement of traditional culturing methods in the field of bacterial identification in clinical diagnostics, as well as in the food sector (Kuipers 1999). Furthermore, Fourier Transform Infrared Spectroscopy (FT-IR) has been described as a new method for rapid and reliable bacterial identification (Sandt et al. 2006). At the same time, proteomic tools, such as mass spectrometry were introduced for the identification of microorganisms (Klaenhammer et al. 2007; Emerson et al. 2008). The biggest impact has been in the clinical sector, to adopt fast and accurate methods to detect microorganisms that cause infectious diseases. The progress in the food area has been lower due to the elevated costs of molecular techniques and the need for a greater technical expertise compared to traditional culturing methods.

2.1 Classical methods

Classical methods of microbial cultures are used for the determination of the total bacterial content or for the detection of the presence of a particular pathogen. With the help of selective media, bacteria can be identified based on phenotypic and biochemical properties. However, in most cases, bacterial identification is carried out using some basic parameters, such as colony morphology, Gram stain, tests for oxidase and catalase and growth conditions, such as the absence of oxygen and the ability to grow on selective media. For a better classification more tests are needed, including the determination of the type of hemolysis, the presence of amino peptidases, urease and coagulase and indole production. There are commercial identification systems, such as API and VITEK, with which several biochemical tests can be performed simultaneously in minimized space and time (Aslanzadeh 2006). However, these techniques do not always allow an accurate determination of the identity of the microbiota. Furthermore, phenotypic assays are designed to identify particular genera and groups of bacteria, but they cannot be applied for a wide range of different bacterial genera that are of interest in seafood safety and quality.

2.2 Bacterial identification by DNA-based methods

Methods based on the analysis of deoxyribonucleic acid (DNA) are being widely used for bacterial identification in medicine and, in recent years, its use in food analyzing laboratories has reached considerable importance. DNA has the advantage that it is not degraded when foods are subjected to physical, chemical, thermal or high pressure treatments, common in food processing industries. Thus, DNA samples can be analyzed even in highly processed and/or degraded foodstuffs.

By means of DNA amplification by PCR techniques, specific DNA fragments are obtained, resulting in specific fingerprints that enable the differentiation of bacterial strains. In this sense, a number of PCR-techniques have been applied to the detection and differentiation of foodborne pathogenic bacteria in foodstuffs, such as multiplex PCR (Lee et al. 2003; Alarcón et al. 2004; Cremonesi et al. 2005), RAPD (Carmen Collado and Hernández 2007), RFLP (Nawaz et al. 2006) and hybridization assays (Hill and Keasler 1991; Ben-Gigirey et al. 2002; Lee et al. 2003). Recently, RTi-PCR (Real Time PCR) has been introduced in the detection and quantification of major foodborne bacteria. Its main advantage is that besides detecting the bacterial species, the amount of DNA of the bacterial species that is intended to be identified is quantified. A number of RTi-PCR procedures have been developed with the purpose of detecting the amounts of *Salmonella, L. monocytogenes, S. aureus* or *Leuconostosc mesenteroides,* amongst others, in food products (Hein et al. 2001; Malorny et al. 2004; Elizaquível et al. 2008).

The most studied bacterial gene, which most bacterial identification methods are based on, is the 16S rRNA gene due to its highly conserved regions that allow the design of universal primers for the simultaneous amplification of most bacterial species (Greisen et al. 1994). At the same time, the 16S rRNA gene has variable regions unique to each bacterial species, reflecting the genetic evolution (Han 2006). Other genes have similar characteristics and are also being considered for bacterial identification, such as the 23S rRNA gene, the region between the 16S and 23S rRNA gene (ITS), the β subunit of RNA polymerase (*rpoB*) and the β subunit of DNA gyrase (*gyrB*). The sequencing of the 16S rRNA gene is the most applied method for bacterial identification in both, clinical diagnosis, and in the food control sector (Kolbert and Persing 1999; Drancourt et al. 2000). In the case of marine products, sequencing of the 16S rRNA gene has been applied in the identification of several bacterial strains isolated from different fish (Kim et al. 2003; Romero and Navarrete 2006). Furthermore, analysis of the 16S rRNA gene has been frequently used to establish phylogenetic relationships for the classification and typing of bacterial strains (Jay 2003; Mignard and Flandrois 2006).

2.3 Bacterial identification by analysis of proteins

In recent years, a variety of identification methods based on protein analysis have been used for the identification of bacterial species in both, the medical field, as well as in food products. Immunological techniques, such as ELISA, where a specific antibody binds to the corresponding antigen conjugated to an enzyme, are used to identify food borne pathogenic bacteria such as *Salmonella* spp., *E. coli* and *L. monocytogenes* (Jay et al. 2005). Other methods, based on the study of proteins, include the techniques of two-dimensional electrophoresis in polyacrylamide gels (2D-PAGE) and protein sequencing. The 2D-PAGE technique divides complex protein mixtures according to their isoelectric points and molecular weights, generating protein maps that are stored in databases and allow the identification of an unknown bacterial species by comparing the electrophoretic patterns (Emerson et al. 2008). However, this method is more often used to characterize mixtures of proteins and isolate proteins of interest for their further characterization and identification. Recently, protein identification has been carried out by mass spectrometry. The development of soft ionization techniques, such as electrospray ionization (ESI) and matrix-assisted laser desorption/

ionization (MALDI), allowed the analysis of large biomolecules, including the identification of whole proteins (top-down) or peptides obtained from the digestion of proteins (bottom-up) (Klaenhammer et al. 2007). At first, mixtures of proteins are separated by electrophoresis or liquid chromatography (HPLC). The identification of proteins is accomplished by comparing the spectra of peptide mass fingerprinting (PMF) or the fragmentation spectra of peptides with the information contained in the databases such as UniProt (www.uniprot.org), using searching programs such as SEQUEST (Eng et al. 1994) or MASCOT (Perkins et al. 1999). In this manner, a number of bacterial peptides and proteins have been identified, providing extensive information of the proteome of many bacterial species, including also, proteins associated with pathogenicity and toxicity that could serve as potential biomarkers for the rapid identification of the corresponding bacterial species.

3 MALDI-TOF MS for Bacterial Identification

3.1 General aspects

Mass spectrometry can be classified according to the source of ionization employed. The ionization by MALDI (Matrix-assisted laser desorption/ ionization) is a very soft ionization technique that allows the analysis of large biomolecules, such as proteins (Demirev et al. 2004). Ions are produced with great sensitivity without fragmentation of the molecules. For this, the sample is mixed with a matrix solution and placed in a well on a stainless steel plate. Upon evaporation of the solvents, the matrix molecules crystallize including the sample molecules. The matrix protects the biomolecules, absorbing the energy applied with a nitrogen laser (337 nm UV). Afterwards, the matrix ions transfer the energy to the sample molecules, resulting in ions of charge 1. Subsequently, an electric field is applied and the ions are accelerated and separated with relation to their *m/z* ratio by a type of flight analyzer (TOF) (van Baar 2000) (Fig. 1).

MALDI-TOF MS emerged as a new tool for bacterial differentiation due to its speed and simplicity together with the potential to analyze intact proteins. Studies showed that when bacterial cells were treated with lysozyme, spectra were not affected, but after a treatment with proteinase the peaks disappeared, leading to the conclusion that the biomolecules detected by MALDI-TOF MS are associated with intracellular proteins (Arnold and Reilly 1999; Conway et al. 2001). The identification of certain proteins indicated that most of them corresponded to ribosomal proteins.

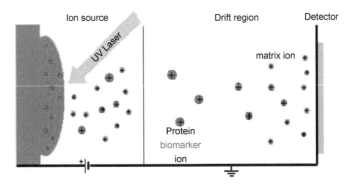

Figure 1. Principle of MALDI-TOF mass spectrometry.
Color image of this figure appears in the color plate section at the end of the book.

MALDI-TOF MS fingerprinting proved to be applicable for bacterial identification at the genus, species and even strain level due to the high specificity of the spectral profiles of soluble and low molecular weight poteins (<15,000 Da) obtained for the corresponding bacterial species (Lay Jr. 2001). In the routine bacterial identification in the clinical sector, it achieved correct species identification in 84–93% of the unknown strains. This is a significantly better result than that obtained by conventional biochemical identification systems or even the sequence analysis of the 16S rRNA gene (Bizzini et al. 2010; van Veen et al. 2010). Some authors conclude that MALDI-TOF MS is one of the latest tools for microbial species identification, forging a revolution in microbial diagnostics with the potential to replace the time-consuming and manpower-intensive conventional procedures that have been used for decades. Several authors agree that, when taking into account the costs of materials and staff, the costs of bacterial identification by MALDI-TOF MS fingerprinting is around two-thirds less than conventional methods and results are achieved in less than 1 h (Seng et al. 2009; van Veen et al. 2010). Furthermore, it has several advantages over other fast methods relying on genomics, such as DNA-microarrays, because fewer steps are necessary to achieve bacterial identification and thus, fewer errors are introduced along the analyzing process.

Another advantage of MALDI-TOF mass fingerprinting is the effortless analysis of the results, since no extensive data processing and statistical analysis is required, as it is the case in other rapid methods for bacterial identification, such as FT-IR and DNA-microarrays. The vast spectral data can be effectively examined by searching common peak masses and cluster analysis. On one hand, characteristic peak patterns can be determined and serve as biomarker proteins for the rapid identification of an unknown spectrum. On the other hand, the construction of a cluster, based on spectral data, allows for the typing of closely related strains, extending conventional typing methods.

Apart from applications of WC-MS in clinical diagnostics, other fields of microbiology are also adopting the technology with success (Welker and Moore 2011). However, most studies of the application of MALDI-TOF MS fingerprinting for bacterial identification discussed the ability of the technique for routine analysis in the clinical sector, highlighting the advantages in comparison to traditional culturing and biochemical methods (Bizzini et al. 2010; Carbonnelle et al. 2010).

For bacterial identification, two different techniques have been described; one is based on identifying ion biomarker masses that are correlated to theoretically determined protein masses deposited in databases and the second approach is based on the comparison of the whole spectral profile to a spectral reference database (fingerprinting) (van Baar 2000). In the first approach, bacterial strains are identified by determining the experimental masses of bacterial proteins by MALDI-TOF MS, which are then compared to theoretical sequence-derived masses, published in proteome databases. In this way, the proteins and consequently the corresponding bacterial species can be identified (Demirev et al. 2004; Fagerquist et al. 2005). A critical challenge of protein database searches is the necessary high mass accuracy, since some proteins have very similar masses. It has to be mentioned that peaks with similar m/z values in mass spectra do not necessarily represent the same protein and therefore, a single peak is not sufficient for a characterization in most cases. Furthermore, identification is limited to well-characterized microorganisms with known protein sequences (Dare 2006).

The second approach allows for the differentiation of bacterial strains, due to the high specific spectral profiles, named "fingerprints". MALDI-TOF MS fingerprinting is the most frequently applied method for bacterial identification. It relies on spectral differences of bacterial strains and the identification of an unknown strain is carried out by comparing its spectral profile with a library of reference spectra (Mazzeo et al. 2006; Giebel et al. 2010). For this purpose, it is not necessary to identify the proteins, but just to determine a number of characteristic peaks that are representative for the corresponding species and/or genus. The mass range of 1,500 to 20,000 Da pronounced high inter-specific variability and at the same time high intra-specific similarity has been found for most bacterial species. In lower mass ranges, e.g., 500–2500 Da, fingerprints of closely related isolates can be very dissimilar, although this mass range has been proposed for identification in early studies (Claydon et al. 1996). This is due to the fact that some bacterial species may produce a high diversity of secondary metabolites and thus, strain-specific mass patterns can be observed (Welker and Moore 2011). In the mass range above 20,000 Da, in general, only a limited number of mass signals are recorded by MALDI-TOF MS of intact bacterial cells, despite the fact that a large diversity of proteins is present in living cells.

To allow the comparison of spectra of reference strains to each other, as well as to unknown strains, the spectral profiles have to be reproducible. A critical challenge of this method is the sensitiveness to changes in sample preparation and analytical parameters. In a number of studies it has been shown that protocol changes and different parameters, such as culture media, growth temperature, salt content, organic solvents for the preparation of cell solutions, matrices, etc. had significant influences on the spectral profiles (Saenz et al. 1999; Williams et al. 2003; Valentine et al. 2005). These authors emphasized the need to establish and use a standardized protocol to carry out spectral comparisons.

3.2 Sample preparation for MALDI-TOF MS analysis of bacterial isolates

In the initial studies, protein fractions were isolated from the bacterial cells prior to analysis, but the focus shortly turned to the analysis of whole cells directly taken from bacterial cultures without any sample pre-treatment, called intact cell mass spectrometry (ICMS). Many authors focused on the optimization of the sample preparation protocol, with the aim to establish a standardized protocol to obtain specific and reproducible spectral profiles in a rapid and labor-saving way.

Nowadays, three different sample preparation protocols are commonly applied. In the first, the bacterial colonies are harvested from the culture plate and placed directly onto the sample plate for analysis. Subsequently, the matrix is applied above the biomass in each well. This method proved to be very fast and less laborious (Bright et al. 2002). However, no reproducible spectra were obtained with this technique due to the difficulty in obtaining a homogeneous mixture of sample and matrix. Furthermore, the proportion of biomass/matrix strongly influenced the quality of the spectra, making it difficult to set the proper amount of biomass (Böhme et al. 2010b). The second technique involves the preparation of a suspension of bacterial biomass. Possible washing steps increase the time and work needed. Also, washing steps can lead to the loss of small and soluble proteins. To avoid this, colonies are directly dissolved in the matrix solution without washing. The suspension is then applied to the sample plate for analysis (Carbonnelle et al. 2007). The third technique is based on the preparation of protein extracts from bacterial biomass in one dissolution/centrifugation step. This method yielded the best spectral profiles compared to the previously described techniques, representing less noise and a higher reproducibility (Böhme et al. 2010b). Moreover, the preparation resulted quick and effortless. Bacterial protein extracts are prepared, collecting a loop-full of biomass and suspending it in 100 µl of a mixture of organic solvents (50% ACN, and 1% TFA) (Fig. 2). The strong acid lyses cells of Gram-negative, as well as Gram-positive strains and the organic solvent extract small soluble proteins. After mixing by vortex and centrifuging the suspension, the supernatant is recovered and mixed with the matrix to conduct the analysis. Nowadays, most applications of MALDI-TOF MS for bacterial identification analyze bacterial cell extract, obtained by just one dilution/centrifugation step.

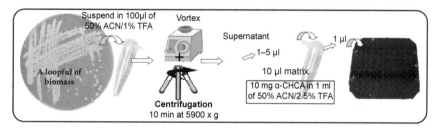

Figure 2. Sample preparation to obtain bacterial protein extracts for MALDI-TOF MS analysis.

3.3 Spectral databases and data analysis

MALDI-TOF MS fingerprinting has been shown to have a great capacity for bacterial identification, being a rapid, cost effective and accurate technique. For bacterial identification, the spectral profile of an unknown bacterial strain is compared to a spectral library of reference strains.

The main drawback of MALDI-TOF MS fingerprinting approach is the lack of public databases. Several private databases have been created, including spectral profiles of more than 500 bacterial strains, such as the Spectral Archive and Microbial Identification System (Saramis™; AnagnosTec GmbH, Potsdam, Germany) (Erhard et al. 2008) and the Microbelynx bacterial identification system (Waters Corporation, Manchester, UK) (Dare 2006). The MALDI Biotyper 2.0 (Bruker Daltonics) searches against an ample database of more than 1800 bacterial species and new spectral profiles are being added on a daily basis (Sauer et al. 2008). However, these spectral libraries are commercial, requiring high charges for access and are conditioned to the corresponding instrument. Furthermore, most studies are targeted at bacterial species with importance in clinical infection diseases and little work has been done in the environmental and food sector.

Thus, a few attempts to start a public database have been achieved. Mazzeo et al. (2006) constructed a library containing spectra of 24 food-borne bacterial species, including *Escherichia, Yersinia, Proteus, Morganella, Salmonella, Staphylococcus, Micrococcus, Lactococcus, Pseudomonas, Leuconostoc* and *Listeria*. Although the spectral profiles and peak mass lists are freely available on the Web (http://bioinformatica.isa.cnr.it/Descr_Bact_Dbase. htm), the library only includes a few bacterial species important in food-borne diseases and food spoilage.

Recently, a more ample reference database with free access has been created by Böhme et al. (2012a), called "SpectraBank" (www.spectrabank. org, www. spectrabank.eu, www.spectrabank.es). At present, the database includes spectral information obtained by MALDI-TOF MS of more than 200 bacterial strains of 56 different bacterial species with interest in food safety and quality and is continuously increasing with new genera, species and strains. In Table 1, bacterial species included in the spectral reference library to date are listed. The spectral reference library consists of a series of mass lists with the corresponding spectrum of each studied strain (Fig. 3). SpectraBank is an open access database that permits other researchers to download all relevant information and perform spectral comparisons with their own data and/or carry out bacterial identification of an unknown strain. It should be emphasized that the compiled reference library of foodborne and spoilage bacterial species can be applied to any type of foodstuffs, since the constituted database may easily be enlarged by further bacterial species and strains that are of interest in the corresponding food product.

The difficulty of developing an 'in house' database lies in the need for particular algorithms to analyze or compare obtained spectra and to carry out searches against the constructed reference library. Peak matching techniques eliminate the subjectivity of visual comparison. Jarman et al. (2003) developed an automated peak detection algorithm to extract representative mass ions from a fingerprint and to compare spectra to fingerprints in a reference library. This algorithm carries out the identification of an unknown spectrum by calculating a degree of matching and is robust to the variability in the ion intensity (Jarman et al. 2003). Other authors developed a software (BGP-database, available on http://

Table 1. Bacterial species included in the spectral database SpectraBank to date.

Bacterial Species	
Acinetobacter baumanii	*Leuconostoc mesenteroides*
Aeromonas hydrophila	*Leuconostoc pseudomesenteroides*
Bacillus amyloliquefaciens	*Listeria innocua*
Bacillus cereus	*Listeria ivanovii*
Bacillus licheniformis	*Listeria monocytogenes*
Bacillus megaterium	*Listeria seeligeri*
Bacillus pumilus	*Listeria welshimeri*
Bacillus subtilis	*Morganella morganii*
Bacillus subtilis subsp. *subtilis*	*Pantoea agglomerans*
Bacillus subtilis subsp. *spizizenii*	*Photobacterium damselae*
Bacillus thuringiensis	*Photobacterium phosphoreum*
Carnobacterium divergens	*Proteus mirabilis*
Carnobacterium gallinarum	*Proteus penneri*
Carnobacterium maltaromaticum	*Proteus vulgaris*
Citrobacter freundii	*Providencia rettgeri*
Clostridium botulinum	*Providencia stuartii*
Clostridium perfringens	*Pseudomonas fluorescens*
Enterobacter aerogenes	*Pseudomonas fragi*
Enterobacter cloacae	*Pseudomonas putida*
Enterobacter hormaechei	*Pseudomonas syringae*
Enterobacter sakazakii	*Raoultella planticola*
Enterococcus faecium	*Serratia liquefaciens*
Enterococcus faecalis	*Serratia marcescens*
Enterococcus gilvus	*Serratia proteamaculans*
Enterococcus mundtii	*Shewanella algae*
Enterococcus sanguinicola	*Shewanella baltica*
Enterococcus malodoratus	*Shewanella putrefaciens*
Enterococcus gallinarum	*Staphylococcus aureus*
Enterococcus casseliflavus	*Staphylococcus epidermidis*
Enterococcus durans	*Staphylococcus pasteuri*
Enterococcus hirae	*Staphylococcus xylosus*
Escherichia coli	*Stenotrophomonas maltophilia*
Hafnia alvei	*Vagococcus salmoninarum*
Klebsiella oxytoca	*Vibrio alginolyticus*
Klebsiella pneumoniae	*Vibrio parahaemolyticus*
Lactococcus garvieae	*Vibrio vulnificus*

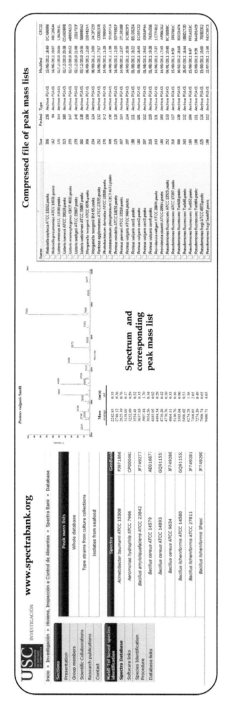

Figure 3. Principles and content of the spectral database SpectraBank.

sourceforge.net/projects/bgp) to analyze and compare spectral profiles, allowing for rapid identification. This software determines the best match between the tested strain and the reference strains of the database, taking into account a possible error of the m/z value (Carbonnelle et al. 2007).

Further bioinformatics programs, such as Statgraphics Plus 5.1 (Statpoint Technologies, inc., Warrenton, USA), offer a variety of functionalities. First, spectral data has to be transformed to a binary table, indicating the presence (1) and absence (0) of a peak mass. Afterwards, various algorithms for cluster analysis can be applied, as well as Principal Component Analysis (Böhme et al. 2011b). In a different study, the BioNumerics 6.0 software (Applied-Maths, Sint-Martens-Latem, Belgium) was used for data analysis and machine learning for bacterial identification by MALDI-TOF MS (De Bruyne et al. 2011).

In further studies, bacterial strains were analyzed in at least quadruplicate and a final consensus peak mass list was created for every studied strain, calculating average values and standard deviations of reproducible peak masses (Böhme et al. 2012a). The freely available web-based application, SPECLUST (http://bioinfo.thep.lu.se/speclust.html) (Alm et al. 2006), was used to determine common peak masses to the four replicates with the aim of extracting only reproducible peak masses that were present in all four spectra of the same strain. Later on, the consensus mass lists can be compared to each other and common peak masses can be defined (Fig. 4). For that, all consensus peak mass lists that are of interest are joined together in a compressed file and this file is then processed with the SPECLUST application. The web program has been found to be very fast, easy to handle and also includes a clustering option. The compressed file of peak mass lists can be modified in a simple manner by deleting or adding new spectral mass lists. In this way, bacterial identification of an unknown strain can be carried out by means of comparing the peak mass list of the unknown spectra to the peak mass lists included in the reference file and carrying out cluster analysis (Fig. 4). In addition, the program allows the rapid determination of specific biomarker peaks and permits the creation of phyloproteomic relations between the studied strains by cluster analysis.

3.3.1 Biomarker Peaks

As mentioned before, some authors described the use of MALDI-TOF MS for bacterial identification as determining the experimental masses of bacterial proteins that are then compared to theoretical sequence-derived masses, published in proteome databases. In this sense, characteristic biomarker

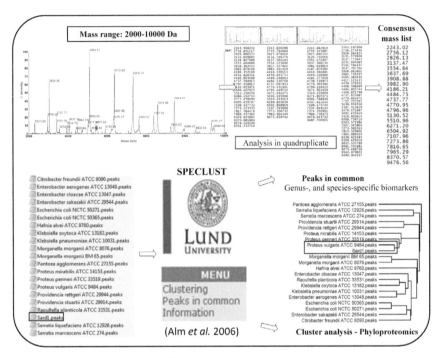

Figure 4. Scheme of spectral data analysis.

peaks have been defined that are specific for a certain genus and/or species and allow the identification of the corresponding bacterial species (Demirev et al. 2004; Fagerquist et al. 2005). In the fingerprinting approach, bacterial identification is not based on a few biomarker peaks, but on the whole spectral profile. In this manner, a more accurate classification can be obtained taking into account intra-specific variability. For example, it has been observed that spectra of strains isolated from different sources varied in some peaks from the spectra of reference strains obtained from type culture collections. Nevertheless, beside spectral variability within a species, some peak masses could be found in common for all strains of the same species and/or genus. Such characteristic, conserved peak masses could serve as biomarker masses for the rapid discrimination of the corresponding species or genus and for the development of other identification approaches, such as a protein array. In addition, such biomarker peaks become more important when working with microbial mixtures, since the presence or absence of unique peaks could lead to a clear indication of the presence of a bacterial species. The detection of biomarker proteins by MALDI-TOF MS has been successfully applied for the identification of two bacterial species isolated from contaminated water, lettuce and cotton cloth (Holland et al. 2000).

However, until now, the application of MALDI-TOF MS fingerprinting for microbial mixtures has not yet been demonstrated.

Böhme et al. (2010a; 2011b) found genus-specific peak masses for the genera *Carnobacterium, Clostridium, Listeria, Proteus, Providencia, Pseudomonas, Serratia, Shewanella, Vibrio* and a number of species-specific peak masses. However, it should be noted that these peaks were assigned as biomarker pattern masses in the context of the studied species. When new species are included, some specific masses may change the designation of specific peaks. In addition, in these studies, a number of characteristic peak masses have been determined for all the studied species, including specific peak masses, as well as common peak masses for a concrete group of bacterial species. The whole unit of such characteristic peak masses represents a peak map, on the base of which fast characterization and identification of an unknown strain can be carried out.

3.3.2 Phyloproteomics

Cluster analysis of MALDI-TOF MS spectral data represents another approach for spectral comparison. Spectral differences can be better visualized by clustering of the peak mass lists than by comparison of the spectra (Böhme et al. 2011b). Conway et al. (2001) introduced the term 'Phyloproteomics' and clustering has been successfully applied for the differentiation and identification of bacterial strains at the genus and species level (Conway et al. 2001; Vargha et al. 2006; Carbonnelle et al. 2007; Dubois et al. 2010; Böhme et al. 2012c). Recently, the construction of phyloproteomic relations has also gained in importance as a typing method, with the aim to classify different strains corresponding to the same species (Siegrist et al. 2007; Teramoto et al. 2007). The combination of various typing methods could be helpful for a better understanding of the epidemiology and health risks caused by foodborne pathogens. The phyloproteomic typing based on spectral data obtained by MALDI-TOF MS fingerprinting represents an approach that extends phenotypic and genotypic approaches, allowing a much more ample classification of bacterial strains with interest in the food producing sector.

When comparing the dendrograms representing phyloproteomic relations revealed by MALDI-TOF MS analysis to phylogenetic trees based on sequences of the 16S rRA gene, a high concordance have been described by a number of authors (Siegrist et al. 2007; Teramoto et al. 2007; Dieckmann et al. 2008; Dubois et al. 2010; Rezzonico et al. 2010). This is not surprising, since the molecules detected by MALDI-TOF MS are generally attributed to ribosomal proteins (Ryzhov and Fenselau 2001).

Böhme et al. (2010a) described the cluster analysis of Gram-negative species with interest in seafood safety and quality and compared the phyloproteomic cluster to the phylogenetic tree obtained by 16S rRNA sequence analysis. Both dendrograms showed a high level of concordance. However, the classification of bacterial strains by MALDI-TOF MS fingerprinting was more discriminating.

4 MALDI-TOF MS as a Reliable Method for Seafood Safety Assessment

4.1 Classification and identification of seafood-borne pathogenic and spoilage bacteria based on MALDI-TOF MS fingerprinting

As already mentioned, most studies in the field of bacterial identification by MALDI-TOF MS fingerprinting, as well as the few private databases, are targeted at human pathogens causing infectious diseases and clinical routine analysis. Nevertheless, these databases also include bacterial species that play an important role in food safety and quality. However, only a few works have been published about MALDI-TOF MS for bacterial identification in the area of environmental bacterial strains and food safety. Table 2 gives an overview of studies based on MALDI-TOF MS analysis of bacterial strains, including bacterial species that are of interest in the seafood safety and quality. These authors proved the applicability of the MALDI-TOF MS approach for bacterial species differentiation with a high percentage of correct species identification (85–95%), but in general in these studies, little spectral information is given to other researchers for further spectral comparisons.

A further study was aimed at the detection of foodborne pathogens and food spoilage bacteria, including genera such as *Escherichia, Yersinia, Proteus, Morganella, Salmonella, Staphylococcus, Micrococcus, Lactococcus, Pseudomonas, Leuconostoc* and *Listeria* (Mazzeo et al. 2006). Mellmann et al. (2008) analyzed a huge number of non-fermenting bacterial strains, including important species related with seafood spoilage, such as *Acinetobacter* spp., *Pseudomonas* spp., *Shewanella* spp. and *Stenotrophomonas* spp. In a different study, the ability of MALDI-TOF MS fingerprinting for the differentiation of bacterial species involved in scombroid fish poisoning has been shown (Fernández-No et al. 2011).

The before mentioned spectral database SpectraBank (www.spectrabank. org) includes the main pathogenic and spoilage bacterial species potentially present in seafood, such as *Acinetobacter, Aeromonas, Bacillus, Carnobacterium, Listeria, Pseudomonas, Shewanella, Staphylococcus, Stenotrophomonas, Vibrio* and genera of the *Enterobacteriaceae* family (Böhme et al. 2012a). In

Table 2. Review of MALDI-TOF MS studies carried out on bacterial species with interest in the seafood safety and quality.

Species	Reference
Acinetobacter baumanii	(Keys et al. 2004)
Aeromonas hydrophila	(Holland et al. 2000; Donohue et al. 2006)
Bacillus cereus	(Krishnamurthy and Ross 1996)
Bacillus pumilus	(Keys et al. 2004)
Bacillus subtilis	(Krishnamurthy and Ross 1996; Wang et al. 1998; Valentine et al. 2005)
Bacillus thuringiensis	(Krishnamurthy and Ross 1996; Wang et al. 1998)
Campylobacter spp.	(Mandrell et al. 2005; Bessède et al. 2011)
Clostridium spp.	(Grosse-Herrenthey et al. 2008)
Cronobacter spp.	(Zhu et al. 2011)
Escherichia coli	(Holland et al. 1996; Arnold and Reilly 1998; Holland et al. 2000; Lin et al. 2005)
Enterobacter cloacae	(Holland et al. 1996)
Enterococcus spp.	(Lin et al. 2005; Giebel et al. 2008; Wieser et al. 2012)
Erwinia spp.	(Sauer et al. 2008)
Klebsiella pneumoniae	(Lynn et al. 1999; Keys et al. 2004)
Listeria spp.	(Williams et al. 2003; Mazzeo et al. 2006; Barbuddhe et al. 2008)
Pantoea spp.	(Wahl et al. 2002; Rezzonico et al. 2010)
Proteus mirabilis	(Holland et al. 1996)
Providencia rettgeri	(Lynn et al. 1999)
Pseudomonas spp.	(Hotta et al. 2010)
Salmonella spp.	(Lynn et al. 1999; Dieckmann et al. 2008)
Serratia marcescens	(Holland et al. 1996)
Shigella flexneri	(Holland et al. 1996; Holland et al. 2000)
Staohylococcus aureus/ Staphylococcus spp.	(Edwards-Jones et al. 2000; Du et al. 2002; Smole et al. 2002; Walker et al. 2002; Bernardo et al. 2002a; Jackson et al. 2005; Lin et al. 2005; Carbonnelle et al. 2007; Liu et al. 2007; Szabados et al. 2010a; Wolters et al. 2010)
Streptococcus spp.	(Cherkaoui et al. 2010; Hinse et al. 2011)
Vibrio parahaemolyticus	(Hazen et al. 2009)
Yersinia enterocolitica	(Valentine et al. 2005; Lasch et al. 2010)

Fig. 5, spectral profiles of some important foodborne pathogens and spoilers are shown, demonstrating the high specificity of the spectral fingerprints obtained by MALDI-TOF MS analysis.

In the following, some species, genera or concrete groups of bacteria with special interest in seafood safety assessment are discussed, based on the studies carried out by Böhme et al. (2010a; 2011b).

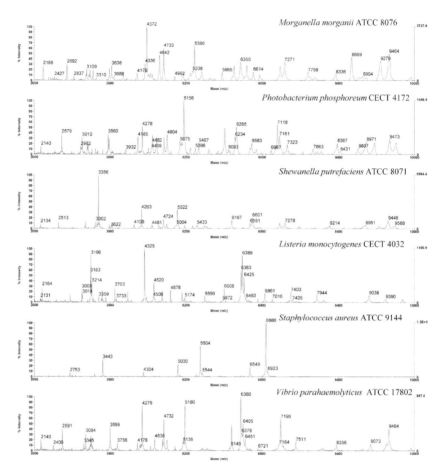

Figure 5. MALDI-TOF MS profiles of some seafood-borne and spoilage bacteria.

Marine species

A group of spectral profiles included species that are typically found in marine environments including *Aeromonas* spp., *Photobacterium* spp., *Shewanella* spp. and *Vibrio* spp. In this group, the spectral profiles of the different genera could be easily distinguished. A number of species-specific peak masses could also be defined for each species. A few peaks were found in common for species of this group, resulting that the strains of these genera grouped together in the cluster analysis. Interestingly, the genus *Photobacterium* showed more similarity to the family *Enterobacteriaceae*.

Enterobacteriaceae

Strains belonging to the *Enterobacteriaceae* family demonstrated profiles very similar to each other. The number and intensities of peaks in a certain mass range were similar among the individual species and the analogue arrangement of peaks made it difficult to differentiate between the individual species. However, a more detailed study of the mass lists, showed certain mass differences of related peaks. The genus *Serratia* formed an exception, showing spectral profiles different from those obtained for other species of *Enterobacteriaceae*. Furthermore, spectra corresponding to the genera *Enterobacter*, *Klebsiella* and *Raoultella* were very similar and no genus-specific peak masses could be found. In addition, the spectra of the mentioned genera had a number of mass ions in common. However, discrimination at the species level was possible due to a number of species-specific peaks. The spectral profile of *Morganella morganii* was analogous to the typical peak arrangement of *Enterobacteriaceae* spectra, but peak masses showed certain differences leading to eight species-specific peaks.

Pseudomonas spp.

Strains corresponding to the genus *Pseudomonas* showed very similar peak patterns in respect to the distribution, intensities and masses of peaks. A number of characteristic peak masses at the genus level could be determined, as well as some specific peak masses which were found for every species. The latter could serve as biomarker peaks, allowing the differentiation of the very close *Pseudomonas* species.

Stenotrophomonas maltophilia

Stenotrophomonas maltophilia has been described as an emerging nosocomial pathogen and important cadaverine producer in fish (Ben-Gigirey et al. 2000). Böhme et al. (2012c) analyzed for the first time, a number of *S. maltophilia* strains isolated from seafood. The spectra could be easily differentiated from all the other studied genera and a number of species-specific peak masses could be determined. Furthermore, these authors observed differences at the strain level, allowing the classification of *S. maltophilia* strains into different groups. The discrimination at the strain level can be of importance for the evaluation of the pathogenicity and spoilage character of *S. maltophilia* strains.

Bacillus spp.

Within the genus *Bacillus*, the species *B. cereus* and *Bacillus thuringiensis* had a number of peak masses in common and therefore could be distinguished from the other *Bacillus* species. It should be mentioned that the species *Bacillus anthracis*, *B. thuringiensis* and *B. cereus* form one group that can be distinguished phylogenetically from other *Bacillus* spp., being more related to the genus *Staphylococcus* than to the other *Bacillus* spp. (Akhtar et al. 2008). At the same time, it is difficult to distinguish species within this group using the common gene markers, such as 16S rRNA and 23S rRNA. Phylogenetic studies of *B. cereus* and *B. thuringiensis* strains resulted in different groups, but could not clearly differentiate between these two species (Chen and Tsen 2002; Bavykin et al. 2004; Kolstø et al. 2009). However, the pathogenic character between these *Bacillus* species differs significantly, so it is important to differentiate these species. Böhme et al. (2011b) studied a number of *B. cereus* and *B. thuringiensis* by MALDI-TOF MS analysis. As a result, a number of common peak masses were found for both species, but also some spectral differences. Furthermore, some peak masses could be determined which were present only in one species but not in the other species. Similarly, the species *Bacillus amyloliquefaciens*, *Bacillus licheniformis* and *Bacillus subtilis* are difficult to differentiate by phylogenetic analysis (Wang et al. 2007). With MALDI-TOF MS analysis, *B. licheniformis* strains could be easily distinguished from the other *Bacillus* species, due to the very characteristic spectral profiles obtained. In the case of *B. amyloliquefaciens*, it could also be distinguished from strains corresponding to *B. subtilis* by specific peak masses. Thus, as a conclusion, the proteomic approach based on MALDI-TOF MS fingerprinting has a higher discriminating potential than genetic analysis of the 16S rRNA gene.

Carnobacterium spp.

Spectral profiles of the genus *Carnobacterium* were very characteristic, since they exhibited very few peak masses, most of them being specific peak masses for the genus or the corresponding species that allow the clear differentiation of *Carnobacterium* spp. strains from other genera.

Clostridium spp. and Listeria spp.

Spectra corresponding to strains of the genera *Clostridium* and *Listeria* exhibited a high number of peaks, thus forming an exception inside the group of Gram-positive strains that generally exhibited a lower number of peaks. Both genera could be easily distinguished from the other genera due to a number of specific peak masses.

Staphylococcus spp.

Claydon et al. (1996) analyzed spectral profiles of three different *Staphylococcus* species and found some peak masses in common in the low mass range (<1,000 Da). However, in further studies spectra of *Staphylococcus* spp. did not show any similarity at the genus level. In this sense, the unambiguous differentiation of *S. aureus* from the other *Staphylococcus* species was easily possible. As shown in Table 2, there exist a great number of studies concerning *S. aureus* due to its pathogenicity and antibiotic resistance. However, although spectral differences at the strain level have been described, no correlation could be found to the antibiotic resistance. Likewise, the differentiation of *S. aureus* strains in different groups based on spectral differences could not be associated to properties related to their pathogenicity, such as toxin production (Böhme et al. 2012b).

4.2 Application of MALDI-TOF MS fingerprinting and the public database SpectraBank for the identification of strains isolated from seafood

MALDI-TOF MS fingerprinting has been successfully applied to clinical routine analysis, achieving >92% of correct species identification, better than conventional biochemical methods (van Veen et al. 2010; Croxatto et al. 2012). Despite the great number of publications in the field of MALDI-TOF MS for clinical routine analysis, only a few works have been reported about MALDI-TOF MS for bacterial identification in the area of food safety and quality in general and in the seafood sector in particular (Mazzeo et al. 2006; Böhme et al. 2011a). In two different studies, the abilities of MALDI-TOF MS fingerprinting in the differentiation of bacterial species involved in scombroid fish poisoning (Fernández-No et al. 2011), as well as for species identification of bacterial strains isolated from commercial seafood products (Böhme et al. 2011a), have been demonstrated.

Furthermore, the potential of the created spectral database SpectraBank to identify isolated strains by means of MALDI-TOF MS analysis has been tested (Böhme et al. 2012c). The 50 bacterial strains of this study were isolated from different fish and processed seafood products, being part of shelf-life studies of pasteurized seafood and histamine-producing strains in fresh fish. For the identification by MALDI-TOF MS fingerprinting, a final peak mass list was created for every strain, extracting representative and reproducible peaks that were present in all of the four replicates of the corresponding sample. Based on the spectral database SpectraBank, species identification of the unknown strains was carried out. For that, the obtained peak mass lists of the isolated strains were compared to the peak mass lists of the reference strains present in the spectral library with the web-tool SPECLUST, determining common peak masses with special attention paid to species-specific peak masses. As a result, the comparison of the peak mass lists and the search for characteristic peak masses resulted in a clear attribution of the spectra of most unknown strains to a certain bacterial species present in the reference library. Figure 6 shows the clear differentiation of the spectra corresponding to the genera *Pseudomonas* and *Serratia*, respectively, as well as the identification of the unknown isolated strains Turb 32 and Proc 7T6, indicating the high number of peak masses in common (▼) with the corresponding reference species, as well as genus- (o) and species-specific (*) peak masses. In a further approach, a dendrogram was constructed by clustering peak mass lists of the isolated strains together with the peak mass lists of all reference strains. In Fig. 7 a part of the cluster analysis is shown, highlighting the grouping of the unknown isolated strains together with the corresponding bacterial reference species.

With the MALDI-TOF MS fingerprinting approach, 76% of the isolated strains were identified at the species level, corresponding to *Bacillus* spp.,

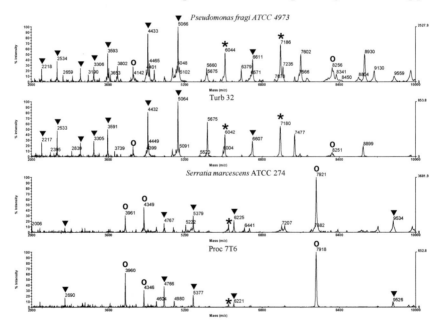

Figure 6. MALDI-TOF MS profiles of two reference strains (*Pseudomonas fragi* and *Serratia marcescens*) and two unknown strains isolated from seafood. Species-specific peaks are indicated by (✱), genus-specific peaks by (**O**) and further common peaks by (▼).

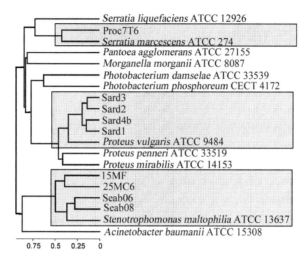

Figure 7. Cluster analysis of spectral data obtained by MALDI-TOF MS for species identification of unknown strains isolated from seafood.

Proteus vulgaris, Pseudomonas spp., *Serratia* spp. and *S. maltophilia* (Table 3). Ten of the remaining strains were correctly classified at the genus level: two of them as *Pseudomonas* spp., three as *Serratia* spp. and five strains as *Bacillus* spp. (dark grey in Table 3). The spectral profile of a further strain that has been later identified as *Enterobacter hormaechei* showed a high similarity to the spectra of the strains *Enterobacter* spp., *Klebsiella* spp. and *Raoultella* spp., but could not be identified at that moment, due to the lack of sufficient reference spectra in the spectral library (light grey in Table 3). There was similar occurrence with a strain that was identified as *Staphylococcus pasteuri* afterwards, but could not be identified by MALDI-TOF MS fingerprinting, since the spectrum could not be related to any reference species or genus present in the spectral library. As mentioned before, within the genus *Staphylococcus,* no spectral similarities between the different species could be observed, making the identification in this case especially difficult. Nevertheless, new spectra can be easily submitted to the reference database; thus, in future, a correct identification of these species is expected. The improvement of correct species identification by adding the information of isolated strains to the spectral library has been demonstrated (Sogawa et al. 2011).

It should also be mentioned that it is of high importance to include spectral information of strains isolated from different food matrices, since isolates can differ significantly from type strains in their phenotypic and proteotypic properties, due to modifications caused by environmental changes.

Table 3. Bacterial species identification of unknown strains isolated from seafood by means of MALDI-TOF MS analysis and 16S rRNA gene sequencing.

Species (MALDI)	Species (DNA)	No. of strains
Bacillus licheniformis	*Bacillus licheniformis*	1
Bacillus megaterium	*Bacillus megaterium*	1
Bacillus amyloliquefaciens	**Bacillus sp.**[b]	3
Bacillus subtilis	**Bacillus sp.**[b]	5
Bacillus sp.[a]	*Bacillus sp.*[a]	5
Carnobacterium maltaromaticum	*Carnobacterium maltaromaticum*	2
-	*Enterobacter hormaechei*	1
Proteus vulgaris	*Proteus vulgaris*	4
Pseudomonas putida	*Pseudomonas putida*	2
Pseudomonas fragi	**Pseudomonas sp.**	6
Pseudomonas fluorescens	**Pseudomonas sp.**	4
Pseudomonas syringae	**Pseudomonas sp.**	1
Pseudomonas sp.	*Pseudomonas sp.*	2
Serratia marcescens	*Serratia marcescens*	1
Serratia sp.	*Serratia proteamaculans*	3
Staphylococcus aureus	*Staphylococcus aureus*	1
-	*Staphylococcus pasteuri*	1
Stenotrophomonas maltophilia	*Stenotrophomonas maltophilia*	7

[a]B. cereus/B. thuringiensis group
[b]B. amyloliquefaciens/B. polyfermenticus/B. subtilis group

4.3 MALDI-TOF MS fingerprinting vs. 16S rRNA sequencing for bacterial species identification

The method of choice for bacterial species identification in most microbiological laboratories is the sequencing of the 16S rRNA gene and matching the sequences against reference sequences deposited in huge DNA databases, as for example the GenBank of the National Center for Biotechnolgy Information (NCBI) (Drancourt et al. 2000; Han 2006; Mellmann et al. 2008). Although, MALDI-TOF MS is less expensive and less laborious, 16S rRNA sequencing remains the 'gold standard' for bacterial species identification and is used to identify critical strains that could not be identified by conventional methods and also to confirm the results obtained by MALDI-TOF MS analysis (Cherkaoui et al. 2010; Zhu et al. 2011). In this section, the proteomic approach based on MALDI-TOF MS fingerprinting is discussed for bacterial species identification of strains isolated from seafood in comparison to the usually applied DNA-based technique of 16S rRNA gene sequencing.

In this sense, it has been demonstrated that species differentiation is not always possible by means of 16S rRNA gene sequencing that to date represents the 'gold-standard' for bacterial identification. 16S rRNA sequencing showed limitations to discriminate at the species level, especially for closely related species, such as *Bacillus* spp., *Pseudomonas* spp. and members of the *Enterobacteriaceae* family. In contrast, with MALDI-TOF MS fingerprinting, a much better percentage of correct species identification was obtained due to the higher discrimination potential, allowing the differentiation and correct identification of close bacterial species and even strains of the same species (Hazen et al. 2009; Zhu et al. 2011; Böhme et al. 2012c).

Besides the application for bacterial species identification, the phyloproteomic approach as a typing method for bacterial classification and comparison of the obtained clusters to phylogenetic relations based on the 16S rRNA sequences is of great interest (Conway et al. 2001; Donohue et al. 2006; Carbonnelle et al. 2007). When comparing the clusters obtained by MALDI-TOF MS analysis to the phylogenetic ones, based on DNA-based classifications, a high degree of concurrence has been observed (Siegrist et al. 2007; Teramoto et al. 2007; Dieckmann et al. 2008; Dubois et al. 2010; Rezzonico et al. 2010). However, a higher discriminating potential was obtained by MALDI-TOF MS fingerprinting compared to DNA-based methods and this has been highlighted for the differentiation of closely related species. The high efficacy of MALDI-TOF MS analysis as a bacterial typing method, obtaining a much better differentiation of closely related bacterial species, has been demonstrated for bacterial species with importance in seafood safety, such as *Cronobacter* spp., *Clostridium* spp., *Listeria* spp., *Pantoea* spp., *Salmonella* spp., *S. aureus* and *Vibrio parahaemolyticus* (Barbuddhe et al. 2008; Dieckmann et al. 2008; Grosse-Herrenthey et al. 2008; Hazen et al. 2009; Dubois et al. 2010; Rezzonico et al. 2010; Szabados et al. 2010b). An exception is for species of *Streptococcus* spp.; for these, the identification by MALDI-TOF MS fingerprinting was difficult and not always possible (Cherkaoui et al. 2010; Risch et al. 2010). In this case, more studies need to be done.

In the study of Böhme et al. (2012c) 50 strains isolated from seafood were identified by the two techniques, MALDI-TOF MS fingerprinting and 16S rRNA gene sequencing. The results are shown in Table 3. Genomic analysis achieved species identification in only 50% of the isolated strains, demonstrating limitations in the differentiation of species corresponding to the genera *Pseudomonas* and *Bacillus*.

Within the genus *Pseudomonas*, only two isolated strains could be clearly identified as *Pseudomonas putida*. However, twelve further strains were also identified as *Pseudomonas*, but could not be assigned to a certain species by genomic analysis. Matching the sequences against sequences of the NCBI database resulted in a list of a number of different *Pseudomonas* species with the same percentage of sequence similarity, such as *Pseudomonas fluorescens, Pseudomonas fragi, Pseudomonas gessardii, Pseudomonas gingeri, Pseudomonas libaniensis, Pseudomonas syringae* and *Pseudomonas veronii*. The genus *Pseudomonas* is a very complex and heterogeneous group and its taxonomy has undergone many changes in the last decades. The diversity at the intra-generic level resulted in an insufficient discriminating into different groups by 16S rRNA sequencing (Moore et al. 1996; Yamamoto et al. 2000; Scarpellini et al. 2004; Mulet et al. 2010). In addition, sequences published in the NCBI database may have been incorrectly assigned to a particular species, thus leading to false identifications and confusing classification results. The same problem has been reported for *Pantoea* species and *Enterobacteriaceae* in general (Rezzonico et al. 2010). In the case of *Bacillus* spp., two isolated strains could be clearly identified as *B. licheniformis* and *B. megaterium*, respectively. However, thirteen further isolates could not be identified at the species level by 16S rRNA sequencing. Nevertheless, these strains were classified into two groups; one group including the species *B. amyloliquefaciens, B. polyfermenticus* and *B. subtilis* (b in Table 3) and the other group corresponding to the species *B. cereus* and *B. thuringiensis* (a in Table 3). Within these groups, identification at the species level was not possible due to the high sequence similarities of the mentioned species. *Bacillus* spp. are known to be difficult to distinguish using common gene markers (Chen and Tsen 2002; Bavykin et al. 2004; Daffonchio et al. 2006; Wang et al. 2007; Kolstø et al. 2009).

However, it should be considered that the pathogenic and spoilage character of the species can differ significantly, making the differentiation of these species at the intra-generic level substantial. *B. cereus* is a known human pathogen causing serious foodborne intoxications (Ghelardi et al. 2002) and *B. subtilis*, as well as some species of the genus *Pseudomonas* are the major cause of the alteration of fresh and processed seafood (Gram and Huss 1996).

In this sense, analysis by MALDI-TOF MS exhibited spectral differences and specificity for all the studied species, allowing the identification of 76% of the isolated strains at the species level. As an exception, a better result has been achieved by 16S rRNA sequencing for five strains, identified as *E. hormaechei, S. proteamaculans* and *S. pasteuri*, respectively. In these cases, species identification by MALDI-TOF MS fingerprinting was not possible due to the lack of reference data in the SpectraBank database, as already mentioned in the previous section. Special attention was given to strains corresponding to the genera *Pseudomonas* and *Bacillus* due to the difficulties in identifying them by genetic tools. The proteomic approach allowed the classification of *Pseudomonas*, as well as *Bacillus* strains, into different groups, confirming the higher discrimination potential of the proteomic approach.

Interestingly, ten of the *Pseudomonas* spp. strains that could not be identified at the species level by 16S rRNA sequencing were assigned to a concrete species by MALDI-TOF MS fingerprinting, as shown in Table 3. The remaining two strains have been classified at the genus level also, by MALDI-TOF MS. When analyzing the phylogenetic relations between the *Pseudomonas* spp. strains studied in the present work, only *P. putida* strains grouped together and were separated from the other species (Fig. 8). In contrast, the other *Pseudomonas* spp. strains could not be classified at the species level, representing very low bootstrap levels. A much better classification of *Pseudomonas* spp. strains was obtained by the proteomic clustering based on the peak mass lists of MALDI-TOF MS analysis (Fig. 8), exhibiting five subclusters corresponding to the species *P. putida*, *P. syringae*, *P. fragi*, *P. fluorescens* and two strains that could not be correlated to any reference species, due to the lack of reference data.

Likewise, in the case of the *Bacillus* spp. strains, of which three could be assigned to the species *B. amyloliquefaciens* and five to the species *B. subtilis*, respectively. An exception formed the *B. cereus/B. thuringiensis* group, for which, species differentiation was not possible with any of the two techniques. Nevertheless, with the proteomic approach, the *Bacillus* spp. strains could be classified into distinct groups, achieving a better

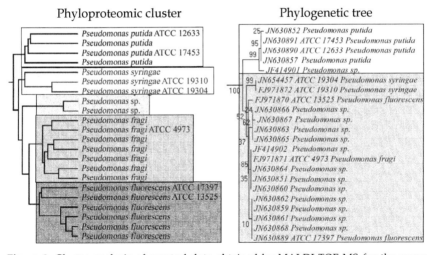

Figure 8. Cluster analysis of spectral data obtained by MALDI-TOF MS for the genus *Pseudomonas*.

differentiation compared to the genomic approach that exhibited 100% sequence similarity for these strains. Furthermore, in the case of *B. subtilis* strains, a differentiation at the subspecies level was obtained for *B. subtilis* subs. *spizizenii*.

There was a similar occurence in the case of *S. maltophilia;* strains for those two groups at the subspecies level were obtained.

Within the *Enterobacteriaceae* family, in general, the 16S rRNA sequences of the different species are very similar to each other and some of the branches created by the phylogenetic clustering showed very low bootstrap levels. In this sense, the species *Proteus penneri* and *Proteus vulgaris* were difficult to distinguish by the genomic approach, but the spectral profiles of *P. vulgaris* obtained by MALDI-TOF MS analysis could be distinguished from the spectra of *P. penneri* by few species-specific peak masses. Likewise, a better classification was obtained by the proteomic approach for the species *Serratia* spp. that were separated in the phylogenetic tree, but grouped together at the genus-level in the proteomic cluster (Böhme et al. 2010a).

In conclusion, MALDI-TOF MS fingerprinting has resulted in a competent bacterial typing tool that could extend phenotypic and genotypic approaches, allowing for a much more ample classification of bacterial strains, related to the pathogenic and spoilage character.

5 Conclusion

Regarding the application to routine bacterial identification, MALDI-TOF MS fingerprinting has been demonstrated to be a fast, reliable and cost-effective technique, forging a revolution in microbial diagnostics, with the potential to replace conventional, time-consuming and man-power-intensive identification procedures in clinical laboratories. The main objective of the present chapter was to emphasize the applicability of the proteomic approach for bacterial species identification in the food control sector also. The methodology and spectral reference database, Spectrabank, have been successfully applied to identify unknown bacterial strains isolated from different seafood products, resulting in a much higher percentage of correct species identification when compared to conventional methods and DNA-based techniques. Since the pathogenic and spoilage character of bacteria in food can differ significantly between species of the same genus, an accurate discrimination at the species-level is important for a prediction of the microbial risk. In this sense, species differentiation is not always possible with conventional phenotypic methods. The 'gold standard' for bacterial species identification in most microbiological laboratories is the sequencing of the 16S rRNA gene. Nevertheless, in some cases, species identification is not possible by 16S rRNA sequencing for closely related bacterial species. In contrast, with MALDI-TOF MS fingerprinting, a much better percentage of correct species identification was obtained due to the higher discrimination potential, allowing the differentiation and correct identification of extremely close bacterial species and even strains of the same species. In this sense, MALDI-TOF MS fingerprinting turned out to be a competent bacterial typing tool that could extend phenotypic and genotypic approaches, allowing for a much more comprehensive classification of bacterial strains and for the deduction of correlations between the proteotype and the pathogenic and spoilage character.

In addition to the higher discriminating potential, MALDI-TOF MS fingerprinting has been described as a rapid, less-laborious and cost-effective method. In comparison to conventional methods, the results were obtained in hours instead of days and the costs were around two-thirds less, when taking into account the cost of materials and staff. Furthermore, it has several advantages over other fast methods, such as DNA-microarrays or FT-IR, because no extensive data processing is required and fewer steps are necessary to achieve bacterial identification and thus, fewer errors are introduced along the analyzing process. It should be mentioned that, although this chapter and the database SpectraBank are focused on bacterial species with importance in seafood-borne diseases and spoilage, the methodology and created reference library can be applied to any other foodstuffs, since the reference database can be easily extended with the addition of further strains and species.

The main drawback of the proteomic approach, when compared to DNA-based methods, is the lack of public databases, since most of the existing spectral libraries are commercial and require high costs for access. Furthermore, most studies are targeted at bacterial species with importance in clinical infection diseases and little work has been done in the environmental and food industry field. Thus, little spectral information is available for bacterial species with impact in food safety, leading to a lack of database entries and consequent misidentification. The creation of the publicly available spectral library, SpectraBank, represents an important advance concerning this matter. Another disadvantage of the MALDI-TOF MS technique is that analysis can only be carried out for isolated colonies, thus it is not applicable for bacteria that need special conditions or long time for cultivation, neither is it applicable for bacterial mixtures. In future, the determination of biomarker proteins could facilitate the identification of food-borne pathogenic and spoilage bacterial species in mixtures, as well as seafood samples, by proteomic tools.

Acknowledgements

This work was funded by project 10PXIB261045PR from Xunta de Galicia, project AGL2010-19646 from the Spanish Ministry of Science and Technology and project IPT-2011-1290-010000 Co financed by the Spanish Ministry of Economy and Competitivity and the European Regional Development Fund 2007–2013 (FEDER). The work of the authors K. Böhme and M. Quintela was supported by project PT-2011-1290-010000.

References

Akhtar, N., M.A. Ghauri, et al. 2008. Biodiversity and phylogenetic analysis of culturable bacteria indigenous to Khewra salt mine of Pakistan and their industrial importance. Brazilian Journal of Microbiology. 39: 143–150.

Alarcón, B., V. García-Cañas, et al. 2004. Simultaneous and Sensitive Detection of Three Foodborne Pathogens by Multiplex PCR, Capillary Gel Electrophoresis, and Laser-Induced Fluorescence. Journal of Agricultural and Food Chemistry. 52(23): 7180–7186.

Alm, R., P. Johansson, et al. 2006. Detection and identification of protein isoforms using cluster analysis of MALDI-MS mass spectra. Journal of Proteome Research. 5(4): 785–792.

Arnold, R.J. and J.P. Reilly. 1998. Fingerprint matching of *E. coli* strains with matrix-assisted laser desorption/ionization time-of-flight mass spectrometry of whole cells using a modified correlation approach. Rapid Communications in Mass Spectrometry. 12(10): 630–636.

Arnold, R.J. and J.P. Reilly. 1999. Observation of *Escherichia coli* ribosomal proteins and their posttranslational modifications by mass spectrometry. Analytical Biochemistry. 269(1): 105–112.

Aslanzadeh, J. 2006. Biochemical profile-based microbial identification systems. Advanced Techniques in Diagnostic Microbiology. Y.-W. Tang and C. W. Stratton. New York, Springer Science+Business Media, LLC: 84–116.

Barbuddhe, S.B., T. Maier, et al. 2008. Rapid identification and typing of *Listeria* species by matrix-assisted laser desorption ionization-time of flight mass spectrometry. Applied and Environment Microbiology. 74(17): 5402–5407.

Barros-Velázquez, J., A. Jiménez, et al. 2002. Speciation of thermotolerant *Campylobacter* isolates involved in foodborne disease by means of DNA restriction analysis and molecular probes. Journal of Agricultural and Food Chemistry. 50(22): 6563–6568.

Bavykin, S.G., Y.P. Lysov, et al. 2004. Use of 16S rRNA, 23S rRNA, and gyrB gene sequence analysis to determine phylogenetic relationships of *Bacillus cereus* group microorganisms. Journal of Clinical Microbiology. 42(8): 3711–3730.

Ben-Gigirey, B., J.M. Vieites Baptista de Sousa, et al. 2000. Characterization of biogenic amine-producing *Stenotrophomonas maltophilia* strains isolated from white muscle of fresh and frozen albacore tuna. International Journal of Food Microbiology. 57(1–2): 19–31.

Ben-Gigirey, B., J.M. Vieites, et al. 2002. Specific detection of *Stenotrophomonas maltophila* strains in albacore tuna (*Thunnus alalunga*) by reverse dot-blot hybridization. Food Control. 13: 293–299.

Bernardo, K., N. Pakulat et al. 2002a. Identification and discrimination of *Staphylococcus aureus* strains using matrix-assisted laser desorption/ionization-time of flight mass spectrometry. Proteomics. 2(6): 747–753.

Bessède, E., O. Solecki, et al. 2011. Identification of Campylobacter species and related organisms by matrix assisted laser desorption ionization–time of flight (MALDI-TOF) mass spectrometry. Clinical Microbiology and Infection. 17(11): 1735–1739.

Bizzini, A., C. Durussel, et al. 2010. Performance of matrix-assisted laser desorption ionization-time of flight mass spectrometry for identification of bacterial strains routinely isolated in a clinical microbiology laboratory. Journal of Clinical Microbiology. 48(5): 1549–1554.

Böhme, K., I. Fernández-No, et al. 2011a. Safety assessment of fresh and processed seafood products by MALDI-TOF mass fingerprinting. Food and Bioprocess Technology. 4(6): 907–918.

Böhme, K., I.C. Fernández-No, et al. 2010b. Comparative analysis of protein extraction methods for the identification of seafood-borne pathogenic and spoilage bacteria by MALDI-TOF mass spectrometry. Analytical Methods. 2(12): 1941–1947.

Böhme, K., I.C. Fernández-No, et al. 2010a. Species differentiation of seafood spoilage and pathogenic Gram-negative bacteria by MALDI-TOF mass fingerprinting. Journal of Proteome Research. 9(6): 3169–3183.

Böhme, K., I.C. Fernández-No, et al. 2011b. Rapid species identification of seafood spoilage and pathogenic Gram-positive bacteria by MALDI-TOF mass fingerprinting. Electrophoresis. 32(21): 2951–2965.

Böhme, K., I.C. Fernández-No, et al. 2012a. SpectraBank: An open access tool for rapid microbial identification by MALDI-TOF MS fingerprinting. Electrophoresis. 33(14): 2138–2142.

Böhme, K., S. Morandi, et al. 2012b. Characterization of *Staphylococcus aureus* strains isolated from Italian dairy products by MALDI-TOF mass fingerprinting. Electrophoresis. 33(15): 2355–2364.

Böhme, K., I.C. Fernández-No, et al. 2012c. Identification of seafood spoilage and pathogenic bacteria: 16S rRNA sequencing vs. MALDI-TOF MS fingerprinting.

Bright, J.J., M.A. Claydon, et al. 2002. Rapid typing of bacteria using matrix-assisted laser desorption ionisation time-of-flight mass spectrometry and pattern recognition software. Journal of Microbiological Methods. 48(2–3): 127–138.

Carbonnelle, E., J.-L. Beretti, et al. 2007. Rapid identification of *Staphylococci* isolated in clinical microbiology laboratories by matrix-assisted laser desorption ionization-time of flight mass spectrometry. Journal of Clinical Microbiology. 45(7): 2156–2161.

Carbonnelle, E., C.c. Mesquita, et al. 2010. MALDI-TOF mass spectrometry tools for bacterial identification in clinical microbiology laboratory. Clinical Biochemistry. 44(1): 104–109.

Carmen Collado, M. and M. Hernández. 2007. Identification and differentiation of *Lactobacillus*, *Streptococcus* and *Bifidobacterium* species in fermented milk products with bifidobacteria. Microbiological Research. 162(1): 86–92.

Claydon, M.A., S.N. Davey, et al. 1996. The rapid identification of intact microorganisms using mass spectrometry. Nature Biotechnology. 14(11): 1584–1586.

Conway, G.C., S.C. Smole, et al. 2001. Phyloproteomics: species identification of *Enterobacteriaceae* using matrix-assisted laser desorption/ionization time-of-flight spectrometry. Journal of Molecular Microbiology and Biotechnology. 3(1): 103–112.

Cremonesi, P., M. Luzzana, et al. 2005. Development of a multiplex PCR assay for the identification of *Staphylococcus aureus* enterotoxigenic strains isolated from milk and dairy products. Molecular and Cellular Probes. 19(5): 299–305.

Croxatto, A., G. Prod'hom, et al. 2012. Applications of MALDI-TOF mass spectrometry in clinical diagnostic microbiology. FEMS Microbiology Reviews. 36(2): 380–407.

Chen, M.L. and H.Y. Tsen. 2002. Discrimination of Bacillus cereus and Bacillus thuringiensis with 16S rRNA and gyrB gene based PCR primers and sequencing of their annealing sites. Journal of Applied Microbiology. 92(5): 912–919.

Cherkaoui, A., J. Hibbs, et al. 2010. Comparison of two matrix-assisted laser desorption ionization-time of flight mass spectrometry methods with conventional phenotypic identification for routine identification of bacteria to the species level. Journal of Clinical Microbiology. 48(4): 1169–1175.

Daffonchio, D., N. Raddadi, et al. 2006. Strategy for identification of *Bacillus cereus* and *Bacillus thuringiensis* strains closely related to *Bacillus anthracis*. Applied and Environmental Microbiology. 72(2): 1295–1301.

Dare, D. 2006. Rapid bacterial characterization and identification by MALDI-TOF mass spectrometry. Advanced Techniques in Diagnostic Microbiology. Y.-W. Tang and C.W. Stratton. New York, Springer Science+Business Media, LLC: 117–133.

De Bruyne, K., B. Slabbinck, et al. 2011. Bacterial species identification from MALDI-TOF mass spectra through data analysis and machine learning. Systematic and Applied Microbiology. 34(1): 20–29.

Demirev, P.A., A.B. Feldman, et al. 2004. Bioinformatics-based strategies for rapid microorganism identification by mass spectrometry. John Hopkins APL Technical Digest. 25(1): 27–37.

Dieckmann, R., R. Helmuth, et al. 2008. Rapid classification and identification of *Salmonellae* at the species and subspecies levels by whole-cell matrix-assisted laser desorption ionization-time of flight mass spectrometry. Applied and Environmental Microbiology. 74(24): 7767–7778.

Donohue, M.J., A.W. Smallwood, et al. 2006. The development of a matrix-assisted laser desorption/ionization mass spectrometry-based method for the protein fingerprinting and identification of *Aeromonas* species using whole cells. Journal of Microbiological Methods. 65(3): 380–389.

Drancourt, M., C. Bollet, et al. 2000. 16S ribosomal DNA sequence analysis of a large collection of environmental and clinical unidentifiable bacterial isolates. Journal of Clinical Microbiology. 38(10): 3623–3630.

Du, Z., R. Yang, et al. 2002. Identification of *Staphylococcus aureus* and determination of its methicillin resistance by matrix-assisted laser desorption/ionization time-of-flight mass spectrometry. Analytical Chemistry. 74(21): 5487–5491.

Dubois, D., D. Leyssene, et al. 2010. Identification of a variety of *Staphylococcus* species by matrix-assisted laser desorption ionization-time of flight mass spectrometry. Journal of Clinical Microbiology. 48(3): 941–945.

Edwards-Jones, V., M.A. Claydon, et al. 2000. Rapid discrimination between methicillin-sensitive and methicillin-resistant *Staphylococcus aureus* by intact cell mass spectrometry. Journal of Medical Microbiology. 49(3): 295–300.

Eja, M., C. Abriba, et al. 2008. Seasonal occurrence of *Vibrios* in water and shellfish obtained from the Great Kwa River Estuary, Calabar, Nigeria. Bulletin of Environmental Contamination and Toxicology. 81(3): 245–248.

Elizaquível, P., E. Chenoll, et al. 2008. A TaqMan-based real-time PCR assay for the specific detection and quantification of *Leuconostoc mesenteroides* in meat products. FEMS Microbiology Letters. 278(1): 62–71.

Emerson, D., L. Agulto, et al. 2008. Identifying and characterizing bacteria in an era of genomics and proteomics. BioScience. 58(10): 925–936.

Eng, J., A. McCormack, et al. 1994. An approach to correlate tandem mass spectral data of peptides with amino acid sequences in a protein database. Journal of The American Society for Mass Spectrometry. 5(11): 976–989.

Erhard, M., U.-C. Hipler, et al. 2008. Identification of dermatophyte species causing onychomycosis and tinea pedis by MALDI-TOF mass spectrometry. Experimental Dermatology. 17(4): 356–361.

Fagerquist, C.K., W.G. Miller, et al. 2005. Genomic and proteomic identification of a DNA-binding protein used in the "Fingerprinting" of *Campylobacter* species and strains by MALDI-TOF-MS protein biomarker analysis. Analytical Chemistry. 77(15): 4897–4907.

Fernández-No, I.C., K. Böhme, et al. 2011. Characterisation of histamine-producing bacteria from farmed blackspot seabream (*Pagellus bogaraveo*) and turbot (*Psetta maxima*). International Journal of Food Microbiology. 151(2): 182–189.

Flick, G.J., M.P. Oria, et al. 2001. Potential hazards in cold-smoked fish: Biogenic amines. Journal of Food Science. 66(s7): S1088–S1099.

Fonnesbech Vogel, B., K. Venkateswaran, et al. 2005. Identification of *Shewanella baltica* as the most important H_2S-producing species during iced storage of danish marine fish. Applied and Environmental Microbiology. 71(11): 6689–6697.

Ghelardi, E., F. Celandroni, et al. 2002. Identification and characterization of toxigenic *Bacillus cereus* isolates responsible for two food-poisoning outbreaks. FEMS Microbiology Letters. 208(1): 129–134.

Giebel, R., C. Worden, et al. 2010. Microbial fingerprinting using matrix-assisted laser desorption ionization time-of-flight mass spectrometry (MALDI-TOF MS): Applications and challenges. Advances in Applied Microbiology. I.L. Allen, S. Sima and M.G. Geoffrey. San Diego, USA, Academic Press. 71: 149–184.

Giebel, R.A., W. Fredenberg, et al. 2008. Characterization of environmental isolates of *Enterococcus* spp. by matrix-assisted laser desorption/ionization time-of-flight mass spectrometry. Water Research. 42(4–5): 931–940.

Gram, L. 1992. Evaluation of the bacteriological quality of seafood. International Journal of Food Microbiology. 16(1): 25–39.

Gram, L. and P. Dalgaard. 2002. Fish spoilage bacteria—problems and solutions. Current Opinion in Biotechnology. 13(3): 262–266.

Gram, L. and H.H. Huss. 1996. Microbiological spoilage of fish and fish products. International Journal of Food Microbiology. 33(1): 121–137.

Greisen, K., M. Loeffelholz, et al. 1994. PCR primers and probes for the 16S rRNA gene of most species of pathogenic bacteria, including bacteria found in cerebrospinal fluid. 32: 335–351.

Grosse-Herrenthey, A., T. Maier, et al. 2008. Challenging the problem of clostridial identification with matrix-assisted laser desorption and ionization-time-of-flight mass spectrometry (MALDI-TOF MS). Anaerobe. 14(4): 242–249.

Han, X. 2006. Bacterial identification based on 16S ribosomal RNA gene sequence analysis. Advanced Techniques in Diagnostic Microbiology, Springer US: 323–332.

Han, X.Y. 2006. Bacterial Identification based on 16S ribosomal RNA gene sequence analysis. Advanced Techniques in Diagnostic Microbiology. Y.-W. Tang and C.W. Stratton. New York, Springer Science+Business Media, LLC: 323–332.

Hazen, T.H., R.J. Martinez, et al. 2009. Rapid identification of *Vibrio parahaemolyticus* by whole-cell matrix-assisted laser desorption ionization-time of flight mass spectrometry. Applied Environmental Microbiology. 75(21): 6745–6756.

Hein, I., D. Klein, et al. 2001. Detection and quantification of the iap gene of *Listeria monocytogenes* and *Listeria innocua* by a new real-time quantitative PCR assay. Research in Microbiology. 152(1): 37–46.

Hill, W.E. and S.P. Keasler. 1991. Identification of foodborne pathogens by nucleic acid hybridization. International Journal of Food Microbiology. 12(1): 67–75.

Hinse, D., T. Vollmer, et al. 2011. Differentiation of species of the *Streptococcus bovis/equinus*-complex by MALDI-TOF Mass Spectrometry in comparison to sodA sequence analyses. Systematic and Applied Microbiology. 34(1): 52–57.

Holland, R.D., F. Rafii, et al. 2000. Matrix-assisted laser desorption/ionization time-of-flight mass spectrometric detection of bacterial biomarker proteins isolated from contaminated water, lettuce and cotton cloth. Rapid Communications in Mass Spectrometry. 14(10): 911–917.

Holland, R.D., J.G. Wilkes, et al. 1996. Rapid identification of intact whole bacteria based on spectral patterns using matrix-assisted laser desorption/ionization with time-of-flight mass spectrometry. Rapid Communications in Mass Spectrometry. 10(10): 1227–1232.

Hotta, Y., K. Teramoto, et al. 2010. Classification of genus *Pseudomonas* by MALDI-TOF MS based on ribosomal protein coding in *S10-spc-alpha* operon at strain level. Journal of Proteome Research. 9(12): 6722–6728.

Huss, H.H. 1994. Assurance of seafood quality. Rome, Food & Agriculture Organization of the United Nations (FAO).

Huss, H.H. 1995. Quality and quality changes in fresh fish. Rome, Food & Agriculture Organization of the United Nations (FAO).

Huss, H.H., A. Reilly, et al. 2000. Prevention and control of hazards in seafood. Food Control. 11(2): 149–156.

ICMSF. 1986. Microorganisms in Foods. 2. Sampling for microbiological analysis: Principles and specific applications. Toronto, Blackwell Scientific Publications.

Iwamoto, M., T. Ayers, et al. 2010. Epidemiology of seafood-associated infections in the United States. Clinical Microbiology and Infection. 23(2): 399–411.

Jackson, K.A., V. Edwards-Jones, et al. 2005. Optimisation of intact cell MALDI method for fingerprinting of methicillin-resistant *Staphylococcus aureus*. Journal of Microbiological Methods. 62(3): 273–284.

Jarman, K.H., D.S. Daly, et al. 2003. A new approach to automated peak detection. Chemometrics and Intelligent Laboratory Systems. 69(1–2): 61–76.

Jay, J.M., M.J. Loessner, et al. 2005. Modern Food Microbiology, Springer Science + Business Media, Inc.

Jay, J.M. 2003. A review of recent taxonomic changes in seven genera of bacteria commonly found in foods. Journal of Food Protection. 66(7): 1304–1309.

Keys, C.J., D.J. Dare, et al. 2004. Compilation of a MALDI-TOF mass spectral database for the rapid screening and characterisation of bacteria implicated in human infectious diseases. Infection, Genetics and Evolution. 4(3): 221–242.

Kim, S.-H., J. Barros-Velázquez, et al. 2003. Identification of the main bacteria contributing to histamin formation in seafood to ensure product safety. Food Science and Biotechnology. 12(4): 451–460.

Klaenhammer, T.R., E. Pfeiler, et al. 2007. Genomics and proteomics of foodborne microorganisms. Food Microbiology: Fundamentals and Frontiers. M.P. Doyle and L.R. Beuchat. Washington, D.C., ASM Press: 935–951.

Kolbert, C.P. and D.H. Persing. 1999. Ribosomal DNA sequencing as a tool for identification of bacterial pathogens. Current Opinion in Microbiology. 2: 299–305.

Kolstø, A.-B., N.J. Tourasse, et al. 2009. What sets *Bacillus anthracis* apart from other *Bacillus* species? Annual Review of Microbiology. 63(1): 451–476.

Krishnamurthy, T. and P.L. Ross. 1996. Rapid identification of bacteria by direct matrix-assisted laser desorption/ionization mass spectrometric analysis of whole cells. Rapid Communications in Mass Spectrometry. 10(15): 1992–1996.

Kuipers, O.P. 1999. Genomics for food biotechnology: prospects of the use of high-throughput technologies for the improvement of food microorganisms. Current Opinion in Biotechnology. 10: 511–516.

Lasch, P., M. Drevinek, et al. 2010. Characterization of *Yersinia* using MALDI-TOF mass spectrometry and chemometrics. Analytical Chemistry. 82(20): 8464–8475.

Lay, J.O., Jr. 2001. MALDI-TOF mass spectrometry of bacteria. Mass Spectrometry Reviews. 20(4): 172–194.

Lee, C.-Y., G. Panicker, et al. 2003. Detection of pathogenic bacteria in shellfish using multiplex PCR followed by CovaLink NH microwell plate sandwich hybridization. Journal of Microbiological Methods. 53(2): 199–209.

Lee, R.J., R.E. Rangdale, et al. 2008. Bacterial Pathogens in Seafood. Improving Seafood Products for the Consumer. T. Børresen, Woodhead Publishing Limited.

Lehane, L. and J. Olley. 2000. Histamine fish poisoning revisited. International Journal of Food Microbiology. 58(1–2): 1–37.

Lhafi, S.K. and M. Kühne. 2007. Occurrence of *Vibrio* spp. in blue mussels (*Mytilus edulis*) from the German Wadden Sea. International Journal of Food Microbiology. 116(2): 297–300.

Lin, Y.-S., P.-J. Tsai, et al. 2005. Affinity capture using vancomycin-bound magnetic nanoparticles for the MALDI-MS analysis of bacteria. Analytical Chemistry. 77(6): 1753–1760.

Liu, H., Z. Du, et al. 2007. Universal sample preparation method for characterization of bacteria by matrix-assisted laser desorption ionization-time of flight mass spectrometry. Applied and Environmental Microbiology. 73(6): 1899–1907.

Lynn, E.C., M.-C. Chung, et al. 1999. Identification of *Enterobacteriaceae* bacteria by direct matrix-assisted laser desorptiom/ionization mass spectrometric analysis of whole cells. Rapid Communications in Mass Spectrometry. 13(20): 2022–2027.

Malorny, B., E. Paccassoni, et al. 2004. Diagnostic real-time PCR for detection of *Salmonella* in food. 70: 7046–7052.

Mandrell, R.E., L.A. Harden, et al. 2005. Speciation of *Campylobacter coli, C. jejuni, C. helveticus, C. lari, C. sputorum,* and *C. upsaliensis* by matrix-assisted laser desorption ionization-time of flight mass spectrometry. Applied and Environmental Microbiology. 71(10): 6292–6307.

Mazzeo, M.F., A. Sorrentino, et al. 2006. Matrix-assisted laser desorption ionization-time of flight mass spectrometry for the discrimination of food-borne microorganisms. Applied and Environmental Microbiology. 72(2): 1180–1189.

Mellmann, A., J. Cloud, et al. 2008. Evaluation of matrix-assisted laser desorption ionization-time-of-flight mass spectrometry in comparison to 16S rRNA gene sequencing for species identification of nonfermenting bacteria. Journal of Clinical Microbiology. 46(6): 1946–1954.

Mignard, S. and J.P. Flandrois. 2006. 16S rRNA sequencing in routine bacterial identification: A 30-month experiment. Journal of Microbiological Methods. 67(3): 574–581.

Moore, E.R.B., M. Mau, et al. 1996. The determination and comparison of the 16S rRNA gene sequences of species of the genus *Pseudomonas* (sensu stricto) and estimation of the natural intrageneric relationships. Systematic and Applied Microbiology. 19(4): 478–492.

Mulet, M., J. Lalucat, et al. 2010. DNA sequence-based analysis of the *Pseudomonas* species. Environmental Microbiology. 12(6): 1513–1530.

Nawaz, M., K. Sung, et al. 2006. Biochemical and Molecular Characterization of Tetracycline-Resistant *Aeromonas veronii* Isolates from Catfish. 72: 6461–6466.

Paludan-Müller, C., P. Dalgaard, et al. 1998. Evaluation of the role of *Carnobacterium piscicola* in spoilage of vacuum- and modified-atmosphere-packed cold-smoked salmon stored at 5°C. International Journal of Food Microbiology. 39(3): 155–166.

Perkins, D.N., D.J.C. Pappin, et al. 1999. Probability-based protein identification by searching sequence databases using mass spectrometry data. Electrophoresis. 20(18): 3551–3567.

Rezzonico, F., G. Vogel, et al. 2010. Application of whole-cell matrix-assisted laser desorption ionization-time of flight mass spectrometry for rapid identification and clustering analysis of *Pantoea* species. Applied and Environmental Microbiology. 76(13): 4497–4509.

Rezzonico, F., G. Vogel, et al. 2010. Whole cell MALDI-TOF mass spectrometry application for rapid identification and clustering analysis of *Pantoea* species. Applied and Environmental Microbiology: AEM.03112-09.

Risch, M., D. Radjenovic, et al. 2010. Comparison of MALDI TOF with conventional identification of clinically relevant bacteria. Swiss Medical Weekly.

Rodriguez, O., J. Barros-Velazquez, et al. 2003. Evaluation of sensory and microbiological changes and identification of proteolytic bacteria during the iced storage of farmed turbot (*Psetta maxima*). Journal of Food Science 68(9): 2764–2771.

Romero, J. and P. Navarrete. 2006. 16S rDNA-based analysis of dominant bacterial populations associated with early life stages of Coho Salmon (*Oncorhynchus kisutch*). Microbial Ecology. 51(4): 422–430.

Ryzhov, V. and C. Fenselau. 2001. Characterization of the protein subset desorbed by MALDI from whole bacterial cells. Analytical Chemistry. 73(4): 746–750.

Saenz, A.J., C.E. Petersen, et al. 1999. Reproducibility of matrix-assisted laser desorption/ionization time-of-flight mass spectrometry for replicate bacterial culture analysis. Rapid Communications in Mass Spectrometry. 13(15): 1580–1585.

Sandt, C., C. Madoulet, et al. 2006. FT-IR microspectroscopy for early identification of some clinically relevant pathogens. Journal of Applied Microbiology. 101(4): 785–797.

Sauer, S., A. Freiwald, et al. 2008. Classification and identification of bacteria by mass spectrometry and computational analysis. PLoS ONE. 3(7): e2843.

Scarpellini, M., L. Franzetti, et al. 2004. Development of PCR assay to identify *Pseudomonas fluorescens* and its biotype. FEMS Microbiology Letters. 236(2): 257–260.

Seng, P., M. Drancourt, et al. 2009. Ongoing revolution in bacteriology: Routine identification of bacteria by matrix-assisted laser desorption ionization time-of-flight mass spectrometry. Clinical Infectious Diseases. 49(4): 543–551.

Siegrist, T.J., P.D. Anderson, et al. 2007. Discrimination and characterization of environmental strains of *Escherichia coli* by matrix-assisted laser desorption/ionization time-of-flight mass spectrometry (MALDI-TOF-MS). Journal of Microbiological Methods. 68(3): 554–562.

Smole, S.C., L.A. King, et al. 2002. Sample preparation of Gram-positive bacteria for identification by matrix assisted laser desorption/ionization time-of-flight. Journal of Microbiological Methods. 48(2–3): 107–115.

Sogawa, K., M. Watanabe, et al. 2011. Rapid identification of microorganisms by mass spectrometry: improved performance by incorporation of in-house spectral data into a commercial database. Analytical and Bioanalytical Chemistry: 1–12.

Szabados, F., J. Woloszyn, et al. 2010a. Identification of molecularly defined *Staphylococcus aureus* strains using matrix-assisted laser desorption/ionization time of flight mass spectrometry and the Biotyper 2.0 database. Journal of Medical Microbiology. 59(7): 787–790.

Szabados, F., K. Becker, et al. 2010b. The matrix-assisted laser desorption/ionisation time-of-flight mass spectrometry (MALDI-TOF MS)-based protein peaks of 4448 and 5302 Da are not associated with the presence of Panton-Valentine leukocidin. International Journal of Medical Microbiology. 301(1): 58–63.

Teramoto, K., H. Sato, et al. 2007. Phylogenetic classification of *Pseudomonas putida* strains by MALDI-MS using ribosomal subunit proteins as biomarkers. Analytical Chemistry. 79(22): 8712–8719.

Valentine, N., S. Wunschel, et al. 2005. Effect of culture conditions on microorganism identification by matrix-assisted laser desorption ionization mass spectrometry. Applied and Environmental Microbiology. 71(1): 58–64.

van Baar, B.L.M. 2000. Characterisation of bacteria by matrix-assisted laser desorption/ionisation and electrospray mass spectrometry. FEMS Microbiology Reviews. 24: 193–219.

van Veen, S.Q., E.C.J. Claas, et al. 2010. High-throughput identification of bacteria and yeast by matrix-assisted laser desorption ionization-time of flight mass spectrometry in conventional medical microbiology laboratories. Journal of Clinical Microbiology. 48(3): 900–907.

Vargha, M., Z. Takáts, et al. 2006. Optimization of MALDI-TOF MS for strain level differentiation of *Arthrobacter* isolates. Journal of Microbiological Methods. 66(3): 399–409.

Wahl, K.L., S.C. Wunschel, et al. 2002. Analysis of microbial mixtures by matrix-assisted laser desorption/ionization time-of-flight mass spectrometry. Analytical Chemistry. 74(24): 6191–6199.

Walker, J., A.J. Fox, et al. 2002. Intact cell mass spectrometry (ICMS) used to type methicillin-resistant *Staphylococcus aureus*: media effects and inter-laboratory reproducibility. Journal of Microbiological Methods. 48(2–3): 117–126.

Wang, L.-T., F.-L. Lee, et al. 2007. Comparison of gyrB gene sequences, 16S rRNA gene sequences and DNA-DNA hybridization in the *Bacillus subtilis* group. International Journal of Systematic and Evolutionary Microbiology. 57(8): 1846–1850.

Wang, Z., L. Russon, et al. 1998. Investigation of spectral reproducibility in direct analysis of bacteria proteins by matrix-assisted laser desorption/ionization time-of-flight mass spectrometry. Rapid Communications in Mass Spectrometry. 12(8): 456–464.

Ward, D.R. 2001. Chapter I: Description of the Situation. Journal of Food Science Suppl. 6(7): 1067–1071.

Welker, M. and E.R.B. Moore. 2011. Applications of whole-cell matrix-assisted laser-desorption/ionization time-of-flight mass spectrometry in systematic microbiology. Systematic and Applied Microbiology. 34(1): 2–11.

Wieser, A., L. Schneider, et al. 2012. MALDI-TOF MS in microbiological diagnostics—identification of microorganisms and beyond (mini review). Applied Microbiology and Biotechnology. 93(3): 965–974.

Williams, T.L., D. Andrzejewski, et al. 2003. Experimental factors affecting the quality and reproducibility of MALDI TOF mass spectra obtained from whole bacteria cells. Journal of the American Society for Mass Spectrometry. 14(4): 342–351.

Wolters, M., H. Rohde, et al. 2010. MALDI-TOF MS fingerprinting allows for discrimination of major methicillin-resistant *Staphylococcus aureus* lineages. International Journal of Medical Microbiology. 301(1): 64–68.

Yamamoto, S., H. Kasai, et al. 2000. Phylogeny of the genus *Pseudomonas*: intrageneric structure reconstructed from the nucleotide sequences of *gyrB* and *rpoD* genes. Microbiology. 146(10): 2385–2394.

Zhu, S., S. Ratering, et al. 2011. Matrix-assisted laser desorption and ionization time-of-flight mass spectrometry, 16S rRNA gene sequencing, and API 32E for identification of *Cronobacter* spp.: A comparative study. Journal of Food Protection. 74(12): 2182–2187.

8

Production and Application of Microbial Transglutaminase to Improve Gelling Capabilities of Some Indonesian Minced Fish

Ekowati Chasanah[1],* and Yusro Nuri Fawzya[2]

1 Introduction

For food processing industries and food scientists, transglutaminase enzyme has been helpful in modifying and meeting various needs such as improving flavor, texture, appearance and function of the modified food. Microbial transglutaminase (MTGase EC 2.3.2.13) is popular in the protein modification of surimi based products and others restructured meat products (Mirzaei 2011). The enzyme catalyzes an acyl transfer reaction, where the γ-carboxyamide groups of glutamine residues in proteins act as acyl donors and some primary amines as acyl acceptors. The reaction produces covalent cross-links, such as ε-(γ-glutamil)-lysin bonds which result in improving functional properties of the food proteins. Microbial enzymes are generally recognized as safe (GRAS), as well as easy and cheaper to produce.

[1] Indonesian Research and Development Center for Marine and Fisheries Product Processing and Biotechnology, Jl. KS Tubun, Petamburan 6, Slipi, Jakarta, Indonesia.
[2] Indonesian Research and Development Center for Marine and Fisheries Product Processing and Biotechnology, Jl. KS Tubun, Petamburan 6, Slipi, Jakarta, Indonesia.
Email: nuri_fawzya@yahoo.com
* Coresponding author: ekowatichasanah@gmail.com

Surimi, prepared from washed minced fish with the addition of cryoprotectant, is now in a great demand in Indonesia and exported overseas. Processed products developed using surimi as raw material are becoming popular and various surimi based products are in the market. Surimi in Indonesia, as in other producing countries, is made of white flesh marine fish; however, due to overfishing in some of Indonesia's waters, the raw material for surimi moved to smaller, low economical marine fish, some of which was once considered bycatch. The quality of surimi's gelling properties is very important for its application in the surimi based products. The role of transglutaminase enzyme is crucial along with the actomyosin in fish meat which is very dependent on the species of fish as surimi raw material.

With the government program of increasing fish production by 353% at the end of the year 2015, fish culture is booming in Indonesia, and fresh water fish is becoming abundant. Therefore, cultured fish can be as alternative raw material for surimi and surimi based products. Research on surimi from Indonesian fresh water fish had been done and the result showed that even though the yield was quite low (approximately 13–33%), fresh water fish such as kissing gourami (*Helostoma temminski*) was reported to have relatively good gelling properties, i.e., elasticity, while some others such as Java carp (*Puntius gonionotus*) and carp (*Osteochillus hasselti* cv.) produced low gelling ability (Heruwati et al. 1995). Surimi from Indonesian catfish (*Pangasius hypophthalmus*) has been reported to have a problem due to the high fat content in the meat, but it has good gelling properties, especially when gelling agents such as calcium lactate and carageenan are added (Suryaningrum et al. 2009).

In the small scale surimi industries, the availablity of suitable water in quality and quantity, which is an important factor in surimi production, is becoming a problem. Consequently, the production of minced fish with minimum washing treatment has increased to meet the needs of small and medium scale processors of surimi based products, including fish balls, *otak-otak, mpek-mpek, siomay,* etc. Since the washing process of minced fish is not as intensive as that for surimi, the remaining fat and water-soluble substances, especially sarcoplasmic protein, are higher than in surimi, thus affecting the gel formation, resulting in lower gel strength of the products (Hall and Ahmad 1997; Hossain et al. 2004). Therefore, some additives need to be introduced into the minced fish to improve the gel forming ability.

The gelling properties of fish mince are dictated by the main component of myosin. Endogenous transglutaminase which presents in fish, depending on the fish species, has been reported to influence the gelling capabilities of fish meat by catalyzing polymerization of the myosin heavy chains (MHC). Actomyosin, its present varied among fish species, controls the crosslinking reaction mediated by transglutaminase (Araki and Seki 1993).

During surimi processing, washing treatment reduces the presence of native transglutaminase (TGase) in fish meat while screw-pressed dewatering was reported not to remove the enzyme. Studies in Threadbream fish (*Nemipterus* spp.) found that endogenous TGase was reduced significantly, even though not completely removed, during the washing process. About 44% of TGase activity was left in Threadbream surimi (Yongsawatdigul et al. 2002), while Araki and Seki (1993) found a slight change in TGase from pollock mince (0.41 unit/g) to pollock surimi (0.33 unit/g). Since the enzyme is found in sarcoplasmic proteins (Folk 1980), it can be removed when washing is too extensive. It seems that there is still controversy regarding the role of sarcoplasmic protein in surimi processing. Kim et al. (2005) found that sarcoplasmic proteins positively contribute to myosin gelation as reported by Morioka et al. (1997); Piyadhammavibon (2008). Morioka et al. (1997) also found that the gelation properties of sarcoplasmic proteins were related to their composition, which varied among fish species, while Kim et al. (2005) assumed that a certain minimum of myofibrillar protein content was required to obtain appropriate gelling properties of surimi.

2 Production of Microbial Transglutaminase

Production of MTGase related to various aspects, including the use of economical media and environmental conditions of fermentation, is still attractive enough to be studied, for reasons such as the utilization of agricultural waste to decrease production cost, and various factors which may increase the enzyme activity or shorten the production time. Miller and Churchill (1986) stated that media contribute almost 30% of the total cost for microbial fermentation. This is due to the commonly used expensive media for nutrient source, such as yeast extract and pepton. As an alternative, the Marine Biotechnology Research Group at the Indonesian Research and Development Center for Marine and Fisheries Product Processing and Biotechnology (IRDCMFPPB) has used liquid waste of tofu simultaneously with hydrolysate of cassava starch to produce MTGase from *Streptoverticillium ladakanum* NRL 3191 (Zilda et al. 2011). By using Fractional Factorial Research Design involving 4 factors, i.e., liquid waste of tofu processing, cassava starch, yeast extract and pepton, an optimum media composition for MTGase production has been developed. The MTGase produced by the media had the highest activity at the 4th day's cultivation, when the isolate was in stationary phase (Fig. 1). The enzyme worked optimally at 55°C and pH 8.

The use of surimi liquid waste in cultivation media of *S. ladakanum* NRL 3191 with/without cassava starch hydrolysate was reported by Chasanah et al. (2011) and Fawzya et al. (2012). The highest MTGase activity has been demonstrated using cultivation media of the hydrolysate of surimi liquid waste simultaneously with the cassava starch hydrolysate for 3 days cultivation (Fig. 2). The optimum temperature and pH of this enzyme activity were 40°C and 6, respectively. When surimi liquid waste and cassava starch hydrolysate were used to substitute pepton and glycerol respectively in a media formula (Fig. 3.), the highest enzyme activity was produced

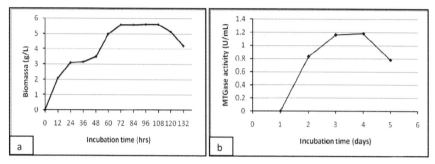

Figure 1. (a) Growth of *S. ladakanum* NRL 3191 in liquid media containing liquid waste of tofu processing and cassava starch hydrolysate (b) MTGase activity produced during cultivation.

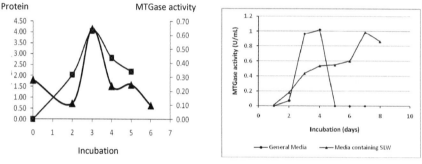

Figure 2. Transglutaminase activity produced by *S. ladakanum* NRL 3191 in liquid media consisting of (a) surimi liquid waste and cassava starch hydrolysate (b) and surimi liquid waste (SLW) substituted for sodium caseinate in comparison to general media.

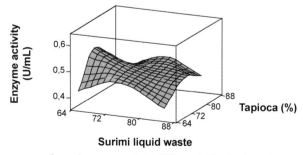

Figure 3. Response surface of enzyme activity: Effect of substitution of pepton with surimi liquid waste (SLW) (35:65)% and glycerol with cassava starch hydrolysate (25:75)%.

from the medium containing 75% surimi liquid waste with 25% pepton and 65% cassava hydrolysate with 35% glycerol after 3 days incubation (Chasanah et al. 2011). The enzyme worked optimally at 35°C and pH 6, whereas surimi liquid waste substituted for sodium caseinate showed a longer production period even though the enzyme activity was competitive with that produced from ordinary (general) media (Fawzya et al. 2011). The enzyme had optimum activity at 50°C and pH 8. It was relatively comparable with the optimum temperature and pH of MTGase activity as reported by Zilda et al. (2011), due to the similar conditions of fermentation of *S. ladakanum* NRL 3191, particularly the temperature of fermentation, i.e., 26°C; whereas the fermentation temperature of *S. ladakanum* NRL 3191 grown by using media containing hydrolysate of surimi liquid waste simultaneously with cassava starch hydrolysate was 30°C. The difference in fermentation temperature resulted in differences in optimum temperature and pH of MTGase activity. Cui et al. (2007) reported that MTGase from *Streptomyces hygroscopicus* exhibited optimum activity at 37–45°C, and pH 6–7, while Ho et al. (2000) found that the optimal temperature and pH of MTGase activity from *S. ladakanum* was 40°C and 5.5, respectively.

3 Application of Transglutaminase

MTGase produced by *S. ladakanum* NRL 3191 using modified media containing local nutrient sources, e.g., surimi and tofu industrial waste, was used in the research on application of MTGase to some Indonesian fish. Before being applied to 'mata goyang' fish (*Priacanthus macracanthus*), the MTGase *S. ladakanum* NRL 3191 produced from modified media was concentrated with ultrafiltration. The commercial MTGase was used as control. Since the commercial MTGase (CMTGase) contained MTGase, salt and sodium caseinate, the application of MTGase produced from the formulated media was done with treatments as presented at Fig. 4. The MTGase treatment showed that the gel forming ability of the minced fish based products was almost as good as that of those with the added commercial MTGase (Fawzya et al. 2012). This enzyme was also able to improve the gel strength, springiness and cohesiveness of the minced fish based products made of *P. macracanthus* compared to those produced from the treatments of 1% NaCl or 1% NaCl and 1% sodium caseinate. The presence of sodium caseinate, as in the commercial MTGase product, does not seem to affect the enzyme activity.

The result of scanning electron microscopy (SEM) on the minced fish based product of *P. macracanthus* showed fewer hollows in the structure of the product surface with increasing concentration of MTGase due to the formation of cross-links [ε-(γ-glutamyl)-lysine] by MTGase (Fig. 5).

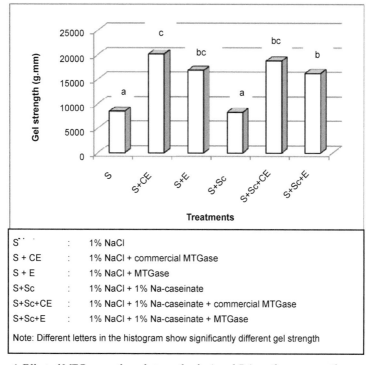

Figure 4. Effect of MTGase on the gel strength of minced *Priacanthus macracanthus*.

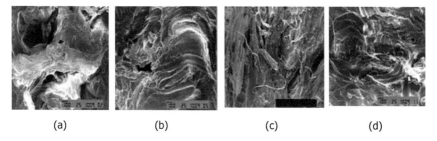

| (a) | (b) | (c) | (d) |

Figure 5. Microstructure of minced fish based product of *P. macracantus* under scanning electron microscope with magnification of 75x; from the treatment of 2% salt with the concentration of MTGase (a) 0%, (b) 0.3%, (c) 0.6% and (d) 1%.

In the second study of MTGase application, 'tongkol' fish (*Euthynnus* spp.), a pelagic fish that is not normally utilized for surimi production was used. This is because the fish, as do other small pelagic fish with higher percentage of dark meat, contains high lipids and sarcoplasmic proteins in the muscle tissue, including proteolytic enzymes, inducing a lower gel-forming ability (Shimizu et al. 1992 in Chaijan and Panpipat 2010). Indonesia is rich in this fish, and we were trying to improve the minced fish based product of *Euthynnus* spp. meat by the addition of 0.5–1% MTGase. The gelling properties of the product after addition of the MTGase was determined including hardness, springiness, cohesiveness, breaking force, deformation and gel strength. Results showed that the addition of 1% MTGase into minced tongkol based products could improve gelling properties of minced fish, being 22 times higher in white meat of tongkol and 13 times higher in red meat of tongkol compared to the minced tongkol without addition of the enzyme (Fig. 6). This study showed that tongkol, both red meat and white meat, can be used as raw material for the minced fish based products with the addition of the MTGase.

Other studies on the application of MTGase for improving gel forming ability of fresh water fish were conducted for Nile tilapia (*Oreochronis niloticus*), 'lele' or catfish (*Clarias* sp.) and 'patin' (*Pangasius* sp.). As stated before, these fresh water fish are becoming abundant due to the strong efforts of accelerating fish culture to support the government program of achieving 353% increment of cultured fish production in the year 2015. The effect of MTGase's addition on the gel forming ability of *Clarias* sp. surimi as well as on minced fish have been studied. The enzyme produced by using media developed by Marine Biotechnology Research Group of the IRDCMFPPB containing surimi liquid waste as well as its hydrolysate form, was added to washed minced fish and the unwashed minced fish (Chasanah et al. 2011). The two treatments used the same concentration of MTGase, i.e., 300 U/kg meat. As control, commercial MTGases was used at the same concentration (300 U/kg meat), carragenan 1% and sodium tripolyphosphate (STPP) 1%. The results showed that MTGase addition significantly increased the gel strength of the products (Fig. 7), with the higher improvement taking place on the unwashed meat (approximately 4 times) as compared to washed meat (approximately 1.7 times).

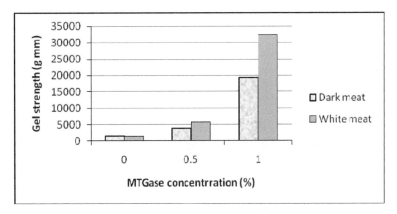

Figure 6. Effect of MTGase on the gel strength of minced *Euthynnus* spp.

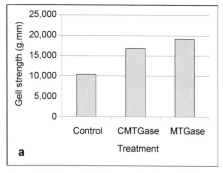 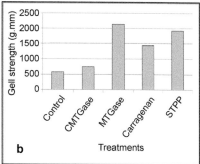

Figure 7. Effect of MTGase and other additives on the gel strength of (a) washed minced-*Clarias* sp. and (b) unwashed minced-*Clarias* sp.

A slight difference in the result was shown on the influence of MTGase on the minced of *Pangasius* sp. The highest gel strength of the product was reached in once-washed minced fish with the addition of 0.6% MTGase (Chasanah et al. 2010) as presented in Fig. 8. The improvement was less than 50%; this might be due to the actomyosin of the fish which is known as having good gelling properties in fish (Suryaningrum et al. 2012). Based on the microstructure appearance of minced *Pangasius* based product (Fig. 9), the holes shown on the natural product without the addition of MTGase were nearly similar to that with added MTGase.

Application of MTGase on the Nile tilapia (*Oreochronis niloticus*) was carried out by using a different source of MTGase and different treatment. A commercial MTGase was added at the concentration levels of 0; 0.5025; 1.005 and 1.675 U/g meat into minced Nile tilapia with and without salt addition.

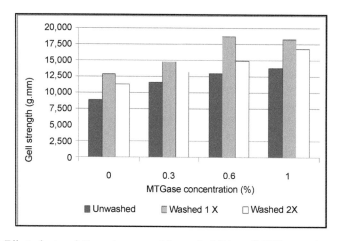

Figure 8. Effect of minced-*Pangasius* sp. washing and addition of MTGase on the gel strength of restructured fish.

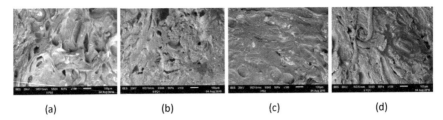

(a)   (b)   (c)   (d)

Figure 9. Microstructure of minced fish based product of *Pangasius* sp. under scanning electron microscope with magnification of 100x; from the treatment of 2% salt with the concentration of MTGase 0% (a), 0.3% (b), 0.6% (c) and 1% (d).

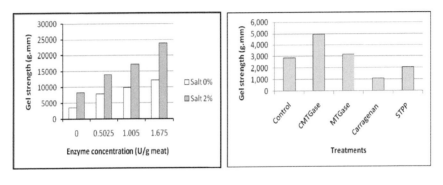

A   B

Figure 10. Effect of commercial MTGase addition to washed minced Nile tilapia (A) and some additives incorporated to unwashed minced Nile tilapia (B) on the gel strength of the fish minced based product.

A significant improvement of gel strength was observed on the minced fish based product with higher salt addition; indicating the synergism effect of both salt and the MTGase.

Table 1 summarized the result of MTGase application on the above fish sample with the enzyme concentration of about 1% compared to those without addition of MTGase (Table 1). The data showed that all samples had relative gel strength more than 100%, showing that the addition of MTGase increased the gel strength, which related to the cross-linking function of MTGase. However, high gel strength improvement resulted from *Euthynnus* spp. which is known to be low in actomyosin, protein that contributes in gel-forming ability. This was similar to the result found by Wu and Corke (2005) indicating that the effect of incorporation of MTGase to the low gluten content wheat flour was much more than the high gluten content flour. Several factors may affect the results including internal and external factors such as fish species related to natural actomyosin, endogenous enzymes, fish

Table 1. Application of MTGase on some Indonesian minced fish.

Fish Species	Gel strength (g.mm)	Relative gel strength (%)*	Treatment	Reference
Nile Tilapia (*Oreochronis niloticus*)	23,950	290	W 2x, S 2%, E 1% (1,675 U/kg meat)**	Fawzya et al. 2011
	3,178	111	UW, S 2%, E 300 U/kg meat	Chasanah et al. 2011
'Patin' (*Pangasius* sp.)	18,237	132	W1x, S 2%, E 1%**	Chasanah et al. 2010
'Lele' (*Clarias* sp.)	2,141	379	UW, S 2%, E300 U/kg meat	Chasanah et al. 2011
	19,181	186	W 2x, S 2%, E 300 U/kg meat	Chasanah et al. 2011
'Tongkol' (*Euthynnus* spp.)	19,278	1285	W 2x, S 1%, E 1% (1,675 U/kg meat)**	Fawzya et al. 2011
'Mata goyang' (*P. macracantus*)	30,000	250	W 2x, S 2%, E 1% (1,675 U/kg meat)**	Fawzya et al. 2011
	16,848	199	W 2x, S 1%, E 300 U/kg meat	Fawzya et al. 2011
	20,150	238	W 2x, S 1%, E 300 U/kg meat**	Fawzya et al. 2011

*Relative to control (without addition of MTGase)
**commercial MTGase containing ingredient other than MTGase
W: washed, UW: unwashed, S: salt, E: enzyme (MTGase)

freshness, additives (salt, etc.), washing treatments and cooking procedures as stated by Lin and Park (1996), Kongpun (1999), Hossain et al. (2004), Kamal et al. (2005), Ramirez et al. (2006).

Acknowledgement

Thank you to Agricultural Research Service, United State Department of Agricultural for providing the *Streptoverticillium ladakanum* NRL 3191 and Prof. Hari Eko Irianto for reviewing this paper.

References

Araki, H. and N. Seki. 1993. Comparison of reactivity of transglutaminase to various fish actomyosins. Nippon Suisan Gakkaishi. 59: 711–716.

Chaijan, M. and W. Panpipat. 2010. Gel-Forming Ability of Mackerel (*Rastrelliger Branchysoma*) Protein Isolate as Affected by Microbial Transglutaminase. Walailak J. Sci. & Tech. 7(1): 41–49.

Chasanah, E., Y.N. Fawzya, A. Poernomo, A.S. Dewi, I. Munifah, G. Patantis and A. Pratitis. 2010. Research on Utilization of microorganism and enzyme to develop tropical catfish surimi based products, bioenergy from seaweed processing waste and nutraceuticals from shrimp waste. Technical Report. Research Centre of Marine and Fishery Product Processing and Biotechnology. Jakarta (In Indonesian).

Chasanah, E., Y.N. Fawzya, A. Poernomo, S. Wibowo, S. Amini, I. Munifah, Pujoyuwono, H.I. Januar and G. Patantis. 2011. Research on Utilization of marine microorganism and potential enzymes to develop food and non food products. Technical Report. Research Centre of Marine and Fishery Product Processing and Biotechnology. Jakarta (In Indonesian).

Cui, L., G. Du, D. Zhang, H. Liu and J. Chen. 2007. Purification and characterization of transglutaminase from a newly isolated *Streptomyces hygroscopicus*. Food Chemistry. 105: 612–618.

Fawzya, Y.N., D.S. Zilda, A. Poernomo, I. Kristiana and H. Nursyam. 2011. Characterization and application of Transglutaminase produced from *Streptoverticillium ladakanum* on the *Priacanthus macracanthus* minced meat. Journal of Marine and Fisheries Post Harvest and Biotechnology (In Indonesian). 6(2): 157–166.

Fawzya, Y.N., T.K.A. Puruhita, Gunawan and G. Patantis. 2011. Effect of salt and transglutaminase on the physical and sensory properties of restructured *Priacanthus macracanthus*. Journal of Marine and Fisheries Post Harvest and Biotechnology (In Indonesian). 4(1): 69–78.

Fawzya, Y.N., E. Chasanah, T.K.A. Puruhita and R. Indrati. 2011. Effect of salt and transglutaminase addition on the characteristics of Nile Tilapia (*Oreochronis niloticus*), minced meat gel. Paper Presented at National Seminar on Aquaculture Technology Inovation Forum at Bali, 19–21 July 2011.

Fawzya, Y.N., A. Pratitis and G. Patantis. 2011. Effect of transglutaminase addition on the gel forming ability of *Euthynnus* spp. minced meat. Proceeding of the National Seminar on Fisheries Product Processing Community. Bogor Agricultural University. Bogor.

Fawzya, Y.N., M.F. Ridwan, W. Trilaksani and D.S. Zilda. 2012. Production and characterization of *Streptoverticillium ladakanum* transglutaminase in media substituted with liquid surimi waste. Proceeding of the National Seminar on Fisheries Product Processing Community. Brawijaya University, pp. 245–254.

Folk, J.E. 1980. Transglutaminases. Annual Review of Biochemistry. 49: 517–531. DOI: 10.1146/annurev.bi.49.070180.002505.

Hall, G.M. and N.H. Ahmad. 1997. Surimi and fish mince products. pp. 74–92. *In*: G.M. Hall (ed.). Fish Processing Technology. Chapman and Hall. Springer US. http://link.springer.com/chapter/10.1007%2F978-1-4613-1113-3_3#.

Heruwati, E.S., Jovita Tri Murtini, Siti Rahayu and Memen Suherman. 1995. Effect of fish species and food additives on the elasticity of fresh water fish surimi. Journal of Marine and Fisheries Post Harvest and Biotechnology (In Indonesian). 1(1): 86–94.

Ho, M.-L., S.-Z. Leu, J.-F. Hsieh and S.-T. Jiang. 2000. Technical approach to simplify the purification method and characterization of microbial transglutaminase produced from Streptoverticillium ladakanum. J. of Food Sci. 65(1): 76–80.

Hossain, M.I., M.M. Kamal, F. Hoque Shikha and M.D. Shahidul. 2004. Effect of washing and salt concentration on the gel forming ability of two tropical fish gpecies. International Journal of Agriculture & Biology. 1560–8530/2004/06–5–762–766. http://www.ijab.org.

Kamal, M., M. Ismail Hossain, M.N. Sakib, F.H. Shikha, M. Neazuddin, M.A.J. Bapary and M.N. Islam. 2005. Effect of salt concentration and cryoprotectants on gel-forming ability of surimi prepared from queen fish (*Chorinemus lysan*) during frozen storage. Pak. J. Biol. Sci. 8: 793–797.

Kim, Y.S., J. Yongsawatdigul, J.W. Park and Thawornchinsombut. 2005. Characteristics of sarcoplasmic proteins and their interaction with myofibrillar proteins. Journal of Food Biochemistry. 29: 517–532.

Kongpun, O. 1999. The Gel Forming Ability of Washed and Unwashed Fish Meat (Lizardfish and Nile tilapia). Kasetsart J. Nat. Sci. 33: 258–269.

Lin, T.M. and J.W. Park. 1996. Extraction of proteins from Pacific whiting mince at various washing conditions. Journal of Food Science. 61(2): 432–438.

Mahta Mirzaei. 2011. Microbial Transglutaminase application in food industry. International Conference on Food Engineering and Biotechnology IPCBEE. 9: 267–271.

Miller, T.L. and B.W. Churchill. 1986. Substrates for Large-Scale Fermentations. *In*: A.L. Demain and L.A. Solomon (eds.). Manual of Industrial Microbiology and Biotechnology. American Society for Microbiology. Washington DC.

Morioka, K., T. Nishimura, A. Obatake and Y. Shimizu. 1997. Relationship between the myofibrillar protein gel strengthening effect and the composition of sarcoplasmic proteins from Pacific mackerel. Fisheries Science. 63(1): 111–114.

Piyadhammavibon, P. 2008. Effect of Sarcoplasmic Proteins from Threadfin bream as Gel Enhancer of Lizardfish Surimi Gel [Thesis]. Suranaree University of Technology. Thailand. http://sutir.sut.ac.th:8080/sutir/bitstream/123456789/3127/2/fulltext.pdf.

Ramirez, J.A., A.D. Angel, G. Velazquez and M. Vazquez. 2006. Production of low-salt restructured fish products from Mexican flounder (Cyclopsetta chittendeni) using microbial transglutaminase or whey protein concentrate as binders. Eur. Food. Res. Technol. 223: 341–345. DOI 10.1007/s00217-005-0210-z. http://www.springerlink.com/content/p3824m33t1t32j10/fulltext.pdf. Diakses 13 Juni 2011.

Shimizu, Y., H. Toyohara and T.C. Lanier. 1992. Surimi Production from Fatty and Dark-Fleshed Fish Species. pp. 181–207. *In*: T.C. Lanier and C.M. Lee (eds.). Surimi Technology. Marcel Dekker, New York.

Suryaningrum, T.D., Diah Ikasari and Syamdidi. 2009. The addition of gelling agents on catfish (*Pangasius hypophthalmus*) surimi processing. Journal of Marine and Fisheries Post Harvest and Biotechnology (In Indonesian). 4(1): 37–47.

Wu, J. and H. Corke. 2005. Quality of dried white salted noodles affected by microbial transglutaminase. J. Sci. Food Agric. 85: 2587–2594. DOI: 10.1002/jsfa.2311.

Yongsawatdigul, J., A. Worratao and J. Park. 2002. Effect of endogenous transglutaminase on threadfin bream surimi gelation. J. Food Sci. 67(9): 3258–3263.

9

Lactic Acid Bacteria in Seafood Products: Current Trends and Future Perspectives

Panchanathan Manivasagan,[1] *Jayachandran Venkatesan*[1] *and Se-Kwon Kim*[1,2,]*

1 Introduction

Lactic acid bacteria constitute a large group of non sporulating, Gram positive catalase and oxidize negative rods and cocci that produce lactic acid as the major metabolite of the carbohydrate fermentation. LAB are aerotolerant anaerobe and generally have complex nutritional requirements especially for amino acids and vitamins. The genera that comprise the LAB are at its core *Lactobacillus, Lactococcus, Leuconostoc, Pediococcus, Streptococcus* as well as the more peripheral *Aerococcus, Carnobacterium, Enterococcus, Oenococcus, Sporolactobacillus, Tetragenococcus, Vagococcus, Weissella* and *Bifidobacterium* genus.

LAB are widespread in most ecosystems and are found in soil, water, plants, and animals. They are responsible for many food fermentation processes, but they are also commonly found in non-fermented foods

---

[1] Marine Biotechnology Laboratory, Department of Chemistry Pukyong National University, Busan 608-737, Republic of Korea.
  Email: manimaribtech@gmail.com
[2] Marine Bioprocess Research Center, Pukyong National University, Busan 608-737, Republic of Korea.
* Corresponding author: sknkim@pknu.ac.kr

such as dairy products, meat products, seafood, fruits, vegetables, cereals, sewage, and the genital, intestinal, and respiratory tracts of humans and animals. LAB are widely used as starter cultures in the food industry for the production of fermented foods, including dairy (yogurt, cheese), meat (sausages), fish, cereals (bread and beverages such as beer), fruit (malolactic fermentation processes in wine production), and vegetables (sauerkraut, kimchi, silage). Most LAB are considered GRAS (generally recognized as safe) by the US Food and Drug Administration (FDA (Silva et al. 2002)).

LAB are used to ensure safety, preserve food quality, develop characteristic new flavors and to improve the nutritional qualities of food. LAB exert strong antagonistic activity against many related and unrelated microorganisms, including food spoilage organisms and pathogenic bacteria such as *Listeria, Clostridium, Staphylococcus,* and *Bacillus* spp. The antagonistic effect of LAB is mainly due to a lowering of the pH of the food, competition for nutrients, and the production of inhibitory metabolites (Stiles 1996).

LAB are able to grow at refrigeration temperatures. They tolerate atmosphere packaging, low pH, high salt concentrations, and the presence of additives such as lactic acid, ethanol, or acetic acid. In this sense, Dalgaard et al. (2003) identified LAB strains found in cooked and brined shrimp stored under a modified atmosphere between 0 and 25°C. The major organism found in the 15–25°C range was *Enterococcus faecalis,* whereas *Carnobacterium divergens* and *Lactobacillus curvatus* were the major organisms in the 0, 5, and 8°C range (Dalgaard et al. 2003). The potential applications of lactic acid bacteria in the biopreservation of seafood products are discussed in the present review.

2 Microbial Ecology of Seafood

Foods are dynamic systems in which changes occur in pH, atmosphere, nutrient composition and microflora over time. Each food product has its own unique flora, determined by the raw materials used, food processing parameters and subsequent storage conditions. Despite the variation in composition of the microflora on newly caught fish, seafood products can be categorized into groups with similar microbial ecology (Gram and Huss 2000). During storage, the microflora changes owing to the different abilities of the microorganisms to tolerate the preservation conditions. Gram-negative, fermentative bacteria (such as *Vibrionaceae*) spoil unpreserved fish, whereas psychrotolerant Gram-negative bacteria (*Pseudomonas* spp. and *Shewanella* spp.) grow on chilled fish (Gram and Huss 2000). CO_2 packing inhibits respiratory organisms and selects for

Photobacterium phosphoreum and lactic acid bacteria (LAB) (Dalgaard 2000). Respiratory Gram-negative bacteria are typically inhibited in fish products preserved by the addition of low levels of NaCl, slight acidification and chill storage in vacuum-packs (e.g., cold-smoked fish). Under these conditions, the microflora typically becomes dominated by LAB (*Lactobacillus* and *Carnobacterium*) with an association of Gram-negative fermentative bacteria such as *P. phosphoreum* and psychrotrophic Enterobacteriaceae (Hansen et al. 1995; Jørgensen et al. 2000; Leroi et al. 1998). Increasing the 'preservation pressure', for example, by acidification or the addition of preservatives like sorbate and benzoate, as in the so-called semi preserved seafood (e.g., marinated herring), allows for growth of lactobacilli and yeasts. Drying or heavily dry-salting of fish eliminates bacterial growth and these products will spoil due to growth of filamentous fungi or insect infestation. Yeasts may grow in heavily wet-salted fish (e.g., barrel-salted herring). Several fish products are subjected to a mild heat treatment (equaling pasteurization) and spore-forming bacteria (*Clostridium* or *Bacillus*) may grow in such products, particularly if unsalted (e.g., products cooked in vacuum pouches, i.e., *sous vide* products).

3 Lactic Acid Bacteria

Although not the most common, it is generally accepted that LAB occur among the normal intestinal microbiota of fish from the first few days onwards (Yang et al. 2007; Ringø 2008). Lactobacilli, notably *Lactobacillus plantarum*, have been found in Atlantic salmon (Ringø et al. 1997), pollock (Schrøder et al. 1980), Arctic char (Ringø and Gatesoupe 1998) and cod (Strøm and Olafsen 1990). Carnobacteria, including *Carnobacterium maltaromaticum* (previously *piscicola*), *divergens, gallinarum* and *inhibens*, have been isolated from all these species, as well as from rainbow trout (Jöborn et al. 1999; Ringø et al. 2001; Huber et al. 2004). Carnobacteria have even been quoted as being the dominant genus in the gastro-intestinal tract of juvenile Atlantic salmon (Ringø et al. 1997) and cod (Seppola et al. 2006). Other authors have also reported the presence of *Leuconostoc mesenteroides, Lactococcus piscium, Vagococcus salmoninarum, Lactobacillus fuchuensis, Streptococcus* spp., and *Weissella* spp. (Wallbanks et al. 1990; Williams et al. 1990; Ringø and Gatesoupe 1998; Liu et al. 2009; Matamoros et al. 2009). Various factors, like salinity of the water or stress, can affect the presence of LAB (Ringø and Strøm 1994). For instance, the number of lactobacilli in the gastro-intestinal tract of Arctic char was smaller in those reared in sea water than in fresh water, while the number of *Leuconostoc* and enterococci remained the same (Ringø and Strøm 1994). Atlantic salmon stressed by daily manipulations experienced a drop in the number of Gram-negative bacteria in their intestinal microbiota and an increase of carnobacteria count (Ringø and Gatesoupe 1998).

Even though most LAB are generally recognized as safe by the US Food and Drug administration, their implication in fish disease has been reported. In marine farmed fish, numerous epizootics linked to streptococci have been notified, beginning in Japan and North America then spreading worldwide (Eldar et al. 1996). These bacteria, reclassified as *Lactococcus garvieae*, are responsible for septicemias, ophthalmias and hemorrhages. *C. maltaromaticum* has been isolated from different diseased fish and its virulence has been clearly established in rainbow trout and striped bass experimentally infected (Baya et al. 1991). Recently, a novel *Weissella* species has been described as an opportunistic pathogen for rainbow trout (Liu et al. 2009).

4 LAB Occurrence in Seafood Products

LAB are not considered as belonging to aquatic environments, but certain genera (i.e., *Carnobacterium, Vagococcus, Lactobacillus, Enterococcus, Lactococcus*) have been found in fish and their surrounding environment (Huss 1995; Stiles 1996; Gänzle et al. 2000; Ringø et al. 2000). Carnobacteria, mainly *C. maltaromaticum* and *C. divergens*, have been reported to be part of the normal intestinal microbial population of many fish species such as Atlantic salmon (*Salmo salar*), wild pike (*Esox lucius*), and wild brown trout (*Salmo trutta* (Ringø et al. 2000; Ringø and Gatesoupe 1998; Gonzalez et al. 2000; Gonzalez et al. 1999)).

LAB have also been isolated in processed seafood products. Bacteriocinogenic strains of the species *C. maltaromaticum* and *Lactobacillus* spp. have been isolated from fish and cold-smoked fish products (Baya et al. 1991; Alves et al. 2005; Duffes, Corre et al. 1999; González-Rodríguez et al. 2002a; González-Rodríguez et al. 2002b; Groth Laursen et al. 2005; Leroi et al. 2003; Pellé et al. 2005). In brined shrimps, Dalgaard et al. (2003) found *C. divergens* and *L. curvatus* in spoilage associations of products stored at refrigeration temperatures (Dalgaard et al. 2003). Enterococci with antilisterial activity have also been isolated from the skin of farmed turbot (Campos et al. 2006) and commercial cold-smoked salmon (Tomé et al. 2008). Wessels and Huss (1996) tested *L. lactis* ssp. *lactis* ATCC 11454 as a protective culture for lightly preserved fish products such as commercial cold-smoked salmon. They found that when salmon slices were inoculated with *L. monocytogenes* at 10 CFU/g and a 300-fold excess of washed lactococci cells, the pathogenic population declined a half log over the first 1.5 days, then increased at a rate slightly lower than that of the controls not inoculated with the lactococci (Wessels and Huss 1996).

LAB tend to grow slowly at refrigeration temperatures (Huis in't Veld 1996). Psychrotrophic LAB are capable of growth at 5°C or below. Psychrotrophic LAB include *Carnobacterium*, *Leuconostoc*, and *Weissella* spp. and some facultatively heterofermentative lactobacilli, such as *L. sakei* and *L. curvatus*. At temperatures below 20°C under anaerobic conditions, psychrotrophic LAB compete with other psychrotrophic spoilage microorganisms such as *L. monocytogenes*. The LAB selection is generally recognized as beneficial for extending the shelf life of fish because their growth can inhibit more potent spoilers by means of antagonistic activities such as the production of organic acids, diacetyl, hydrogen peroxide, CO_2, and bacteriocins (Daeschel 1989; Lindgren and Dobrogosz 1990). In this sense, several studies have been carried out on the antilisterial activity of many bacteriocinogenic LAB strains isolated from seafood.

5 Seafood Fermentation by LAB

Unlike in dairy, meat, cereal and plant products, LAB are not often used for fermenting seafood products. There are some traditional fermented products, like anchovies placed in a barrel with salt and sugar for a curing process, and fish sauces which are very popular in Asiatic countries. Recently in Japan, fish sauce production has increased dramatically (quadruple the amount produced five years ago) as the fishing industry tries to reduce waste by making full use of fish materials. Many fish sauces are made of small sea water fish with long term fermentation of more than a year. Halophilic LAB, mainly *Staphylococcus* spp. and *Tetragenococcus* spp., were isolated as dominant bacteria during fermentation (Taira et al. 2007) but their real role is quite controversial. Much of the degradation is essentially due to the presence of endogenous enzymes in fish. However, it is clear that certain salt tolerant strains of *Tetragenococcus halophilus* and *T. muriaticus* contribute to lowering the pH value and reduce the risk of putrefaction of the fermenting sauce, especially when an extra source of carbohydrate is added, like soy sauce *koji* (Uchida et al. 2005). On the other hand, this positive effect can be balanced by the fact that *T. halophilus* is a histamine producer, and levels of 1000 ppm have been recorded in some fish sauces (Satomi et al. 2008).

Traditional lightly salted fermented fish products are widespread in South-East Asia. They are typically composed of fresh water fish species, salt (2–7%), a carbohydrate source (rice, millet, sugar or fruit) and spices (garlic, ginger, chilli, pepper). The mixture is tightly packed in banana leaves or plastic bags and left to ferment for two to five days at 30°C before being consumed as the main course or as a snack. In these kind of products, Paludan-Müller et al. (1999) have shown that LAB isolated from fish raw material were dominated by *L. lactis* subsp. *lactis* (Paludan-Müller et al. 1999). However, the LAB responsible for the fermentation were mostly of vegetable origin, dominated by aciduric heterofermentative *Lactobacillus* spp. and homofermentative *L. plantarum/pentosus*. There have also been some attempts to develop new fermented marine products, like salted salmon fillets inoculated with different commercial lactobacilli starters and with sugar added (Morzel et al. 1997). Nevertheless, this research is still at the laboratory stage and the products, to our knowledge, have not been commercialized.

Fish is a source of high quality proteins, essential minerals, vitamins and polyunsaturated fatty acids. Some recent works, therefore, are concerned with the development of new products that ideally, should retain all the nutritional properties of fish but not its typical odor so that they can be included in meat-based preparations. Glatman et al. (2000) have inoculated yellowfin tuna with high level of *L. plantarum, Pediococcus pentosaceus* and *L. mesenteroides* and they showed that a strain of *L. mesenteroides* could effectively ferment tuna flesh, making it lose its character and develop new odors and texture close to those of meat, and, in addition, reduce the level of histamine. The best results were obtained after 5 weeks of fermentation at 8°C (Gelman et al. 2000).

The silage of fish has been investigated since the 1950s and should have provided a commercial alternative for some of the by-products of the fish industry. Different strains of *L. plantarum, L. acidophilus, Pediococcus halophilus* and *P. acidilactici* have been tested, with some success, for their ability to lower the pH of fish flesh rapidly. However, the low level of sugar in the matrix requires a carbohydrate supplement, such as starch, malt or molasses to be added. Nevertheless, in the anaerobic conditions favoring the LAB fermentation, the risk of contamination by *Bacillus* and *Clostridium* cannot be avoided. Paradoxically in the presence of air, yeasts and moulds may then degrade the product. An avenue of research for developing these products is to look for psychrotrophic strains suitable for the very rapid fermentation of fish. Yoon et al. (1997) described the characteristics of LAB for the preparation of silage from tuna viscera (Yoon et al. 1997). Successful studies have been performed on the biotransformation of salmon (Bower and Hietala 2008) and tuna waste (Vijayan et al. 2009) as novel aquafeed ingredients for fish.

6 Natural Antimicrobial Compounds Produced by LAB

The preservative action of starter strains in seafood and beverage systems is attributed to the combined action of a range of antimicrobial metabolites produced during the fermentation process (Caplice and Fitzgerald 1999). These include many organic acids such as lactic, acetic and propionic acids produced as end products; these provide an acidic environment unfavorable for the growth of many pathogenic and spoilage microorganisms. Acids are generally thought to exert their antimicrobial effect by interfering with the maintenance of cell membrane potential, inhibiting active transport, reducing intracellular pH and inhibiting a variety of metabolic functions. They have a very broad mode of action and inhibit both Gram-positive and Gram-negative bacteria as well as yeast and moulds (Caplice and Fitzgerald 1999). The selection of protective cultures is relevant in order to biopreserve and improve the functional safety of food products, mainly through inhibition of spoilage and/or pathogenic bacteria. The present investigated potential applications of lactic acid bacteria (LAB) is in the biopreservation of fish and shellfish products. For this purpose, a collection of 84 LAB strains isolated from sea bass (*Dicentrarchus labrax*) and sea bream (*Sparus aurata*) was identified and characterized for its inhibitory activities against the most relevant seafood-spoilage and pathogenic bacteria potentially present in commercial products (Bourouni et al. 2012).

Other examples of secondary metabolites produced by LAB which have antagonistic activity include the compound reuterin (Axelsson et al. 1989; Chung et al. 1989) and the recently discovered antibiotic reuterocyclin (Gänzle et al. 2000; Hoeltzel et al. 2000), both of which are produced by strains of *Lactobacillus reuteri*. Reuterin is an equilibrium mixture of monomeric, hydrated monomeric and cyclic dimeric forms of β-hydroxypropionaldehyde. It has a broad spectrum of activity and inhibits fungi, protozoa and a wide range of bacteria including both Gram-positive and Gram-negative microorganisms. This compound is produced by stationary phase cultures during anaerobic growth on a mixture of glucose and glycerol or glyceraldehyde. Consequently, in order to use reuterin-producing *L. reuteri* for biopreservation in a seafood product, it would be beneficial to include glycerol with the strain. This approach was used to extend the shelf-life of herring fillets stored at 5°C and involved dipping the fish in a solution containing 1×10^9 CFU/ml of *L. reuteri* and 250 mM glycerol (Lindgren and Dobrogosz 1990). Results demonstrated that after 6 days of storage, there were approximately 100-fold-less Gram-negative bacteria in the *L. reuteri* samples than in the untreated control.

More recently, the first antibiotic produced by a LAB was discovered (Gänzle et al. 2000; Hoeltzel et al. 2000). Reuterocyclin is a negatively charged, highly hydrophobic antagonist, and structural elucidation revealed it to be a novel tetramic acid. The spectrum of inhibition of the antibiotic is confined to Gram-positive bacteria including *Lactobacillus* spp., *Bacillus subtilis, B. cereus, E. faecalis, S. aureus* and *Listeria innocua*. Interestingly, inhibition of *E. coli* and *Salmonella* is observed under conditions that disrupt the outer membrane, including truncated LPS, low pH and high salt concentrations. Since it is well known that nisin can kill Gram-negative bacteria under conditions which disturb the outer membrane (Stevens et al. 1992), it is likely that there are similarities in the mode of action of nisin and this novel antibiotic.

7 Bacteriocin-producing LAB in Seafood Safety

The deterioration of fish is generally caused by gram-negative microorganisms and as such, few attempts have focused on evaluating the potential of LAB bacteriocins in such products. However, in vacuum-packed seafood, spoilage organisms such as *C. Botulinum* and *L. monocytogenes* can cause problems. The inhibition of these spoilage organisms by bacteriocins has been studied (Degnan et al. 1994; Einarsson and Lauzon 1995; Nilsson et al. 1997). However, up to recently, little work had focused on incorporating live bacteriocin-producing cultures into the product but rather on addition of the bacteriocin as a concentrated preparation. In the last number of years, different groups tested the growth and bacteriocin production of *Carnobacterium divergens* V41 and *Carnobacterium piscicola* V1 in a simulated cold-smoked fish system at 4°C. In co-culture, these strains were very effective inhibiting *L. monocytogenes* as early as day 4 (Duffes, Leroi et al. 1999). The same group later demonstrated the inhibition of *L. monocytogenes* by *Carnobacterium* strains on sterile and commercial vacuum-packed cold-smoked salmon stored at 4 and 8°C. *C. piscicola* V1 was bactericidal against *L. monocytogenes* at the two temperatures, whereas *C. divergens* V41 had a bacteriostatic effect. *C. piscicola* SF668 delayed *L. monocytogenes* growth at 8°C and had a bacteriostatic effect at 4°C (Duffes, Corre et al. 1999). Importantly, the use of these bacteriocin-producing strains did not have any effect on the quality of the end-product. Recently, Nilsson et al. (1999) also demonstrated that a bacteriocin-producing *Carnobacterium* could be used as a biopreservative in vacuum-packed, cold-smoked salmon stored at 5°C. *C. piscicola* (A9b) initially caused a 7-d lag phase of *L. monocytogenes*, followed by a significant reduction in numbers from 10^3 CFU/ml to below 10 CFU/ml after 32 d of incubation (Nilsson et al. 1999). More recently, *C. divergens* V41 has been shown to have potential as a biopreservative for refrigerated cold-smoked salmon given that the bacteriocin divercin V41 was produced under harsh culture conditions which included variations in temperature, NaCl and glucose concentration (Connil et al. 2002).

8 Bacteriocins Produced by LAB

Bacteriocins are ribosomally synthesized, extracellularly released bioactive peptides or peptide complexes with a bactericidal or bacteriostatic effect on other closely and non-closely related species. In all cases, the producer cell exhibits specific immunity to the action of its own bacteriocin. Bacteriocins are found in all major bacterial lineages and constitute a diverse family of proteins showing a wide range of antimicrobial activity. Bacteriocins were first discovered by A. Gratia (Gratia 1925). He was involved in searching for ways to kill bacteria, which also resulted in the development of antibiotics and the discovery of bacteriophages, all within a span of a few years. He called his first discovery a *colicine* because it killed *Escherichia coli*. Some of the bacteriocins of LAB are well studied because these bacteria are used in the food industry.

The classification of bacteriocins has been reviewed by several authors (Klaenhammer 1988; Ennahar et al. 1999; Cleveland et al. 2001). Based on Klaenhammer's classification of bacteriocins, they are classified, based on common elements, into four well defined classes (Klaenhammer 1988):

Class I bacteriocins (lantibiotics): These are post-translationally modified to contain amino acids such as lanthionine and B-methyllanthionine and several dehydrated amino acids. Lantibiotics are further divided into two subgroups, A and B, based on structural features and their mode of killing.

>**Type A lantibiotics-** Leader peptides are cleaved by a serin proteinase. They kill the target cell by depolarizing the cytoplasmic membrane. Type A are linear and larger than type B lantibiotics; they range in size from 21 to 38 amino acids. Nisin is the best-known and best-studied Gram-positive bacteriocin and has been approved as a food additive.

>**Type B lantibiotics-** have a more globular secondary structure and are smaller than type A (the largest is 19 amino acids in length). Leader peptides are cleaved by an ABC-transporter. Type B lantibiotics function through enzyme inhibition. Thus, mersacidin interferes with cell wall biosynthesis.

Class II LAB bacteriocins (small heat-stable non-lantibiotics): These are small, heat-stable, non-lanthionine-containing peptides, not post-translationally modified, ranging in size from 30 to 60 amino acids, and they are usually positively charged at neutral pH. They are divided into subgroups:

Class IIa (pediocin-like bacteriocins with a strong antilisterial effect)- is the largest group, and its members are distinguished by a conserved amino-terminal sequence (YGNGVXaaC) and a shared strong inhibitory activity against *Listeria*. Because of their effectiveness against the food pathogen *Listeria*, class IIa bacteriocins are currently attracting much interest as potential natural and non-toxic food preservatives. Like type A lantibiotics, class IIa bacteriocins act through the formation of pores in the cytoplasmic membrane. Examples include pediocin PA-1 and pediocin AcH (*Pediococcus acidilactici*), sakacins A and P (*L. sakei*), leucocin A (*Leuconostoc gelidum*), enterocins A and P (*E. faecium*), and carnobacteriocin (*Carnobacterium* sp.).

Class IIb (bacteriocins whose activity depends on the complementary activity of two peptides)- These form pores composed of two different proteins in the membrane of their target cells. This group includes lacticin F and lactococcin G.

Class IIc (*sec*-dependent secreted bacteriocins)-includes bacteriocins that are sec-dependent, such as acidocin B and enterocin P, which can be included in both class IIa and Class IIc.

Class IId- Includes other bacteriocins such as lactocin A and B.

Class III bacteriocins: These are large heat-labile proteins. They include helveticins J and V and lactacin B.

Class IV bacteriocins (cyclic peptides): These require lipid or carbohydrate moieties for their activity. Little is known about the structure and function of this proposed class. The group has yet to be confirmed by purification and biochemical characterization. These bacteriocins include leuconocin S and lactocin 27.

Episodes of adverse food reactions related to chemical food preservatives have raised the concern of consumers. Owing to their peptide structure, the bacteriocins of LAB are considered good biopreservatives, as it is assumed that they are degraded by the proteases of the gastrointestinal tract. Most bacteriocins have good thermostability, and hence, can survive the thermal processing cycle of foods, whereas others can function at both low pH and low temperature and could therefore be useful in acid foods and cold-processed or cold-stored products. Bacteriocins may also find applications in minimally processed refrigerated foods, e.g., vacuum- and modified-atmosphere-packaged refrigerated meats and ready-to-eat meals, which lack the multiple impediments to the growth of pathogenic and spoilage bacteria conferred by traditional preservation techniques (O'Keeffe and Hill 2000). Furthermore, the specific and bactericidal actions of certain bacteriocins against bacteria of special concern, e.g., *Listeria* spp. and *Clostridium* spp., have increased the interest in these compounds (O'Keeffe and Hill 2000; Riley and Wertz 2002).

The major drawbacks of the use of bacteriocins as preservatives in foods are (1) their hydrophobic nature, although they are relatively small molecules and so they can easily diffuse into the water phase of food products, (2) the specific environment of the food, (3) the poor solubility, (4) uneven distribution of the bacteriocin molecules, (5) sensitivity to food enzymes, and (6) the negative impact of high levels of salt or other added ingredients that affect the production or activity of the bacteriocin (O'Keeffe and Hill 2000).

9 Biopreservation by Lactic Acid Bacteria

A mixed population of microorganisms that interact together generally colonizes food. The interactions have been classified on the basis of effects like competition, commensalism, mutualism, ammensalism or neutralism (Viljoen 2001). In food, more than one type of interaction may occur simultaneously leading to the specificity of the final product. While microbial antagonism has currently been observed, people thousands years ago used naturally occurring yeast, moulds and bacteria cultured foods with improved preservation properties. In 1962, Jameson, in studies concerning growth of *Salmonella*, reported the suppression of growth of all microorganisms on the food when the total microbial population achieved the maximum population density characteristic of the food (Jameson 1962). The same 'Jameson effect' has been reported for *Staphylococcus aureus* in seafood (Ross 1996) and *Listeria monocytogenes* in CSS (Gimenez and Dalgaard 2003). Among the different microorganisms naturally present in food and responsible for pathogens inhibition, LAB are currently listed. The acidification process due to the lactic acid production is one of the most commonly described effects. However, other mechanisms may be involved, such as production of inhibitory molecules, redox modification, competition for substrate, etc. In CSS, Mejlholm et al. (2005) have modeled the antagonism effect of naturally occurring LAB against *L. monocytogenes* (Mejlholm et al. 2005). The model is entirely empirical and it does not include assumptions about mechanisms of the microbial interaction that take place. Since the nineties, many studies have been conducted on the selection of bacteria with antimicrobial properties that could be inoculated at high level in seafood in order to inhibit the growth of undesirable micro-organisms. This technology is termed biopreservation (Françoise 2010).

LAB are good candidates for this technology as they produce a wide range of inhibitory compounds (organic acids, hydrogen peroxide, diacetyl and bacteriocins). In addition, they often have the GRAS status granted by the US-FDA and benefit from the healthy image associated with dairy products (Rodgers 2001). Although a great deal of work has been done on the selection of bacteria exhibiting antimicrobial properties in liquid medium and the number of bacteriocins characterized is increasing every day, very few commercial applications have appeared in seafood products. A major hurdle is that these products are not fermented and the selected LAB strain should not change their organoleptic and nutritional qualities. Many bacteria that gave promising results in liquid medium proved to be ineffective in products, either because they were poorly established in the environmental conditions (Wessels and Huss 1996), or because they produced unpleasant odors (Nilsson et al. 1999). Nevertheless, since the importance of LAB in semi-preserved fish has been highlighted, research into this subject has intensified to prevent growth of pathogenic and spoiling bacteria (Françoise 2010).

10 Nisin as a Seafood Preservative

The potential hazard of botulism in both vacuum-packed and modified atmosphere packed fish led to work at the Torry Research Station in the UK, where application of nisin by spray to fillets of cod, herring and smoked mackerel inoculated with *C. botulinum* Type E spores resulted in a significant delay in toxin production at 10 and 26°C. Another problem in smoked fish is growth of the psychroduric pathogen *L. monocytogenes*, especially in fresh and lightly preserved products. Nisin is an effective antilisterial agent in smoked salmon, especially when packed in a carbon dioxide atmosphere.

Nisin, at 25 mg/kg in combination with a reduced heat process that does not cause product damage of lobster meat, achieved a *Listeria* kill significantly better than either heat or nisin alone. Washing crabmeat with nisin reduced levels of *L. monocytogenes* (Delves-Broughton 2005).

11 Carnobacteriocins as a Seafood Preservative

C. maltaromaticum produces different carnobacteriocins (Nettles and Barefoot 1993; Quadri et al. 1994). These have been identified as heat-resistant, stable over a wide range of pH, and having a bactericidal mode of action (Schillinger et al. 1993; Jack et al. 1996). They are mainly effective against microorganisms such as other LAB and *L. monocytogenes* (Stoffels et al. 1992; McMullen and Stiles 1996). Some authors have studied the use of bacteriocins from *C. maltaromaticum* as biopreservatives for different food systems (Mathieu et al. 1994; Jack et al. 1996; Schöbitz et al. 1999).

Carnocin UI49 is a potential biopreservative produced by *C. piscicola*. Einarsson and Lauzon (1995) evaluated the biopreservation of brined shrimp by nisin combined with carnocin UI49, with a negative result on the extension of shelf life (Einarsson and Lauzon 1995). Stoffels et al. (1993) purified carnocin UI49 and nisin Z and tested their activity against various Gram-positive bacteria, including *Listeria* spp. They found that the *Listeria* strains were weakly sensitive to carnocin UI49, whereas when in combination, lower concentrations of nisin were needed to inhibit growth (Stoffels et al. 1993).

12 Enterocins as a Seafood Preservative

Enterococci (*E. faecium*) were approved as probiotics and are currently being used as such. *E. faecium* P13 and other *E. faecium* strains isolated from dry-fermented sausages produce enterocin P, a bacteriocin with strong inhibitory activity against *L. monocytogenes*. Enterocin P (EntP) is a pediocin-like bacteriocin with an N-terminal signal peptide, which may allow it to be secreted via the sec pathway. EntP is synthesized as a pre-peptide consisting of a 27-amino acid signal peptide and the 44-amino acid mature bacteriocin and has been shown to dissipate the membrane potential of energized cells and to form specific potassium ion conducting pores in the cytoplasmic membranes of target cells (Cintas et al. 1997).

A drawback in the use of enterocins is that they may be produced by enterococcal species carrying antibiotic resistance genes and/or genes coding for potential virulence factors (Eaton and Gasson 2001; Franz et al. 2003). In this sense, strains should be selected for their sensitivity to antibiotics. Owing to their broad antimicrobial spectra, EntP produced by food origin strains of *E. faecium* (Cintas et al. 1997), especially those isolated from seafood (Campos et al. 2006; Arlindo et al. 2006), may be of interest as antimicrobials in the food industry. Besides its antimicrobial activity, enterocin P shows some additional properties such as protease sensitivity, thermal stability, activity over a broad range of pH values, and the maintenance of antimicrobial activity after freeze-thawing, lyophilization, and chilled storage, all of which make it a potentially remarkable food preservative (Cintas et al. 1997). Enterocin-P-producing *Enterococcus* isolated from seafood (Campos et al. 2006; Arlindo et al. 2006) has shown great potential as an inhibitor of undesirable microflora.

13 Bacteriocin-producing LAB and the Biopreservation of Ready-to-eat Seafoods

Over the past decade, the recurrence of listeriosis outbreaks has focused the attention of bacteriocin researchers on this organism. *L. monocytogenes*, responsible for food-related listeriosis, has the ability to grow in vacuum-packaged food at low temperatures and is relatively tolerant to salt and low pH, making it difficult to control its growth in food. *L. monocytogenes* shows resistance to traditional food preservation methods and has the ability to grow at near-freezing temperatures. Cases of human listeriosis have been reported from the consumption of lightly preserved, ready-to-eat fish products (Ericsson et al. 1997; Miettinen et al. 2003).

The consumer acceptance and demand for minimally processed refrigerated foods has been increasing over the past several years. However, the microbiological safety of these foods is of concern, owing to the possible presence of non-proteolytic strains of *C. botulinum* able to grow at 4°C, and post-processing contamination with other psychrotrophic pathogens such as *L. monocytogenes*. *L. monocytogenes* may be found in cold-smoked salmon, typically in low numbers (Cortesi et al. 1997; Jørgensen and Huss 1998). *L. monocytogenes* is a common microorganism in the environment, being a source of contamination in the production of cold-smoked products such as cold-smoked salmon where there is no listericidal step. Bacteriocins can act against *Listeria* spp., and hence, they could be of value where *Listeria* contamination is a potential problem. Government agencies and industry have focused on introducing critical control points to prevent the growth of *L. monocytogenes* in ready-to-eat fish products. Bacteriocins have been discovered in a wide range of habitats and conditions, such as fresh and cured meats and seafoods, milk and milk products, spoiled salad dressing, and soybean paste. Several authors have studied different bacteriocin-producing LAB to extend the shelf life of ready to eat products.

LAB peptides with antilisterial activity may be promising candidates for the biopreservation of cold-smoked and ready-to-eat seafood stored at 5°C. Several studies have been carried out on the biopreservation of aquatic food products. Katla et al. (2003) studied the effect of some bacteriocin-producing strains against *L. monocytogenes* along the complete storage period, finding that sakacin-P-producing or the non-producing *L. sakei* had a bacteriostatic effect on *L. monocytogenes* (Katla et al. 2003). However, those authors observed a bactericidal effect on *L. monocytogenes* when the sakacin-P-producing *L. sakei* culture was added to vacuum-packed cold-smoked salmon together with sakacin P. Moreover, the sakacin P produced by *L. sakei* (Tichaczek et al. 1992; Holck et al. 1994) exhibits strong activity against *L. monocytogenes* (Aasen et al. 2003; Eijsink et al. 1998).

14 Conclusions

Antimicrobial packaging may play an important role in reducing the risk of pathogen development as well as extending the shelf life of seafoods. However, any innovative preservation technique would not substitute for good quality seafood materials and good manufacture practices. In this sense, it is important to bear in mind that careful monitoring of the whole seafood process will lessen the chances for bacteria to grow. Studies of new food-grade bacteriocins as preservatives and the development of suitable systems of bacteriocin treatment of plastic films for food packaging are important for a better preservation of seafood products.

As the most innovative procedure, active and intelligent food packaging focuses on on-command preservative releasing packaging systems as well as on the development of noninvasive microbial growth sensors to monitor the sterility of food products. This is a promising technique that should ensure the good quality of the product throughout its shelf-life.

Acknowledgements

This research was supported by a grant from Marine Bioprocess Research Center of the Marine Biotechnology Program funded by the Ministry of Oceans and Fisheries, Republic of Korea.

References

Aasen, I.M., S. Markussen, T. Møretrø, T. Katla, L. Axelsson and K. Naterstad. 2003. Interactions of the bacteriocins sakacin P and nisin with food constituents. Int. J. Food Microbiol. 87(1): 35–43.

Alves, V.F., E.C.P. De Martinis, M.T. Destro, B.F. Vogel and L. Gram. 2005. Antilisterial activity of a *Carnobacterium piscicola* isolated from Brazilian smoked fish (Surubim *Pseudoplatystoma* sp.) and its activity against a persistent strain of *Listeria monocytogenes* isolated from Surubim. J. Food Protection. 68(10): 2068–2077.

Arlindo, S., P. Calo, C. Franco, M. Prado, A. Cepeda and J. Barros-Velázquez. 2006. Single nucleotide polymorphism analysis of the enterocin P structural gene of *Enterococcus faecium* strains isolated from nonfermented animal foods. Mol. Nut. Food Res. 50(12): 1229–1238.

Axelsson, L.T., T.C. Chung, W.J. Dobrogosz and S.E. Lindgren. 1989. Production of a broad spectrum antimicrobial substance by *Lactobacillus reuteri*. Microbial Ecol. Health Dis. 2(2): 131–136.

Baya, A.M., A.E. Toranzo, B. Lupiani, T. Li, B.S. Roberson and F.M. Hetrick. 1991. Biochemical and serological characterization of *Carnobacterium* spp. isolated from farmed and natural populations of striped bass and catfish. Appl. Environ. Microbiol. 57(11): 3114–3120.

Bourouni, O.C., M. El Bour, P. Calo-Mata, A. Boudabous and J. Barros-Velàzquez. 2012. Discovery of novel biopreservation agents with inhibitory effects on growth of food-borne pathogens and their application to seafood products. Res. Microbiol. 163: 44–54.

Bower, C.K. and K.A. Hietala. 2008. Acidification methods for stabilization and storage of salmon by-products. J. Aquatic Food Product Technol. 17(4): 459–478.

Campos, C.A., Ó. Rodríguez, P. Calo-Mata, M. Prado and J. Barros-Velázquez. 2006. Preliminary characterization of bacteriocins from *Lactococcus lactis*, *Enterococcus faecium* and *Enterococcus mundtii* strains isolated from turbot (*Psetta maxima*). Food Res. Int. 39(3): 356–364.

Caplice, E. and G.F. Fitzgerald. 1999. Food fermentations: Role of microorganisms in food production and preservation. Int. J. Food Microbiol. 50(1): 131–149.

Chung, T.C., L. Axelsson, S.E. Lindgren and W.J. Dobrogosz. 1989. *In vitro* studies on reuterin synthesis by *Lactobacillus reuteri*. Microbial Ecol. Health Dis. 2(2): 137–144.

Cintas, L.M., P. Casaus, L.S. Håvarstein, P.E. Hernandez and I.F. Nes. 1997. Biochemical and genetic characterization of enterocin P, a novel sec-dependent bacteriocin from *Enterococcus faecium* P13 with a broad antimicrobial spectrum. Appl. Environ. Microbiol. 63(11): 4321–4330.

Cleveland, J., T.J. Montville, I.F. Nes and M.L. Chikindas. 2001. Bacteriocins: safe, natural antimicrobials for food preservation. Int. J. Food Microbiol. 71(1): 1–20.

Connil, N., L. Plissoneau, B. Onno, M.-F. Pilet, H. Prevost and X. Dousset. 2002. Growth of *Carnobacterium divergens* V41 and production of biogenic amines and divercin V41 in sterile cold-smoked salmon extract at varying temperatures, NaCl levels, and glucose concentrations. J. Food Protection. 65(2): 333–338.

Cortesi, M.L., T. Sarli, A. Santoro, N. Murru and T. Pepe. 1997. Distribution and behavior of Listeria monocytogenes in three lots of naturally-contaminated vacuum-packed smoked salmon stored at 2 and 10°C. Int. J. Food Microbiol. 37(2-3): 209–214.

Dalgaard, P. 2000. Fresh and lightly preserved seafood. Shelf life Eval. Foods. 139: 110–141.

Dalgaard, P., M. Vancanneyt, N.E. Vilalta, J. Swings, P. Fruekilde and J.J. Leisner. 2003. Identification of lactic acid bacteria from spoilage associations of cooked and brined shrimps stored under modified atmosphere between 0°C and 25°C. J. Appl. Microbiol. 94(1): 80–89.

Degnan, A.J., C.W. Kaspar, W.S. Otwell, M.L. Tamplin and J.B. Luchansky. 1994. Evaluation of lactic acid bacterium fermentation products and food-grade chemicals to control *Listeria monocytogenes* in blue crab (*Callinectes sapidus*) meat. Appl. Environ. Microbiol. 60(9): 3198–3203.

Delves-Broughton, J. 2005. Nisin as a food preservative. Food Australia. 57(12): 525–532.

Duffes, F., C. Corre, F. Leroi, X. Dousset and P. Boyaval. 1999. Inhibition of Listeria monocytogenes by *in situ* produced and semipurified bacteriocins of *Carnobacterium* spp. on vacuum-packed, refrigerated cold-smoked salmon. J. Food Protection. 62(12): 1394–1403.

Daeschel, M.A. 1989. Antimicrobial substances from lactic acid bacteria for use as food preservatives. Food Technology. 43(1): 164–167.

Duffes, F., F. Leroi, P. Boyaval and X. Dousset. 1999. Inhibition of *Listeria monocytogenes* by *Carnobacterium* spp. strains in a simulated cold smoked fish system stored at 4°C. Int. J. Food Microbiol. 47(1): 33–42.

Eaton, T.J. and M.J. Gasson. 2001. Molecular screening of *Enterococcus* virulence determinants and potential for genetic exchange between food and medical isolates. Appl. Environ. Microbiol. 67(4): 1628–1635.

Eijsink, V.G.H., M. Skeie, P.H. Middelhoven, M.B. Brurberg and I.F. Nes. 1998. Comparative studies of class IIa bacteriocins of lactic acid bacteria. Appl. Environ. Microbiol. 64(9): 3275–3281.

Einarsson, H. and H.L. Lauzon. 1995. Biopreservation of brined shrimp (*Pandalus borealis*) by bacteriocins from lactic acid bacteria. Appl. Environ. Microbiol. 61(2): 669–676.

Eldar, A., C. Ghittino, L. Asanta, et al. 1996. *Enterococcus seriolicida* is a junior synonym of *Lactococcus garvieae*, a causative agent of septicemia and meningoencephalitis in fish. Cur. Microbiol. 32(2): 85–88.

Ennahar, S., K. Sonomoto and A. Ishizaki. 1999. Class IIa bacteriocins from lactic acid bacteria: antibacterial activity and food preservation. J. Biosci. Bioeng. 87(6): 705–716.

Ericsson, H., A. Eklöw, M.L. Danielsson-Tham et al. 1997. An outbreak of listeriosis suspected to have been caused by rainbow trout. J. Clin. Microbiol. 35(11): 2904–2907.

Françoise, L. 2010. Occurrence and role of lactic acid bacteria in seafood products. Food Microbiol. 27(6): 698–709.

Franz, C.M.A.P., M.E. Stiles, K.H. Schleifer and W.H. Holzapfel. 2003. Enterococci in foods—a conundrum for food safety. Int. J. Food Microbiol. 88(2): 105–122.

198 *Seafood Science: Advances in Chemistry, Technology and Applications*

Gänzle, M.G., A. Höltzel, J. Walter, G. Jung and W.P. Hammes. 2000. Characterization of reutericyclin produced by *Lactobacillus reuteri* LTH2584. Appl. Environ. Microbiol. 66(10): 4325–4333.

Gelman, A., V. Drabkin and L. Glatman. 2000. Evaluation of lactic acid bacteria, isolated from lightly preserved fish products, as starter cultures for new fish-based food products. Innovative Food Sci Emerging Technologies. 1(3): 219–226.

Gimenez, B. and Paw Dalgaard. 2003. Modelling and predicting the simultaneous growth of Listeria monocytogenes and spoilage micro-organisms in cold-smoked salmon. J. Appl. Microbiol. 96(1): 96–109.

Glatman, L., V. Drabkin and A. Gelman. 2000. Using lactic acid bacteria for developing novel fish food products. J. Sci. Food Agri. 80(3): 375–380.

González-Rodríguez, Maria-Nieves, José-Javier Sanz, Jesús-Ángel Santos, A. Otero and Maria-Luisa García-López. 2002. Numbers and types of microorganisms in vacuum-packed cold-smoked freshwater fish at the retail level. Int. J. Food Microbiol. 77(1): 161–168.

González-Rodríguez, M.N., J.J. Sanz, J.A. Santos, A. Otero and M.L. García-López. 2002. Foodborne pathogenic bacteria in prepackaged fresh retail portions of farmed rainbow trout and salmon stored at 3°C. Int. J. Food Microbiol. 76(1): 135–141.

Gonzalez, Cesar-Javier, Teresa-Maria Lopez-Diaz, M. Prieto and Andres Otero. 1999. Bacterial microflora of wild brown trout (*Salmo trutta*), wild pike (*Esox lucius*), and aquacultured rainbow trout (*Oncorhynchus mykiss*). J. Food Protection. 62(11): 1270–1277.

Gonzalez, C.J., J.P. Encinas, M.L. García-López and A. Otero. 2000. Characterization and identification of lactic acid bacteria from freshwater fishes. Food Microbiol. 17(4): 383–391.

Gram, L. and H.H. Huss. 2000. Fresh and processed fish and shellfish. Microbiological Safety Quality Food. 1: 472–506.

Gratia, A. 1925. Sur un remarquable exemple d'antagonisme entre deux souches de colibacille. Comptes Rendus des Seances de la Societe de Biologie et de ses Filiales. 93: 1040–1041.

Groth, L., Birgit, L. Bay, I. Cleenwerck, et al. 2005. *Carnobacterium divergens* and *Carnobacterium maltaromaticum* as spoilers or protective cultures in meat and seafood: phenotypic and genotypic characterization. Sys. Appl. Microbiol. 28(2): 151–164.

Hansen, L.T., T. Gill and H.H. Hussa. 1995. Effects of salt and storage temperature on chemical, microbiological and sensory changes in cold-smoked salmon. Food Res. Int. 28(2): 123–130.

Hoeltzel, A., M.G. Gaenzle, G.J. Nicholson, W.P. Hammes and G. Jung. 2000. The first low molecular weight antibiotic from lactic acid bacteria: Reutericyclin, a new tetramic acid. Angewandte Chemie International Edition. 39(15): 2766–2768.

Holck, A.L., L. Axelsson, K. Hühne and L. Kröckel. 1994. Purification and cloning of sakacin 674, a bacteriocin from *Lactobacillus sake* Lb674. FEMS Microbiol. Lett. 115(2): 143–149.

Huber, I., B. Spanggaard, K.F. Appel, L. Rossen, T. Nielsen and L. Gram. 2004. Phylogenetic analysis and *in situ* identification of the intestinal microbial community of rainbow trout (*Oncorhynchus mykiss*, Walbaum). J. Appl. Microbiol. 96(1): 117–132.

Huis in't Veld, J.H. 1996. Microbial and biochemical spoilage of foods: an overview. International Journal of Food Microbiology. 33(1): 1–18.

Huss, Hans Henrik. 1995. Quality and quality changes in fresh fish. FAO fisheries technical paper (348).

Jack, R.W., J. Wan, J. Gordon, et al. 1996. Characterization of the chemical and antimicrobial properties of piscicolin 126, a bacteriocin produced by *Carnobacterium piscicola* JG126. Appl. Environ. Microbiol. 62(8): 2897–2903.

Jameson, J.E. 1962. A discussion of the dynamics of Salmonella enrichment. J. Hyg. (*Lond*). 60(2): 193–207.

Jöborn, A., M. Dorsch, J.C. Olsson, A. Westerdahl and S. Kjelleberg. 1999. *Carnobacterium inhibens* sp. nov., isolated from the intestine of Atlantic salmon (Salmo salar). Int. J. Sys. Evol. Microbiol. 49(4): 1891–1898.

Jørgensen, L.V. and H.H. Huss. 1998. Prevalence and growth of *Listeria monocytogenes* in naturally contaminated seafood. Int. J. Food Microbiol. 42(1): 127–131.

Jørgensen, L.V., H.H. Huss and P. Dalgaard. 2000. The effect of biogenic amine production by single bacterial cultures and metabiosis on cold-smoked salmon. J. Appl. Microbiol. 89(6): 920–934.

Katla, T., K. Naterstad, M. Vancanneyt, J. Swings and L. Axelsson. 2003. Differences in susceptibility of *Listeria monocytogenes* strains to sakacin P, sakacin A, pediocin PA-1, and nisin. Appl. Environ. Microbiol. 69(8): 4431–4437.

Klaenhammer, T.R. 1988. Bacteriocins of lactic acid bacteria. Biochimie. 70(3): 337–349.

Leroi, F., N. Arbey, J.-J. Joffraud and F. Chevalier. 2003. Effect of inoculation with lactic acid bacteria on extending the shelf-life of vacuum-packed cold smoked salmon. Int. J. Food Sci. Technol. 31(6): 497–504.

Leroi, F., J.-J. Joffraud, F. Chevalier and M. Cardinal. 1998. Study of the microbial ecology of cold-smoked salmon during storage at 8°C. Int. J. Food Microbiol. 39(1): 111–121.

Lindgren, S.E. and W.J. Dobrogosz. 1990. Antagonistic activities of lactic acid bacteria in food and feed fermentations. FEMS Microbiol. Lett. 87(1): 149–163.

Liu, J.Y., A.H. Li, C. Ji and W.M. Yang. 2009. First description of a novel *Weissella* species as an opportunistic pathogen for rainbow trout *Oncorhynchus mykiss* (Walbaum) in China. Veterinary Microbiol. 136(3-4): 314–320.

Matamoros, S., F. Leroi, M. Cardinal, et al. 2009. Psychrotrophic lactic acid bacteria used to improve the safety and quality of vacuum-packaged cooked and peeled tropical shrimp and cold-smoked salmon. J. Food Protection. 72(2): 365–374.

Mathieu, F., M. Michel, A. Lebrihi and G. Lefebvre. 1994. Effect of the bacteriocin carnocin CP5 and of the producing strain *Carnobacterium piscicola* CP5 on the viability of *Listeria monocytogenes* ATCC 15313 in salt solution, broth and skimmed milk, at various incubation temperatures. Int. J. Food Microbiol. 22(2): 155–172.

McMullen, L.M. and M.E. Stiles. 1996. Potential for use of bacteriocin-producing lactic acid bacteria in the preservation of meats. J. Food Protection. 64–71.

Mejlholm, O., N. Bøknæs and P. Dalgaard. 2005. Shelf life and safety aspects of chilled cooked and peeled shrimps (*Pandalus borealis*) in modified atmosphere packaging. J. Appl. Microbiol. 99(1): 66–76.

Miettinen, H., A. Arvola, T. Luoma and G. Wirtanen. 2003. Prevalence of *Listeria monocytogenes* in, and microbiological and sensory quality of, rainbow trout, whitefish, and vendace roes from Finnish retail markets. J. Food Protection. 66(10): 1832–1839.

Morzel, M., G.F. Fitzgerald and E.K. Arendt. 1997. Fermentation of salmon fillets with a variety of lactic acid bacteria. Food Res. Int. 30(10): 777–785.

Nettles, C.G. and S.F. Barefoot. 1993. Biochemical and genetic characteristics of bacteriocins of food-associated lactic acid bacteria. J. Food Protection. 56(4): 338–356.

Nilsson, L., L. Gram and H.H. Huss. 1999. Growth control of *Listeria monocytogenes* on cold-smoked salmon using a competitive lactic acid bacteria flora. J. Food Protection. 62(4): 336–342.

Nilsson, L., H.H. Huss and L. Gram. 1997. Inhibition of *Listeria monocytogenes* on cold-smoked salmon by nisin and carbon dioxide atmosphere. Int. J. Food Microbiol. 38(2): 217–227.

O'Keeffe, T. and C. Hill. 2000. Bacteriocins. pp. 183–197. *In*: Robinson (ed.), *Encyclopaedia of Food Microbiology*.

Paludan-Müller, C., H.H. Huss and L. Gram. 1999. Characterization of lactic acid bacteria isolated from a Thai low-salt fermented fish product and the role of garlic as substrate for fermentation. Int. J. Food Microbiol. 46(3): 219–229.

Pellé, E., X. Dousset, H. Prévost and D. Drider. 2005. Specific molecular detection of *Carnobacterium piscicola* SF668 in cold smoked salmon. Lett. Appl. Microbiol. 40(5): 364–368.

Quadri, L.E., M. Sailer, K.L. Roy, J.C. Vederas and M.E. Stiles. 1994. Chemical and genetic characterization of bacteriocins produced by *Carnobacterium piscicola* LV17B. J. Biological Chem. 269(16): 12204–12211.

Riley, M.A. and J.E. Wertz. 2002. Bacteriocins: evolution, ecology, and application. Annual Rev. Microbiol. 56(1): 117–137.

Ringø, E., H.R. Bendiksen, M.S. Wesmajervi, R.E. Olsen, P.A. Jansen and H. Mikkelsen. 2000. Lactic acid bacteria associated with the digestive tract of Atlantic salmon (*Salmo salar* L.). J. Appl. Microbiol. 89(2): 317–322.

Ringø, E. 2008. The ability of carnobacteria isolated from fish intestine to inhibit growth of fish pathogenic bacteria: a screening study. Aquaculture Res. 39(2): 171–180.

Ringø, E. and F.J. Gatesoupe. 1998. Lactic acid bacteria in fish: A review. Aquaculture. 160(3): 177–203.

Ringø, E., R.E. Olsen, Ø. Øverli and F. Løvik. 1997. Effect of dominance hierarchy formation on aerobic microbiota associated with epithelial mucosa of subordinate and dominant individuals of Arctic charr, *Salvelinus alpinus* (L.). Aquaculture Res. 28(11): 901–904.

Ringø, E. and E. Strøm. 1994. Microflora of Arctic charr, *Salvelinus alpinus* (L.): gastrointestinal microflora of free-living fish and effect of diet and salinity on intestinal microflora. Aquaculture Res. 25(6): 623–629.

Ringø, E., M.S. Wesmajervi, H.R. Bendiksen, et al. 2001. Identification and characterization of carnobacteria isolated from fish intestine. Sys. Appl. Microbiol. 24(2): 183–191.

Rodgers, Svetlana. 2001. Preserving non-fermented refrigerated foods with microbial cultures—A review. Trends Food Sci. Technol. 12(8): 276–284.

Ross, T. 1996. Indices for performance evaluation of predictive models in food microbiology. J. Appl. Microbiol. 81(5): 501–508.

Satomi, M., M. Furushita, H. Oikawa, M. Yoshikawa-Takahashi and Y. Yano. 2008. Analysis of a 30 kbp plasmid encoding histidine decarboxylase gene in *Tetragenococcus halophilus* isolated from fish sauce. Int. J. Food Microbiol. 126(1): 202–209.

Schillinger, U., M.E. Stiles and W.H. Holzapfel. 1993. Bacteriocin production by *Carnobacterium piscicola* LV 61. Int. J. Food Microbiol. 20(3): 131–147.

Schöbitz, R., T. Zaror, O. León and M. Costa. 1999. A bacteriocin from *Carnobacterium piscicola* for the control of *Listeria monocytogenes* in vacuum-packaged meat. Food Microbiol. 16(3): 249–255.

Schrøder, K., E. Clausen, A.M. Sandberg and J. Raa. 1980. Psychrotrophic *Lactobacillus plantarum* from fish and its ability to produce antibiotic substances. Adv. Fish Sci. Technol. 148: 480–483.

Seppola, M., R.E. Olsen, E. Sandaker, P. Kanapathippillai, W. Holzapfel and E. Ringø. 2006. Random amplification of polymorphic DNA (RAPD) typing of carnobacteria isolated from hindgut chamber and large intestine of Atlantic cod (*Gadus morhua* L.). Sys. Appl. Microbiol. 29(2): 131–137.

Silva, J., A.S. Carvalho, P. Teixeira and P.A. Gibbs. 2002. Bacteriocin production by spray-dried lactic acid bacteria. Lett. Appl. Microbiol. 34(2): 77–81.

Stevens, K.A., B.W. Sheldon, N.A. Klapes and T.R. Klaenhammer. 1992. Effect of treatment conditions on nisin inactivation of gram-negative bacteria. J. Food Protection. 55(10): 763.

Stiles, Michael, E. 1996. Biopreservation by lactic acid bacteria. Antonie Van Leeuwenhoek. 70(2): 331–345.

Stoffels, G., I.F. Nes and A. Guomundsdóttir. 1992. Isolation and properties of a bacteriocin-producing *Carnobacterium piscicola* isolated from fish. J. Appl. Microbiol. 73(4): 309–316.

Stoffels, G., H.-G. Sahl and Á. Gudmundsdóttir. 1993. Carnocin U149, a potential biopreservative produced by *Carnobacterium piscicola*: large scale purification and activity against various Gram-positive bacteria including *Listeria* sp. Int. J. Food Microbiol. 20(4): 199–210.

Strøm, E. and J.A. Olafsen. 1990. The indigenous microflora of wild-captured juvenile cod in net-pen rearing. Microbiol. Poecilotherms. 181–185.

Taira, W., Y. Funatsu, M. Satomi, T. Takano and H. Abe. 2007. Changes in extractive components and microbial proliferation during fermentation of fish sauce from underutilized fish species and quality of final products. Fisheries Sci. 73(4): 913–923.

Tichaczek, P.S., J. Nissen-Meyer, I.F. Nes, R.F. Vogel and W.P. Hammes. 1992. Characterization of the Bacteriocins Curvacin A from *Lactobacillus curvatus* LTH1174 and Sakacin P from *L. sake* LTH673. Sys. Appl. Microbiol. 15(3): 460–468.

Tomé, E., V.L. Pereira, C.I. Lopes, P.A. Gibbs and P.C. Teixeira. 2008. *In vitro* tests of suitability of bacteriocin-producing lactic acid bacteria, as potential biopreservation cultures in vacuum-packaged cold-smoked salmon. Food Control. 19(5): 535–543.

Uchida, M., J. Ou, Bi-Wen Chen, et al. 2005. Effects of soy sauce koji and lactic acid bacteria on the fermentation of fish sauce from freshwater silver carp Hypophthalmichthys molitrix. Fisheries Sci. 71(2): 422–430.

Vijayan, H., I. Joseph and R.P. Raj. 2009. Biotransformation of tuna waste by co-fermentation into an aquafeed ingredient. Aquaculture Res. 40(9): 1047–1053.

Viljoen, B.C. 2001. The interaction between yeasts and bacteria in dairy environments. Int. J. Food Microbiol. 69(1): 37–44.

Wallbanks, S., A.J. Martinez-Murcia, J.L. Fryer, B.A. Phillips and M.D. Collins. 1990. 16S rRNA sequence determination for members of the genus Carnobacterium and related lactic acid bacteria and description of *Vagococcus salmoninarum* sp. nov. Int. J. Sys. Bacteriol. 40(3): 224–230.

Wessels, S. and H.H. Huss. 1996. Suitability of *Lactococcus lactis* subsp. *lactis* ATCC 11454 as a protective culture for lightly preserved fish products. Food Microbiol. 13(4): 323–332.

Williams, A.M., J.L. Fryer and M.D. Collins. 1990. *Lactococcus piscium* sp. nov. a new *Lactococcus* species from salmonid fish. FEMS Microbiol. Lett. 68(1): 109–113.

Yang, G., B. Bao, E. Peatman, H. Li, L. Huang and D. Ren. 2007. Analysis of the composition of the bacterial community in puffer fish *Takifugu obscurus*. Aquaculture. 262(2): 183–191.

Yoon, H.-D., D.-S. Lee, C.-I. Ji and S.-B. Suh. 1997. Studies on the utilization of wastes from fish processing I-characteristics of lactic acid bacteria for preparing skipjack tuna viscera silage. J.-Korean Fisheries Soc. 30: 1–7.

10

Feeding Trial of Red Sea Bream with Dioxin Reduced Fish Oil

T. Honryo

1 Introduction

Polychlorinated dibenzodioxins (PSDDs), polychlorinated dibenzofurans (PCDFs) and polychlorinated biphenyls (PCBs) constitute a group of persistent environmental chemicals known as the 'dioxins' and these are emergent especially in imperfect combustion of chlorine materials. A number of dioxin or furan congeners, as well as co-planner PCBs (Co-PCBs) have been included as the toxic dioxins. The toxic impacts of dioxins on living organism are complicated and variously dependent on assortment, isomeric form and congeners, therefore TEQ (toxic equivalent) is converted by TEF (toxic equivalency factors) to calculation, consideration and discussion. The dioxins are of concern due to their carcinogenic and teratogenic potential on humans and background exposure may occur through the diet, skin absorption and aspiration.

In Japan, the Ministry of Health, Labour and Welfare (http://www.mhlw.go.jp/) has defined the TDI (Tolerable Daily Intake) of dioxins as 4 pg-TEQ/kgbw/day, which means that this amount of daily and per body weight exposure would not affect human health on a lifetime basis. The Bureau of Social Welfare and Public Health in Tokyo Metropolitan Government reported that the dioxin exposure for residents in Tokyo has been far below levels of TDI since 1998; however, 99% of this exposure is

Oshima Station, Fisheries Laboratory of Kinki University, Wakayama 649-3633 Japan.
 Email: t.honryo@kindaisuiken.jp

accounted from food intake. In addition, the evaluation of food contamination revealed that daily dietary intakes of dioxins from fish and shellfish (Group 10, 90%) dominate, followed by meat and eggs (Group 11, 6.7%) from the total diet study (TDS) of 14 food group composites (Tsutsumi et al. 2008). It is considered that Co-PCBs contribute to over 70% of dioxin intake and reducing this amount from Group 10 is the most essential for human health (Mastuda 2009). Even the level of TDI has undergone a reduced level of prescription in Japan. It is important to monitor the dioxin contamination especially from the food group 10, fish and shellfish, for risk assessments.

The dioxins discharged into aquatic environments accumulate in phyto- and zooplanktons and transfer to the predators, although bioaccumulation patterns of dioxins varied greatly among congeners and depended on the metabolic capabilities of creatures (Naito et al. 2003). The survey of natural fish and shellfish cumulative dioxins, carried out by Fisheries Agency of Japan (http://www.jfa.maff.go.jp/), elucidated that the TEQ differed depending on its captured place even in the same species. This profile indicated that habitat and prey types have a strong effect on natural fish and shellfish accumulation. Moreover, Hannu et al. (2003) reported that dioxins (PCDD/Fs and PCBs) in Baltic herring varied depending on the age, catchment area and fat percentage. On the other hand, fed food is considered as the main factor of dioxin accumulation in the case of cultured fish and shellfish because the aquaculture generally takes place at fixed locations. The survey carried out by the Fisheries Agency of Japan, which is mentioned above, showed that, in general, cultured fish contained higher amounts of accumulated dioxins than natural fish. Moreover, Hites et al. (2004) indicated that concentrations of organochlorine in a total of 700 salmon (farmed and wild) and fish feed collected from around the world showed significantly higher contaminants in farmed salmon than in wild. They pointed out that these differences are most likely a function of their diets. For the reasons mentioned above, reducing the dioxin content in cultured fish by adapting feed materials, may add value in aquaculture-products, making them more safe and secure.

In this study, we investigated by using red sea bream *Pagrus major* (RSB) which is one of the most developed in its aquaculture techniques and widely cultured in the west coast of Japan (Kumai 2000). Focusing on the lipid source of the artificial diets, a trial was conducted over 90 days in order to compare the fish oil used heretofore in the pellet (MD oil) with dioxins-reduced fish oil (CT oil) developed by *Ueda-oils and Fats MFG Co Ltd* (http://www.uedaoil.co.jp/), with its total TEQ reduced approximately by 1/6 of its dioxins contents. The formula of the artificial diets are shown in Table 1. The effect of replacement of normal fish oil as a lipid source by dioxin-reduced fish oil on RSB (571.7±96.5 g BW, 25.9±1.1 cm BL) was explored.

Table 1. Formula of the artificial test diets.

	MD	CT
Ingredients (%)		
Vitamin-free Casein	55.0	55.0
Fish oil (MD)	15.0	-
Fish oil (CT)	-	15.0
Mineral mixture[a]	8.0	8.0
Starch	8.0	8.0
Cellulose	5.5	5.5
Vitamin mixture[a]	5.0	5.0
CMC	2.0	2.0
Feeding stimulant[b]	1.0	1.0

[a]Halver 1957 mineral and vitamin mixture.
[b]Sodium asparatate, 0.309%; Alanine, 0.309%; L-glutamic acid, 0.381%.

In this study, the dioxin was defined as compounded by congeners and isomeric forms of PCDDs, PCDFs and Co-PCBs. The chemical analysis was carried out by Kobelco Research Institute, INC (http://www.kobelcokaken.co.jp/) and the TEF prescript by WHO 2009 contributed to the TEQ calculation of both MD and CT oils, muscle and liver contents. The values under limits of detection were represented as ND on the fish oil in experimental diets, muscle and liver of RSB.

2 The Dioxin Contents in Experimental Fish Oil

The contents of toxic equivalents (pg-TEQ/g) of PCDDs, PCDFs, Co-PCBs and total toxic equivalents in fish oil as an ingredient of an experimental diet are shown in Table 2. MD oil is made mainly from Clupeoidei fish Menhaden *Brevoortiatyrannus* which is harvested in the Gulf of Mexico and widely used as food for pets, i.e., dogs and cats, as well as fish. CT oil, mainly made from MD oil mixed with a small amount of Anchoveta oil, has controlled its dioxin TEQ contents as 3 pg-TEQ/g. This admixture was based on consideration of manufacturing cost. CT oil was some dioxin-reduced treated by basely activated-carbon treatment. As a result of dioxin-reduced treatments, CT oil contained 1/6 of total TEQ contents compared with MD oil which is 99% of PCDDs and 63% of Co-PCBs were successfully reduced.

Table 2. The contents of toxic equivalents (pg-TEQ/g) of PCDDs, PCDFs, Co-PCBs and total toxic equivalents in fish oil as an ingredient of experimental diet.

	No.	total TEQ(pg-TEQ/g)	PCDDs(pg-TEQ/g)	PCDFs (pg-TEQ/g)	Co-PCBs (pg-TEQ/g)
oil	CT	3.0	0.000065	ND*	3
	MD	17.7	7.6	1.9	8.2

*ND=Concentration below limits of detection.

3 Results

3.1 Growth performances

Only a 1 fish dead at the CT group during the experimental duration, hence there was no difference among the group in survival. The body weight (g), body length (cm), CF, Hepatosomatic index (%), total feed intake (g) and other parameters were shown in Table 3. Only the significantly differences was detected in the HSI which decline from initial value. Weight gain (MD;10.8 ± 2.9, CT;11.5 ± 2.3%), FR (MD;3.2 ± 1.2, CT;2.5 ± 0.7%) and FE (MD;34.8 ± 12.2, CT;43.1 ± 9.5%) showed any differences among the groups, therefore dioxin-reduced fish oil was not influenced on growth performance on RSB.

Table 3. Growth performances of red sea bream fed the experimental diet.

		Body weight(g)	Body length(cm)	CF	feed intake/ fish(g)	Weight gain(%)	FR(%)	FCE(%)	HSI(%)
Initial		571.1±96.5	25.9±1.1	32.8±1.3	-	-	-	-	1.2 ± 0.3[a]
Final	MD	892.5±108.1	30.6±0.9	30.9±1.0	926.3 ± 44.6	10.8 ± 2.9	3.2 ± 1.2	34.8 ± 12.2	0.5 ± 0.2[b]
	CT	949.6±92.7	30.7±1.3	32.9±3.5	875.1 ± 89.1	11.5 ± 2.3	2.5 ± 0.7	43.1 ± 9.5	0.6 ± 0.2[b]

Values are expressed by mean ± S.D. of triplication and three more specimens as accession, respectively.
Different superscript letters represent a significance difference ($p<0.05$, One-way ANOVA with Bonfferoni correction).

3.2 Dioxins contents in fish

The contents of toxic equivalents (pg-TEQ/g) in the muscle and liver of the experimental fish at initial and final day were shown in Table 4. At the beginning of experiment, total TEQ contents of the fish muscle was 0.137

Table 4. The contents of total TEQ(pg-TEQ/g) in the muscle and the liver of the experimental fish on initial and final day.

Muscle	No,	amount pools	total TEQ(pg-TEQ/g)	PCDDs(pg-TEQ/g)	PCDFs(pg-TEQ/g)	Co-PCBs(pg-TEQ/g)
Initial	Initial	9	0.137	ND*	0.017	0.12
Final	CT	9	0.071 (0.0610.082)[a]	0.000012 (ND-0.000036)[a]	0.004 (0.003-0.0063)[a]	0.067 (0.058-0.076)[a]
	MD	9	0.348 (0.272-0.456)[b]	0.14 (0.11-0.19)[b]	0.032 (0.022-0.046)[b]	0.177 (0.14-0.22)[b]
Liver						
Initial	Initial	3	0.37	ND*	0.02	0.35
Final	CT	18	0.338	ND*	0.018	0.32
	MD	18	1.49	0.6	0.15	0.74

*ND=Concentration below limits of detection.
The data are shown by average and range in parenthesis.
Different superscript letters represent a significance difference ($p<0.05$, One-way ANOVA) among CT and MD.

pg-TEQ/g included 88% of Co-PCBs. And at the end of experiment, total TEQ contents of the fish muscle in MD group showed at an average of 0.348 pg-TEQ/g (0.272–0.456 pr-TEQ/g) included 51% of Co-PCBs. In contrast, total TEQ contents of fish muscle in CT group showed at an average of 0.071 pg-TEQ/g (0.061–0.082 pr-TEQ/g) included 94% of Co-PCBs which was significantly lower than that of MD group ($P<0.05$, $n=3$). All ingredients of CT group such as PCDDs, PCDFs and Co-PCBs were significantly lower than MD group. Furthermore, about 50% of total TEQ contents of fish muscle in CT group were decreased compared with initial contents.

Obvious difference was observed in liver total TEQ contents among the groups. In MD group, total TEQ contents increased about 4 times (1.49 pg-TEQ/g) than the initial contests (0.37 pg TEQ/g), on the other hands, there was no increment in CT group (0.338 pg-TEQ/g). Moreover, PCDDs content was below the limit of detection and over 50% of Co-PCBs content was reduced than MD group in CT group.

The target fish species of this study, RSB inhabits almost all cost side of Japanese mainland and its aquaculture amount was 69,700 tons which is the second largest production scale in Japan. Japan Fisheries Agency reported that total TEQ contents on cultured RSB which harvested at the northwest coast of Kyusyu-island was 0.55–0.61 pgTEQ/g, and at the west coast of Seto Inland Sea was 0.44 pgTEQ/g. These contents were similar with RSB in this present study fed the MD diet (0.348 pgTEQ/g). On the other hands, the total TEQ contents in fish fed CT diet composed of dioxins-reduced fish oil showed 0.071 pgTEQ/g that remarkably lower contents. Particularly, Co-PSBs contents greatly decreased in CT group (0.071 pgTEQ/g) compared with cultured RSB at the northwest coast of Kyusyu-island (0.43~0.52 pgTEQ/g) and at the west coast of Seto Inland Sea (0.29 pgTEQ/g). It is thought that the accumulation amount of dioxins in cultured fish derived from mainly the bait in many cases from the artificial pellet. Therefore, adapting dioxins-reduced fish oil examined in this study enable cultured fish to decrease its dioxin content. In addition, this dioxin-reduced fish oil (CT oil) has already been established its mass production techniques. Even though the manufacture cost shows a bit rise in price because of additional processing, it may achieve the practical use of this oil in consideration of offering added-value products.

3.3 Impact for RSB aquaculture

In general classification, there are two types of fish and shellfish; wild-caught and cultured, it seems adversity to reduce the dioxins contents from wild-caught fish and shellfish. Since implementation of the law regulation towards dioxins discharge, estimated discharge amount of dioxins to the environment dramatically decreased in Japan (Ministry of Environment,

Japan. http://www.env.go.jp/en/). However, dioxins have been finally accumulated and pervaded in sea either directs or indirect route through the effusion of industrial wastage and/or living drainage included preliminary emissions. Due to inherent hydrophobic and persistence profiles of dioxins, bulks of substance are prone to remain in the form of residue in aquatic environment such as water, sediment and suspended solids for prolong period without decomposition. It is well known that accumulation and excretory function of living organism differences in depending on dioxin homologs and congeners (Opperhuizen and Sjim 1990), the dioxins are generally considered to be biomagnified along with trophic level of biota. Japan Ministry of Agriculture, Forestry and Fisheries reported on 2009 that the dioxins contents in fish and shellfish remained at the same levels. In addition, the dioxins contamination of seawater is characterized by low diffusibility and regional pollution (Kubota et al. 2002) which means that is anticipated to be contaminated and accumulated fish and shellfish depending on habitat environment such as costal industry, riparian environment or harvested place. Therefore, there is concern that the effect of regional food contamination of dioxins on human through the fish, shellfish or fish-oil-derived productions. Current main protein source as fish meal or lipid source as fish oil of pellet for the aquaculture fish composed by basely wild-caught raw materials, and the cultured fish fed these materials contained dioxins are possibly distributed and consumed all over the world. Accordingly, it is necessary to monitor the dioxins contents in market fish and shellfish by many organizations. On the other hands, the approach to reduce the dioxins content in cultured fish and shellfish is also important. For example, Kawashima et al. (2009) reported the applicable methods of removal dioxins from fish oil by counter current super critical CO_2 extraction and activated carbon treatment. Moreover, Koshio 2009 reported the effect of dioxins-reduced fish oil on *Seriolaquinqueradiata* which investigated the growth and dioxin contents in fish. Results of feeding trial, dioxins-reduced fish oil in EP pellet showed any negative impacts on growth and reduced dioxins contents in fish that is coincident with this present study. Besides, replacement of fish oil with plant oil such as palm oil could effectively reduce the dioxin contents in rainbow trout *Oncorhynchusmykiss* (Oo et al. 2007).

Japanese Ministry of Health, Labour and Welfare (http://www.mhlw.go.jp/) has been defined the TDI as 4 pg-TEQ/kgbw/day. In contrast, WHO (World Health Organization) introduced lower TDI as 1–3 pg-TEQ/kgbw/day. And United States Environmental Protection Agency (EPA) recommended the limits of monthly and/or weekly intake of fish and shellfish amount. It is reported that total dioxin exposure for the Japanese from fish and shellfish domains 60–90% due to the Japanese tend to consume fish and shellfish more than meat, in general (Takayama et al. 1991). The effect of replacement of fish oil as lipid source in artificial diet for RSB which used heretofore (MD group) and dioxin-reduced fish oil (CT group) was investigated in this study. Basely active-carbon treatment and adjustment of ingredient material fish species; it enables CT oil to reduce 1/6 of dioxin contents and contributes significant lower accumulation in cultured RSB. Positive effects on accumulation dioxin content in fish were much clear in liver; i.e., Co-PCBs reduced over 50% in CT group (0.32 pg-TEQ/g) comparing with MD group (0.74 pg-TEQ/g). Furthermore, we confirmed any negative impact on growth and survival between the groups, therefore this study demonstrated the efficacy of dioxin-reduced fish oil in RSB aquaculture.

Thus, it is obvious that replacement of fish oil with dioxins-reduced fish oil or adapting dioxins-reduced fish oil reflects dioxins contents on cultured fish, even though it is comparatively difficult to reduced dioxins contents in wild-caught fish and shellfish.

Overall, the issues remain to be solved in this study are the lipid composition or lipid class analysis, according to lipophilicity profile of dioxins. The involvement with lipid contents in fish and dioxins accumulation amount are unignorable, also it is well known that additional effect of taurine in food for fish fed the diet without fish-derived protein source induced modification of lipid accumulation dynamics. The dioxins which were discharged to environment finally effused and accumulated in sea, then thought to be biomagnified via food chain. Hence, it is adversity to reduce the dioxins content in artificially cultured fish fed the pellet composed by naturally derived fish oil. On the other hands, this study elucidated that adapting dioxins-reduced fish oil in pellet enables the cultured fish to remarkably reduce its dioxins contents especially Co-PSBs which were considered as the highest contribution rate of exposures without any negative impacts on growth and survival. In conclusion, this study indicated the possibility that supply and produce the additional high-value cultured fish that is guarantee more safety and complacency.

Acknowledgement

The author would like to thank Mr. Hiroshi Yamamoto, Mr. Naoki Kaze, Mr. Hideki Hirose and Mr. Atsunori Kumanishi (Ueda Oils and Fats, MFG Co., Ltd.) for providing materials as well as kind support, and the students of Kinki University; Mr. Takao Nagai, Mr. Ryusei Ono and Ms. Ayako Ishii who help this feeding trial at Oshima Experimental station, Fisheries Laboratory of Kinki University.

References

Halver, J.E. 1957. Nutrition of salmonid fishes -III. Water soluble vitamin requirements of chinook salmon. J. Nutr. 62: 225–243.

Hannu, K., T. Vartiainen, R. Parmanne, A. Hallikainen and J. Koistinen. 2003. PCDD/Fs and PCB in Baltic herring during the 1990s. Chemosphere. 50: 1201–1216.

Hites, R., A.F. Jeffery, O.C. David, C.M. Hamilton, A.B. Knuth and J.S. Zchwager. 2004. Global Assessment of Organic Contaminants in Farmed Salmon. Science. 303: 226–229.

Kawashima, A., S. Wanatabe, R. Iwakiri and K. Honda. 2009. Removal of dioxins and dioxin-like PCBs from fish oil by counter current super critical CO_2 extraction and active carbon treatment. Chemosphere. 75: 788–794.

Kosio, S. 2009. Aquaculture Magazine. 46(7): 60–61 (in Japanese). Tokyo, Japan.

Kubota, A., M. Someya, M. Watanabe and S. Tanabe. 2002. Contamination status of PCBs (including coplanar congeners), polychlorinated dibenzo-*p*-dioxins (PCDDs) and dibenzofurans (PCDFs) in sediments from Uwa Sea, Japan. Nippon Suisan Gakkaishi. 68: 695–700 (in Japanese with English abstract).

Kumai, H. Aquaculture of marine fish. 2000. Soubunsya, Tokyo, Japan.

Mastuda R. 2009. The research on food contaminations by hazardous chemical substance (dioxins). Available at; http://www.mhlw.go.jp/topics/bukyoku/iyaku/syoku-anzen/dioxin/sessyu08/dl/sessyu08b.pdf (in Japanese).

Naito, W., J. Jin, Y. Kang, M. Yamamuro, S. Masunaga and J. Nakanishi. 2003. Dynamics of PCDDs/DFs and coplanar-PCBs in an aquatic food chain of Tokyo Bay. Chemosphere. 53: 347–362.

Oo, N.A., S. Satoh and N. Tsuchida. 2007. Effect of replacements of fishmeal and fish oil on growth and dioxin contents of rainbow traut. Fisheries Science. 73: 750–759.

Opperhuizen, A. and D.T.H.M. Sjim. 1990. Bioaccumulation and biotransformation of polychlorinated dibenzo-*p*-dioxins and dibenzofurans in fish. Environmental Toxicology and Chemistry. 9: 175–186.

Takayama, K., H. Miyata, O. Aozasa, M. Mimura and T. Kashimoto. 1991. Dietary intake of dioxin-related compounds through food in Japan. J. Food Hyg. Soc. Jpn. 32: 525–532.

Tsutsumi, T., Y. Amakura, T. Yanagi, Y. Kono, M. Nakamura, T. Nomura, K. Sasaki, T. Maitani and R. Matsuda. 2008. Dietary exposure to dioxins in Japan: Nationwide total diet study 2002–2006. Organohalogen Compounds. 70: 2313–2316.

11

Chitosan as Bio-based Nanocomposite in Seafood Industry and Aquaculture

Alireza Alishahi,[1] Jade Proulx[2] and Mohammed Aider[3,4,] *

1 Introduction

Because consumers are increasingly conscious of the relationship between diet and health, the consumption of marine-based foods has been growing continuously. Consumers identify seafood as nutritious, complete foods that are an excellent source of high-quality proteins and valuable lipids with high amounts of polyunsaturated fatty acids (PUFAs). These compounds are well known to enhance human health, namely by reducing the risk of cardiovascular disease, coronary disease, and hypertension. Additionally, marine-based food products are easily digested and constitute an excellent source of essential minerals. To further strengthen its healthy connotation, seafood has recently been recognized as a functional food and a source

[1] Department of Fisheries, The University of Agriculture and Natural Science of Gorgan, Gorgan, Iran.
[2] Department of Food Science and Agricultural Chemistry, Macdonald Campus of McGill University, Macdonald Campus, 21 111 Lakeshore, Ste. Anne de Bellevue, Que., Canada H9X-3V9.
[3] Institute of Nutrition and Functional Foods (INAF), Université Laval, Quebec, Qc, G1V 0A6, Canada.
[4] Department of Soil Sciences and Agri-Food Engineering, Laval University, Quebec, Qc, G1V 0A6, Canada.
* Corresponding author: mohammed.aider@fsaa.ulaval.ca

of nutraceuticals. Functional foods, first identified in Japan in 1980, are defined as foods demonstrating a beneficial effect on one or more targeted functions of the human organism (Ross 2000). Marine-based functional foods, or nutraceuticals, include omega-3 fatty acids, chitin and chitosan, fish protein hydrolysates, algal constituents, carotenoids, antioxidants, fish bone, shark cartilage, and taurine (Kadam and Prabhasankar 2010).

Despite the aforementioned desirable properties, seafood products are highly susceptible to quality deterioration, mainly due to lipid oxidative reactions involving PUFAs. These reactions are catalyzed by the presence of high concentrations of heme and nonheme proteins that contain iron and other metal ions within their structures (Decker and Haultin 1992). Marine-based food quality is also highly susceptible to autolysis, bacterial contamination, and loss of protein functionality (Jeon et al. 2002). Additionally, there has been a growing concern with seafood consumption due to contamination with various hazardous materials, such as refinery and industrial wastes, and heavy metals (Kadam and Prabhasankar 2010). Another negative perception associated with seafood relates to the intensive farming practices of the aquaculture industry, which have recently gained much attention due to the worldwide depletion of wild fish and shellfish stocks. Not only that, intensive farming is a major cause of stress in fish, which is the most important factor affecting their immune systems (Ledger et al. 2002).

To address the aforementioned problems, the use of chitosan-based nanocomposites as a protective material is very promising. Chitosan is naturally found in the cell walls of fungi of the class zygomycetes, in the green algae Chlorella, in yeast, in protozoa in insect cuticles, and especially in the exoskeletons of crustaceans. Chitosan is a deacetylated derivative of chitin, the second most abundant polysaccharide in nature after cellulose. Chitin was first discovered in 1811 by the French scientist Henri Braconnot, who isolated it from mushroom. It was then isolated from insect cuticles in 1820 (Bhatnagar and Sillanpa 2009). In 1859, Rouget reported finding chitosan after boiling chitin in potassium hydroxide, a treatment that rendered the material soluble in organic acids. Hoppe-Seyler then named this new material 'chitosan' in 1894 (Khor 2001). Chemically, chitosan is a high-molecular weight, linear, polycationic heteropolysaccharide consisting of two monosaccharides, N-acetyl-glucosamine and D-glucosamine linked by β-$(1{\rightarrow}4)$ glycosidic bonds. The relative amounts of these two monosaccharides in chitosan vary considerably. Chitosan can thus have degrees of deacetylation ranging from 75% to 95%, molecular weights ranging from 50 to 2,000 kDa, and have varying viscosities and pKa values (Tharanathan and Kittur 2003). Chitosan has three functional moieties on its backbone: an amino group on C2, as well as a primary and secondary hydroxyl groups on the C3 and C6 positions, respectively. These functional groups play important roles in the various functionalities of chitosan. The amino group is the most important one, especially in acidic conditions where its protonation makes it interact with negatively charged molecules (or sites). Chitosan polymers can also interact with metal cations through their amino groups, hydroxyl ions, and coordination bonds. Commercially, chitosan is produced from chitin by exhaustive alkaline deacetylation, involving boiling chitin in concentrated alkali for several hours. Because this process almost never goes to completion, chitosan is classified as a partially N-deacetylated derivative of chitin (Kumar 2000). In the industry, the biocompatibility, biodegradability, nontoxicity, and bioadhesion of chitosan make it a valuable compound for pharmaceutical (Dias et al. 2008), cosmetic (Pittermann et al. 1997), medical (Carlson et al. 2008), food (Shahidi et al. 1999; No et al. 2007; Kumar 2000), textile (El Tahlawy et al. 2005), wastewater treatment (Chi and Cheng 2006), paper finishing, photographic paper (Kumar 2000), and agricultural applications (Hirano 1996).

Although there are several existing reviews on the use of chitosan in food applications (Alishahi and Aider 2012; No et al. 2007; Shahidi et al. 1999), the use of chitosan-based nanocomposites in seafood—especially its novel application as a nanocarrier for bioactive compounds for shelf life extension—has not yet been reported. Other recent findings include a study on the use of chitosan nanoparticles for stability enhancement of vitamin C in a rainbow trout diet (Alishahi et al. 2011a,b), and a growing interest in the mechanical and barrier properties of edible films and coatings containing chitosan nanoparticles. Effluent treatment of seafood processing plants using chitosan-based nanocomposites to recover protein might also present new potential applications (Alishahi et al. unpublished data).

As a whole, nanotechnology refers to a vast area of technological activity focused on the engineering and manipulation of objects at the nano-scale between 1 and 100 nm. The properties of materials at this scale can be very different from conventional materials. This is due to nanomaterials having an increased relative surface area, and to quantum effects dominating the behavior of matter at the nano-scale. These factors can change or enhance properties such as strength, reactivity, and electrical conductivity (Siegrist et al. 2008). On the other hand, food nanotechnology refers to the application of nanoscience to the food sector. Food nanotechnology includes a wide range of potential applications: alterations to the properties of foods with nano-additives and nano-ingredients, improved delivery, quality and safety of food and the development of new food packaging (Buzby 2010). Figure 1 shows general nanotechnology applications in the food industry, where it is made obvious that sectors involving highly perishable goods such as aquaculture and the seafood industry would highly benefit from nanotechnology. Hence, this review attempts to survey the applications of chitosan-based nanocomposites in various fields of marine-based products.

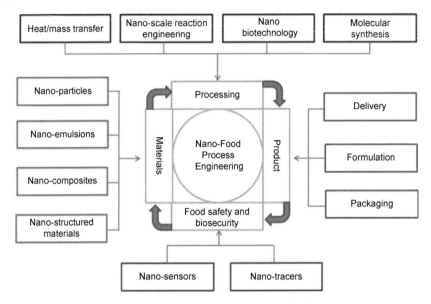

Figure 1. Application of nanotechnology in food science.

2 Nanoencapsulation

Currently, the value of functional foods and bioactive compounds is increasing due to consumer awareness and consciousness. However, many of these compounds are very sensitive to environmental factors, such as oxygen, light, and temperature. In addition, when incorporated into food and drug delivery systems, these bioactive components are often hydrolyzed due to the harsh acidic conditions in the gastrointestinal tract (Alishahi et al. 2011a,b). Schep et al. (1999) stated that many oral delivery systems for bioactive aquaculture compounds meet three major barriers when passing through the gastrointestinal tract: luminal and membrane bound enzymes from the host, immunological cells from enterocytes and underlying connective tissue, and physical barriers from epithelial cells. Based on these considerations, the encapsulation of bioactive compounds and functional foods could be a promising way to overcome these problems. Encapsulation is a process in which thin films, generally consisting of polymeric materials, are applied to small solid particles, liquids, or gas droplets (Figs. 2 and 3).

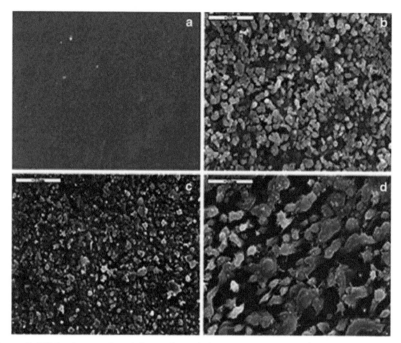

Figure 2. SEM micrographs of the surface of chitosan and chitosan–gelatin blend films at different concentrations: (a) chitosan, (b) chitosan and gelatin (1%), (c) chitosan and gelatin (2%), and (d) chitosan and gelatin (4%). Adapted from Vargas et al. (2009) with permission.

This method is used to entrap active components and release them under controlled conditions (Deladino et al. 2008). In the food industry, encapsulation is used on numerous materials such as vitamins, minerals, antioxidants, colorants, enzymes, and sweeteners (Shahidi and Han 1993). Chitosan can act as an encapsulating agent due to its nontoxicity, biocompatibility, mucus adhesiveness, and biodegradability (Alishahi et al. 2011a,b; Kumar 2000). Recently, Alishahi et al. (2011a,b) showed that a chitosan/vitamin C nanoparticle system successfully increased the shelf life and delivery of vitamin C in rainbow trout over a storage period of 20 days (Fig. 4).

Moreover, the controlled release behavior of vitamin C, both *in vitro* and *in vivo*, showed that vitamin C was gradually released in the rainbow trout gastrointestinal tract over 48 hours, and that the vitamin C was thus successfully protected from the harsh acidic and enzymatic conditions of the gastrointestinal tract by the chitosan nanoparticles. The shelf life of vitamin C was also significantly ($p<0.05$) increased by 20 days at ambient temperature in rainbow trout feed. In comparison, the control

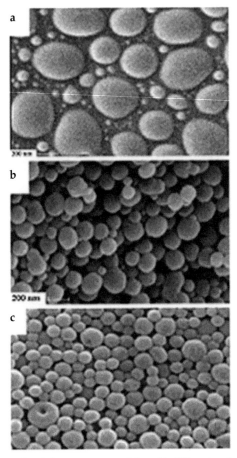

Figure 3. SEM micrographs of chitosan nanoparticles with different molecular weights: (a) 450 kDa, (b) 110 kDa, and (c) 65 kDa. Adapted from Alishahi et al. (2011a,b) with permission.

feed—which only contained uncapsulated vitamin C—lost significant vitamin C content in only a few days. Finally, the authors also showed that chitosan nanoparticles containing vitamin C significantly (p<0.05) induced the nonspecific immunity system of the rainbow trout compared with the control.

Rajeshkumar et al. (2009) demonstrated that chitosan nanoparticles could be used to encapsulate DNA, which was then beneficially incorporated into shrimp feed to protect them from the white spot syndrome virus. Their results showed that these nanoparticles increased the survival rates of shrimp with white spot syndrome for 30 days post-treatment. Likewise, Rajeshkumar et al. (2008) incorporated chitosan nanoparticles

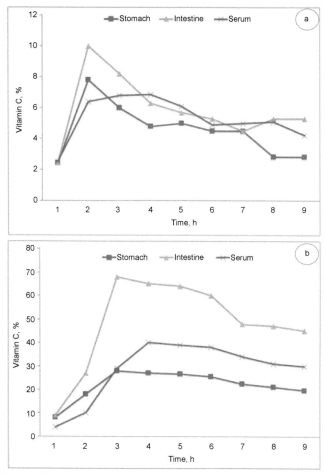

Figure 4. Controlled release of vitamin C in rainbow trout. (a) rainbow trout fed with feed supplemented with vitamin C, and (b) fish fed with feed supplemented with chitosan nanoparticles loaded with vitamin C. Adapted from Alishahi et al. (2011a,b).

Color image of this figure appears in the color plate section at the end of the book.

containing a DNA vaccine into Asian sea bass (*Lates calcarifer*) feed. Their results indicated that sea bass orally vaccinated with the chitosan–DNA (pV AOMP38) complex showed moderate protection against experimental Vibrio anguillarum infection. Similarly, Tian et al. (2008) reported that chitosan microspheres loaded with plasmid vaccine were used to orally immunize Japanese flounder (*Paralichthys olivaceus*). They explained that the release profile of DNA from chitosan microspheres in phosphate-buffered saline buffer (pH 7.4) was extended up to 42 days after intestinal imbibitions.

Finally, Aydin and Akbuga (1996) showed that salmon calcitonin-loaded chitosan beads could be prepared by gelling the cationic chitosan with an anionic counterpart, providing a controlled release property.

Shark liver oil was efficiently encapsulated in calcium alginate beads coated with chitosan to mask its unpleasant taste (Peniche et al. 2004). The chitosan coating controlled the capsules' permeability and prevented leakage. Chitosan/calcium alginate capsules loaded with shark liver oil were initially resistant to the acidic environment of the stomach, but after 4 hours at intestinal pH (7.4), the capsule wall weakened; it was thus easily disintegrated by the mechanical and peristaltic movements of the gastrointestinal tract. Likewise, Klinkesorn and McClements (2009) stated that the encapsulation of tuna oil droplets with chitosan affected their physical stability and digestibility when they were evaluated in an *in vitro* digestion model using pancreatic lipase. The amount of free fatty acids released from the emulsions decreased as the concentration of chitosan was increased, but was unaffected by the chitosan's molecular weight. These results showed that chitosan reduced the amount of free fatty acids released from the emulsion. This may be attributable to a number of different physiological mechanisms: the formation of a protective chitosan coating around the lipid droplets, the direct interaction of chitosan with lipase, or the fatty acids binding with the chitosan. Additionally, the authors showed that pancreatic lipase was able to digest chitosan and release glucosamine; this has important implications for the utilization of chitosan coatings for the encapsulation, protection and delivery of omega-3 fatty acids. They suggested that encapsulation with chitosan could be used to protect emulsified polyunsaturated lipids from oxidation during storage while allowing the capsules to release functional lipids after they are consumed. Industrially, tuna oil encapsulation with chitosan using an ultrasonic atomizer was shown to be a promising technique for the near future (Klaypradit and Huang 2008).

3 Packaging Film

Edible films and coatings are thin layers of edible material applied on food products for conservation, distribution and marketing purposes. Some of their functions are to protect the product from mechanical, physical, chemical, and microbiological damage. Their use in food applications—especially for highly perishable products like seafood—is based on factors such as cost, availability, functional attributes, mechanical properties (flexibility, tension), optical properties (brightness and opacity), gas barrier properties, microbial resistance, structural resistance to water and sensory acceptability (Falguera et al. 2011). Nanotechnology is now applied with great results in many research areas, one of them being polymer research, where ultrafine particles in the nanometer size order called nanoparticles (Hosokawa et al. 2008) form nanobiocomposite films when combined with natural polymers. Recently, the employment of chitosan nanoparticle in edible film proved to enhance the aforementioned characteristics; nanocomposites exhibit increased barrier properties, increased mechanical strength, and improved heat resistance compared to their neat polymers and conventional composites (Sinha Ray and Okamoto 2003; Sinha Ray et al. 2006; Sorrentino et al. 2007). The research and development of nanobiocomposite materials for food applications is expected to grow with the advent of new polymeric materials containing inorganic nanoparticles, although it is not widely spread yet (Restuccia et al. 2010; Sorrentino et al. 2007).

Traditionally, mineral fillers such as clay, silica and talc have been incorporated in film preparation at a 10–50% w/w ratio in order to reduce costs or to improve performance (Rhim and Ng 2007). So far, clays have been the most important functional inorganic fillers to be added to edible films. According to Rhim and Ng (2007), the nanometer-size dispersion of polymer-clay nanocomposites exhibits the larger-scale improvement in mechanical and physical properties compared with pure polymer or conventional composites. Both proteins (Shotornvit et al. 2009) and polysaccharides (Casariego et al. 2009; Tang et al. 2008) have been combined with nano-clay particles to form films. Other nanoparticles such as tripolyphosphate-chitosan (De Moura et al. 2009) have also been added to biopolymers to obtain films. These nanoparticles are able to improve moisture barrier properties (Casariego et al. 2009; De Moura et al. 2009; Shotornvit et al. 2009) and restrict microbial growth (Shotornvit et al. 2009). In this way, Rhim et al. (2006) found that the use of nanoparticles could potentially help develop new natural, biopolymer-based, biodegradable packaging materials. Their experiments showed that different nanoparticles improved the physical properties of chitosan-based nanocomposite films, and could potentially improve its antimicrobial activity.

Xu et al. (2005) demonstrated that exfoliated nano-structures were formed after the addition of small amounts of MMT-Na to the chitosan matrix. Intercalation together with some exfoliation occurred when the amount of MMT-Na was increased to 5% wt. Other observations were that the surface roughness increased with the addition of a small amount of nanoclay, and that micro-scale composites (tactoids) were formed when Cloisite 30B was added to the chitosan matrix. While the chitosan film's tensile strength increased with the addition of small amounts of MMT-Na, it did not increase significantly when Cloisite 30B was added. Elongation-at-break decreased with the addition of clays, but not with the addition of MMT-Na. Melting temperature (Tm) and onset temperature of thermal degradation (Td) of chitosan composite films increased when MMT-Na was added; Tm and Td did not change significantly with the addition of Cloisite 30B. Generally, addition of MMT-Na improved the film's mechanical and thermal properties more than Cloisite 30B. Most interestingly, AFM images (Figs. 5 and 6) of chitosan nanocomposite film showed unique traits compared with the plain mood.

Likewise, Li et al. (2009), stated that chitosan-nanocomposite films mixed well with cellulose whiskers. The results indicated that the nanocomposites exhibited good miscibility, and strong interactions occurred between the whiskers and the matrix. With increasing the whisker content from 0 to 15–20 wt %, the tensile strength of the composite films in dry and wet states increased from 85 to 120 MPa and 9.9 to 17.3 MPa, respectively. Furthermore, the nanocomposite films displayed excellent thermal stability and water

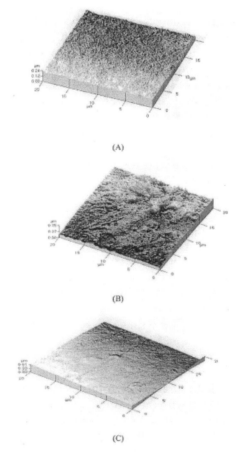

(A)

(B)

(C)

Figure 5. AFM topographic images of (A) chitosan and its nanocomposite films with (B) 3 wt % MMT-Na and (C) 5 wt % MMT-Na. Adapted from Xu et al. (2005) with permission.

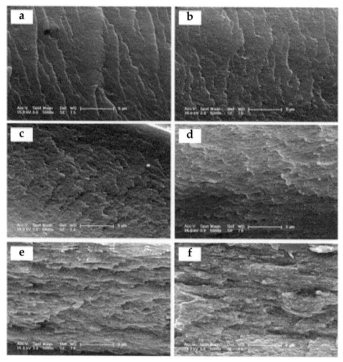

Figure 6. SEM images of the (a) fracture surfaces of the CH film and the nanocomposites of (b) CW-5, (c) CW-10, (d) CW-20, (e) CW-25, and (f) CW-30 (The scale bar is 5 lm.). Adapted from Li et al. (2009) with permission.

resistance with the incorporation of cellulose whiskers. SEM images of the films are shown in Fig. 5. The chitosan film, which was prepared from chitosan alone, exhibited a smooth surface. The chitosan-whisker mixes with different concentration of whisker, CW-5, CW-10, and CW-20, displayed a homogeneous and dense structure, indicating a good dispersion level of whiskers for the composite film. It is difficult to observe the exact positions of the dispersed filler in the chitosan matrix due to its small particle size. However, as the whisker content increased, the fractured surface of CW-30 became relatively rough, suggesting a more brittle rupture. Moreover, some white points, corresponding to the large aggregates of the fillers, could be observed at the fracture surface of CW-30.

Further, Vimala et al. (2010) created a novel film based on chitosan nanocomposites that showed even better applied characteristics. As shown in Fig. 7 and Fig. 8, the film exhibited a porous structure, thus improving mechanical and barrier strength. This study describes the preparation of porous chitosan–silver nanocomposite films in an easy, three-step process. The chitosan and polyethylene glycol employed were biocompatible in

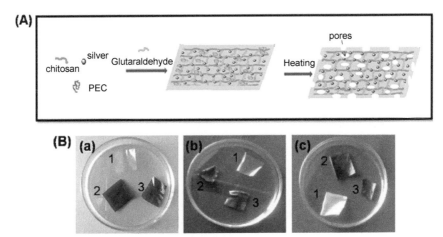

Figure 7. (A) Schematic illustration of preparation of porous chitosan–silver nanocomposite (PCSSNC) films. (B) (1) chitosan, (2) chitosan–silver nanocomposite, and (3) porous chitosan–silver nanocomposite films, (a–c) different composition of chitosan:PEG, 1:1, 2:1, and 5:1, respectively. Adapted from Vimala et al. (2010) with permission.

Color image of this figure appears in the color plate section at the end of the book.

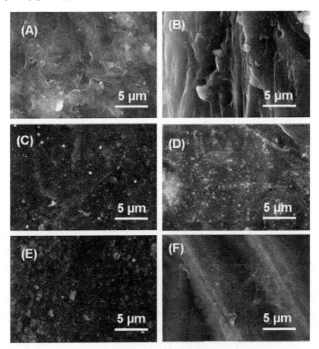

Figure 8. Scanning electron micrographs of (A and B) chitosan, (C and D) chitosan–silver nanocomposite, and (E and F) porous chitosan–silver nanocomposite films. Bar size indicates 5 μm. Adapted from Vimala et al. (2010) with permission.

nature, and not toxic. These polymers not only help in reducing the metal ions into nanoparticles, but also provide excellent stability for a sustained release of nanoparticles for antibacterial applications. The developed porous nanocomposite film exhibited superior antibacterial properties, and better mechanical properties than the chitosan and chitosan–silver nanocomposites.

4 Environmental Impact; Freshwater and Marine Organisms (Fish)

Given the increasing production of nanomaterials (NMs) of all types, the potential for their release in the environment and the subsequent effects on ecosystem health are becoming an increasing concern. Regulatory agencies in particular should address this issue by determining the fate and behavior of manufactured NMs in the environment. Do they retain their nominal nanoscale size, original structure, and reactivity in aquatic and soil/ sedimentary systems? Do they associate with other colloidal and particulate constituents? Is their effect on aquatic and sedimentary biota different from that of larger particles of the same material? Do biota, such as biofilms and invertebrates, modify their behavior? What are the effects of solution and physical (e.g., flow) conditions? Answers to these questions, among others, will guide the setting of regulatory guidelines that will provide adequate protection to ecosystems while permitting nanotechnology advantages to be fully developed.

In freshwater fish, Smith et al. (Smith et al. 2007) studied the toxicity of single-wall carbon nanotube (SWCN) on rainbow trout. Single walled carbon nanotube exposure caused a dose-dependent rise in ventilation rate, gill pathologies (edema, altered mucocytes, hyperplasia), and mucus secretion with SWCN precipitation on the gill mucus. It also caused statistically significant increases in $Na^+ K^+$ -adenosine triphosphatase activity in the gills and intestine, but not in the brain. A thiobarbituric acid (TBA) test was carried out to assess relative amounts of lipid peroxidation products, and thus to detect any effect SWCN might have on oxidative stress. Results demonstrated dose-dependent, statistically significant decreases in lipid peroxidation products in the gill, brain, and liver. Overall, the authors concluded that SWCN are a respiratory toxicant in trout, and that observed cellular pathologies suggest cell cycle defects, neurotoxicity, and blood-borne factors that possibly mediate systemic pathology (Smith et al. 2007).

In addition, the effect of numerous other metal NPs was assessed on rainbow trout (Federici et al. 2007) and zebrafish (Lee et al. 2007). Results indicate that effects will depend on the type of material being studied, with copper (Griffite et al. 2007) and silver (Lee et al. 2007) more likely to result in toxic effects at lower concentrations than titanium dioxide (Federici et al. 2007).

To date, few ecotoxicological studies on marine invertebrates and fish have been reported. Overall trends for nanoparticle effects on marine creatures have been depicted in Fig. 8. The information is limited to a few reports, and we are far from having reached a general consensus about NP absorption, distribution, metabolism, excretion, toxic effects on body systems, and toxicity to different groups of marine organisms. Moore (Moore 2006) argues that marine bivalves such as *Mytilus edulis* might take up NPs using endocytosis, and demonstrated that polyester NPs were taken up by mussels' endosomes and lysosomes. Such modes of uptake may be especially relevant to marine organisms where aggregation of NPs will occur mostly on their surface. Moore (Moore 2006) also raised concerns about NPs acting as delivery vehicles for other chemicals via the endocytosis pathway. Some evidence also exists that colloidal metals have a different bioavailability from aqueous metals for uptake by oysters (Guo et al. 2002). It may also be possible for the surfaces of marine organisms to promote NP formation. Scarano and Morelli (Scarano and Morelli 2003) noted that stable nanocrystals form on marine phytoplankton when exposed to cadmium. This raises the possibility that metal NP exposure to appropriate conditions for crystal formation at the surface of the organism could lead to subsequent uptake, even when manufactured NPs are not present in the original polluting material (Fig. 9).

Chitosan nanoparticle showed somewhat the same trend in freshwater fish rather than in rainbow trout (Alishahi et al. 2011a). When the fish was completely immersed with the chitosan nanoparticle, the fish died after 8 hours. However, when the nanoparticle was put in the feed, it did not cause any mortality. This is because, in the immersion state, chitosan nanoparticles have a positive charge causing them to stick to the gills' epithelial cells. An interesting observation is that when put in the feed, chitosan nanoparticles not only did not cause mortality, but also provoked the innate immunity system, and increased the feed's residence time into the digestive tract. In doing so, it enhanced the absorption rate of the feed at the intestine's epithelial surface (Alishahi et al. 2011b). Overall, chitosan nanoparticles have no severe side effect on animal and human cells because of their natural biocompatibility. This strengthens chitosan-based nanocomposites' potential in seafood and marine-based products applications, compared with other previously mentioned nanoparticles.

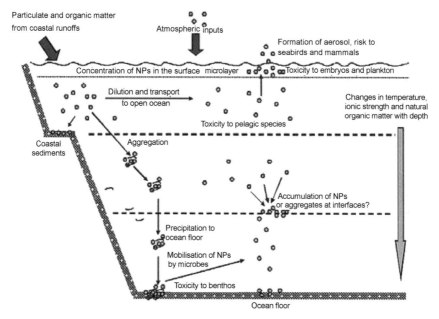

Figure 9. Schematic diagram outlining the possible fate of nanoparticles (NPs) in the marine environment and the organisms at risk of exposure. Adapted from Klaine et al. (2008) with permission.

5 Impact on Human Health

There are three different ways in which nanoparticles can enter organisms: inhalation, skin penetration, and ingestion. However, there is growing scientific evidence reporting that free nanoparticles can cross cellular barriers, and that exposure to some of these nanoparticles may lead to oxidative damage and inflammatory reactions (Chaudhry et al. 2008; Bouwmeester et al. 2009; Wiesner et al. 2011).

In the case where nanomaterials could be used in food packaging, many people fear the risk of indirect exposure due to potential migration of nanoparticles from the packaging material. However, the only cases where such leakage occurred are almost all exclusively related to workers in nanomaterial producing factories. For these workers, wearing personal protection gear such as gloves, glasses, and masks with high efficiency filters is recommended.

As cited in previous segments of this review, chitosan is a non-toxic, biodegradable, biocompatible, and eco-friendly polymer. In consequence, its utilization in food nanotechnology—especially in seafood and aquatic products with high spoilage rate—should be safe. Moreover, Alishahi et al. (2011a,b) and many other authors have proved that no inflammatory response incurred after the use of chitosan nanoparticles in food, drug delivery systems, as well as human and animal digestive tracts. In fact, it was shown that chitosan NPs could be given as an immunostimulants since they augmented the innate immunity system and the growth rate of rainbow trout (*Oncorhynchus mykiss*) (Alishahi et al. 2011b).

6 Regulation Issues

As food nanotechnology is increasingly used, its safety gains a lot of attention. Despite the rapid commercialization of nanotechnology, it is not subject to any specific regulations anywhere in the world. Most regulatory agencies remain in information-gathering mode, lacking the legal and scientific information necessary to oversee, adequately, the exponential market growth of nanotechnology (Delgado 2010; Mansour 2006). International agencies such as the Food and Agriculture Organization (FAO), the World Health Organization (WHO), the Organization for Economic Cooperation and Development (OECD) and the European Food Safety Authority (EFSA) identified a number of research areas that need to be further investigated in order to improve the basic scientific knowledge needed to support such regulatory work:

- Development of reliable measurement methods, reference materials, and material characterization;
- Review and development of test methods for human health, safety and the environment;
- Development of exposure information throughout the life-cycle of nanomaterials;
- Review of existing risk-assessment methods;
- Risk management for workers' protection;
- Establishment of new infrastructures, and network creation for existing ones to examine health, safety and environmental aspects of nanomaterials (Blasco and Pico 2011).

On the other hand, certain countries are trying to include nanomaterials in their current regulations. For example in the United States, the Food and Drug Administration (FDA) requires manufacturers to demonstrate that foods and their ingredients are not dangerous to health, but there are no specific rules for nanoparticles because the agency regulates products, not technologies. Nevertheless, the FDA expects that many nanotechnology products will come under its jurisdiction; the FDA regulates a wide range of products including foods, cosmetics, drugs, devices, and veterinary products, some of which contain nanomaterials or are produced nanotechnologically (FDA 2007).

The European Commission (EC) aims to reinforce nanotechnology regulation and, at the same time, enhance support for collaborative research and development on the potential effect of nanotechnology on human health and the environment via toxicological and ecotoxicological studies. The EC is performing a regulatory inventory, covering EU regulatory frameworks that are applicable to nanomaterials. The purpose of this inventory is to examine, and where appropriate, to propose adaptations of EU regulations in relevant sectors (EC 2009).

7 Public Perception

Studies examining public perception of food nanotechnology—and more interestingly seafood nanotechnology—in the U.S. and Europe are very limited (Cobb and Macoubrie 2004; Lee et al. 2005; Scheufele et al. 2009). Furthermore, there is evidence of a growing nanotechnology-related knowledge gap based on two surveys in the U.S. in 2004 (Scheufele and Lewenstein 2005) and 2007 (Scheufele et al. 2009); there is an escalation in food nanotechnology awareness for highly educated respondents, and a decrease in food nanotechnology awareness for the least informed respondents (Corely and Scheufele 2010). Surveys have also examined public perceptions of nanotechnology, finding, for example, that the U.S. public perceives fewer potential benefits and, in most cases, more risks from nanotechnology than U.S. scientists (Scheufele et al. 2007), and that the U.S. public is more optimistic toward nanotechnology that the European public (Gaskell et al. 2005).

A handful of recent studies by Siegrist and colleagues have examined potential public response to food nanotechnology. Their outcomes identified factors that are likely to impact public acceptance of food-related nanotechnologies. Their results suggest that individuals are more likely to accept food nanotechnologies and perceive higher levels of naturalness in nanotechnology food products (Siegrist et al. 2009) when they have higher levels of trust in the food industry (Siegrist et al. 2007; Siegrist 2008; Siegrist et al. 2008). Their results also suggest that individuals are less likely to accept food nanotechnologies when their level of trust in the food industry is lower (Siegrist et al. 2008). Their work has also shown that individuals' perceived benefits and risks influence their degree of acceptance (Siegrist 2008), and that nano-derived food packaging is perceived of as more beneficial than nano-derived foods (Siegrist et al. 2007, 2008). The research reveals valuable insights into public perceptions of food nanotechnology and suggests that trust in the nano-related industries may be an important factor for the successful introduction of nanotechnological foods on the market.

Despite the aforementioned points, in Japan, the US and many southeast countries of Asia, chitosan-based nanocomposites are increasingly utilized in active and intelligent packaging, bioactive compound delivery systems, and functional food introduction in marine-based products (Shahidi 2006).

References

Alishahi, A., A. Mirvaghefi, M. Rafie-Tehrani, H. Farahmand, S.A. Shojaosadati and F.A. Dorkoosh. 2011a. Shelf life and delivery enhancement of vitamin C using chitosan nanoparticles. Food Chemistry. 126: 935–940.

Alishahi, A., A. Mirvaghefi, M.R. Tehrani, H. Farahmand, S. Koshio and F. Dorkoosh. 2011b. Chitosan nanoparticle to carry vitamin C through the gastrointestinal tract and induce the non-specific immunity system of rainbow trout (Oncorhynchus mykiss). Carbohydrate Polymer. 86: 142–146.

Alishahi, A. and M. Aïder. 2012. Applications of Chitosan in the Seafood Industry and Aquaculture: A Review. Food and Bioprocess Technology. 5(3): 817–830.

Aydin, Z. and J. Akbuga. 1996. Chitosan beads for delivery of salmon calcitonin: preparation and release characteristics. International Journal of Pharmaceutics. 131: 101–103.

Bhatnagar, A. and M. Sillanpa. 2009. Application of chitin and chitosan-derivatives for water and waste water—a short review. Advance in Colloid and Interface Science. 55: 9479–9488.

Blasco, C. and Y. Pico. 2011. Determining nanomaterials in food. Trends in Analytical Chemistry. 30: 84–100.

Bouwmeester, H., S. Dekkers, M.Y. Noordam, W.I. Hagens, A.S. Bulder, C. Heer, E.C.G. Sandra, S.W.P. Wijnhoven, H.J.P. Marvin and A.J.A.M. Sips. 2009. Review of health safety aspects of nanotechnologies in food production. Regulation in Toxicology and Pharmaceutics. 53: 52–62.

Buzby, J.C. 2010. Nanotechnology for food applications. More questions than answers. Journal of Consumer Affairs. 44: 528–545.

Carlson, R.P., R. Taffs, W.M. Davison and P.S. Steward. 2008. Anti-biofilm properties of chitosan coated surfaces. Journal of Biomaterial Science Polymer. 19: 1035–1046.

Casariego, A., B.W.S. Souza, M.A. Cerqueira, J.A. Teixeira, L. Cruz, R. Dıaz and A.A. Vicente. 2009. Chitosan/clay films' properties as affected by biopolymer and clay micro/nanoparticles' concentrations. Food Hydrocolloid. 23: 1895–1902.

Chaudhry, Q., M. Scotter, J. Blackburn, B. Ross, A. Boxall and L. Castle. 2008. Applications and implications of nanotechnologies for the food sector. Food Additive Conta, part A. 25: 241–258.

Chi, F.H. and W.P. Cheng. 2005. Use of chitosan as coagulant to treat waste water from milk processing plant. Journal of Polymer Environment. 14: 411–417.

Cobb, M. and J. Macoubrie. 2004. Public perceptions about nanotechnology. Risks, benefits and trust. Journal of Nanoparticle Research. 6: 395–405.

Corley, E.A. and D.A. Scheufele. 2010. Outreach gone wrong? When we talk nano to the public, we are leaving behind key audiences. The Scientist. 24: 22–37.

Decker, E.A. and H.O. Haultin. 1992. Lipid oxidation in muscle foods via redox ion. pp. 33–54. *In*: A.J. Angelo (ed.). Lipid oxidation in food Washington, D.C. American Chemistry Society.

Delgado, G.C. 2010. Economics and governance of nanomaterials: Potential and risks. Technology Society. 32: 137–144.

Deladino, L., P.S. Anbinder, A.S. Navarro and M.N. Martino. 2008. Encapsulation of natural antioxidants extracted from Ilex paraguariensis. Carbohydrate Polymer. 71: 126–134.

De Moura, M.R., R.J. Avena-Bustillos, T.H. McHugh, J.M.M. Krochta and L.H.C. Mattoso. 2008. Properties of novel hydroxypropyl methyl cellulose films containing chitosan nanoparticles. Journal Food Science. 73: 31–37.

Dias, F.S., D.C. Querroz, R.F. Nascimento and M.B. Lima. 2008. Simple system for preparation of chitosan microspheres. Quimica Nova. 31: 160–163.

EC (European Community), 2009. Scientific Opinion. The Potential Risks Arising from Nanoscience and Nanotechnologies on Food and Feed Safety. Scientific Opinion of the Scientific Committee (Question No EFSA-Q-2007-124a). Adopted on 10 February 2009. The EFSA Journal. 958: 1–39.

El Tahlawy, K.F., M.A. El Benday, A.G. Elhendawy and S.M. Hudson. 2005. The antimicrobial activity of cotton fabrics treated with different cross linking agents and chitosan. Carbohydrate Polymer. 60: 421–430.

Falguera, V., J.P. Quintero, A. Jimenez, J.A. Munoz and A. Ibarz. 2011. Edible films and coatings: structures, active functions and trends in their use. Trends in Food Science and Technology. 22: 292–303.

Falguera et al. http://www.sciencedirect.com/science/article/pii/S0924224411000318.

Federici, G., B.J. Shaw and R.D. Handy. 2007. Toxicity of titanium oxide nanoparticles to rainbow trout (Oncorhynchus mykiss): Gill injury, oxidative stress, and other physiological effects. Aquatic Toxicology. 84: 415–430.

FDA, 2007. Nanotechnology. A Report of the U.S. Food and Drug Administration Nanotechnology Task Force July. 25: 1–38.

Gaskell, G., T. Ten Eyck, J. Jackson and G. Veltri. 2005. Imaging nanotechnology. Cultural support for technological innovation in Europe and the United States., public understanding of science. 14: 81–90.

Griffith et al. 2007.

Guo, L., P.H. Santschib and S.M. Ray. 2002. Metal partitioning between colloidal and dissolved phases and its relation with bioavailability to American oysters. Marine Environment Science. 54: 49–64.

Hosokawa, Nogi, Makio and Yokoyama. 2008.

Hirano, S. 1996. Chitin biotechnology applications. Biotechnology Annual Review. 2: 237–258.

Jeon, Y.J., J.Y.V.A. Kamil and F. Shahidi. 2002. Chitosan as an edible invisible film for quality preservation of herring and Atlantic cod. Journal of Agriculture and Food Chemistry. 50: 67–78.

Kadam, S.U. and P. Prabhasankar. 2010. Marine foods as functional ingredients in bakery and pasta products. Food Research International. 43: 1975–1980.

Khor, E. 2001. The relevance of chitin. pp. 1–8. *In*: Chitin: Fulfilling a biomaterial promise. New York: Elsevier.

Klaine, S.J., P.J.J. Alvarez, G.E. Batley, T.F. Fernandez, R.D. Handy, D.Y. Lyon, S. Mahendra, M.J. Mclaughlin and J.R. Lead. 2008. Nanomaterials in the environment: Behavior, fate, bioavailability and effects. Environment and Toxicological Chemistry. 27: 1825–1851.

Klaypradit, W. and Y.W. Huang. 2008. Fish oil encapsulation with chitosan using ultrasonic atomizer. LWT-Food Science and Technology. 41: 1133–1139.

Klinkesorn, U. and D.J. McClements. 2009. Influence of chitosan on stability and lipase digestibility of lecithin-stabilized tuna oil-in-water emulsions. Food Chemistry. 14: 1308–1315.

Kumar, M.N.V. 2000. A review of chitin and chitosan applications. Reactive and Functional Polymer. 46: 1–27.

Ledger, R., I.G. Tucker and G.F. Walker. 2002. The metabolic barrier of the lower intestinal tract of salmon to the oral delivery of protein and peptide drugs. Journal of Control Release. 85: 91–103.

Lee, C.J., D.A. Scheufele and B.V. Lewenstein. 2005. Public attitudes toward emerging technologies. Examining the interactive effects of cognitions and effect on public attitudes toward nanotechnology. Science Communication. 27: 240–265.

Li, Q., J. Zhou and L. Zhang. 2009. Structure and properties of the nanocomposite films of chitosan reinforced with cellulose whiskers. Journal of Polymer Science: Part B. 47: 1069–1077.

Mansour, M. 2010. The emerging regulatory framework for nanomaterials. Nanomaterial and Nanotechnology in Biological Medicine. 2: 269–312.

Moore, M.N. 2006. Do nanoparticles present ecotoxicological risks for the health of the aquatic environment? Environment International. 32: 967–976.

No, H.K., S.P. Meyers, W. Prinyawiwatkul and Z. Xu. 2007. Applications of chitosan for improvement of quality and shelf life of foods: a review. Journal of Food Science. 72: 87–100.

Peniche, C., I. Howland, O. Carrillo, C. Zaldivar and W. Arguelles-Monal. 2004. Formation and stability of shark liver oil loaded chitosan/calcium alginate capsules. Food Hydrocolloid. 18: 865–871.

Pittermann, W., V. Horner and R. Wachter. 1997. Food applications of high molecular weight chitosan in skin care applications. *In:* R.A.A. Muzzarelli and M.G. Peter (eds.). Chitin handbook, Grottammare, Italy: European Chitin Society. 361 p.

Rajeshkumar, S., V.D. Ishaq Ahmed, V. Parameswaran, R. Sudhakaran, V. SarathBabu and A.S. Sahl Hameed. 2008. Potential use of chitosan nanoparticles for oral delivery of DNA vaccine in Asian sea bass (Latescalcarifer) to protect from Vibrio anguillarum. Fish and Shellfish Immunology. 25: 47–56.

Rajeshkumar, S., C. Venkatesan, M. Sarathi, V. Sarathbabu, J. Thomas and K. AnverBasha. 2009. Oral delivery of DNA construct using chitosan nanoparticles to protect the shrimp from white spot syndrome virus (WSSV). Fish and Shellfish Immunology. 26: 429–437.

Restuccia, D., U.G. Spizzirri, O.I. Parisi, G. Cirillo, M. Curcio, F. Iemma, F. Puoci, G. Vinci and N. Picci. 2010. New EU regulation aspects and global market of active and intelligent packaging for food industry applications. Food Control. 21: 1425–1435.

Rhim, J., S. Hong, H. Park and K.W. Perry. 2006. Preparation and characterization of chitosan-based nanocomposite films with antimicrobial activity. Journal of Agricultural and Food Chemistry. 54: 5814–5822.

Rhim, J.W., S.I. Hong, H.M. Park and P.K.W. Ng. 2006. Preparation and characterization of chitosan-based nanocomposite films with antimicrobial activity. Journal of Agriculture and Food Chemistry. 54: 5814–5822.

Rhim, J.W. and P.K.W. Ng. 2007. Natural biopolymer-based nanocomposite films for packaging applications. Critical Review in Food Science and Nutrition. 47: 411–33.

Ross, S. 2000. Functional foods: the food and drug administration perspective. American Journal of Clinical Nutrition. 71: 1735s–1738s.

Scarano, G. and E. Morelli. 2003. Properties of phytochelatin-coated CdS nanocrystallites formed in a marine phytoplanktonic alga (Phaeodactylum tricornutum, Bohlin) in response to Cd. Plant Scienec 165: 803–810.

Schep, L.G., I.G. Tucker, G. Young, R. Ledger and A.G. Butt. 1999. Controlled release opportunities for oral peptide delivery in aquaculture. Journal of Control Release. 59: 1–14.

Scheufele, D.A., E.A. Corley, T.J. Shih, K.E. Dalrymple and S.S. Ho. 2009. Religious beliefs and public attitudes to nanotechnology in Europe and the U.S. nature nanotechnology. 4: 91–94.

Scheufele, D.A. and B.V. Lewenstein. 2005. The public and nanotechnology. How citizens make sense of emerging technologies. Journal of Nanoparticle Research. 7: 659–667.

Shahidi, F., J.Y.V.A. Kamil and Y.J. Jeon. 1999. Food applications of chitin and chitosan. Trends in Food Science and Technology. 10: 37–51.

Shahidi, F. and X. Han. 1993. Encapsulation of food ingredients. Critical Review in Food Science and Nutrition. 33: 501–547.

Shahidi, F. 2006. Functional Foods: Their Role in Health Promotion and Disease Prevention. Journal of Food Science. 69: 146–149.

Shotornvit, R., J.W. Rhim and S.I. Hong. 2009. Effect of nano-clay type on the physical and antimicrobial properties of whey protein isolate/clay composite films. Journal of Food Engineering. 91: 467–493.

Siegrist, M., N. Stampfli, H. Kastenholz and C. Keller. 2008. Perceived risks and perceived benefits of different nanotechnology foods and nanotechnology food packaging. Appetite. 51: 283–290.

Siegrist, M., M.E. Cousin, H. Kastenholz and A. Wiek. 2007. Public acceptance of nanotechnology foods and food packaging. The influence of affect and trust. Appetite. 49: 459–466.

Siegrist, M. 2008. Factors influencing public acceptance of innovative food technologies and products. Trend in Food Science and Technology. 19: 603–608.

Siegrist, M., N. Stampfli and H. Kastenholz. 2009. Acceptance of nanotechnology foods. A conjoint study examining consumer's willingness to buy. Britain Food Journal. 111: 660–668.

Sinha Ray, S. and M. Okamoto. 2003. Polymer/layered silicate nanocomposites: a review from preparation to processing. Progress in Polymer Science. 28: 1539–1641.

Sinha Ray, S., S.Y. Quek, A. Easteal and X.D. Chen. 2006. The potential use of polymerclay nanocomposites in food packaging. International Journal of Food Engineering. 2: 1–11.

Smith, C.J., B.J. Shaw and R.D. Handy. 2007. Toxicity of single walled carbon nanotubes to rainbow trout (Oncorhynchus mykiss): Respiratory toxicity, organ pathologies, and other physiological effects. Aquatic Toxicology. 82: 94–102.

Sorrentino, A., G. Gorrasi and V. Vittoria. 2007. Potential perspectives of bio-nanocomposites for food packaging applications. Trends in Food Science and Technology. 18: 84–95.

Tang, X., S. Alavi and T.J. Herald. 2008. Effects of plasticizers on the structure and properties of starch-clay nanocomposite films. Carbohydrate Polymers. 74: 552–558.

Tharanathan, R.N. and F.S. Kittur. 2003. Chitin-the undisputed biomolecule of great potential. Critical Review in Food Science Nutrition. 43: 61–87.

Tian, J., J. Yu and X. Sun. 2008. Chitosan microspheres as candidate plasmid vaccine carrier for oral immunization of Japanese flounder (Paralichthy solivaceus). Veterinary Immunity and Immunology. 126: 220–229.

Vargas, M., A. Albors, A. Chiralt and C. Gonzalez-Martinez. 2009. Characterization of chitosan-oleic acid composite films. Food Hydrocolloid. 23: 536–547.

Vimala, K., Y. Murali Mohana, K. Samba Sivudua, K. Varaprasada, S. Ravindraa, N. Narayana Reddya, Y. Padmab, B. Sreedharc and K. MohanaRaju. 2010. Fabrication of porous chitosan films impregnated with silver nanoparticles: A facile approach for superior antibacterial application. Colloid Surface B: Biointerfaces. 76: 248–258.

Weiss, J., P. Takhistov and D.J. McClements. 2006. Functional materials in food nanotechnology. Journal of Food Science. 71: R107–R116.

Wiesner, M.R. and J.Y. Bottero. 2011. A risk forecasting process for nanostructured materials, and nanomanufacturing. Critical Review in Physics. 12: 659–668.

Xu, H.H.K. and C.G.J. Simson. 2005. Fast setting calcium phosphate–chitosan scaffold: mechanical properties and biocompatibility. Biomaterial. 26: 1337–1348.

12

Recent Developments in Quality Evaluation, Optimization and Traceability System in Shrimp Supply Chain

Imran Ahmad,[1,2,]* Chawalit Jeenanunta[1] and Athapol Noomhorm[2]

1 Introduction

1.1 Shrimp Industry

Dramatic increase in commercial shrimp farming has been recorded in the decades following the 1970s, particularly due to the market demand of the EU, USA, and Japan. Currently, approximately 80% of farmed shrimp is cultured in Asia, including China and Thailand. Latin America, especially Brazil, is also a major producer. Nevertheless, Thailand is the top exporter. Most farmed shrimp belongs to the family *Penaeidae* which includes Pacific white shrimp and Black tiger shrimp, comprising 80% of all farmed shrimp.

[1] Sirindhorn Institute of Technology, Thammasat University, Pathum Thani, Thailand.
[2] Asian Institute of Technology, Pathum Thani, Thailand.
  Emails: himranz.ahmad@gmail.com, chawalit@siit.tu.ac.th,
  athapol.noomhorm@gmail.com
* Corresponding author

1.2 General Considerations for Shrimp Supply Chain

A set of approaches is employed to efficiently integrate suppliers, manufacturers, warehouses and transportation modes, so that merchandise is produced and distributed at the required quantities, to the right locations, and the right time, in order to minimize system wide costs while satisfying service level requirements. The integration of quality and safety control measures at each stage in supply chain management affects the cost and quality of frozen shrimp. The supply chain starts from cultivation practice at farm level, moving on to harvesting, and post-harvest, transportation, processing, and marketing. Shrimp is harvested from farm and delivered to factory for freezing into different types of product. At farm level, farmers can be grouped together or use the services of middle men to manage handling and selling to a factory. For transportation, the logistic cost of handling depends on the type of vehicle. The facilities of each vehicle such as cold storage truck and air supply truck are an additional cost and also have an effect on the quality of shrimp. The adoption of appropriate post-harvest methods is very important; among them, the temperature is the most important factor influencing the rate of spoilage process (Le 1997). Then, shrimp is processed in the factory for freezing in terms of frozen raw, frozen cooked, frozen breaded shrimp and others. The methods of freezing in shrimp process include Individual Quick Frozen (IQF) and Block Freezing methods. After processing, the finished frozen products should be stored at frozen temperature of about –18°C. This cold chain must be used for distribution and logistics. The temperature of transportation and storage should be inspected constantly until the shrimp is delivered to customers. It can be recognized that there are many issues concerning shrimp processing such as freezing methods, storage time, and the temperature of storage room that affect the quality of shrimp.

A generic form of supply chain of shrimp is depicted in Fig. 1. In practice, the supply chain is multifaceted due to the involvement of numerous interlinked activities performed by multiple types of players located in different regions of the globe.

For agricultural commodities, the focus of supply chain management is unequivocally on finding the most effective and efficient way of adding value, with the aim of generating cross-functional solutions to the many complex problems associated with meeting consumer requirements effectively, and at minimal cost. The production of raw materials for

Figure 1. The generic value chain of sea food.

processing requires breeding, production, storage and distribution, the procurement of other inputs and the management of a number of distinct 'production functions'. Agricultural commodities are subject to grading and further preparation for certain markets. Although the functions of a supply chain may be universal, their implementation varies within and outside the organization. For an organization, the real challenge of supply chain management is to perform all the functions, from procurement to distribution of finished goods, to the competitive advantage. Further details of main functions of processing and handling are given in next section.

Issues in fresh shrimp supply chain are, therefore, improving efficiency, product quality, farm-to-plate traceability, and increased supply and demand for sustainably harvested fish. The current practice of dividing the activities of the supply chain into multiple parts results in issues such as outdated preservation and handling practices, non-negotiable price fixing by processors and unsynchronized flows of supply and demand for fresh shrimp. Trends in the aquaculture industry currently focus on sustainability at every level of farming, harvesting and processing. Consumers are increasingly aware of sustainable practices and demand for fair and environment-friendly treatment of produce, labor and other supply chain activities. However, the information needed to determine the sustainability of a particular seafood product is usually unavailable at the point of purchase—for both consumers and wholesale seafood buyers. It is estimated that about 30 percent of information (such as species, origin, or catch method) may actually be false, if it does accompany a seafood product at all. All actors in the supply chain agree that traceability is critical for supply chain security, safeguarding sustainability, preventing fraud, and guaranteeing product safety. However, a major portion of seafood lacks information regarding species, origin, farm inputs and catch method in the supply chain. A number of hurdles prevent companies from embarking on traceability options such as, (i) investing in enabling services and systems, (ii) verification systems. Moreover, small farmers lack in the required level of education to implement a record keeping system—a pre-requisite for certification process.

1.3 Perishability management

The perishable nature of seafood and products adds additional complexity to the supply chain distribution process. The postmortem deteriorating process of perishable products cannot be stopped or reversed, but it can be slowed down. Various postharvest strategies are employed to slow the rate of deterioration in order to maximize product's shelf life; among them, monitoring and controlling low temperature is of foremost importance. Generally speaking, foods stored above 41°F provide a rich growth environment for both spoilage organisms and pathogens, resulting in greater risk of both economic loss and an outbreak of food borne illness. Therefore, managing perishable food requires the management of temperature, including refrigerated trucks, containers, warehousing, packaging, display cases; personal training; and additional information about the 'condition' of the products in order for the associated processes to be managed properly. All stages of perishable product movement are important for all members in the supply chain because the product's characteristic is related to both quality and safety. Therefore, forging vertical partnerships with its supply chain and sharing information among members can reduce some risks in the product quality and safety. Furthermore, effective supply chain gives several benefits for customers such as less out of stock (lost sale), fresh product with a longer shelf life, better availability, and potential cost savings.

2 Handling and Distribution

Farmed shrimp is transported to the processing utility after harvesting, either submerged in iced water with air-supply or in crushed ice. Both catch and handling methods have significant impact on the final quality of shrimp. This section compares both handling methods in terms of physical quality and describes key handling activities in the shrimp supply chain.

2.1 Chilling

Shelf-life of shrimp stored in ice has been reported as 4 days; however at −3°C it was 6 days, and at −10°C it was 9 weeks. The *k-value* (rate of deteriorative reaction) and TVB-N are good freshness indices for shrimp shelf-life studies, especially during frozen storage (Yamakata and Low 1995). Vongsawasdi (1996) assessed that during ice storage, giant freshwater shrimp are susceptible to a textural problem called mushiness. It is believed that the cause of the phenomena is mainly due to the activities of enzymes released from hepatopancreas after the death of the shrimp. Garrido et al. (2000) reported that the shrink is the natural product weight loss due to seepage or draining of fluids from seafood tissue. Shrimp have been reported to lose up to 18 percent of the original product weight during 36 hours of ice storage in retail establishments; the shrimp stored on ice at 38°F were judged to have a potential acceptable shelf-life of 17 days; the texture remained firm throughout most of the study. It changed to slightly soft and shrimp got slimy towards the 17th day on ice storage.

Ahmad et al. (2013) reported loss of quality in white shrimps (*Litopenaeus vannamei*) during a cold supply chain (0–8°C for 96 hours). After 84 hours of storage in variable temperature conditions, the level of TVB-N was still within the acceptable range (≤25 mg/100 gm N), but samples were unacceptable due to high microbial growth (>7.5).

2.2 Super-chilling

Commercially, icing plays a major role in slowing down bacteria and enzymatic activities in seafood. The quality changes of northern shrimp (*Pandalus borealis*), stored in different cooling conditions (ice, liquid-ice or salt-water ice at either –1.5°C or 1.5°C) showed the total volatile nitrogen (TVB-N) level decreased during the first day of storage in liquid ice, whereas, the TVB-N level increased with the time of storage in ice or salt-water ice. Total viable counts (TVC) showed that bacteria grew most quickly in shrimp stored in ice and in salt-water ice, followed by those in liquid ice at 1.5°C and –1.5°C, respectively, throughout the storage period. Liquid ice storage at –1.5°C gave the longest shelf-life of shrimp based on sensory analysis (Zeng et al. 2005).

2.3 Freezing

Storage of shrimp in crushed ice is suitable only during transport. For long storage, freezing or other alternative preservation methods are employed during distribution and retail sale. In a period of 3 months of storage at freezing temperature, the total volatile basic nitrogen (TVB-N), pH and cooking loss values are reported to rise gradually. The quality of products also depends on the method and mode of freezing. A comparison of Individual Quick Freezing (IQF) and Block Freezing (BF) methods showed IQF offered better quality retention due to effective heat removal from the product body (Saranakomkul 1999). It was found that total volatile basic nitrogen (TVB-N) and trimethylamine (TMA-N) values rose gradually, while TMA-O decreased with time. High freezing rates are achieved with IQF in which sub-freezing temperatures of –45°C are used. However, maintaining such low temperatures throughout the supply chain is expensive and non-practical. Therefore, storage of products at –18°C or lower is recommended (Morrison 1993).

There are three basic freezing methods available for freezing shrimps: air-blast freezing where a continuous stream of cold air is passed over the product, secondly, plate or contact freezing where the product is placed in direct contact with the freezer plates on which a cold fluid is passed, and thirdly, spray or immersion freezing where the product is placed in direct contact with a fluid refrigerant (Garthwaite 1997).

During the freezing and storage of products, quality changes (such as chemical reactions) can occur, some of which may be desirable, some not desirable (Mallikarjunan and Hung 1997). It has been established that shrimp has the best quality right after harvesting, and after freezing pH, TVB, TVC score were increased. After 3 months of frozen storage at –18°C, pH and TVC did not change significantly, but TVB increased (Sripongpankul 1992; Aran et al. 1995; Ahmad et al. 2013).

In conclusion, if shrimp is frozen in a conventional freezer in a block of ice, the texture, flavor and overall quality of the product will not change for approximately 3–6 months. If prawns are individually quick frozen (IQF), in a freezer which uses either gaseous nitrogen or carbon dioxide to quick freeze, shrimp shelf life in terms of texture, flavor, and overall quality of the product will be approximately 10 months.

Effect of freezing rate

Rapid freezing forms smaller ice crystals, and smaller ice crystals usually mean less damage. However, even slight temperature fluctuations during frozen storage or handling cause thawing and refreezing. The more freeze/thaw cycles a product undergoes, the larger the ice crystals become. As the water in a product thaws and refreezes, the crystal formation ruptures physical structures, promoting water migration, syneresis, emulsion breakdown, and loss of viscosity. Several cycles often cause significant deterioration in flavor and texture (Lyn 1996).

2.4 Thawing

Thawing is an important operation in seafood handling. Thawing is conventionally carried out by leaving products at room temperature or immersing in warm water. The conventional water immersion thawing has the disadvantages of long thawing times and the potential of cross-contamination (John 2002). Recently, direct heating (ohmic heating) and dielectric heating (microwave) have become popular. Ohmic thawing has been reported to be more economical. It helps retain sensory and other quality attributes. Moreover, there is less chance of contamination. Both technologies require a high initial cost but low operational costs. Still, due to limited availability and the risk of over cooking of products, processors are hesitant to convert their processes to modern thawing techniques. Different thawing methods have a significant effect on the quality attributes; for example, texture, color, cooking loss and TVC. Similarly, repeated freezing and thawing reduce shrimp quality in terms of texture scores and moisture loss.

3 Processing

Vongsawasdi (1996) stated the effects of cooking time on giant fresh water shrimp: percent cooking loss increased with cooking time. Shrimp cooked for 5 min. were moderately firm and very juicy, thus obtaining the highest acceptability in sensory evaluation. Erdogdu et al. (2000) reported that the cook-related yield losses for different sizes of white shrimp (*Penaeus vannamei*) showed that increase in the internal temperature results in higher yield losses.

Ready-to-eat (RTE) shrimps are gaining popularity due to convenience in serving and preparation. The products are prepared by deveining, peeling and marinating prior to cooking. The most common preservation technique is freezing. However, other combinatorial preservation hurdles are also being explored. Microbiological safety and sensorial quality are the two major challenges in RTE seafood.

Various strategies are employed to ensure microbiological safety of ready-to-eat shrimps and products. Products stored at low temperature range of 3–12°C coupled with modified packaging atmosphere were acceptable after 15 days of storage. This strategy is useful when strict temperature control is difficult or freezing is non-practical or undesirable, such as during processing, transportation, retail display, or home use; additional antimicrobial hurdles may be necessary to ensure safety (Rutherford et al. 2007).

Irradiation is another candidate in the process of using hurdles to prepare stable, ready to eat shrimp, which included reduced water activity (0.85 ± 0.02), packaging and γ-irradiation (2.5 kGy). The ready to eat shrimp has been reported as microbiologically safe and sensorially acceptable even after 2 months of storage at ambient temperature (Kanatt et al. 2006).

Kaur et al. (2012) studied the effect of high-pressure (100, 270, and 435 MPa for 5) processing on quality and shelf life of black tiger shrimp. They found no significant changes in TVB-N and TMA-N levels of shrimp after processing; however, these significantly increased with storage. Whiteness index increased with pressure intensity while hardness showed a decreasing trend during storage. Lower TVC was observed as compared to the untreated sample. With the optimum pressure level of 435 MPa, a shelf life of 15 days was possible which, without HPP, would be 4–5 days. Although significant results were obtained with the application of HPP, practical aspects are yet to be considered owing to the very high cost of HPP for 15 days of shelf-life as compared to freezing, which can be achieved with much less effort and cost.

4 Quality Evaluation

Seafood, such as shrimps, contains high amount of non-protein nitrogen and are highly susceptible to spoilage and deterioration caused by autolysis and the growth of postmortem microbial population. In order to fully assess quality in terms of freshness, both internal quality parameters as well as external appearance are quantified. Physical appearance of products is basically a manifestation of internal quality. Consumers usually make purchase decisions based on sensorial appearance unbeknownst to the onset of a deteriorative reaction. Therefore, for consumers' safety and product integrity, products are simultaneously evaluated by chemical methods, physical properties measurement, microbial analysis and sensorial appeal.

4.1 Quality evaluation by chemical methods

Chemical indicators have been used to replace more time consuming microbiological methods. Such objective methods should, however, correlate with sensory quality evaluations and the chemical compound to be measured should increase or decrease with the level of microbial spoilage or autolysis (Botta 1995). Total volatile basic nitrogen (TVB-N), for example, is one of the most widely used measurements of seafood quality. It is a general term which includes the measurement of trimethylamine (produced by spoilage bacteria), dimethylamine (produced by autolytic enzymes during frozen storage), ammonia (produced by the deamination of amino-acids and nucleotide catabolites) and other volatile basic nitrogenous compounds associated with seafood spoilage. TVB-N values are not linearly related to the length of time (days) the species being evaluated was stored in ice and it cannot be used to predict the storage life of that species (Botta 1995).

4.2 Physical quality evaluation

Color, texture, water holding capacity are among the parameters of physical quality. Conventionally, physical quality evaluation is based on destructive methods; however, with the advancement in instrumentation, some parameters such as color can be measured inline without prior preparation.

Color

The system used for color measurement is referred to as Hunter color parameters (L^*, a^*, and b^*), and the total change in color (ΔE) is defined as the difference between each parameter at initial and final storage stage and root of the square sum of L^*, a^*, b^*. The translucent appearance of fresh shrimp turns darker with time, which is quantified in terms of the b*-value (yellowness) and L*-value (lightness), during shrimp storage. Color is measured at the time of harvest, prior to cooking, and after cooking to determine the change. The change in color is dependent on sample composition and storage temperature. A zero order relationship has been used to describe the change in b-value over a period of time (Tsironi et al. 2009). Likewise, Ahmad et al. (2013) reported color degradation in cold chain by a zero order reaction with R^2 values >0.9.

Texture

Texture is one of those quality attributes which affect product acceptability and consumer perception (Monaco et al. 2007). Texture depends on the sample's physico-chemical properties and human perception. Bourne (1978) has published extensively on various aspects of texture, its measurement and interpretation of data with relation to the structure of foods. The TPA (texture profile analysis) method based on compression of samples with the texture analyzer is used to objectively evaluate texture with a cylindrical probe of 35 mm diameter. TPA comprises of two consecutive 40% compressions of the sample at a crosshead speed of 12 mm/min.

One of the earlier works done on texture measurement of shrimp showed that it was influenced by pH, cooking time, and length of storage but not by the method of thawing (Ahmed et al. 1972). However, recent works have reported contradictory results (Qingzhu 2003). Vongsawasdi (1996) assessed that during ice storage, giant freshwater shrimp are susceptible to textural condition called mushiness mainly due to the activities of enzymes released from hepatopancreas after the death of the shrimp. Storage temperature was shown to be a crucial parameter in determining the rate of toughening of precooked, frozen lobster. Significant textural changes occurred after 3–4 months at –12°C but only after 9–10 months at –27°C. Texture largely depends on the structural component which is affected by the amount and size of ice crystals during cold storage. There are some quality changes during seafood freezing and frozen storage period.

Qingzhu (2003) reported there were no obvious regular trends in the changes in the springiness, resilience and cohesiveness of the shrimp stored under different conditions (–1.5 to 1.5°C). Springiness and cohesiveness decreased at the beginning of storage and increased again later. Bourn (1978) noted that a large set of data with sufficient number of replications should be considered if data varies too often without any apparent trend.

In general, various studies reported only minor changes in texture. In a storage method comparison study of 3 months in terms of IQF and block frozen shrimp qualities of black tiger shrimp, the texture of raw shrimp did not change over storage. In another study of shrimp quality, storage for a period 3 months at –18°C also revealed no change in texture over time.

However, Vongsawasdi (1996) and more recently, Ahmad et al. (2013), reported significant change in TPA parameters. Vongsawasdi (1996) demonstrated that variation in geometry of testing probe, sample dimensions and temperature and cross head speed of texture analyzer has significant impact on texture analysis. One of the drawbacks of TPA is that the tests are highly dependent on testing parameters and are not comparable. Using data analysis techniques such as multiple linear regression, the number of variables obtained in TPA, hardness (H), springiness (S), gumminess (G), cohesiveness (Co) and chewiness (Ch), can be reduced. Hardness (H) has been reported the most significant parameter (P<0.05) among the others. Traditionally, the Arrhenius relationship has been applied to well defined chemical and microbial reactions; however, due to inconsistency in textural data, it has been rarely applied to model changes in textural properties. Ahmad et al. (2013) reported the application of kinetic parameters, reaction rates (k, $min.^{-1}$) and activation energy (E_a, KJ. Mole^{-1}.$^\circ$K) as the TPA parameters. Furthermore, the possibility of using Artificial Networks has also been explored (Ahmad et al. 2013).

4.3 Microbiological methods

The aim of microbiological examinations of seafood products is to evaluate the total microbial load and possible presence of pathogens, major public health concerns. Total viable count (TVC) does not give any information about eating quality and freshness but, it is a good indication of processing and handling standards maintained during the supply chain, such as temperature abuse. TVC or Standard Plate Count (SPC) is used to determine the total number of aerobic bacteria present at mesophylic temperatures (30°C–37°C). Total numbers of bacteria in terms of total viable count (TVC) grow very quickly in shrimp throughout the storage period. A normal range of 10^2–10^7 CFU (colony forming units)/g in shrimp samples are usually an acceptable standard. The number varies in skin, gills and intestine.

In addition to the assessment of total microbial load, knowledge of Spoilage Specific Organism (SSO) and level of population, i.e., minimum spoilage level and environmental conditions in which the SSO grows, is essential. It has been noted that assessment of SSO in laboratory conditions is not necessarily applicable in real foods (Baranyi et al. 1995). The error of predictive models can be as high as 53%. Therefore, validation of developed models in field trials is crucial. *Pseudomonas* sp. is the most common spoilage organism occurring in all meat types and seafood during supply chain distribution. Its occurrence largely depends on farm conditions, handling conditions, and storage temperature.

4.4 Sensory methods

Despite the subjective nature of the sensory methods of quality evaluation, the industry heavily relies on them due to the potential advantage of their being simple to conduct, cheap and readily providing results for decision making. However, care must be practiced as they are based on the assessment of individuals, their likes and dislikes. The subjectivity and bias can be reduced significantly by proper training and the use of proper descriptors and structured scaling. Nevertheless, developments are underway for instruments capable of measuring parameters such as texture and other rheological properties, and microscopic methods combined with image analysis are being employed to assess structural changes (Wu and Sun 2012; Pathare et al. 2013) and 'the artificial nose' is now used to evaluate odor profile (Botta 1995). Sensory methods relying on trained assessors, i.e., objective sensory methods are required for inspection/grading and use in quality control for the evaluation of freshness and the determination of remaining shelf life of seafood.

4.5 Rapid methods of quality evaluation

Modern instrumentation and data analysis techniques have warranted increased reliability of indirect estimation of quality. With conventional sensory, microbial and chemical analysis, simple and rapid tests based on image analysis, electronic nose, near infra-red (NIR) spectroscopy are being explored to assess seafood quality. The major challenge in using these instruments is careful calibration and the establishment of interrelationship between these methods and traditional methods.

4.5.1 Image analysis

In order to replace tedious length and weight measurements and growth studies for determining age classes, which are routinely conducted by an electronic slide caliper, an image analysis setup is used. An image capturing set-up includes a camera to grab high resolution images and a system to calibrate pixels with the length. The pixel areas of shrimp can be measured automatically and efficiently and they can provide precise and robust estimates of length. Pan et al. (1993) demonstrated that weight of shrimp can be estimated through machine vision which is an important parameter for the grading process. The traditional weight prediction of shelled shrimp is not accurate enough because it ignores shrimp thickness. A multivariate prediction model containing area, perimeter, length, and width was established using an image analysis set-up. An artificial neural network (ANN) was used for the shrimp weight predication which yielded an average relative error of 2.67%. Other applications of image analysis include separation of diseased and damaged shrimps, color sorting and general grading.

4.5.2 Application of E-nose

Evaluation of shrimp freshness in terms of chemical and microbial indicators using an electronic nose (Cyranose 320™) is reported by Salvi (2003). The experiments were set up to measure changes in quality of shrimp during storage at different temperatures (0, 5, 12, 20°C and at the room temperature of about 28°C). The data obtained from the e-nose was trained using ANN to predict (TVBN) and total plate count (TPC). TVBN and TPC appeared to be related and their relationships with time and temperature were established. It is reported that the electronic nose was able to classify samples with different levels of freshness in a particular storage condition using canonical discriminate analysis (CDA) and principle component analysis (PCA). Evidently, even small changes in TVBN and TPC was taken into account by the electronic nose.

In another study, the electronic nose has shown good correlation to TVB-N measurements and sensory analysis (Olafsdottir et al. 2004). In conclusion, an electronic nose has the potential to be used as a new rapid technique to monitor freshness and spoilage of seafood. However, e-nose technology has not made inroads in industry due to tedious calibration of chemical sensors, contamination of sensitive sensor with other volatiles, the need for trained workmanship and difficulty in data analysis.

Shrimp patterns obtained from electronic nose with different freshness levels were determined by conventional chemical method (TVBN value). The prediction of TVBN value from e-nose patterns was carried out by multiple linear regression (MLR), partial least square regression (PLS-R) and artificial neural networks (ANN). In the classification of electronic nose patterns based on arbitrary limits of TVBN values and Euclidean distance, considerable overlapping was detected among the classes. Finally, a three layer back propagation neural network was able to classify patterns with overall success rating of about 93.54% based on two classes corresponding to different TVBN ranges (above and below 35 mg N/100g) (Salvi 2003).

4.5.3 Spectroscopy

As shrimp processing such as cooking and processing can lead to denaturation of proteins, optimization of shrimp processing is, therefore, an essential task to ensure minimal moisture loss and to ensure a high quality of the final product. Traditional analytical methods for physicochemical measurements are, however, time-consuming, sample-destructive and expensive, and are not poorly suited for on-line process monitoring and optimization. Several studies show the potential of using Near Infra-Red Spectroscopy and its variants for on-line process control during the processing of shrimp. The basic principle of these methods is to find and develop strong correlations to various physicochemical properties, such as muscle pH, moisture content, water holding capacity and other important quality factors during the processing. NIR reflectance measurements are performed directly on intact, peeled shrimp muscle over the wavelength range 800 to 2500 nm. Gudjonsdotter et al. (2011) reported a high correlation coefficient of moisture content and water holding capacity with NIR reflectance. Studies have shown the usefulness of NIR spectroscopy in applications such as freshness of seafood, fat content determination, rancidity assessment and evaluation of other physico-chemical properties that are complex, time-consuming, and expensive.

5 Optimization and Predictive Modeling of Quality Parameters

The continuous change in the quality from the time the fresh products leave growers to the time the product reach consumers offers a difficult challenge to quantify. The spoilage process of perishable products begins as soon as they are harvested. Methods are available to reduce the impact on products' quality, but the methods cannot maintain the natural freshness. The most crucial time of a fresh product's life is the time it spends on vehicles since the loss in quality during transportation directly affects the commercial value of the product. Distributors struggle to minimize distribution costs but often without considering the loss due to quality deterioration, as the traditional view of reducing the distribution cost is not valid when perishability is in question.

Tarantilis and Kiranoudis (2001) described that the time period between preparation and trading of fresh products was crucial for both producers and traders due to decline in their quality and loss of sale efficiency. Sloof et al. (1996) divided the quality of perishable product into 3 sub-models, the first one describing the quality of the assignment by customers, the next one describing the physiological behavior of the product, and the last one describing the change in the environment of the product. The environment model describes physical processes, whereas the product model describes complex biochemical processes. The acceptability of a product is an evaluation of the assigned quality in the context of extrinsic properties of the product such as its price in relation to other available products.

A number of approaches have been considered to improve the supply chain of fresh products which indirectly delivers a better quality. Wilson (1996) proposed central purchase which gives additional benefits to retailers in terms of economies of scale, reduction in number of suppliers and thus vehicles, and shorter routes. Tarantilis and Kiranoudis (2001) formulated the distribution of fresh milk as a heterogeneous fixed fleet vehicle routing problem. They provided practical solutions for the improvement of operational performance. However, the problem remained a highly complex problem which could be solved by a threshold-accepting algorithm. An open multi-depot approach was used to solve the distribution problem of fresh meat by Tarantilis and Kiranoudis (2002). Supply Chain Management (SCM) in food industry should be an integration of logistics and product quality. This concept is called Quality Controlled Logistics (QCL). Van der Vorst et al. (2005) attempted to fulfill the need for theory building and developing tools and methods for successful SCM practices. More recently, Osvald and Stirn (2008) attempted to include perishability as part of the overall distribution costs by developing an algorithm for the distribution of fresh vegetables. The problem was formulated as a vehicle routing problem with time windows and time-dependent travel-times (VRPTWTD) where the travel-times between two locations depends on both the distance and on the time of day.

5.1 Quality Loss in Raw and Frozen Shrimp

In the case of raw and frozen shrimp, overall quality and shelf life of whole processed shrimp is the composite result of the following changes that occur simultaneously: color fading (Chandrasekaran 1994; Ghosh and Nerkar 1991); lipid oxidation (Riaz and Qadri 1990); denaturation of protein (Bhobe and Pai 1986); sublimation and recrystallization of ice (Londahl 1997); increase in volatile basic nitrogen (Riaz and Qadri 1990; Yamagata and Low 1995); and reduced water binding capacity, as well as microbial spoilage (Bhobe and Pai 1986). The rates of these changes depend on storage temperature. Therefore, it is important to prevent temperature fluctuations during transportation and storage, and to avoid thawing and re-freezing to maintain the quality of frozen shrimp. The highest shrimp quality can be obtained if shrimp is frozen immediately after harvest (Fennema et al. 1975). Tsironi et al. (2009) investigated the effect of variable storage conditions on shelf life and quality characteristics of frozen shrimp. Color change, microbial growth, bio-chemical changes were modeled by apparent zero order equations and showed high dependence on temperature. Gormley et al. (2002) compared the effect of fluctuating and constant frozen storage temperature regimes on quality parameters of selected products and found a straight line relationship of quality loss with time. They found a significant difference in the final quality of products undergoing constant and fluctuating storage temperatures. The temperature dependence of quality deterioration was adequately modeled by the Arrhenius equation $Q_t = Q_0 e^{-kt}$, where Q_t and Q_0 are values of quality indication at time $= t$

and at time = 0, respectively. k is a slope of the rate of quality loss. In the secondary level of modeling, the dependence of parameter k on operating temperature is determined. A number of k values (rates of reaction at different temperatures) yield activation energy (E_a), the energy required to overcome a deteriorating reaction. E_a values for changes in bio-chemical properties of shrimp have been reported (Tsironi et al. 2009), however, these are rarely reported for physical changes. For instance, Ahmad et al. (2013) has reported activation energy (–10.03 to –31.15 KJ/mole, °K) required for textural changes in shrimp stored above freezing temperatures (0–8°C) in terms of textural profile analysis (TPA). The scarcity of k-values in literature is mainly due to the difficulty in data collection in sub-freezing temperatures.

5.2 Total Distribution Cost Models for Fresh Shrimp

In the traditional supply chain, shrimp is transported alive from farms to processing centers. Transportation from a farm to a factory determines the freshness, which is one of the important factors of commercial value; this largely depends upon the speed of transportation, conditioning during transport (forced air or iced) and on timely delivery of the load. Then, the load is processed through the means of chilling, freezing, canning, semi-cooking or ready-to-eat by manufacturer. The manufacturer then transports the processed shrimp to distribution centers (Noomhorm et al. 2008).

Under assumptions that loss of the goods' quality during delivery is a linear function of time and that arrival time is defined by the time of delivery, Ahmad et al. (2009) proposed mathematical models to represent various components of shrimp logistics to provide the basis for cost and, eventually, the loss in quality minimization procedure: (i) Transportation costs model (TC_{fp}) between farms and processing plants is comprised of minimizing loading and unloading costs, vehicle cost and the distance cost between farm and processing plant (ii) The transport cost from processing plant to distributor (TC_{pd}) will be based on the cost for loading & unloading, vehicle cost and distance (iii) Loss in quality is directly proportional to commercial or saleable quantity, i.e., *Quality loss* $\propto$ *(1 - commercial value)*. It means that if the load can be sold at its full anticipated price, this factor would be zero. However, if it is sold at 80% of the price due to the loss of quality, the cost of quality loss will be 20% (iv) However, it is quite difficult to define the actual rate of deterioration of the quality of a load over time, as it is depended on storage conditions and the load's condition after harvest. The cost of quality loss can be expected from the time spent in loading and the number of loadings. By minimizing the term of $(T_i)(L_i)$, the total costs of quality loss can actually be reduced (v) Since time spent on a vehicle is directly related to the quality loss, therefore, a term 'Total Penalty Cost'

(TC_p) is introduced; if the goods are supplied within the required time, the penalty factor will be zero (vi) Finally, the total distribution costs (TDC) model for the fresh products (in this case, raw & fresh shrimp) is given as the objective function of minimization for various conditions. The total costs model includes transportation costs, the cost of quality loss and penalty cost due to the delay in transportation. All models (i–vi) are given in box 1 and their explanations in box 2. The efficiency of the distribution process of perishable goods, where the amount of time a food spends on the vehicle is the most significant and crucial factor, represents an important problem in fresh produce distribution. Therefore, the loss of quality is directly proportional to the absolute loss in saleable quantity.

Box 1. Transportation cost minimization models (*modified from Ahmad et al. 2009*).

$$Min \ TC_{fp} = min \sum_{f=1}^{n}\sum_{p=1}^{m}\sum_{i=1}^{k} \left[L_f + \left(D_{fp}\right)\left(V_{fpi}\right) + U_p \right](T) \tag{i}$$

$$Min \ TC_{pd} = min \sum_{p=1}^{m}\sum_{d=1}^{c}\sum_{i=1}^{k} \left[L_p + \left(D_{pd}\right)\left(V_{pdi}\right) + U_d \right](T) \tag{ii}$$

$$TC_{QL} = \sum_{i=1}^{k}(T_i)\,(L_i)\,(CQL) \tag{iii}$$

$$TC_P = \sum_{i=1}^{k} |AAT_i - RAT_i|\,(DCF) \tag{iv}$$

$$TDC = TC_{fp} + TC_{pd} + TC_{QL} + TP_c \tag{v}$$

Box 2. Explanation of model parameters.

f	= farm number, f = 1, 2, ..., n
p	= plant number, p = 1, 2, ..., m
i	= type of vehicle, i = 1, 2, ..., k
L_f	= supplier's cost of loading per tonne at a farm (f)
D_{fp}	= distance from a farm to a plant (kilometer, km)
V_{fpi}	= vehicle cost per ton-km from a farm to a plant (including empty return)
U_p	= supplier's cost of unloading per tonne at plant (p)
T	= total weight of the product (tonne)
L_p	= supplier's cost of loading per tonne at plant (p)
D_{pd}	= distance from plants to DCs (kilometre)
V_{pd}	= vehicle cost (rent or depreciation) per tonne-km from plant to DC (including empty return)
U_d	= supplier's cost of unloading per tonne at destination (d)
TC_{QL}	= total costs of quality loss
T_i	= the time spent in loading (hour) in vehicle i
L_i	= the number of loadings in vehicle i
CQL	= cost of quality loss per hour
TP_c	= total penalty costs
AAT_i	= actual arrival time (hour)
RAT_i	= required arrival time (hour)
DCF	= delay cost factor (per hour)

5.3 Generic modeling

The last decade has witnessed a number of new trends in modeling optimization problems. The use of computational-based optimization algorithms is one such example. Originally, the use of stochastic searching methods evolved for solving non-differentiable, non-continuous and multi-modal optimization problems based on Darwin's natural selection principle (Enitan and Adeyemo 2011). Jindal and Chauhan (2002), and Enitan and Adeyemo (2011) reviewed applications of evolutionary algorithms such as Artificial Neural Networks (ANN) in food processing. Unlike regression, the ANN does not depend on distributional assumptions of input and target variables, i.e., no exact mathematical functional form of the model are required. This greatly simplifies the modeling procedure, especially for a phenomenon where no direct physical relationship among variables can be established. Therefore, ANN has the capability to accurately predict target variables without having to understand the underlying physical or chemical phenomena. Geeraerd et al. (1998) applied artificial neural networks as a non-linear modular modeling technique to describe bacterial growth in chilled food products and reported promising results. Commonly used multilayered feed forward neural network architecture and a nonlinear, sigmoid transfer function are considered sufficient for solving a nonlinear problem. There can be only one neuron (a quality indicator such as storage time) in the input layer. The nodes in each layer accept the input values, and successive layers of nodes receive inputs from previous layers (hidden layers). The number of neurons in the hidden layers affects the output error function during the network training. Too few hidden neurons may limit the ability of the neural network to model the process, and too many can result in mimicking the noise present in the dataset (Shmueli et al. 2007). In frozen seafood industry, prediction of freezing and thawing times are of significant importance. Mittal and Zhang (2000) used a neural network to predict freezing and thawing times for a number of food products including shrimps. In their study, product dimensions and thermal properties were used as input variables. In this way, a complex problem that is based on heat transfer principles could be solved with convenience. Similarly, Goni et al. (2008) predicted food freezing and thawing times using ANN and genetic algorithm (GA).

Ahmad et al. (2014) applied a genetic algorithm (GA) to optimize an Artificial Neural Network (ANN) model to predict end-of-storage quality parameters of frozen shrimp (*Litopenaeus vannamei*). Input (independent) variables were freezing rate, thawing rate, storage time, and product dimensions. The output or dependent variables were physical properties such as CIE Colour L*a*b* values, and textural properties (hardness, cohesiveness and chewiness). In comparison with Multiple Linear

Regression (MLR) and ANN trained with back propagation (BP) algorithm, the GANN model showed much better prediction, in terms of RMSE and highest R^2.

In situations where a categorical response of network is desired, such as dividing goods into different classes or grades based on quality, evolutionary algorithms are an excellent alternative. Classification techniques such as K-nearest neighbor, naïve Bayes, decision tree and support vector machines (SVM) have been explored (Ahmad et al. 2010, 2014).

6 Monitoring Temperature During Distribution of Shrimp

The data generated through a monitoring system can be used to develop empirical quality models and to establish temperature dependency of most of the quality deteriorating reactions. In addition, empirical models are needed for developing quality assignment models representing a homogeneous group of consumers for their quality perception (Sloof et al. 1996). White (2007) reported that the temperature in 30% of trips from the supplier to the distribution center in temperature-controlled shipment rises above the specified temperature. Therefore, monitoring temperature fluctuations is important.

A number of data collection strategies are used in cold chain management. The most basic solution for data collection is the use of data loggers with temperature measuring sensors. However, these data logger devices are cumbersome to use, require hard drives for data storage and prone to damage. Moreover, these monitoring devices are not suitable for translating data into useful information. For smart supply chain solutions, the data capturing and acquisition system should be able to provide meaningful conclusions instantly.

6.1 TTI

Time-Temperature Integrators (TTI), or 'smart labels' as commonly called, respond to time-temperature dependent change. The TTI method is an indirect estimation of quality with the help of a temperature sensitive chemical sensor that is calibrated with the underlying deteriorative reaction (Giannakourou et al. 2006). In order to predict the shelf-life of frozen products, suitable TTIs can be calibrated with the temperature dependent change for the real distribution of frozen foods. Wells and Singh (1988) and Giannakourou and Taoukis (2002), demonstrated the use of TTI for prediction of quality level and development of an optimized inventory management and stock rotation system on Least Shelf Life First Out (LSFO) issuing policy. TTIs have a considerable role in building consumer confidence; however, this technology has not been fully adopted in the context of supply chain management where information sharing across multiple actors is needed.

6.2 RFID

The rise of Radio Frequency Identification (RFID) Technology has provided the much needed visibility in a supply chain. In addition, the technology has potentially added newer dimensions in monitoring a cold chain in an economical way. The so called 'active' RFIDs which are enabled with temperature sensors and storage memory are replacing data loggers. With these tags attached, when a shipment undergoes real temperature fluctuation scenarios, the recorded data is extracted at the destination and with suitable temperature dependent shelf-life models, the remaining shelf-life is calculated (Ruiz-Garcia 2008). There are, however, a number of issues such as low reading range and frequency band compatibility, for which research is under way. Ogasawara and Yamasaki (2006) demonstrated application of embedded temperature sensors RFID tags for effective risk management throughout transportation processes.

7 Supply Chain Security, Traceability and Technological Requirements

7.1 Supply chain security

A strategy to measure competency of food supply chain security has been proposed by defining capabilities and translating them into competencies. The proposed model was demonstrated by implemention in frozen shrimp supply chain (grower, processor, wholesale/export, and supermarket) Ahmad and Komolavanij (2010). The results were comparable with the world class benchmark of supply chain security (Fig. 2). The approach provided a step towards standardization of supply chain security processes by defining competencies and understanding their impact on security performance. The benchmarking data does not exist for every segment; therefore, a well-planned comprehensive study is needed to measure performance metrics and benchmark the supply chain security of frozen shrimp industry from an Asian perspective.

The overall gap between the international benchmark reported by Helferich and Cook (2003) and the mean of scores (+1 S.D.) given by survey respondents was –0.11, which is quite close. However, comparison of each capability reveals that Thai shrimp industry greatly lacks in Process Management (–0.91), Management Technology (–0.88) and Communication

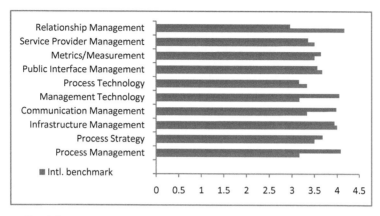

Figure 2. Capability comparison of Thai shrimp industry and international benchmark of food supply chain security. Ahmad and Komolavanij (2010).

Color image of this figure appears in the color plate section at the end of the book.

Management (–0.65). This shows that the Thai Shrimp Industry is not able to effectively use information management systems. On the other hand, the Thai Shrimp Industry greatly surpassed the tnternational benchmark in Relationship Management (+1.21). The integrated supply chain security technological requirements include: timely access for data verification, efficient verification process, scalability, validity of ID numbers, termination of IDs, information exchange protocols, heuristics and process logic.

7.2 Traceability

It is customary that consumers hold food companies and their management responsible for delivering safe food. Moreover, mass distribution of products without online traceability leads to delayed responsiveness and costly recalls.

Traceability is a verifiable method of conducting product identification from the growers, through all the steps in the supply chain, up to the retailers and eventually the customers. Traceability comprises two components, tracking and tracing. Tracking involves monitoring a product and all its inputs through all steps and agents along the supply chain, whereas tracing involves following a product from any point in the supply chain back to its origin. There are many reasons for tracing a food product back to their source. The most important, of course, is to find the source and origin of the food. In case of a safety breach, the SC actors should be able to locate and retrieve the entire product as quickly as possible. Governments worldwide are making stricter new regulations that require all players in the food supply chain to know more about where their product came from, how their product was processed, which chemicals or pesticides were used, when their product was harvested, slaughtered, or processed, and maybe what these animals were fed. The U.S. Bio-terrorism Act 2003, the EU General Food Law 178/2002–2005, and the Japan Food Sanitation Law 2003, ISO 22000:7.5.3 (traceability) and Mandatory Country of Origin Labeling (COOL) are some examples of legislations that are re-shaping the fresh food industry. Traceability is not about introducing a new procedure within a supply chain (such as HACCP and other benchmarking procedures), but is about extracting all the information from the existing processes.

Case study: Designing of traceability system of frozen shrimp supply chain

Upon receiving of shipment at the manufacturer's facilities, a pre-installed electronic trace back system enables manufacturers to learn everything they need to know about how, when, and where the shrimp was grown. Manufacturers will also use the traceability software to manage the processes within the processing plant, while recording vital information that can be traced by anybody in the supply chain. Use of the traceability software will vastly improve the efficiency of the plant. The historical data being collected could also help tremendously in forecasting. Similarly, retailers will also be able to gain access to all the information collected up to the point that the shrimp arrives at their stores. That information would be vital in conducting a quick and targeted recall if the need arises. Without the traceability software providing them with all the detailed and precise information, the impact of a full scale recall could be very costly. All those seemingly separate functions of the traceability software work together to provide the most comprehensive tracing and tracking system; almost any food industry will find it hard to remain competitive without a functional trace and track system in place, in the not too distant future. The fundamental basis for a traceability system is to track systems in place, tracing products through farming, processing and logistics, addressing both inputs and activities.

Traditionally, the record keeping is done manually, using paper checklists. With the rise of affordable Radio Frequency Identification (RFID) tags, information can be stored. In fact, currently, RFID is the backbone of the traceability system. It is an automatic identification method, relying on storing and remotely retrieving data using devices called RFID tags or transponders. An RFID tag is an object that can be applied to or incorporated into a product, animal, or person for the purpose of identification and tracking using radio waves. Some tags can be read from several meters away and beyond the line of sight of the reader. Most RFID tags contain at least two parts. One is an integrated circuit for storing and processing information, modulating and demodulating a (RF) signal, and other specialized functions. The second is an antenna for receiving and transmitting the signal. Chipless RFID allows for discrete identification of tags without an integrated circuit, thereby allowing tags to be printed directly onto assets at a lower cost than traditional tags. The data from the farm, logistics by truck and the processing is recorded using the RFID system. At the factory or farm receiving the raw material, data is put in an RFID card on the particular lot. In the processing line, shrimp from different sources is also received; therefore the RFID card will be updated in the basket for each lot with the recording data for identification of that lot. The RFID card on the shrimp basket will be updated in every step of shrimp processing until the process is completed. Finally, the RFID card with that shrimp lot is passed through the packing process. The RFID card is patched with the package, pallet or container, depending on the designed system.

RFID written database structure and information retrieval/management software are the core requirements. Though commercial software such as 'OpsSmart™' suite of applications (FXA Co. Ltd.), GTNet (IBM), Oracle, Mettler Toledo and MES (Light House) are available they come with a hefty price tag often not affordable or suitable for medium to small business owners. However, commercial packages provide additional flexibility such as data integration into company's existing ERP or management system.

In order to provide information in public domain for free access to RFID practitioners to construct a database for shrimp traceability system, Wannaiampikul (2008) has outlined step-by-step instructions (Fig. 3). Database software such as Microsoft Access can be used for this purpose.

Create 2 input forms:

1. For the farmers to input the farm data and
2. For factory personnel to input processing and packing information

At farm level:

- Farm name, code, ID number, telephone number, and address.
- Details of pond on each farm, including the status of the pond such as the date of use, area, production capacity, density, farm system, source of water and water system.
- Feed and substance that are used in cultured shrimp: production lot, FDA number, production date, expiry date, quantity, invoice number, purchasing number and receiving date are recorded.

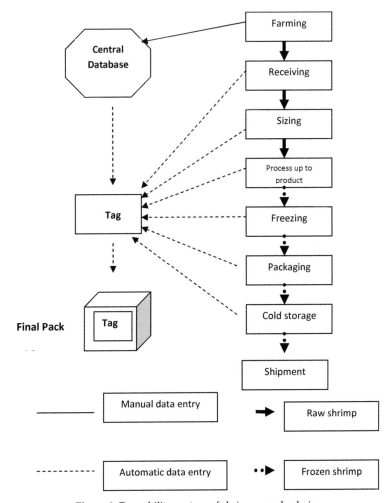

Figure 3. Traceability system of shrimp supply chain.

- Batch information: all data about culturing is recorded in this page. The details of shrimp breeding include the original country, laboratory's name and result, hatchery's name and quantity. The record of resources used is also stored in the program. The customer can be informed about the chemicals which the farmer puts in the pond and their quantity.
- Results of water inspection and sampling size: This is the most important information for inspection of quality in the process of shrimp farming.
- The result from the laboratory is reported: When shrimp are cultured in ponds, farmers have to inspect many kinds of quality of shrimp. It is

not only the water, but also the health of shrimp which are inspected. Farmers may send shrimp samples, which include the drug residue, for inspection to the Department of Fisheries (DOF) or a private laboratory.

- Harvesting details can be recorded.
- Buyers' information such as contract farming with factory or open bidding.

At factory level:

- Raw material receipt that describes on receiving, details such as shrimp information, supplier details and item details. They include the vehicle plate number, time of arriving and unloading, farm name, batch code, quantity and others. The most important information is the shrimp raw material lot code because it identifies the shrimp lot and it is given as an ID number.
- Raw material analysis, i.e., inspection report of shrimp raw material.
- Sizing process refers to tabulating the numbers in various sizes, weight of each size and product name. In addition, total weight before and after sizing is recorded automatically.
- De-head process: data of date, time and weight. The start—end time and date are recorded automatically, and also the total weight of shrimp before and after this process.
- Peeling: date, time and weight. The start—end time and date are recorded automatically, and also total weight of shrimp before and after this process.
- Dipping, soaking, marinating process includes the details of raw material used in this process, including lot number. The duration of this process and weight of shrimp are automatically recorded.
- Cooking process describes the temperature range, the time duration of boiler and fryer machines among others. In the cooking process, some raw materials are used and mentioned, including lot number.
- Inspection of cooking process in which the sampling is recorded every time.
- The freezing process describes the type of freezer machine, the temperature, the duration, and other details.
- The inspection of freezing process refers to the core temperature of shrimp sample and machine. The sampling time is recorded every time.
- The packing process page that has remarks about the packing details. The package type and lot number are recorded and also the sealing machine code and name. The product and packaging details are described.

- During the packing process, the inspection of finished goods is recorded. This page is called FG lab analysis. A shrimp package is taken at random from the production line for inspection in terms of chemical and microorganism analysis. This template shows the result of the analysis. In addition, each importing country has different standard regulations.
- After the packing process, every pack of shrimp has to pass through a metal detector. This page is for the metal detection process. Testing details are recorded with sampling time. This is the last CCP in HACCP.
- Physical inspection: This template shows the result of inspection. A shrimp package is taken at random from the production line for inspection of physical characteristics such as uniformity, weight, number of shrimp per pack, and the percentage of defect. All these are the customers' requirements that the factory has to follow.
- The shipment process describes shipment details. It includes the port destination and loading, time of departure and arriving, packing, customer name and code, freight type, container number, temperature and total cartons. Besides this, the expiry date and lot number are also shown.

Framework of technology selection

One important decision for chain actors is to select the best system to suit the supply chain from several possible temperature monitoring and traceability alternatives. A number of monitoring solutions are commercially available for the shrimp industry; however, the complex nature of supply chain demands a suitable solution that is technologically appropriate and financially affordable. Raab (2011) outlined three dimensions on which appropriate selection of technology is made. These dimensions are based on technological, organizational and functional requirements of a supply chain. For requirement determination, a scoring method (0–100) was employed where the preference level of a heterogeneous system was determined by weight coefficients.

Summary

This chapter reviews key activities of a modern supply chain of frozen and raw shrimp which includes postharvest processing quality evaluation with an emphasis on non-destructive and rapid methods of estimation. In addition to conventional mathematical modeling techniques, the use of evolutionary algorithms for the prediction of final quality level at the retailer's end is also reviewed. Traceability and temperature monitoring are the two crucial activities of a modern supply chain; an easy to develop and implementation system is detailed for the benefit of practitioners and the knowledge of others. Two case studies: comparison of two handling methods and; traceability system are given. Finally, criterion for appropriate technology selection on technological, organizational and functional level is discussed.

References

Ahmad, I. and S. Komolavanij. 2010. Competence measurement of Thai frozen shrimp industry for implementation of supply chain security. GCMM2010, King Mongkut's University of Technology North Bangkok Press, Bangkok, Thailand (http://www.gcmm2010.org).

Ahmad, I., C. Jeenanunta and S. Komolavanij. 2013. Characterization of Quality Degradation during Chilled Shrimp (Litopenaeus vannamei) Supply Chain. International Food Research Journal. 20(4): 1833–1842.

Ahmad, I., C. Jeenanunta, P. Chanvarasuth and S. Komolavanij. 2014. Prediction of physical quality parameters of frozen shrimp (Litopenaeus vannamei): An Artificial Neural Networks and Genetic Algorithm Approach. Food and Bioprocess Technology. 7: 1433–1444.

Ahmad, I., S. Komolavanij and P. Chanvarasuth. 2010. Prediction of raw milk microbial quality using data mining techniques. Agricultural Information Research. 19(3): 64–70. www.jstage.jst.go.jp on 23/10/2012.

Ahmad, I., S. Komolavanij and P. Chongphaisal. 2009. Quantification of quality loss during distribution of fresh products—a proposed study of frozen shrimp industry in Thailand. Proceedings of International Symposium on Scheduling 2009, 256–259. The Japan Society of Mechanical Engineers. Japan.

Ahmed, E.M., E.C. Fluck and R.A. Dennison. 1972. Textural properties of irradiated tomatoes. J. Texture Stud. 3: 115–121.

Aran H-Kittikun, P. Singrat and S. Chaiwatcharakul. 1995. Effect of post harvesting techniques on the quality of frozen shrimp (*Penaeus monodon*). Prince of Songkla University. 51 p.

Baranyi, J., T.P. Robinson, A. Kaloti and B.M. Mackey. 1995. Predicting growth of Brochotrix thermosphacta at changing temperatures. Intl. J. of Food Microbiology. 27(1): 6–75.

Bhobe, A.M. and J.S. Pai. 1986. Study of the properties of frozen shrimps. Journal of Food Science and Technology. 23: 143–147.

Botta, J.R. 1995. Evaluation of Seafood Freshness Quality VHC Publishers, Inc., New York. 180 p.

Bourne, M.C. 1978. Texture Profile Analysis. Food Technology. 32(7): 62–72.

Chandrasekaran, M. 1994. Methods for preprocessing and freezing of shrimps: A critical evaluation. Journal of Food Science and Technology. 31(6): 441–452.

Enitan, A.M. and J. Adeyemo. 2011. Food processing optimization using evolutionary algorithms. African Journal of Biotechnology. 10(72): 16120–16127. DOI: 10.5897/AJB11.410.

Erdoğdu, F. and M.O. Balabana. 2000. Thermal processing effects on the textural attributes of previously frozen shrimp. Journal of Aquatic Food Product Technology. 9: 1–4.

Fennema, O.R., M. Karel and D.B. Lund. 1975. Principles of food sciences Part II. Physical principles of food preservation. New York: Marcel Dekker, Inc.

Garrido, L.R., R.A. Benner, P. Ross and W.S. Otwell. 2000. Assessing product quality, shelf-life and consumer acceptance for freshwater, farm raised shrimp (*Litopenaeus vannamei*) University of Florida, Aquatic Food Products Lab.

Garthwaite, G.A. 1997. Chilling and freezing of fish. pp. 93–118. *In*: G.M. Hall (ed.). Fish Processing Technology. 2nd ed. Blackie Academic and Processional, London.

Geeraerd, A.H., C.H. Herremans, C. Cenens and J.F. Van Impe. 1998. Application of artificial neural networks as a non-linear modular modeling technique to describe bacterial growth in chilled food products. International Journal of Food Microbiology. 44: 49–68.

Ghosh, S. and D.P. Nerkar. 1991. Preventing discoloration in small dried shrimps. Fleischwirtschaft. 71: 789.

Giannakourou, M.C., P.S. Taoukis and J.E. Nychas. 2006. Monitoring and Control of the Cold Chain. *In*: D.-W. Sun (ed.). Handbook of Frozen Food Processing and Packaging. CRC Press, Taylor & Francis Group.

Giannakourou, M.C. and P.S. Taoukis. 2002. Systematic application of time–temperature integrators as tools for control of frozen vegetable quality. Journal of Food Science. 67(6): 2221–2228.

Goni, S.M., S. Oddone d, J.A. Segura, R.H. Mascheroni and V.O. Salvadori. 2008. Prediction of foods freezing and thawing times: Artificial neural networks and genetic algorithm approach. Journal of Food Engineering. 83: 164–178.

Gormley, R., T. Walshe, K. Hussey and F. Butler. 2002. The effect of fluctuating vs. constant frozen storage temperature regimes on some quality parameters of selected food products. LWT Food Science and Technology. 35: 190–200.

Gudjónsdóttir, M., Á. Jónsson, A.B. Bergsson, S. Arason and T. Rustad. 2011. Shrimp Processing Assessed by Low Field Nuclear Magnetic Resonance, Near Infrared Spectroscopy, and Physicochemical Measurements—The Effect of Polyphosphate Content and Length of Prebrining on Shrimp Muscle. Journal of Food Science. 76(4): 357–367.

Helferich, O.K. and R.L. Cook. 2003. Securing the Supply Chain (Chicago: Council of Logistics Management).

Jindal, V.K. and V. Chauhan. 2002. Neural network approach to modeling in food processing operations. *In*: Joseph Irudayaraj (ed.). Food Processing Operations Modeling: Design and Analysis. Marcel Dekker Inc.

John, S.R., M.O. Balaban and D.A. Luzuriaga. 2002. Comparison of Quality Attributes of Ohmic and Water Immersion Thawed Shrimp Journal of Aquatic Food Product Technology. 11(2): 3–11.

Kanatt, S.R., S.P. Chawla, R. Chander and A. Sharma. 2006. Development of shelf-stable, ready-to-eat (RTE) shrimps (*Penaeus indicus*) using γ-radiation as one of the hurdles. LWT - Food Science and Technology. 39(6): 621–626.

Kaur, B.P., N. Kaushik, P.S. Rao and O.P. Chauhan. 2012. Effect of high-pressure processing on physical, biochemical, and microbiological characteristics of black tiger shrimp (*Penaeus monodon*). Food and Bioprocess Technology. 6(6): 1390–1400.

Le Ha, N. 1997. Changes in freshness of shrimp (*Penaeus simisulcatus*) during storage at various temperatures. M.S. thesis, Asian Institute of Technology, Thailand.

Londahl, G. 1997. Technological aspects of freezing and glazing shrimp. INFOFISH International. 3: 49–56.

Lynn A. Kuntz. 1996. Food product design. http://www.foodproductdesign.com/archive.

Mallikarjunan, P. and Y.-C. Hung. 1997. Physical and Ultra structural measurements. pp. 313–339. In Erickson, Quality in Frozen Food. Chapman and Hall, USA.

Mittal, G.S. and J. Zhang. 2000. Prediction of freezing time for food products using a neural network. Food Research International. 33: 557–562.

Monaco, R.D., R. Cavella and P. Masi. 2007. Journal of Texture Studies. 39(2): 129–149.

Morrison, C.R. 1993. Fish and shellfish. pp. 197–235. *In*: C.P. Mallett (ed.). Frozen Food Technology. Blackie Academic and Processional, London.

Noomhorm, N., S. Wannaiampikul and I. Ahmad. 2008. Development of a Supply Chain Management System for Frozen Shrimp in Thailand. A report submitted to Asian Productivity Organization (APO) Tokyo, Japan.

Ogasawara, A. and K. Yamasaki. 2006. A temperature-managed traceability system using RFID tags with embedded temperature sensors. NEC Technical Journal. 1,2: 82–86.

Olafsdottir, G., P. Nesvadba, C.D. Natale, M. Careche, J. Oehlenschlager, S.V. Tryggvadottir et al. 2004. Multisensor for fish quality determination. Trends in Food Science and Technology. 15: 86–93.

Onyuksel, H. and I. Rubinstein. 2001. Materials and methods for making improved micelle compositions. U.S. Patent # 6,217,886.

Osvald, Ana and Lidija Zadnik Stirn. 2008. A vehicle routing algorithm for the distribution of fresh vegetables and similar perishable food. Journal of Food Engineering. 85(2): 285–295.

Pan, B.S. and W.T. Yeh. 1993. Biochemical and morphological changes in grass shrimp (*Penaeus monodon*) muscle following freezing by air blast and liquid nitrogen methods. Journal of Food Biochemistry. 17: 147–160.

Pathare, P.B., U.L. Opara and A. Fahad Al-Said. 2013. Colour measurement and analysis in fresh and processed foods: a review. Food Bioprocess Technology. 6: 36–60/DOI.

Qingzhu, Z. 2003. Quality indicators of Northern shrimp (*Pandalus borealis*) stored under different cooling conditions. The United Nations University. Tokyo, Japan: UNU-Fisheries Training Programme.

Raab, von V. 2011. Assessment of novel temperature monitoring systems for improving cold chain management in meat supply chains. Doctoral Dissertation. University of Bonn.

Riaz, M. and R.B. Qadri. 1990. Time–temperature tolerance of frozen shrimp 2. Biochemical and microbiological changes during storage of frozen glazed shrimps. Tropical Science. 30(4): 343–356.

Ruiz-Garcia, L. 2008. Development of monitoring applications for refrigerated perishable goods transportation. *In*: Ingenieria Rural. Madrid: Universidad Politécnica de Madrid.

Rutherford, T.J., D.L. Marshall, L.S. Andrews, P.C. Coggins, M.W. Schilling and P. Gerard. 2007. Combined effect of packaging atmosphere and storage temperature on growth of Listeria monocytogenes on ready-to-eat shrimp. Food Microbiology. 24(7–8): 703–710.

Salvi, A.D. 2003. Study of quality deterioration of shrimp by chemical, microbial and electronic nose analysis. AIT Master's Thesis. Asian Institute of Technology, Thailand.

Saranakomkul, P. 1999. Quality changes in black tiger shrimp in relation to water salinity, cooking time and freezing method. Master's Thesis, Asian Institute of Technology, Thailand.

Shmueli, G., N.R. Patel and P.C. Bruce. 2007. Data mining for business intelligence. Wiley.

Sloof, M., L.M.M. Tijskens and E.C. Wilkinson. 1996. Concepts for modeling the quality of perishable products. Trends in Food Science and Technology. 7: 165–171.

Sripongpankul. 1992. Quality changes of black tiger shrimp (*Penaeus monodon F.*) during Harvesting and Frozen Storage. Master thesis. Prince of Songkla University. 98 p.

Tarantilis, C.D. and C.T. Kiranoudis. 2001. A meta-heuristic algorithm for the efficient distribution of perishable foods. Journal of Food Engineering. 50(1): 1–9.

Tarantilis, C.D. and C.T. Kiranoudis. 2002. Distribution of fresh meat. Journal of Food Engineering. 51(1): 85–91.

Tsironi, T., E. Dermesonlouoglou, M. Giannakourou and P. Taoukis. 2009. Shelf life modeling of frozen shrimp at variable temperature conditions. LWT - Food Science and Technology. 42: 664–671.

Van der Vorst. 2007. Proceedings of the 2005 Winter Simulation Conference. pp. 1658–1667.

Vongsawasdi, P. 1996. Effects of handling and preservation methods on qualities of giant freshwater prawns. Ph.D. thesis, Asian Institute of Technology, Thailand.

Wannaiampikul, S. 2008. Development of a supply chain management system for frozen shrimp in Thailand AIT Master's Thesis, Asian Institute of Technology, Thailand.

Wells, John Henry and R. Paul Singh. 1989. A quality-based inventory issue policy for perishable foods. Journal of Food Processing and Preservation. 12(4): 271–292.

White, J. 2007. How Cold Was It? Know the Whole Story. Frozen Food Age. 56(3): 38–40. *In*: L. Ruiz-Garcia (2008). Development of monitoring applications for refrigerated perishable goods transportation. Ingenieria Rural. Madrid: Universidad Politécnica de Madrid.

Wilson, N. 1996. Supply chain management: a case study of a dedicated supply chain for bananas in the UK grocery market. British Food Journal. 1(2): 28–35.

Wu, D. and D. Sun. 2012. Colour measurements by computer vision for food quality control—A review. Trends in Food Science & Technology, Inpress, 1-16. 10.1007/s11947-012-0867-9.

Yamakata, M. and L.K. Low. 1995. Banana Shrimp, *Penaeus merguiensis*, Quality changes During Iced and Frozen Storage. J. Food Sci. 60(4): 721–726.

Zeng, Q.Z., K.A. Thorarinsdottir and G. Olafsdottir. 2005. Quality changes of shrimp (Pandalus borealis) stored under different cooling conditions. J. Food Science. 70(7): 459–466.

13

Anti-aging & Immunoenhancing Properties of Marine Bioactive Compounds

Ranithri Abeynayake and *Eresha Mendis**

1 Introduction

Emerging research evidence regarding the impact of diet on human health beyond the basic nutrition has aroused the curiosity of consumers. Marine based bioactive compounds, in particular, are believed to provide a number of health benefits. The marine ecosystem covers more than 70% of the earth's surface but represents 95% of the biosphere with phenomenal biodiversity (Faulkner 1995). Therefore, marine bioactive compounds can be derived from a number of sources including marine plants, microorganisms and by-products of the fish industry. Many marine organisms live in complex, competitive and aggressive habitats exposed to extreme conditions and in adapting to new environmental surroundings, they produce a wide variety of biologically active secondary metabolites which cannot be found in other organisms. While the effect of these compounds on the human body may be very small over relatively short periods, they could contribute significantly to health when they are consumed throughout one's life as a part of the daily diet (Biesalski et al. 2009).

Department of Food Science & Technology, Faculty of Agriculture, University of Peradeniya, Peradeniya (KY 20400), Sri Lanka.
  Emails: aranithri@yahoo.com, ereshamendis@yahoo.com
* Corresponding author

Japanese females have the longest average lifespan with better health than other elderly people in the world; this is believed to be due to the anti-aging and immunoenhancing properties of their diet, rich in marine foods. Therefore, many scientific research studies have been carried out to find the anti-aging and immunoenhancing mechanisms of marine derived bioactive compounds. Among millions of bioactive compounds, fatty acids, proteins, polysaccharides and antioxidants such as carotene are considered to be highly beneficial.

2 Mechanisms and Theories of Aging Process

Aging is a unique biological process for each and every living organism. Accumulation of damaged and defective cellular components, loss of cell or organ physiological functions, failure of physical activities and loss of memory occur during the aging process. Most of the aging related changes cause the development of diseases such as cardiovascular disease, type II diabetes, neurodegenerative diseases, cognitive decline and depression, etc. As a result of scientific investigations carried out to discover the secret of aging, a number of theories have been suggested but none of these theories alone provides the means to explain the secret of aging. Genetic mutation and imbalance of the signaling pathways are believed to have a great impact on the aging process. Genes with genetic mutation regulate aging through acting on cellular stress response, energy and metabolism control, growth modulation, gene dysregulation, genetic stability and nutrition sensing. The coordinated action of signaling networks systematically modulates homeostasis and functions that respond to stress, damage, nutrition and temperature. Imbalance of these signaling pathways has been noted in various organs and tissues during age related pathology (Pan et al. 2012).

3 Immune System, Immunoenhancement and Stimulation

Immunity is the resistance of an organism to avoid infection, disease or other unwanted biological invasion. The immune system is the most diverse and complex system in the body; it is a network of cells, tissues and organs that work together to defend the body against attacks by microbes such as bacteria, parasites, viruses and fungi. The function of the immune system depends on its ability to distinguish between the body's own cells and foreign cells. The immune system starts to react as soon as it recognizes the attack of an antigen. Anything that can trigger an immune response is called an antigen. An antigen can be a microbe or a part of a microbe such as a molecule.

Many exogenous factors like bioactive compounds have been found to enhance the activity of the immune system. In the past few decades, the discovery of metabolites with immunoenhancing properties from marine sources has increased significantly. The compounds responsible for immunoenhancement are known as immunostimulants.

4 Sources of Marine Bioactive Compounds

The importance of the marine ecosystem as a source of bioactive compounds is growing rapidly. Sea weeds, marine micro algae or by products of fish processing can be the root of marine bioactive compounds.

Marine algae are one of the richest sources of bioactive compounds. Algae can be classified into macro algae (seaweed) and microalgae, based on the size. The biodiversity of microalgae is enormous and they represent an almost untapped resource. Marine macro algae are classified into three main categories based on their pigmentation. Phaeophyta or brown seaweeds are predominantly brown due to the presence of the carotenoid fucoxanthin and the primary polysaccharides present include alginates, laminarins, fucans and cellulose (Haugan and Liaaenjensen 1994; Goni et al. 2002). Chlorophyta or green seaweeds are dominated by chlorophyll a and b, with ulvan being the major polysaccharide component (Robic et al. 2009). The principal pigments found in rhodophyta or red seaweeds are phycoerythrin and phycocyanin and the primary polysaccharides are agars and carrageenans (McHugh 2003). By products of fish processing refer to tissues that remain after the fish muscle has been removed and include heads, frames, viscera, and skin. By products can be used to make fish meal and fish oil. Fish, fish oils, fish meal and by products of fish processing industry are considered rich sources of bioactive compounds.

5 Role of Marine Bioactive Compounds in Anti-aging and Immunoenhancing Mechanisms

5.1 Carotenoids

Chemically, carotenoids are polyunsaturated hydrocarbons containing 40 carbon atoms per molecule and variable numbers of hydrogen atoms. Oxygenated derivatives of carotenoids are known specifically as xanthophyll. Among the other carotenoid sources, seaweeds and microalgae are known to have a greater impact on anti-aging and immunoenhancing mechanisms (Table 1).

Table 1. Macro and micro algal sources of carotenoids.

Carotenoid	Molecular Structure	Source	Reference
β-carotene		*Dunaliella salina* *Chlorella ellipsoidea* *Chlorella vulgaris*	(Rabbani et al. 1998) (Plaza et al. 2009) (Plaza et al. 2009)
α-carotene		*Dunaliella salina*	(Ocean White 2003)
Lutein		*Chlorella ellipsoidea* *Chlorella vulgaris*	(Plaza et al. 2009) (Plaza et al. 2009)
Astaxanthin		*Haematococcus pluvialis* crustacean shells	(Tripathi et al. 1999) (Cha et al. 2008)
Cryptoxanthin		*Dunaliella salina*	(Ocean White 2003)
Canthaxanthin		*Haematococcus pluvialis*	(Demming-Adams and Adams 2002)
Lutein		*Haematococcus pluvialis* *Scenedesmus almeriensis*	(Demming-Adams and Adams 2002) (Abe et al. 2005)
Zeaxanthin		*Dunaliella salina*	(Yokthongwattana et al. 2005)

Carotenoids in anti-aging and age related disease prevention: In the last decades, many laboratory and epidemiological studies have been conducted which suggest that intake of carotenoids is inversely related to aging and age related diseases. Although the mechanisms are not well understood, the action of carotenoids is believed to be due to its antioxidant effect. The conjugated double bonds of carotenoids show antioxidant properties in reducing oxidative stress created by reactive oxygen species. Astaxanthin possesses powerful antioxidant properties as it contains two additional oxygenated groups on each ring structure compared with other carotenoids.

The anticancer activity of marine carotenoids has been widely observed. In particular, β-carotene, astaxanthin, cantaxanthin and zeaxanthin have been shown to have positive impacts (Gradelet et al. 1998; Nishino et al. 2002). DNA oxidation due to oxidative stress is believed to be one of the major risk factors of cancers. In this sense, the anticancer activity of marine carotenoids may be due to its antioxidant ability to prevent DNA oxidation.

Cardiovascular diseases are the most commonly found disease among elderly populations and atherosclerosis is considered the most common cause. Low density lipoproteins oxidation induced by oxidative stress plays a key role in the pathogenesis of atherosclerosis. Numerous research studies have shown that astaxanthin, cantaxanthin, lutein, α-carotene and β-carotene exerted significant anti-atherogenic properties due to the ability to suppress oxidative stress (Palozza et al. 2008; Vílchez et al. 2011).

In addition, a marine carotenoid enriched diet has been found to diminish the risk of degenerative diseases such as Parkinson's and Alzheimer's (Guerin et al. 2003). Neurodegeneration is the progressive loss of structure or function of neurons. Many neurodegenerative diseases are caused by genetic mutations. Therefore, the protective action of marine carotenoids due to their antioxidant activity on DNA oxidation can be considered.

Carotenoids in immunoenhancement: Many studies have demonstrated the activity of marine β-carotenoids upon immunoenhancement. In early days the immunoenhancing role of β-carotene was believed to be due to its provitamin A activity in maintaining non-specific host defence. In humans, four carotenoids; β-carotene, α-carotene and β-cryptoxanthin have provitamin A activity.

However, subsequent studies have shown the specific action of carotenoids without provitamin A activity, carotenoids such as in lutein, canthaxanthin and astaxanthin. In fact, these non-provitamin A carotenoids were as active and at times more active than β-carotene in enhancing immune response in animals and humans (Chew and Park 2004). Though the mode of action is not clear, antioxidant properties of conjugated double bonds of carotenoids may be effective in immunoenhancement through neutralizing singlet oxygen and other reactive oxygen species.

During the non-specific immune responses to infection, free radicals and reactive oxygen species are used by certain white blood cells against bacterial invaders. Neutrophils are a major class of white blood cells which use reactive oxygen species and free radicals to kill phagocytized bacteria. However, effectiveness and efficiency of the immune system can be depleted as a result of injury to white blood cells as well as neighboring cells and tissues caused by the excessive production of free radicals. Scientists have found that the β-carotene protects neutrophils from free radical damage without suppressing its activity on bacteria (Weitberg et al. 1985).

The cells involved in specific immune responses can also be affected by lipid peroxides and other oxidative products formed due to oxidative reactions (Gurr 1983). β-carotene and canthaxanthin together inhibited the loss of macrophage receptors following exposure to reactive oxygen intermediates (Gruners et al. 1986). At the same time, carotenoids can protect lipids from oxidation and reduce peroxide generation.

Apart from the antioxidant activity β-carotene enhances many aspects of immune function including T and B lymphocyte proliferation and secretion of factors required for communication between the cells responsible for immune functions.

5.2 Proteins, peptides and amino acids

Amino acids contain an amine group, a carboxylic acid group and a side chain that is specific to each amino acid. Peptides are short polymers of amino acid monomers linked by peptide bonds between the carboxyl and amino groups of adjacent amino acid residues. Typically peptides contain fewer than 50 monomer units. Proteins are biochemical compounds consisting of one or more polypeptide folded into a globular or fibrous form, facilitating a biological function.

Proteins, peptides and amino acids in anti-aging and age related disease prevention: Anti-aging and age related disease preventive actions of proteins, peptides and amino acids isolated from marine sources have been widely studied. It has been reported that several proteins, peptides and amino acids have the antioxidant property to suppress the activity of reactive oxygen species. On the other hand, oxidative stress created by free radicals is believed to have a strong relationship with aging and age related diseases. Therefore, proteins, peptides and amino acids are considered effective in anti-aging and prevention of age related diseases.

Polypeptides with 3–20 amino acids are found to have functional properties. However their activities are based on their amino acid composition and sequence (Pihlanto-Leppälä 2000; Kim and Wijesekara 2010). It has already been shown that bioactive peptides isolated from fish protein hydrolysates, algal fucans, and galactans possess anticoagulant, anticancer, hypocholesterolemic (Moskowitz 2000) and antioxidative properties (Jun et al. 2004; Rajapakse et al. 2005). In addition, purified peptides extracted from *Chlorella vulgaris* have shown protective effects against cellular damage (Sheih et al. 2009). On the other hand, fish protein has been reported to decrease cholesterol which influences the lipid metabolism of human subjects in serum and liver (Hosomi et al. 2009). The major proteins in *Spirulina platensis* and *Porphyridium* are known to have antioxidant and anticancer effects (Plaza et al. 2009).

The basic unit of proteins and peptides is amino acid. In the intestinal tract, proteins and peptides are broken down to amino acids which are distributed to the whole body after absorption by the intestinal tube wall. Therefore, it is suggested that proteins and peptides are absorbed as amino acids and the amino acids inhibit age related diseases due to their antioxidant properties. In particular, these amino acids are considered to be a leading part of the antioxidant property. Researchers have found that phenylalanine, histidine, and tryptophan in fish product peptides may scavenge HO• (Dean et al. 1997). Therefore, regular intake of marine peptides and proteins might contribute to the prevention of aging and age related diseases by regulating the balance of oxidative stress.

Proteins, peptides and amino acids in immunoenhancement: There has not been much research on the immunoenhancing properties of marine derived bioactive proteins, peptides or amino acids. Anyhow, the immunoenhancing properties of protein hydrolysate isolated from *Chlorella vulgaris,* which is a small green one celled algae, have received considerable attention. Though the specific functions are little known, *Chlorella* is believed to stimulate the production of interferons, macrophages and T cells, thus functioning as an immune stimulant (Tse 2000).

5.3 Fatty acids

Fatty acids are dietary nutrients important for the healthy functioning of the body. Unlike other fatty acids that can be created in the body, ω-6 and ω-3 fatty acids can only be obtained through one's diet. Considerable research activities have focused on the relationship between increased ω-3 fatty acid intake and anti-aging and immunoenhancing activities. Oily fish, fish oil and microalgae contain varying amounts of polyunsaturated ω-3 fatty acids known as eicosapentaenoic acid (EPA, 20:5n-3) and docosahexaenoic acid (DHA, 22:6n-3). Microalgae are being developed as a commercial source.

Fatty acids in anti-aging and age related disease prevention: Experimental studies have shown the relationship between higher intakes of DHA and EPA from marine sources and inhibition of tumor growth. EPA and DHA may act either through the same or different mechanisms, but differential efficacy could exist. Several molecular mechanisms, whereby ω-3 fatty acids

Table 2. Sources of marine ω-3 fatty acids.

Fatty Acid	Molecular Structure	Source	Reference
EPA		Microalgae *Porphyridium cruentum* *Chlorella minutissima* Fish and fish oil Cod liver, herring, mackerel, salmon, menhaden, sardine	(Fuentes et al. 1999) (Rasmussen and Morrissey 2007)
DHA		Microalgae *Crypthecodinium cohnii* Fish and fish oil Cod liver, herring, mackerel, salmon, menhaden, sardine	(Wikipedia 2012)
γ-linolenic acid		Microalgae *Arthrospiraplatensis* *Arthrospira maxima*	(Wikipedia 2012)

may modify the carcinogenic process, have been proposed. These include suppression of arachidonic acid derived eicosanoid biosynthesis, influences on transcription factor activity, gene expression and signal transduction pathways, alteration of estrogen metabolism, increased or decreased production of free radicals and reactive oxygen species and mechanisms involving insulin sensitivity and membrane fluidity (Larsson et al. 2004).

At the same time, numerous epidemiological and experimental studies have conclusively shown that higher intakes of marine EPA and DHA are associated with reduced risk of cardiovascular disease. The cardiovascular protective effects of DHA contained in fish lipid have been reported to be due to suppression of cholesterol secretion from the liver to the plasma (Garg et al. 1998).

Moreover, DHA derived metabolites have been found to promote resolution and protect neural cells in brain tissue from neurodegeneration (Bazan 2005). At the same time, marine DHA intake has been reported to reduce the incidence of Alzheimer's disease which is characterized by a decline in cognitive function. The presence of extracellular amyloid peptide deposits and intracellular neurofibrillary tangles in the brain are considered the main features of the disease. Although the cause of the disease is not well understood, increased inflammation (Akiyama et al. 2000) and oxidative stress (Markesbery 1997) are considered the key contributing factors. Therefore, the preventive activity of DHA can be recommended due to its ability to suppress neuroinflammation and oxidative stress.

Fatty acids in immunoenhancement: Immunoenhancing activities of marine fatty acids are lacking in evidence. The hot water extract of *Spirulina* was discovered to enhance immunity by increasing the phagocytic activity of macrophages and stimulating the natural killer cells. The immunoenhancing properties of *Spirulina* may also be due to activation and mobilization of T and B cells due to its impact in the production of cytokines and antibodies (Schwartz and Shklar 1987). The immunoenhancing property of *Spirulina* is believed to be due to γ-linolenic acid which is a precursor of prostaglandins, leukotrienes and thromboxans (Burtin 2003).

5.4 Oligosaccharides and polysaccharides

Significant efforts have been made to find and confirm the bioactivity of saccharides. Polysaccharides and oligosaccharides are long carbohydrate molecules of repeated monomer units joined together by glycosidic bonds. Typically, oligosaccharides contain two to ten monomer units while polysaccharides contain more than ten monosaccharide units.

Fucoidan is a sulfated, water soluble, branched polysaccharide with L-fucose monomer units. The basic structure of 3-linked, preponderantly 4-sulfated fucoidan from *Eckloniakurome* (Nishino and Nagumo 1991) is shown in Fig. 1. Fucoidans are found both intercellularly and in the cell wall of brown algae, such as *Hizikiafusiforme, Fucusvesiculosus, Eiseniabicyclics, Fucusserratus, Fucusdistichus, Himanthalialorea, Bifurcaria bifurcate* and *Eckloniakurome* (Berteau and Mulloy 2003).

Chitin is one of the major structural polysaccharides of crustacean shells and shellfish wastes and built from *n*-acetyl-glucosamine monomer. The structure of the chitin molecule, showing two of the N-acetylglucosamine units that repeat to form long chains in β-1,4 linkage are given in Fig. 2.

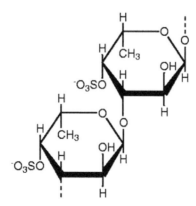

Figure 1. The basic structure of 3-linked, preponderantly 4-sulfated fucoidan.

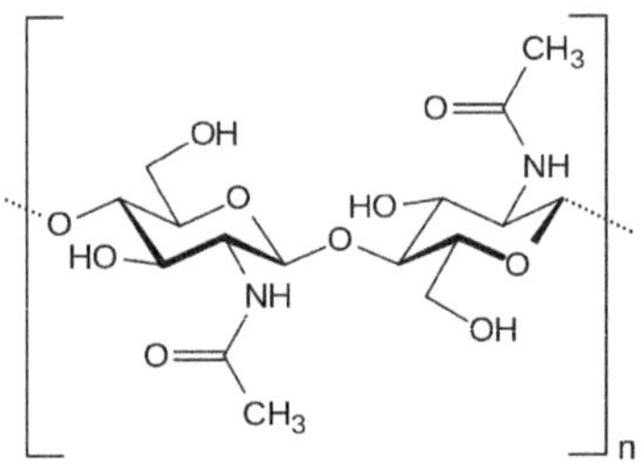

Figure 2. Structure of the chitin molecule, showing two of the N-acetylglucosamine units.

In practice, chitosan is produced commercially by deacetylation of chitin. Chitosan is a linear cationic polysaccharide composed of randomly distributed β-(1-4)-linked D-glucosamine (deacetylated unit) and N-acetyl-D-glucosamine (acetylated unit). The proportion of the two monosaccharide units in chitosan depends on the alkaline treatment. Chitosan has three types of reactive functional groups, an amino/acetamido group as well as both primary and secondary hydroxyl groups at the C-2, C-3 and C-6 positions, respectively. Chito-oligosaccharides are chitosan derivatives which can be generated via chemical or enzymatic hydrolysis of chitosan.

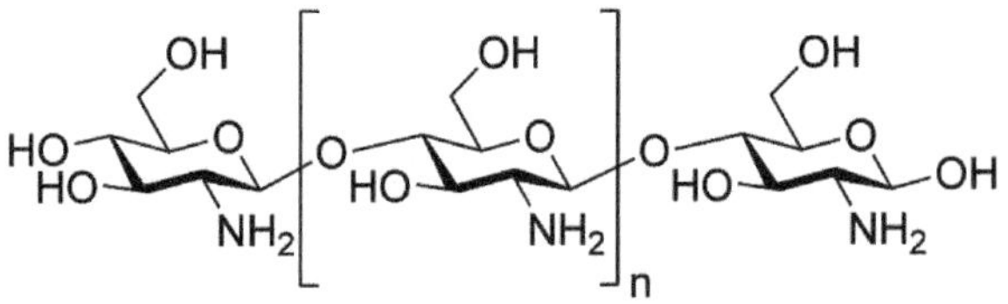

Figure 3. Chemical structure of chitosan.

Laminaran and sodium alginate are polysaccharides with proven functional health benefits. Laminaran is a β-glucan found widely in brown algae. It is a β-1,3/1,6 glucan which is considered the most active form of β-glucans.

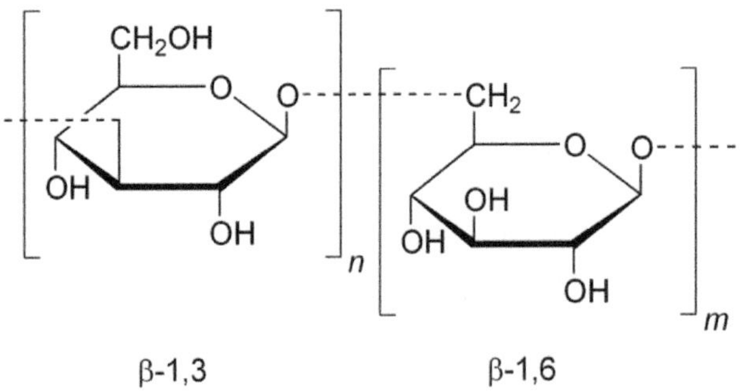

β-1,3   β-1,6

Figure 4. Chemical structure of laminaran.

Sodium alginate is the sodium salt of alginic acid and is extracted from brown sea weeds by the use of dilute alkali. Basically, sodium alginate composed of β-D-mannuronic acid monomers linked by glycosidic bonds.

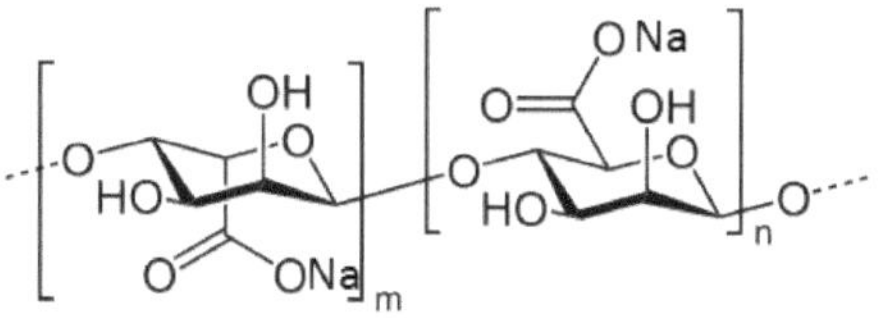

Figure 5. Chemical structure of sodium alginate.

Oligosaccharides and polysaccharides in anti-aging and age related disease prevention: Marine polysaccharides such as fucoidans and chitosan and chito-oligosaccharides present a real potential in anti-aging mechanisms. Fucoidans are reported to display physiological and biological activities including antithrombotic, antitumor and antioxidant (Li et al. 2008) activities with therapeutic potential increasing with the degree of sulfation (Berteau and Mulloy 2003).

In the medical field, chitosan has been developed as new physiological material due to its antitumor and hypocholesterolemic properties (Jeon et al. 2000). These functions have been limited by its high molecular weight and highly viscous nature, resulting in its low solubility in acid free aqueous media. Recently, these oligosaccharides have been the subject of increased attention in terms of their high solubility and their antioxidant, anticancer, hypocholesterolemic, hypoglycemic and anticoagulant properties (Wijesekara and Kim 2010).

Oligosaccharides and polysaccharides in immunoenhancement: Great attention has been paid to immunoenhancing properties of marine derived polysaccharides such as fucoidan, glucan, alginate, chitin and chitin derivatives, etc.

Marine algae are considered a rich source of polysaccharides, having a number of health benefits including immunoenhancement. β-1,3-glucan of *Chlorella* (Spolaore et al. 2006) and 1→3:1→6-β-D-glucans, produced from laminaran, have been found to have immunostimulating activities in animals and plants (Kuznetsova et al. 1994; Chertkov et al. 1999). Preparations containing 1→3:1→6-β-D-glucans, laminaran and fucoidan, are marketed for their beneficial properties on the immune system. High molecular weight fucoidans of *Okinawa mozuku* are immunostimulators as they increase the proportion of murine cytotoxic T cells (Shimizu 2005). Fucoidan from *Fucus vesiculosus* has immunostimulating and maturing effects on dendritic cells which are powerful antigen presenting cells (Kima and Joo 2008). The marine brown alga, *Endarachne binghamiae,* which is a rich source of sodium alginate, exhibited strong immunostimulation activity for macrophage and T cell proliferation (Rang and Huiting 2005). Although the immunoenhancing mechanisms of alginates and their hydrolysates are still unclear, it is proposed that the molecular size and molecular conformational characters are highly related to their bioactivity (Ji et al. 2011).

Recent advances demonstrate the health benefits of oligosaccharides including enhanced immunity. Chitin and chitin derivatives stimulate macrophages to produce cytokines by interacting with different cell surface receptors such as macrophage mannose receptor, toll-like receptor-2 (TLR-2), C-type lectin receptor Dectin-1, and leukotriene 134 receptor (BLT1) (Lee 2009). Chitosan showed an immunoenhancing effect by enhancement of antibody response. 70%-DD chitosan, in particular, was an immune regulator that could activate macrophages and natural killer cells and improve the delayed type hypersensitive reaction (Nishimura et al. 1984).

6 Perspectives

Since antiquity, due to their functional properties, marine sources have been used as a part of daily diet and also as a medicinal remedy in some parts of the world. The longest average lifespan and better health of Japanese females are also believed to be due to the anti-aging and immunoenhancing properties of their diet, rich in marine foods. As a result, scientific investigations have been carried out to find the exact molecules responsible for the anti-aging and immunoenhancing activities. But it is still difficult to explain how exactly these molecules perform their activity. Therefore, further studies are needed in order to investigate the detailed molecular mechanisms.

Although marine food has been found to possess a number of health benefits, including anti-aging and immunoenhancement, its intake is not adequate due to its cost and lack of availability. At the same time, many individuals do not like the taste and odor of marine food. Foods fortified with marine bioactive compounds could play an important role in meeting the requirements for optimal benefits. Technological innovations in the field of food science should be applied in the development of food products fortified with bioactive compounds, having anti-aging and immunoenhancing properties.

References

Abe, K., H. Hattor and M. Hiran. 2005. Accumulation and antioxidant activity of secondary carotenoids in the aerial microalga *Coelastrellastriolata* var. *multistriata*. Food Chem. 100: 656–661.

Akiyama, H., S. Barger, S. Barnum, B. Bradt, J. Bauer, G.M. Cole, N.R. Cooper, P. Eikelenboom, M. Emmerling, B.L. Fiebich, C.E. Finch, S. Frautschy, W.S.T. Griffin, H. Hampel, M. Hull, G. Landreth, L.-F. Lue, R. Mrak, I.R. MacKenzie, P.L. McGeer, M.K. O'Banion, J. Pachter, G. Pasinetti, C. Plata-Salaman, J. Rogers, R. Rydel, Y. Shen, W. Streit, R. Strohmeyer, I. Tooyoma, F.L. Van Muiswinkel , R. Veerhuis, D. Walker, S. Webster, B. Wegrzyniak, G. Wenk and T. Wyss-Coray. 2000. Inflammation and Alzheimer's disease. Neurobiology of Aging. 21(3): 383–421.

Bazan, N.G. 2005. Neuroprotectin D1 (NPD1): a DHA-derived mediator that protects brain and retina against cell injury-induced oxidative stress. Brain Pathology. 15(2): 159–166.

Berteau, O. and B. Mulloy. 2003. Sulfated fucans, fresh perspectives: structures, functions, and biological properties of sulfated fucans and an overview of enzymes active toward this class of polysaccharide: A review. Glycobiology. 13: 30R.

Biesalski, H. K., L.O. Dragsted, I. Elmadfa, R. Grossklaus, M. Müller, D. Schrenk, P. Walter and P. Weber. 2009. Bioactive compounds: Definition and assessment of activity. Nutrition. 25: 1202–1205.

Burtin, P. 2003. Nutritional value of seaweeds. EJEAFChe. 2: 498–503.

Cha, K.H., S.Y. Koo and D.U. Lee. 2008. Antiproliferative effects of carotenoids extracted from *chlorella ellipsoidea* and *chlorella vulgaris* on human colon cancer cells. J. Agric. Food Chem. 56: 10521–10526.

Chertkov, K.S., S.A. Davydova, T.A. Nesterova, T.N. Zviagintseva and L.A. Eliakova. 1999. Efficiency of polysaccharide translam for early treatment of acute radiation illness. Radiats Biol. Radioecol. 39: 572–577.

Chew, B.P. and J.S. Park. 2004. Carotenoid action on the immune response. J. Nutr. 134(1): 257S–261S.

Dean, R.T., S. Fu, R. Stocker and M.J. Davies. 1997. Biochemistry and pathology of radical-mediated protein oxidation. Biochem. J. 324: 1–18.

Demming-Adams, B. and W.W. Adams. 2002. Antioxidants in photosynthesis and human nutrition. Science. 298: 2149–2153.

Faulkner, D.J. 1995. Chemical Riches from the Ocean. Chem. Brit. 31: 680–684.

Fuentes, M.M.R., J.L.G. Sánchez, J.M.F. Sevilla, F.G.A. Fernández, J.A.S´. Pérez and E.M. Grima. 1999. Outdoor continuous culture of *Porphyridium cruentum* in a tubular photobioreactor: Quantitative analysis of the daily cyclic variation of culture parameters. J. Biotechnol. 70: 271–288.

Garg, M.L., A.A. Wierzbicki, A.B. Thomson and M.T Clandinin. 1998. Fish oil reduces cholesterol and arachidonic acid content more efficiently in rats fed diets containing low linoleic acid to saturated fatty acid ratios. Biochim. Biophys. Acta. 962(3): 337–344.

Goni, I., L. Valdivieso and M. Gudiel-Urbano. 2002. Capacity of edible seaweeds to modify *in vitro* starch digestibility of wheat bread. Nahrung. 46: 18–20.

Gradelet, S., A.M. Le Bon, R. Berges, M. Suschetet and P. Astorg. 1998. Dietary carotenoids inhibit aflatoxin B1-induced liver preneoplastic foci and DNA damage in the rat: Role of the modulation of aflatoxin B1 metabolism. Carcinogenesis. 19: 403–411.

Gruners, S., H. Volk, P. Falck and R. Vonbaehr. 1986. The influence of phagocytic stimuli on the expression of HLA-DR Antigens; role of reactive oxygen intermediates. Eur. J. Immunol. 6: 212–215.

Guerin, M., M.E. Huntley and M. Olaizola. 2003. *Haematococcus* astaxanthin: Applications for human health and nutrition. Trends Biotech. 21: 210–216.

Gurr, M.I. 1983. The role of lipids in the regulation of the immune system. Prog. Lipid Res. 22: 257–287.

Haugan, J.A. and S. Liaaenjensen. 1994. Algal Carotenoids 54. Carotenoids of Brown-Algae (Phaeophyceae). Biochem. Syst. Ecol. 22: 31–41.

Hosomi, R., K. Fukunaga, H. Arai, T. Nishiyama and M. Yoshida. 2009. Effects of dietary fish protein on serum and liver lipid concentrations in rats and the expression of hepatic genes involved in lipid metabolism. J. Agric. Food Chem. 57: 9256–9262.

Jeon, Y.J., F. Shahidi and S.K. Kim. 2000. Preparation of chitin and chitosan oligomers and their application in physiological functional foods. Food Rev. Int. 16: 159–176.

Ji, J., L.C. Wang, H. Wu and H.M. Luan. 2011. Bio-function summary of marine oligosaccharides. International Journal of Biology. 3(1): 74–76.

Jun, S.Y., P.J. Park, W.K. Jung and S.K. Kim. 2004. Purification and characterization of an antioxidative peptide from enzymatic hydrolysate of yellow fin sole (*limanda aspera*) frame protein. Eur. Food Res. Technol. 219: 20–26.

Kim, S.K. and I. Wijesekara. 2010. Development and biological activities of marine-derived bioactive peptides: A review. J. Funct. Foods. 2: 1–9.

Kima, M.H. and H.G. Joo. 2008. Immunostimulatory effects of fucoidan on bone marrow-derived dendritic cells. Immunol. Lett. 115: 138–143.

Kuznetsova, T.A., N.V. Krylova, N.N. Besednova, V.N. Vasil'eva, T.N. Zviagintseva, S.V. Krashevskii and L.A. Eliakova. 1994. The effect of translam on the natural resistance indices of the irradiated organism. Radiats Biol. Radioecol. 34: 236–239.

Larsson, S.C., M. Kumlin, A. Wolk, M. Ingelman-Sundberg and A. Wolk. 2004. Dietary long-chain n–3 fatty acids for the prevention of cancer: A review of potential mechanisms. Am J. Clin. Nutr. 79(6): 935–945.

Lee, C.G. 2009. Chitin, chitinases and chitinase-like proteins in allergic inflammation and tissue remodeling. Yonsei Med. J. 50: 22–30.

Li, B., F. Lu, X.J. Wei and R.X. Zhao. 2008. Fucoidan: Structure and bioactivity. Molecules. 13: 1671–1695.

Markesbery, W.R. 1997. Oxidative stress hypothesis in Alzheimer's disease. Free Radical Biology and Medicine. 23(1): 134–147.

McHugh, D.J. 2003. [FAO] Food and Agriculture Organization of the United Nations. A Guide to the Seaweed Industry. FAO Technical Paper in Fisheries 441, Rome: 118 p.

Moskowitz, R.W. 2000. Role of collagen hydrolysate in bone and joint disease. Semin. Arthritis Rheum. 30: 87–99.

Nishimura, K., S. Nishimura, N. Nishi, I. Saiki, S. Tokura and I. Azuma. 1984. Immunological activity of chitin and its derivatives. Vaccine. 2(1): 93–99.

Nishino, H., M. Murakoshi, T. Li, M. Takemura, M. Kuchide, M. Kanazawa, X. Mou, S. Wada, M. Masuda, Y. Ohsaka, S. Yogosawa, Y. Satomi and K. Jinno. 2002. Carotenoids in cancer chemoprevention. Cancer Metast. Rev. 21: 257–264.

Nishino, T. and T. Nagumo. 1991. Structural characterization of a new anticoagulant fucan sulfate from the brown seaweed *Ecklonia kurome*. Carbohyd. Res. 211: 77–90.

Ocean White. 2003. *Dunaliella Salina Algae (Algae Extract)* [online]. Available at <http://www.oceanwhite.com/glossary/glossary.html> [Accessed 6 August 2012].

Palozza, P., E. Barone, C. Mancuso and N. Picci. 2008. The protective role of carotenoids against 7-keto-cholesterol formation in solution. Mol. Cell. Biochem. 309: 61–68.

Pan, M.H., C.S. Lai, M.L. Tsai, J.C. Wu and C.T. Ho. 2012. Molecular mechanisms for anti-aging by natural dietary compounds. Mol. Nutr. Food Res. 56(1): 88–115.

Pihlanto-Leppälä. 2000. A. Bioactive peptides derived from bovine whey proteins: Opioid and ace-inhibitory peptides. Trends Food Sci. Technol. 11: 347–356.

Plaza, M., M. Herrero, A. Cifuentes and E. Ibáñez. 2009. Innovative natural functional ingredients from microalgae. J. Agric. Food Chem. 57: 7159–7170.

Rabbani, S., P. Beyer, J. von Lintig, P. Hugueney and H. Kleinig. 1998. Induced β-carotene synthesis driven by triacylglycerol deposition in the unicellular alga *Dunaliellabardawil*. Plant Physiol. 116: 1239–1248.

Rajapakse, N., W.K. Jung, E. Mendis, S.H. Moon and S.K. Kim. 2005. A novel anticoagulant purified from fish protein hydrolysate inhibits factor xiia and platelet aggregation. Life Sci. 76: 2607–2619.

Rang, H. and L. Huiting. 2005. Immunological Properties of the Marine Brown Alga *Endarachne binghamiae* (Phaeophyceae). International Journal of Applied Science and Engineering. 3(3): 167–173.

Rasmussen, R.S. and M.T. Morrissey. 2007. Marine biotechnology for production of food ingredients. Adv. Food Nutr. Res. 52: 237–292.

Robic, A., C. Gaillard, J.F. Sassi, Y. Lerat and M. Lahaye. 2009. Ultrastructure of Ulvan: A polysaccharide from green seaweeds. Biopolymers. 91: 652–664.

Schwartz, J. and G. Shklar. 1987. Regression of experimental hamster cancer by beta carotene and algae extracts. J. Oral Maxillofac Surg. 45(6): 510–515.

Sheih, I.C., T.K. Wu and T.J. Fang. 2009. Antioxidant properties of a new antioxidative peptide from algae protein waste hydrolysate in different oxidation systems. Bioresour. Technol. 100: 3419–3425.

Shimizu, J., U.Wada-Funada, H. Mano, Y. Matahira, M. Kawaguchi and M. Wada. 2005. Proportion of murine cytotoxic T cells is increased by high molecular-weight fucoidan extracted from *Okinawa mozuku* (Cladosiphon okamuranus). J. Health Sci. 51: 394–397.

Wikipedia. 2012. *Spirulina (dietary supplement)* [online]. Available at <http://en.wikipedia.org/wiki/Spirulina_(dietary_supplement)> [Accessed 25 July 2012].

Spolaore, P., C. Joannis-Cassan, E. Duran and A. Isambert. 2006. Commercial applications of microalgae. J. Biosci. Bioeng. 101: 87–96.

Tripathi, U., R. Sarada, S.R. Rao and G.A. Ravishankar. 1999. Production of astaxanthin in haematococcuspluvialis cultured in various media. Bioresour. Technol. 68: 197–199.

Tse, P. 2000. The detoxification, immunostimulation and healing property of chlorella. Proc. World convention of traditional medicine & acupuncture, Singapore. 18–19.

Vílchez, C., E. Forján, M. Cuaresma, F. Bédmar, I. Garbayo and J.M. Vega. 2011. Marine Carotenoids: Biological Functions and Commercial Applications: A review. Mar. Drugs 9: 319–333.

Weitberg, A.B., S.A. Weitzman, E.P. Clark and T.P. Stossel. 1985. Effects of antioxidants on oxidant-induced sister chromatid exchange formation. Clin. Invest. 75: 1835–1841.

Wijesekara, I. and S.K. Kim. 2010. Angiotension-i-converting enzyme (ace) inhibitors from marine resources: Prospects in the pharmaceutical industry. Mar. Drugs. 8: 1080–1093.

Wikipedia. 2012. *Docosahexaenoic acid* [online]. Available at: <http://en.wikipedia.org/wiki/Docosahexaenoic_acid> [Accessed 25 July 2012].

Yokthongwattana, K., T. Savchenko, J.E.W. Polle and A. Melis. 2005. Isolation and characterization of a xanthophyll-rich fraction from the thylakoid membrane of *dunaliellasalina* (green algae). Photochem. Photobiol. Sci. 4: 1028–1034.

14

Arsenic in Seaweed: Presence, Bioavailability and Speciation

Cristina García Sartal, María Carmen Barciela Alonso and
*Pilar Bermejo Barrera**

1 Introduction

The ubiquity of arsenic in the environment is associated with geological and anthropogenic sources. Drinking water is the main route of arsenic exposure. The population of arsenic-rich areas is among the most affected by chronic health disorders related to arsenic. For this reason, the Bangladeshi and Indian governments have established national regulations for arsenic in drinking water (50 µg L^{-1}) (Bhattacharya et al. 2007), and the WHO and US-EPA have established a recommended value of 10 µg L^{-1}, and a maximum permissible arsenic value of 50 µg L^{-1} in drinking water (WHO 2001; Smedley and Kinniburgh 2002; Mandal and Suzuki 2002). In areas without natural or anthropogenic arsenic contamination, food is the main route of arsenic exposure for humans, with seafood being the main contributor to this metalloid intake (Devesa et al. 2008).

Seaweed (macroalgae) and microalgae are present in most people's daily diet, either directly as raw or processed food, indirectly through by-products used in food and pharmaceutical industry such as gelling, stabilizing and thickening agents, or as a source of colours, flavours and textures added to food products (García-Sartal et al. 2012a). Seaweeds satisfy the nutritional requirements in humans and are catalogued as 'health food', since they are rich in soluble dietary fibre (higher fibre content than

Department of Analytical Chemistry, Nutrition and Bromatology, Faculty of Chemistry, University of Santiago de Compostela, 15782 Santiago de Compostela, Spain.
*Corresponding author: pilar.bermejo@usc.es

in some fruits), in proteins containing all essential amino acids, minerals, vitamins, antioxidants, polyunsaturated fatty acids (with low caloric value), and are also a source of bioactive compounds (Gupta and Abu-Ghannam 2011; Dawczynski et al. 2007a). In general, seaweeds have shown the following properties in *in vivo* studies: cancer preventive properties, antiobesity, antidiabetic, antihypertensive, antihyperlipidemic, antioxidant, anticoagulant, anti-inflammatory, immunomodulatory, antiestrogenic, thyroid stimulating, neuroprotective, antiviral, antifungal, antibacterial and tissue healing properties (Mohamed et al. 2012).

Although seaweeds are a source of essential minerals for humans, they also might present a risk for human health, because they are high accumulators of non-essential elements, some of them widely recognized for their high toxicity, such as arsenic, among others. Moreda-Piñeiro et al. (Moreda-Piñeiro et al. 2012) have recently reviewed the significance of the presence of trace and ultratrace elements in seaweed. Depending on the concentration of metals in the environment and their bioavailability ratio from seaweed, seaweed can accumulate metals at levels several thousand times higher than those found in the surrounding seawater.

Because of the increasingly widespread consumption of these marine vegetables in Western countries, different European regulations concerning toxic elements in seaweed have been established. The EU has set maximum levels for As, Cd, Hg and Pb in food products derived from seaweed used as additives in the food industry (Commission Directive 2008/84/EC 2008; Commission Directive 2009/10/EC 2009). However, there are no specific regulations for limits of metals in raw seaweed. Eleven macroalgae consisting of five Phaeophyta (*Ascophylum nodosum, Fucus vesiculosis, Fucus serratus, Himanthalia elongata, Undaria pinnatifida*), four Rhodophyta (*Porphyra umbilicalis, Palmaria palmata, Cracilaria verrucosa, Chondrus crispus*) and two Chlorophyta (*Ulva* spp., *Enteromorpha* spp.) as well as two microalgae (*Spirulina* sp., *Odontella aurita*) have been authorised as vegetables and dressings/flavourings by the French government. France has established the following limits in edible seaweed: Pb, I and Sn <5 µg g^{-1}, Cd <0.5 µg g^{-1}, Hg <0.1 µg g^{-1}, and inorganic As <3 µg g^{-1}, expressed on a dry weight basis (Moreda-Piñeiro et al. 2012; Besada et al. 2009; Almela et al. 2006). In July 2004, the Food Standards Agency of the UK established a warning on the consumption of Hijiki (*Hijikia fusiforme*), because this seaweed contains high amounts of inorganic arsenic (Nakajima et al. 2006), although no toxic episodes have been described after its ingestion.

The knowledge of the total element concentration in seaweed is not enough to evaluate its toxic effects in humans. Toxicological effects vary among different species of the same element, and frequently, the elemental bioavailability is dependent on its chemical form. Nomenclature, abbreviated names and molecular formulas of these relevant arsenic species are shown in Table 1.

Table 1. Nomenclature, abbreviated name and molecular formula of the main arsenic species cited in this review.

Nomenclature	Abbreviated name	Molecular formula
Arsenite (arsenous acid)	AsIII	$As(OH)_3$
Arsenate (Arsenic acid)	AsV	$AsO(OH)_3$
Monomethylarsonous acid	MMAIII	$CH_3As(OH)_2$
Monomethylarsonic acid	MMAV	$CH_3AsO(OH)_2$
Dimethylarsinous acid	DMAIII	$(CH_3)_2AsOH$
Dimethylarsinic acid	DMAV	$(CH_3)_2AsO(OH)$
Arsenobetaine	AB	$(CH_3)_3As^+CH2COO^-$
Arsenocholine	AC	$(CH_3)_3As^+CH_2CH_2OH$
Trimethylarsine oxide	TMAO	$(CH_3)_3AsO$
Tetramethylarsonium ion	Me_4As^+ (TETRA)	$(CH_3)_4As^+$
Trimethylarsine	TMAIII	$(CH_3)_3As$
Dimethylarsinoyl ethanol	DMAE	$(CH_3)_2AsOCH_2CH_2OH$
Dimethylarsinoyl acetic acid	DMAA	$(CH_3)_2AsOCH_2COOH$
5-dimethylarsinoyl-ß-ribofuranose	Hydrogen-ribose, H-sugar	
3-[5′-deoxy-5′-(dimethylarsinoyl)-ß-ribofuranosyloxy]-2-hydroxypropylene glycol	Glycerol-ribose, OH-sugar	
3-[5′-deoxy-5′-(dimethylarsinoyl)-ß-ribofuranosyloxy]-2-hydroxypropane sulfonic acid	Sulfonate-ribose, SO_3-sugar	
3-[5′-deoxy-5′-(dimethylarsinoyl)-ß-ribofuranosyloxy]-2-hydroxypropyl hydrogen sulfate	Sulfate-ribose, SO_4-sugar	
3-[5′-deoxy-5′-(dimethylarsinoyl)-ß-ribofuranosyloxy]-2-hydroxypropyl-2,3-hydroxypropyl phosphate	Phosphate-ribose, PO_4-sugar	

This chapter reviews the information published over the last twelve years about arsenic speciation in seaweed, its bioavailability and its excretion into urine after seaweed consumption. Information about total arsenic level, analytical trends, sample preparation and species identification in seaweed, gastrointestinal fluids and urine are discussed throughout this review.

2 Total Arsenic Determination and Arsenic Speciation in Seaweed Samples: Extraction, Analytical Determination and Chemical Distribution

Arsenic concentrations are typically around 1.5 µg L^{-1} in open seawater and 4 µg L^{-1} in estuarine water, occurring mostly as inorganic forms (Smedley and Kinniburgh 2002). However, marine organisms can bioaccumulate inorganic arsenic from the surrounding seawater by factors of up to 100,000, and biotransform it mainly into organoarsenic compounds. While seaweed accumulates arsenic directly from the water, marine animals could incorporate arsenic from the surrounding water and from their diet of seaweed (Craig 2003).

The ability of arsenic to be bioaccumulated varies greatly between the three algal divisions established; Chlorophyta (or green seaweed), Phaeophyta (or brown seaweed) and Rhodophyta (or red seaweed). Brown seaweeds, in particular, are known to contain higher amounts of arsenic than red and green seaweeds (Dawczynski et al. 2007). Geographical origin and harvesting time, other than family divisions, are some of the more influential factors in seaweed metal content (Larrea-Marín et al. 2010).

Arsenosugars are the main arsenic compounds found in seaweed. Arsenosugar biosynthesis in seaweed involves no clear and complex mechanisms. Arsenate is the main arsenic species in seawater and its absorption metabolism by seaweed is similar to that reported for phosphate. It is assumed that seaweeds produce arsenosugars as end products of a detoxifying mechanism. Craig et al. suggested that absorbed arsenate is sequentially reduced and methylated until DMA(V) formation, with S-adenosylmethionine (SAM), acting as a methyl group donor. Finally, DMA(V) is reduced to DMA(III) and a glycosylation is produced due to the addition of the adenosyl group from SAM, resulting in dimethylarsinoylribosides (Craig 2003; Murray et al. 2003). Other authors have suggested that arsenosugar-phospholipids (lipid-soluble arsenic species found in seaweed) are synthetized by seaweed, and arsenosugars are the degradation products of the excess of arsenosugar-phospholipids. Other findings have suggested that the phosphate content in seawater is related to the formation of arsenosugars, and arsenosugar-phospholipids are formed in phosphate-rich environments (García-Salgado et al. 2012a).

Only about 1% of the absorbed inorganic arsenic remains in seaweed as inorganic arsenic (Raber et al. 2000).

Arsenic toxicity depends on arsenic speciation. Inorganic arsenic is generally more toxic than organic forms for living organisms. Arsenite (As(III)) is usually more toxic than arsenate (As(V)). Unlike methylated pentavalent arsenic species (DMA(V) and MMA(V)), methylated trivalent forms (DMA(III) and MMA(III)) are even more toxic and genotoxic than their precursor compounds. LC_{50} values for inorganic arsenic and dimethylated tri- and pentavalent arsenic species were calculated in a recent study. Results suggested that toxicity decreases as follows: DMA(III) > As(III) > As(V) > DMA(V) (Sharma and Sohn 2009). On the other hand, organic arsenic species such as arsenobetaine, arsenocholine and arsenosugars are regarded as non-toxic, because mutagenic, cytotoxic and immunotoxic effects have not been observed. A synthetic trivalent arsenosugar showed cytotoxicity and DNA nicking at the 500–600 µM level; however, this action was not found in the synthetic pentavalent counterpart (Borak and Hosgood 2007). Although arsenosugars have been considered inert and harmless compounds for humans, it has been found that both arsenosugars and inorganic arsenic share their capacity to be metabolized into the same compound, DMA(V). This arsenic specie is known for its carcinogenicity in humans. Thus, arsenosugar safety has been questioned (Andrewes et al. 2004).

The differences in toxicity and mobility among arsenic species and the efforts to understand the arsenic metabolic pathway have spurred the development of analytical methods for its determination. Fast and high resolution separation techniques, sensitive detection systems as well as optimized extraction methodologies are required.

2.1 Sample treatment for total analysis and arsenic speciation in seaweed samples

Most of the cited literature does not give information about sample pre-treatment, either because seaweeds under study are commercially available as dehydrated products (sun dried, oven baked or flame dried) or because this part of the information is omitted. A recent review (Rubio et al. 2010) suggests that sample handling is one of the most critical steps, because epiphytes and epifauna which might live in symbiosis with algae should be removed prior to arsenic extraction. These microorganisms might add small amounts of arsenic or arsenic species which are not naturally found in seaweed.

Most protocols for fresh seaweed pre-treatments consist of rinsing a sample with tap water (Raber et al. 2000), deionized water (Tukai et al. 2002) or both (Van Hulle et al. 2002; Karthikeyan and Hirata 2004), to remove interfering epiphytes, epifauna, salts or sediment remains, followed by freezing and further freeze-drying for water elimination. Alternatively, water removal has been performed by applying heat in an oven at 50–60°C for 2 days (Van Hulle et al. 2002; Wuilloud et al. 2006). Instead of a dry weight basis, other authors expressed their results based on the fresh sample weight analysed (Kohlmeyer et al. 2003). Finally, dried seaweed is ground to a fine powder in either a mortar with a pestle, a mixer, an ultra-centrifugal mill, or a stainless steel grinding mill. Few studies of the reviewed literature report data on particle size after powdering. Han et al. (2009) used a 0.5 mm ring sieve, which was bigger than that used by Kuehnelt et al. (2001) (0.25 mm sieve). García-Salgado et al. (2006, 2008) used a particle size of 125 µm in their studies. Sample powdering is a controversial issue because some authors consider that the destruction of tissues could alter the distribution of arsenic species. However, studies performed by Rubio et al. (2010) suggest that homogeneity and quantitative extraction are favoured by the rupture of cells and membranes.

Sample treatment for total arsenic determination depends on the analytical technique used. Different sample pretreatments for total arsenic determination are shown in Table 2. As it can be seen, the sample was digested by either using ultrapure nitric acid and hydrogen peroxide or only nitric acid with microwave energy, when total arsenic determination was performed either by inductively coupled plasma with mass spectrometry (ICP-MS) or atomic emission spectroscopy (ICP-AES) detection. However, when hydride generation was used, the presence of nitric acid produced matrix interference affecting hydride formation (Kumar and Riyazuddin 2010). An alternative, which has been proposed by different authors, is the removal of the residual HNO_3 by heating or diluting the digested sample with HCl before its introduction into the hydride generation system.

Moreda-Piñeiro et al. (2007) have developed a sample treatment method for total arsenic determination using acetic acid, based on pressurized liquid extraction (PLE). This methodology offered accurate and repeatable results, in a shorter time (7 min) than that needed by the traditional microwave assisted acid digestion processes (60 min).

Extraction procedures are usually performed in speciation studies. Extraction is a critical step since it can affect both qualitative and quantitative results. Hence, the integrity of the arsenic species must be ensured when speciation studies are performed and high extraction efficiencies are desirable. The different extraction procedures used in arsenic speciation studies in seaweed samples are shown in Table 3. Solvent extraction, microwave-assisted extraction and pressurized liquid extraction (PLE) are the main extraction techniques found in the literature for this purpose.

Solvent extraction is assisted by sonication (ultrasound bath, ultrasonic probe) or mechanical shaking. Deionized water (García Salgado 2006; Pedersen and Francesconi 2000; Nischwitz et al. 2006; Nischwitz and

Table 2. Total arsenic determination in seaweed samples.

Seaweed	Digestion procedure	Arsenic detection	Total arsenic concentration	CRM	Ref.
Fucus spiralis and *Halidrys siliquosa* seaweed	Approximately 100 mg of seaweed powder were mixed with 0.30 mL (30 g Mg(NO$_3$)$_2$ + 50 g MgO + 500 mL of DIW). The mixture was dried at 80°C overnight and heated in a muffle furnace at 200°C for 1 h, 300°C for 1 h and 500°C for 8 h. Residue was dissolved in 2.5 mL of 6M HCl and mixed with 2.5 mL of DIW.	HG-AAS	31.5 ± 0.9 and 21.3 ± 0.6 µg g^{-1} in *Fucus* sp. and *Halidrys siliquosa*, respectively		(Raber et al. 2000)
Fucales and *laminares*, *Hizkia fusiforme*, *Himanthalia*, *Palmaria palmata*, *Porphyra umbilicalis*, *Sargassum lacerifolium*, *Undaria pinnatifida* seaweed	Total arsenic determinations: 0.5 g of sample was digested overnight with 5 mL of 65% HNO$_3$, followed by addition of 35 mL of water and digestion at 90°C in a water bath. Inorganic arsenic (i-As) extraction: Separation of arsenic as AsCl$_3$ by distillation from organoarsenic compounds, following the method of Whyte and Englar (1983).	GFAAS	Total arsenic concentration: 3.5–134 µg g^{-1} i-As: 1.9–62 µg g^{-1} (i-As constituted between 36–74% of the t-As)	*Laminaria digitata* (a reference seaweed) Total As: 145 µg g^{-1}, inorganic As: 61 µg g^{-1}	(McSheehy and Szpunar 2000)
Laminaria digitata and *Fucus vesiculosis* seaweed	Mineralisation in a muffle oven following the method of Penrose et al. (1975).	HG-AAS	43 and 140 µg g^{-1} dry mass for *Laminaria d.* and *Fucus v.*, respectively	BCR-279 (*Ulva lactuca*): 2.89–3.29 µg g^{-1}	(Pedersen and Francesconi 2000)
Hizkia fusiforme seaweed	Microwave assisted acid digestion: 0.2 g of algal material with 5 mL of nitric acid and 0.5 mL of 33% hydrogen peroxide. Digests were diluted to 50 mL.	ICP-MS	87 ± 5 mg As kg^{-1}		(Kuehnelt et al. 2001)

Hizikia fusiforme and Laminaria seaweed	Microwave assisted acid digestion: 0.2 g of sample with 4 mL of HNO$_3$ and 1 mL of H$_2$O$_2$ (180°C, 30 min). The digests were filled to 15 mL with DIW.	ICP-MS	51.2 ± 1.1 and 49.5 ± 1.0 µg g^{-1} for Hizikia and Laminaria, respectively		(Kohlmeyer et al. 2002)
Phacophyta, Rhodophyta, Chlorophyta seaweed	Microwave assisted acid digestion: 0.07 g of sample with 1 mL of HNO$_3$. The digests were filled to 10 mL with DIW.	ICP-MS	13.0 µg g^{-1} for *Padina fraseri* (Phaeophyta), 5.47 µg g^{-1} for *Laurencia* sp. (Rhodophyta) and 5.77 µg g^{-1} for *Cladophora subsimplex* (Chlorophyta)	NIES-09 (*Sargasso*): 106–124 µg g^{-1} BCR-279 (*Ulta lactuca*): 2.89–3.29 µg g^{-1}	(Tukai et al. 2002)
Laminaria japonica, Porphyra crispate and *Eucheuma denticulatum* seaweed	Samples were irradiated at a thermal neutron flux of 1.2 * 10^{12} n cm^{-2}s^{-1} for 2.5 h. The γ-rays of the ^{76}As (t$_{1/2}$ 26.3 h) isotope were measured with a multi-channel analyser Ge(Li) detector.	INAA	43.2 ± 0.4, 30.1 ± 1.3 and 5.6 ± 0.1 µg g^{-1}, for *Laminaria, Porphyra* and *Eucheuma* respectively	NIST SRM 1571 (Orchard Leaves): 8–12 µg g^{-1}	(Van Hulle et al. 2002)
Hizikia fusiforme, Laminaria, Porphyra and a roasted seaweed	The same as previously reported (Kohlmeyer et al. 2002) (0.2 g of sample + 4 mL of HNO$_3$ + 1 mL of H$_2$O$_2$ (180°C, 30 min))	ICP-MS	Between 50–110 µg g^{-1} in brown seaweed (*Hizikia fusiforme* and *Laminaria*). And between 30–40 µg g^{-1} for Phorphyra and a roasted seaweed	BCR-279 (*Ulva lactuca*): 2.89–3.29 µg g^{-1}	(Kohlmeyer et al. 2003)

Table 2. contd....

Table 2. contd.

Seaweed	Digestion procedure	Arsenic detection	Total arsenic concentration	CRM	Ref.
Phorphyra seaweed	Microwave assisted acid digestion: 0.5 g of sample with 4 mL of HNO_3:H_2O_2 (3:1 (v/v)) for 3–4 min at 750 W, three times. The final solution was diluted to 25 mL with DIW.	ICP-MS	2.1–21.6 µg As g^{-1}	NIST SRM 1571 (Orchard Leaves): 8–12 µg g^{-1}	(Wei et al. 2003)
Five red seaweed 'Zicai' and five Brown seaweed 'Haidai'	Microwave assisted acid digestion: 0.1 g of dried sample with 3 mL of HNO_3 and 2 mL H_2O_2 for 5 min at 315 W. The final solution was diluted to 10 mL with DIW.	ICP-MS	1.7–38.7 µg g^{-1}	BCR-CRM 627 (Tuna fish): 4.5–5.1 µg g^{-1}	(Li et al. 2003)
Sargassum piluliferum and *Melanosiphen intestinalis* seaweed	Microwave assisted acid digestion: 200 mg (or 100 mg) powdered samples with 5.0 mL of HNO_3 and 1.0 mL (or 2 mL) H_2O_2. Digests were diluted to 50 mL with DIW.	ICP-MS	133 ± 7 µg g^{-1} for *S. piluliferum* and 14.2 ± 1.6 µg g^{-1} for *M. intestinalis*	DORM-2 (dogfish muscle), with AB (16.4 ± 1.1 µg g^{-1}) and TeMAs$^+$ (0.248 ± 0.054 µg g^{-1}) certified values	(Karthikeyan and Hirata 2004; Hirata and Toshimitsu 2005)
Canadian Kelp	Microwave assisted acid digestion.	ICP-MS	18.9 ± 0.8 µg g^{-1} for "old" Canadian kelp and 39.6 ± 2.4 µg g^{-1} for "new" Canadian kelp		(Nischwitz et al. 2006; Nischwitz and Pergantis 2006)
Chlorella vulgaris, *Hizikia fusiformis* and *Laminaria digitata* seaweed	Microwave assisted acid digestion: 250 mg of sample with 10 mL of HNO_3 (70% (v/v)) (30 min, 225 psi and 210°C). Digests were diluted to 25 mL with DIW.	ICP-AES	88 ± 6 µg g^{-1}, 41 ± 4 µg g^{-1} and 39 ± 3 µg g^{-1} for *Hizikia*, *Laminaria* and *Chlorela*, respectively	NIES-09 (*Sargasso*): 106–124 µg g^{-1}	García Salgado et al. 2006a and 2006b; García Salgado et al. 2008

Hizikia fusiforme seaweed	Microwave acid digestion: *Hijiki* sample with 5 mL of HNO_3 (60%) and 2 mL H_2O_2 (30%), Sulfuric acid (1.5 mL, 96%) was added to the degraded acid solution and the beaker was covered with a glass dish. Digestion was performed on a hotplace (<230°C) until sulfur trioxide fumes appeared. After cooling, 0.1 mL of ammonium hydrogencitrate (25% (w/w) was added. Digested solution was neutralized with ammonium hydroxide. Hydrochloric acid (4 mL, 25% (w/w)), ascorbic acid (2 mL, 20% (w/w)) and potassium iodine (2 mL, 20% (w/w)) were added to the sample solutions. Digests were diluted to 100 mL.	HG-AAS	Total arsenic concentrations in *Hizikia* from Japan, South Korea and China ranged between 41.7–46.7, 65.6–79.8 and 36.0–48.6 µg g⁻¹, respectively. Around 28.2–58.8% of the t-As in seaweed was removed after soaking, and 49.3–60.5% was removed after cooking		(Ichikawa et al. 2006)
Iridaea cordata, Ascoseira mirabilis, Adenocystis utricularis, Desmarestia menziesii and *Gigartina skottbergii* seaweed	Acid digestion in a CEM microwave (maximum temperature 200°C, maximum pressure 350 psi): 0.25 g of seaweed with 5 mL of ultrapure HNO_3, 2 mL of DIW and 1 mL of ultrapure HF. After cooling, digests were diluted to 50 mL with DIW.	ICP-MS	8.4–29.3 µg g⁻¹	NIST SRM 1571 (Orchard Leaves): 8–12 µg g⁻¹	(Wuilloud et al. 2006)
Ascophyllum nodosum, Laminaria digitata, Fucus vesiculosus and kelp	Microwave assisted acid digestion: 150–200 mg of sample with 5 mL of HNO_3. Digests were diluted to 50 mL with DIW.	ICP-MS	25–83 µg g⁻¹	BCR-CRM 627 (Tuna fish): 4.5–5.1 µg g⁻¹	(Nischwitz and Pergantis 2006)

Table 2. contd....

Table 2. contd.

Seaweed	Digestion procedure	Arsenic detection	Total arsenic concentration	CRM	Ref.
Porphyra (Nori), *Palmaria* (Dulse), *Undaria pinnatifida* (Wakame), *Himanthalia elongata* (Sea spaghetti), *Laminaria ochroleuca* (Kombu) *and Ulva rigida* (Sea lettuce) seaweed	Pressurized Liquid Extraction: 0.5 g of sample was mixed with 2.5 g of inert diatomaceous earth (DE). The mixture was introduced into Dionex standard 11 mL stainless steel extraction cell containing cellulose filters. PLE conditions: 0.75 M acetic acid, 25°C extraction temperature, 5 min extraction time, 10.3 MPa pressure, and one extraction step. Samples were finally diluted to 25 mL with DIW.	ICP-OES	48.8 ± 5.0, 8.2 ± 0.2, 19 ± 0.3, 14.7 ± 0.2, 0.79 ± 0.01 and 13.2 ± 0.2 µg g^{-1} in Kombu, Sea spaghetti, Nori, Dulse, Sea lettuce and Wakame, respectively	NIES-03 (*Chlorella kessleri*): non-certified As value NIES-09 (*Sargasso*): 106–124 µg g^{-1}	(Moreda-Piñeiro et al. 2007)
Sargassum fulvellum, *Sargassum piluliferum*, *Undaria pinnatifida*, Kelp, *Hizikia fusiforme* and *Pelvetia wrightii* seaweed	100 mg of powdered samples were acid digested as previously reported (Karthikeyan and Hirata 2004) (5.0 mL of HNO$_3$ and 1.0 mL of H$_2$O$_2$).	ICP-MS	14.5–288.0 µg g^{-1}		(Hirata and Toshimitsu 2007)
Hijiki, Kombu, Arame, Wakame and *Nori* seaweed	Total As determination: Samples (0.5 g dry weight) were digested with 5 mL of concentrated HNO$_3$ by using a microwave-assisted system. Digested samples were made up to volume (10 mL) with water. Inorganic As determination: Arsenic was converted into As(III) after dissolution of samples in concentrated hydrochloric acid. As(III) was extracted into chloroform. The arsenic was back extracted into dilute hydrochloric acid.	ICP-MS	t-As in seaweed: 95–124 µg g^{-1}, 28–32 µg g^{-1}, 29–42 µg g^{-1}, 19–75 µg g^{-1} and 18–32 µg g^{-1} for *Hijiki, Arame, Wakame, Kombu* and *Nori*, respectively. I-As in seaweed: only detected in Hijiki samples (73–96 µg g^{-1}). Mean t-As after seaweed		(Rose et al. 2007)

Seaweed	Digestion	Technique	Results	CRM	Reference
			preparation: 5.4, 1.1, 0.4 and 0.3 µg g⁻¹ for *Hijiki, Arame, Wakame* and *Kombu*, respectively. I-As in prepared seaweed: 5.1–23 µg g⁻¹ in *Hijiki* samples. I-As in soaking water: 0.4–4.0 µg g⁻¹ in *Hijiki* samples		(Jiang et al. 2009)
Laminaria japonica, Porphyra tenera and *Sarcodia montagneana* seaweed	Microwave assisted acid digestion: 200 mg of sample + 4 mL HNO_3 + 1 mL H_2O_2 (digestion program: 120°C, 0.5MPa, 5 min and 140°C, 1 MPa, 10 min). Digests were heated in a 50 mL beaker to remove the residual HNO_3. Then, 10 mL of reducing agent (10% ascorbic acid + 10% thiourea) was added and samples were acidified with H_2SO_4 to a final concentration of 1M.	EC-HG-AFS	11.54 ± 0.53, 11.32 ± 0.58 and 16.25 ± 0.81 µg g⁻¹ in *Laminaria japonica, Porphyra tenera* and *Sarcodia montagneana*, respectively	GBW08517 (*Laminaria japonica* Aresch): 11.5–16.3 µg g⁻¹. GBW10023 (*Porphyra crispata*): 21–33 µg g⁻¹	
Sargassum fusiforme seaweed	Pre-digestion overnight at 25°C: 0.2 mg of sample, 8 mL of HNO_3 and 2 mL of H_2O_2. Digestion in an ETHOS 1 microwave oven according to the US EPA method 3052. Digested solution was diluted with DIW to 25 g and filtered (0.45 µm).	ICP-MS	65.3–90.3 µg g⁻¹ in *Sargassum* samples	TORT-2 (Lobster hepatopancreas): 19.8–22.4 µg g⁻¹	(Han et al. 2009)

Table 2. contd....

Table 2. contd.

Seaweed	Digestion procedure	Arsenic detection	Total arsenic concentration	CRM	Ref.
Porphyra umbilicalios (Nori), *Palmaria palmata* (Dulse), *Undaria pinnatifida* (Wakame), *Himanthalia elongata* (Sea spaghetti), *Laminaria ochrouleca* (Kombu) and *Ulva rigida* (Sea lettuce) seaweed	Microwave assisted acid digestion: 0.2 g of sample with 8 mL of 69% HNO$_3$ and 2 mL of 33% H$_2$O$_2$ (maximum temperature 190°C). Digests were diluted up to 25 mL with DIW.	HG-AFS	t-As concentrations in Wakame, Sea spaghetti, Kombu, Dulse, Nori and Sea lettuce seaweed ranged from 27.9–45.5, 10.5–16.0, 147.3–383.8, 3.1–7.3, 10.7–24.8, 2.7–6.0 µg g^{-1}, respectively	NIES-09 (*Sargasso*): 106–124 µg g^{-1}	(García-Sartal et al. 2010)
Himanthalia elongata (Sea spaghetti), *Laminaria ochrouleca* (Kombu), *Undaria pinnatifida* (Wakame), *Porphyra umbilicalis* (Nori), *Eisenia arbórea* (Arame), *Hizikia fusiformis* (Hijikia), *Fucus vesiculosus* (Fucus) and *Laminaria digitata* (Laminaria) seaweed	Microwave assisted acid digestion: 250 mg of sample with 10 mL of HNO$_3$/H$_2$O$_5$ (1:1). The digestion method applied a temperature slope of 200°C over 15 min with a hold time of 10 min. Digests were diluted up to 50 mL with DIW.	ICP-AES	23–126 µg g^{-1}	NIES-09 (*Sargasso*): 106–124 µg g^{-1}	(García Salgado et al. 2012b)

Abbreviations and acronyms: CRM (certified reference material), Ref (references); DIW (de-ionized water); HG (hydride generation); AAS (atomic absorption spectroscopy); GFAAS (graphite furnace atomic absorption spectrometry); i-As (inorganic arsenic); t-As (total arsenic); ICP-MS (inductively coupled plasma—mass spectrometry); INAA (instrumental neutron activation analysis); ICP-AES (inductively coupled plasma —atomic emission spectroscopy); ICP-OES (inductively coupled plasma—optical emission spectroscopy); EC (electrochemical); AFS (atomic fluorescence spectrosocopy)

Table 3. Arsenic speciation in seaweed samples.

Seaweed	Arsenic species extraction procedure	Arsenic species detection	Chromatographic conditions	Results	Ref.
Fucus spiralis and *Halidrys siliquosa* seaweed	About 0.1 mg of seaweed powder was extracted with water/methanol (1 + 1, 1.0 mL) in a sonicator for 40 s. After centrifugation (5,000 g), supernatant was collected. This procedure was repeated twice more, and supernatants were combined. After evaporation to dryness at 30°C, dry residue was redissolved in 1 mL of water and filtered (0.45 µm)	HPLC-ICP-MS	Anion-exchange ▪ Column: PRP-X100 (Hamilton Company) ▪ Mobile Phase: 20 mM $(NH_4)H_2PO_4$, pH: 5.6 ▪ Flow rate: 1.5 mL.min⁻¹ ▪ Vinj: 100 µL ▪ Tª: 40°C	About 61% of the arsenic in *Fucus* sp. and 68% in *Halidrys siliquosa* were extracted. Most of the arsenic extracted from *Fucus* sp. was present in the form of sulfate-ribose (9.7 ± 0.2 µg g⁻¹, 56%), sulfonate-ribose (4.6 ± 0.3 µg g⁻¹, 26%) and phosphate ribose (1.95 ± 0.02 µg g⁻¹, 11%). In *Halidrys siliquosa*, sulfate-ribose was the main arsenic specie (6.16 ± 0.07 µg g⁻¹, 55%), followed by sulfonate-ribose (1.89 ± 0.04 µg g⁻¹, 17%), and phosphate-ribose (1.10 ± 0.06 µg g⁻¹, 10%)	(Raber et al. 2000)
Kelp powder	About 0.500 g of sample was extracted with water/methanol (1 + 1, v/v, 5 mL). Mixture was sonicated (20 min, 25°C) and centrifuged (20 min). Residue was extracted twice and supernatants combined. After evaporation to dryness, residue was redissolved in 10 mL of DIW and filtered (0.45 µm)	HPLC-ICP-MS	Reversed-phase ▪ Column: Discovery C_{18} ▪ Mobile Phase: 5 mM tetrabutylammonium hydroxide pH: 6.2 ▪ Flow rate: 0.7 mL.min⁻¹	Relative amounts of glycerol, sulfate, sulfonate and phosphate riboses and DMA were 31, 12.5, 49, 4.9 and 2.6%, respectively	(Wangkarn and Pergantis 2000)

Table 3. contd....

Table 3. contd.

Seaweed	Arsenic species extraction procedure	Arsenic species detection	Chromatographic conditions	Results	Ref.
Sargassum lacerifolium seaweed	Ultrasound extraction of 1 g of seaweed with water/methanol (1 + 1, 20 mL) for 3 h. After centrifugation (2500 rpm, 20 min), supernatants were removed and reserved. The extraction was repeated once more with water/methanol (1 + 9, v/v), and the residue washed three times with the extracting mixture. Solvent was removed in combined extracts and washings by means of a rotary evaporator (40°C). The residue was redissolved in 10 mL of water	HPLC-ICP-MS	**Size-exclusion** • Column: Superdex peptide HR 10/30 (Pharmacia Biotech) • Mobile Phase: 1% aq. Acetic acid pH: 3.0 • Flow rate: 0.6 mL.min^{-1} • Vinj: 100 µL **Anion-exchange** • Column: Supelcosil SAX1(Supelco) • Mobile Phase: 5–16 mM phosphate buffer pH: 6.0 • Flow rate: 1.0 mL.min^{-1} • Vinj: 100 µL	Phosphate, sulphate and sulfonate riboses have been identified in *Sargassum* by AE-HPLC-ICP-MS	(Szpunar et al. 2000)
Fucales and *laminares*, *Hizikia fusiforme*, *Himanthalia*, *Palmaria palmata*, *Porphyra umbilicalis*, *Sargassum lacerifolium*, *Undaria pinnatifida* seaweed	The same as previously reported (Szpunar et al. 2000) (ultrasound assisted extraction: 1 g of sample with water/methanol (1 + 1, 20 mL) for 3 h)	HPLC-ICP-MS	**Size-exclusion** • Column: Superdex peptide HR 10/30 (Pharmacia Biotech) • Mobile Phase: 1% aq. Acetic acid pH: 3.0 • Flow rate: 0.6 mL.min^{-1} • Vinj: 100 µL **Anion-exchange** • Column: Supelcosil SAX1(Supelco) • Mobile Phase: 5–30 mM phosphate buffer pH: 6.0 • Flow rate: 1.0 mL.min^{-1}	SEC-HPLC-ICP-MS chromatogram of an extract of *Hizikia* seaweed shows three resolved peaks (fractions II, III, IV), the first one preceeded by a shoulder (fraction I). Fraction I was confirmed by spiking with sulfate, sulfonate and phosphate riboses. Fraction III matched the elution volumes of MMA and DMA. Finally, fraction IV was identified as As(V). Phosphate and sulphate riboses have been identified in fraction	(McSheehy and Szpunar 2000; McSheehy et al. 2000)

Sample	Preparation	Technique	Method	Results	Reference
			• Injection volume: 100 μL	I of SEC, phosphate, sulfate and sulfonate riboses in fraction II of SEC, MMA and DMA in fraction III of SEC, and As(V) in fraction IV of SEC by AE-HPLC-ICP-MS	
Laminaria digitata and *Fucus vesiculosis* seaweed	A portion of 200 mg of powder seaweed was extracted with 5 mL of water with a sonication probe. Mixture was centrifuged (3000 g, 5 min) and filtered (0.45 μm)	LC/MS	Anion-exchange • Column: PRP-X100 (Hamilton Company) • Mobile Phase: 20 mM NH_4HCO_3 (10% MeOH) pH: 10.3 • Flow rate: 0.4 mL.min^{-1} • Vinj: 10 μL	About 93% of the arsenic in *Laminaria* and 82% in *Fucus* was extracted. Phosphate and sulfate riboses were identified in the extracts of *Laminaria* and *Fucus*. Sulfonate ribose was found in *Fucus*. The concentration of arsenosugars in extracts accounted for 1.25 and 3.96 μg g^{-1} for *Laminaria* and *Fucus*, corresponding to 78 and 86% of the t-As in the extract, respectively	(Pedersen and Francesconi, 2000)
Seaweed A and *Sargassum muticum* seaweed	Arsenic species were extracted via pressurized liquid extraction (PLE). Optimized conditions are as follows: 3 mL PLE cell with 30/70 (w/w) MeOH/H₂O, a pressure of 500 psi, ambient temperature, 1 min heat step, 1 min static step, 3 cycles, 90% vol. flush and 120 s purge	HPLC-ICP-MS	Anion-exchange • Column: PRP-X100 (Hamilton Company) • Mobile Phase: 20 mM $(NH_4)_2CO_3$ pH: 9.0 • Flow rate: 1.0 mL.min^{-1} • Vinj: 100 μL	PLE extraction efficiencies were 25.6% and 50.5% for seaweed A and *Sargassum*, respectively. Glycerol (3.9%), phosphate (2.6%), sulfate (15.4%) and sulfonate (8.2%) riboses, as well As(III) (8.1%), DMA (7.5%) and As(V) (54.4%) were detected in seaweed A. Glycerol (1.9%), phosphate (8.3%), sulfate (45.1%) and sulfonate (15.8%) riboses, as well as DMA (1.0%) and As(V) (27.8%) were detected in *Sargassum*. Results are expressed as relative area percentage	(Gallagher et al. 2001)

Table 3. contd....

Table 3. contd.

Seaweed	Arsenic species extraction procedure	Arsenic species detection	Chromatographic conditions	Results	Ref.
Hizikia fusiforme seaweed	Three extraction methods were described. Approximately 10 mL of 1.5 M orthophosphoric acid, 10 mL of water or 10 mL of methanol/water (9 + 1) were added to 500 mg of seaweed. A cross-shaped rotor was used to extract the arsenic by rotating the vials containing the mixture (45 rpm, 14 h, 25°C). The acid extracts were neutralized with 25% aqueous ammonia, filled to 50 mL, centrifuged and filtered. Water extracts were centrifuged and filtered. Methanolic extracts were centrifuged. Methanolic residues were washed three times with 10 mL of methanol/water (9 + 1). Combined methanolic supernatants were rotaevaporated and the residues were dissolved in 10 mL of water and filtered	HPLC-ICP-MS	Anion-exchange • Column: Hamilton PRP-X100 (Hamilton Company) • Mobile Phase: 20 mM NH$_4$H$_2$PO$_4$ pH: 5.6 or 6.0 Cation-exchange • Column: Supelcosil LC-SCX • Mobile Phase: 20 mM pyridine buffer pH: 2.6	About 76% of the total arsenic was extracted with orthophosphoric acid; 65% of the total arsenic was extracted with water and only 33% when methanol/water (9 + 1) was used as extractant solution. Glycerol ribose and As(V) were best extracted with orthophosphoric acid. DMA and sulfate-ribose were slightly better extracted with orthophosphoric acid and water, respectively. No differences in the extractability of phosphate and sulfonate riboses were observed. MMA, AB, AC, TMAO and TETRA were below the detection limit (50 µg As kg^{-1})	(Kuehnelt et al. 2001)

Seaweed	Method	Extraction	HPLC conditions	Results	Reference
Hizikia fusiforme and *Laminaria* seaweed	HPLC-ICP-MS	Sample (0.2 g) was weighed and methanol/water (20 mL, 3:1 v/v) was added. The mixture was homogenized in an UltraTurrax® T25 basis dispersion unit (2 min). Samples were centrifuged (3500 rpm) and supernatant was separated. The residue was washed twice with 10 mL of methanol/water (10 mL, 1:1 v/v) and mechanical shaken. Combined supernatants were filled with water to 50 mL. The extracts were filtered (0.22 µm) and diluted 100-fold	Anion-exchange - Guard column: IonPak AG7 (Dionex) - Column: IonPak AS7 (Dionex) - Mobile Phase: 0.5–50 mM HNO_3 + 0.05 mM benzene-1,2-disulfonic acid dipotassium salt + 0.5% MeOH pH: 1.8–3.4 - Flow rate: 1.0 mL.min⁻¹	AC, AB, TMAO, TMAs and As(III) were absent in both seaweed. MMA (1.75 µg.g⁻¹) DMA (1.32 µg g⁻¹), As(V) (12.3 µg g⁻¹), sulfate ribose (31.4 µg g⁻¹), sulfonate ribose (4.17 µg g⁻¹) and two unknown species were identified and quantified in *Hizikia*. MMA (2.74 µg.g⁻¹), DMA (0.27 µg g⁻¹), As(V) (13.7 µg g⁻¹), sulfonate ribose (28.2 µg g⁻¹), sulfate ribose (1.37 µg g⁻¹) and an unknown peak were identified and quantified in *Laminaria*	(Kohlmeyer et al. 2002)
Phaeophyta, Rhodophyta, Chlorophyta seaweed	HPLC-ICP-MS	A sample of 0.05 g, 0.07 g and 0.08 g were weighed for Phaeophyta, Rhodophyta and Chlorophyta, respectively. Sample was mixed with aqueous methanol (56%, 66% and 78% (v/v) for Phaeophyta, Rhodophyta and Chlorophyta, respectively). Samples were extracted in a microwave oven (5 min, 60°C) followed by centrifugation (4000 rpm, 20 min). Extraction was	Anion-exchange - Column: Hamilton PRP-X100 (Hamilton Company) - Mobile Phase: 20 mM ammonium phosphate buffer pH: 9.2 - Flow rate: 1.5 mL.min⁻¹ - T°: 45°C Cation-exchange - Column: Supelcosil LC-SCX - Mobile Phase: 20 mM pyridine buffer pH: 2.6 - Flow rate: 1.5 mL.min⁻¹ - T°: 45°C	Arsenic extraction efficiencies were 88 ± 9%, 90 ± 14, 88 ± 6% for Phaeophyta, Rhodophyta and Chlorophyta seaweed, respectively. Phosphate and glycerol riboses and trace quantities of sulfonate ribose were the main arsenic species found in Phaeophyta. Sulfonate ribose was the dominant specie in Rhodophyta samples, and traces of glycerol, sulfate and phosphate riboses have also been found. Only two arsenoribosides have been found in Chlorophyta samples (glycerol and phosphate	(Tukai et al. 2002)

Table 3. contd....

Table 3. contd.

Seaweed	Arsenic species extraction procedure	Arsenic species detection	Chromatographic conditions	Results	Ref.
	repeated and supernatants combined. An aliquot, 4 mL, was evaporated to dryness (45°C) by centrifugal evaporation. Residues were redissolved in 0.8 mL of DIW and filtered			riboses), the presence of glycerol ribose being dominant	
Laminaria japonica, Porphyra crispata and *Eucheuma denticulatum* seaweed	Portions of 0.1–0.2 g were extracted with 6 mL of aqueous MeOH (50% (v/v)). Samples were ultrasonicated (15 min) and centrifuged (9000 rpm, 15 min). This procedure was repeated five times more, and supernatants were combined. Combined fractions were left overnight in an oven (50°C). Residue was diluted with DIW. For HPLC-ESI-MS analysis, extracts were subjected to a purification step with diethyl ether	HPLC-ICP-MS and HPLC-ESI-MS	ICP-MS chromatographic conditions: Anion-exchange ▪ Column: Hamilton PRP-X100 (Hamilton Company) ▪ Mobile Phase: (A) 20 mM $(NH_4)_2HPO_4$, pH: 7.0; (B) 20 mM NH_4HCO_3, pH: 7.5 ▪ Flow rate: 1.0 mL.min^{-1} ▪ Injection volume: 50 µL Cation-exchange ▪ Column: Dionex Ionpac CS-10 ▪ Mobile Phase: 20 mM pyridine pH: 2.4 ▪ Flow rate: 1.0 mL.min^{-1} ▪ Injection volume: 50 µL ESI-MS chromatographic conditions: Anion-exchange ▪ Column: Hamilton PRP-X100 (Hamilton Company)	Extraction efficiencies were 76.4%, 69.8% and 25.0% for *Laminaria japonica, Porphyra crispate* and *Eucheuma denticulatum* seaweed, respectively. *Porphyra crispata* contains glycerol and phosphate riboses and trace quantities of DMA and sulfonate ribose. High amounts of DMA were found in *Laminaria* (16.8% of the total extractable arsenic). Glycerol and phosphate riboses were also detected, as well as a negligible amount of sulfate ribose	(Van Hulle et al. 2002)

Species	Sample preparation	Instrument	Conditions	Results	Reference
			• Mobile Phase: 20 mM NH_4HCO_3, pH: 7.7 in 20% MeOH • Flow rate: 1.0 mL.min^{-1} • Vinj: 50 µL	In the *Porphyra* sample, the phosphate ribose (26.5 µg g^{-1}) and glycerol ribose (4.1 µg g^{-1}) were the main arsenic species found. DMA (0.2 µg g^{-1}) was also detected. In *Hizikia*, As(V) (25.6 µg g^{-1}), sulfate ribose (10.2 µg g^{-1}) and DMA (2.8 µg g^{-1}) were the main arsenical found. As(V) (12 µg g^{-1}) and phosphate ribose (2.0 µg g^{-1}) were the main arsenic species detected in roasted seaweed	(Kohlmeyer et al. 2003)
Hizikia fusiforme, Porphyra and a roasted seaweed	The same as previously reported (Kohlmeyer et al. 2002)	HPLC-ICP-MS	The same conditions as previously reported (Kohlmeyer et al. 2002)		
Porphyra seaweed	One gram of sample was extracted with water/methanol (1 + 1, v/v, 20 mL). Mixture was sonicated (15 min) and centrifuged (15 min). The procedure was repeated twice more and supernatants were collected. After rotary evaporation, residue was redissolved with a minimal amount of DIW	HPLC-ICP-MS	<u>Anion-exchange</u> • Column: Hamilton PRP-X100 (Hamilton Company) • Mobile Phase: 20 mM NH_4HCO_3, pH: 10.2 • Flow rate: 1.0 mL.min^{-1} • Vinj: 50 µL <u>Cation-exchange</u> • Column: Ionpac CS-10 (Dionex) • Mobile Phase: 20 mM pyridinium in water pH: 2.4 • Flow rate: 1.0 mL.min^{-1} • Vinj: 50 µL	Arsenic extraction efficiencies were higher than 90%. Two arsenosugars have been identified in *Phorphyra* seaweed: phosphate and glycerol riboses, being higher in the quantity of arsenosugar-PO_4 than that for arsenosugar-OH in most of the samples. The average ratio found for the two riboses was 1.64:1 (arsenosugar PO_4:OH), being lower in Dalian *Phorphyra* (0.079:1)	(Wei et al. 2003)

Table 3. contd....

Table 3. contd.

Seaweed	Arsenic species extraction procedure	Arsenic species detection	Chromatographic conditions	Results	Ref.
Five red seaweed "Zicai" and five Brown seaweed "Haidai"	Around 0.5 g of seaweed was placed in a centrifuge tube and 10 mL of MeOH: H_2O (1:1 (v/v)) was added. Tubes were sonicated and centrifuged (10 min). Extraction was repeated four times more. Combined extracts were rotary evaporated (30°C) to dryness and residue was dissolved in 10 mL of DIW	HPLC-ICP-MS	The same as previously reported (Wei et al., 2003). Mobile phase pH for cation-exchange chromatographic conditions was pH 2.0	Extraction efficiencies were 87–96%. Arsenosugars were detected in all seaweed extracts (1.5–33.8 $\mu g\ g^{-1}$). Arsenosugars -OH and -PO_4 were the major arsenic species found in red seaweed. However, arsenosugars -OH, -PO_4 and -SO_3, as well as DMA were the arsenic species found in brown 'Haidai'	(Li et al. 2003)
Fucus serratus and *H. fusiforme* seaweed	A portion of sample (100 mg) was extracted with 5.0 mL of Milli-Q® water and shaken overnight. After centrifugation, the supernatant was collected and filtered (0.22 µm).	HPLC-HG-ICP-MS	Anion-exchange • Column: PRP-X100 (Hamilton) • Mobile Phase: 10 mM $NH_4H_2PO_4$, pH: 6.0 • Flow rate: 1.5 mL.min^{-1} • Vinj: 20 µL • T°: 40°C	Glycerol-, phosphate-, sulfonate- and sulfate- sugars have been quantified in *F. serratus* seaweed (0.098, 0.082, 0.58 and 0.39 mean As content in µm, respectively). Glycerol-, phosphate-, sulfonate- and sulfate- sugars, as well as As(V) have been quantified in *H. fusiforme* seaweed (3.57, 1.27, 1.26, 12.3 and 31.1 mean As content in µg, respectively)	(Schmeisser et al. 2004)
Sargassum piluliferum and *Melanosiphon intestinalis* seaweed	Around 100 mg of seaweed was placed in a centrifuge tube and 10 mL of MeOH: H_2O (1:1 (v/v)) was added. Tubes were sonicated and centrifuged (15 min). Extraction was repeated. Combined extracts were diluted 20 times with Milli-Q® water	HPLC-ICP-MS	Anion-exchange • Guard column: IonPak AG4A (Dionex) • Column: IonPak AS4A (Dionex) • Mobile Phase: (A) 0.4 mM nitric acid pH: 3.3; (B) 50 mM nitric acid pH: 1.3 • Flow rate: 1.2 mL min^{-1} • Vinj: 100 µL	74% extraction efficiency percentage was found for *S. piluliferum*, and 48%, in the case of *M. intestinalis*. As(V) (113 $\mu g\ g^{-1}$) was dominant in *S. piluliferum*. As(III), MMA and DMA were also found in the range 2.00–12.6 µg As g^{-1}. Two unknown peaks (possibly phosphate and sulfonate riboses) eluted from the column.	(Karthikeyan and Hirata 2004)

Sargassum piluliferum and *Melanosiphen intestinalis*, *Ecklonia cava* Kjellman and *Myelophycus simplex* seaweed	The same as previously reported (Karthikeyan and Hirata 2004)	HPLC-ICP-MS	As(V), MMA, DMA, As(III), AB, TMAO, AC and TeMAs± separation: • Column: Capcell Pack C$_{18}$ • Mobile Phase: (A) 5 mM 1-butanesulfonic acid sodium salt; (B) 0.3 mM hexanesulfonic acid sodium salt; (C) 2 mM malonic acid + 0.5% methanol • Flow rate: 1.0 mL.min^{-1} • Vinj: 10 µL Arsenosugars separation: • Column: PRP-X100 (Hamilton Company) • Mobile Phase: 20 mM NH$_4$H$_2$PO$_4$ pH: 5.2 • Flow rate: 1.0 mL.min^{-1} • Vinj: 10 µL	As(III), DMA, MMA and an unknown peak were detected in *M. intestinalis*. 72% extraction efficiency percentage was found for *S. piluliferum*, and 51%, in the case of *M. intestinalis*. As(V) (112.4 and 47.4 µg g^{-1}) was dominant in *S. piluliferum* and *M. intestinalis*, respectively. As(III), MMA, TeMAs$^+$ and DMA were also found in the range 1.05–6.31 µg As g^{-1}. Phosphate, sulfate and sulfonate riboses have been quantified in the four seaweeds under study. Phosphate, sulfonate and sulfate riboses were between 0.18–0.61 µg As g^{-1}, 0.42–9.59 µg As g^{-1} and 1.15–2.90 µg As g^{-1}	(Hirata and Toshimitsu 2005)
Canadian and Icelandic Kelp powder	Around 1 g of Kelp was extracted with 20 mL of deionised water. Samples were shaken (10 min) and centrifuged (2000 g, 15 min). Finally, supernatants were filtered (0.45 µm)	HPLC-ESI-MS/MS and HPLC-ICP-MS	ICP-MS chromatographic conditions: Anion-pairing Reversed-phase: • Column: Discovery C$_{18}$ • Mobile Phase: 5 mM tetrabutylammonium hydroxide in 5% MeOH$_{(aq)}$ pH: 7.5 • Flow rate: 1.0 mL.min^{-1} • Vinj: 50 µL	Extraction efficiencies were 86–92% for the Canadian Kelp samples. Glycerol, phosphate, sulfonate and sulfate riboses, as well as, their Thio-analogues, have been identified in 'new' Kelp samples. Quantification of DMThioAsSugarPhosphate (0.25 ± 0.06 µg g^{-1}), DMThioAsSugarSulfonate (2.6 ± 0.5 µg g^{-1}) and	(Nischwitz et al. 2006)

Table 3. contd....

Table 3. contd.

Seaweed	Arsenic species extraction procedure	Arsenic species detection	Chromatographic conditions	Results	Ref.
			ESI-MS chromatographic conditions: Anion-exchange ■ Column: Hamilton PRP-X100 (Hamilton Company) ■ Mobile Phase: / A) 20 mM NH_4HCO_3 pH: 10; (B) 20 mM NH_4HCO_3, 40% MeOH pH: 10; ■ Flow rate: 1.0 mL.min^{-1} (24% split postcolumn) ■ Vinj: 20 µL	DMThioAsSugarSulfate (3.3 ± 0.5 µg g^{-1}) in "new" kelp samples has been performed. These thio-arsenosugars constituted about 8% of the total As eluted in 'new' kelp	
Sargassum fulvellum, Chlorella vulgaris, Hizikia fusiformis and *Laminaria digitata* seaweed	0.2 g of seaweed and 5 mL of DIW were sonicated for 30 s and then centrifuged (14,000 g, 10 min). The extraction process was repeated 2-3 times more. Mixed supernatants were rotaevaporated (60°C) and the residue was dissolved in 4 mL of DIW	HPLC-ICP-AES	Anion-exchange ■ Column: PRP-X100 (Hamilton Company) ■ Mobile Phase: 17 mM phosphate buffer pH: 5.5 ■ Flow rate: 1.0 mL.min^{-1} ■ Injection volume: 100 µL	Recoveries were found between 51–88%. As(V) was identified and quantified in *Sargassum* (38 ± 2 µg g^{-1}), *Hizikia* (46 ± 2 µg g^{-1}) and *Chlorella* (9 ± 1 µg g^{-1}). DMA was found in *Chlorella* (13 ± 1 µg g^{-1}). *Laminaria* seaweed chromatogram showed an unknown peak close to the void volume	(García Salgado et al. 2006b)
Sargassum fulvellum, Chlorella vulgaris, Hizikia fusiformis and *Laminaria digitata* seaweed	Microwave assisted extraction: 200 mg of seaweed and 8 mL of DIW (90°C, 5 min). Extracts were centrifuged (14,000 g, 10 min) and diluted up to 10 mL with DIW. The extraction process was repeated three times more. The final extract was centrifuged (14,000 g, 10°C) and filtrated (0.20 µm)	HPLC-HG-ICP-MS	The same as previously described (Salgado et al. 2006)	Recovery percentages ranged between 64–106% for the seaweed under study. As(III) (6.1± 0.4 µg g^{-1}), As(V) (11.8 ± 0.5 µg g^{-1}), MMA (2.2 ± 0.3 µg g^{-1}) and DMA (15 ± 2 µg g^{-1}) have been quantified in *Chlorella* seaweed. As(V) has been identified in *Hizikia*. (62 ± 2 µg g^{-1}) and in *Sargasso* (72 ± 3 µg g^{-1}). Arsenic species were not observed in *Laminaria*	(García Salgado et al. 2006a)

Seaweed	Sample preparation	Method	Conditions	Results	Reference
Hizikia fusiforme seaweed	*Hijiki* (5 g) was soaked in 20 mL of water (3 h, 25°C). The soaked *Hijiki* was separated by filtration. Separated *Hijiki* was homogenized and arsenic compounds were extracted by sonication (15 min) with water	HPLC-ICP-MS	▪ Column: Inertsil As ▪ Mobile Phase: 10 mM sodium 1-butanesulfonate, 4 mM tetramethylammonium hydroxide, 4 mM malonic acid, 0.5% MeOH ▪ Flow rate: 0.20 mL.min⁻¹ ▪ Vinj: 5 µL ▪ Tª: 40°C	As(V) (25.0–69.0 µg g⁻¹), As(III) (0.7–12.6 µg g⁻¹) DMA (0.2–3.1 µg g⁻¹) and arsenosugars (0.4–1.4 µg g⁻¹) were found in *Hizikia*	(Ichikawa et al. 2006)
Iridaea cordata, Ascoseira mirabilis, Adenocystis utricularis, Desmarestia menziesii and *Gigartina skottbergii* seaweed	A portion of sample (1 g) was mixed with 50 mL MeOH:H₂O (1:1). The mixture was shaken in a vortex (2 min) and then sonicated (3 h, 30°C). Extracts were centrifuged (3600 rpm, 30 min). The collected supernatant was successively filtered (0.45 µm and 0.2 µm) before chromatographic analysis	HPLC-ICP-MS and HPLC-ESI-ITMS	Anion-exchange ▪ Column: IonPac AS7 (Dionex) ▪ Mobile Phase: (A) 25 mM ammonium bicarbonate pH: 10; (B) DIW ▪ Flow rate: 0.8 mL.min⁻¹ ▪ Vinj: 5 µL Cation-exchange ▪ Column: PRP-X200 (Hamilton) ▪ Mobile Phase: 4 mM pyridine pH: 2.4 ▪ Flow rate: 0.8 mL.min⁻¹ ▪ Vinj: 5 µL	Extraction efficiencies were within 83–108%. Relative amount of each arsenic species was: 0.7–8.4%, 1.3–15.7%, 0.4–0.8%, 2.0–10.1%, 67.4%, 6.6–18.9%, 0.8–78.6% and 3.9–78.9% for As(III), As(V), MMA, DMA, arsenosugar-H, arsenosugar-OH, arsenosugar-SO₃ and arsenosugar-PO₄, respectively	(Wuilloud et al. 2006)
Ascophyllum nodosum, Laminaria digitata, Fucus vesiculosus and kelp	Algal powder (0.25 g) was soaked in 5 mL of DIW and shaken for 30 min. Supernatant was collected and the residue was extracted again in the same way.	HPLC-ICP-MS and HPLC-ES-MS/MS	Anion-exchange ▪ Column: PRP-X100 (Hamilton) ▪ Mobile Phase: 20 mM HH₄HCO₃ in 3% MeOH pH: 10 ▪ Flow rate: 1.0 mL.min⁻¹ ▪ Vinj: 20 µL	Extraction efficiencies for arsenic ranged from 59.1–74%. Arsenosugars -OH (2.4–5.0 µg g⁻¹), -PO₄ (0.8–7.3 µg g⁻¹), -SO₃ (2.0–27.5 µg g⁻¹) and -SO₄ (0.6–11.3 µg g⁻¹) and DMA (0.2–0.6 µg g⁻¹) were quantified in all extracts of the samples under study by HPLC-ICP-MS. Arsenate (7.1–15.2 µg g⁻¹) was only present in *Laminaria* seaweed samples	(Nischwitz and Pergantis 2006)

Table 3. contd....

Table 3. contd.

Seaweed	Arsenic species extraction procedure	Arsenic species detection	Chromatographic conditions	Results	Ref.
			ESI-MS chromatographic conditions: The same as previously reported (Nischwitz, Kanaki and Pergantis , 2006) For combined cation-anion exchange method, 10 mM ammonium acetate (pH 3.0), 20 mM NH_4HCO_3 in 10 % MeOH (pH: 10.0) and DIW were the mobile phase	DMAThioAsSugar –OH, -PO$_4$, SO$_3$ and –SO$_4$, as well as arsenobetaine, DMAsSugarAminoSulfonate, TMAsSugarSulfate, DMAA, and DMAsSugarManitol have been also detected in those seaweed samples	(Hirata and Toshimitsu 2007)
Sargassum fulvellum, *Sargassum piluliferum*, *Undaria pinnatifida*, Kelp, *Hizikia fusiforme* and *Myelophycus simplex* seaweed	The same as previously reported (Karthikeyan and Hirata 2004)	HPLC-ICP-MS	The same as previously reported (Hirata and Toshimitsu 2005). The mobile phase selected for arsenosugars separation was 20 mM NH_4HCO_3 pH: 8.4	Extraction efficiencies were from 9.3% to 94.6%. Phosphate ribose was found in all the seaweed under study (1.09– 9.66 µg g⁻¹). Sulfonate ribose was not detected in *Undaria pinnatifida* (0.17–14.4 µg g⁻¹). Sulfate ribose (1.17–11.9 µg g⁻¹) was not detected in *Undaria pinnatifida*, Kelp and *Myelophycus simplex* seaweed. Arsenosugars accounted for 6.0– 34.9% of the total extracted arsenic	
Sargassum fusiforme seaweed	Microwave assisted extraction: 0.2 g of sample + 10 mL of 50% (v/v) methanol-pure water at 70ºC for15 min. The suspension was centrifuged (9,000 rpm, 15 min). This procedure was repeated four times more	HPLC-ICP-MS	Anion-exchange • Column: PRP-X100 (Hamilton Company) • Mobile Phase: $(NH_4)_2CO_3$	The As(V) and DMA contents amount to 15–35 and 0.6–18 µg g⁻¹ (dry mass) of the total extractable As. Other unknown arsenic species eluted from the column. Unknown species were tentatively identified as arsenosugars	(Han et al. 2009)

Sample	Extraction	Technique	Method	Results	Reference
Undaria pinnatifida, Hizikia fusiformis and *Monostroma nitidum wittrock* seaweed	<u>Alkaline digestion for CT-HGAAS analysis</u> A 2 M NaOH solution (5 mL) was mixed with 0.05 g of sample. The mixture was heated at 90–95°C for 3 h. <u>Arsenic species extraction for HPLC-HGAFS analysis</u> A 0.25 M H_3PO_4 solution (10 mL) containing 0.5 g of powdery sample was shaken in a stoppered centrifuge tube for 12 h.	CT-HGAAS and HPLC-HGAFS	<u>HPLC-HGAFS method</u> <u>Anion-exchange</u> • Column: PRP-X100 (Hamilton) • Mobile Phase: 25 mM KH_2PO_4 pH: 5.8 • Flow rate: 1.1 mL.min⁻¹ • Vinj: 200 µL	<u>Arsenic species by HPLC-HGAFS</u> As(III) (0.64 ± 0.05 µg g⁻¹ and 0.78 ± 0.03 µg g⁻¹), DMA (1.03 ± 0.02 and 0.91 ± 0.02 µg g⁻¹) and arsenosugars –OH (7.65 ± 0.36 and 9.97 ± 0.50 µg g⁻¹), -PO_4 (9.81 ± 0.52 and 5.45 ± 0.14 µg g⁻¹) and-SO_3H (27.4 ± 1.6 and 8.43± 0.34 µg g⁻¹) have been quantified in *Laminaria* sp. and *Undaria pinnatifida*, respectively. MMA (0.27 ± 0.07 µg g⁻¹) has been detected in *Undaria pinnatifida* seaweed. As(V) (78.2 ± 0.4 µg g⁻¹) was the main arsenic specie found in *Hizikia fusiformis*. Arsenosugar –OH (10.6 ± 0.1 and 2.22 ± 0.08 µg g⁻¹) has also been quantified in *Hizikia fusiformis* and *Monostroma nitidum wittrock*, respectively. <u>Arsenic species by CT-HGAAS</u> Inorganic arsenic accounted for 0.36 ± 0.05, 0.21 ± 0.03, 79.9 ± 0.3 and 0.66 ± 0.04 µg g⁻¹ and dimethylated arsenic was 45.9 ± 0.4, 23.5 ± 0.4, 11.3 ± 0.5 and 3.32 ± 0.18 µg g⁻¹ for *Laminaria* sp., *Undaria pinnatifida, Hizikia* and *Monostroma* respectively.	(Geng et al. 2009)
Himanthalia elongata (Sea spaghetti), *Laminaria ochrouleca* (Kombu), *Undaria*	The same as previously reported (García Salgado et al. 2006)	HPLC-(UV)-HG-AFS	<u>Anion-exchange</u> • Column: Hamilton PRP-X100 (Hamilton Company) • Mobile Phase: (A) 5 mM phosphate buffer pH: 9.0; (B) 20 mM phosphate	Extraction efficiencies ranged from 49–98%. DMA (0.48 ± 0.04 µg g⁻¹), As(V) (7.0 ± 0.1 µg g⁻¹), glycerol sugar (1.5 ± 0.1 µg g⁻¹), and phosphate	(García-Salgado et al. 2012a)

Table 3. contd....

Table 3. contd.

Seaweed	Arsenic species extraction procedure	Arsenic species detection	Chromatographic conditions	Results	Ref.
pinnatifida (Wakame), *Porphyra umbilicalis* (Nori), *Eisenia arbórea* (Arame), *Hizikia fusiformis* (Hijika), *Fucus vesiculosus* (Fucus) and *Laminaria digitata* (Laminaria) seaweed			buffer pH: 9.0; (C) phosphate buffer pH: 9.0 ■ Flow rate: 1.0 mL.min⁻¹ ■ Vinj: 100 µL Cation-exchange ■ Column: PRP-X200 (Hamilton) ■ Mobile Phase: 2.5 mM pyridine pH: 2.65 ■ Flow rate: 1.0 mL.min⁻¹ ■ Injection volume: 100 µL	sugar (2.04 ± 0.01 µg g⁻¹) were quantified in Arame seaweed. DMA (0.55 ± 0.07 µg g⁻¹), As(V) (11 ± 1 µg g⁻¹) and the four arsenosugars OH-sug (2.6 ± 0.1 µg g⁻¹), PO_4-sugar (0.8 ± 0.1 µg g⁻¹), SO_3-sugar (0.6 ± 0.1 µg g⁻¹) and SO_4-sugar (7.2 ± 0.1 µg g⁻¹) were detected in Fucus seaweed. Sea spaghetti showed As(V) (2.0 ± 0.1 µg g⁻¹), OH-sugar (4.5 ± 0.4 µg g⁻¹), PO_4-sugar (0.11 ± 0.01 µg g⁻¹) and SO_3-sugar (4.2 ± 0.9 µg g⁻¹) arsenic species. DMA, As(V), Gly-sug PO_4-sug, SO_3-sug and SO_4-sug were found in Hijiki seaweed (0.44 ± 0.06, 50.3 ± 0.4, 1.05 ± 0.03, 0.4 ± 0.1, 0.7 ± 0.1 and 2.7 ± 0.4 µg g⁻¹, respectively). DMA (0.36-0.40 µg g⁻¹) was detected in only two samples of Kombu, however, As(V), Gly-sug and PO_4-sug were detected in the three Kombu samples analysed (11-32, 3.1-11 and 1.9-22 µg g⁻¹, respectively). As(V) (77 ± 3 µg g⁻¹), Gly-sug (10.2 ± 0.7 µg g⁻¹) and PO_4-sug (3.5 ± 0.1 µg g⁻¹) were detected in Lamianria. As(V) (2.2-4.5 µg g⁻¹), Gly-sug (2.68-14.3 µg g⁻¹) and PO_4-sug (1.5-10.10 µg g⁻¹) were detected in two samples of Wakame.	

Sample	Extraction	Method	Chromatography	Results	Reference
Undaria pinnatifida (Wakame) and *Hizikia fusiformis* (Hijiki) seaweed	Sample (1.0 g) was shaken overnight with a mixture of chloroform:methanol (2:1 v/v, 25 g). Supernatant was collected and washed with bicarbonate solution (1% w/v; 2*20 mL). The chloroform layer was separated, evaporated and redissolved in chloroform/acetone (1 : 1 v/v, 1 mL). Around 500 mL of solution was applied to a 'plug' of silica packed into a Pasteur pipette. The silica (5*1 mL) was washed with chloroform/acetone, 1 : 1, methanol (3*1 mL) and methanol containing 1% aqueous ammonia (10*1 mL). Five fractions were combined, evaporated and redissolved in methanol	HPLC-ES-MS and HPLC-ICP-MS	Reverse-phase: • Column: Zorbax Eclipse XDB-C8 • Mobile Phase: (A) 10 mM acetic acid pH 6.0; (B) MeOH	DMA was only quantified in one sample of Wakame (0.025 ± 0.007 µg g^{-1}). Gly-sug (1.02–1.6 µg g^{-1}) and PO$_4$-sug (13–20.1 µg g^{-1}) were quantified in two Nori seaweed. As(V) (70 ± 1 µg g^{-1}) was the main arsenic specie found in Sargasso (DMA (0.9 ± 0.1 µg g^{-1}), Gly-sug (1.0 ± 0.2 µg g^{-1}), PO$_4$-sug (1.4 ± 0.2 µg g^{-1}) and SO$_4$-sug (7 ± 2 µg g^{-1})). The lipid fraction accounted for 6.7% and 1.6% of the total arsenic in Wakame and Hijiki, respectively. Three arsenic-hydrocarbons, five arsenosugar-phospholipids and two unknown compounds have been quantified in Wakame seaweed. Three arsenic-hydrocarbons, six arsenosugar-phospholipids and two unknown compounds have been quantified in Hijiki seaweed.	(García-Salgado et al. 2012a)

Table 3. contd....

Table 3. contd.

Seaweed	Arsenic species extraction procedure	Arsenic species detection	Chromatographic conditions	Results	Ref.
Undaria pinnatifida (Wakame), *Laminaria ochroleuca* (Kombu), *Porphyra umbilicalis* (Nori) and *Ulva rigida* (Sea lettuce) seaweed	Ultrasound assisted extraction: 0.05 g of kombu, (or 0.1 g of wakame and nori and 0.2 g of sea lettuce) + 20 mM ammonium acetate (pH: 7.4), for 1 h at 25°C. Samples were centrifuged (1,000 g, 30 min) and supernatant collected and filtered (0.45 µm)	HPLC-ICP-MS and HPLC-ESI-MS-MS	Anion-exchange: • Column: Hamilton PRP-X100 (Hamilton Company) • Mobile Phase: 20 mM NH_4HCO_3 pH: 9.0 + 1% MeOH • Flow rate: 1.0 mL.min^{-1} • Vinj: 50 µL Size-exclusion: • Column: TSK gel G 3000 PWxL • Mobile phase: 10 mMNH$_4$Ac (pH 8.5) • Flow rate: 0.5 mL.min^{-1} • Vinj: 40 µL	Arsenic extraction efficiencies were between 40–61%. Arsenosugar glycerol (8.12±0.25 and 3.17±0.09), arsenosugars phosphate (3.56±0.12 and 5.35±0.14) and arsenosugars sulfonate (22.8±0.0 and 5.68±0.09) were quantified in Kombu and Wakame seaweed samples, respectively. However, only arsenosugar glycerol (4.30±0.09 and 0.56±0.03) and arsenosugars phosphate (26.2±0.4 and 0.29±0.01) were quantified in Nori and Sea lettuce seaweed samples, respectively. DMA was a minor arsenic specie found in the four seaweed extracts. And AB was only identified in Sea lettuce	(García Sartal et al. 2012c)

Abbreviations and acronyms: HPLC (high performance liquid chromatography); ICP-MS (inductively coupled plasma—mass spectrometry); SEC (size exclusion chromatography); AE (anion exchange chromatography); ESI-MS (electrospray ionisation—mass spectrometry); LC (liquid chromatography); PLE (pressurized liquid extraction); DIW (deionized-water); HG (hydride generation); ITMS (ion trap mass spectrometry);CT (cryogenic trap); AAS (atomic absorption spectroscopy); AFS (atomic fluorescence spectroscopy); UV (ultraviolet radiation); Vinj (injection volume); T° (temperature)

Pergantis 2006; Ichikawa et al. 2006) and water/methanol mixtures are the most common extractant solution used. Water/methanol 1:1 ratio (Raber et al. 2000; Van Hulle et al. 2002; Karthikeyan and Hirata 2004; Wuilloud et al. 2006; Han et al. 2009; Wangkarn and Pergantis 2000; Szpunar et al. 2000; Wei et al. 2003) and 1:3 water/methanol ratio (Kohlmeyer et al. 2002) have been used for the extraction of arsenic species. The extraction time varies from a few seconds to several hours. The most commonly selected temperature was 25°C.

Supernatants containing the arsenic species are collected after centrifugation. The extraction strategy often continues with at least two consecutive extractions of the algal pellet. The extracting solution selected for successive extractions occasionally varies in relation to the initial solution used. Methanol:water (9:1, v/v) is the commonly selected solution in successive extractions (Szpunar et al. 2000; McSheehy and Szpunar 2000). Finally, supernatants are combined and often evaporated to dryness, further redissolved, filtered and directly analysed. Due to the high water content (90%) in seaweed, different authors recommend the use of 100% MeOH instead of aqueous methanol as an extracting solution when fresh seaweed is analysed (McSheehy et al. 2000). Kuehnelt et al. (2001) have compared the extraction efficiencies and the arsenic species distribution in the arsenic extraction procedure from *Hijiki fuziforme* and DORM-2 certified reference material, using 1.5 M orthophosphoric acid and water or methanol:water (9 + 1) solutions. Around 76% of the total arsenic in *H. fuziforme* was extracted with orthophosphoric acid and lower extraction percentages were achieved with water (65%) and aqueous methanol (33%). The authors concluded that the extraction procedure should be optimized with respect to the target arsenic compound and to the kind of seaweed in the sample. Therefore, glycerol ribose, DMA and arsenic acid were better extracted with orthophosphoric acid from *H. fuziforme* seaweed. However, sulfate ribose was slightly better extracted with water than with aqueous methanol or orthophosphoric acid. Beside methanol-water mixtures, other extracting solutions have been used for the arsenic species extraction from seaweed samples. Geng et al. (2009) performed the extraction of arsenic species by using a 0.25 M H_3PO_4 solution when the samples were analysed by HPLC-HG-AFS. They used alkaline digestion (2 M NaOH extracting solution) when samples were subjected to CT-HG-AAS for speciation studies. The decomposition of proteins, and also the transformation of hydride-inactive species into hydride-active species, can be favourably carried out when the sample is subjected to an alkaline treatment. Arsenosugar-phospholipids and chloroform-methanol extracting solutions with mechanical shaking were used for the extraction of lipid-soluble compounds (García-Salgado et al. 2012a).

The use of methanol or methanol:water mixtures as extracting solutions has been discussed by Francesconi et al. (2004). The authors criticize the extended use of methanol to extract naturally-occurring arsenic species. Even though these are mainly organic species, their polarity and water-solubility is well recognized. Thus, the authors proposed water as the best solvent for the extraction of arsenic species, particularly arsenosugars.

The aggressive and destructive nature of microwave assisted extraction has probably limited its applicability in the extraction of arsenic species from seaweed samples (Niegel and Matysik 2010). Tukai et al. (2002) evaluated the effect of temperature, extraction time, solvent composition and sample mass on the extraction efficiency (using microwave energy) by using a chemometric approach. The authors found that the solvent composition and the sample mass were the most significant factors, and these were highly dependent on the type of seaweed (Phaeophyta, Rhodophyta or Chlorophyta). In another study developed by García Salgado et al. (2006a) the arsenic species were extracted with water by using microwave energy. The extraction efficiencies obtained with this method (78–98%) were better than those obtained by focused sonication (65%). The stability of arsenic species during the extraction procedure was evaluated by spiking As(III), As(V), MMA and DMA to the seaweed material. Results suggested that the arsenic species remained stable during the extraction procedure, since 93–115% of the spiked arsenic standards were recovered. Chao Han et al. (Han et al. 2009) observed that temperature was one of the most important parameters to be optimized in microwave assisted extraction, and variations in solvent composition (aqueous methanol ranging from 10–90%) and sample mass (0.1–0.3 g) led to similar extraction efficiencies. These authors found that the efficiency of the extraction increased at temperatures close to the boiling point of the solvent (70°C), probably due to the thermal stability of the analytes.

Extraction can also be performed using pressurized liquid extraction (PLE). High extraction efficiency, reproducibility, low solvent consumption, ease of handling and high degrees of automation (Moreda-Piñeiro et al. 2007) are the main advantages of this technique. Parameters such as solvent mixtures, applied pressure, temperature and static time must be optimized for more quantitative extraction efficiency (Niegel and Matysik 2010). Gallagher et al. (2001) used a 3 mL PLE cell for the extraction of arsenicals from ribbon kelp. The authors concluded that the pressure applied did not affect the arsenicals' distribution and had little effect on the arsenic recovery percentage. The recovery percentages and the arsenic distribution were the same, using one and five minutes as static time. On the other hand, extraction temperature had a pronounced effect on extraction efficiency, increasing at higher temperatures. Another determining parameter was the solvent composition. Results indicated that 100% pure water extracted more arsenical species from the kelp sample. However, the sample had a tendency to swell upon hydration when pure water was used as an extractant, and a cell blockage may be produced; as a solution, 30% aqueous methanol was used.

There is little literature on arsenic species' stability in seaweed or seaweed extracts. The stability of the arsenical is influenced by the sample matrix and the concentration of arsenic species, therefore these parameters should be included in the stability studies. García Salgado et al. (2008) have assessed the stability of the total arsenic and arsenic species in *Hijiki* seaweed and NIES no 9 *Sargasso*, as well as in their aqueous extracts. A stable behaviour for at least 12 months was observed in those samples when they were stored in polystyrene containers at 20°C. *Sargasso* aqueous extracts were stable for at least 15 days and *Hijiki* aqueous extracts for 10 days when they were stored in polystyrene containers at 18 and 4°C, respectively.

2.2 Total arsenic determination

Atomic absorption spectroscopy (AAS) coupled to a graphite furnace (GF) (McSheehy and Szpunar 2000; Raber et al. 2000) or a hydride generation system (HG) (Raber et al. 2000; Pedersen and Francesconi 2000; Ichikawa et al. 2006; Raber et al. 2009), atomic fluorescence spectroscopy (AFS) (Jiang et al. 2009; García-Sartal et al. 2010) with hydride generation devices and inductively coupled plasma coupled to either mass spectrometry (ICP-MS) (Tukai et al. 2002; Karthikeyan and Hirata 2004; Wuilloud et al. 2006; Kohlmeyer et al. 2003; Han et al. 2009; Kuehnelt et al. 2001; Nischwitz et al. 2006; Nischwitz and Pergantis 2006; Wei et al. 2003; Kohlmeyer et al. 2002; Li et al. 2003; Hirata and Toshimitsu 2005, 2007; Rose et al. 2007) or optical emission spectrometry (ICP-OES) (García Salgado et al. 2006a and 2006b; García Salgado et al. 2008; Moreda-Piñeiro et al. 2007; García Salgado et al. 2012) systems, are the analytical techniques commonly used in the determination of total arsenic in seaweed samples, as can be seen in Table 2.

Most studies reported in Table 2 are part of a more comprehensive arsenic speciation research in seaweed, and detection and quantification limits data of total arsenic are omitted in most of them. Detection and quantification limits of 0.1 and 0.3 $\mu g\ g^{-1}$, respectively, were obtained by García-Sartal et al. (2010) for total arsenic determination in acid digested samples by HG-AFS. Moreda-Piñeiro et al. (2007) have reported 0.01 $\mu g\ g^{-1}$ and 0.05 $\mu g\ g^{-1}$ for LOD and LOQ, respectively, in total arsenic determination in seaweed samples by ICP-OES. Detection limits of 0.02 $\mu g\ g^{-1}$ and 0.3 $\mu g\ g^{-1}$ were reported by Rose et al. (2007) for total arsenic and inorganic arsenic determination in seaweed samples by ICP-MS.

Gallium (Kuehnelt et al. 2001), iridium (Wei et al. 2003), rhenium (Li et al. 2003), yttrium (Han et al. 2009) or germanium, indium and iridium

(Wuilloud et al. 2006) have been used as internal standards for ICP-MS signal drift correction. Some authors have monitored the ion intensity at m/z 77 (Se+, ArCl+) and m/z 82 (Se+) (Van Hulle et al. 2002) to check possible chloride interferences (Van Hulle et al. 2004) when ICP-MS detectors were used for total arsenic determination.

The influence of the concentration of both hydrochloric acid and sodium tetrahydroborate used as carrier and reductant solutions, respectively, on the hydride generation has been evaluated by García-Sartal et al. for total arsenic determination by HG-AFS. 1.25% (m/v) NaBH$_4$ and 0.75 M HCl solutions were selected as optimal concentrations (García-Sartal et al. 2010). Jiang et al. proposed the addition of a reducing agent (10% ascorbic acid + 10% thiourea) and H$_2$SO$_4$ after the acid digestion for total arsenic determination in seaweed samples by electrochemical hydride generation AFS (Jiang et al. 2009). Ichikawa et al. added HCl (25% (w/w)), ascorbic acid (20% (w/w)) and KI (20% (w/w)) to digested seaweed samples for total arsenic determination by HG-AAS (Ichikawa et al. 2006). Alternatively, Raber et al. (Raber et al. 2000) redissolved the algal residue obtained after mineralization and heating in a muffle, using 6 M HCl, before arsenic quantification by HG-AAS.

2.3 Arsenic speciation in seaweed samples

Most analytical methods for arsenic speciation in seaweed samples (Table 3) are based on hyphenated techniques by using high-performance liquid chromatography (HPLC) coupled to sensitive arsenic specific or non-specific detectors, such as ICP-MS, ICP-AES, ESI-MS or by coupling hydride generation systems (HG) with ICP-MS, AAS and AFS systems.

Due to the polar behaviour and pH dependence of the oxo-arsenic compounds, chromatographic separation techniques based on anion-exchange and ion-pairing mechanisms have been used the most for arsenic speciation studies. Size-exclusion, reversed phase or the parallel use of two or three chromatographic modes have also been reported in arsenic speciation studies in seaweed samples. The coupling of size exclusion and anion exchange chromatography improves the peak resolution and reduces the peak overlap due to the simplification of the matrix (Szpunar et al. 2000). Ion-exchange chromatography is the separation technique selected by the majority of researchers. Anion-exchange chromatography has been used in more applications than cation-exchange chromatography. However, the combination of both anion- and cation-exchange chromatography for the analysis of cationic and anionic arsenic species is common. Cationic species and As(III) which elute in or close to the void volume by anion-exchange, are well resolved when cation-exchange is employed. Tukai et al. (2002) have used both anion and cation-exchange separation columns for arsenic speciation studies. DMA, As(V), glycerol-, phosphate-, sulfonate- and sulfate-riboses were resolved within 15 min when an anion-exchange column was used. Under these conditions, As(III) and MA showed the same retention time at the optimum pH (9.2) and glycerol-ribose eluted close to the void volume of the column. Alternatively, a method based on cation-exchange separations was used to determine these species. Kuehnelt et al. (2001) separated AB, AC, TMAO, TETRA and glycerol-ribose by cation-exchange chromatography. As(III), As(V), MMA, DMA, phosphate-, sulfate- and sulfonate-riboses were determined by anion-exchange by using 20 mM $NH_4H_2PO_4$ as mobile phase at two different pH (5.6 and 6.0). Wuilloud et al. (2006) separated arsenosugar-H and arsenosugar-OH in Antarctic seaweed extracts using cation exchange chromatography. These species eluted in the void volume by anion-exchange chromatography. Van Hulle et al. (2002) used both chromatographic modes to discard new arsenic species eluting from the column. Discrepancies between chromatographic modes were not observed. In a study developed by Li et al. (2003), AB and arsenosugars-OH coeluted from the column when anion-exchange chromatography was used. On the other hand, cation-exchange chromatography allowed the separation of arsenosugar-OH from AB, as well as AC, DMA and phosphate-ribose.

In general, as it can be seen in Table 3, phosphate buffer (($NH_4)_2HPO_4$, $NH_4H_2PO_4$, KH_2PO_4) and carbonate buffer (NH_4HCO_3, ($NH_4)_2CO_3$) adjusted to a pH ranging from 5.2 to 7.0 and from 7.0 to 10.3, respectively, are the mobile phases most widely used in anion-exchange chromatography. Phosphate buffers are preferred over carbonate mobile phases when hydride generation devices are coupled to the detection system. Because of their volatility, carbonate buffers are the common mobile phases selected when the chromatographic system is coupled with mass spectrometric techniques for arsenic speciation.

The separation of cationic and anionic species in gradient mode using nitric acid together with or without an ion-pairing reagent (benzene-1, 2-disulfonic acid dipotassium salt (DBSA)) (Karthikeyan and Hirata 2004; Kohlmeyer et al. 2003; Kohlmeyer et al. 2002) on an anion-exchange column has also been described. On the other hand, pyridine buffers adjusted to lower pH (2.4–2.65) are commonly used when cation-exchange separations are performed.

Ion pair-reversed phase-HPLC separation systems can be used to separate anionic, cationic and neutral arsenic species. Wangkarn and Pergantis (2000) developed a separation procedure based on narrow-bore IC-RP-HPLC to overcome the shortcomings of the conventional separation technique. Lower analytical times and lower column equilibrium times were needed with this technique. Furthermore, the volume of solvent and sample were reduced, as well as the amount of waste. In recent years, Todolí and Grotti (2010) have confirmed the advantages of narrow bore HPLC with

ICP-MS for fast determination of arsenosugars in algal extracts. The authors investigated the behaviour of naturally occurring arsenosugars on an anion-exchange microbore column, and the use of high efficiency nebulizer (HEN) or PFA micronebulizer together with a low inner volume cyclonic spray as an external interface. Improvement in sensitivity (20–50%) was obtained with the HEN nebulizer. The sensitivity among arsenic species varied to a maximum of 10%, suggesting that arsenosugars could be quantified using standard solutions of other arsenic species.

Mobile phases containing organic solvents, such as methanol, increase the ICP-MS signals of elements with ionization energies ranging from 9 to 11 eV. The influence of methanol on the arsenic species separation in seaweed has been studied by Raber et al. (2000). The authors corroborated that at 3% methanol (v/v) concentration in the mobile phase, the arsenic species sensitivity was enhanced by a factor of 2.5. The retention times of some arsenic species such as arsenous acid, dimethylarsinic acid and methylarsonic acid remain constant at higher methanol concentrations. However, the retention time of phosphate-, sulfonate- and sulfate-riboses were shorter, and co-elution between species was produced. These authors also observed that the retention times of sulfate-ribose and sulfonate-ribose were reduced by about 20% and 10%, respectively, when the separation was performed at 55°C.

Other authors have taken advantage of the ICP-MS multielemental detection to monitor arsenic and sulfur from oxo-arsenic and thio-arsenic compounds in a single chromatographic run, using the separation methods described above. However, strong retention and broad peaks of thio-arsenicals have been found using an anion-exchange column. Separation methods based on anion-pairing reversed-phase chromatography coupled to ICP-MS have been developed to determine thio-arsenic compounds. Nischwitz et al. (2006) and Raml et al. (2006) found that the addition of methanol to the mobile phase shortened the elution time of the thio-compounds when a Hamilton PRP-X100 column was coupled to an ESI-MS detection system. However, high amounts of methanol (40%) were not compatible with the ICP-MS detection technique. Alternatively, anion-pairing reversed-phase with a mobile phase containing 5 mM TBAH (pH: 7.5) and 5% MeOH was adequate for efficient separation of DMThioAsSugarGlycol, DMThioAsSugarPhosphate, DMThioAsSugarSulfonate and DMThioAsSugarSulfate.

Hydride generation coupled with atomic absorption and atomic fluorescence spectroscopy is relatively inexpensive and easy to maintain in a laboratory. This technique can be used as detectors for arsenic speciation analysis. The main advantage of this technique is the ability to measure arsenic species in high-chloride content matrices, such as seaweed, without polyatomic interference ion, $^{40}Ar^{35}Cl^+$, formation (Kirby et al. 2004). The main drawback of the technique is the inability of certain species such as AB, AC and arsenosugars to form arsines under normal conditions, thus requiring a previous decomposition step to generate the volatile hydride (Wangkarn and Pergantis 2000). A study developed by Schmeisser et al. (2004) confirmed, for the first time, that arsenosugars are hydride-active species. In this study, arsenosugars were separated and detected by HPLC-HG-ICP-MS without a previous decomposition step. The authors concluded that the type of hydride generation system and the conditions employed (NaBH$_4$ and HCl concentrations) were crucial in the arsenosugars activity.

Arsenic speciation studies in *Sargasso*, *Hizikia*, *Laminaria* and *Chlorella* by HPLC-ICP-AES and HPLC-HG-ICP-AES have been performed by García Salgado et al. (2006a and 2006b). The authors found that As(V) was the only arsenic species found in both, *Sargasso* and *Hizikia* seaweeds, with and without the use of a hydride generation system. An unknown arsenic peak was found in *Laminaria* by HPLC-ICP-AES; however, no arsenic traces were observed when the analysis was performed with a previous vapour stage. This finding suggests that this compound is not hydride-active. The authors attributed this peak to a possible arsenosugar. As(III), As(V), DMA and MMA were identified in *Chlorella* by HPLC-HG-ICP-AES; however, As(III) and MMA were not detected without a previous hydride generation step. The explanation to these findings in *Chlorella* could be attributed to an increase in sensitivity provided by the hydride generation system, or matrix interferences which could cause the retention of the arsenic compounds on the chromatographic column. Geng et al. (2009) compared two methods based on HPLC-HG-AFS and CT-HGAAS for arsenic speciation studies in seaweed. The authors clearly observed the peaks of arsenosugars in the extracts of those samples when a photooxidation step was applied prior to the analysis by HG-AFS. On the other hand, arsenosugars were not detected when the sample was subjected to an alkaline digestion and CT-HGAAS detection. On the contrary, high amounts of DMA were quantified in the alkaline extracts. The authors concluded that arsenosugars are transformed into hydride-active DMA during the alkaline digestion, because the sum of the amounts of DMA and three arsenosugars quantified by the HPLC-HG-AFS method was approximately equal to the amount of DMA quantified by the CT-HGAAS method. Kirby et al. (2004) analysed a *Durvillea potatorum* Tasmanian kelp by HPLC-ICP-MS and HPLC-HG-ICP-MS. Results suggested that the four arsenosugars (OH-, PO$_4$-, SO$_3$- and SO$_4$-riboses) were not hydride active, since they were not detected by HPLC-HG-ICP-MS. As(III) could only be quantified with a previous arsine-forming step which increased the sensitivity of the technique. Furthermore, there was close agreement between DMA, MMA and As(V) concentrations quantified by both techniques.

Arsenic speciation differs between seaweeds; however a speciation pattern could be elucidated from the data shown in Tables 2 and 3. In general, it can be said that brown seaweeds are higher arsenic accumulators than red seaweeds, and the latter accumulate more arsenic than green seaweeds (Sanchez-Rodriguez et al. 2001).

2.4 Identification of arsenic species in seaweed samples

The lack of structural information when ICP-MS is used and the limited availability of arsenic standards have led to the development of molecular mass spectrometric methodologies in arsenic research. Electrospray spectrometry, single or in tandem, has been used for structural verification of arsenosugar compounds. Table 4 shows the arsenic species identified by mass spectrometry techniques in seaweed samples, as well as the main fragments found in those published studies.

Pedersen and Francesconi (2000) identified and quantified the major arsenosugars (phosphate, sulfonate and sulfate riboses) in the extracts of two brown seaweeds using HPLC-ESI/MS in the positive mode. Two commonly fragments, m/z 237 and m/z 75, provided the necessary evidence for a structural assignment. Szpunar and McSheehy et al. (Szpunar et al. 2000; McSheehy and Szpunar 2000; McSheehy et al. 2000; McSheehy et al. 2001) have developed an arsenosugar characterization method based on pneumatically assisted electrospray tandem mass spectrometry in the CID MS mode. The authors suggested that the direct infusion of the methanolic extract of the sample produces signal suppression by the matrix and a considerable baseline noise due to the presence of other compounds. The coupling with chromatographic devices allowed the simplification of the sample matrix and the elimination of the interfering compounds. Size-exclusion chromatography was selected over other separation techniques because it has a high tolerance to the matrix and allows the fractionation of the arsenosugar compounds. Wei et al. (2003) have confirmed the presence of PO_4- and OH-riboses in *Porphyra* seaweed by direct infusion into the ESI-MS-MS system of the fractions collected after an anion and cation exchange in chromatographic columns.

Different studies reported that arsenobetaine, the main arsenic species found in fish and crustaceans, is absent in seaweed samples (Van Hulle et al. 2002; Kohlmeyer et al. 2002). However, Nischwitz and Pergantis (2005) have confirmed the presence of arsenobetaine in *Ascophyllum nodosum*, *Laminaria digitata* and *Fucus vesiculosus*, as well as in fresh marine algae *Ulva lactuca*, *Padina pavonica* and some red seaweeds. Anion and cation exchange separation methods coupled to electrospray tandem mass spectrometry in the selected reaction monitoring (SRM) mode offered a full chromatographic separation and sensitive detection for the individual identification of

Table 4. Arsenic species identification by mass spectrometry.

Seaweed	Identification technique	Specie	Characteristic fragments (m/z)	Ref.
Sargassum lacerifolium seaweed	ESI-MS/MS	• 3-[5′-deoxy-5′-(dimethylarsinoyl)-β-ribofuranosyloxy]-2-hydroxy-propanesulfonic acid: sulfonate ribose • 3-[5′-deoxy-5′-(dimethylarsinoyl)-β-ribofuranosyloxy]-2-hydroxypropyl hydrogen sulfate: sulphate ribose • 3-[5′-deoxy-5′-(dimethylarsinoyl)-β-ribofuranosyloxy]-2-hydroxypropyl 2,3-hydroxy propyl phosphate: phosphate ribose	• 393.1, 374.8, 236.5 • 408.7, 328.9, 237.1 • 483.4, 328.9, 237.1	(Szpunar et al. 2000)
Fucales and *laminares, Hizikia fusiforme, Himanthalia, Palmaria palmata, Porphyra umbilicalis, Sargassum lacerifolium, Undaria pinnatifida* seaweed	ESI-MS/MS	• 3-[5′-deoxy-5′-(dimethylarsinoyl)-β-ribofuranosyloxy]-2-hydroxypropyl hydrogen sulfate: sulphate ribose • 3-[5′-deoxy-5′-(dimethylarsinoyl)-β-ribofuranosyloxy]-2-hydroxypropyl 2,3-hydroxy propyl phosphate: phosphate ribose	• 408.9 • 483.0	(McSheehy and Szpunar 2000)
Laminaria digitata and *Fucus vesiculosis* seaweed (Pedersen and Francesconi 2000)	LC/MS	• 3-[5′-deoxy-5′-(dimethylarsinoyl)-β-ribofuranosyloxy]-2-hydroxy-propanesulfonic acid: sulfonate ribose • 3-[5′-deoxy-5′-(dimethylarsinoyl)-β-ribofuranosyloxy]-2-hydroxypropyl hydrogen sulfate: sulphate ribose • 3-[5′-deoxy-5′-(dimethylarsinoyl)-β-ribofuranosyloxy]-2-hydroxypropyl 2,3-hydroxy propyl phosphate: phosphate ribose	• 393 • 409 (only identified in *Fucus*) • 483 m/z 75 and 237 fragments confirmed the presence of those arsenosugars in both seaweed samples	(Pedersen and Francesconi 2000)
Laminaria japonica, Porphyra crispata and *Eucheuma denticulatum* seaweed	AE-HPLC-ESI-MS	• 3-[5′-deoxy-5′-(dimethylarsinoyl)-β-ribofuranosyloxy]-2-hydroxy-propylene glycol: glycerol ribose • 3-[5′-deoxy-5′-(dimethylarsinoyl)-β-ribofuranosyloxy]-2-hydroxy-propanesulfonic acid: sulfonate ribose • 3-[5′-deoxy-5′-(dimethylarsinoyl)-β-ribofuranosyloxy]-2-hydroxypropyl hydrogen sulfate: sulphate ribose • 3-[5′-deoxy-5′-(dimethylarsinoyl)-β-ribofuranosyloxy]-2-hydroxypropyl 2,3-hydroxy propyl phosphate: phosphate ribose • DMA	• 329.1, 236.9, 195.0, 97.0 • 393.0, 237.0, 195.0, 97.0 • 409.1, 329.0, 237.0, 195.0, 97.1 • 483.0, 329.0, 236.9, 195.0, 97.2 • 138.9, 121.0, 91.0	(Van Hulle et al. 2002)

Table 4. contd....

Table 4. contd.

Seaweed	Identification technique	Specie	Characteristic fragments (m/z)	Ref.
Phorphyra seaweed	ESI-MS/MS	▪ 3-[5′-deoxy-5′-(dimethylarsinoyl)-β-ribofuranosyloxy]-2-hydroxy-propylene glycol: glycerol ribose ▪ 3-[5′-deoxy-5′-(dimethylarsinoyl)-β-ribofuranosyloxy]-2-hydroxypropyl 2,3-hydroxy propyl phosphate: phosphate ribose	▪ 329, 311, 237, 195, 97 ▪ 483, 465, 391, 329, 277, 237, 195	(Wei et al. 2003)
Canadian and Icelandic Kelp powder	HPLC-ESI-MS/MS	▪ DMThioAsSugarGlycol ▪ DMThioAsSugarPhosphate ▪ DMThioAsSugarSulfonate ▪ DMThioAsSugarSulfate	▪ 345, 327, 253, 97 ▪ 499, 253, 97 ▪ 409, 253, 97 ▪ 425, 345, 253, 97	(Nischwitz et al. 2006)
Granulated Kelp seaweed	HPLC-ES-MS/MS	▪ DMAA ▪ DMAsSugarMannitol	▪ 181.0, 164.1, 162.9, 138.9, 118.9, 118,1 ▪ 418.9, 236.7, 218.8, 195.2, 175.2, 175.2, 167.2, 97.0	(Nischwitz and Pergantis 2006)
BCR-279 (*Ulva lactuca*) reference material	ESI-MS	▪ Dimethylarsinoyl fatty acid (arsenolipid)	▪ 437, 419.1, 149.0, 122.9	(Hsieh and Jiang 2012)
Undaria pinnatifida (Wakame) and *Hizikia fusiformis* (Hijiki) seaweed	HPLC-MS	▪ Arsenosugar-phospholipids	▪ Several compounds with the following molecular mass: 944, 956, 982, 958, 988, 956, 1012, 1014, 1042 and 1070	(García-Salgado et al. 2012a)
Undaria pinnatifida (Wakame), *Laminaria ochroleuca* (Kombu), *Porphyra umbilicalis* (Nori) and *Ulva rigida* (Sea lettuce) seaweed	HPLC-ESI-MS/MS	▪ 3-[5′-deoxy-5′-(dimethylarsinoyl)-β-ribofuranosyloxy]-2-hydroxy-propylene glycol: glycerol ribose ▪ 3-[5′-deoxy-5′-(dimethylarsinoyl)-β-ribofuranosyloxy]-2-hydroxypropyl 2,3-hydroxy propyl phosphate: phosphate ribose ▪ 3-[5′-deoxy-5′-(dimethylarsinoyl)-β-ribofuranosyloxy]-2-hydroxy-propanesulfonic acid: sulfonate ribose ▪ Arsenobetaine	▪ 329.1, 311.1, 237.0, 218.9, 194.997.2 ▪ 483.0, 465.1, 391.0, 329.1, 237.0 ▪ 393.0, 375.1, 295.1, 236.9, 218.9, 194.9 ▪ 179, 161, 137, 105	(García-Sartal et al. 2012c)

Abbreviations and acronyms: HPLC (high performance liquid chromatography); LC (liquid chromatography); ESI-MS (electrospray ionisation—mass spectrometry); ESI-MS/MS (electrospray ionisation tandem mass spectrometry); AE (anion exchange chromatography) arsenobetaine in the algal extracts by monitoring the product ion mass spectra of m/z 179. The presence of AB in Sea lettuce seaweed has also been confirmed by García-Sartal et al. (García Sartal et al. 2012c).

Nischwitz et al. (2006) were the first to identify arsinothioyl-sugars in aqueous kelp extracts. The authors prepared the thioarsenosugar standards through the reaction of their arsinoyl-sugar analogues with H_2S. The four thioarsenosugars examined (OH-, PO_4-, SO_3- and SO_4-thioarsenosugar) showed the formation of two common product ions with m/z 253 and m/z 97. Additionally, the product ion m/z 345 was only found in the case of DMThioAsSugarSulfate, which coincides with the $[M + H]^+$ molecular ion for OH-thioarsenosugar (m/z 345). Furthermore, sulfate and sulfonate thioarsenosugars exhibit the same transition in the SRM mode (409 → 97). In order to avoid misinformation, standards and seaweed extracts were first separated by anion-exchange and then characterized with ESI-MS/MS. Sulfonate and sulfate thioarsenosugars were identified in the kelp extracts by AE-ES-MS-MS and further confirmed by anion pairing-reversed phase (AP-RP) coupled to an ICP-MS detection system. Moreover, other arsenic species such as OH-, SO_4-, SO_3-riboses and PO_4-, OH-thioarsenosugars were tentatively identified in that kelp sample. The authors also discussed the possibility of oxoarsenosugars conversion into thioarsenosugars between the time of harvesting and drying of the kelp.

Most of the arsenic compounds identified in seaweed samples are water-soluble arsenic species. Nowadays, arsenic speciation of the lipid-soluble fraction is gaining popularity, because arsenolipids from seaweed are presumably the forerunners of the arsenolipids found in fish. Hsieh and Jiang (2012) have tentatively identified an arsenolipid by direct infusion of the sample extract into an ESI-MS system. This sample extract was obtained from a seaweed reference material (BCR-279). Therefore, the CRM seaweed sample was subjected to a microvawe extraction procedure using MeOH:ammonium carbonate (1:1) as an extracting solution. The liquid phase was separated (discarded) and the residue was treated with 1% (v/v) HNO_3. The new liquid phase obtained corresponded with the arsenolipid fraction. This fraction was then injected into the ESI-MS system for identification. The compound spectrum showed a major peak at m/z 437, coinciding with previous studies for fish samples (Amayo et al. 2011). Recently, García-Salgado et al. (2012a) have identified five and six arsenosugar-phospholipids in Wakame and Hijiki seaweed, respectively, as well as a new arsenic-hydrocarbon in Wakame (m/z 388). Arsenosugar-phospholipid with m/z 959 was the main lipid-soluble arsenic species found in both kinds of seaweed. Other peaks in the mass spectra before and after m/z 959 indicated the presence of saturated and unsaturated homologous series of arsenosugar-phospholipids.

2.5 Effects of cooking on arsenic speciation in seaweed

Edible seaweeds are commonly consumed as cooked products. Boiling, baking, frying, toasting and roasting are some of the cooking treatments to which seaweed are subjected before their consumption. These treatments are performed either to enhance the palatability of seaweed, or as part of an industrial conservation process. Loss of water and soluble constituents such as proteins, vitamins and polar arsenic species, as well as interconversion of arsenic species are some of the phenomena which may be produced during the thermal processing (Devesa et al. 2008).

Devesa et al. (2001) performed a kinetic study using standards of AB, DMA, MMA, TMAO, TMA$^+$ and AC. The authors studied the possible transformation of the mentioned standards (main arsenic species found in seafood), after subjecting them to typical cooking and industrial processing temperatures (85–190°C), over various periods of time (15, 22, 30, 37 and 44 min), and to various pH values (4.5, 5.5, 6.5 and 8.0). In their study, the authors concluded that arsenic species were stable up to 120°C. A partial decomposition of AB into TMAO (1.5%) at 150 °C, and into TMAO (2.1–16%) and TMA$^+$ (0.1–2%) at 160°C or above was observed. AC was also slightly transformed into TMAO (<1.1%) at temperatures ranging between 150–180°C, and into DMA (0.1–0.2%) at temperatures above 170°C. The authors emphasized that the presence of other compounds in the samples could enhance or impede transformations, thus this study should be extrapolated to real samples with caution. Ichikawa et al. (2006) observed that 88.7–91.5% of the total arsenic was removed from *H. fusiforme* seaweed as a result of a soaking pre-treatment (28.2–58.8% of the total arsenic was removed) and cooking in water for 20 min at 90°C (49.3–60.5% of the total arsenic was removed). Even so, inorganic arsenic (As(V)) was the main arsenic species found in the cooked seaweed and changes in arsenic speciation were not appreciated. Other studies performed with the same kind of seaweed (*Hijiki*) have reported that high amounts of water soluble inorganic arsenic are retained in the seaweed (61–73% of inorganic arsenic in cooked seaweed) even after subjecting it to a soaking process (Rose et al. 2007).

García-Sartal et al. (2012b) estimated that aproximately 69% of arsenic in Kombu, 50% in Wakame, 71% in Nori and 34% in Sea Lettuce was removed from the matrix during the cooking procedure, being released into the cooking water. The authors concluded that the arsenic distribution in cooked seaweed and cooking water was quite similar. OH-ribose was the main arsenic specie found in Wakame and Sea lettuce, SO$_3$-ribose for Kombu seaweed, and PO$_4$-ribose in the case of Nori.

In order to evaluate the arsenosugar stability after heat treatment, Wei et al. (2003) subjected extracts of *Porphyra* containing PO$_4$-ribose and OH-ribose to a cooking process (oil bath for 10 min at 100°C). Results suggested that no changes in arsenic speciation were produced during short-term heating.

3 Arsenic Bioaccessibility and Bioavailability in Seaweed

Ingestion, inhalation and cutaneous contact are the main routes of exposure to a particular environmental contaminant, either organic or inorganic.

Food provides essential elements for the normal development of the metabolic activity. However, it can also be a source of non-essential elements, such as toxic heavy metals (Intawongse and Dean 2006). Food ingestion is the main route to arsenic exposure in humans, especially through consumption of marine products, including seaweed (Moreda-Pineiro et al. 2011).

On the other hand, the knowledge of the total content of a contaminant in food is insufficient when assessing its potential adverse health effects. The total amount of the contaminant does not necessarily reflect the bioavailable quantity, since only a portion of the total content is absorbed and participates in body functions (Versantvoort et al. 2004).

It is important to differentiate between two biological terms: bioaccessibility and bioavailability. Bioaccessibility is defined as the fraction of a given element that is released from the food matrix after ingestion, and solubilized in the intestinal lumen. However, the term bioavailability describes the fraction of the solubilized element which is absorbed in the intestinal tract and reaches the circulatory system (Caussy 2003). Bioavailability is a broader term which schematically consists of: 1) Ingestion; 2) Release of the element from the surrounding matrix in the gastrointestinal tract (bioaccessibility); 3) Element absorption through the intestinal epithelium; and 4) 'First-pass' effect described as the metabolic transformation that occurs in the liver (Caussy 2003; Brandon et al. 2006; Caussy et al. 2003). Consequently, element bioaccessibility itself does not provide a direct measure of element bioavailability, since not all soluble element species are absorbable (Laparra et al. 2007).

Bioaccessibility of a compound is mainly affected by dietary constituents of the matrix (Brandon et al. 2006); however, bioavailability and absorption of an inorganic nutrient are dependent on factors related to the element or host, such as element properties (chemical form, molecular weight, solubility or lipophilicity of the element in the small intestine), capacity of the element to be modified by gastrointestinal juices or dietary components, gut flora, hepatic and renal function, possible presence of infection in the body, the physical stage of the individual (age, pregnancy), among others (Fairweather-Tait 1999; Fairweather-Tait and Hurrell 1996). It is believed that once the element is released from the matrix, the matrix itself does not affect the absorption of the element, although a competition between food constituents and the element to be transported through the intestinal epithelium might take place (Versantvoort et al. 2004). Matrix influence was observed by Laparra et al. (2007), who have subjected organoarsenical standards and a dogfish muscle reference material (DORM-2) to *in vitro* gastrointestinal digestion. They found that AB standard was soluble in gastrointestinal solution (67.5–100% bioaccessibility), and only 1.7% of AB was transported by Caco-2 cells. However, 12% of AB in DORM-2 reference material was bioavailable after transportation by Caco-2 cells (intestinal epithelia model).

In vivo or *in vitro* approaches can be used to assess the bioavailability and bioaccessibility of an element. While *in vivo* techniques use living beings to evaluate the bioavailability of a contaminant, *in vitro* approaches are developed on a laboratory level by simulating the human gastrointestinal digestion.

3.1 In vivo assays for bioavailability assessment

In vivo bioavailability assessment of metals is mainly performed by chemical, radioisotope or stable isotope balances (Van Campen and Glahn 1999). Arsenic *in vivo* studies are based on chemical balances which represent the difference between intake and excretion. He and Zeng (2010) correlated the total urinary, fecal and blood arsenic to the total arsenic ingested in food and beverages to assess arsenic bioavailability by chemical balances. Primates, swine, dogs, rabbits and rodents are the animals most commonly selected when developing an *in vivo* method (Rees et al. 2009). Primates are more suitable, since they are physiologically and behaviourally more similar to humans but their high cost makes them inaccessible to the great majority of researchers (Roberts et al. 2002). The use of rodents as an animal model has two drawbacks. High arsenic doses are needed to be detected in blood and arsenic distribution in rodents is different to that in humans (Caussy 2003). Swines seem to be the best option since young specimens metabolize and excrete arsenic following human pattern excretion, and furthermore, they are monogastric and omnivorous like humans are (Juhasz et al. 2006).

Ichikawa et al. (2010) have developed an *in vivo* method in mice to assess the absorption, metabolism, excretion and accumulation of arsenic compounds after ingestion of *Hijikia fusiforme* seaweed. They observed a small arsenic accumulation in mice tissues suggesting that 66–92% of arsenic was excreted in urine and feces within three days after *Hijikia* administration. In other work, the authors concluded that arsenate from *Hijikia* seaweed was excreted as DMA after consumption of this seaweed by mice (Ichikawa et al. 2006). Arsenic *in vivo* studies in humans and sheep after seaweed or synthetically prepared arsenosugar ingestion are described in more detail in the urinary arsenic speciation section.

The strict protocols followed in animal and human testing, the high cost associated with the method, as well as ethical restrictions are the main shortcomings of the *in vivo* studies. Another drawback is the large degree of variability in the physiology of individuals and the variability in the results obtained, caused by interactions between the element under study and other components of the diet (Parada and Aguilera 2007).

3.2 In vitro assays for bioavailability assessment

Alternatively, *in vitro* methods are also designed to evaluate metal bioavailability. These types of testing offer simplicity, speed, low cost, ease of handling, and precise and reproducible results on a small scale, being often preferred over *in vivo* assays (Moreda-Pineiro et al. 2011). Some considerations must be taken into account when performing gastrointestinal *in vitro* studies in order to reproduce faithfully, the human gastrointestinal digestion. Temperature, agitation, pH, gastrointestinal juice compositions, among others, are some of the variables that may affect the proper development of the proceedings. Therefore, the following points should be considered (Dean 2007):

- Simulation of the human body temperature (37°C).
- Human rhythmic muscular contractions must be simulated with mechanical shaking.
- Reproduction of the gastric conditions (hydrochloric acid, pH: 1.0–4.0 and pepsin enzyme): stomach acidity and pepsin enzyme produced denaturation of proteins and release of metals bound to weak oxides, sulfides and carbonates from the diet.
- Reproduction of the small intestine conditions (pH: 4.0–7.5 and intestinal juices). Fat, polysaccharides and proteins from diet are digested with intestinal juices (trypsin, pancreatin and amylase), bile salts and bicarbonate. Nutrient absorption takes place in the small intestine.

Static and dynamic *in vitro* methodology has been developed to assess nutrient bioavailability and bioaccessibility (Parada and Aguilera 2007). While the food digest is sequentially exposed to simulated mouth and gastrointestinal conditions in static *in vitro* assays, the digest is gradually moving through the simulated digestive tract in dynamic methods (Oomen et al. 2002).

The static methods include those that determine the bioaccessibility of a compound as the maximum soluble concentration in simulated gastrointestinal solution, such as the German DIN model and the RIVM model (from the National Institute for Public Health and the Environment); and those that only or only in simulate gastric conditions such as SBET (Simple bioaccessibility extraction test) (Oomen et al. 2002). Bioavailability has been assessed by dialyzability (Miller et al. 1981) and Caco-2 cell culture (Polarized human colon carcinoma cells) (Van Campen and Glahn

1999) static techniques. Both methods measure the bioavailable ratio as the fraction of a compound which can dialyze through a semi-permeable dialysis membrane or through a microporous support containing Caco-2 cells, respectively. More complex dynamic *in vitro* methods have been developed for bioavailability and bioaccessibility assessment, including TIM (TNO Gastrointestinal Model) and SHIME (Simulator of the Human Intestinal Microbial Ecosystem) methods (Van Campen and Glahn 1999; Torres-Escribano et al. 2011). The stomach, small intestine and even the large intestine are represented by different computer-controlled reaction compartments in those dynamic methods (Torres-Escribano et al. 2011).

Several *in vitro* gastrointestinal methods have been applied to measure arsenic bioavailability and bioaccessibility from seaweed samples. The approaches are based on extraction techniques which simulate human gastric or gastrointestinal digestion (temperature, pH, enzyme and chemical composition). Tables 5 and 6 describe the *in vitro* procedures found in literature for arsenic bioaccessibility and bioavailability assessment in seaweed samples, respectively.

Total arsenic (t-As) and total inorganic arsenic (i-As) bioaccessibility has been assessed by Laparra et al. (2003) in edible seaweed (*H. fusiforme*, *Porphyra* sp. and *Enteromorpha* sp.) before and after being cooked. The authors found that the effect of cooking increased the total and inorganic arsenic bioaccessibility. They suggested that the heat treatment might favour a denaturation of proteins. This affects the As(III)-protein bonds, and therefore, inorganic arsenic solubilisation in cooked seaweed is facilitated due to the greater accessibility of gastrointestinal enzymes. A later work developed by the same research group (Laparra et al. 2004) confirmed a reduction of 30–42% in the arsenic concentration of cooked *H. fusiforme* with respect to the raw product by solubilisation in the cooking water. The authors also observed that the cooking procedure enhanced the t-As and i-As bioaccessibility percentages. As(III) was the main inorganic arsenic species present in either the gastric and gastrointestinal solutions of raw *Hizikia*, whereas in the cooked sample, As(III) and As(V) predominance was dependent on the batch analysed. These findings suggested that arsenic solubilisation is mainly influenced by gastric pH, with the stomach being the limiting region of the gastrointestinal tract for arsenic bioaccessibility. Almela et al. (2005) have confirmed an increase in As(V) concentration in methanolic extracts of cooked *Hizikia Fusiforme*, compared with their content in the raw sample. They have also found arsenosugars bioaccessibility percentages (>80%) in *Hizikia fusiforme*, *Undaria pinnatifida* and *Porphyra* sp., which exceeded the inorganic arsenic bioaccessibility previously reported (49–89%). These results suggest that the cooking procedure does not affect the arsenic bioavailability. The authors did not observe arsenosugar degradation during the *in vitro* gastrointestinal digestion, which confirms

Table 5. Arsenic *in vitro* bioaccessibility testing in seaweed.

Seaweed	*In vitro* bioaccessibility assay	Arsenic species	Results	Ref.
H. fusiforme, Porphyra sp. and *Enteromorpha* sp. seaweed	Gastric juices: 1 g of pepsin in 10 mL of 0.1 M HCl. Intestinal juices: 0.2 g of pancreatin and 1.25 g of bile salts in 50 mL of 0.1M NaHCO$_3$. Gastric digestion: 90–160 mL of water was added to 5 g of seaweed. The solution was adjusted to pH: 2.0. Then, sample was filled up to 100–170 mL with water after adding the gastric solution. Gastric digest was incubated for 2 h in a shaking water bath (37°C). Intestinal digestion: Gastric digest pH was raised to 5.0 and intestinal juices were added. Then, intestinal digest was incubated for 2 h at 37°C. Soluble arsenic was recovered after centrifugation at 15000 rpm for 30 min and 4°C.	Total arsenic ([]](t-As) and total inorganic arsenic (i-As) in raw and cooked seaweed.	Total arsenic bioaccessibility: 32.0–67.2% in raw seaweed and 65.7–79.9% in cooked seaweed. Inorganic arsenic bioaccessibility: 48.6–77.2% in raw seaweed and 72.6–87.9% in cooked seaweed.	(Laparra et al. 2003)
Hizikia fusiforme seaweed	The same as previously reported (Laparra et al. 2003)	t-As, i-As, As(III) and As(V) in raw and cooked seaweed.	Total arsenic bioaccessibility: 55.6–84.3% and 61.9–74% in raw and cooked seaweed, respectively. In raw seaweed, As(III) bioaccessible concentration ranged between 18.4–45.3 µg g^{-1} (36–94%) and As(V) was found between 2.2-26.1 µg g^{-1} (5–51%). In cooked seaweed, bioaccessible As (III) and As(V) were in the ranges 7.1–25.4 µg g^{-1} (20–70%) and 7.5.23.8 µg g^{-1} (21–68%), respectively.	(Laparra et al. 2004)

Table 5. contd....

Table 5. contd.

Seaweed	*In vitro* bioaccessibility assay	Arsenic species	Results	Ref.
Kelp powder, *Hizikia fusiforme, Undaria pinnatifida* and *Porphyra* sp. seaweed	The same as previously reported (Laparra et al., 2003, Laparra et al. 2004)	t-As and glycerol, phosphate, sulfonate and sulfate riboses in raw and cooked seaweed.	Total arsenic bioaccessibility: 12.5–69.9% in raw seaweed and 27.8–53.8% in cooked seaweed. Glycerol ribose bioaccessibility ranged between 98–119 % in raw seaweed and 81% in cooked seaweed. Bioaccessibility in phosphate ribose was found to be between 89–120% and 93–101% in raw and cooked seaweed respectively. Sulfonate ribose bioaccessibility ranged between 81–120 % and 114% in raw and cooked seaweed, respectively. Sulfate ribose bioaccessibility was found to range between 84–127% and 109% in raw and cooked seaweed, respectively.	(Almela et al. 2005)
Fucus sp. seaweed	Gastric juices: 30.03 g of glycine in 800 mL DIW (pH: 1.5, adjusted with 1M HCl) and diluted to 1 L. Intestinal juices: 0.0175 g of bile salts + 0.005 g pancreatin. Gastric digestion: 10 mL of gastric fluid were added to 1 ± 0.0005 g of seaweed. The mixture was shaken on a platform shaker (150 rpm, 37°C) for 1 h. Intestinal digestion: Gastric digest pH was adjusted to 7.0 with 50% (m/v) NaOH and the intestinal juices were added. Samples were shaken for 4 h and finally filtered (0.45 µm hydrophilic PVDF filter).	t-As, i-As (As(III) and As(V)), DMA, MMA and arsenosugars (glycerol, phosphate, sulfonate and sulfate riboses).	Total arsenic bioaccessibility: 68–77% and 78–79% in gastric and gastrointestinal solutions, respectively. Total inorganic arsenic bioaccessibility: 54 ± 12% and 45 ± 7% in gastric and gastrointestinal solutions, respectively. And 1.2–1.9% of total bioaccessible arsenic was found as DMA and MMA. Arsenosugars were found between 43–46% of the bioaccessible arsenic.	(Koch et al. 2007)

| IAEA-140/TM (*Fucus* sp.) seaweed reference material | Static method (SM): Same procedure as previously reported (Laparra et al. 2003) Dynamic method (TIM-1): Author have followed the method described by Minekus et al. (Minekus et al. 1995). | t-As (as well as Cd, Hg and Pb). | Total arsenic bioaccessibility: 72% and 47% after applying a static and a dynamic *in vitro* method, respectively. | (Torres-Escribano et al. 2011) |

Total arsenic bioaccessibility % was calculated as:

$$Total\ BAc\% = \frac{Total\ bioaccessible\ arsenic\ in\ seaweed}{Total\ arsenic\ in\ seaweed} \times 100$$

Individual arsenic specie bioaccessibility % was calculated as:

$$As\ specie\ BAc\% = \frac{Bioaccessible\ As\ specie\ concentration\ in\ seaweed}{Total\ As\ specie\ concentration\ in\ seaweed} \times 100$$

Abbreviations and acronyms: DIW (de-ionized water); PVDF (polyvinyldene fluoride)

Table 6. Arsenic *in vitro* bioavailability testing in seaweed.

Seaweed	*In vitro* bioavailability assay	Arsenic species	Results	Ref.
Laminaria japonica seaweed	Saliva composition: • Inorganic materials: 10 mL of 189.6 g L⁻¹ KCl, 10 mL of 20 g L⁻¹ KSCN, 10 mL of 88.8 g L⁻¹ NaH$_2$PO$_4$, 10 mL of 57 g L⁻¹ Na$_3$PO$_4$, 1.7 mL of 175.3 g L⁻¹ NaCl, 1.8 mL of 40 g L⁻¹ NaOH • Organic materials: 8 mL of 25 g L⁻¹ urea • Bioenzymes: 145 mg of α-amylase, 15 mg of uric acid, 50 mg of mucin • pH: 6.5 ± 0.2 Gastric juice composition: • Inorganic materials: 15.7 mL of 175.3 g L⁻¹ NaCl, 3.0 mL of 88.8 g L⁻¹ NaH$_2$PO$_4$, 9.2 mL of 89.6 g L⁻¹ KCl, 18 mL of 22.2 g L⁻¹ CaCl$_2$·2H$_2$O, 10 mL of 30.6 g L⁻¹ NH$_4$Cl, 8.3 mL of 37% (g g⁻¹) HCl • Organic materials: 10 mL of 65 g L⁻¹ glucose, 10 mL of 2 g L⁻¹ glucuronic acid, 3.4 mL of 25 g L⁻¹ urea, 10 mL of 33 g L⁻¹ glucoseamine hydrochloride • Bioenzymes: 1 g of bovine serum albumin, 1 g of pepsin, 3 g of muncin • pH: : 1.30 ± 0.07 Duodenal composition • Inorganic materials: 40 mL of 175.3 g L⁻¹ NaCl, 40 mL of 84.7 g L⁻¹ NaHCO$_3$, 6.3 mL of 89.6 g L⁻¹ KCl, 9 mL of 22.2 g L⁻¹ CaCl$_2$·2H$_2$O, 10 mL of 8 g L⁻¹ KH$_2$PO$_4$, 0.18 mL of 37% (g g⁻¹) HCl, 10 mL of 5 g L⁻¹ MgCl$_2$ • Organic materials: 4 mL of 25 g L⁻¹ urea • Bioenzymes: 1 g of bovine serum albumin, 1 g of pancreatin, 0.5 g of lipase • pH: : 7.8 ± 0.2	t-As (as well as, V, Cr, Mn, Fe, Ni, Cu, Zn, Se, Cd and Pb)	Contents of affinity-monolayer liposome metals (AMLMs) and water-soluble metals (WSMs) in the simulated stomach or intestine are given under different conditions: gastric and intestinal acidity, semibionic digestion (excluding the digestive enzymes) and whole bionic digestion (including the digestive enzymes). Results are given as the ratio of the concentrations of AMLMs to WSMs (D_{aw} parameter). D_{aw} for As was 0.59 and 9.00, 0.45 and 4.75, 3.25 and 0.43 in stomach and intestine under acidity, semibionic and whole bionic digestion conditions, respectively. D_{aw} values >1 suggests that AMLMs are greater than their WSMs.	(Li et al. 2011)

		(García-Sartal et al. 2011)
	Total arsenic bioavailability percentages were found between 14.0–17.0% in raw and 7.4–15.3% in cooked seaweed, respectively.	
	t-As in raw and cooked seaweed	

Bile composition

- Inorganic materials: 30 mL of 175.3 g L^{-1} NaCl, 68.3 mL of 4.7 g L^{-1} NaHCO$_3$, 4.2 mL of 89.6 g L^{-1} KCl, 10 mL of 22.2 g L^{-1} CaCl$_2$:2H$_2$O, 0.2 mL of 37% (g g^{-1}) HCl
- Bioenzymes: 1.8 g of bovine serum albumin, 6 g of bile
- pH: : 8.0 ± 0.2

In vitro bionic digestion procedure:

About 4.0 g of seaweed was mixed with 5 mL of saliva. Chewing was simulated by oscillation (5 min, 37°C). A quantity of 150 mL of gastric juices was added and gastric digest was stirred for 3 h at 37°C. Finally, 150 mL of duodenal and 60 mL of bile solutions were added. Intestinal digest was stirred for 8 h at 37°C. Around 25 mL of chyme (gastric or gastrointestinal) was mixed with a multilayer liposome, vibrated (37°C) in N$_2$ atmosphere, frozen (–71°C, 20 min) and thawed at 37°C. Monolayer liposome bond metals (AMLMs) and water soluble metals (WSMs) could be separated by using suction filtration with 0.45 and 0.22 μm membranes, respectively.

| *Porphyra umbilicalis* (Nori), *Laminaria ochroleuca* (Kombu), *Undaria pinnatifida* (Wakame) and *Ulva rigida* (Sea lettuce) seaweed | Gastric juices: Pepsin 6.0% (m/v) dissolved in 0.1 M HCl
Intestinal juices: Pancreatine 0.4% (m/v) + bile salts 2.5 % (m/v) dissolved in 0.1 M sodium hydrogen carbonate
Gastric digestion: a volume of 20 mL of DIW was added to 20 g of seaweed and the pH was adjusted to 2.0 with 6 M HCl. Then, 0.15 g of gastric solution was added. Gastric digest was shaken (150 rpm, 37°C) for 2 h. The enzymatic digestion was stopped by placing the flasks in an ice-bath.
Intestinal solution: A quantity of 5 mL of intestinal juices was added to the gastric digest. Dialysis membranes filled with 20 mL of a 0.15 N PIPES solution (pH: 7.5) were placed inside the flasks. The intestinal digestion was completed by shaking (150 rpm, 37°C) for 2 h. The enzymatic digestion was stopped by placing the flasks in an ice-bath. | |

Table 6. contd....

Table 6. contd.

Seaweed	In vitro bioavailability assay	Arsenic species	Results	Ref.
Porphyra umbilicalis (Nori), *Laminaria ochroleuca* (Kombu), *Undaria pinnatifida* (Wakame) and *Ulva rigida* (Sea lettuce) seaweed	The same as previously reported (García-Sartal et al. 2011)	t-As and glycerol, phosphate and sulfate riboses species	Total arsenic bioavailability percentages were found between 11–16% for the seaweed under study, being the arsenosugar percentage in dialysates ranging between 93–120%, except for Sea lettuce (41%).	(García Sartal et al. 2012c)
Cooked *Porphyra umbilicalis* (Nori), *Laminaria ochroleuca* (Kombu), *Undaria pinnatifida* (Wakame) and *Ulva rigida* (Sea lettuce) seaweed	The same as previously reported (García-Sartal et al. 2011)	t-As, inorganic As, DMA, AB and glycerol, phosphate and sulfate riboses species	Glycerol sugar and sulfonate sugar were the most bioavailable species for cooked Kombu and Wakame. Phosphate sugar showed to be in highest proportion in the dialyzable fraction of Nori. AB and glycerol sugar became more bioavailable in the Sea Lettuce sample.	(García Sartal et al. 2012b)

Total arsenic bioavailability % was calculated as:

$$Total\ BAv\% = \frac{Total\ bioavailable\ arsenic\ in\ seaweed}{Total\ arsenic\ in\ seaweed} \times 100$$

Individual arsenic specie bioavailability % was calculated as:

$$As\ specie\ BAv\% = \frac{Bioavailable\ As\ specie\ concentration\ in\ seaweed}{Total\ As\ specie\ concentration\ in\ seaweed} \times 100$$

Abbreviations and acronyms: PIPES (Piperazine-NN-bis (2-ethane-sulfonic acid) disodium salt); DIW (de-ionised water) the integrity of arsenosugars even after their having been subjected to strong enzymatic and acid gastrointestinal conditions. However, Gamble et al. (2002) found a slow hydrolytic degradation of arsenosugars isolated from ribbon kelp and which had been subjected to simulated gastric juices. DMA (main urinary arsenic species after seaweed ingestion) was not found as a degradation product suggesting that methylated arsenic species are not generated due to the stomach conditions. However, a common degradation product (m/z: 254) for the three arsenosugars under study (m/z: 392, 408 and 482) was tentatively identified as another arsenosugar by ESI-MS/MS.

The bioaccessibility of i-As (As(III) and As(V)), methylated arsenic species (MMA and DMA) and arsenosugars (glycerol, phosphate, sulfonate and sulfate riboses) from *Fucus* sp. harvested in an arsenic-contaminated marine area has been evaluated by Koch et al. (2007). These authors found that approximately 1.2–1.9% of total bioaccessible arsenic was attributed to the methylated arsenic species, and 43–46% to arsenosugars.

A static (SM) and a dynamic (TIM-1) *in vitro* gastrointestinal digestion method have been compared by Torres Escribano et al. (2011) for As, Cd, Pb and Hg bioaccessibility assessment in food certified reference materials, including a IAEA-140/TM (*Fucus* sp.) reference material. The high bioaccessibility percentages obtained (near to or greater than 50%) suggested that arsenic was present in this seaweed under easily solubilizable chemical forms, such as water-soluble arsenosugar species. Significant differences were found in the concentration of metals, (including As) in the bioaccessible fraction of *Fucus* sp. when it was evaluated by a static and a dynamic method. Bioaccessible arsenic concentration in the seaweed CRM was 28.4–33.6 and 18.5–21.8 $\mu g \ g^{-1}$ on a dry weight basis by a static and dynamic *in vitro* approach, respectively.

As previously mentioned, a loss of arsenic from seaweed into the cooking water has been confirmed by García-Sartal et al. (García Sartal et al. 2012b; García-Sartal et al. 2011). The authors assessed arsenic bioavailability in seaweed by an *in vitro* method based on absorbable arsenic dialysis through a semi-permeable membrane of 10 kDa molecular weight cut-off. Dialyzability percentages found in raw seaweed (Kombu, Wakame and Nori), were statistically comparable with those found in cooked seaweed, suggesting that the bioavailable arsenic species remain in the sample after cooking. The exception was Sea lettuce seaweed which showed lower bioavailability percentages when cooked. The authors also confirmed, by HPLC-ICP-MS in parallel with ESI-MS/MS, that the relative arsenic species distribution in seaweed-buffered extracts and dialysates was very similar (García Sartal et al. 2012c). They found that approximately 11–16% of the total arsenic in the raw seaweed samples was bioavailable, with arsenosugars comprising 93–120% of the dialyzable arsenic fraction for

Kombu, Wakame and Nori, and only 41% in the case of Sea lettuce. The arsenic species present in the dialysate fractions of four types of cooked seaweed (Kombu, Wakame, Nori and Sea lettuce) were determined in a recent study developed by these authors (García Sartal et al. 2012b). Results suggested that the acidity of the solution, the enzymes used and the dialysis through the membrane did not affect the arsenic speciation. Furthermore, arsenic species distribution showed that OH-ribose was the arsenic species more soluble or dialyzable in Kombu, Wakame and Nori cooked seaweed, but AB and As(III) were the arsenic species more dialyzable in the case of cooked Sea lettuce.

Metal bioavailability, including As, V, Cr, Mn, Fe, Ni, Cu, Zn, Se, Cd and Pb from *Laminaria japonica* seaweed has been studied by Li et al. (2011). After seaweed *in vitro* digestion, metals were absorbed through a monolayer liposome membrane as coordinated complexes. These authors concluded that metal bioavailability was affected by metal species, gastrointestinal acidity and composition of the gastric and intestinal juices. Therefore, intestinal acidity, semibionic intestinal digestion (excluding enzymes) and bionic digestion in the stomach (including enzymes) conditions enhanced arsenic absorption through the liposomal membrane.

4 Urinary Arsenic Speciation after Seaweed Consumption

Arsenic metabolism depends on the chemical form in which it is consumed and on the animal species (Suzuki et al. 2002). It has been demonstrated that humans, dogs and mice are able to methylate inorganic arsenic for further urinary excretion. However, marmoset monkeys, chimpanzees and guinea pigs seem to use other kinds of mechanisms, such as protein binding, for arsenic elimination from the organism (Vahter 2002). Arsenic speciation studies in urine can give us valuable information about the arsenic biotransformation in the human body.

When arsenic enters the human body, its bioavailable fraction reaches the bloodstream. Subsequently, it is metabolized and urine is the main pathway for arsenic elimination from the organism (Crecelius 1977; Ma and Le 1998).

Because of the rapid metabolism of this element and its short half-life (approximately 4 days), total urinary arsenic or the sum of inorganic arsenic (As(III) and As(V)), MMA(V) and DMA(V) excreted have been used as biomarkers of human exposure (Hughes 2006; Navas-Acien et al. 2011). Their presence in urine give us information about the potential health risks for humans exposed to inorganic arsenic.

Different authors proposed a detoxifying mechanism in humans which involves *in vivo* inorganic arsenic methylation processes in the liver. The organic arsenic species formed, such as DMA(V) and MMA(V) are less toxic, less reactive with tissue constituents and easily excretable. The proposed methylation mechanism alternates two-electron reduction steps of pentavalent arsenic [As(V) and MMA(V)] to trivalent forms [As(III) and MMA(III)] (acting glutathione as a reducing agent) with oxidative additions of a methyl group from a methyl donor compound (S-adenosylmethionine) to finally be transformed into DMA(V) (Le et al. 2000b; Xie et al. 2006). Other authors questioned the proposed metabolic pathway after finding MMA(III) (Le et al. 2000b; Le et al. 2000a) and DMA(III) (Le et al. 2000a) presence in human urine. New biomethylation processes suggest the formation of DMA(III) from DMA(V) which could be metabolized into TMAO(V). The presence of TMAO below the detection limit in urine samples could be explained as a result of its possible metabolization to trimethylarsine [TMA(III)], a volatile species that can be exhaled while breathing (Le et al. 2000a). The non-toxicity of methylated arsenic species has been called into question. Several studies have suggested that the methyl donor compound S-adenosylmethionine is required for both arsenic and DNA methylation. Arsenic might be genotoxic, altering the DNA methylation (Le et al. 2000b; Feldmann et al. 1999). Furthermore, it has been shown that MMA(III) and DMA(III) produce enzyme inhibition, cell toxicity and clastogenicity (Hsueh et al. 2002), being more toxic than the inorganic forms. Therefore, biomethylation must be considered as a bioactivation rather than a detoxification mechanism.

On the other hand, recently, Hayakawa et al. (2005) elucidated an arsenic methylation pathway in which DMA(III) is a DMA(V) precursor formed as an intermediate between MMA(V) and DMA(V). Their findings reaffirm the former theory that the methylation of inorganic arsenic into a less toxic DMA(V) is a detoxification process in the human body.

Taking into account this biomethylation process, there is an apparent pattern of excretion under which around 80% of the inorganic arsenic is detoxified and excreted mainly as DMA(V) (60–70%), 10–20% as MMA(V) and 10–30% as inorganic arsenic (Ruiz-Navarro et al. 1998; Ritsema et al. 1998; Sur et al. 1999; Samanta et al. 2000; Soleo et al. 2008).

On the other hand, it has been found that the intake of a single meal of seafood, including seaweed, increases by over 1000 µg L^{-1} the total arsenic content in urine (Vahter 1994). As previously mentioned, seaweed and seafood contain high concentrations of organoarsenic compounds such as arsenosugars or arsenobetaine. Unlike arsenosugars, arsenobetaine is considered a very stable compound and is excreted unchanged into the urine. The As-C bond in the arsenobetaine structure might confer stability to this organoarsenic specie, hindering its interaction with other biomolecules (Feldmann and Krupp 2011). However, a recent study questions this established theory, suggesting two postulates: either arsenobetaine is accumulated, to be later released from tissues, or arsenobetaine is generated in the inorganic arsenic or DMA(V) metabolism (Newcombe et al. 2010). Nevertheless, arsenosugars are mainly metabolized to DMA, among other species. DMA is a common metabolite in the urine from humans exposed to inorganic or organic arsenic. For this reason, to obtain a reliable assessment of inorganic arsenic exposure, subjects under study must be interviewed as to their food habits and refrain from eating seafood before their urine is sampled (Ma and Le 1998; Navas-Acien et al. 2011; Lintschinger et al. 1998). The similarities in the metabolism of inorganic arsenic and arsenosugars after finding DMA(V) and other more toxic common urinary metabolites (trivalent methylated species) have been recognized. Therefore, it is important to elucidate the metabolism of arsenosugars to rule out toxic effects, even though arsenosugars have not shown cytotoxicity in macrophages at trace level (Sakurai 2002).

It is known that arsenosugars are absorbed by humans and subsequently excreted into the urine but their metabolism is more complicated than the metabolism of inorganic arsenic. Raml et al. (2005) have suggested an arsenosugar degradation mechanism in which the ingested arsenosugar could be thiolated before further degradation. They analysed the urine of a human volunteer who ingested a synthetically prepared glycerol-sugar. The authors found thio-arsenic species in the samples under study. They suggested a first degradation of ingested arsenosugar into DMAE, DMAA and their thio-counterparts, followed by a further degradation of these compounds into DMA and thio-DMA, respectively. Other degradation mechanisms may also be possible. In another alternative mechanism proposed by these authors, the thio-DMAA and thio-DMAE may result from the intermediate-forming thio-arsenosugar or from the ingested arsenosugar via DMAA or DMAE, respectively. Feldmann and Krupp (2011) proposed a metabolic transformation pathway for a DMA(V)-sugar. The authors suggested that the ingested arsenosugar was transformed into DMA(V), via DMAE and DMAA. They also suggested that thiolated and trivalent intermediate species of the precursor arsenosugar and the three metabolites formed might occur.

Thio-DMA presence has been confirmed in urine samples from wild sheep which live on the beaches of North Ronaldsay (Orkney Islands, UK) and feed on arsenosugar-rich seaweed (Hansen et al. 2004). On the other hand, Raml et al. (2007) corroborated, for the first time, the presence of thio-DMA in the urine of Bangladeshi people exposed to inorganic arsenic via drinking water, suggesting that this metabolite is not exclusive to arsenosugar metabolism. They have also observed the slow conversion of thio-DMA into DMA(V) over time.

It has been demonstrated that around 85–95% of the arsenic ingested as arsenosugars is excreted in urine within 90 hours. However, findings suggest that the excretion rate is inter-individual dependent. Raml et al.

(2009) developed a study with six volunteers who ingested a synthetically prepared oxo-arsenosugar. The authors classified the volunteers into two groups; low and high-excretors. Only 3.9–15% of the ingested arsenic was excreted by low-excretors within 90 hours, mainly as DMA, oxo-arsenosugar (unchanged) and thio-arsenosugar. DMA, thio-DMAA, thio-DMAE and an unknown thio-arsenic compound were the main arsenic species found in the urine of the high-excretors group; however, unchanged ingested arsenosugar was not found in their samples. The differences between both groups might be due to biotransformation enzymes in the liver or differences in gut flora. The high-excretors may be able to metabolize the arsenosugar to a bioaccessible form which could be biotransformed in the liver and further excreted in the urine. Feces might be the route of arsenic excretion in the case of the low-excretors.

Arsenic toxic effects are associated with low methylation capacity in humans, thus, with higher arsenic concentration in tissues (Vahter 2000). However, harmful health effects were not noticed in a group of seaweed-eating sheep from Northern Scotland with high levels of arsenic in urine and blood (two times higher than unexposed control sheep) and in liver, kidney, muscle and wool (100 times higher than in control sheep) (Feldmann et al. 2000). Similarly, in order to obtain a biological response or clinical disorder, some biological parameters such as serum aspartic acid transaminase (AST), glutathione (GSH), glutathione peroxidase (GPX), lipid peroxidation and uric acid have been analysed after controlled seafood consumption. Results showed that these parameters were not affected by prolonged seafood intake in the 16 volunteers under study (Choi et al. 2010).

4.1 Factors affecting arsenic biotransformation in humans

As discussed above, arsenic epidemiological studies are based on the methylation capacity of individuals. The inter-individual variation found in arsenic methylation suggests that there are some factors affecting the arsenic biotransformation pathway. Age, sex, smoking status, body mass index (BMI) and nutritional and dietary status are some of the factors that might be influential in arsenic metabolism (Tseng 2009).

An assay of arsenic metabolites in urine belonging to 479 adults has concluded that age and gender factors affect the methylation of inorganic arsenic. The older volunteers showed the lowest methylation capacity, while women exhibited a higher methylation capacity than men (Tseng et al. 2005). Estrogen might be responsible for more efficient methylation of arsenic in pregnant women (Tseng 2009). It has been found that smokers excrete higher MMA(V) percentages than DMA(V) in urine; this is likely, due to the competition of the chemical constituents of cigarettes for enzymes involved in the methylation mechanism to DMA (Tseng 2009). Body mass index as an indicator of obesity has been positively correlated with high DMA and low MMA excretion in urine. This fact could be related to a higher nutritional status, increased human adipose tissue or changes in the gene expression profile induced by obesity (Gomez-Rubio et al. 2011). Brima et al. (2007) have suggested that fasting affects the arsenic methylation into the body, increasing the percentage of MMA(V) excreted. They concluded that the most toxic arsenic species are removed from the body during fasting. Low protein and nutrient intake have been associated with higher MMA excretion, and consequently, low biomethylation efficiency (Mitra et al. 2004).

On the other hand, factors such as ethnicity or genetic profile have been evaluated. Similar arsenic metabolism patterns have been found between individuals belonging to the same population or family (Tseng 2009). However, inter-ethnic differences have also been observed. This fact could be explained as a result of the difference in dietary habits between different populations, since diet may enhance the activity of particular enzymes across generations resulting in genetic polymorphism (Brima et al. 2006), or due to other factors such as duration and dosage exposure, or lifestyle (Tseng 2009).

4.2 Sampling, storage and sample treatment of urine samples

Appropriate sampling and storage are essential prerequisites to keep the integrity of the arsenic species in the sample and to obtain reliable results in speciation studies.

In the arsenic speciation studies of urine samples, volunteers under study must avoid the ingestion of arsenic-rich food (e.g., seafood, mushrooms, rice) for at least 4 days before the beginning of the experiment. The purpose is to ensure that all the arsenic metabolites present in the urine come from no other sources, except from seaweed. A common practice in this kind of research is to collect the first morning urine sample of the fourth day, before the seaweed ingestion. This sample will be used to evaluate the arsenic background, as a blank.

There are three common urine collection methods: a 24 hour sampling, a first morning void (FMV) and a spot urine sampling. The first method is probably the most reliable but it entails a complex logistic. Arsenic speciation studies have been carried out using FMV and spot urine samples. A significant correlation between them was found suggesting that both kinds of sampling can be used to assess arsenic exposure (Rivera-Núñez et al. 2010).

The influence of several factors, including storage temperature, time and addition of acid or preservatives to the urine samples, has been studied. Results suggested that strong acidification with hydrochloric acid produces changes in the arsenic speciation, causing the conversion of As(V) to As(III). Additives, including sodium azide, benzoic acid, benzyltrimethylammonium chloride and cetylpyridinium chloride, have been evaluated for sample preservation. The addition of these agents did not show any improvement in arsenic species recoveries when compared with untreated samples.

The stability of the arsenic species with at different temperatures was studied by Feldmann et al. (1999). Three storage temperatures (25, 4 and –20°C) were tested. No substantial changes in arsenic speciation were observed in samples kept at low temperature (4 and –20°C) for up to 2 months. However, longer storage times (4 and 8 months), affected the arsenic species' distribution, which varied depending on the sample matrix.

More recently, trivalent arsenic metabolites, MMA(III) and DMA(III), have been detected in human urine. The instability of these trivalent forms with the temperature during handling and storage steps was demonstrated by Gong et al. (2001). The authors observed the oxidation of MMA(III) (>90%) to MMA(V) in a five-month period at 4 or –20°C, and within a week at room temperature. They also showed that DMA(III) was completely oxidized to DMA(V) in just one day at the mentioned storage conditions.

In speciation studies, urine samples are subjected to a simple treatment before their introduction into the separation system. Most treatments consist of a sample dispersion by using ultrasonic wave energy (Hsueh et al. 2002), followed by a sample filtration step. Sep-Pak C_{18} columns (Hsueh et al. 2002) or filtration through a 0.45 or 0.22 µm nylon membranes has been used in urine treatments (Raml et al. 2005; Hansen et al. 2003a). Centrifugation followed by supernatant filtration through a 0.22 µm PVDF syringe filter has also been used for urinary arsenic speciation (Van Hulle et al. 2004). Other authors have extracted the organoarsenic species to a phenol phase to avoid the interference of inorganic salts (Francesconi et al. 2002).

4.3 Arsenic species determination and identification in urine samples

Urinary arsenic speciation research has been performed by coupling a variety of liquid chromatographic separation techniques with an arsenic-specific detector. Tables 7 and 8 show different methods used for total arsenic determination and arsenic speciation in urine samples using different detectors. ICP-MS is the most common detector used for arsenic determination in urine samples; however, due to arsenic's ability to generate volatile hydrides, hydride generation (HG) combined with AAS and AFS has been used as an alternative technique offering high sensibility and sensitivity.

Table 7. Total urinary arsenic determination.

Matrix ingested	Digestion procedure	Arsenic detection	Total arsenic concentration	Certified reference material	Ref.
Synthetically prepared arsenosugar glycerol	Acid digestion assisted by microwave energy: 2.00 mL of urine + 1.00 mL of water + 2.00 mL of HNO$_3$. Digests were diluted up to 25.0 mL with water.	ICP-MS	Background As: 1.25 µg of As or 0.35 µg As/g of dry weight of urine Maximum As concentration at 27 hours: 194 µg of As or 29.6 µg As/g of dry weight of urine		(Francesconi et al. 2002)
Laminaria seaweed	Urine samples were diluted 40-fold with 0.14 M HNO$_3$ and 3% MeOH was added to improve the signal intensity.	ICP-MS	Urine samples from 5 volunteers with normalised concentrations between 20–120 µg As/g of creatinine		(Van Hulle et al. 2004)
Synthetically prepared arsenosugar glycerol	Acid digestion assisted by microwave energy: 1.00 mL of urine + 2.00 mL of water + 2.00 mL of HNO$_3$. Digests were diluted up to 10.0 mL with water.	ICP-MS	Background As: < LOD (<3 µgL^{-1}) Maximum As concentration at 9 hours: 221 µg of As	NIES CRM no. 18 (Human Urine): 126–148 µg L^{-1} DORM-2 (dogfish muscle): 16.9–19.1 µg g^{-1}	(Raml et al. 2005)
Synthetically prepared arsenosugar glycerol	Microwave-assisted acid digestion 2.00 mL of HNO$_3$ and 2 mL of water were added to 1 mL of urine sample. Samples were mineralized by using microwave energy from an Ultraclave III, and finally were diluted up to 50.0 mL with water. Urine preparation without mineralization 200 µL of urine were diluted up to 10 mL with water containing 100 µL of HNO$_3$ and 500 µL of MeOH.	ICP-MS	Background As: 2.9–12.0 µg As/L normalised concentration for volunteers 1–6. Maximum As concentration: 25–532 µg As/L normalised concentration for volunteers 1–6, excreted within 24 hours after ingestion.	NIES CRM no. 18 (Human Urine): 126–148 µg As L^{-1}	(Raml et al. 2009)

| Kelp, dried laver, flatfish and conch | Urine samples were reduced by adding HCl, ascorbic acid and KI. Arsine gas was produced by reaction of the mixture with borohydrate and sodium peroxide. | HG-AAS | Background As: 1.4 µg As/g creatinine Maximum As concentration: 29.6 µg As/g creatinine at 24 h after seafood ingestion. Seafood intake during six consecutive days increased total urinary As up to higher levels (48.0 µg As/g creatinine). | | (Choi et al. 2010) |

Abbreviations and acronyms: ICP-MS (inductively coupled plasma—mass spectrometry); HG-AAS (hydride generation—atomic absorption spectroscopy)

Table 8. Urinary arsenic speciation after seaweed consumption.

Matrix ingested	Chromatographic conditions	As species detection	As species determined	Ref.
Laminaria digitata	• Anion-exchange Column: PRP-X100 (Hamilton Company) Mobile Phase: 30 mM H$_3$PO$_4$ pH: 6.0 Flow rate: 1.3 mL.min⁻¹ Vinj: 20 µL • Cation-exchange Column: Supelcosil SCX (Supelco) Mobile Phase: 20 mM pyridine pH: 9.0 Flow rate: 1.0 mL.min⁻¹ Vinj: 20 µL	HPLC-ICP-MS	As(III): <LOD, arsenosugar glycerol: <1–44 µg L⁻¹, DMA(V): 457–9116 µg L⁻¹, MMA(V): <1–259 µg L⁻¹, As(V): <LOD, TMA⁺: <1–56 µg L⁻¹, an unidentified specie: <1–144 µg L⁻¹	(Feldmann et al. 2000)
Synthetically prepared arsenosugar glycerol	• Anion-exchange Column: PRP-X100 (Hamilton Company) Temperature: 40°C Mobile Phase: 20 mM NH$_4$HCO$_3$ pH: 9.0 + 3% MeOH Flow rate: 1.5 mL.min⁻¹ Vinj: 5–20 µL • Cation-exchange Column: Ionospher-C (Chrompack Denmark ApS,) Temperature: 40°C Mobile Phase: 10 mM pyridine pH: 3.0 + 3% MeOH Flow rate: 1.5 mL.min⁻¹ Vinj: 5–20 µL	HPLC-ICP-MS	DMA(V): 67%, DMAE: 5%, TMAO: 0.5% and possibly MMA(III): 20%. Percentages are referred to the total arsenic found in urine sample at 27 hours after ingestion.	(Francesconi et al. 2002)
Seaweed *Laminaria hyperborea* and *Laminaria digitata*	• Anion-exchange Column: PRP-X100 (Hamilton Company) Mobile Phase: 30 mM acetic acid pH: 5.3 Flow rate: 1.0 mL.min⁻¹	HPLC-ICP-MS	DMAE, DMA(V) and DMAA	(Hansen et al. 2003a)

Sample	Conditions	Technique	Species	Reference
Seaweed *Laminaria hyperborea* and *Laminaria digitata*	• Anion-exchange Column: PRP-X100 (Hamilton Company) Mobile Phase: 30 mM H_3PO_4 pH: 5.3 or 6.0 • Cation-exchange Column: Supelcosil LC-SCX (Supelco) Mobile Phase: 20 mM pyridine pH: 2.5, 3.0 or 7.0	HPLC-ICP-MS	DMA(V) (60 ± 22% of intake), DMAE, TMA+, MMA(V), As(V) and seven unknown compounds	(Hansen et al. 2003b)
Laminaria	• Anion-exchange Column: PRP-X100 (Hamilton Company) Mobile Phase: Method 1: 20 mM $(NH_4)_2HPO_4$ pH: 7.0 + 3% MeOH Method 4: 30 mM ammonium acetate pH: 6.0 + 3% MeOH Flow rate: 1.0 mL.min^{-1} Vinj: 50 µL • Cation-exchange Column: Dionex Ionpac CS-10 (Dionex) Mobile Phase: 20 mM pyridine pH: 2.7 Flow rate: 1.0 mL.min^{-1} Vinj: 50 µL	HPLC-ICP-MS	DMA(V), MMA(V), DMAE	(Van Hulle et al. 2004)
Synthetically prepared arsenosugar glycerol	• Anion-exchange Method A Column: PRP-X100 (Hamilton company) Mobile Phase: 20 mM $NH_4H_2PO_4$ pH: 6.0 + 3% MeOH Flow rate: 1.5 mL.min^{-1} Temperature: 40°C Method B Column: PRP-X100 (Hamilton company) Mobile Phase: 20 mM NH_4HCO_3 pH: 9.0 + 3% MeOH Flow rate: 1.5 mL.min^{-1} Temperature: 40°C • Cation-exchange	HPLC-ICP-MS	DMAE: 6%, Unchanged arsenosugar: 1%, DMA: 28%, DMAA: 4%, Thio-DMAE: 17%, Thio-DMAA: 31%, and TMAO and Thio-DMA at trace level. Percentages are referred to the total arsenic found in urine sample at 9 hours after ingestion.	(Raml et al. 2005)

Table 8. contd....

Table 8. contd.

Matrix ingested	Chromatographic conditions	As species detection	As species determined	Ref.
	Method C Column: Zorbax 300 SCX (Agilent) Mobile Phase: 20 mM pyridine pH: 2.6 Flow rate: 1.5 mL.min⁻¹ Temperature: 30°C Method D Column: Ionospher 5C (Varian) Mobile Phase: 10 mM pyridine pH: 3.0 + 3% MeOH Flow rate: 1.5 mL.min⁻¹ Temperature: 40°C Method E Column: Shodex NN-614 (Showa Denko) Mobile Phase: 6 mM NH_4NO_3 + 5 mM HNO_3 pH: 2.2 Flow rate: 0.4 mL.min⁻¹ Temperature: 25°C Method F Column: Atlantis (Waters) Mobile Phase: 20 mM $NH_4H_2PO_4$ pH: 3.0 Flow rate: 1.0 mL.min⁻¹ Temperature: 30°C			
Hijiki seaweed	• Anion-exchange Column: Gelpack GL-IC-A15 (Showadenko) Mobile Phase: 5 mM HNO_3;6 mM NH_4NO_3: 1.5 mM PDCA Flow rate: 1.0 mL.min⁻¹ Vinj: 50 µL • Cation-exchange Column: Shodex Rspak NN-614 (Hitachi Resin) Mobile Phase: 3 mM NaH_2PO_4 pH: 6.0 Flow rate: 0.8 mL.min⁻¹ Vinj: 50 µL	HPLC-ICP-MS	As(V), As(III), MMA(V), DMA(V) and several unidentified arsenic peaks	(Nakajima et al. 2006)

Synthetically prepared arsenosugar glycerol	• Anion-exchange Column: PRP-X100 (Hamilton company) Mobile Phase: 20 mM $NH_4H_2PO_4$ pH: 6.0 + 3% MeOH Flow rate: 1.5 mL.min^{-1} Temperature: 40°C • Cation-exchange Column: Chrompack Ionospher C (Varian) Mobile Phase: 10 mM pyridine pH: 3.0 + 3% MeOH Flow rate: 1.5 mL.min^{-1} Temperature: 40°C • Reversed phase Method A Column: Waters Atlantis dC18 Mobile Phase: 20 mM $NH_4H_2PO_4$ pH: 3.0 Flow rate: 1.0 mL.min^{-1} Temperature: 30°C Method B Column: Waters Atlantis dC18 Mobile Phase: 20 mM NH_4HCO_2 pH: 3.0 Flow rate: 1.0 mL.min^{-1} Temperature: 30°C	HPLC-ICP-MS	DMA (40–46%), thio-DMAA (15–19%), thio-DMAE (5–9%), DMAE, DMAA, TMAO, thio-DMA, TMAS, unchanged arsenosugar, thio-arsenosugar and a pair of thio- and oxo-unknown compounds	(Raml et al. 2009)
Kelp, dried laver, flatfish and conch	• Anion-exchange Column: PRP-X100 (Hamilton company) Mobile Phase: 20 mM NH_4HCO_3/20 mM $(NH_4)_2SO_4$ Flow rate: 1.5 mL.min^{-1} Temperature: 25°C Vinj: 100 µL	HPLC-ICP-MS	DMA (V), AsC, and inorganic arsenic [As(III) and As(V)] and MMA(V) detected only at a limited level	(Choi et al. 2010)

Abbreviations and acronyms: HPLC (high performance liquid chromatography); ICP-MS (inductively coupled plasma-mass spectrometry); PDCA (2,6 Pyridinedicarboxylic acid); Vinj (injection volume).

The main drawback in hydride generation systems is their inability to distinguish between trivalent and pentavalent methylated species. While pentavalent As species are detected after forming the hydride at low acidic conditions (pH<1), higher pH conditions (pH>5) are required in the case of trivalent As species. Consequently, two separation methods of the same sample are needed for the determination of both species (Gong et al. 2001; Burguera and Burguera 1997).

Arsenic species identification by using HPLC separation techniques is based on the retention time of arsenic compounds. Arsenic species are identified by comparing their retention times with those of individually injected standards. Spiking samples with standards is another useful alternative. Analytes in samples can be identified by matching retention times with the spiked standards without influence from the sample matrix. Following this identifying methodology, As(III), As(V), MMA(V), DMA(V), AB or TMAO have been found in urine (Francesconi et al. 2002; Verdon et al. 2009).

The major drawback in urinary arsenic speciation is that there are no commercial standards available for all the arsenic species present in urine. Some authors use lab-made standards to solve this problem. MMA(III) and DMA(III) have been synthesized by hydrolysis of CH_3AsI_2 and $(CH_3)_2AsI$, respectively (Xie et al. 2006). Thio-DMAA and thio-DMAE have been synthesized as crystals after dissolving their oxo analogues in water, and bubbling H_2S through the solution (Raml et al. 2005, 2007). Thio-DMA was prepared by the addition of aqueous H_2S and ethyl acetate to a solution of dimethylarsinate (pH: 3.0). Similarly, thio-MMA was formed after MMA solution was added to a saturated H_2S solution (Raml et al. 2007). DMAA and DMAE have been prepared as previously reported by Francesconi et al. (1989) and Edmonds et al. (1982). Trimethylarsine sulfide (TMAS) has been synthesized by the addition of saturated aqueous H_2S solution to a solution of TMAO (Raml et al. 2009). However, the use of uncharacterized standards may lead to wrong analyte assignments in samples. For example, Hansen et al. (2004) assumed that the lab-made standard synthesized by the reduction of the methylated pentavalent arsenic species with sodium-metabisulfite $(Na_2S_2O_5)$/sodium thiosulfate $(Na_2S_2O_3)$ reagent was DMA(III). However, HPLC-ICP-MS coupled simultaneously to HPLC-ES-MS and ES-Q-TOF-MS techniques revealed that the supposed DMA(III) was actually thio-DMA. This fact suggests that the DMA(III) assignment made in other studies may be an error.

Molecular specific techniques such as electrospray tandem mass spectrometry (ESI-MS-MS) or electrospray single mass spectrometry (ESI-MS) offers molecular and structural information in the absence of standards. These techniques are a supporting tool for ICP-MS in arsenic species identification. However, it has been shown that HPLC-ESI-MS-MS cannot be used to quantify arsenic species in urine samples, since electrospray ionisation is compound dependent and produces a reduction in the analytical signal due to the sample matrix (Hansen et al. 2003a).

Taking into acount the advantages of organic mass spectrometry, Hansen et al. (2003a) have identified DMAE, DMA(V) and DMAA in urine samples from sheep fed on seaweed. Urine samples were analysed by parallel HPLC-ICP-MS and HPLC-ESI-MS in positive ion mode. Each peak eluted from the column in the ICP-MS system was assigned to a spectrum originating from the ESI-MS at the same retention time, using a fragmenter voltage of 100 V. In this way, peak 1 with a dominant presence of m/z 167 was identified as the main $[M+H]^+$ ion of DMAE. Peak 2 was assigned as DMA(V), as the mass spectrum was dominated by m/z 139 ($[M+H]^+$ ion), m/z 277 ($[2M+H]^+$ ion) and m/z 259 ($[2M+H-18]^+$ ion). Finally, peak 3 at the same retention time as the dominant ions with m/z 361 and m/z 181 was assigned to DMAA. The presence of these organo arsenic species was corroborated when m/z 91 was found as dominant ion in all three mass spectra after increasing the fragmenter voltage up to 240 V. An m/z 91 is a common breakdown product of organo arsenic species due to the formation of either AsO^+ or CH_3AsH^+.

Recently, Raml et al. (2009) have elucidated possible structures for an oxo/thio pair found in human urine by HPLC-ESI-MS. Two main peaks were found in the spectra ($[M+H]^+$ with m/z 211 and $[M+H]^+$-$H_2O]$ with m/z 193). The compound was formulated as $Me_2As(O)$-$(C_4H_9O_2)$. A possible candidate with the elucidated molecular formula and its thio-analogue was then synthesized by the authors. Mass spectra data of the synthesized compounds was in accordance with that found in urine; however. the chromatographic properties were different. This data indicated that the formulation was likely to be correct but not the synthesized isomer.

Table 9 shows the dominant and common fragments obtained by MS for the arsenic species found in urine samples after seaweed ingestion.

Table 9. Arsenic species identification in urine samples.

Arsenic specie	Dominant fragments	Common fragments
DMAE	m/z 167	
DMA(V)	m/z 139, 277, 259	m/z 121, 109, 107, 89
DMAA	m/z 361, 181	m/z 163, 121, 109, 107, 105, 89
TMAS	m/z 153, 107	

4.4 Results normalization

Intake of liquids, diet and perspiration are some of the causes that affect urinary dilution and, accordingly, the concentration of urine constituents. These factors produce intra-individual variations that must be normalized for results to be compared (Carrieri et al. 2000).

Urinary creatinine concentration and specific gravity are the most common corrections. Adjustment of the results is done on the basis of creatinine measurements and urine density or dried weight of fresh urine samples, respectively. Specific gravities higher than 1.03 or below 1.01 and creatinine concentrations outside the range 0.5–3 g.L^{-1}, are inadequate and may lead to false results (AIHA 2014; Rivera-Núñez et al. 2010). Parameters outside the range are indicative of liver or kidney disease, or are characteristic of people who handle chemicals containing arsenic (Soleo et al. 2008).

Creatinine has been determined by reversed phase HPLC with UV detection (Van Hulle et al. 2004). Creatinine can also be analysed by photometric techniques by using commercial kits such as Metra Creatinine Assay Kit (Quidel Corporation, USA) (Brima et al. 2007) or creatinasa and N-(3-sulfpuropyl)-3-methoxy-5-methylaniline from Pure Auto CRE-N kit (Daiichi Pure Chemicals Co. Ltd, Japan) (Nakajima et al. 2006). On the other hand, a pycnometer has been used for specific gravity measurements to provide an estimate of dissolved material and particle concentration in urine. The density of the urine was calculated based on the density of water as a reference (Rivera-Núñez et al. 2010).

5 Conclusion

Arsenosugars are the main arsenic species found in seaweed. Different techniques based on solvent extraction, microwave energy or accelerated solvent extraction have been developed for the extraction of arsenic species from seaweed. Extraction efficiencies obtained were dependent on the extraction technique, type of seaweed and arsenic species present in the seaweed sample. Ion exchange chromatography coupled with ICP-MS was the separation and detection technique selected by a vast majority of researchers. Further investigation of non-extractable or non-chromatographable arsenic species should be carried out in those seaweed samples with low recovery percentages.

Simple, inexpensive and reproducible *in vitro* techniques which simulate human gastrointestinal digestion have been developed to determine the bioavailable and bioaccessible fraction of arsenic from seaweed. It has been observed that arsenic is mainly solubilized in the stomach compartment due to the acidic pH. Arsenosugars, the main arsenic species found in seaweed, are soluble and easily absorbable under gastrointestinal conditions. These are probably the most bioavailable arsenic species for human body functions.

On the other hand, DMA(V) is the main arsenic specie excreted in urine after seaweed ingestion. Arsenosugars from seaweed are transformed into different metabolites in the liver. DMA(V) is the most toxic specie and the main product formed. Due to the difference in toxicity between the ingested arsenosugar and its metabolite (DMA(V)), arsenosugar safety has been questioned.

References

AIHA, 2014. Biological Monitoring: A practical field Manual. American Industrial Higiene Association (AIHA). Fairfax, VA: AIHA, Press.

Almela, C., M. Jesús Clemente, D. Vélez and R. Montoro. 2006. Total arsenic, inorganic arsenic, lead and cadmium contents in edible seaweed sold in Spain. Food and Chemical Toxicology. 44(11): 1901–1908.

Almela, C., J.M. Laparra, D. Velez, R. Barbera, R. Farreand R. Montoro. 2005. Arsenosugars in raw and cooked edible seaweed: Characterization and bioaccessibility. Journal of Agricultural and Food Chemistry. 53(18): 7344–7351.

Amayo, K.O., A. Petursdottir, C. Newcombe, H. Gunnlaugsdottir, A. Raab, E.M. Krupp and J. Feldmann. 2011. Identification and quantification of arsenolipids using reversed-phase HPLC coupled simultaneously to high-resolution ICPMS and high-resolution electrospray MS without species-specific standards. Analytical Chemistry. 83(9): 3589.

Andrewes, P., D.M. DeMarini, K. Funasaka, K. Wallace, V.W.M. Lai, H. Sun, W.R. Cullen and K.T. Kitchin. 2004. Do arsenosugars pose a risk to human health? The comparative toxicities of a trivalent and pentavalent arsenosugar. Environmental Science & Technology. 38(15): 4140–4148.

Besada, V., J.M. Andrade, F. Schultze and J.J. González. 2009. Heavy metals in edible seaweeds commercialised for human consumption. Journal of Marine Systems. 75(1-2): 305–313.

Bhattacharya, P., A.H. Welch, K.G. Stollenwerk, M.J. McLaughlin, J. Bundschuh and G. Panaullah. 2007. Arsenic in the environment: Biology and Chemistry. Science of The Total Environment. 379(2-3): 109–120.

Biological Monitoring: A Practical Field Manual. 2004. American Industrial Hygiene Association, Fairfax, VA.

Borak, J. and H.D. Hosgood. 2007. Seafood arsenic: Implications for human risk assessment. Regulatory Toxicology and Pharmacology. 47(2): 204–212.

Brandon, E.F.A., A.G. Oomen, C.J.M. Rompelberg, C.H.M. Versantvoort, J.G.M. van Engelen and A.J.A.M. Sips. 2006. Consumer product *in vitro* digestion model: Bioaccessibility of contaminants and its application in risk assessment. Regulatory Toxicology and Pharmacology. 44(2): 161–171.

Brima, E.I., P.I. Haris, R.O. Jenkins, D.A. Polya, A.G. Gault and C.F. Harrington. 2006. Understanding arsenic metabolism through a comparative study of arsenic levels in the urine, hair and fingernails of healthy volunteers from three unexposed ethnic groups in the United Kingdom. Toxicology and Applied Pharmacology. 216(1): 122–130.

Brima, E.I., R.O. Jenkins, P.R. Lythgoe, A.G. Gault, D.A. Polyaand P.I. Haris. 2007. Effect of fasting on the pattern of urinary arsenic excretion. Journal of Environmental Monitoring. 9(1): 98–104.

Burguera, M. and J.L. Burguera. 1997. Analytical methodology for speciation of arsenic in environmental and biological samples. Talanta. 44(9): 1581–1604.

Carrieri, M., A. Trevisan and G.B. Bartolucci. 2000. Adjustment to concentration-dilution of spot urine samples: correlation between specific gravity and creatinine. International Archives of Occupational and Environmental Health. 74(1): 63–67.

Caussy, D. 2003. Case studies of the impact of understanding bioavailability: arsenic. Ecotoxicology and Environmental Safety. 56(1): 164–173.

Caussy, D., M. Gochfeld, E. Gurzau, C. Neagu and H. Ruedel. 2003. Lessons from case studies of metals: Investigating exposure, bioavailability, and risk. Ecotoxicology and Environmental Safety. 56(1): 45–51.

Choi, B., S. Choi, D. Kim, M. Huang, N. Kim, K. Park, C. Kim, H. Lee, Y. Yum, E. Han, T. Kang, I. Yu and J. Park. 2010. Effects of repeated seafood consumption on urinary excretion of arsenic species by volunteers. Archives of Environmental Contamination and Toxicology. 58(1): 222–229.

Commission Directive 2008/84/EC, 2008, of 27 August 2008 laying down specific purity criteria on food additives other than colours and sweeteners. Official Journal of the European Communities. vol. 20/09/2008, pp. L253/1–253/75.

Commission Directive 2009/10/EC 2009. of 13 February 2009 amending Directive 2008/84/EC laying down specific purity criteria on food additives other than colours and sweeteners. Official Journal of the European Communities. vol. 14/2/2009, pp. L44/62–L44/78.

Craig, P. 2003. Organometallic Compounds in the Environment, John Willey & Sons Ltd, The Atrium, Southern Gate, Chinchester.

Crecelius, E.A. 1977. Changes in the chemical speciation of arsenic following ingestion by man. Environmental Health Perspectives. 19: 147.

Dawczynski, C., R. Schubert and G. Jahreis. 2007. Amino acids, fatty acids, and dietary fibre in edible seaweed products. Food Chemistry. 103(3): 891–899.

Dean, J.R. 2007. Bioavailability, Bioaccessibility and Mobility of Environmental Contaminants, John Wiley & Sons Ltd, The Atrium, Southern Gate, Chichester, West Sussex PO19 8SQ, England.

Devesa, V., A. Martinez, M.A. Suner, V. Benito, D. Velez and R. Montoro. 2001. Kinetic study of transformations of arsenic species during heat treatment. Journal of Agricultural and Food Chemistry. 49(5): 2267–2271.

Devesa, V., D. Velez And R. Montoro. 2008. Effect of thermal treatments on arsenic species contents in food. Food and Chemical Toxicology. 46(1): 1–8.

Edmonds, J., K. Francesconi and J. Hansen. 1982. Dimethyloxarsylethanol from anaerobic decomposition of brown kelp (Ecklonia radiata): A likely precursor of arsenobetaine in marine fauna. Cellular and Molecular Life Sciences. 38(6): 643–644.

Fairweather-Tait, S. and R.F. Hurrell. 1996. Bioavailability of minerals and trace elements. Nutrition Research Reviews. 9(1): 295–324.

Fairweather-Tait, S.J. 1999. The importance of trace element speciation in nutritional sciences. Fresenius' Journal of Analytical Chemistry. 363(5-6): 536–540.

Feldmann, J. and E.M. Krupp. 2011. Critical review or scientific opinion paper: Arsenosugars—A class of benign arsenic species or justification for developing partly speciated arsenic fractionation in foodstuffs? Analytical and Bioanalytical Chemistry. 399: 1735–1741.

Feldmann, J., V.W.M. Lai, W.R. Cullen, M. Ma, X. Lu and X.C. Le. 1999. Sample preparation and storage can change arsenic speciation in human urine. Clinical Chemistry. 45(11): 1988.

Feldmann, J., K. John and P. Pengprecha. 2000. Arsenic metabolism in seaweed-eating sheep from Northern Scotland. Fresenius' Journal of Analytical Chemistry. 368(1): 116–121.

Francesconi, K.A., J.S. Edmonds and R.V. Stick. 1989. Accumulation of arsenic in yelloweye mullet (Aldrichetta forsteri) following oral administration of organoarsenic compounds and arsenate. The Science of the Total Environment. 79(1): 59–67.

Francesconi, K.A. and D. Kuehnelt. 2004. Determination of arsenic species: A critical review of methods and applications, 2000–2003. Analyst. 129(5): 373–395.

Francesconi, K.A., R Tanggaard, C.J. McKenzie and W. Goessle. 2002. Arsenic metabolites in human urine after ingestion of an arsenosugar. Clinical Chemistry. 48(1): 92–101.

Gallagher, P.A., J.A. Shoemaker, X. Wei, C.A. Brockhoff-Schwegel and J.T. Creed. 2001. Extraction and detection of arsenicals in seaweed via accelerated solvent extraction with ion chromatographic separation and ICP-MS detection. Fresenius' Journal of Analytical Chemistry. 369(1): 71–80.

Gamble, B.M., P.A. Gallagher, J.A. Shoemaker, X. Wei, C.A. Schwegel and J.T. Creed. 2002. An investigation of the chemical stability of arsenosugars in simulated gastric juice and acidic environments using IC-ICP-MS and IC-ESI-MS/MS. Analyst. 127(6): 781–785.

García Salgado, S., M.A. Quijano Nieto and M.M. Bonilla Simón. 2006a. Determination of soluble toxic arsenic species in alga samples by microwave-assisted extraction and high performance liquid chromatography–hydride generation–inductively coupled plasma-atomic emission spectrometry. Journal of Chromatography A. 1129(1): 54–60.

García Salgado, S., M.A. Quijano Nieto and M.M. Bonilla Simon. 2008. Assessment of total arsenic and arsenic species stability in alga samples and their aqueous extracts. Talanta, 75(4): 897–903.

García Salgado, S., M.A. Quijano Nieto and M.M. Bonilla Simon. 2006b. Optimisation of sample treatment for arsenic speciation in alga samples by focussed sonication and ultrafiltration. Talanta. 68(5): 1522–1527.

García Sartal, C., M.C. Barciela Alonso and P. Bermejo Barrera. 2012a. Handbook of marine macroalgae: Biotecnology and applied phycology. Wiley-blackbell. 512–531.

García Sartal, C., M.C. Barciela-Alonso and P. Bermejo-Barrera. 2012b. Effect of the cooking procedure on the arsenic speciation in the bioavailable (dialyzable) fraction from seaweed. Microchemical Journal. 105: 65–71.

García Sartal, C., S. Taebunpakul, E. Stokes, M.C. Barciela-Alonso, P. Bermejo-Barrera and H. Goenaga-Infante. 2012c. Two-dimensional HPLC coupled to ICP-MS and electrospray ionisation (ESI) MS/MS for investigating the bioaccessibility of arsenic species from edible algae. Analytical and Bioanalytical Chemistry. 402: 3359.

García-Salgado, S., G. Raber, R. Raml, C. Magnes and K.A. Francesconi. 2012a. Arsenosugar phospholipids and arsenic hydrocarbons in two species of brown macroalgae. Environmental Chemistry. 9: 63.

García-Salgado, S., M.A. Quijano and M.M. Bonilla. 2012b. Arsenic speciation in edible alga samples by microwave-assisted extraction and high performance liquid chromatography coupled to atomic fluorescence spectrometry. Analytica Chimica Acta. 714(0): 38–46.

García-Sartal, C., M.C. Barciela-Alonso, P. Herbello-Hermelo and P. Bermejo-Barrera. 2010. A sensitive, simple and safe method for total arsenic determination in edible seaweed by HG-AFS. Atomic Spectroscopy. 31(1): 27–33.

García-Sartal, C., V. Romarís-Hortas, M.d.C. Barciela-Alonso, A. Moreda-Piñeiro, R. Dominguez-Gonzalez and P. Bermejo-Barrera. 2011. Use of an *in vitro* digestion method to evaluate the bioaccessibility of arsenic in edible seaweed by inductively coupled plasma-mass spectrometry. Microchemical Journal. 98: 91–96.

Geng, W., R. Komine, T. Ohta, T. Nakajima, H. Takanashi and A. Ohki. 2009. Arsenic speciation in marine product samples: Comparison of extraction–HPLC method and digestion–cryogenic trap method. Talanta. 79(2): 369–375.

Gomez-Rubio, P., J. Roberge, L. Arendell, R.B. Harris, M.K. O'Rourke, Z. Chen, E. Cantu-Soto, M.M. Meza-Montenegro, D. Billheimer and Z. Lu. 2011. Association between body mass index and arsenic methylation efficiency in adult women from southwest US and northwest Mexico. Toxicology and Applied Pharmacology. 252: 176.

Gong, Z., X. Lu, W.R. Cullen and X.C. Le. 2001. Unstable trivalent arsenic metabolites, monomethylarsonous acid and dimethylarsinous acid. Journal of Anaytical Atomic Spectrometry. 16(12): 1409–1413.

Gupta, S. and N. Abu-Ghannam. 2011. Recent developments in the application of seaweeds or seaweed extracts as a means for enhancing the safety and quality attributes of foods. Innovative Food Science & Emerging Technologies. 12(4): 600–609.

Han, C., X. Cao, J. Yu, X. Wang and Y. Shen. 2009. Arsenic speciation in Sargassum fusiforme by microwave-assisted extraction and LC-ICP-MS. Chromatographia. 69(5-6): 587–591.

Hansen, H.R., A. Raab and J. Feldmann. 2003a. New arsenosugar metabolite determined in urine by parallel use of HPLC-ICP-MS and HPLC-ESI-MS. Journal of Analytical Atomic Spectrometry. 18(5): 474–479.

Hansen, H.R., A. Raab, M. Jaspars, B.F. Milne and J. Feldmann. 2004. Sulfur-containing arsenical mistaken for dimethylarsinous acid [DMA (III)] and identified as a natural metabolite in urine: major implications for studies on arsenic metabolism and toxicity. Chemical Research in Toxicology. 17(8): 1086–1091.

Hansen, H.R., A. Raab, K.A. Francesconi and J. Feldmann. 2003b. Metabolism of arsenic by sheep chronically exposed to arsenosugars as a normal part of their diet. 1. Quantitative intake, uptake, and excretion. Environmental Science and Technology. 37(5): 845–851.

Hayakawa, T., Y. Kobayashi, X. Cui and S. Hirano. 2005. A new metabolic pathway of arsenite: Arsenic–glutathione complexes are substrates for human arsenic methyltransferase Cyt19. Archives of Toxicology. 79(4): 183–191.

He, Y. and Y. Zheng. 2010. Assessment of *in vivo* bioaccessibility of arsenic in dietary rice by a mass balance approach. Science of The Total Environment. 408(6): 1430–1436.

Hirata, S. and H. Toshimitsu. 2005. Determination of arsenic species and arsenosugars in marine samples by HPLC–ICP–MS. Analytical and Bioanalytical Chemistry. 383(3): 454–460.

Hirata, S. and H. Toshimitsu. 2007. Determination of arsenic species and arsenosugars in marine samples by HPLC-ICP-MS. Applied Organometallic Chemistry. 21(6): 447–454.

Hsieh, Y.J. and S.J. Jiang. 2012. Application of HPLC-ICP-MS and HPLC-ESI-MS procedures for arsenic speciation in seaweeds. Journal of Agricultural and Food Chemistry. 60: 2083.

Hsueh, Y.M., M.K. Hsu, H.Y. Chiou, M.H. Yang, C.C. Huang and C.J. Chen. 2002. Urinary arsenic speciation in subjects with or without restriction from seafood dietary intake. Toxicology Letters. 133(1): 83–91.

Hughes, M.F. 2006. Biomarkers of exposure: a case study with inorganic arsenic. Environmental Health Perspectives., 114(11): 1790.

Ichikawa, S., M. Kamoshida, M.H. Ken'ichi Hanaoka, T. Maitani and T. Kaise. 2006. Decrease of arsenic in edible brown algae, Hijikia fusiforme, by the cooking process. Applied Organometallic Chemistry. 20(9): 585–590.

Ichikawa, S., S. Nozawa, K. Hanaoka and T. Kaise. 2010. Ingestion and excretion of arsenic compounds present in edible brown algae, Hijikia fusiforme, by mice. Food and Chemical Toxicology. 48(2): 465–469.

Intawongse, M. and J.R. Dean. 2006. *In-vitro* testing for assessing oral bioaccessibility of trace metals in soil and food samples. TrAC Trends in Analytical Chemistry. 25(9): 876–886.

Jiang, X., W. Gan, S. Han, H. Zi and Y. He. 2009. Design and application of a novel integrated electrochemical hydride generation cell for the determination of arsenic in seaweeds by atomic fluorescence spectrometry. Talanta. 79(2): 314–318.

Juhasz, A.L., E. Smith, J. Weber, M. Rees, A. Rofe, T. Kuchel, L. Sansom and R. Naidu. 2006. *In vivo* assessment of arsenic bioavailability in rice and its significance for human health risk assessment. Environmental Health Perspectives. 114(12): 1826.

Karthikeyan, S. and S. Hirata. 2004. Ion chromatography-inductively coupled plasma mass spectrometry determination of arsenic species in marine samples. Applied Organometallic Chemistry. 18(7): 323–330.

Kirby, J., W. Maher, M. Ellwood and F. Krikowa. 2004. Arsenic species determination in biological tissues by HPLC-ICP-MS and HPLC-HG-ICP-MS. Australian Journal of Chemistry. 57(10): 957–966.

Koch, I., K. McPherson, P. Smith, L. Easton, K.G. Doe and K.J. Reimer. 2007. Arsenic bioaccessibility and speciation in clams and seaweed from a contaminated marine environment. Marine Pollution Bulletin. 54(5): 586–594.

Kohlmeyer, U., E. Jantzen, J. Kuballa and S. Jakubik. 2003. Benefits of high resolution IC-ICP-MS for the routine analysis of inorganic and organic arsenic species in food products of marine and terrestrial origin. Analytical and Bioanalytical Chemistry. 377(1): 6–13.

Kohlmeyer, U., J. Kuballa and E. Jantzen. 2002. Simultaneous separation of 17 inorganic and organic arsenic compounds in marine biota by means of high-performance liquid chromatography/inductively coupled plasma mass spectrometry. Rapid Communications in Mass Spectrometry. 16(10): 965–974.

Kuehnelt, D., K.J. Irgolic and W. Goessler. 2001. Comparison of three methods for the extraction of arsenic compounds from the NRCC standard reference material DORM-2 and the brown alga, Hijikia fusiforme. Applied Organometallic Chemistry. 15(6): 445–456.

Kumar, A.R. and P. Riyazuddin. 2010. Chemical interferences in hydride-generation atomic spectrometry. TrAC Trends in Analytical Chemistry. 29(2): 166–176.

Laparra, J.M., D. Velez, R. Barbera, R. Montoro and R. Farre. 2007. Bioaccessibility and transport by Caco-2 cells of organoarsenical species present in seafood. Journal of Agricultural and Food Chemistry. 55(14): 5892–5897.

Laparra, J.M., D. Velez, R. Montoro, R. Barbera and R. Farre. 2004. Bioaccessibility of inorganic arsenic species in raw and cooked Hizikia fusiforme seaweed. Applied Organometallic Chemistry. 18(12): 662–669.

Laparra, J.M., D. Velez, R. Montoro, R. Barbera and R. Farre. 2003. Estimation of arsenic bioaccessibility in edible seaweed by *in vitro* digestion method. Journal of Agricultural and Food Chemistry. 51(20): 6080–6085.

Larrea-Marín, M., M. Pomares-Alfonso, M. Gómez-Juaristi, F. Sánchez-Muniz and S.R. de la Rocha. 2010. Validation of an ICP-OES method for macro and trace element determination in Laminaria and Porphyra seaweeds from four different countries. Journal of Food Composition and Analysis. 23(8): 814–820.

Le, X.C., X. Lu, M. Ma, W.R. Cullen, H.V. Aposhian and B. Zheng. 2000a. Speciation of key arsenic metabolic intermediates in human urine. Analytical Chemistry. 72(21): 5172–5177.

Le, X.C., M. Ma, W.R. Cullen, H.V. Aposhian, X. Lu and B. Zheng. 2000b. Determination of monomethylarsonous acid, a key arsenic methylation intermediate, in human urine. Environmental Health Perspectives. 108(11): 1015.

Li, S.X., L.X. Lin, F.Y. Zheng and Q.X. Wang. 2011. Metal bioavailability and risk assessment from edible brown alga Laminaria japonica, using biomimetic digestion and absorption system and determination by ICP-MS. Journal of Agricultural and Food Chemistry. 59: 822–828.

Li, W., C. Wei, C. Zhang, M. Van Hulle, R. Cornelis and X. Zhang. 2003. A survey of arsenic species in Chinese seafood. Food and Chemical Toxicology. 41(8): 1103–1110.

Lintschinger, J., P. Schramel, A. Hatalak-Rauscher, I. Wendler and B. Michalke. 1998. A new method for the analysis of arsenic species in urine by using HPLC-ICP-MS. Fresenius' Journal of Analytical Chemistry. 362(3): 313–318.

Ma, M. and X.C. Le. 1998. Effect of arsenosugar ingestion on urinary arsenic speciation. Clinical Chemistry (Washington, D.C.). 44(3): 539–550.

Mandal, B.K. and K.T. Suzuki. 2002. Arsenic round the world: A review. Talanta. 58(1): 201–235.

McSheehy, S., M. Marcinek, H. Chassaigne and J. Szpunar. 2000. Identification of dimethylarsinoyl-riboside derivatives in seaweed by pneumatically assisted electrospray tandem mass spectrometry. Analytica Chimica Acta. 410(1-2): 71–84.

McSheehy, S., P. Pohl, R. Lobinski and J. Szpunar. 2001. Investigation of arsenic speciation in oyster test reference material by multidimensional HPLC-ICP-MS and electrospray tandem mass spectrometry (ES-MS-MS). Analyst. 126(7): 1055–1062.

McSheehy, S. and J. Szpunar. 2000. Speciation of arsenic in edible algae by bi-dimensional size-exclusion anion exchange HPLC with dual ICP-MS and electrospray MS/MS detection. Journal of Analytical Atomic Spectrometry. 15(1): 79–87.

Miller, D.D., B.R. Schricker, R.R. Rasmussen and D. Van Campen. 1981. An *in vitro* method for estimation of iron availability from meals. American Journal of Clinical Nutrition. 34(10): 2248.

Minekus, M., P. Marteau, R. Havenaar and J. Huis in't Veld. 1995. A multicompartmental dynamic computer-controlled model simulating the stomach and small intestine. Atla. 23: 197–209.

Mitra, S.R., D.N.G. Mazumder, A. Basu, G. Block, R. Haque, S. Samanta, N. Ghosh, M.M.H. Smith, O.S. Von Ehrenstein and A.H. Smith. 2004. Nutritional factors and susceptibility to arsenic-caused skin lesions in West Bengal, India. Environmental health perspectives. 112(10): 1104.

Mohamed, S., S.N. Hashim and H.A. Rahman. 2012. Seaweeds: A sustainable functional food for complementary and alternative therapy. Trends in Food Science & Technology. 23(2): 83–96.

Moreda-Piñeiro, A., E. Peña-Vázquez and P. Bermejo-Barrera. 2012. Significance of the Presence of Trace and Ultratrace Elements in Seaweed. Handbook of Marine Macroalgae. ed. Se-Kwon Kim, John Wiley & Sons, Ltd, UK, pp. 116–170.

Moreda-Pineiro, J., C. Moscoso-Perez, P. Lopez-Mahia, S. Muniategui-Lorenzo, D. Prada-Rodriguez, A. Moreda-Pineiro, V. Romaris-Hortas and P. Bermejo-Barrera. 2011. *In vivo* and *in vitro* testing to assess the bioaccessibility and the bioavailability of arsenic, selenium and mercury species in food samples. TrAC Trends in Analytical Chemistry. 30: 324–345.

Moreda-Piñeiro, J., E. Alonso-Rodríguez, P. López-Mahía, S. Muniategui-Lorenzo, D. Prada-Rodríguez, A. Moreda-Piñeiro and P. Bermejo-Barrera. 2007. Development of a new sample pre-treatment procedure based on pressurized liquid extraction for the determination of metals in edible seaweed. Analytica Chimica Acta. 598(1): 95–102.

Murray, L.A., A. Raab, I.L. Marr and J. Feldmann. 2003. Biotransformation of arsenate to arsenosugars by Chlorella vulgaris. Applied Organometallic Chemistry. 17(9): 669–674.

Nakajima, Y., Y. Endo, Y. Inoue, K. Yamanaka, K. Kato, H. Wanibuchi and G. Endo. 2006. Ingestion of Hijiki seaweed and risk of arsenic poisoning. Applied Organometallic Chemistry. 20(9): 557–564.

Navas-Acien, A., K.A. Francesconi, E.K. Silbergeld and E. Guallar. 2011. Seafood intake and urine concentrations of total arsenic, dimethylarsinate and arsenobetaine in the US population. Environmental Research. 111(1): 110–118.

Newcombe, C., A. Raab, P.N. Williams, C. Deacon, P.I. Haris, A.A. Meharg and J. Feldmann. 2010. Accumulation or production of arsenobetaine in humans? Journal of Environmental Monitoring. 12(4): 832–837.

Niegel, C. and F. Matysik. 2010. Analytical methods for the determination of arsenosugars—A review of recent trends and developments. Analytica Chimica Acta. 657(2): 83–99.

Nischwitz, V. and S.A. Pergantis. 2005. First report on the detection and quantification of arsenobetaine in extracts of marine algae using HPLC-ES-MS/MS. Analyst. 130(10): 1348–1350.

Nischwitz, V., K. Kanaki and S.A. Pergantis. 2006. Mass spectrometric identification of novel arsinothioyl-sugars in marine bivalves and algae. Journal of Analytical Atomic Spectrometry. 21(1): 33–40.

Nischwitz, V. and S.A. Pergantis. 2006. Improved arsenic speciation analysis for extracts of commercially available edible marine algae using HPLC-ES-MS/MS. Journal of Agricultural and Food Chemistry. 54(18): 6507–6519.

Oomen, A.G., A. Hack, M. Minekus, E. Zeijdner, C. Cornelis, G. Schoeters, W. Verstraete, T. Van de Wiele, J. Wragg, C.J.M. Rompelberg, A.J.A.M. Sips and J.H. Van Wijnen. 2002. Comparison of five *in vitro* digestion models to study the bioaccessibility of soil contaminants. Environmental Science and Technology. 36(15): 3326–3334.

Parada, J. and J.M. Aguilera. 2007. Food microstructure affects the bioavailability of several nutrients. Journal of Food Science. 72(2): R21–R32.

Pedersen, S.N. and K.A. Francesconi. 2000. Liquid chromatography electrospray mass spectrometry with variable fragmentor voltages gives simultaneous elemental and molecular detection of arsenic compounds. Rapid Communications in Mass Spectrometry. 14(8): 641–645.

Penrose, W.R., R. Black and M.J. Hayward. 1975. Limited arsenic dispersion in sea water, sediments, and biota near a continuous source. Journal of the Fisheries Board of Canada. 32(8): 1275–1281.

Arsenic in Seaweed: Presence, Bioavailability and Speciation 349

Raber, G., K.A. Francesconi, K.J. Irgolic and W. Goessler. 2000. Determination of arsenosugars in algae with anion-exchange chromatography and an inductively coupled plasma mass spectrometer as element-specific detector. Fresenius' Journal of Analytical Chemistry. 367(2): 181–188.

Raber, G., S. Khoomrung,M.S. Taleshi, J.S. Edmonds and K.A. Francesconi. 2009. Identification of arsenolipids with GC/MS. Talanta. 78(3): 1215–1218.

Raml, R., W. Goessler and K.A. Francesconi. 2006. Improved chromatographic separation of thio-arsenic compounds by reversed-phase high performance liquid chromatography-inductively coupled plasma mass spectrometry. Journal of Chromatography A. 1128(1): 164–170.

Raml, R., W. Goessler, P. Traar, T. Ochi and K.A. Francesconi. 2005. Novel thioarsenic metabolites in human urine after ingestion of an arsenosugar, 2′, 3′-Dihydroxypropyl 5-Deoxy-5-dimethylarsinoyl-β-d-riboside. Chemical Research in Toxicology. 18(9): 1444–1450.

Raml, R., G. Raber, A. Rumpler, T. Bauernhofer, W. Goessler and K.A. Francesconi. 2009. Individual variability in the human metabolism of an arsenic-containing carbohydrate, 2′,3′-Dihydroxypropyl 5-deoxy-5-dimethylarsinoyl-Î²-D-riboside, a naturally occurring arsenical in seafood. Chemical Research in Toxicology. 22(9): 1534–1540.

Raml, R., A. Rumpler, W. Goessler, M. Vahter, L. Li, T. Ochi and K.A. Francesconi. 2007. Thio-dimethylarsinate is a common metabolite in urine samples from arsenic-exposed women in Bangladesh. Toxicology and Applied Pharmacology. 222(3): 374–380.

Rees, M., L. Sansom, A. Rofe, A.L. Juhasz, E. Smith, J. Weber, R. Naidu and T. Kuchel. 2009. Principles and application of an in vivo swine assay for the determination of arsenic bioavailability in contaminated matrices. Environmental Geochemistry and Health. 31: 167–177.

Ritsema, R., L. Dukan, T.R. Navarro, W. van Leeuwen, N. Oliveira, P. Wolfs and E. Lebret. 1998. Speciation of arsenic compounds in urine by LC–ICP MS. Applied Organometallic Chemistry. 12(8-9): 591–599.

Rivera-Núñez, Z., J.R. Meliker, A.M. Linder and J.O. Nriagu. 2010. Reliability of spot urine samples in assessing arsenic exposure. International Journal of Hygiene and Environmental Health. 213(4): 259–264.

Roberts, S.M., W.R. Weimar, J.R.T. Vinson, J.W. Munson and R.J. Bergeron. 2002. Measurement of arsenic bioavailability in soil using a primate model. Toxicological Sciences. 67(2): 303–310.

Rose, M., J. Lewis, N. Langford, M. Baxter, S. Origgi, M. Barber, H. MacBain and K. Thomas. 2007. Arsenic in seaweed—Forms, concentration and dietary exposure. Food and Chemical Toxicology. 45(7): 1263–1267.

Rubio, R., M.J. Ruiz-Chancho, J.F. López-Sánchez, R. Rubio and J.F. López-Sánchez. 2010. Sample pre-treatment and extraction methods that are crucial to arsenic speciation in algae and aquatic plants. TrAC Trends in Analytical Chemistry. 29(1): 53–69.

Ruiz-Navarro, M.L., M. Navarro-Alarcón, H. Lopez Gª-de la Serrana, V. Pérez-Valero and M.C. López-Martinez. 1998. Urine arsenic concentrations in healthy adults as indicators of environmental contamination: Relation with some pathologies. The Science of the Total Environment. 216(1-2): 55–61.

Sakurai, T. 2002. Biological effects of organic arsenic compounds in seafood. Applied Organometallic Chemistry. 16(8): 401–405.

Salgado, S.G., Q. Nieto and M.M. Bonilla Simon. 2006. Optimisation of sample treatment for arsenic speciation in alga samples by focused sonication and ultrafiltration. Talanta. 68(5): 1522–1527.

Salgado, S.G., M.A.Q. Nieto and M.M.B. Simón. 2008. Assessment of total arsenic and arsenic species stability in alga samples and their aqueous extracts. Talanta. 75(4): 897–903.

Samanta, G., U.K. Chowdhury, B.K. Mandal, D. Chakraborti, N.C. Sekaran, H. Tokunaga and M. Ando. 2000. High performance liquid chromatography inductively coupled plasma mass spectrometry for speciation of arsenic compounds in urine. Microchemical Journal. 65(2): 113–127.

Sanchez-Rodriguez, I., M. Huerta-Diaz, E. Choumiline, O. Holguin-Quinones and J. Zertuche-Gonzalez. 2001. Elemental concentrations in different species of seaweeds from Loreto Bay, Baja California Sur, Mexico: implications for the geochemical control of metals in algal tissue. Environmental Pollution. 114(2): 145–160.

Schmeisser, E., W. Goessler, N. Kienzl and K.A. Francesconi. 2004. Volatile analytes formed from arsenosugars: determination by HPLC-HG-ICPMS and implications for arsenic speciation analyses. Analytical Chemistry. 76(2): 418–423.

Sharma, V.K. and M. Sohn. 2009. Aquatic arsenic: Toxicity, speciation, transformations, and remediation. Environment International. 35(4): 743–759.

Smedley, P.L. and D.G. Kinniburgh. 2002. A review of the source, behaviour and distribution of arsenic in natural waters. Applied Geochemistry. 17(5): 517–568.

Soleo, L., P. Lovreglio, S. Iavicoli, A. Antelmi, I. Drago, A. Basso, L. Di Lorenzo, M.E. Gilberti, G. De Palma and P. Apostoli. 2008. Significance of urinary arsenic speciation in assessment of seafood ingestion as the main source of organic and inorganic arsenic in a population resident near a coastal area. Chemosphere. 73(3): 291–299.

Sur, R., J. Begerow and L. Dunemann. 1999. Determination of arsenic species in human urine using HPLC with on-line photooxidation or microwave-assisted oxidation combined with flow-injection HG-AAS. Fresenius' Journal of Analytical Chemistry. 363(5): 526–530.

Suzuki, K.T., B.K. Mandal and Y. Ogra. 2002. Speciation of arsenic in body fluids. Talanta. 58(1): 111–119.

Szpunar, J., S. McSheehy, K. Polec, V. Vacchina, S. Mounicou, I. Rodriguez and R. Lobinski. 2000. Gas and liquid chromatography with inductively coupled plasma mass spectrometry detection for environmental speciation analysis—advances and limitations. Spectrochimica Acta, Part B: Atomic Spectroscopy. 55B(7): 779–793.

Todolí, J.L. and M. Grotti. 2010. Fast determination of arsenosugars in algal extracts by narrow bore high-performance liquid chromatography–inductively coupled plasma mass spectrometry. Journal of Chromatography A. 1217(47): 7428–7433.

Torres-Escribano, S., S. Denis, S. Blanquet-Diot, M. Calatayud, L. Barrios, D. Vélez, M. Alric and R. Montoro. 2011. Comparison of a static and a dynamic *in vitro* model to estimate the bioaccessibility of As, Cd, Pb and Hg from food reference materials, Fucus sp. (IAEA-140/TM) and Lobster hepatopancreas (TORT-2). Science of the Total Environment. 409(3): 604–611.

Tseng, C.H. 2009. A review on environmental factors regulating arsenic methylation in humans. Toxicology and Applied Pharmacology. 235(3): 338–350.

Tseng, C., Y. Huang, Y. Huang, C. Chung, M. Yang, C. Chen and Y. Hsueh. 2005. Arsenic exposure, urinary arsenic speciation, and peripheral vascular disease in blackfoot disease-hyperendemic villages in Taiwan. Toxicology and Applied Pharmacology. 206(3): 299–308.

Tukai, R., W.A. Maher, I.J. McNaught and M.J. Ellwood. 2002. Measurement of arsenic species in marine macroalgae by microwave-assisted extraction and high performance liquid chromatography-inductively coupled plasma mass spectrometry. Analytica Chimica Acta. 457(2): 173–185.

Vahter, M. 2000. Genetic polymorphism in the biotransformation of inorganic arsenic and its role in toxicity. Toxicology Letters. 112: 209–217.

Vahter, M. 1994. What are the chemical forms of arsenic in urine, and what can they tell us about exposure? Clinical Chemistry. 40(5): 679.

Vahter, M. 2002. Mechanisms of arsenic biotransformation. Toxicology. 181-182: 211–217.

Van Campen, D.R. and R.P. Glahn. 1999. Micronutrient bioavailability techniques: Accuracy, problems and limitations. Field Crops Research. 60(1-2): 93–113.

Van Hulle, M., C. Zhang, B. Schotte, L. Mees, F. Vanhaecke, R. Vanholder, X.R. Zhang and R. Cornelis. 2004. Identification of some arsenic species in human urine and blood after ingestion of Chinese seaweed Laminaria. Journal of Analytical Atomic Spectrometry. 19(1): 58–64.

Van Hulle, M., Zhang, C., Zhang, X. and Cornelis, R. (2002) Arsenic speciation in Chinese seaweeds using HPLC-ICP-MS and HPLC-ES-MS. Analyst. 127(5): 634–640.

Verdon, C.P., K.L. Caldwell, M.R. Fresquez and R.L. Jones. 2009. Determination of seven arsenic compounds in urine by HPLC-ICP-DRC-MS: a CDC population biomonitoring method. Analytical and Bioanalytical Chemistry. 393(3): 939–947.

Versantvoort, C.J.M., E. Van de Kamp and C.J.M. Rompelberg. 2004. Development and applicability of an *in vitro* digestion model in assessing the bioaccessibility of contaminants from food. RIVM, report number 320102002/2004. RIVM, Bilthoven.

Wangkarn, S. and S.A. Pergantis. 2000. High-speed separation of arsenic compounds using narrow-bore high-performance liquid chromatography on-line with inductively coupled plasma mass spectrometry. Journal of Analytical Atomic Spectrometry. 15(6): 627–633.

Wei, C., W. Li, C. Zhang, M. Van Hulle, R. Cornelis and X. Zhang. 2003. Safety evaluation of organoarsenical species in edible porphyra from the China Sea. Journal of Agricultural and Food Chemistry. 51(17): 5176–5182.

WHO. 2001. Environmental health criteria 224, arsenic and arsenic compounds. Inter-organization Programme for the Sound Management of Chemicals, Geneva.

Whyte, J.N.C. and J.R. Englar. 1983. Analysis of inorganic and organic-bound arsenic in marine brown algae. Botanica Marina 26: 159–164.

Wuilloud, R.G., J.C. Altamirano, P.N. Smichowski and D.T. Heitkemper. 2006. Investigation of arsenic speciation in algae of the Antarctic region by HPLC-ICP-MS and HPLC-ESI-Ion Trap MS. Journal of Analytical Atomic Spectrometry. 21(11): 1214–1223.

Xie, R., W. Johnson, S. Spayd, G.S. Hall and B. Buckley. 2006. Arsenic speciation analysis of human urine using ion exchange chromatography coupled to inductively coupled plasma mass spectrometry. Analytica Chimica Acta. 578(2): 186–194.

15

Application of Bacterial Fermentation in Edible Brown Algae

Sung-Hwan Eom and *Young-Mog Kim* *

1 Introduction

Seaweeds have used in various dishes including appetizers, casseroles, muffins, pilafs, and soups in Korea and Japan (Eom et al. 2011). In addition, seaweeds are rich in minerals and dietary fiber and are thus typically used as a health food. Among them, many brown algae exhibit various biological activities against tumors (Noda et al. 1989), Alzheimer's disease (Jung et al. 2010), inflammatory diseases (Shibata et al. 2003), and allergic diseases (Shibata et al. 2002). Despite its key biological activities, brown algae have largely been used for dried, frozen, or salted processed products with low-value seaweeds (Heo and Jeon 2005). The typical odor of seaweeds is also one of the major obstacles to developing highly processed seaweed products. Therefore, in order to attract consumers, new seaweed-processing technologies are required to remove the seaweed odor while keeping or enhancing its biological activities (Bae and Kim 2010; Eom et al. 2010; Song et al. 2011).

Fermentation is a very interesting process used to increase the nutritional quality and remove undesirable compounds (Martha et al. 2005; Wijesinghe et al. 2012). In addition, fermentation enhances the nutrient

Department of Food Science and Technology, Pukyong National University, Busan, Republic of Korea.
* Corresponding author: ymkim@pknu.ac.kr

content of foods through the biosynthesis of vitamins, essential amino acids, and proteins. Several products of microbial fermentation are also incorporated in food as additives and supplements such as antioxidants, flavors, colorants, preservatives, and sweeteners (Couto and Sanroman 2006). Mould fermented foods play an important role, especially in Asian countries where the production process for many foods includes a fungal fermentation (Geisen and Farber 2002). Microorganisms play a central role in the production of a wide range of primary and secondary metabolites. The edible yeast *Candida utilis* is generally recognized as a safe substance by the Food and Drug Administration (Miura et al. 1998) and is an industrially important microorganism. It is used in the production of several biologically useful materials such as glutathione, and certain amino acids and enzymes (Kondo et al. 1995). Furthermore, fermentation improves micronutrient bioavailability and aids in the degradation of anti-nutritional factors.

In this chapter, we report on the application of microbial fermentation on brown algae processing and its effectiveness. The present investigation is aimed at the fermented brown algae and its potential application as functional and nutritional ingredients.

2 Application of Bacteria to Ferment Brown Algae

2.1 Brown algae

Marine algae have become an important source of pharmacologically active metabolites. They are also widely distributed and abundant throughout the coastal areas of many countries. They are a source of useful secondary metabolites such as agar, carrageenan and alginate with interesting pharmaceutical properties (Taskin et al. 2001). Among marine algae, brown algae have been reported to contain high phlorotannin contents. Phlorotannins are phenolic compounds that consist of polymers of phloroglucinol (1,3,5–tryhydroxybenzene) units and are formed in the acetate-malonate pathway in marine algae. Furthermore, these phlorotannins are highly hydrophilic components with a wide range of molecular sizes (126 Da–650 kDa) (Ragan and Glombitza 1986; Wijesekara and Kim 2010).

Furthermore, several phlorotannins purified from brown seaweeds such as *Ecklonia cava, E. kurome, E. stolonifera, Eisenia aborea, E. bicyclis, Ishige okamurae,* and *Pelvetia siliquosa* contain medicinal and pharmaceutical benefits and have shown strong anti-oxidant, anti-inflammatory, anti-viral, anti-tumor, anti-diabetes and anti-cancer properties (Cha et al. 2011; Eom et al. 2011; Gupta and Abu-Ghannam 2011; Kim et al. 2009). Collectively, phlorotannins can be used for functional ingredients in the pharmaceutical industries.

2.2 Microbial strains to ferment brown algae

Fermentation is a biochemical reaction that splits complex organic compounds into relatively simple components. Also, microbial fermentation results in either the break-down or production of bioactive compounds (Achinewhu et al. 1998; Adewusi et al. 1999). Fermentation not only enhances the nutrient content of foods through the biosynthesis of vitamins, essential amino acids and proteins, but also improves protein quality and fiber digestibility. It also improves the micronutrient bioavailability and aids in degrading anti-nutritional factors (Achinewhu et al. 1998; Adewusi et al. 1999; Bae and Kim 2010).

Eom et al. (2011) isolated microbial strains from Korean traditional fermented foods. Three strains of *Lactobacillus* sp. (LB-1, LB-2 and LB-3) were isolated from Kimchi through an enrichment culture using MRS broth (Difco, Detroit, MI), two *Bacillus* sp. strains were isolated from Meju using Mannitol-Egg-Yolk-Polymyxine-Agar (Difco), and three yeast strains (YM-1, YM-2 and YM-3) were isolated from Meju using YM broth (Difco) to ferment edible brown algae *Eisenia bicyclis*. Three yeast strains were identified as *Candida utilis* by API 20 C AUX (Bio-Mérieux, Marcy l' Etoile, France). *C. utilis* is an industrially important microorganism which has been approved as a safe substance by the U.S. Food and Drug Administration (FDA) (Boze et al. 1992; Ichi et al. 1993). Lactic acid bacteria are known to release various enzymes into the intestinal lumen that exert synergistic effects on digestion, alleviating symptoms of intestinal malabsorption (Parvez et al. 2006).

Seo et al. (2012) reported that fermentation by *Aspergillus oryzae* can reduce off-flavors from sea tangle (*Laminaria japonica*) extract since the typical odor of seaweed is an impediment to the consumption of seaweed products. *A. oryzae* is one of the most important microorganisms used in biotechnology. It has been in use already, for many decades. to produce extracellular enzymes and citric acid. In fact, citric acid and many *A. oryzae* enzymes are considered GRAS (Generally Regarded As Safe) by the U.S. FDA (Schuster et al., 2002).

In addition, Cha et al. (2011) reported that the sea tangle fermented by *Saccharomyces cerevisiae* can apply as a suitable model system of alcohol metabolism. The yeast *S. cerevisiae* is an attractive microorganism due to the very well established molecular biology and fermentation techniques, together with its GRAS status (FDA, USA) (Moreira dos Santos et al. 2004).

3 Antioxidant Activity of Fermented Brown Algae by Bacteria

Enterococcus faecium was characterized for their ability to utilize brown algae *Sargassum* sp. fermentation (Shobharani et al. 2013). By the observation of drop in pH and increase in lactic acid content, the fermentation process in seaweed broth was confirmed. Further, the antioxidant activity during fermentation period was correlated with the increase in polyphenols content as compared with the control. Fermentation may cause disruption of seaweed cell walls, resulting in increased polyphenol content in the fermented sample. Polyphenols are known to be effective antioxidant compounds because of their hydroxyl group that aids in scavenging free radicals (Jimenez-Escrig et al. 2001).

Eom et al. (2011) reported that *C. utilis* YM-1 strain for the fermentation of brown algae *E. bicyclis* extracts is a candidate strain to develop value-added food ingredients. Additionally, the total phenol (TP) content, DPPH radical scavenging activity, and free amino acid analysis content measured strongly indicate that the optimum level of fermentation by *C. utilis* YM-1 was achieved after one day of fermentation. They examined the possibility of using fermentation to develop value-added food ingredients from *E. bicyclis*. Microbial fermentation by *C. utilis* YM-1 resulted in enhanced biological activity, including an increased TP content and antioxidant activity.

According to Wijesinghe et al. (2012), the brown algae *Ecklonia cava* processing by-product fermented by *C. utilis* was evaluated its potential antioxidant activity. The extract from the *Ecklonia cava* processing by-product fermented for one day exhibited the highest TP contents and was also found to be the strongest antioxidant. The one day extract of fermented *Ecklonia cava* processing by-product strongly enhanced cell viability against H_2O_2-induced oxidative damage in the Vero cell line. This sample also exhibited good protective properties against H_2O_2-induced cell apoptosis as was demonstrated by a decreased quantity of sub-G1 hypodiploid cells and decreased apoptotic body formation in the flow cytometry analysis.

Kang et al. (2012b) reported on the antioxidant effects of fermented sea tangle (FST) on healthy volunteers with high levels of γ-glutamyltransferse (γ-GT) by a randomized, double-blind and placebo-controlled clinical study. Forty-eight participants were divided into a placebo group and an FST group that received FST (1.5 g/day) for 4 weeks. Serum γ-GT, malondialdehyde (MDA), catalase (CAT), superoxide dismutase (SOD), and glutathione peroxidase (GPx) activities were determined before and after the trial. FST exhibited potent antioxidant effects. FST significantly ameliorated serum γ-GT levels, which is an oxidative marker, and effectively decreased serum MDA levels compared to those in the placebo group. Additionally, FST significantly augmented the serum antioxidant enzyme activities, including SOD and CAT, compared to those in the placebo group. These findings suggest that FST may be helpful for reducing oxidative stress, and further present a possible mechanism underlying the clinical prophylactic advantages of this supplement.

Eom et al. (2010) evaluated the effective use of sea tangle (*Laminaria japonica*) fermented by *Saccaromyces cerivisiae* SC-2. The antioxidant activity of DPPH radical scavenging activity, superoxide radical scavenging activity, nitric oxide scavenging activity, and xanthine oxidase inhibition assay were enhanced as compared to the control group. In addition, inhibition of LPS-induced NO production in RAW 264.7 macrophages was enhanced by *S. cerivisiae* SC-2 fermentation. Therefore, this study suggests that sea tangle fermented by *S. cerivisiae* SC-2 can be a good source of a natural anti-inflammatory agent.

Edible seaweed, *Hizikia fusiforme,* fermented by *Lactobacillus brevis* BJ-20 was also evaluated with reference to anti-oxidant and anti-inflammatory activities (Song et al. 2011). Antioxidant activity was evaluated by assaying DPPH radical, hydroxyl, superoxide and alkyl scavenging activity. Fermented *H. fusiforme* extracts suppress the production of pro-inflammatory cytokines as TNF-α and the expression of iNOS in HepG2 cells more than the control group.

4 Conclusion

Marine organisms provide a promising source of antioxidant, anti-inflammatory effect, and low cytotoxicity, of which can be used to develop potential pharmacological agents (Kang et al. 2012a; Yoon et al. 2009). Recently, increasing consumer knowledge of the link between safe and health has raised the demand for novel health promotion and functional food products. Hence, edible brown algae fermented by generally regarded as safe bacteria should be considered as potent anti-oxidative and anti-inflammatory agents due to their confirmed efficacy and excellent safety profiles.

References

Achinewhu, S.C., L.I. Barber and I.O. Ijeoma. 1998. Physicochemical properties and garification (gari yield) of selected cassava cultivars in Rivers State, Nigeria. Plant Foods Hum. Nutr. 52: 133–140.

Adewusi, S.R.A., T.V. Ojumu and O.S. Falade. 1999. The effect of processing on total organic acids content and mineral availability of simulated cassava-vegetable diets. Plant Foods Hum. Nutr. 53: 367–380.

Bae, H.N. and Y.M. Kim. 2010. Improvement of the functional qualities of sea tangle extract through fermentation by *Aspergillus oryzae*. Fish Aquat. Sci. 13: 12–17.

Boze, H., G. Moulin and P. Galzy. 1992. Production of food and fodder yeasts. Crit. Rev. Biotechnol. 12: 65–86.

Cha, J.Y., H.J. Yang and Y.S. Cho. 2011. Effect of fermented sea tangle on the alcohol dehydrogenase and acetaldehyde dehydrogenase in *Saccharomyces cerevisiae*. J. Microbiol. Biotechnol. 21: 791–795.

Couto, S.R. and M.A. Sanroman. 2006. Application of solid-state fermentation to food industry—a review. Food Eng. 76: 291–302.

Eom, S.H., B.J. Lee and Y.M. Kim. 2010. Effect of yeast fermentation on the antioxidant and anti-inflammatory activity of sea tangle water extract. Kor. J. Fish Aquat. Sci. 43: 117–124.

Eom, S.H., Y.M. Kang, J.H. Park, D.U. Yu, E.T. Jeong, M.S. Lee and Y.M. Kim. 2011. Enhancement of polyphenol content and antioxidant activity of brown alga *Eisenia bicyclis* extract by microbial fermentation. Fish Aquat. Sci. 14: 192–197.

Geisen, R. and P. Farber. 2002. New aspects of fungal starter cultures for fermented foods. Appl. Microbiol. 2: 13–29.

Gupta, S. and N. Abu-Ghannam. 2011. Recent developments in the application of seaweeds or seaweed extracts as a means for enhancing the safety and quality attributes of foods. Innov. Food Sci. Emerg. Tech. 12: 600–609.

Heo, S.J. and Y.J. Jeon. 2005. Antioxidant effect and protecting effect against cell damage by enzymatic hydrolysates from marine algae. Food Ind. Nutr. 10: 31–41.

Ichi, T., S. Takenaka, H. Konno, T. Ishida, H. Sato, A. Suzuki and K. Yamazuki. 1993 Development of a new commercial-scale airlift fermentor for rapid growth of yeast. J. Ferment. Bioeng. 75: 375–379.

Jiménez-Escrig, A., I. Jiménez-Jiménez, R. Pulido and F. Saura-Calixto. 2001. Antioxidant activity of fresh and processed edible seaweeds. J. Sci. Food Agr. 81: 530–534.

Jung, H.A., S.H. Oh and J.S. Choi. 2010. Molecular docking studies of phlorotannins from *Eisenia bicyclis* with BACE1 inhibitory activity. Bioorg. Med. Chem. Lett. 20: 3211–3215.

Kang, S.M., S.J. Heo, K.N. Kim, S.H. Lee, H.M. Yang, A.D. Kim and Y.J. Jeon. 2012a. Molecular docking studies of a phlorotannin, dieckol isolated from *Ecklonia cava* with tyrosinase inhibitory activity. Bioorg. Med. Chem. Lett. 20: 311–316.

Kang, Y.M., B.J. Lee, J.I. Kim, B.H. Nam, J.Y. Cha, Y.M. Kim, C.B. Ahn, J.S. Choi, I.S. Choi and J.Y. Je. 2012b. Antioxidant effects of fermented sea tangle (*Laminaria japonica*) by *Lactobacillus brevis* BJ20 in individuals with high level of γ-GT: A randomized, double-blind, and placebo-controlled clinical study. Food Chem. Toxicol. 50: 1166–1169.

Kim, K.L., M.S. Park, K.S. Lim, D.H. Lee and Y.M. Kim. 2010. Antiviral activity of seaweed extracts against feline calicivirus. Fish Aquat. Sci. 13: 96–101.

Kondo, K., T. Saito, S. Kajiwara, M. Takagi and N. Misawa. 1995. A transformation system for the yeast *Candida utilis*: use of a modified endogenous ribosomal protein gene as a drug-resistant marker and ribosomal DNA as an integration target for vector DNA. J. Bact. 177: 7171–7177.

Martha, J.F., L. Miranda, R. Doblado and C. Vidal-Valverde. 2005. Effect of germination and fermentation on the antioxidant vitamin content and antioxidant capacity of *Lupinus albus* L. var. Multolupa. Food Chem. 92: 211–220.

Miura, Y., K. Kondo, T. Saito, H. Shimada, P.D. Fraser and N. Misawa. 1998. Production of the carotenoids lycopene, β-carotene, and astaxanthin in the food yeast *Candida utilis*. Appl. Environ. Microbiol. 64: 1226–1229.

Moreira dos Santos, M., V. Raghevendran, P. Kötter, L. Olsson and J. Nielsen. 2004. Manipulation of malic enzyme in *Saccharomyces cerevisiae* for increasing NADPH production capacity aerobically in different cellular compartments. Metabolic engineering. 6: 352–363.

Noda, H., H. Amano, K. Arashima, S. Hashimoto and K. Nisizawa. 1989. Studies on the antitumour activity of marine algae. Nippon Suisan Gakkaishi. 55: 1259–1264.

Parvez, S., K.A. Malik, S.A. Kang and H.Y. Kim. 2006. Probiotics and their fermented food products are beneficial for health. J. Appl. Microbiol. 100: 1171–1185.

Ragan, M.A. and K.W. Glombitza. 1986. Phlorotannins, Brown Algal Polyphenols. pp. 129–241. *In*: F.E. Round and D.J. Chapman (eds.). Progress in Phycological Research, 4th edn. Biopress Ltd.: Bristol.

Schuster, E., N. Dunn-Coleman, J. Frisvad and P. Van Dijck. 2002. On the safety of *Aspergillus niger*—a review. Appl Microbiol. Biotechnol. 59: 426–435.

Seo, Y.S., H.N. Bae, S.H. Eom, K.S. Lim, I.H. Yun, Y.H. Chung, M.S. Lee, Y.B. Lee and Y.M. Kim. 2012. Removal of off-flavors from sea tangle (*Laminaria japonica*) extract by fermentation with *Aspergillus oryzae*. Bioresour. Technol. 121: 475–479.

Shibata, T., K. Fujimoto, K. Nagayama, K. Yamaguchi and T. Nakamura. 2002. Inhibitory activity of brown algal phlorotannins against hyaluronidase. Int. J. Food Sci. Technol. 37: 703–709.

Shibata, T., K. Nagayama, R. Tanaka, K. Yamaguchi and T. Nakamura. 2003. Inhibitory effects of brown algal phlorotannins on secretory phospholipase A2s, lipoxygenases and cyclooxygenases. J. Appl. Phycol. 15: 61–66.

Shobharani, P., P.M. Halami and N.M. Sachindra. 2013. Potential of marine lactic acid bacteria to ferment *Sargassum* sp. for enhanced anticoagulant and antioxidant properties. J. Appl. Microbiol. 114: 96–107.

Song, H.S., S.H. Eom, Y.M. Kang, J.D. Choi and Y.M. Kim. 2011. Enhancement of the antioxidant and anti-inflammatory activity of *Hizikia fusiforme* water extract by lactic acid bacteria fermentation. Korean J. Fish Aquat. Sci. 44: 111–117.

Taskin, E., M. Ozturk, O. Kurt. 2001. Antibacterial activities of some marine algae from the Aegean Sea (Turkey). Afr. J. Biotechnol. 6: 2746–2751.

Wijesekara, I. and S.K. Kim. 2010. Angiotensin-I-converting enzyme (ACE) inhibitors from marine resources: prospects in the pharmaceutical industry. Mar Drugs. 8: 1080–1093.

Wijesinghe, W.A.J.P., W.W. Lee, Y.M. Kim, Y.T. Kim, S.K. Kim, B.T. Jeon, J.S. Kim, M.S. Heu, W.K. Jung, G. Ahn, K.W. Lee and Y.J. Jeon. 2012. Value-added fermentation of *Ecklonia cava* processing by-product and its antioxidant effect. J. Appl. Phycol. 24: 201–209.

Yoon, N.Y., T.K. Eom, M.M. Kim and S.K. Kim. 2009. Inhibitory effect of phlorotannins isolated from *Ecklonia cava* on mushroom tyrosinase activity and melanin formation in mouse B16F10 melanoma cells. J. Agric. Food Chem. 57: 4124–4129.

16

Production, Handling and Processing of Seaweeds in Indonesia

Hari Eko Irianto[1], and *Syamdidi[2]*

1 Introduction

Indonesia is an archipelago located between the continents of Asia and Australia, as well as the Pacific and Indian oceans. Consequently, both oceans have affected the diversity of seaweed species. Indonesia possesses a largely coral reef coastline, 95.181 km in length, and occupies an area of 7.1 million km^2 (Ministry of Marine Affairs and Fisheries 2010). The archipelago consists of more than 17,508 islands (Working Group of Geostrategy and National Resilience 2012). Kalimantan is the largest island, followed by Papua, Sulawesi, Sumatera and Java islands. Many islands have coastlines environmentally suitable for seaweed culture.

Seaweed is abundant in Indonesia, of which around 550 species of them have been already identified (AntaraNews 2012). Since 2003, the Indonesian Ministry of Marine Affairs and Fisheries (MOMAF) started taking the opportunity for economic development by introducing seaweed as a main commodity of fisheries products in addition to tuna and shrimps (Ministry of Marine Affairs and Fisheries 2010). Seaweed production of Indonesia

[1] Research Center for Fisheries Management and Conservation, Jl.Pasir Putih II, Gedung Balitbang KP II - Komplek Bina Samudra, Ancol Timur, Jakarta Utara 14430, Indonesia.
[2] Research and Development Center for Marine and Fisheries Product Processing and Biotechnology, Jl. Petamburan VI, Jakarta 10260, Indonesia.
* Corresponding author: harieko_irianto@yahoo.com

reached 5,170,201 tons in 2011 (Ministry of Marine Affairs and Fisheries 2013) and is expected to be 10 million tons in 2014, while seaweed export is supposed to achieve US$ 230 million in 2013 (Tribunnews.com 2013). Most of the seaweed produced by Indonesia through farming is mainly *Eucheuma* sp. and *Gracilaria* sp.

As the global demand for seaweed and its derivative products sharply increases, seaweed production in Indonesia should be supported by the availability of appropriate postharvest technologies and activities, including harvesting technology, organic substrate extraction technology, processing technology and seaweed products marketing. To support this, the Indonesian MOMAF has encouraged the development of seaweed processing industries.

2 Seaweed Farming in Indonesia

Seaweed industries have been facing the same problem over years, i.e., lack of raw material supply in terms of quality and quantity due to both inconsistent supply and inconsistent quality prior to the introduction of commercial farming. Seaweed farming programs were conducted by involving financiers for providing financial support and also the buyers, thus stimulating the cultivation of *Eucheuma spinosum* in tropical Southeast Asian countries such as Singapore, Philippines and Indonesia. Successful field cultivation ensures yields at predictable quantity and quality and possibly also reduces transport costs. So far, the results have been very encouraging and the Philippine experience has become a model of such efforts (Doty 1973).

Indonesian seaweed culture at laboratory scale was started in 1966 (Soegiarto 1968) but the progress of seaweed research and development was very slow due to political and economic instability. Later on, in 1972, the Indonesian government, represented by Indonesian National Institute of Oceanology (NIO) and the Indonesian Marine Fisheries Research Institute (MFRI), was trying to set up outdoor seaweed culture at several places, i.e., Pari Islands, Jakarta Bay (Soegiarto et al. 1975), Tanjung Benoa, Bali (Sulistijo and Atmadja 1980), Samaringa Islands, Central Sulawesi (Mubarak 1975), North Tanimbar Island and Maumere (Mubarak 1974), Riau (Mubarak 1976) and Aru Islands, Moluccas (Mubarak 1978). The results showed that *Eucheuma spinosum* farming in Terora-Tanjung Benoa, Bali, produced the best yield with a monthly production volume around 15–20 tons dry weight in 1979 (Soegiarto et al. 1990).

Currently, Indonesia seaweed culture for *Eucheuma* sp. adopts three methods; i.e., off bottom, floating rack and long line. In addition, culture for *Gracilaria* sp. is suitable using the bottom method. The use of these methods varies and is dependent on location and water conditions. Currently,

E. cottonii culture has spread widely in Indonesian waters, including Bali, Sulawesi, West Nusa Tenggara, Moluccas, Papua and Riau Islands, while *Gracilaria* sp. has been cultured in Java Coast, South Sulawesi, Central Sulawesi, Southeast Sulawesi, Lombok Island and Sumbawa Island (Anggadiredja et al. 2011).

3 Seaweed Utilization

Only three out of four classes of seaweed are economically utilized in Indonesia, those are red algae (*Rhodophyceae*), green algae (*Chlorophyceae*) and brown algae (*Phaephyceae*). Based on the phycocolloids which can be extracted, seaweed can be divided into three groups; those are agarophyte, carragenophyte and alginophyte. Agar and carrageenan are extracted from agarophyte and carrageenophyte seaweeds respectively, in which both seaweeds are actually red algae. Meanwhile alginophyte seaweed can be used to produce alginate (Directorate General of Aquaculture 2005).

Anggadiredja et al. (2006) showed that 61 species of seaweed from 27 genera have already been used as food and 21 species as traditional herbal medicine for people living along the coastal line. The usage of seaweed in Indonesia can be seen in Table 1.

Red algae are the most popular seaweed which is used as a food, pharmaceutical and industrial usages. From 17 genera of red algae, 34 species have already been used for many purposes. Moreover, 23 species are able to be cultured, i.e., 6 species from genus *Eucheuma*, 3 species from genus *Gelidium*, 10 species from genus *Gracilaria* and 4 species from genus *Hypnea*. Genera *Eucheuma*, *Gracilaria* and *Gelidium* are commonly found and cultured in Indonesian waters. However, only genera producing agar (agarophytes) and carrageenan (carrageenophytes) are commercially cultured to support seaweed industries in Indonesia and to also fulfill worldwide demands (Kordi 2011).

Green algae commonly found in shallow water along the coastal line. *Ulva* called as 'salada' in Indonesia is one of the most common green algae scattered in Indonesian waters. Green algae, which are abundant in tropical water, can be easily found above tidal zones having sufficient sunshine. Indonesian waters has 12 genera of green algae which are *Caulerpa, Ulva, Valonia, Dictyosphaera, Halimeda, Chaetomorpha, Codium, Udotea, Tydemania, Bernetella, Burgesenia, Neomeris* (Juana 2009).

Brown algae grow rapidly in cold waters. However, there are several genera found along the coastal line in Indonesia such as *Cystoseira* sp., *Dictyopteris* sp., *Dictyota, Hormophysa, Hydroclathrus, Padina, Sargassum,* and *Turbinaria* (Juana 2009).

Table 1. Seaweed utilization in Indonesia.

Seaweed Genera	Location in Indonesia	Utilization
Chlorophyceae		
Acanthopora specifera	Kangean islands, Lombok, Flores, Sumbawa and Alor island	Pickle, Salad
Bostrychia radicans	West and South Java Island, South Lampung, Damar Island and Tanibar Islands	Salad, soup
Caloglosa leprierii	Solor, Alor and Wetar	Salad, soup
Caloglosa leprieurii	South Kalimantan Islands, Alor, Wetar	Soup, vermifuges
Catenella nipae	South Kalimantan, Aru Island and North Papua	Salad, soup
Catenella impudica	North Java Island, Madura Island, Kangean Islands	Salad, soup
Corallopsis salicorni	Bali, Maumere Bay, Solor Islands, Riau Islands, Tanimbar Islands	Salad, soup, pickle
Eucheuma edule	Riau Islands, Seribu Islands, Madura, Kei Islands, Tanimbar	Agar, carrageenan
Eucheuma gelatine	Sumba Island, Alor, Kei Islands, Tanimbar Islands	Agar, carrageenan, goiter, anti asthma, anti bronchitis
Eucheuma horridium	Kei Islands, Tanimbar Island, Rote Island, Sumba Island	Soup, carrageenan
Eucheuma muricatum	Riau Island, Bangka and Belitung Islands, Seribu Islands, Flores, Sumba and Tanimbar Islands	Agar, carrageenan
Eucheuma spinosum (*E. denticulatum*)	Cultured and scattered across Indonesian coast	Iota-carageenan, pickle, soup, agar, salad
Eucheuma cottonii (*Kappaphyrus alvarezi*)	Cultured and scattered across Indonesian coast	Iota-carageenan, pickle, soup, agar, salad
Gelidium amansii	Alor Islands, Tanimbar Islands, Maluku Islands	Agar, anti stomach ache
Gelidium rigidium	Scattered across Indonesian coast	Agar
Gelidium latifolium	Bengkulu, Lampung, Southern part of Java Island, West Nusa Tenggara Islands	Agar, anti stomachache
Gracilaria confervoides	Scattered across Indonesian coast	Agar, salad
Gracilaria crassa	West Java, South Sulawesi	Salad, soup

Glacilaria blodgetti	West Java, East Java, Lombok and Sumba Islands	Salad, soup
Gracilaria arcuata	West Java, Lombok, Sumbawa, Sumba, Sawu Island	Salad, soup
Gracilaria verucosa	West Sumbawa, Sawu Island, South Sulawesi, Southeast Sulawesi	Agar, anti stomach ache, anti mumps, anti bladder disease
Gracilaria eucheumoides	South Lampung, South Java Island, Southeast Celebes, South and Southeast Moluccas	Agar, salad, anti stomach ache, anti mumps, anti bladder disease
Gracilaria lichenoides	Scattered across Indonesian coast	Pickle, agar, salad, candied agar
Gracilaria gigas	Scattered across Indonesian coast	Pickle, agar, salad, candied agar
Gracilaria taenoides	Riau Islands, Bangka and Belitung Islands, Lampung	Pickle, agar, salad, candied agar
Grateloupia filicina	West and South Java Island, South Lampung, Seribu Islands	Pickle, agar, salad, candied agar, fermivuges
Halymenia durvilliae	South and Southeast Celebes, Ambon and Seram Islands, Papua, East Nusa Tenggara, Lombok, Sumbawa and Halmahera	Pickle, salad, candied agar
Hypnea cenomyce	Riau Island, South Kalimantan and Sulu Island	Salad, soup
Hypnea cervicornis	Riau Islands, Bali, Tawi-tawi	Agar sweetener, salad
Hypnea divacirata	Riau Islands, Moluccas Islands	Salad, soup, candied agar
Hypnea musciformis	Scattered across Indonesian coast	Candied agar
Lurencia obtusa	Lingga Island, Riau Islands, Bangka Island	Salad, soup, anti stomach ache
Porphyra atropurpurae	Halmahera Island and Kei Island	Candied agar, soup, anti mumps, anti bladder disease, anti malnutrition
Rhodymenia palmata	Scattered across Indonesian coast	Salad, soup
Sarcodia montagneana	Riau, Lingga Island, Bangka Island, South Java Island, Lombok, Flores, Ambon and Seram Islands	Soup, salad, pickle
Chlorophyceae		
Acetabularia mayor	West Java, Seribu Islands	Anti scrofula
Caulerpa peltata	Bangka Island, Seribu Islands, Southeast Sulawesi	Soup, salad
Caulerpa racemosa laeferens	Bali, Kei Island, Seram Island, Damar Islands	Salad, soup, pickle
Caulerpa racemosa plavifera	Scattered across Indonesian coast	Salad, soup

Table 1. contd....

Table 1. contd.

Seaweed Genera	Location in Indonesia	Utilization
Caulerpa racemosa uinifera	Riau Islands, Sulawesi, Buru Island and Rote Island	Salad, soup, pickle
Caulerpa serrulata	East Kalimantan, Sulawesi, Moluccas, Southern Papua	Salad, soup, pickle
Caulerpa sertularoides	Scattered across Indonesian coast	Salad, soup, pickle
Caulerpa crasa	Seribu Islands, Southern Lampung	Salad, candied agar
Caulerpa javanica	Southern Java Island, Seribu Islands, Ambon and Seram Island	Salad, soup
Codium tenue	Sulu, Ambon, Halmahera	Salad, soup, fermivuges
Codium tamentosum	Scattered across Indonesian coast	Salad, soup, fermivuges
Enteromorpha compressa	Bengkulu, Lampung, West Java, Lombok, Sumba and Flores	Salad, soup, anti mumps, anti cough, antipyretic, anti bronchitis
Enteromorpha intestinalis	Northern Java, Lampung, South and Southeast Sulawesi	Salad, soup, anti mumps, sun block
Enteromorpha prolifera	Seribu Islands, Southern Lampung, Bali, Lombok, Flores	Soup, anti pyretic, anti cough
Ulva lactuca	Sulawesi, Lombok, Sulu, Kei, Sumba, Banda, Solor, West Java, Southern Lampung	Salad, soup, anti pyretic, anti ulcer, anti bladder disease, nosebleed
Phaeophyceae		
Dictyota apiculata	South and Southeast Sulawesi	Salad
Hydroclathrus clathratus	Kalimantan, Java, Sumbawa	Salad, pickle
Padina australis	Riau Islands, Southern Lampung, Southern Java, Sumbawa, Sumba, Ambon, Tanimbar, Kai, Aru, South and Southeast Sulawesi, Lombok, Flores	Soup, candied agar
Sargassum aquifolium	Scattered across Indonesian coast	Alginate, soup, candied agar, anti bladder disease, anti mumps, cosmetic
Sargassum polycystum	Scattered across Indonesian coast	Alginate, candied agar, salad, anti pyretic, anti mumps
Sargassum siliqosum	Southern Java, South and Southeast Sulawesi, Aru Islands, Kei, Tanibar	Alginate, candied agar, salad, anti pyretic, anti mumps
Turbinaria ornata	Scattered across Indonesian coast	Salad, soup
Turbinaria conoides	Scattered across Indonesian coast	Salad, soup

Source: Anggadiredja et al. (2006)

4 Seaweed Harvesting and Handling

Before 1980, seaweed was harvested from nature and seaweed culture was started from 1980 (Sugiarto and Sulustijo 1990). Seaweed production through harvesting of wild seaweed is unpredictable because it depends on the season, monsoons and environmental conditions (Trono 1990). Hence, to ensure the availability and continuity of raw material supply, seaweed culture was developed gradually to increase seaweed production. The first seaweed culture was *E. spinosum* in 1982 and then *E. cottonii* in 1984 (Anggadiredja et al. 2011).

The Directorate General of Aquaculture (2005) recommended the following important activities which should be considered to obtain the accepted quality of seaweed as raw material for further processing (a) **Harvesting** should be carried out at the right age and when the weather is sunny; (b) **Sortation** is to remove unwanted materials such as coral, mud and others; (c) **Washing** is to remove mud and salts attached to seaweeds; (d) **Drying** should be carried out directly under sunlight at open space; and (e) **Packing and storaging** of dried seaweed is performed using plastic sacks.

Seaweed is normally harvested after 45 days if cultured on the beach, but harvested after 60–75 days if cultured in ponds (Kordi 2011). Directorate General of Aquaculture (2005) suggested that *Gracillaria* is harvested after 90 days cultivation, and the next harvesting can be performed after 60 days. *Eucheuma* is harvested after 45 days cultivtion. *Sargassum,* growing wild in nature, can be harvested when the length of its tallus is over 50 cm. Washing is introduced after either harvesting or drying. Washing for *Eucheuma* is performed by soaking in 0.5–3.0% KOH solution for 2–3 hours. *Gracillaria* is normally washed using fresh water until the seaweed is free from unwanted materials.

Harvested *Sargassum* can be washed using fresh water and then soaked in inorganic acid solution or base solution. Soaking in HCl solution increases alginate content and frees mineral salts. The use of a base solution such as $Ca(OH)_2$ is to remove mineral salts and to bind alginate, resulting in calcium alginate. NaOH solution is used to remove protein and non alginate substances. The use of NaOH must be done carefully because excessive usage will result in breaking the chain of alginate so that the alginate produced has a low viscosity (Irianto and Giyatmi 2009). In addition, soaking can also be performed using 0.1% KOH solution for 60 minutes to avoid a degradation of alginates (Yunizal 2009), as well as improved yield with lower moisture content conpared to the sample without soaking in 0.1% KOH solution. Alginate contents of *Sargassum* with and without soaking in 0.1% KOH solution are 23.17% and 16.7% respetively (Fateha 2010).

Drying seaweed normally takes 3–4 days in order to reach a moisture content of 32–35% for *Eucheuma*, 18–22% for *Gracillaria* and 20% for *Sargassum* (Directorate General of Aquaculture 2005). Drying is done on a rack or on the ground, given a base to avoid contamination with foreign material. In addition, during drying, seaweed is protected from dew and rain. In order to ensure faster drying, seaweed is turn upside down after certain periodic intervals. The yield at the end of drying is about 10%. For a long storage period, dried seaweed should have 17% moisture content or less (Irianto and Giyatmi 2009).

5 Seaweed Processing

Seaweed processing technology was introduced to extract organic substances including agar, carrageenan and alginate. Some researchers have developed extraction technologies but these may need some improvements and modification depending on some of the following factors, i.e., seaweed species, habitat, nature or wild crops, and cultivation period. The facts show that the organic composition of seaweed varies from one to another. Experience and knowledge of seaweed characteristics will lead us to overcome the problems arising during the extraction process.

Methods of seaweed extraction are presented in Table 2. These methods were used to extract organic substances with some modification, if needed. In the case of Indonesia, not all methods can be applied due to lack of adequate equipment and knowledge, particularly for processors living along the coastal line. The commonly used methods are namely (a) pressing method and freezing—thawing method for agar production; (b) alcohol method for refined carrageenan, particularly iota and kappa carrageenan; (c) pressing + KCl method only for kappa carrageenan; (d) mild alkali or strong alkali extraction method for semi refined carrageenan; and (e) CaCl$_2$ methods for alginate extraction (Anggadiredja et al. 2011).

Table 2. Methods of seaweed processing.

Product	Method
1. Agar	- Freezing-thawing - Pressing

2. Refined Carrageenan
  - Iota-carrageenan
  - Kappa-carrageenan
	- Alcohol
	- Pressing
	- Pressing + KCl

3. Semirefined carrageenan (powder and chips)	- Mild alkali - Strong alkali
4. Alginate	- CaCl$_2$ - Green Cold - Hecter

Source: Anggadiredja et al. (2011)

5.1 Agar processing

Red algae such as *Gelidium, Gracilaria, Gigartina* and *Rhodymenia* have been revealed as a source of agar. While *Gracilaria* is naturally available across Indonesian waters, it is also widely cultured by the farmers along the coastal line, particularly in Sulawesi dan Java. The other three are harvested from nature. However, *Gracilaria* was commonly used as a source of agar because of its greater availability compared with the other three genuses. However, the gel strength of agar made from *Gracilaria* was lower than that made from *Rhodymenia*. The gel strength of agar made from *Gracilaria* was around 30 g/ cm^2 (Utomo et al. 2003) and gel strength of that made from *Rhodymenia* reached 56.83 g/cm^2 (Chasanah 2004). Agar processors usually combine *Gracilaria* and *Rhodymenia* together to produce agar with a better gel strength.

The term 'agar' is originally from Indonesia. The study of seaweed producing agar in Indonesia was started around 1921. At that time, Hofstede studied the agar content from various seaweeds and assessed them in order to establish the large scale agar industry in Indonesia. He found that among the seaweed studied, only a few contained agar and its availability in nature was not abundant. A few years later, an improved method was applied to determine the actual agar content and the result showed that seaweeed containing agar had sufficient amount to justify the establishment of large scale industry (Zaneveld 1955).

The production of agar in Indonesia had been started before World War II. In 1930, the first agar factory was established in Kudus, Central Java, then followed by others such as those in Jakarta, Surabaya and Makassar (Sulistijo 2002). Currently, there are 11 agar factories in Indonesia. Even though Indonesia is one of the main seaweed producing countries worldwide, the domestic demand of agar is fulfilled from imports. The demand of agar in Indonesia is around 9,850 ton per year (Anon 2012), and 90% of that is supplied from imports. Agar was imported from China, Malaysia, Canada, France, Germany and Spain. The Indonesian import volume of agar in 2011 was 903.86 tons (Antara News 2012).

Research on agar processing from red algae in Indonesia was directed towards increasing the production in order to meet domestic needs in terms of quantity and quality. The quality standards of agar, issued as Indonesian National Standard (SNI), are shown in Table 3.

It is common that an industry combines at least two species of seaweed to produce a high gel strength agar. *Gracilaria*, which is known as a source of agar, has low gel strength if used solely, so, combining it with other seaweeds would improve gel strength. Combining *Gracilaria* with *Gelidium* at the ratio of 1:1 (w/w) can increase the gel strength from 514 g/cm^2 to 835 g/cm^2 (Suryaningrum et al. 1994). The research also showed that the higher proportion of *Gelidium* used in the mixture, the higher the gel strength of agar will be.

Table 3. Quality standard of agar sheet (SNI 01-4105-1996).

Parameter	Unit	Value
a. Sensory		
- Minimum		7 (9 scale)
- Yeast		None
b. Chemistry		
- Moisture content	%w/w	15
- Acid insoluble ash content	%w/w	0.5
c. Metal Content (max)		
- Pb	mg/kg	2.0
- Cu	mg/kg	20.0
- Zn	mg/kg	100.0
- Sn	mg/kg	40.0
- Hg	mg/kg	0.5
- As	mg/kg	1.0
d. Physical content		
- Gel strength	g/cm^2	150

Source: SNI (1996)

In the agar pocessing as shown in Fig. 1, dried seaweeds are first bleached by soaking in 1% $CaOCl_2$ solution for an hour and then washed with fresh water until pH neutral. Agar extraction is carried out by boiling in water at 80–90°C for two hours and subsequently filtering, using a 150 mesh vibrator. The filtrate obtained is placed in pans and left overnight for gel formation. The gel is cut into 1 cm thick sheets, wrapped with cotton fabric, and then arranged in the presses. Pressing is conducted by putting a burden on the top of the presses until the agar sheets become thin enough. The agar sheets are sundried and unwrapped later on to obtain dried agar sheets. Agar powder can be produced by milling dried agar sheets (Darmawan et al. 2006).

Aji et al. (2003) suggested that agar extraction is performed in two steps. The first extraction is carried out by boiling seaweed with water and dried seaweed in the ratio of 14:1 at pH 6–7 and 85–95°C. The second extraction is performed on the residue of the filtering process of the first extraction with a water and dried seaweed ratio of 6:1. However, when a mixture of *Gracilaria* harvested from nature and ponds is used as raw material, the extraction is done with water and dried mixed *Gracilaria* in the ratio of 12:1 at 80–85°C and pH 4.5 for two hours boiling. Gel formation is done by the addition of KCl or KOH solution into the filtrate. Murdinah and Sinurat (2011) informed that the KCl amount added for gel formation is 3% of the dried seaweed amount. KCl is dissolved in water prior to its addition into the agar filtrate.

5.2 Carrageenan processing

The same as agar, carrageenan is also a product obtained from red algae. Three predominant seaweeds which contain carrageenan are *E. denticlatum* (*spinosum* as a trade name), *Kappaphycus alvarezii* (cottonii as a trade name) and *K. striatum* (sacol as a trade name) (Anggadiredja et al. 2011). *K. alvarezii* is a major commodity of seaweed exports from Indonesia and the Philippines.

Carrageenans are hydrocolloids that are commonly used as ingredients in many products, such as in the dairy industry (Doty 1973). Only two types of carrageenan, kappa and iota, are in the international market since lambda carrageenan is unavailable commercially. Lambda carrageenan is a non-gelling carrageenan.

Indonesia has several carrageenan industries, i.e., 3 refined carrageenan factories and 11 semi-refined carrageenan factories but those are still insufficient to fulfill the domestic demand. In 2007, Indonesia only contributed 13% of the total worldwide carrageenan supply (Hakim et al. 2011). The bulk of its domestic demand was fulfilled through import from Korea, China, Philippines, United Stated, Nederland, France and Denmark. The total import volume of carrageenan in 2011 was 1,320.82 tons (Antara News 2012).

The organic composition of seaweeds differs from one to the other either by division or species. It can also vary geographically, according to depth, and even from one part of a thallus to another. Therefore, it is not surprising that the published data on the organic content of seaweeds varies significantly. Seaweeds are important sources of chemical substances for various industrial or medical purposes in Indonesia, among other important resources from the sea.

Suryaningrum et al. (1991) studied the effect of the cultivation period on the quality of carrageenan extracted from *E.* cottonii and *E. spinosum* collected from Seribu Islands, Jakarta. The results showed that seaweed harvested after 7 weeks cultivation would have a higher yield and gel strength compared to the one harvested before 7 weeks.

In Indonesia, carragenophyte is processed into at least three products, i.e., alkali treated cottonii (ATC), semirefined carrageenan (SRC) and refined carrageenan (RC).

ATC is prepared by first boiling seaweed in 6–8% KOH solution at 80–85°C for 2–3 hours. The volume of KOH solution used is 3–4 times the seaweed weight. After boiling, seaweed is soaked and washed several times using water until neutral pH of the wash water is achieved. The seaweed is then cut into pieces about 5 cm in length and sundried for 2–3 days. ATC chips are the product from this process. However, ATC powder can be obtained through a further process of milling and sieving (Aji 2003).

SRC is a produced through a process of extraction carragenophyte. Extraction is carried out by boiling seaweed at 90–95°C for two hours and then filtering to separate most cellulose. KCl is added to the filtrate obtained to form gel. Further processing is the same as that applied to produce dried agar sheets or agar powder described above. Besides dried seaweed, SRC can also be processed using ATC as raw material. SRC powder is characteristically white or slightly yellowish, odourless, tasteless, and 100 mesh in size. In addition, SRC has 8–12% moisture content, 18–23% ash content and >550 g/cm^2 gel strength (Wibowo 2006). Basmal et al. (2005) suggested that carrageenan extraction is performed by boiling *E. cottonii* in 12% KOH solution with dried seaweed and KOH soluton ratio of 1:8.

For RC processing, seaweed extracts from two times of extraction using 3.5% KOH solution are passed through a vibrating filter. The addition of 0.03M KCl into filtrate is carried out while heating it at 90°C for 30 minutes. The filtrate is then poured into isopropyl alcohol (IPA) with the filtrate and IPA ratio of 1:2. Carrageenan will settle in IPA, and is then separated after 12 hours of drying in the sun. The dried carrageenan is ground and screened with a 60 mesh sieve to produce carrageenan powder (Suryaningrum et al. 1991).

5.3 Alginate processing

Alginate is a polysaccharide produced from brown algae extraction. *Sargassum* sp. and *Turbinaria* sp. are sources of alginate which are found in abundance on the Indonesian coast. Wild crops of *Sargassum* and *Turbinaria* are available during the year along the south coast of Java Island and Sumatera waters. Alginate in form of sodium salt is the most widely used in foods.

The properties of sodium alginate extracted from several species of brown algae can be seen in Table 4. Sodium alginate contents and the viscosity value of brown algae were affected by seaweed species and harvesting location. A lower viscosity value is demonstrated by *T. conoides* harvested from Gili Petagan, i.e., 134 cps, whereas a higher viscosity value is shown by *S. polycystum*, i.e., 503.7 cps (Rasyid 2009). Mushollaeni and Rusdiana (2011) found that yields of sodium alginate extrated from *S. duplicatum*, *S. crassifolium* and *Padina* sp. are 30.5%, 30.3% and 16.93% correspondingly, with the highest viscosity exhibited by *S. crassifolium*, i.e., 39 cps. Susanto et al. (2001) also extracted alginate from *Padina* sp., in which yield, viscosity, ash content, and moisture content of alginate salt were 30.51%, 122 cps, 30.51% and 11.21% respectively. Rasyid (2007) noted that sodium alginate contents of *Padina australis* collected from several places

Table 4. Properties of sodium alginate extracted from various brown seaweeds.

No	Species/Sampling Location	Sodium Alginate Content (%)	Moisture Content (%)	Viscosity Value (cps)
1.	*Turbinaria conoides*/Gili Petagan	19.96	17.43	134
2.	*Sargassum polycystum*/Batuampar	18.05	16.85	503.7
3.	*Sargassum* sp./Batuampar	16.89	15.12	143.5
4.	*Turbinaria ornata*/Gili Bedil	17.05	15.86	335
5.	*Sargassum polycystum*/Sumbawa Island	18.12	16.92	390
6.	*Sargassum* sp./Sumbawa Island	17.25	16.82	285
7.	*Turbinaria decurrens*/Sumbawa Island	13.17	16.34	335
8.	*Turbinaria conoides*/Pari Island	25.65	14.96	560
9.	*Turbinaria decurrens*/Barranglompo Island	20.30	15.03	560
10.	*Turbinaria decurrens*/Otangala Island	30.19	15.20	680
11.	*Sargassum polycystum*/Pameungpeuk	28.60	15.20	1,500
12.	*Sargassum echinocarphum*/Pari Island	24.32	15.12	3,000

Source: Rasyid (2009)

were in the range of 4.79%–27.07% with viscosity values of 37 cps–125 cps. Meanwhile, the sodium alginate content of *S. echinocarphum* is 17.07% yield and the alginate has 14.97% moisture content and 6,100 cps viscosity (Rasyid 2010).

Technology for alginate extraction has been developed by Indonesian researchers, i.e., Basmal et al. (1998), Tazwir et al. (2000), Wikanta et al. (2000), Yunizal et al. (2000) and Basmal et al. (2002). Subaryono et al. (2010) successfully improved the viscosity of alginate, approaching values similar to carrageenan. Prior to alginate extraction, dried seaweed is soaked in 1% HCl solution for an hour and subsequently washed with fresh water until pH neutral. Alginate extraction is carried out by heating seaweed in 2% Na_2CO_3 solution at 60–70°C for an hour, and then macerating and further heating for another an hour. Macerated seaweed is filtered using a vibrating filter equipped with a 150 mesh sieve. The filtrate obtained is added with NaOCl as bleaching agent amounting to 4% of filtrate volume. Alginic acid is separated by the addition of 10% HCl solution until pH 2.8–3.2 is reached. Conversion of alginic acid into sodium alginate is performed by the addition of 10% NaOH solution until pH 7 is obtained. Separation of sodium alginate is carried out by slowly pouring this solution into isopropyl alcohol and then sundrying it for 12 hours. Finally, dried sodium alginate is ground and sieved using 60 mesh sieve (Yunizal 2000; Subaryono et al. 2010). Subaryono and Apriani (2010) proved that alginate quality can be improved with the decantation treatment of alginate filtrate up to three hours prior to the formation of alginic acid, particularly in terms of reducing total water insoluble matter from 9.46% to 1.70% and increasing alginate viscosity from 131 cP to 746 cP. Yulianto (2007) used 2%, 5% and 10% NaOH for convertion of alginic acid into sodium alginate, and resulted in sodium alginate with viscosity values of 13,558.67 cps, 5,262.66 cps and 3,736.05 cps respectively. So, the use of higer concentration of NaOH will produce sodium alginate with lower viscosity.

Zailanie et al. (2001) noted that part of the plant used, i.e., top, leaves, lower end and whole, of *S. filipendula*, affects yield and viscosity during extraction of alginate. Higher yield and viscosity are obtained from lower end part. Raw material conditions, i.e., fresh and dried, also affect the yield and viscosity of extracted sodium aginate: fresh seaweed produces higher yield and viscosity of sodium alginate.

However, even with having abundant sources and sufficient technology, Indonesia still has no alginate factory. It is allegedly since raw material is still dependent on wild crops. Actually, *Sargassum* sp. can be cultured. One report showed that *S. polycystum* grew about 2.34 cm per week through culture (Kalangi 2001). Researchers from the Indonesia Research and Development Center for Marine and Fisheries Product Processing and Biotechnology have sucessfully cultured *S. filipendula* using a floating rack method, carried out in coastal waters at Binuangan-Pandeglang District, Banten Province.

6 Conclusion

The Indonesian seaweed industry is now growing fast since the government has a concern as to its economic value. Seaweed culture has increased, making Indonesia the biggest producer of cultured seaweed. Indonesia contributes about 50% of the world's total production of *E. cottonii*.

However, seaweed processing activities do not correspond to the seaweed production. Most of the seaweeds are exported as raw material to the Philippines and other countries and after that their processed products are exported back to Indonesia. It is an irony that Indonesia does not get any special benefits from the seaweed industry, particularly added value and economic benefits.

The government of Indonesia is encouraging the developmet of the seaweed processing industry. Several agar as well as semi refined and refined carrageenan factories have been estabished in some provinces. Currently, Indonesia is the largest agar producer in the world. Up to now, 14 carrageenan companies can be found throughout Indonesia and perhaps in future, the production capacity will increase to fulfill domestic demand and minimize import through setting up new seaweed processing factories.

References

Aji, N., F. Ariyani and T.D. Suryaningrum. 2003. Technology of Seaweed Utilization. Research Center for Marine and Fisheries Product Processing and Socioeconomic. Jakarta (in Indonesian).

Anggadiredja, J.T., M.A. Widodo, A. Arfah, A. Zatnika, S. Kusnowirjono and I. Indrayani. 2011. Strategy of Seaweed industry Development and Its Sustainable Utilization. Jakarta: BPPT Press (in Indonesian).

Anggadiredja, J.T., A. Zatnika, H. Purwoto and S. Istini. 2006. Seaweed: Culturing, Processing and Marketing Potential Fisheries Commodity. Jakarta: Penebar Swadaya (in Indonesian).

Anon. 2012. Highly Culturable, Ministry of Marine Affairs and Fisheries Raise Seaweed Production at North Coast of Java. Retrieved 20 September 2012 from http://www.lensaindonesia.com/2012/04/07/kkp-pacu-produksi-rumput-laut-di-pantura.html.

AntaraNews. 2012. Carrageenan Import Reached 1,300 tons. Retrieved 20 September 2012 from http://makassar.antaranews.com/berita/38849/impor-karaginan-nasional-capai-13-juta-ton (in Indonesian).

Basmal, J., T.D. Suryaningrum and Y. Yennie. 2005. Effects of concentration and ratio of potassium hydroxide solution to seaweed on the quality of carrageenan sheet. Jurnal Penelitian Perikanan Indonesia. 11(8): 29–38 (in Indonesian).

Basmal, J., T. Wikanta and Tazwir. 2002. Effect of potassium hydroxide and sodium carbonate on the quality of alginate. Jurnal Pasca Panen dan Bioteknologi Kelautan dan Perikanan. 8(6): 45–52 (in Indonesian).

Basmal, J., Yunizal and J.T. Murtini. 1998. Effect of volume and extraction time of sodium alginate in sodium carbonate solution. Paper presented at the Forum Komunikasi I. Ikatan Fikologi Indonesia. Serpong, 8th September 1999. pp. 119–126 (in Indonesian).

Chasanah, D.S.N. 2004. Quality of agarose extracted from *Rhodymenia ciliata using DAEA-celulose*. Bachelor Thesis. Bogor Agricultural University, Bogor (in Indonesian).

Darmawan, M., Syamdidi and E. Hastarini. 2006. Processing of bacto agar from red seaweed (*Rhodymenia ciliata*) by alkali pre-treatment. Jurnal Pasca Panen dan Bioteknologi Kelautan dan Perikanan. 1(1): 9–18 (in Indonesian).

Directorate General of Aquaculture. 2005. Seaweed Profiles of Indonesia. Ministry for Marine Affairs and Fisheries. Jakarta (in Indonesian).

Doty, M.S. 1973. Farming the Red Seaweed, Eucheuma, for Carrageenans. Micronesica. 9: 59–73.

Fateha. 2010. Post-harvest handling technique of brown seaweed, Sargassum filipendula used as raw material of alginate. Bul. Tek. Lit. Akuakultur. 6(1): 69–73 (in Indonesian).

Hakim, A.R., S. Wibowo, F. Arfini and R. Peranginangin. 2011. Effect of medium extract ratio, temperature of precipitation and potassium chlorideconcentration on quality of carrageenan. Jurnal Pasca Panen dan Bioteknologi Kelautan dan Perikanan. 6(1): 1–11 (in Indonesian).

Irianto, H.E. and S. Giyatmi. 2009. Processing technology of fisheries products. Penerbit Universitas Terbuka. Jakarta (in Indonesian).

Juana, K.R.S. 2009. Biologi Laut: Ilmu Pengetahuan Tentang Biota Laut. Jakarta: Djambatan (in Indonesian).

Kalangi, S.M. 2001. Growth and nutritious content of brown seaweed *Sargassum polycystum* C. A. Agard 1824 in Tasik Ria, Minahasa District, North Sulawesi. Retrieved 15 January 2008, from http://digilib.bi.itb.ac.id/go.php?id=saptunsrat-gdl-res-2001-kalangi2c-1936-coklat (in Indonesian).

Kordi, M.G.H. 2011. The Secret of Seaweed Culture in Bay and Pond. Jakarta: Lily Publisher (in Indonesian).

Ministry of Marine Affairs and Fisheries. 2010. Marine and Fisheries in Figures. Center of Data, Statistics and Information. Jakarta.

Ministry of Marine Affairs and Fisheries. 2013. Marine and Fisheries Statistics 2011. Center of Data Statistics and Information. Jakarta.

Ministry of Marine Affairs and Fisheries. 2010. Strategic planning—Ministry of Marine Affairs and Fisheries 2012–2014. Ministry of Marine Affairs and Fisheries. Jakarta (in Indonesian).

Mubarak, H. 1974. Report of *Eucheuma* survey in Moluccas and East Nusa Tenggara waters July–November 1974. Lap. Penel. Perik. Laut. 1: 1–29 (in Indonesian).

Mubarak, H. 1975. Research on Eucheuma spinosum (Rhodophyta, Gigartinales) culture in Samaringa Island, Menur, Central Sulawesi Islands, Last July–early October 1975. LPPL 1/75 - PL.053/75:78-101 (in Indonesian).

Mubarak, H. 1976. Report on the culture of Eucheuma spinosum in Teland Island, Riau. LPPL, Jakarta (in Indonesian).

Mubarak, H. 1978. Seaweed, utilization, source and culture opportunity. Lembaga Oseanologi Nasional—LIPI, Jakarta. 61 p. (in Indonesian).

Murdinah and E. Sinurat. 2011. Improvement of agar-agar functional properties by the addition of various gums. Jurnal Pasca Panen dan Bioteknologi Kelautan dan Perikanan. 6(1): 91–100 (in Indonesian).

Mushollaei, W. and E. Rusdiana. 2011. Characterization of sodium alginate from *Sargassum* sp., *Turbinaria* sp. and *Padina* sp. J. Teknol. dan Industri Pangan. XXII(1): 26–32.

National Standardization Board. 1995. SNI 01-4105-1996. Agar sheet. Jakarta (in Indonesian).

Rasyid, A. 2010. Extraction of sodium alginate from brown algae Sargassum echinocarphum. Oseanologi dan Limnologi di Indonesia. 36(3): 393–400 (in Indonesian).

Rasyid, A. 2009. Quality comparison of sodium alginate from several brown algae. Oseanologi dan Limnologi di Indonesia. 35(1): 57–64 (in Indonesian).

Rasyid, A. 2007. Extraction of sodium alginate from Padina australis. Oseanologi dan Limnologi di Indonesia. 33(2): 271–279 (in Indonesian).

Soegiarto, A. 1968. Seaweed resources from Indonesian waters. NAS-LIPI Workshop on Food, Indonesian Professional Paper, Working Group IV, paper 5. 21 pp. Jakarta, 27–30 May 1968.

Soegiarto, A. and Sulustijo. 1990. Utilization and farming of seaweeds in Indonesia. pp. 9–19. *In*: I.J. Dogma, Jr., G.C. Trono, Jr. and R.A. Tabbada (eds.). Culture and use of algae in Southeast Asia: Proceedings of the Symposium on Culture and Utilization of Algae in Southeast Asia, 8–11 December 1981, Tigbauan, Iloilo, Philippines. Tigbauan, Iloilo, Philippines: Aquaculture Dept., Southeast Asian Fisheries Development Center.

Soegiarto, A., Sulustijo and W.S. Atmadja. 1975. Growth of marine algae Eucheuma spinosum at various depths. The Fourth Biological Seminar-Second National Biological Congress, Yogyakarta 10–12 Jul 1975. 13 p. (in Indonesian).

Subaryono, R. Peranginangin, D. Fardiaz and F. Kusnandar. 2010. Formation of alginate gel extracted from Sargassum filipendula and Turbinaria decurrens using $CaCO_3$ and Gluco-d-lactone (GDL). Jurnal Pasca Panen dan Bioteknologi Kelautan dan Perikanan. 5(1): 43–55 (in Indonesian).

Subaryono and S.N.K. Apriani. 2010. Effect of filtrate decantation in alginate extraction process from Sargassum sp. On the quality of the product. Jurnal Pasca Panen dan Bioteknologi Kelautan dan Perikanan. 5(2): 165–173 (in Indonesian).

Sulistijo. 2002. Research on Seaweed Culture in Indonesia. Paper presented at the Speech on Research Professor Inauguration. LIPI. Jakarta (in Indonesian).

Sulistijo and W.S. Atmadja. 1980. Annual report of National Oceanography Institute—LIPI, Jakarta (in Indonesian).

Suryaningrum, T.D.T., Soekarto and Manulang. 1991. Quality assessment of farmed seaweed: *Eucheuma cottonii* dan *Eucheuma spinosum*. Jurnal Penel. Pasca Panen Perikanan. 68: 13–24 (in Indonesian).

Suryaningrum, T.D.T., S. Wibowo, A. Irawaty and A.N. Assik. 1994. Use of sodium tripolyphosphat in agar extraction from cultured seaweed Gracilaria sp. Jurnal Penel. Pasca Panen Perikanan. 81: 1–11 (in Indonesian).

Susanto, T., S. Rakhmadiono and Mujianto. 2001. Characterisation of alginate extracted from *Padina* sp. J. Teknol. Pertanian. 2(2): 96–109 (in Indonesian).

Tazwir, S. Nasran and Yunizal. 2000. Extraction technique of alginic acid from brown seaweed (*Phaeophyceae*). Paper presented at the Seminar Hasil Perikanan. Sukamandi, 21–22 September 2000 (in Indonesian).

Tribunnews.com. 2013. Seaweed export is projected US$ 230 million this year. Retrieved 18 April 2013 from http://www.tribunnews.com/2013/01/16/ekspor-rumput-laut-tahun-ini-diproyeksi-230-juta-dolar-as (in Indonesian).

Trono, G.C. 1990. Seaweed resources in the developing countries of Asia: production and socioeconomic implications. Aquaculture Department, Southeast Asia Fisheries Development Center. Tigbauan, Iloilo, Philippines.

Utomo, B.S.B., D.S. Zilda, E.S. Heruwati, D.S. Lestari, Subaryono, Murdinah and Suryanti. 2003. The development of phycocolloid products as a substitution of gelatin—Technical Report. Research Center for Marine and Fisheries Product Processing and Socio-economics. Jakarta. 42 pp. (in Indonesian).

Wibowo, S. 2006. Seaweed Industry in Indonesia. pp. 254–295. *In*: F. Cholik, S. Moeslim, E.S. Heruwati, T. Ahmad and A. Jauzi (eds.). 60 Tahun Perikanan Indnesia. Msyarakat Perikanan Indonesia. Jakara (in Indonesian).

Wikanta, T., J. Basmal and Yunizal. 2000. Effect of packaging and ambient temperature storage time on the physicochemical of sodium alginate. Paper presented at the Seminar Hasil Penelitian Perikanan. Sukamandi, 21–22 September 2000 (in Indonesian).

Working Group of Geosrategy and National Resilience. 2012. Concept of National Resilience. Lemhannas RI. Jakarta (in Indonesian).

Yulianto. 2007. Effect of sodium hydroxyde concentration on viscosity of sodium alginate extracted from *Sargassum duplicatum* J.G. AGARD (PHAEOPHYTA). Jurnal Oseanologi dan Limnologi di Indonesia. 33(2): 295–306 (in Indonesian).

Yunizal, Tazwir, J.T. Murtini T. Wikanta. 2000. Research on brown seaweed (*Sargassum filipendula*) post harvest handling using potassium hydroxide. Octopus. 4(1): 49–56 (in Indonesian).

Yunizal. 2004. Alginate Processing Technology. Jakarta: Pusat Riset Pengolahan Produk dan Sosial Ekonomi Kelautan dan Perikanan (in Indonesian).

Zailanie, K., T. Susanto and B.W. Simon. 2001. Extrantion and purification of alginate from Sargassum filipendula: Study on the of effect of plant's part, extraction period and isopropanol concentration. J. Teknol. Pertanian. 2(1): 10–27 (in Indonesian).

Zaneveld, J.S. 1955. Economic marine algae of tropical South and East Asian and their utilization. IPFC Special Publication No. 3. 35 p.

17

Food Applications of By-Products From the Sea

C. Senaka Ranadheera[1] and Janak K. Vidanarachchi[2],*

1 Introduction

Consumption of seafood is not novel and it has been consumed by human beings for thousands of years. Seafood represents a diverse group of marine fish, crustaceans, mollusks, algae, seaweed and marine micro-organisms (Rasika et al. 2013). Large amounts of underutilized by-products are generated globally from the sea food processing industry every year (Senevirathne and Kim 2012). The fish processing industry produces more than 60% by-products as waste, which includes skin, head, viscera, trimmings, liver, frames, bones, and roes (Chalamaiah et al. 2012). Important ingredients and food products can be obtained from marine derived by-products, such as fish oil, antifreeze compounds, collagen and gelatin, carotenoids, enzymes, chitin and chitosan, nutraceutical compounds, fish protein hydrolysate, fish sauce and surimi (Fig. 1). These ingredients, extracted from seafood processing by-products, exhibit potential nutritional and dietary uses other than their various applications in many industries including pharmaceutical, agricultural and cosmetic industries. Therefore, seafood processing by-products can be considered as valuable natural resources, with many applications and various functionalities. Most of these compounds are not found in terrestrial animals or the quality may

[1] Department of Animal & Food Sciences, Faculty of Agriculture, Rajarata University of Sri Lanka, Anuradhapura, Sri Lanka.
[2] Department of Animal Science, Faculty of Agriculture, University of Peradeniya, Peradeniya, Sri Lanka.
* Corresponding author: janakvid@pdn.ac.lk

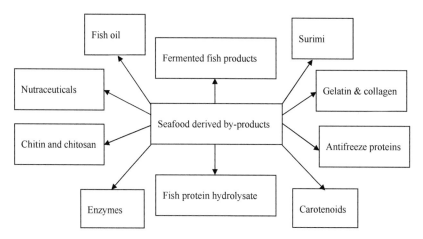

Figure 1. Important foods/food ingredients obtained from seafood processing by-products.

be superior compared to their terrestrial counterparts. Proper utilization of by-products obtained from seafood processing industries is also important for economical as well as ecological reasons in the worldwide context of marine resource depletion (Cudennec et al. 2012).

The potentials and applicability of sea food processing by-products in the food industry are not yet fully explored. Fishy odor and undesirable changes in physico-chemical properties, such as development of rancid flavours and oxidative non-stability during storage, can be considered some major drawbacks of their use of ingredients in the food industry. The addition of odor-masking compounds such as β-cyclodextrin and flavoring agents (vanilla and apple flavors) and the use of novel techniques such as microencapsulation have been successfully utilized in order to improve the sensory attributes of these useful ingredients (Serfert et al. 2010). This chapter's emphasis is mainly on the various by-products derived from the seafood industry and their present and future potential utilizations in the food industry as important food ingredients or food products.

2 By-products from the Sea, Isolation Procedures and their Applications in Food Industry

2.1 Fish oil

Livers of most finfish and shell fish species are considered as by-products in the fisheries industry. Some fish species such as cod (*Gadus morhua*) contain as high as 69% stored lipids in the liver. Although other marine by-products such as head, backbones, viscera, skin, cut-offs and trimmings contain less amounts of fat, these parts can also be successfully utilized to extract fish oil. Fish oil extracted from these tissues of marine organisms possesses important essential fatty acids such as omega-3 fatty acids.

2.1.1 Applications of fish oil in food industry

Dairy foods fortified with fish oil are excellent food products for the delivery of heart health omega-3 long-chain polyunsaturated fatty acids. Fish oils from many marine organisms have been successfully utilized in dairy foods without altering their sensory properties. Strawberry flavored yogurt containing microencapsulated salmon oil (2% wt/vol) demonstrated satisfactory acidity and syneresis (whey separation) during a one month of storage period at 4°C. Although salmon oil incorporated yogurt had more yellow color compared to the control yogurt, the authors have concluded that salmon oil can be successfully utilized in producing strawberry flavored yogurt (Estrada et al. 2011). Another study investigated the effect of incorporation of different oils (butter, fish and oxidized fish) on the sensory characteristics of a savory yogurt. Yogurts were each manufactured at low (1.1–1.2% total fat; 0.43% added oil (wt/wt)) or high (1.6% total fat; 1% added oil (wt/wt)) levels of fish oil, with high levels of fish oil targeted to deliver 145 mg of docosahexaenoic acid + eicosapentaenoic acid/170 g of yogurt. In a preliminary study, untrained panelists (n = 31), using triangle tests, did not discriminate between low levels of fish and butter oils in unflavored yogurts but could discern yogurt with oxidized fish oil, even at the low level (Rognlien et al. 2012). It seems likely that low levels of fish oil can be incorporated into fermented dairy products such as yogurt without a negative effect on their sensory properties, because oxidation and the 'fishy' flavor of fish oil may limit the higher level of fortification (Ye et al. 2009). However, the higher level of fish oil in cheese (71 mg of docosahexaenoic and eicosapentaenoic fatty acids per serving size of (28 g) of cheese compared to 18 and 35 g) was responsible for notable fishy flavor only during the early stages of Cheddar cheese ripening and the fishy flavor decreased as a function of age and became non-significant compared with other samples at 3 months of storage (Martini et al. 2009). Techniques such as encapsulation of fish oil in different matrices may help to reduce unpleasant flavors of fish oil in dairy products. Ye et al. (2009) demonstrated that a fish oil emulsion made with a milk protein complex (encapsulated fish oil) was useful in elevating the fortification level of omega-3 long-chain polyunsaturated fatty acids in processed cheese products without negative effects on sensory properties.

Many studies have examined the possibility of manufacture of fish oil enriched meat products, such as sausage. In a study conducted by Cáceres et al. (2008), pre-emulsified fish oil in caseinates and water was incorporated in a Spanish bologna-type sausage in order to have fish oil levels of 1–6% in the final product. Physico-chemical parameters of the sausage, such as texture, during 90 days at refrigerated storage were not significantly different from the control (no fish oil) sausage samples. The use of deodorized fish oil in partial substitution of pork back fat in producing sausage was successful in terms of sensory acceptability of fish oil incorporated sausage while improving its nutritional quality (Valencia et al. 2006). The use of encapsulated fish oil was also successful and helped in maintaining quality characteristics such as firmness and sensory characteristics of Dutch-style fermented sausages (Josquin et al. 2012).

It is also possible to enhance the lipid profile of livestock and aquatic animal derived food products such as milk (Baer et al. 2001; Tsiplakou and Zervas 2013), flesh (Al-Souti et al. 2012; Hallenstvedt et al. 2010; Zhong et al. 2011) and egg (García-Rebollar et al. 2008) through incorporation of fish oil into animal feed. Incorporation of 5% (wt/wt) fish oil into pig feed has demonstrated no negative effects on the sensory attributes of pork (Hallenstvedt et al. 2010). Similarly, cows fed with a ration enriched with Menhaden fish oil demonstrated good sensory properties in their milk as well as in butter made from the same milk (Baer et al. 2001). Thus, low quality by-products obtained from seafood industries could be effectively utilized in feed formulations for food animals.

Fish oil can be enriched into other food products such as energy bars (Horn et al. 2009) and pasta (Verardo et al. 2009) without major quality or sensory defects.

2.2 Fish enzymes

Digestive enzymes extracted from seafood processing by-products such as fish viscera, liver, stomach, intestine and trimmings and aquatic invertebrates are considered valuable ingredients in the food industry. Chymotrysin, tryptophan hydroxylase, transglutaminase, uricase, alcohol dehydrogenase, pepsin aminopeptidase, elastase, chitinase, cathepsin and trypsin are examples of some of these seafood derived enzymes with potential applications in the food industry (Kim and Mendis 2006; Rasika et al. 2013; Shahidi and Kamil 2001).

2.2.1 Food applications of fish enzymes

Fish enzymes isolated from marine by-products can be utilized to improve the eating qualities of meat through enzymatic hydrolysis of collagen and other proteins in meat. Salt-fermented shrimp sauce prepared from processing by-product (head, shell, and tail) of southern rough shrimp

(*Trachypena curvirostris*) as a meat tenderizer was successful in the case of pork (Kim et al. 2005). Trypsin-like, chymotrypsin-like, carboxypeptidase A and B-like and leucine-aminopeptidase-like activities were detected in the crude extract of the whole digestive tract from the brown shrimp *Penaeus californiensis* (Vega-Villasante et al. 1995). Seafood by-products derived enzymes have been extensively used in the extraction of collagen/gelatin from marine derived sources. For example, Pepsin from the stomach of albacore tuna, skipjack tuna, and tongol tuna has been used in the extraction of collagen from Threadfin bream (*Nemipterus* spp.) skin (Nalinanon et al. 2008). In general, gelatin and collagen-derived hydrolysates and peptides are obtained by enzymatic proteolysis. Enzymatic extracts from fish viscera have also been used to obtain bioactive hydrolysates from the skin and bones of different fish species (Gómez-Guillén et al. 2011). Fish enzymes have been used for ripened fish products and manufacturing of fish sauce. Gastric proteases, including trypsine-like enzymes and cathepsin B from fish and aquatic invertebrates, are more likely to be the most important enzymes in fish sauce production which result in high amounts of soluble proteins during fermentation, adding distinctive flavors to fish sauce (Shahidi and Kamil 2001).

Fish enzyme extracts have been employed in cheese production as a coagulant. Cheddar-type cheeses were manufactured using enzymes extracted from the crustacean *Munida* or chymosin as coagulant and cheeses manufactured with the *Munida* extracts had a higher extent of degradation of β-casein than cheeses made using chymosin as coagulant (Rossano et al. 2005). Further, *Munida* enzymes were found to degrade the chymosin-derived β-casein fragment f193–209, one of the peptides associated with bitterness in cheese, revealing their possible application in cheese industry to lower the unpleasant bitter flavour in some cheeses (Rossano et al. 2011). In addition to this, fish enzymes extracted from seafood by-products are generally less expensive than commercial enzymes used in the cheese industry.

Furthermore, a recent study (Ktari et al. 2012) found that the marine-derived enzymes such as trypsin extracted from zebra blenny (*Salaria basilisca*) viscera have potential application as a detergent additive. Thus enzymes purified from seafood processing by-products such as fish digestive systems could potentially be utilized in manufacturing food grade enzymes. Fish enzymes extracted from seafood processing by-products can be used to hydrolyze the shells of marine crustaceans and extract carotenoproteins (proteins bound with carotene).

2.3 Gelatin and collagen

Collagen, gelatin's parent molecule is a fibrous protein found abundantly in all multi-cellular animals including marine organisms; it is the main component of their connective tissues. Gelatin is mainly obtained from the heat denaturation of collagen, and this soluble protein compound has been widely utilized in biomedical, pharmaceutical, cosmetic and food industrial applications (Rasika et al. 2013). Collagen isolated from marine organisms may possess substantial differences in physico-chemical properties in comparison to that of land animal derived collagen (skins, hides, bones, tendons and cartilages), and thus have special applications in the food industry (Karim and Bhat 2009). Solid waste consisting with whole fish, fish muscles, skin, bone and trace amount of scale fragments and refiner discharge from marine industries such as surimi processing could be utilized as initial material for obtaining collagen/gelatin for industrial applications (Gómez-Guillén et al. 2011). The molecular structure of gelatin is illustrated in the Fig. 2.

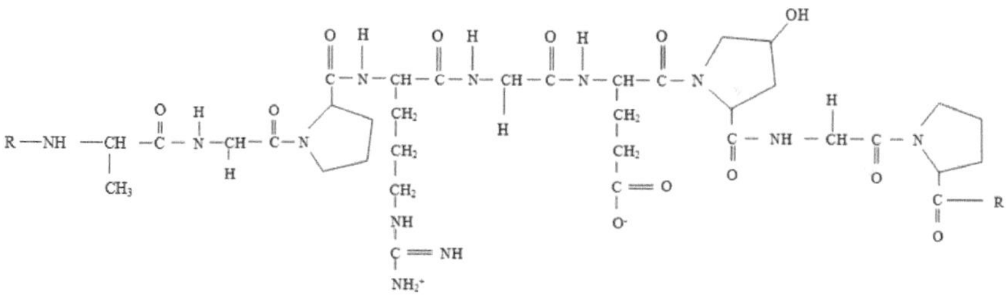

Figure 2. Molecular structure of gelatin. Adapted from Dilan et al. (2013).

2.3.1 Use of collagen and gelatin in food industry

In the food industry, gelatin is utilized in confections, low-fat spreads, dairy, desserts, baked products and meat products to provide specific product properties/functional properties such as chewiness, texture, creaminess, fat reduction, stabilization, emulsification, gelling, water binding and mouth feel (Rasika et al. 2013). Further, gelatin has been identified as a suitable material in developing edible films due to its abundance and biodegradability (Jongjareonrak et al. 2006). Collagen and gelatin films (edible) obtained from fishery by-products such as fish skins provide useful applications in sausage manufacturing (Jongjareonrak et al. 2006). Fishery by-products such as megrim and cod skins have been utilized to produce gelatin with a good gel strength value (Gómez-Guillén et al. 2002), providing wide applications in the food industry. Fish gelatin has been used in microencapsulation of vitamins, colorants and other pharmaceutical additives such as azoxanthine. It is also possible to microencapsulate food flavors such as vegetable oil, lemon oil, garlic flavor, apple flavor or black pepper with warm water fish gelatin. Furthermore, soft gel capsules produced from fish gelatin have been widely used as nutrition supplements (Karim and Bhat 2009).

Even though, there is a growing interest in marine derived gelatin, food applications of marine derived fish gelatin is still relatively lower than that of conventional gelatin isolated from land animals such as cattle and pigs (Rasika et al. 2013).

2.4 Carotenoids

Carotenoids are present in organisms ranging from microorganisms to plants and higher animals, including marine organisms. Carotenoides which are responsible for the yellow to red colour/appearance in organisms are the most widespread pigment in nature and can be categorized into two major groups: xanthophylls (which contain oxygen) and carotenes (which are pure hydrocarbons with no oxygen). Astaxanthin, canthaxanthin, lutein, zeaxanthin and tunaxanthin are the major carotenoids present in aquatic animals, especially in their skins, shells and exoskeletons (Shahidi and Metusalach 1998).

2.4.1 Utilization of carotenoids in the food industry

Carotenoids can be effectively utilized as natural colourants in the food industry. Carotenoids from marine by-products may be useful in preventing the deterioration of foods during processing and storage due to their antioxidant properties. Further, carotenoids possess nutraceutical value in the functional food industry and have been known to have certain therapeutic properties such as prevention of cardiovascular diseases and cancers. Caratenoids also serve as the precursor for vitamin A (Otles and Cagindi 2008; White et al. 2003).

2.5 Chitin and chitosan

Chitin is a glucose-based, branched polysaccharide and it is considered as the second most common biopolymer in nature, mainly found in the exoskeleton (cuticles and peritrophic membranes) of crustaceans & insects, and cell walls of some microorganisms such as bacteria and fungi. Chitosan is the alkaline deacetylated (to varying degrees) form of chitin, which, unlike chitin, is soluble in acidic solutions (Je and Kim 2012; Merzendorfer

2011; Shahidi et al. 1999). The hemical structures of chitin and chitosan are illustrated in Fig. 3.

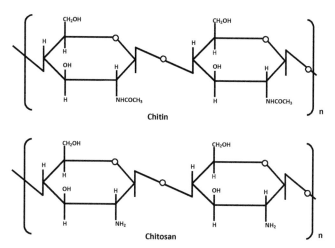

Figure 3. Chemical structures of chitin and chitosan. Adapted from Ghanbarzadeh and Almasi (2013).

2.5.1 Use of chitin and chitosan in food industry

Chitin, chitosan and their derivatives can be considered as natural materials with antioxidant and antimicrobial agents in the food industry (Je and Kim 2012; Vidanarachchi et al. 2010). The poor solubility of chitin may be the major limiting factor in its utilization in the food industry. Chitin is highly hydrophobic and is insoluble in water and most organic solvents. It is soluble in hexafluoroisopropanol, hexafluoroacetone and chloroalcohols in conjugation with aqueous solutions of mineral acids and dimethylacetamide containing 5% lithium chloride. Chitosan is soluble in dilute acids such as acetic acid and formic acid (Ravi Kumar 2000). In the dietary food area, lipase, amylase, 3-hydroxy-3-methylglutaryl CoA reductase, glucokinase and the enzymes of prostaglandin synthesis are involved in the oral administration of chitosan (Muzzarelli 1997). It has been reported that wound dressings made of chitin and chitosan fibers have applications in wastewater treatment. In addition, the removal of heavy metal ions by chitosan through chelation has received much attention (Muzzarelli 1997; Ravi Kumar 2000). Thus, chitosan fibres may have future applications in purification of drinking water. More research is needed to verify this phenomenon.

The *N*-acetylglucosamine moiety (monomeric unit of chitin) present in human milk promotes the growth of bifidobacteria, which suppress the growth and activities of other types of microorganism and generates the lactase required for the digestion of milk lactose. Since cow milk contains only a limited amount of the N-acetylglucosamine moiety, hence some infants fed cow's milk, and also some elderly people, may have problems with indigestion. Incorporation of chitinous materials into whey based chicken feed have shown better feed utilization as well as improvement in intestinal microflora, especially bifidobacteria (Ravi Kumar 2000; Spreen et al. 1984). *Bifidobacterium* are probiotic microorganisms and are extremely important in maintaining gut health. Therefore, chitinous materials extracted from seafood processing by-products may have future potential applications in the dairy industry. Hypercholesterolemic effects of the chitosan and chitooligosaccharides have also recently gained attention, as this property can be applied in the nutraceutical and biomedical fields (Vidanarachchi et al. 2010). Nanopowdered chitosan concentrations (0.3 to 0.5%, vol/vol) could be effectively used to produce a cholesterol-reduced yogurt without significantly adverse effects on the physicochemical, microbial, and sensory properties of yogurt (Seo et al. 2009). Vidanarachchi et al. (2010) have also described the value of marine derived chitin, chitosan and their oligosachchrides in prebiotic applications, enhancement of calcium absorption and toxin binding effects with respect to food applications.

Chitosan and its derivatives have potential applications in the meat industry as well. Fresh pork sausage containing 0.5% and 1% chitosan have shown inhibition of the growth of undesirable microorganisms in pork during 28 days of storage at 4°C. Further, addition of chitosan significantly decreased the rate of lipid oxidation in fresh pork sausage and increased the sensory properties (Soultos et al. 2008).

Antimicrobial effect of chitin and its derivatives has gain wider attention recently due to possible applications in controlling food borne diseases. Chitin isolated from shrimp (*Parapenaeus longirostris*) shell waste exhibited a bacteriostatic effect on Gram-negative bacteria, *Escherichia coli* ATCC 25922, *Vibrio cholerae*, *Shigella dysenteriae*, and *Bacteroides fragilis*. Chitosan exhibited a bacteriostatic effect on all the bacteria tested, except *Salmonella typhimurium*. The oligomers exhibited a bactericidal effect on all these bacteria (Benhabiles et al. 2012). Martinez-Camacho et al. (2010) examined the antifungal effect of chitosan prepared from shrimp waste and commercial chitosan against *Aspergillus niger* and both have inhibited the growth of fungus by 47.26% and 56.16% compared to the cellophane control. These results suggest the potent use of chitosan extracted from seafood processing wastes in food preservation as a natural antibacterial agent. Chitosan also showed a bacteriostatic effect against *Bacillus subtilis* and was helpful in the extension of the shelf life of fresh noodles up to 14 days at 4°C (Huang et al. 2007).

2.6 Nutraceuticals and functional foods

Marine foods can be considered as valuable sources of many healthy food ingredients and biologically active compounds with nutraceutical properties (Vidanarachchi et al. 2012). Various marine organisms such as crab, shrimp, prawn, squid, lobster, cuttle fish and various fish species including salmon, cod, shark, Alaska pollock and mackerel are abundant sources of nutraceutical compounds such as chitin and chitosan, chitoooligosaccharide, omega-3 oils, sulfated polysaccharides, carotenoids, vitamin and minerals, bioactive peptides and taurine (Freitas et al. 2012; Kadam and Prabhasankar 2010; Rasika et al. 2013; Samaranayaka and Li-Chan 2008). Seafood by-products can be utilized to extract nutraceuticals/functional ingredients with potential applications in the food industry (Table 1).

These compounds may protect food products against oxidative degradation or possess potential applications in treating diseases such as cardio vascular diseases, neurodegenerative diseases, Alzheimer's disease, cystic fibrosis, inflammatory disorders and in obesity control. These compounds have also been identified for their antimicrobial activities, anti-carcinogenic properties and immunomodulatory activities. In addition, these compounds may be responsible for reduction of lipid absorption, brain function, growth and function of the body and maintaining teeth and bone strength (Freitas et al. 2012; Kadam and Prabhasankar 2010). Potential

Table 1. Seafood by-product-derived nutraceuticals/functional ingredients having potential applications in food industry.

Compound	Main sources
Polysaccharides	
1. Chitin & chitosan	Crustacean exoskeleton
2. N-acetyl glucosamines	Chitin & chitosan from marine organisms
Bioactive peptides	
1. Bioactive peptides from marine-derived collagen/gelatin	cold water and warm water marine fish skin, scales, bones, cartilage, stomach and soft connective tissues
Polyunsaturated fatty acids	
1. Oleic and palmitoleic acid	Marine fish and crustaceans
2. Linolenic acid	*Arthrospira*
3. Arachidonic acid	*Porphyridium*
4. Eicosapentaenoic acid	*Nannochloropsis, Phaeodactylum, Nitzschina*
5. Docosahexaenoic acid	*Crypthecodinium, Schizochytrium*
Carotenoids	Marine crustaceans
Vitamin & Minerals	Fish liver and fish bone

Adapted and modified (Morganti and Morganti 2008; Ranadheera and Vidanarachchi 2013; Sachindra et al. 2005).

health and nutritional benefits of different marine derived nutraceuticals/ functional ingredients that can be possibly isolated from seafood by-products are shown in Table 2.

Table 2. Health and nutritional benefits of marine-derived nutraceuticals/functonal ingredients.

Compound	Sources	Potential health/nutritional effect
Chitin chitosan Chitoooligosaccharide	Crab, shrimp, prawn, squid, lobster, cuttlefish	Prevention of inflammatory disorders Antimicrobial activity Reduction of lipid absorption Hypocholesterolemic effects
Omega-3 oils	Salmon, cod liver oil	Prevention of cardio vascular diseases Reduction of heart rate and platelet aggregation Hypocholesterolemia Antihypertension Brain function in children Anti-inflammatory effect
Polysaccharides	Carageenan Alginate Fucoidans	Anti-carcinogenic effect Antiviral effect Antihypertensive effect Effective probiotic carriers Antihypertensive effect
Carotenoids	Astaxanthin Fucoxanthin B-carotene	Antioxidants and anticancer agent Prevent neurodegenerative diseases, radical scavenging activities, anti-obese activities Radical scavengers and anti-carcinogenic effect
Vitamin and minerals	Marine fish, fish bone	Growth and physiology of body Teeth and bones strength and antitumor agent
Shark cartilage	Shark	Anti-carcinogenic agent
Bioactive peptides	Fish protein, Alaska pollock backbone Atlantic cod stomach	Obesity control and Ca-binding activity Immunomodulatory activities
Taurine	Cod, mackerel	Prevent cardiovascular diseases, Alzheimer's disease and cystic fibrosis

Adapted and modified (Freitas et al. 2012; Kadam and Prabhasankar 2010; Ranadheera and Vidanarachchi 2013; Vidanarachchi et al. 2012).

2.6.1 Utilization of marine nutraceutical compounds in the food industry

The food applications of marine nutraceutical compounds have already been discussed under the above sections, fish oil, chitin and chitosan and carotenoids, in detail because these compounds possess important nutraceutical values. Other than these uses, omega-3 fatty acids derived from fish and fish wastes can be used to produce fish oil capsules and these products can be used in bakery and confectionary products in order to improve their nutritional value (Freitas et al. 2012). Chitin, chitosans and chitooligosaccharide derivatives may have potential applications as gelling agents, emulsifying agents, food preservatives and sources of dietary fibre in the functional food industry. Further, proteins, bioactive peptides and amino acids from fish and fish wastes can be utilized as stabilizing and thickening agents, protein replacements and gelling agents.

Nowadays, foods containing fish and fish by-products derived oils rich in omega-3 fatty acids, chitin, chitosan, etc. are commercially available in developed countries such as United States, Japan and some European countries (Freitas et al. 2012; Kadam and Prabhasankar 2010). 'Omega bread' available in Denmark (a white bread enriched with omega-3 fatty acids in the form of gelatin coated fish oil), marine oil incorporated pasta and omega-3 fatty acid incorporated dairy foods such as yogurts are good examples of functional food products (Kadam and Prabhasankar 2010; Verardo et al. 2009).

2.7 Antifreeze proteins

Yeasts and some other types of microorganisms can be considered suitable sources of antifreeze proteins; however, the most promising source is fish by-products (body fluids/blood and skin tissues) due to abundant availability and relatively easy isolation procedures (Ferraro et al. 2010). Marine derived antifreeze proteins have been isolated from body fluids of cold sea water fish such as cod. These compounds prevent fish from freezing (Harding et al. 2003). Antifreeze protein (AFP) compounds have also been extracted from many other marine fish species such as right eye flounder, longhorn sculpin, shorthorn sculpin, winter flounder, sea raven, Atlantic herring, Atlantic eel pout, longhorn sculpin, snail fish, cunner and northern notothenioids (Evans and Fletcher 2004; Fletcher et al. 2001; Jorgensen et al. 2008). AFPs found from the marine sources can be categorized into two types: glycoproteins and non-glycoproteins, based on the presence or absence of carbohydrates. Based on the structural diversity, non-glycoproteins have been further classified as Type I, Type II, Type III, and Type IV antifreeze proteins. Both glycoprotein and non-glycoprotein antifreeze compounds can lower the freezing point of an aqueous solution via a non-colligative mechanism, through binding to ice crystal surfaces and inhibiting further crystal growth (Evans and Fletcher 2004; Harding et al. 2003). Therefore, these compounds possess potential applications in food industry. Molecular structures of different types of antifreeze proteins are illustrated in Fig. 4.

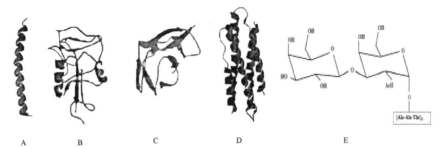

Figure 4. Molecular structures of different types of antifreeze proteins: (A) Type I; (B) Type II; (C) Type III; (D) Type IV; (E) AFGP. Adapted from Raskia et al. (2013).

2.7.1 Use of marine derived antifreeze compounds in food industry

Antifreeze proteins from natural sources may have limited applications in the food industry due to the high cost compared to synthesized antifreeze protein compounds. In particular, antifreeze proteins find large application in frozen food industry including ice cream and frozen meat manufacture (Ferraro et al. 2010). Antifreeze proteins may improve the quality of frozen foods, hence allowing for the maintenance of their natural texture, reduction of cellular damage and loss of nutrients, all of which contribute to preserve their nutritional value (Li and Sun 2002). Antifreeze proteins have a great potential to avoid or minimize defects such as formation of large ice crystals during freezing and dripping during thawing in frozen meat products due to their ability to inhibit ice recrystalization (Rasika et al. 2013). Marine derived antifreeze compounds can be successfully utilized in cryopreservation of fruits and vegetables and surimi production. Detailed views on utilization of antifreeze proteins from marine sources in food industry can be found in a recent article by Rasika et al. (2013).

2.8 Surimi and related products

Surimi can be described as moist frozen concentrated myofibrillar protein of fish flesh. Marine fish species such as Alaska pollock (*Theragrachalcogramma*), white croaker (*Argyrosomusargentatus*), Pacific whiting (*Merlucciusproductus*), hoki (*Macruronusnovaezealandiae*) and blue whiting (*Sillagoparvisquamis*) (Chen et al. 1997; Mahawanich et al. 2010) are some of the common fish species that have been used in traditional surimi production. Due to the unavailability of these traditional fish species in recent years, other sources such as fresh water fish species (Luo et al. 2006; Mahawanich et al. 2010) and marine underutilized fish such as *Sardinops pilchardusts* (Bentis et al. 2005), by-catch species and by-products such as trimmings, cut offs, frames resulting from seafood processing industries (Venugopal and Shahidi 1995) have been utilized in manufacturing surimi and surimi based products.

2.8.1 Food applications of surimi

Deboned, minced and washed fish flesh known as surimi is used in the manufacture of seafood imitation products such as crab legs (Martin-Sanchez et al. 2009). Surimi is an ideal source in the production of sausage (with meat or only with surimi) with higher consumer acceptability (Venugopal and Shahidi 1998). Murphy et al. (2004) have evaluated the effect of added surimi (0–40%), fat (5–30%) and water (10–35%) on the physical, textural and sensory characteristics of fresh breakfast pork sausages. They found that the incorporation of surimi up to 25.3% is possible without major changes in physico-chemical and sensory properties. Incorporation of varying proportions of sardine surimi (0%, 6·5%, 13% and 20%) as an ingredient in Bologna sausage containing various levels of fat (4·8%, 10·6% and 20·8%) resulted in no alteration in the fat and water binding properties and the rheological characteristics of the meat product (Cavestany et al. 1994).

2.9 Fish protein hydrolysate and fish protein concentrate

Fish protein hydrolysates are breakdown products of enzymatic conversion of fish proteins into smaller peptides that generally contain 2–20 amino acids. Any stable and wholesome product with higher protein and nutrients other than the fresh fish can be considered as fish protein concentrate. Both fish protein hydrolysate and concentrate can be prepared from any type of fish or fishery waste such as underutilized body parts. Recently, fish protein hydrolysates have attracted the attention of food biotechnologists due to the availability of large quantities of raw material for the process, as also the presence of high protein content with good amino acid balance and bioactive peptides with antioxidant, antihypertensive, immunomodulatory and antimicrobial peptides (Chalamaiah et al. 2012; Córdova Murueta et al. 2007).

2.9.1 Food applications of fish protein hydrolysate and fish protein concentrate

Nutritionally valuable muscle proteins can be recovered from mechanically deboned fish waste such as fish frames and cutoffs. These fish proteins can be hydrolyzed enzymatically to recover protein biomass into the marketable and acceptable form otherwise discarded as processing waste (Kim and Mendis 2006; Venugopal et al. 1996). Although fish protein hydrolysates from marine fish by-products have been mainly explored for their ability in providing biologically active peptides, the nutritional and functional properties of fish protein hydrolysate may further enhance their potential application in the food industry (Benkajul and Morrissey 1997; Klompong et al. 2007). Fish protein hydrolysate, which is obtained through hydrolysis of tuna waste, can be used as an ingredient in food industries to provide functional effects such as whipping, gelling, and texturing properties (Herpandi et al. 2011). Protein hydrolysate from fish processing co-products have been found to posses desirable physico-chemical properties such as emulsifying, foaming, oil and water binding capacities and many important bio-activities, including anti-oxidative, anti-hypertensive, anti-microbial and anti-anemia properties with potential applications in food and nutritional as well aspharmaceutical products (He et al. 2013). Blue whiting, cod, plaice and salmon hydrolysates were found as significant growth inhibitors and demonstrated antiproliferative activities on human cancer cells *in vitro* due to lipids and or other bioactive trace compounds present in fish protein hydrolysate (Picot et al. 2006). Therefore, there is also a good potential for fish protein hydrolysate to be used as a component of food formulation or for direct human consumption.

2.10 Fermented fish products/fish sauce

Fermented fish products are processed using fin fish, shell fish and crustaceans of fresh water or marine origin with salt to cause fermentation and thereby, to prevent putrefaction. Fish sauce is the well known traditional fermented fish product which commonly used as a condiment and a seasoning agent throughout the world, particularly in South-East Asia. Seafood processing by-products such as fish intestines, viscera, squid head, skin & fins and fish & shrimp processing wastes have been utilized in manufacturing fish sauces with acceptable sensory properties (Dissaraphong et al. 2006; Kim et al. 2005; Xu et al. 2008). Aerobic as well as anaerobic fermentation can be applied to produce fish sauce. In general, fermentation under anaerobic conditions was found to develop fish sauce with a better quality of aroma during the manufacturing process (Sanceda et al. 1994).

2.10.1 Food applications of fermented fish products

Fish sauce can be used for direct human consumption. Fish by-products such as male Arctic capelin and Atlantic cod intestines have been utilized as raw materials for the production of high value fish sauce for human consumption. In this study, supplementation of minced capelin with 5–10% enzyme-rich cod pyloric caeca, resulted in a good recovery of fish sauce protein (60%) after 6 months of storage (Gildberg 2001). A concentrate of tryptic enzymes with high enzymatic activity has been obtained by ultrafiltration of fish sauce made from cod viscera. The concentrate could be stored for 8 months at 3°C or spray-dried with only minor activity loss.

These enzyme preparations isolated from fish sauce have been successfully applied in the maturation of herring fillets (Gildberg and Xian-Quan 1994). Klomklao et al. (2006) examined the effects of the addition of spleen of skipjack tuna (*Katsuwonus pelamis*), at levels of 0%, 10% and 20%, on the liquefaction and characteristics of fish sauce produced from the sardine (*Sardinella gibbosa*) with different salt concentrations (15%, 20% and 25%) during fermentation for 180 days. Sardine with 25% spleen and 15% salt added exhibited the greatest protein hydrolysis, particularly at the early stages, suggesting the combined effects of autolysis and spleen proteinase. Fish sauce samples containing 20% salt, without and with 10% spleen addition had similar acceptability to commercial fish sauce. Another study concluded that the tuna viscera stored at room temperature for up to 8 hours could be used for the production of fish sauce with no detrimental effect on the quality (Dissaraphong et al. 2006), demonstrating the possibility of processing fish by-products into a useful food ingredient with no additional cost for refrigerated storage for raw materials.

Fermentation of shrimp waste for 168 hours using symbiotic lactic acid bacteria *Streptococcus thermophilus, Lactobacillus acidophilus* and *Lactobacillus bulgaricus* facilitated the removal of calcium and protein, with 91.3% calcium, 97.7% protein and 32.3% carotenoid. These extracted ingredients could be possibly used in the food industry (Duan et al. 2011). Squid processing by-products have also been utilized in producing low salt fish sauce without any strong or unpleasant flavours, and the result of amino acids analysis suggested that glutamic acid was the most prominent in sauce samples, demonstrating the nutritional value of fish sauce made from squid processing by-products (Xu et al. 2008).

3 Conclusion

In recent years, there is an increasing trend for utilization of seafood processing by-products in food industry as ingredients or food for direct human consumption. By-products obtained from fish and marine invertebrates, including skin, scales head, viscera, trimmings, liver, frames, bones, shells and roes can be used to extract/produce important food ingredients and products such as fish oil, antifreeze compounds, collagen and gelatin, carotenoids, enzymes, chitin and chitosan, nutraceutical compounds, fish protein hydrolysate, fish sauce and fermented products and surimi. However, the potential and applicability of these ingredients and products in the food industry have not been extensively studied and hence advanced, cost effective and value-added processing technologies need to be developed for the production of high quality materials which possess specific functionalities for food product applications.

References

Al-Souti, A., J. Al-Sabahi, B. Soussi and S. Goddard. 2012. The effects of fish oil-enriched diets on growth, feed conversion and fatty acid content of red hybrid tilapia, *Oreochromis* sp. Food Chemistry. 133(3): 723–727.

Baer, R.J., J. Ryali, D.J. Schingoethe, K.M. Kasperson, D.C. Donovan, A.R. Hippen et al. 2001. Composition and properties of milk and butter from cows fed fish oil. Journal of Dairy Science. 84(2): 345–353.

Benhabiles, M.S., R. Salah, H. Lounici, N. Drouiche, M.F.A. Goosen and N. Mameri. 2012. Antibacterial activity of chitin, chitosan and its oligomers prepared from shrimp shell waste. Food Hydrocolloids. 29(1): 48–56.

Benkajul, S. and M.T. Morrissey. 1997. Protein hydrolysates from Pacific whiting solid wastes. Journal of Agricultural and Food Chemistry. 45: 3423–3430.

Bentis, C.A., A. Zotos and D. Petridis. 2005. Production of fish-protein products (surimi) from small pelagic fish (*Sardinops pilchardusts*), underutilized by the industry. Journal of Food Engineering. 68(3): 303–308.

Cáceres, E., M.L. García and M.D. Selgas. 2008. Effect of pre-emulsified fish oil—as source of PUFA n–3—on microstructure and sensory properties of mortadella, a Spanish bologna-type sausage. Meat Science. 80(2): 183–193.

Cavestany, M., F. Jiménez Colmenero, M.T. Solas and J. Carballo. 1994. Incorporation of sardine surimi in Bologna sausage containing different fat levels. Meat Science. 38(1): 27–37.

Chalamaiah, M., B. Dinesh kumar, R. Hemalatha and T. Jyothirmayi. 2012. Fish protein hydrolysates: Proximate composition, amino acid composition, antioxidant activities and applications: A review. Food Chemistry. 135(4): 3020–3038.

Chen, H.-H., E.M. Chiu and J.-R. Huang. 1997. Color and gel-forming properties of horse mackerel (*Trachurus japonicus*) as related to washing conditions. Journal of Food Science. 62(5): 985–991.

Córdova Murueta, J.H., M.d.l.Á. Navarrete del Toro and F. García Carreño. 2007. Concentrates of fish protein from bycatch species produced by various drying processes. Food Chemistry. 100(2): 705–711.

Cudennec, B., T. Caradec, L. Catiau and R. Ravallec. 2012. Chapter 31—Upgrading of sea by-products: potential nutraceutical applications. *In*: K. Se-Kwon (ed.). Advances in Food and Nutrition Research: Academic Press. 65: 479–494.

Dissaraphong, S., S. Benjakul, W. Visessanguan and H. Kishimura. 2006. The influence of storage conditions of tuna viscera before fermentation on the chemical, physical and microbiological changes in fish sauce during fermentation. Bioresource Technology. 97(16): 2032–2040.

Duan, S., Y.X. Zhang, T.T. Lu, D.X. Cao and J.D. Chen. 2011. Shrimp waste fermentation using symbiotic lactic acid bacteria. *In*: J. Zeng, T. Li, S. Ma, Z. Jiang and D. Yang (eds.). Advanced Materials Research. 194-196: 2156–2163.

Estrada, J.D., C. Boeneke, P. Bechtel and S. Sathivel. 2011. Developing a strawberry yogurt fortified with marine fish oil. Journal of Dairy Science. 94(12): 5760–5769.

Evans, R.P. and G.L. Fletcher. 2004. Isolation and purification of antifreeze proteins from skin tissues of snailfish, cunner and sea raven. Biochimica et Biophysica Acta (BBA)—Proteins and Proteomics. 1700(2): 209–217.

Ferraro, V., I.B. Cruz, R.F. Jorge, F.X. Malcata, M.E. Pintado and P.M.L. Castro. 2010. Valorisation of natural extracts from marine source focused on marine by-products: A review. Food Research International. 43(9): 2221–2233.

Fletcher, G.L., C.L. Hew and P.L. Davies. 2001. Antifreeze proteins of teleost fishes. Annu. Rev. Pysiol. 63: 359–390.

Freitas, A.C., D. Rodrigues, T.A.P. Rocha-Santos, A.M.P. Gomes and A.C. Duarte. 2012. Marine biotechnology advances towards applications in new functional foods. Biotechnology Advances. 30(6): 1506–1515.

García-Rebollar, P., P. Cachaldora, C. Alvarez, C. De Blas and J. Méndez. 2008. Effect of the combined supplementation of diets with increasing levels of fish and linseed oils on yolk fat composition and sensorial quality of eggs in laying hens. Animal Feed Science and Technology. 140(3–4): 337–348.

Ghanbarzadeh, B. and H. Almasi. 2013. Biodegradable polymers. *In*: R. Chamy and F. Rosenkranz (eds.). Biodegradation-Life of Science.

Gildberg, A. 2001. Utilisation of male Arctic capelin and Atlantic cod intestines for fish sauce production—evaluation of fermentation conditions. Bioresource Technology. 76(2): 119–123.

Gildberg, A. and S. Xian-Quan. 1994. Recovery of tryptic enzymes from fish sauce. Process Biochemistry. 29(2): 151–155.

Gómez-Guillén, M.C., B. Giménez, M.E. López-Caballero and M.P. Montero. 2011. Functional and bioactive properties of collagen and gelatin from alternative sources: A review. Food Hydrocolloids. 25(8): 1813–1827.

Gómez-Guillén, M.C., J. Turnay, M.D. Fernández-Díaz, N. Ulmo, M.A. Lizarbe and P. Montero. 2002. Structural and physical properties of gelatin extracted from different marine species: A comparative study. Food Hydrocolloids. 16(1): 25–34.

Hallenstvedt, E., N.P. Kjos, A.C. Rehnberg, M. Øverland and M. Thomassen. 2010. Fish oil in feeds for entire male and female pigs: Changes in muscle fatty acid composition and stability of sensory quality. Meat Science. 85(1): 182–190.

Harding, M.M., P.I. Anderberg and A.D.J. Haymet. 2003. 'Antifreeze' glycoproteins from polar fish. European Journal of Biochemistry. 270(7): 1381–1392.

He, S., C. Franco and W. Zhang. 2013. Functions, applications and production of protein hydrolysates from fish processing co-products (FPCP). Food Research International. 50(1): 289–297.

Herpandi, N.H., A. Rosma and W.A. Wan Nadiah. 2011. The tuna fishing industry: A new outlook on fish protein hydrolysates. Comprehensive Reviews in Food Science and Food Safety. 10(4): 195–207.

Horn, A.F., N.S. Nielsen and C. Jacobsen. 2009. Additions of caffeic acid, ascorbyl palmitate or γ-tocopherol to fish oil-enriched energy bars affect lipid oxidation differently. Food Chemistry. 112(2): 412–420.

Huang, J.-r., C.-y. Huang, Y.-w. Huang and R.-h. Chen. 2007. Shelf-life of fresh noodles as affected by chitosan and its Maillard reaction products. LWT—Food Science and Technology. 40(7): 1287–1291.

Je, J.-Y. and S.-K. Kim. 2012. Chapter 7—Chitosan as potential marine nutraceutical. *In*: K. Se-Kwon (ed.). Advances in Food and Nutrition Research: Academic Press. 65: 121–135.

Jongjareonrak, A., S. Benjakul, W. Visessanguan, T. Prodpran and M. Tanaka. 2006. Characterization of edible films from skin gelatin of brownstripe red snapper and bigeye snapper. Food Hydrocolloids. 20(4): 492–501.

Jorgensen, S.K.K., S. Keskin, D. Kitsios, P.W. Commerou, B. Nilsson and B. Vincents. 2008. Antifreeze Proteins: The Applications of Antifreeze Proteins in the Food Industries: Roskilde University.

Josquin, N.M., J.P.H. Linssen and J.H. Houben. 2012. Quality characteristics of Dutch-style fermented sausages manufactured with partial replacement of pork back-fat with pure, pre-emulsified or encapsulated fish oil. Meat Science. 90(1): 81–86.

Kadam, S.U. and P. Prabhasankar. 2010. Marine foods as functional ingredients in bakery and pasta products. Food Research International. 43(8): 1975–1980.

Karim, A.A. and R. Bhat. 2009. Fish gelatin: properties, challenges, and prospects as an alternative to mammalian gelatins. Food Hydrocolloids. 23(3): 563–576.

Kim, J.-S., F. Shahidi and M.-S. Heu. 2005. Tenderization of meat by salt-fermented sauce from shrimp processing by-products. Food Chemistry. 93(2): 243–249.

Kim, S.-K. and E. Mendis. 2006. Bioactive compounds from marine processing byproducts—A review. Food Research International. 39(4): 383–393.

Klomklao, S., S. Benjakul, W. Visessanguan, H. Kishimura and B.K. Simpson. 2006. Effects of the addition of spleen of skipjack tuna (*Katsuwonus pelamis*) on the liquefaction and characteristics of fish sauce made from sardine (*Sardinella gibbosa*). Food Chemistry. 98(3): 440–452.

Klompong, V., S. Benjakul, D. Kantachote and F. Shahidi. 2007. Antioxidative activity and functional properties of protein hydrolysate of yellow stripe trevally (*Selaroides leptolepis*) as influenced by the degree of hydrolysis and enzyme type. Food Chemistry. 102(4): 1317–1327.

Ktari, N., H. Ben Khaled, R. Nasri, K. Jellouli, S. Ghorbel and M. Nasri. 2012. Trypsin from zebra blenny (*Salaria basilisca*) viscera: Purification, characterisation and potential application as a detergent additive. Food Chemistry. 130(3): 467–474.

Li, B. and D.-W. Sun. 2002. Novel methods for rapid freezing and thawing of foods—A review. Journal of Food Engineering. 54(3): 175–182.

Luo, Y.K., H.X. Shen and D.D. Pan. 2006. Gel-forming ability of surimi from grass carp (*Ctenopharyngodon idellus*): Influence of heat treatment and soy protein isolate. Journal of the Science of Food and Agriculture. 86: 687–693.

Mahawanich, T., J. Lekhavichitr and K. Duangmal. 2010. Gel properties of red tilapia surimi: Effects of setting condition, fish freshness and frozen storage. International Journal of Food Science & Technology. 45(9): 1777–1786.

Martin-Sanchez, A.M., C. VNavarro, J.A. Perez-Alvarez and V. Kuri. 2009. Alternatives for efficient and sustainable production of surimi: A review. Comprehensive Reviews in Food Science and Food Safety. 8(4): 359–374.

Martínez-Camacho, A.P., M.O. Cortez-Rocha, J.M. Ezquerra-Brauer, A.Z. Graciano-Verdugo, F. Rodriguez-Félix, M.M. Castillo-Ortega et al. 2010. Chitosan composite films: Thermal, structural, mechanical and antifungal properties. Carbohydrate Polymers. 82(2): 305–315.

Martini, S., J.E. Thurgood, C. Brothersen, R. Ward and D.J. McMahon. 2009. Fortification of reduced-fat Cheddar cheese with n-3 fatty acids: Effect on off-flavor generation. J. Dairy Sci. 92(5): 1876–1884.

Merzendorfer, H. 2011. The cellular basis of chitin synthesis in fungi and insects: Common principles and differences. European Journal of Cell Biology. 90(9): 759–769.

Morganti, P. and G. Morganti. 2008. Chitin nanofibrils for advanced cosmeceuticals. Clinics in Dermatology. 26(4): 334–340.

Murphy, S.C., D. Gilroy, J.F. Kerry, D.J. Buckley and J.P. Kerry. 2004. Evaluation of surimi, fat and water content in a low/no added pork sausage formulation using response surface methodology. Meat Science. 66(3): 689–701.

Muzzarelli, R.A.A. 1997. Human enzymatic activities related to the therapeutic administration of chitin derivatives. Cellular and Molecular Life Sciences CMLS. 53(2): 131–140.

Nalinanon, S., S. Benjakul, W. Visessanguan and H. Kishimura. 2008. Tuna pepsin: Characteristics and its use for collagen extraction from the skin of threadfin bream (*Nemipterus* spp.). Journal of Food Science. 73(5): C413–419.

Otles, S. and O. Cagindi. 2008. Carotenoids as natural colorants. pp. 51–70 In: C. Socaciu (ed.). Food colorants chemical and functional properties. New York: CRC Press.

Picot, L., S. Bordenave, S. Didelot, I. Fruitier-Arnaudin, F. Sannier, G. Thorkelsson et al. 2006. Antiproliferative activity of fish protein hydrolysates on human breast cancer cell lines. Process Biochemistry. 41(5): 1217–1222.

Ranadheera, C.S. and J.K. Vidanarachchi. 2013. Potentials and applicability of marine derived nutraceutical in dairy induetry. Agro Food Industry hi-tech. 24(3): 22–26.

Rasika, D.M.D., C.S. Ranadheera and J.K. Vidanarachchi. 2013. Applications of marine derived peptides and proteins in the food industry. pp. 545–587. In: S.K. Kim (ed.). Marine Proteins and Peptides: Biological Activities and Applications. Oxford: Wiley-Blackwell.

Ravi Kumar, M.N.V. 2000. A review of chitin and chitosan applications. Reactive and Functional Polymers. 46(1): 1–27.

Rognlien, M., S.E. Duncan, S.F. O'Keefe and W.N. Eigel. 2012. Consumer perception and sensory effect of oxidation in savory-flavored yogurt enriched with n-3 lipids. Journal of Dairy Science. 95(4): 1690–1698.

Rossano, R., M. Larocca, A. Lamaina, S. Viggiani and P. Riccio. 2011. The hepatopancreas enzymes of the crustaceans *Munida* and their potential application in cheese biotechnology. LWT—Food Science and Technology. 44(1): 173–180.

Rossano, R., P. Piraino, A. D'Ambrosio, O.F. O'Connell, N. Ungaro, P.L.H. McSweeney et al. 2005. Proteolysis in miniature cheddar-type cheeses manufactured using extracts from the crustacean *Munida* as coagulant. Journal of Biotechnology. 120(2): 220–227.

Sachindra, N.M., N. Bhaskar and N.S. Mahendrakar. 2005. Carotenoids in crabs from marine and fresh waters of India. LWT—Food Science and Technology. 38(3): 221–225.

Samaranayaka, A.G.P. and E.C.Y. Li-Chan. 2008. Autolysis-assisted production of fish protein hydrolysates with antioxidant properties from Pacific hake (*Merluccius productus*). Food Chemistry. 107(2): 768–776.

Sanceda, N.G., M.F. Sanceda, V.S. Encanto, T. Kurata and N. Arakawa. 1994. Sensory evaluation of fish sauces. Food Quality and Preference. 5(3): 179–184.

Senevirathne, M. and S.-K. Kim. 2012. Chapter 32—Utilization of seafood processing by-products: Medicinal applications. *In*: K. Se-Kwon (ed.). Advances in Food and Nutrition Research: Academic Press. 65: 495–512.

Seo, M.H., S.Y. Lee, Y.H. Chang and H.S. Kwak. 2009. Physicochemical, microbial, and sensory properties of yogurt supplemented with nanopowdered chitosan during storage. Journal of Dairy Science. 92(12): 5907–5916.

Serfert, Y., S. Drusch K. Schwarz. 2010. Sensory odour profiling and lipid oxidation status of fish oil and microencapsulated fish oil. Food Chemistry. 123(4): 968–975.

Shahidi, F., J.K.V. Arachchi and Y.-J. Jeon. 1999. Food applications of chitin and chitosans. Trends in Food Science & Technology. 10(2): 37–51.

Shahidi, F. and Y.V.A.J. Kamil. 2001. Enzymes from fish and aquatic invertebrates and their application in the food industry. Trends in Food Science & Technology. 12(12): 435–464.

Shahidi, F. and B.J.A. Metusalach. 1998. Carotenoid pigments in seafoods & aquaculture. Critical Review in Food Science and Nutrition. 38(1): 1–67.

Soultos, N., Z. Tzikas, A. Abrahim, D. Georgantelis and I. Ambrosiadis. 2008. Chitosan effects on quality properties of Greek style fresh pork sausages. Meat Science. 80(4): 1150–1156.

Spreen, K.A., J.P. Zikakis and P.R. Austin. 1984. The effect of chitinous materials on the intestinal microflora and the utilization of whey in monogastric animals. pp. 57–75. *In*: Z. John (ed.). Chitin, Chitosan, and Related Enzymes: Academic Press.

Tsiplakou, E. and G. Zervas. 2013. The effect of fish and soybean oil inclusion in goat diet on their milk and plasma fatty acid profile. Livestock Science. 155(2–3): 236–243.

Valencia, I., D. Ansorena and I. Astiasarán. 2006. Nutritional and sensory properties of dry fermented sausages enriched with *n*-3 PUFAs. Meat Science. 72(4): 727–733.

Vega-Villasante, F., H. Nolasco and R. Civera. 1995. The digestive enzymes of the Pacific brown shrimp *Penaeus californiensis*—II. Properties of protease activity in the whole digestive tract. Comparative Biochemistry and Physiology Part B: Biochemistry and Molecular Biology. 112(1): 123–129.

Venugopal, V., S.P. Chawla and P.M. Nair. 1996. Spray dried protein powder from threadfin beam: Preparation, properties and composition with FPC type B. Journal of Miuscle Foods. 7: 55–58.

Venugopal, V. and F. Shahidi. 1995. Value-added products from underutilized fish species. Critical Reviews in Food Science and Nutrition. 35(5): 431–453.

Venugopal, V. and F. Shahidi. 1998. Traditional methods to process underutilized fish species for human consumption. Food Reviews International. 14(1): 35–97.

Verardo, V., F. Ferioli, Y. Riciputi, G. Iafelice, E. Marconi and M.F. Caboni. 2009. Evaluation of lipid oxidation in spaghetti pasta enriched with long chain n–3 polyunsaturated fatty acids under different storage conditions. Food Chemistry. 114(2): 472–477.

Vidanarachchi, J.K., M.S. Kurukulasuriya and S.K. Kim. 2010. Chitin, chitosan and their oligosachcharides in food industry. pp. 543–560. *In*: S.K. Kim (ed.). Chitin, Chitosan, Oligosaccharides and Their Derivatives: Biological Activities and Applications. New York, USA: CRC Press.

Vidanarachchi, J.K., M.S. Kurukulasuriya, A. Malshani Samaraweera and K.F.S.T. Silva. 2012. Applications of marine nutraceuticals in dairy products. *In*: K. Se-Kwon (ed.). Advances in Food and Nutrition Research: Academic Press. 65: 457–478.

White, D.A., R. Ørnsrud and S.J. Davies. 2003. Determination of carotenoid and vitamin A concentrations in everted salmonid intestine following exposure to solutions of carotenoid *in vitro*. Comparative Biochemistry and Physiology Part A: Molecular & Integrative Physiology. 136(3): 683–692.

Xu, W., G. Yu, C. Xue, Y. Xue and Y. Ren. 2008. Biochemical changes associated with fast fermentation of squid processing by-products for low salt fish sauce. Food Chemistry. 107(4): 1597–1604.

Ye, A., J. Cui, A. Taneja, X. Zhu and H. Singh. 2009. Evaluation of processed cheese fortified with fish oil emulsion. Food Research International. 42(8): 1093–1098.

Zhong, W., S. Zhang, J. Li, W. Huang and A. Wang. 2011. Effects of dietary replacement of fish oil by conjugated linoleic acid on some meat quality traits of Pacific white shrimp *Litopenaeus vannamei*. Food Chemistry. 127(4): 1739–1743.

18

Mining Products from Shrimp Processing Waste and Their Biological Activities

Asep Awaludin Prihanto, * *Rahmi Nurdiani* and
Muhamad Firdaus

1 Introduction

Aquatic foods constitute a major portion of the protein supply in the world. Seafood (fish, crustaceans, mollusk, etc.) supplies have increased gradually at an average rate of 3.2 percent per year during the period 1961–2009 (FAO 2012). Interestingly, food fish supply growth outpaces population growth (at annual rates of 2.6% and 1.6%, respectively). Mass production of seafood by maximizing fish capture and aquaculture largely contributed to the escalation of fish food supply. Among other regions, Asia and the Pacific were taken into account as the world's largest producers of fish. Both regions supplied 94.2 million tones, including 47.9 million tons from capture fisheries and 46.3 million tons from aquaculture (Lymer et al. 2008). Indonesia is currently the fourth largest producer of capture fisheries with over 6.2 million people directly engaged in fishing and fish farming.

Shrimp has been considered as the 'prima donna' of Indonesian fisheries and aquaculture since the 1980's. Although shrimp aquaculture has encountered several diseases, the production of shrimp has increased

Department of Fishery Product Technology, Faculty of Fisheries and Marine Science, Brawijaya
University, Veteran St, Malang, Indonesia.
Emails: asep_awa@ub.ac.id; rahmi_nurdiani@ub.ac.id; muhamadfir@ub.ac.id
* Corresponding author

considerably. In 1983, the volume of shrimp production was only 27,600 tons (FAO 1990) whereas in 2008, Indonesia was able to produce 410,000 tons of shrimp, making the country the third biggest shrimp producer in the world (ANTARA 2010). Among other seafood products, shrimp is a commodity with relatively high economic value. From 2002 to 2011, Indonesian shrimp has contributed more than 17% of the total volume of exports of fisheries products (Ministry of Marine Affairs and Fisheries 2012). The volume and value of exports of Indonesian shrimp are presented in Fig. 1.

The increasing production of shrimp is associated with increasing processing activity that eventually amplifies post processing waste stream. Shrimp processing generates more than about 40% of waste in average of its total weight depending on the species (Sachindra et al. 2005b). The waste contains many valuable compounds that, after appropriate processing, can add substantially to overall profitability (Stevens et al. 1998). A better economic use of the shrimp processing waste would minimize the pollution problem and at the same time maximize the profits of the processor. The concept of maximum utilizing of fish waste is known as zero waste management concept.

Zero waste concepts bring us to reduce, to segregate and to process all waste. It could be done with reducing, recycling and proper waste management. Segregation of waste can be arranged by optimizing and improving the production process in order to maximize the output. Waste management could be done through minimizing hazards, making it safer for the environment. Meanwhile, recycling waste may produce new

Figure 1. Volume and value of Indonesian shrimp export from 2002 to 2011 (Ministry of Marine Affairs and Fisheries 2012).

Color image of this figure appears in the color plate section at the end of the book.

goods with economic value. Valorization of the waste requires knowledge of the characteristics of the waste, so that we are able to determine exactly, the follow-up effort from zero waste conception in accordance with its character.

It must be noted that some effort to optimize shrimp waste processing has been done; however, in improper ways. For example, the production of chitosan, using high concentration of acid and bases, will eventually lead to new environmental problems. Therefore, alternative treatment in accordance with the principles of zero waste system should be always taken into account. This chapter has been written in the aim of promoting shrimp waste valorization. The potential biological activities of shrimp processing waste which may promote good human health are also described here.

2 Nutritional Value and Characteristic of Shrimp Processing Waste

Shrimp consists of 25–30% head, 50–60% meat, and 10–15% shell (including tail). Most of the shrimp processing wastes are shell, head, and tail. About 35–45% by weight of shrimp raw material is discarded as waste. Balogun and Akegbejo-Samson (1992) observed similar yields of waste from four different species of shrimps *(Macrobrachium rosenbergii, Palaemon serratus, Panaeus notialis* and *Parapenaeopsis atlantica)*. It was stated that head, shell and the edible flesh constitute about 33%, 15% and 51%, respectively, of the total weight. Interestingly, shrimp sex did not significantly affect the yield. The yield of *Farfantepenaeus paulensis* is also similar where muscle, head, shell, tail are 60, 28, 8, 4%, respectively (Sanchez-Camargo et al. 2011a).

The amount of the shrimp waste or discards depends on the species, the size of the raw material and the products resulting from processing. Shrimp products are commonly marketed as head on, head off or peeled forms. The head on shrimp product refers to frozen shrimp in one piece, without beheading. This product is a high value commodity with high demand in the international market. Head off shrimp, on the other hand, are frozen after their heads are separated, yet not peeled. Peeled shrimp products are peeled frozen shrimp after being beheaded. Peeled shrimp products can be divided into three kinds: Peeled Undevined (PUD), i.e., peeled shell, halved and flesh intact without removing tails; Peeled and Devined (PND), i.e., peeled shell, halved lengthwise backs, intestines and tails removed; Tail On Peeled (PTO) is almost the same as the PND, but the tail is not removed.

Generally, industrial shrimp processing includes washing, grading, cutting, cooking, freezing and packaging processes (Fig. 2). Shrimps are usually supplied in frozen or chilled forms. Hence, the washing step is

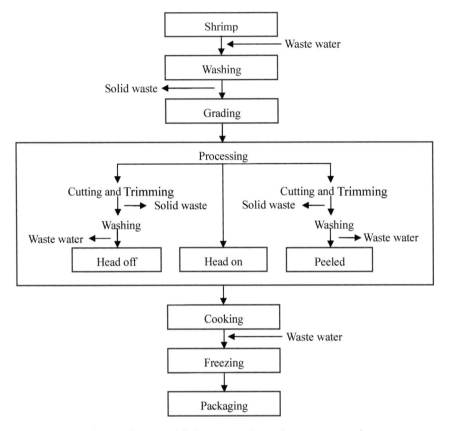

Figure 2. Lay out of shrimp processing and waste generated.

applied to clean the raw materials (RM) as well as to remove other unwanted materials. This step results in a lot of waste water. The next step is grading, which is aimed at classifying the RM based on size and freshness. The RM will then be processed in accordance with the desired end product (head on, head off or peeled). Peeling and cutting shrimp heads usually take place in this process. In terms of waste, the 'head on' product processing generates almost no solid waste. However, it still produces liquid waste. On the other hand, 'peeled shrimp' processing produces large amounts of both solid and liquid waste, especially in PND form. The processes of cooking and freezing produce only waste water.

Almost all shrimp waste-value added projects are focused only on solid waste. It is because solid shrimp waste is commonly known to contain high nutritional value. The nutrient values of the shrimp waste are associated with the nutritional value of the raw materials. There is evidence that the chemical composition of shrimp waste depends on several factors, for instance season and species (Rosa and Nunes 2003). They mentioned that in the winter, protein, fat and ash contents of red and pink shrimps tend to be lower than those in the summer. On the other hand, Karakoltsidis et al. (1995) found that the protein content of Mediterranean shrimp was at a lower level during fall in comparison to spring and winter. Contents of several shrimp waste by species are provided in Table 1. It is evident that the proximate composition of shrimp waste should be considered as one of the best indicators of its potential nutritive value.

In contrast to solid waste, there is still limited research conducted for liquid shrimp processing waste. Liquid waste contains less nutritional value. Moreover, waste water may contain harmful compounds. Shrimp processing liquid waste retains higher organic content than normal domestic waste water. This organic content is related to organic debris or dissolved material released during washing. Even though no toxicological substances were found in waste water, high amounts of organic material in solid and waste water may trigger a decomposition process and become water pollutant.

Organic-rich waste water can be determined by two main parameters, namely Chemical Oxygen Demand (COD) and Biochemical Oxygen Demand (BOD). Another characteristic is Total Suspended Solid (TSS). No reliable data is available for these parameters. The nutritional value of shrimp waste water varies. It depends on the processing efficiency, machinery application and processing plant. For example, in Louisiana,

Table 1. Proximate Values of Shrimp Processing Waste (Head and Shell).

Parameter	P.i	P. nb	M.v	P.s	P. nt
Crude protein	32.5	36.6	41.7	58.07	49.91
Carbohydrate	1.5	1.6	3.74	1.94	4.63
Lipid	9.8	10.3	1.45	2.71	5.40
Fibre	7	9.6	18.84	8.33	12.55
Ash	26.6	28.5	32.03	25.04	25.46
Mineral	**mg/100gr**	**mg/100gr**	**mg/hg**	**mg/hg**	**mg/hg**
Calcium	55*	57.9	1,470	9,460	9,780
Magnesium	30*	31.5	140	450	70
Zinc	ND	47.6	172	313	308
Nickel	ND	-	87	ND	130
Phosphorus	91.5	95.7	7,170	21,280	13,580
Iron	20*	22.2	610	1,890	910
Copper	4.8*	5.7	133	67	74
Cobalt	-	ND	10.9	16.5	58
Sodium	38.6	41.9	10,594	22, 222	15,963
Potassium	33.2	35.3	17,336	34, 444	11,346

Adapted from Ravichandran et al. (2009); Adeyeye et al. (2004); Adeyeye et al. (2008).
Note: *predicted value
P.i—*Penaeus indicus*, M.v—*Macrobrachium vollenhovenii*, P.s—*Palaemon* species A, P.nl—*Penaeus notialis*, P.nb—*Penaeus notibilis*.

USA, different shrimp processing plants generated different qualities of waste water. In term of BOD and COD, plant A produced waste water with 1877 mg/L and 3874 mg/L while plant B generated waste water with 1184 mg/L and 1536 mg/L, respectively (Mauldin and Szabo 1974). A study by Overcash and Pal (1980), showed different BOD and COD values produced by several seafood plants. They found 490 mg/L of BOD and 790 mg/L of COD in shrimp processing waste water, while from crab processing waste water, higher level of COD (1,100mg/L) and BOD (1,400mg/L) were observed. The main characteristic of seafood processing waste water is its high BOD levels (Islam et al. 2004). A high level of BOD is correlated to the potency of waste water as a pollutant which may cause water quality problems. Seafood–related companies should establish effective waste water treatment before releasing the waste to the environment. This effort may increase production costs, though there are several cheap ways which could minimize the cost (Cardoch et al. 2000).

3 Products from Shrimp Waste

3.1 Chitin

Chitin, polysaccharides contain N-acetylglucosamine (GlcNAc) and D-glucosamine which are linked by β 1→4 glycoside bonds. In nature, chitin bonded with protein, pigment, lipid and inorganic materials (e.g. calcium carbonate). A divergent sources of chitin reflected its broad distribution. It is known as the second largest polysaccharide on earth after cellulose. In aquatic ecosystems, chitin is recognized as the most abundant polymer (Aluwihare et al. 2005). About 10^{11} metric tons of chitin are produced annually from aquatic environments (Keyhan and Roseeman 1999). It highlights chitins' role as an importance polysaccharide in the marine food web (Souza et al. 2011).

This polymer can be found not only in the exoskeleton of crustaceans (crab, shrimp, lobster, krill, etc.) but also in insects (cockroache, cricket, mealworm, etc.), fungi (Basidiomycetes, Ascomycetes, and Phycomycetes), mushroom, yeast, and diatom. In the fungi kingdom, 2.6 to 26.2% of the dry weight of the fungal mycelium is chitin. Chitin is a primer material for crustacean shells (Kumar 1999). Hence, among other organisms in the aquatic environment, crustaceans especially shrimp have gained great attention.

The high interest in chitin production is associated with the increased spacious application of this product and its derivates. There are many types of chitin production techniques, yet they can be divided into two categories. The first is a chemical method. As the name suggests, this method uses chemical compounds such as acids and bases to produce chitin. The biological method is another method for chitin production that involves microorganisms in the process.

A combination of biological and chemical methods has now become more popular and can easily be found in recent academic papers. This method becomes a promising method. partly due to the increasing of environmental awareness. Chitin production by the chemical method will lead to new problems since the disposal of acids and bases is not considered an environmentally friendly process. On the other hand, applying only biological methods will produce chitin with poor quality (Beaney et al. 2005). Therefore, the combination of these methods is appealing.

The main steps of chitin manufacturing are demineralization and deproteinization (Fig. 3). During demineralization, high acidic substances such as HCl are commonly used while other acids (CH_3COOH, $HCOOH$, etc.) are also used. The types and strength of the solvent resulted in demineralization to different degrees. It is well known that the surface area of raw material is correlated with process efficiency. The raw material surface area should be wide enough to ensure that it makes contact with the solvent. Process efficiency is also related to the duration of extraction. The duration of extraction is affected by the strength of the bond of each component. Pretreatment should be taken into account if we want to reduce the amount of chemical compounds used as well as to minimize the production cost. Aye and Stevens (2004) proposed that physical pretreatments (drying, grinding and sieving) and chemical pretreatment (soaking in acid water) will double the process efficiency.

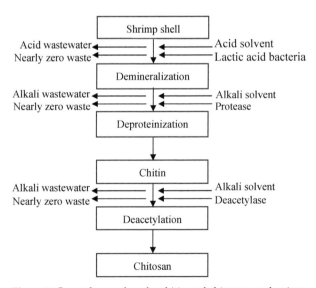

Figure 3. General procedure for chitin and chitosan production.

Nowadays, there is no specific biological method for demineralization that has been used. However, the fermentation of shrimp shell with lactic acid bacteria may trigger the demineralization stage (Duan et al. 2012; Zhang et al. 2012; Arbia et al. 2013).

Deproteinization is performed mostly by alkali solvent. NaOH is regularly used for this process, although the use of other solvents such KOH showed significant results. Complete deproteinization can be achieved at relatively high temperatures ranging from 60 to 100°C. High temperature will assist in the weakening of the bond between protein and chitin, even though it will cause undesirable deacetylation and depolymeration (Kjartansson et al. 2006).

Since the usage of chemical compounds could create environmental problems, an alternative deproteinization method using protease enzyme, which is more environmentally friendly, has been publicized (Valdes-Peña et al. 2010; Ali et al. 2011). Unfortunately, some improvement is needed as its effectiveness is less than the chemical method. Low deproteinization degrees have been noted and longer deproteinization time was needed. Hence, raising the fermentation temperatures would increase the degree of deproteinization.

Adour et al. (2008) suggested an exceptional idea to develop chitin production, by fermenting raw materials using protease-producing thermophilic lactic acid bacteria (LAB). We can perform demineralization and deproteinization in one step using LAB in the method to make it efficient. This concept is not only environmentally friendly but is also efficient. Chitin production by combining demineralization and deproteinization using lactic acid-producing bacteria in relatively high temperature fermentation will be thriving in the next decade.

The processing steps of chitin production can be done in any order or sequence. It will not negatively affect chitin quality. However, if we want to recover another product (astaxanthin or protein hydrolisate), the sequence comes into consideration. For instance, if we want to recover protein, the deproteinization step should be done before demineralization (Benhabiles et al. 2013).

3.2 Chitosan

Chitosan is a cationic polysaccharide, linear polyelectrolyte composed of glucosamine and N-acetyl-glucosamine units linked by β-(1→4) glycosidic bonds. Chitosan has a large molecular weight, is non-toxic, soluble in acid at room temperature, insoluble in organic solvents (such as methanol), capable of binding water, and capable of forming a coating (Alasalvar and Taylor 2002). Chitosan can be produced through deacetylation of chitin. Conversion of chitin to chitosan occurs through amides hydrolysis using bases (Fig. 4). Chitin, which acts as an amide, is hydrolyzed by NaOH as a base. It is initiated by the addition reaction which occurs when the -OH group is attached to $NHCOCH_3$. It causes CH_3COO-group elimination to produce chitosan amide.

The production of chitosan is generally carried out by a thermo chemical process using a strong alkali (typically 40–50%, w/w, NaOH) at high temperatures. The results have not been satisfactory because the quality of chitosan produced varies. In addition, alkali waste water with potential environmental hazards is produced (Tsigos et al. 2000). Immersing chitin in strong sodium hydroxide solution at room temperature before heating will assist optimum deacetylation process (Kurita 2001). Chitosan processing effectiveness is highly dependant on its degree of deacetylation (DD). DD is used to express the proportion of glucosamine monomer residues in chitin. High DD was most often achieved in high temperatures.

At the present time, it is difficult to find a fixed procedure to produce high quality chitosan. The physicochemical characteristics of raw materials and products as a result of each step will influence the optimum condition of later procedures. Hence, it is difficult to find an appropriate procedure to meet all standards for high chitosan quality. Benjakul and Wisitwuttikul (1994) confirmed that a 60 minute deacetylation time was too long to achieve an effective process. Performing deacetylation in longer periods of time will not significantly affect the deacetylation degree. They suggested that the optimum deacetylation process should be conducted for 45 minutes at 100°C using 50% NaOH. With regard to duration and temperature for the deacetylation process, high chitosan viscosity can be achieved after 60 min at 105°C (Bajaj et al. 2011). Meanwhile, Weska et al. (2007) mentioned that chitosan with molecular weight 150 kDa and DD 90% was obtained after 90 min deacetylation at 130°C.

Deacetylation is also conducted by biological methods using enzyme degrading acetyl. Chitin deacetylase (EC 3.5.1.41) is an enzyme that catalyzes the reaction of hydrolysis chitin, a β-1,4-linked N-acetylglucosamine polymer, resulting in glucosamine polymer or chitosan. This enzyme was first discovered in the fungus *Mucor rouxii* (Araki and Ito 1975). Since then, several different fungal and bacterial chitin deacetylases (CDA) have been identified, purified and characterized. Unfortunately, deacetylation using CDA indicated lower DD than alkali treatment. CDA from *M. rouxii* and

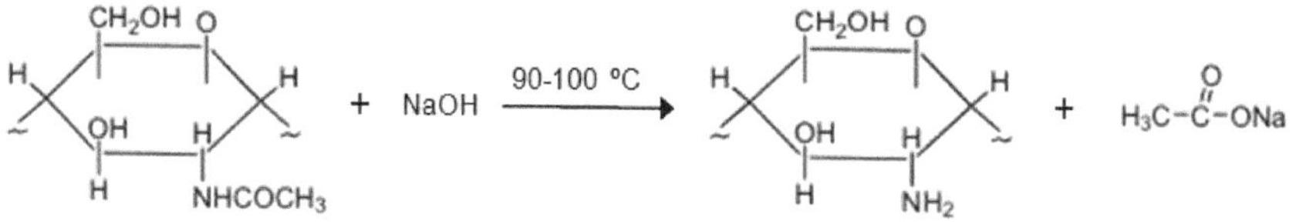

Figure 4. Conversion reaction of chitin to chitosan.

Mortierella sp. resulted in DD less than 50% (Martinou et al. 1995; Kim et al. 2008). The value is considered low compared to the DD of chitosan produced by chemical treatment which ranges from 56% to 99% with an average of 80% (No and Meyers 1995).

The color of chitin and chitosan is associated with the carotenoid pigment. In order to remove it, ab additional step called decolorization is required. Decolorization can be done using reagents or other treatments. No et al. (1989) prepared a white colored crawfish chitin by extraction with acetone, followed by bleaching with 0.315% (v/v) sodium hypochloride solution. Another treatment using ozone was proposed by Seo et al. (2007). They observed that ozonation for 5 min markedly increased the whiteness of chitosan; however, further ozonation resulted in the development of yellowness. This study showed that ozone can be used to remove pigments of chitosan without chemical use in a shorter time and with less cost. Liu et al. (2012), on the other hand, treated chitin with 1% potassium permanganate solution for 1 hr for the purpose of decolorization. Lightly brown chitin was produced.

3.3 Chitooligosaccharides

Unlike chitin or chitosan, chitooligosaccharides (COS), 2-mino-β-1,4-glucose polymer, are water soluble. This characteristic makes COS more reliable for application in pharmaceutical and medicinal purposes. COS have been reported to have abundant biological activities; for example, antimicrobial, antifungal, antioxidant, and immunostimulant effects (Kim and Rajapakse 2005). COS are obtained from chitin or chitosan through a depolymerization process. This depolymerization occurred due to the breaking of the β-glycosidic bond and resulted in polymers with short length and low molecular weight. Jeon and Kim (2000) presumed that the short chain lengths and free amino groups in D-glucosamine (GlcN) units of COS are the main reason for its solubility. The molecular weights of COS are about 10 kDa or less. Several factors such the degree of deacetylation of raw materials, temperature and reaction time affected the molecular weight of oligo-chitosan molecules (Srijanto et al. 2006).

Chemical and enzymatic processes have been used to produce chitosan oligomers. Chemical methods of COS production mainly use acid as a mediator. HCl and HNO_2 are the main chemical compounds for depolymerizing chitosan (Kikkawa et al. 1990; Tommeraas et al. 2001). Since the chemical process has several disadvantages, this method has gradually been abandoned. Aside from its environmental issues, the acid hydrolysis of chitosan usually produced various non-specific oligosaccharides (Bosso et al. 1986). Hence, the enzymatic process to produce COS has become more popular. The enzymatic process is also preferred because environmentally friendly. Chitosanase is the most ideal enzyme used in the enzymatic process because of its ability to recognize and hydrolysis beta-(1->4)-linkages between D-glucosamine residues. However this approach also has some drawbacks, such the high cost of chitosanase enzyme. Using cheap or bulk enzymes seems to be the reasonable way out.

Various enzymes, such as cellulases, pectinases, lipases and proteases as well as chitosanases, have been used to prepare COS (Kittur et al. 2003; Lee et al. 2008; Lin et al. 2009). The effectiveness of using bulk enzymes, nevertheless, still has flaws because it requires a high concentration compared to chitosanase. Hence several breakthroughs have been developed, including mixing several enzymes in COS production and using immobilized chitosanase to avoid its high cost (Zhang et al. 1999). The effective large-scale productions of oligosaccharides with immobilized enzyme combined with process improvement through a dual reactor system which consisted of an ultrafiltration membrane reactor and a column reactor packed with an immobilized enzyme was successfully done by Jeon and Kim (2000).

3.4 Astaxanthin

Astaxanthin is a carotenoid substance which is widely present in microalgae, yeast, *salmon*, trout, krill, crab and shrimp (Johnson and Lewis 1979; Shahidi and Synowiecki 1991; Torrissen et al. 1995; Perez-Galves et al. 2008; Okada et al. 2009). The astaxanthin molecule has two asymmetric benzenoid rings at the beginning and end of the molecule (Fig. 5). This compound is well-known because of its biological activities, especially its high antioxidant activity. Furthermore, recent studies revealed that antioxidant activity of astaxanthin was much higher than β-carotene, α tocopherol and lycopene (Kurashige et al. 1990; Naguib 2000).

Astaxanthin is a dominant pigment in shrimp. A study of four different shrimp species (*Penaeus vannamei*, *P. monodon*, *P. japonicus*, *Metapenaeus monoceros*) showed that the amount of astaxanthin present was more than the other presented carotenoids (Latscha 1989). Among all the body parts of shrimp, the carapace contains the highest known concentration of the carotenoid. The total carotenoid content in the carapace is almost six

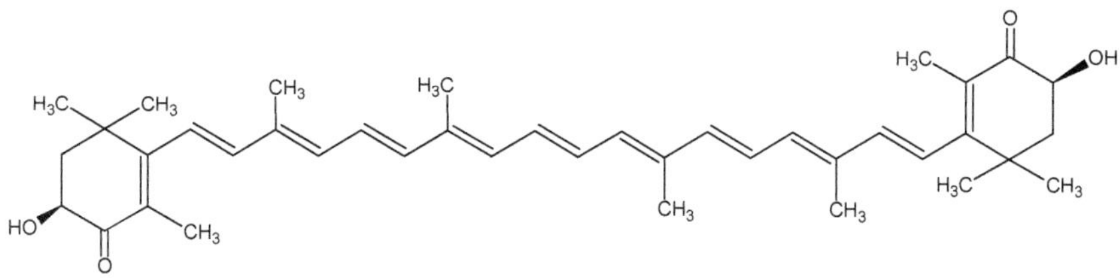

Figure 5. Chemical structure of astaxanthin.

times higher than that in meat and the head (Sachindra and Mahendrakar 2005a). Astaxanthin has an important role in the pigmentation process. It is also of importance in the shrimp pigmentation process while cooking. The visible red color when shrimp is cooked is due to the denaturation of carotenoprotein, which releases the individual astaxanthin prosthetic group. Other pigments existing in shrimp shell are phoenicoxanthin, dihydroxy-piradixanthin, 3,3'-dihydroxy- carotene, and traces of lutein, zeaxanthin, canthaxanthin, echinenone, and β-carotene (Teruhisa and Tadashi 1972).

Extraction of the astaxanthin can be carried out through different methods. The process starts with grinding the raw material (shrimp waste). Initial grinding will assist shrimp waste to easily come in contact with the solvent. Particle size has a significant effect on the yield of extracted pigment. Chen and Meyers (1982) suggested that 14 mm or a smaller particle size is the best range to obtain high yields. The deproteinization can be done by either chemical or biological methods. Demineralization is not the main step. However, several methods have been introduced in order to reduce minerals, such as by involving lactic acid bacteria in protein hydrolysis (Armenta-Lopez et al. 2002; Pacheco et al. 2009).

Since it is known that carotenoids are soluble in water and fat, numerous researches reported chemical methods for astaxanthin extraction which have been successful applied. Plant-derived oil, organic solvents and super critical CO_2 with co solvents such ethanol are commonly used in extracting astaxanthin by chemical methods. Sachindra and Mahendrakar (2005a) showed that carotenoids from shrimp waste (*Penaeus indicus*) could be extracted using several plant-derived oils. Sunflower oil produced a high yield (26,3 µg/gr) compared to other plant-derived oils (groundnut oil, gingelly oil, mustard oil, soya oil, coconut oil, rice bran oil). However, flaxseed oil is the most effective oil to extract astaxanthin from shrimp (*Litopenaeus setiferus*) waste with a maximum yield of 4.83 mg/100gr (Pu et al. 2010). Other plant-derived oils have also been used for extraction of pigments from crustacean wastes, including palm oil (Handayani et al. 2008), and soybean oil (60 mg/100g oil) (Chen and Meyers 1982).

Organic solvents such as acetone, ethanol, and hexane have been applied to extract carotenoids. Among other solvent extraction methods, the ethanol extraction method delivered better results with relatively high yields, even though some researchers were still not satisfied with the results. Hence, they tried to explore this method with new approaches with the same solvent. As a result, they found a method to increase the yield of astaxanthin extraction by incorporating two solvents (Sachindra et al. 2006). The combination of isopropyl alcohol (IPA) and hexane (50:50) provided a high yield of astaxanthin. It is about four times higher than the yield of hexane itself. Organic solvents were more effective in extracting carotenoid pigments than vegetable oil (Guillou et al. 1995).

Yamaguchi et al. (1986) succeeded in separating oil containing astaxanthin and its esters from krill by supercritical CO_2 extraction. Integrating supercritical CO_2 extraction with co solvent can enhance the extraction efficiency. Extracting astaxanthin in crawfish (*Procambarus clarkii*) by supercritical CO_2 method with co solvent ethanol was introduced by Charest et al. (2011). The method has significantly increased the yield. It was noted that temperature and moisture also affected the yield. On the other hand, temperature itself will have no effect. Interestingly, in this method, the size of raw material failed to have an impact on extraction effectiveness.

Carotenoid extractions using plant-derived oil or organic solvents have been proved a successful way to obtain pigments with a good quality (low chitin and ash contents). However, the products are not stable and easily react with oxidants due to the lack of protein in astaxanthin. It is understandable because protein-pigment interaction increases carotenoid stability (Spinelli et al. 1974). Incorporating enzyme from *Bacillus licheniforrnis* in carotenoid extraction successfully recovered 94.2% astaxanthin in total, with 91.9% astaxanthin presented in the top layer (Lee et al. 1999). This recovery was better than that using bovine trypsin which only recovered 80% (Simpson and Haard 1985).

Fascinatingly, astaxanthin can be isolated not only from shrimp solid waste, but also from waste water. Stepnowski et al. (2004) reported that astaxanthin can be recovered from seafood-waste water. The only weak point of optimizing waste water is the additional treatment or tools for collecting the material from waste water. Nevertheless, considering the fact that waste water was almost missed by the researchers, waste water optimization still offers huge opportunities.

3.5 Protein and amino acid

Protein derived from shrimp waste gained significant attention due to the high content of amino acids. It was suggested that protein from shrimp waste can be used to replacing protein from fish. With appropriate treatment, the quantity of protein will increase to about five times higher than that from unhydrolyzed shrimp waste (Brasileiro et al. 2012). Shrimp head silage and hydrolysate with high protein content can be used to replace fish flour in tilapia feed. This replacement resulted in non significant difference in tilapia growth (Cavalheiro et al. 2007; Leal et al. 2010). Shrimp meal also can be used as a part of protein source in broiler diets (Khempaka et al. 2011). Animals are the main object for shrimp waste product application. However, considering its high and abundant amino acid content, protein hydrolysate derived from shrimp waste has a bright prospect as a protein source not only for feed but also for food products or even as human protein supplement.

Producing protein hydrolysate from shrimp waste can be conducted through various processes; for instance, autolytic hydrolysis, acidic hydrolysis, alkaline hydrolysis, and enzymatic hydrolysis (Kristinsson and Rasco 2000). Autolytic hydrolysis incriminated complex endogenous enzymes in the raw material itself. Shrimp heads hydrolyzed using endogenous enzymes showed strong autolysis capacity (AC) for releasing threonine, serine, valine, isoleucine, tyrosine, histidine and tryptophan. However, they have a relatively low AC for releasing cystine and glycine (Cao et al. 2008). Nevertheless, because of its endogenous enzyme complexity, the result of this process is difficult to control.

Shrimp waste hydrolysis assisted by hydrochloric acid, sulfuric acid, sodium hydroxide is an example of acid or alkaline hydrolysis. Hydrolyzing shrimp waste using acetic acid produced a good quality protein which was proved by its capability in replacing fish flour as an ingredient in tilapia feed (Cavalheiro et al. 2007). Alkaline hydrolysis is a less preferable approach to generate protein hydrolysate due to unwanted side effects: significant amounts of amino acids (cysteine, serine and threonine) were reduced due to racemization of L-amino acid to D-amino acid and betha elimination (Kristinsson and Rasco 2000).

Among all others, enzymatic hydrolysis seems to be the most promising hydrolysis method. Low toxicity and eco-friendly compounds are the highlights of this method. Exogenous enzymes are added to assist the hydrolysis process. Several protease enzymes such alcalase, trypsin or protamex are widely used (Dey and Dora 2012). Some researchers preferred protamex since it is well known to be able to produce less bitter hydrolysates. Randriamahatody et al. (2011) noted that different enzymes produced different nutritional compounds. In addition, different species also generated different amino acids. Amino acid profiles derived from shrimp waste are shown in Table 2.

The presence of comprehensive amino acids in shrimp waste hydrolysate will broaden the application of this product. Both essential and non-essential amino acids are found in shrimp waste hydrolysate. Glutamic acid has been confirmed as the most dominant amino acid in the shrimp waste (Ibrahim et al. 1999; Ruttanapornvareesakul et al. 2006).

3.6 Enzymes

Only few enzymes can be found in shrimp waste. The enzymes such as phenoloxidase, alkaline phosphatase, Lipoxigenase (LOX), β-N acetyl-glucosaminidase, chitinases, cathepsin, hyaluronidase and several proteases can be found in heads of shrimp and thawing water (Simpson et al. 1987;

Table 2. Amino acid profiles derived from shrimp waste (gr/100 gr dry matter).

Amino acid	M.a	P.s	P.b	T.c	P.e	M.e	P.m
Arginine	5.4	5.2	5.74	7.02	4.35	6.27	5.17
Histidine	1.1	1.1	1.84	2.69	2.82	2.74	2.62
Isoleucine	1.8	1.9	2.77	3.44	2.45	2.82	2.80
Leucine	3.0	3.0	4.58	5.73	4.90	5.64	5.67
Lysine	3.1	2.8	5.74	7.23	5.07	5.07	6.02
Methionine	1.0	1.0	1.25	0.87	2.42	2.10	2.122
Phenylalanine	2.4	2.4	15.85	20.59	3.69	4.67	4.50
Threonine	2.2	2.3	3.25	4.88	3.43	3.54	3.71
Tyrosine	1.8	1.8	2.05	1.95	3.54	2.79	3.79
Valine	2.5	2.6	4.63	6.00	3.15	3.54	3.46
Alanine	3.8	3.8	4.03	5.33	5.13	5.76	6.42
Aspartic acid	4.6	4.8	6.64	8.82	6.36	7.99	8.50
Glutamic acid	5.9	6.0	9.20	11.43	9.77	11.45	11.81
Glycine	3.6	3.4	4.73	6.08	5.54	6.69	6.81
Proline	3.7	3.2	2.92	4.29	4.42	4.18	5.83
Serine	2.2	2.1	3.36	4.62	3.62	3.68	4.05

Adopted from Ngoan et al. (2000); Heu et al. (2003); Ruttanapornvareesakul et al. (2006).
Notes: M.a—*Metapenaeus afinis*, P.s—*Penaeus semisulcatus*, P.b—*Pandalus borealis*, T.c—*Trachypena curvirostris*, P.e—*Pandalus eous*, M.e—*Metapenaeus endeavouri*, P.m—*Penaeus monodon*.

Kuo and Pan 1992; Le Boulay et al. 1996; Savagon and Sreenivasan 1978; García-Carreño et al. 1997; Benjakul et al. 2005). Among other enzymes, alkaline phosphatase (AP) and phenoloxidase (PO) are probably the only two extensively explored enzymes that have been successfully isolated from shrimp waste. They are known to be widely distributed in nature. The hepatopancreas (head region) is a good source of these enzymes (Chander and Thomas 1999; Benjakul et al. 2005).

Alkaline phosphatase (EC 3.1.3.1) is an enzyme that hydrolyzes phosphomonoesters from several organic molecules. This enzyme could be separated from shrimps' hepatopancreas by fractional technique using several alkaline proteases (Olsen et al. 1991). Alkaline phosphatase is considered a high value enzyme because of its capability to remove phosphate groups from enormous biomolecules, such as DNA and RNA. This enzyme plays an important role in genetic engineering technology, especially in DNA manipulation. The important use of alkaline phosphatase is removing 5' phosphate from vectors (plasmid and bacteriophage) to prevent self ligation due to the tendency of restricted DNA to adhere after being cut by enzyme restriction. Another function of alkaline phosphatase is enhancing the efficacy of DNA labeling by radioactive phosphate.

The shrimp alkaline phosphatase (SAP) becomes more noteworthy due to its unique properties compared to other alkaline phosphatase (calf alkaline phosphatase, Bacillus alkaline phosphatase, etc.). Several shrimp species, including *Penaeus monodon* and Jawala Shrimp (*Acetes indicus*) were reported to be good candidates for alkaline phosphatase (Lee and Chuang 1991; Chander and Thomas 1999). Jawala Shrimp (*Acetes indicus*), an underutilized shrimp, is a potential source of this enzyme. The stability of the enzyme can be maintained at 40°C. Above this temperature, not only the activity but also the stability of enzyme will dramatically decrease. Alkaline phosphatase has optimum activity at pH 9.5. 1,10 phenanthroline or EDTA were strong inhibitor for this enzyme. The inhibition was restored by the addition of Zn^{2+}, Mn^{2+} or Mg^{2+} (Chander and Thomas 1999).

Phenoloxidase (PO) (*EC: 1.14.18.1*) is a redox enzyme which is capable of oxidizing phenols to quinones, which leads to the spontaneous formation of melanin as an end product. For crustaceans, PO plays an important role as natural defense against microbial infections (Liu et al. 2007). This enzyme was also suggested for its involvement in blackening of shrimp. PO spreads in almost all part of the shrimp bodies. However, several reports showed that different species will differently distribute PO. Nakagawa and Nagayama (1981), revealed that monophenol oxidase activity was found in the gills, while diphenol oxidase activity was found both in the hemolymph and gill. For *Cambarus clarkii* and *Penaeus japonicas*, PO can only be found in hemolymph. Meanwhile, the carapace and abdomen-exoskeleton of *Parapenaeus longirostris* are rich sources of PO (Zamorano et al. 2008). The content of PO in the carapace was about six times higher than that found in the abdomen-exoskeleton. Similar results were also shown by the Montero et al. (2001) study. They found that a carapace is the main source of polyphenoloxidase in *Penaeus japonicus*.

Solid waste is a potential source of PO in contrast to waste water from shrimp processing. However, since the waste water released from shrimp processing contains shrimp blood, it can be assumed that phenoloxidase might present because PO can be isolated from blood (haemolymph, hemocytes) (Aspan and Soderhall 1991; Gollas-Galvan et al. 1991).

3.7 Lipids

Fatty acid from shrimp waste can be isolated by several methods, for instance solvent extraction, super-critical fluid extraction and super-critical fluid extraction with a co solvent. Supercritical CO_2 has succeeded in extracting lipid from shrimp waste. The effects of the extraction conditions (pressure and temperature) were observed by Sánchez-Camargo et al. (2011a). They found that the pressure and temperature had no significant effect on the yield of lipid extracted with supercritical CO_2. The traditional soxhlet methods with hexane solvent were compared. It was suggested that supercritical CO_2 generated higher efficiency than hexane extraction. The efficiency of lipid extraction could be increased by performing autolysis approach. The autolysis would break down the bond between lipid and protein, resulting in liberation of lipid and protein. Hence, the lipid extraction would be more efficient (Senphan and Benjakul 2012).

Fatty acids derived from shrimp waste were dominated by unsaturated fatty acids (Sánchez-Camargo et al. 2011b). Fatty acid composition of Red spotted shrimp (*P. paulensis*) lipids waste was determined (Table 3). Twenty three fatty acids were uncovered with total percentages of saturated and polyunsaturated fatty acids of 32.31% and 61.36%, respectively. Hence,

Table 3. Fatty acid profiles of the extracted shrimp waste (g/100 g) (Sanchez-Camargo et al. 2011b).

Fatty acid		*P. paulensis*
Saturated		
C12:0	Lauric	0.16
C14:0	Myristic	1.96
C15:0	Pentadecanoic	1.19
C16:0	Palmitic	16.17
C17:0	Margaric	2.23
C18:0	Stearic	9.4
C20:0	Arachidic	0.7
C22:0	Behenic	-
C24:0	Lignoceric	0.5
Total		32.31
Unsaturated		
C16:1	Palmitoleic	5.2
C17:1	Margaroleic	0.98
C18:1	trans-Elaidic	0.49
C18:1	Oleic	16.85
C18:2	trans-Linoelaidic	0.19
C18:2	ω-6 Linoleic	2
C18:3	trans-Linolenic	0.56
C18:3	ω-3 Linolenic	0.43
C18:4	ω-3 Stearidonic	0.16
C20:1	Gadoleic	3.25
C20:4	ω-6 Arachidonic	5.81
C20:5	ω-3 Eicosapentaenoic–EPA	11.69
C22:5	ω-3 Decosapentaenoic	1.51
C22:6	ω-3 Docosahexaenoic-DHA	12.24
Total		61.36

the total amount of fatty acids found in shrimp (*P. paulensis*) waste was 93.67%. Palmitic and stearic acids were two fatty acids that were highly represented in the saturated form. These fatty acids accounted for 16.17% and 9.40%, respectively. Among unsaturated fatty acids (UFAs), oleic acid, EPA and DHA were the major fatty acids with amounts of 16.85%, 11.69% and 12.24%, respectively. Even though there is proof that the ω-6 family can create inflammatory and carcinogenic eicosanoids, the high amounts of UFAs indicated their potencies for promoting good human health and animal feed enrichment.

4. Biological Activities of Shrimp Waste Derived Products

4.1 Antimicrobial

Chitin, chitosan and their derivates are widely known as shrimp waste derived products with antimicrobial activities. Chitosan has higher antimicrobial activity than chitin (Tareq et al. 2013). The antimicrobial activity of chitin and chitosan is the consequence of complex interaction among various factors. For example, the antimicrobial properties of chitosan depend on its molecular weight, particle size, degree of deacetylation (DD) during processing, and the type of bacteria used in the antimicrobial assay. The higher degree of deacetylation and higher concentration of chitosan cause increased antibacterial activities (Chen et al. 2002). Chitosan performed increasing antimicrobial activity along with increasing DD against bacteria and fungi (Tipparat and Riyaphan 2008; Xiao-Fang et al. 2008).

Chitosan generally showed stronger bactericidal effects for gram positive bacteria (GPB) than for gram-negative bacteria (GNB). Lower molecular weight chitosan is more effective against GNB whereas high molecular weight chitosan is effective against GPB (Omura et al. 2002; Kim et al. 2007). This would suggest that the antibacterial mode of action is dependent upon the host microorganism. The size of the chitosan also showed significant effect on bacterial inhibition activity. The smaller size resulted in a better inhibition on both GNB and GPB (Katas et al. 2011).

There are three mechanisms describing the actions of chitin, chitosan and their derivate as antimicrobial:

1) Chitosan disrupts bacterial cell membranes and the permeability of outer membranes through interaction between positively charged chitin/chitosan molecules and negatively charged microbial cell membranes. Xiao-Fang et al. (2010) proposed that electrostatic interaction between $-NH_3^+$ groups of chitosan and carbonyl or phosphoryl groups of phospholipid components of cell membranes is the main cause of cell membrane damage in gram negative bacteria.

2) mRNA and protein synthesis inhibition through penetration of chitosan into the cells of microorganisms which will adhere to DNA (Sudarshan et al. 1992).

3) Interfering essential nutrients influx by creating an external barrier to chelating metals (Helander et al. 2001).

4.2 Antiviral

Chitosan is known to have an anti-viral infection for plant and animal cells. There was a similar tendency between antiviral, antibacterial and anticancer activities of chitosan. Higher antiviral activities are indicated by reducing chitosan molecular mass. The study conducted by Davydova et al. (2011) on tobacco mosaic virus (TMV) revealed that, reducing chitosan molecular weight from 500–130 to 15–17 kDa will significantly enhance its antiviral activity. He also found that chemically processed chitosan had higher activity levels than enzymatically processed chitosan. In another study, 2.2 and 1.2 kDa chitosan performed higher antiviral activity than 30 kDa chitosan against bean mild mosaic virus (Kulikov et al. 2006). Easy penetration of small chitosan molecules through plants' integuments is a possible reason of this phenomenon.

Inhibition of bacteriophage replication was reported by Kochkina and Chirkov (2000). Interestingly, they found that *Bacillus thuringiensis*, the host of the phage, was not inhibited by chitosan derivatives. The mechanism for this inhibition was almost certainly because of (1) reduction of the viability of cultured bacterial cells, (2) the infectivity neutralization of mature phage particles and (c) blockage of the replication of the virulent phage. Nevertheless, the exact mechanism of phage suppressions remains unknown.

Several papers on anti-human immunodeficiency virus (HIV) activity of chitin derived products were published in the past decades. Chitosan can interact with the envelope glycoprotein of HIV called gp 120. The interaction of glycosaminoglycans in chitin derivates (sulfated chitin) with gp 120 of the AIDS virus depends on the sites of sulfation (Nishimura 1998). This chitin derivate showed other benefits as a low toxic compound. N-carboxymethylchitosan-N-O-sulfate, another chitin derivate, inhibits HIV-human CD4 receptor binding cells. It also inactivates HIV-1 reverse transcriptase through a competitive inhibition (Sosa et al. 1991). N-carboxymethylchitosan-N-O-sulfate can also inhibit a human virus called Hepatitis B virus (HBV). Chitosan oligosaccharides incorporate with 10–23 DNAzyme to form complex nanoparticles, a promising compound in the application in anti-HBV gene therapy (Miao et al. 2012).

4.3 Antioxidant

Molecules or compounds with the ability to eliminate free radicals or disrupt the oxidation reactions in a system by reacting with them to yield harmless products are called antioxidants. Numerous studies have shown the antioxidant activities of shrimp waste products. Chitosan, chitosan derived products and astaxanthins are well known for their antioxidant activity. Antioxidative properties of various chitin and chitosan extracts are gaining great interest in the food industry.

In the food industry, ferrous ion is considered the most pro-oxidant in food. Hence, the research to find a compound that has ferrous ion-chelating properties is emerging in the past decades. The ability of chitosan as as antioxidant mainly comes from the presence of amino groups, which contain lone pairs that help to form the chitosan Fe^{2+} complex (Guzman et al. 2003).

Low molecular weight chitosan showed better antioxidant activity than high molecular weight chitosan. Abd El-Rehim et al. (2012) indicated that the addition of –COOH, –CONH-, and $–SO_3H$ functional groups into chitosan structure will improve not only its water solubility but also its antioxidant activity. In addition, they also proved that, the lower the molecular weight of chitosan, the higher the antioxidant activity. This result corroborated the findings of Xing et al. (2005), that chitosan with 9 kDA molecular weight has antioxidant activity about three times higher than that of 760 kDA chitosan.

Astaxanthin was determined to have the highest antioxidant activity in comparison to other common antioxidants. Naguib (2000), Goto et al. (2001) and Camera et al. (2009) measured the antioxidant activity of astaxanthin compared to other well-known antioxidants. They found that astaxanthin possessed a higher antioxidant activity than lutein, licopene, α and β-carotene, α-tocopherol and canthaxanthin. Carotenoids structures affected their reaction with free radicals, while their reactivity depends on the terminal rings (Goto et al. 2001). However Mortensen et al. (1997) stated that the mechanism and rate of free radical scavenging was dependent on the nature of the free radicals. The phenolic antioxidant in shrimp shell could play a role in providing hydrogen or electrons to a free radical (Seymour et al. 1996). High antioxidant activity of astaxanthin from shrimp waste is due to the synergistic effect of astaxanthin and PUFAs which are present in the astaxanthin monoester and diester fraction (Sindhu and Sherief 2011).

4.4 Anti inflammation

Every cell damage and injury will trigger the body's self defence called inflammation. Localizing and repairing injured cell or tissue is the main purposes of this process. The inflammation process is important and brings many advantages in order to initiate the healing process for the recovery of damaged tissue; however, the side effects of this process caused a severe response in patients with vasodilatation of blood vessels resulting in redness or swelling.

Pain and increase in body temperature are the main symptoms of inflammation (Spector and Willoughby 1963). Anti-inflammation agents have two principal mechanisms to reduce the effects of inflammation. The first is by preventing the generation of mediators for inflammation reactions, such as amine groups (histamine), peptides (brady-kinin and interleukin-1) and lipids groups (prostaglandins, leukotrienes). The second is through their antagonist activity for its pharmacological effect (Vane and Botting 1987).

Chitosan was known to have an anti-inflammatory activity. Kim et al. (2002) presented that the water soluble form of chitosan succeeded in suppressing the secretion and expression of pro-inflammatory cytokines, TNF-α and IL-6 and nitric-oxide synthase (NOS), on human astrocytoma cells. Chou et al. (2003) explained other mechanisms of chitosan as anti-inflammatory. Chitosan inhibited and reduced prostaglandin E2 (PGE2) overproducing, cyclooxygenase-2 (COX-2) protein expression and the pro-inflammatory cytokines. Chitosan derivates, chitosan oligosaccharides also reduced inflammatory reactions by inhibiting PGE2 through blocking of COX-2 protein expression. Two different molecular weights of chitosan oligosaccharides (10 kDa < MW < 20 kDa and 1 kDa < MW < 3 kDa) performed no different results as anti-inflammatory agents (Yang et al. 2010).

Three chemically modified chitins successfully reduced inflammatory disease in chitosan-induced pneumonia. They are phosphated chitin (P-chitin), phosphated–sulfated chitin (PS-chitin), and sulfated chitin (S-chitin). All of them presented anti-inflammatory effects; however, the order of their effectiveness is P-chitin, PS-chitin and S-chitin (Miyatake et al. 2003)

Astaxanthin from shrimp shell also showed its suppression on nitric oxide, PGE2, interleukin-1β (IL-1β), NF-κB activation and TNF-α production (Ohgami et al. 2003; Lee et al. 2003b). Other researches with different cases such gastric inflammation and oxidative stress-induced inflammatory diseases such as diabetic cardiac muscle damage in mice also confirmed the anti-inflammation activity of astaxanthin (Bennedsen et al. 1999; Aoi et al. 2003; Naito et al. 2004).

4.5 Anti-cholesterol

Several researches revealed that plasma and liver cholesterol concentrations decreased when animals or human underwent chitosan diet (LeHoux and

Grondin 1993; Bokura and Kobayashi 2003; Monek and Bartonova 2005; Jaffer and Sampalis 2007). The hypocholesterolemic action of orally administered chitosan was first reported by Sugano et al. (1978), and has been confirmed later by several investigators with an increase of the excretion of colic acid into faeces in rats and in humans (Kobayashi et al. 1979; Maezaki et al. 1993). Rats fed with chitosan, wheat bran and combinations of both experienced not only hyperlipidemia, through decrement in lipid parameters in serum and liver, but also an enhancement of antioxidant enzymes (Lamiaa and Barakat 2011). This finding suggested that the dietary effect of chitosan had the most pronounced effect on the animal's nutritional status. Hence, it has the potency as a functional food for reducing cardiovascular disease risk factors by lowering cholesterol and preventing atherosclerosis.

The mechanism of chitosan as a hypolipidemic agent can be explained as follows. Chitosan triggers upregulation of LDL mRNA receptor expression in the liver (Xu et al. 2007). This reaction will repair liver function, which results in increasing bile salt and bile acid excretions. A study by Fukada et al. (1991) showed that chitosan affected the metabolism of intestinal bile acids. The enhancement of bile acid excretion will significantly reduce cholesterol concentrations in plasma due to the increase of 7α-hydroxylase enzyme (Li et al. 2004).

There is lack of evidence that astaxanthin could lower high cholesterol. Astaxanthin increased HDL while decreasing both triglycerides and non-esterified fatty acids in the blood. Moreover, astaxanthin also inhibits LDL oxidation (Iwamoto et al. 2000). This inhibition mechanism is considered as the secondary effect of astaxanthin since it is well known as an antioxidant.

4.6 Antidiabetes

Two types of diabetes mellitus (DM) have been classified, namely type 1 (insulin-dependent) and type 2 non-insulin-dependent). Unlike type 1, which is mainly due to genetic disorder, DM Type 2 is caused by insulin resistance where the human life style behavior is the main triggering factor (Lin et al. 2012). Hence, DM type 2 is more common than DM type 1 and is projected to increase dramatically in the next decades (Wild et al. 2004; Shaw et al. 2010). It is known that obesity causes insulin resistance which leads to hyperinsulinemia. Therefore, improving both lipid and glucose metabolism will also reduce the obesity-related DM type 2.

Oral administration of low molecular chitosan has a significant impact on reducing DM indicators. Hayashi and Ito (2002) found that KK-Ay mice, genetically obese with type 2 DM, when treated with low molecular chitosan, exhibited hyperglycemia, hyperinsulinemia and hypertriglyceridemia improvement. Chitosan increased the activities of hepatic hexokinase and the intestinal disaccharidases including sucrose and maltase (Yao et al. 2008). The amount of chitosan diet has to be enough to achieve a positive effect on DM because under 0.2% chitosan daily intake in drinking water will not improve obese-type diabetes of KK-Ay mice. It seemed that the anti-diabetic action of low molecular chitosan is not due to D-glucosamine, but may be related to its derivates, oligosaccharides, which are formed during administration until absorbed in the intestine (Hayashi and Ito 2002).

Reducing hyperglycemia and hyperlipidemia using chitooligosaccharides is reported by Katiyar (2011). Reduction in serum lipids (triglycerides, total cholesterol, low density lipoprotein and very low density lipoprotein cholesterol), urea, creatinine and various enzymes of glucose metabolism (serum glutamate oxaloacetate transaminase and glutamate pyruvate transaminase and alkaline phosphatases) was found in studies on diabetic mice. The exact mechanism of COS as an antidiabetic is not so clear. However, a presumed mechanism of the antidiabetic action of COS is related to the induction of glucose-inducible insulin secretion (Lee et al. 2003a). COS also improves metabolic disorders of obesity through inducing adiponectin. Therefore, Kumar et al. (2009) indicated that COS was able to serve as controlling diet intake, body weight gain, blood glucose and lipid profiles of insulin resistant, genetically modified mice.

Astaxanthin from shrimp waste has a hypoglycemic effect. Higher doses of astaxanthin resulted in a higher hypoglycemic effect. 10 mg/kg of astaxanthin taken by oral administration has suppressed hypoglycemia on alloxan-induced diabetic mice (Wang et al. 2012). Astaxanthin increases glucose uptake as well as modulates the circulation of lipid metabolites and adiponectin (Hussein et al. 2007). Astaxanthin prevented the diabetic activity through the ROS scavenging effect in the mitochondria of mesangial cells (Manabe et al. 2008). Hence, the antidiabetic activity of astaxanthin correlated with its antioxidant activity.

4.7 Anticancer

One of the leading causes of death in the world is cancer. Surgery and chemotherapeutic agents are the main treatments of this disease. Since surgery inflicted post trauma and chemotherapeutic agents have severe side effects, use of a natural compound which is safer and as effective is the way out of this problem. In recent decades, research on natural products with respect to natural therapeutic agents for cancer has become the trend. Numerous papers reported new anticancer compounds isolated from marine resources, as very promising (Jimeno et al. 2004).

The group of shrimp waste derived products with anticancer activity includes chitin, chitosan, chitooligosaccharides, pigments and peptides (Qin et al. 2002a). Chitosan acts as an anticancer agent by enhancing the immune system through increasing natural killer (NK) cell activity and other mechanisms (Maeda and Kimura 2004). Water soluble chitosan with 21 kDa molecular weight showed its inhibiting activity against sarcoma. Its derivate, a medium molecular chitosan oligosaccharide (1.5 to 5.5 kDa), also successfully inhibited sarcoma and carcinoma (Jeon and Kim 2002). Water solubility and molecular weight were considered as the two important parameters for their anti-tumor activity (Qin et al. 2002b).

Astaxanthin showed anticancer activity by enhancing the immune system; it improved antitumor immune responses by inhibiting lipid peroxidation (Kurihara et al. 2002). Astaxanthin also suppressed inflammatory mediators, such as tumor necrosis factor alpha (TNF-α), prostaglandin E-2 (PGE-2), interleukin 1B (IL-1b) and nitric oxide (NO) (Lee et al. 2003). Another new anticancer compound from shrimp waste was peptide hydrolysate. Peptide hydrolysates from shrimp shell (*Peneaus setiferus*) with molecular weight less than 10 and 10–30 kDa significantly inhibited the growth of colon (Caco-2) and liver (HepG2) cancers. This study is considered as the first to announc that peptides from shrimp waste conduct anti cancer activity (Kannan et al. 2011).

4.8 Nerve regeneration

Biodegradable and non-biodegradable materials are usually used for bridging nerve stumps and conduits for regenerating and recovering from traumatic nerve injuries. The drawbacks of non-biodegradable materials, such as silicone and polytetrafluoroethylene (PTFE), are their toxicity. They may also cause severe foreign body reactions (Wang et al. 2011). Moreover, after such application, a surgery to remove such non biodegradable materials from our body will be required. This removal process will obviously be inconvenient and may make the patient suffer even more. In comparison to silicon as a conduit, chitosan performed better recovery on nerve function with respect to recovery time (Raisi et al. 2001). Hence, biodegradable materials such polyphosphoesters, chitosan, or casein are the best choice. Using chitosan tubes to bridge the nerve defects is preferable because of its successful application in nerve regeneration therapy (Shen et al. 2010; Li et al. 2010).

It is known that biodegradable materials have good biocompatibility with Schwann cells; they are nontoxic and capable of promoting peripheral nerve defect regeneration (Xu et al. 2011). Chitosan repairs mitochondrial membranes and nerve cell membranes at the same time. Chitosan also repairs the damaged cell membranes, which could prevent the leakage of large molecules from damaged spinal cord cells. The leakage of Lactate dehydrogenase (LDH) from cells is an indicator of wholeness of the plasma membrane cell. The chitosan treated cell showed lower amounts of LDH.

It has been proved that chitosan can maintain membrane and contain membrane damage. Chitosan also reduces harmful reactive oxygen species (ROS), which means it also prevents mitochondrial damage. Mitochondrial damages will produce ROS as a releasing compound in the ATP generating process. ROS levels also decreased after applying chitosan to the damaged tissue. Hence, Cho et al. (2010) suggested that chitosan can be used to induce fusion and seal disrupted plasma membranes in traumatic spinal cord injury.

4.9 Wound healing

Chitin and chitosan have been reported to be able to accelerate wound healing. Complex mechanisms, such as coagulation, inflammation, matrix synthesis and deposition, angiogenesis, fibroplasia, epithelization, contraction and remodelling are numerous mechanisms involving in wound healing (Alemdaroglu et al. 2006). At the first stage, wound healing consists of two important stages; i.e., inflammatory and the formation of new tissues. In this step, chitosan accelerates the inflammatory steps by expediting inflammatory cell infiltration. Chitosan also enhances the production of collagen; consequently, hypertrophic scar diminishes (Ueno et al. 1999).

Glucosamines containing chitin serves as a good substrate for keratinocytes and fibroblasts. This will lead to detraction of hyperproliferation of granulation tissue. Hence, chitosan can be used in preventing scar formation in the restored wound which commonly appears after the natural wound healing process (Antonov et al. 2008). In the early stages of wound healing, chitosan will play its role as (1) an enhancer of polymor-phonuclear (PMN) cell infiltration in the wound area; (2) a fibroblast movement activator in the wound area and in the formation of thick fibrin; (3) a macrophage stimulator; and (4) fibroblast stimulator and collagen producer. Increasing cell proliferation was observed in the epidermal growth factors treated with chitosan (Alemdaroglu et al. 2006). This process will accelerate cell recovery to promote better and faster epithelisation during the wound healing process (Kiyohara et al. 1993).

Topical formulations to recover wound skin in the rabbit ear model, based on water soluble chitin, showed an excellent result (Han 2005). Histological examination showed that granulation tissue was covered by abundant fibroblasts after chitin treatment. Epithelisation and granulation increased in comparison to the control. Experiments on chitin (dibutyryl chitin) application in human, chick and mouse fibroblasts by various methods exhibited brisk results (Muzzarelli et al. 2005). Muzzarelli (2009) suggested that the main biochemical effects of treating wound or scar tissue with chitins and chitosans were fibroblast activation, cytokine production, giant cell migration and stimulation of collagen synthesis, especially type IV.

5 Future Trends

Chitin, chitosan and chitooligosaccharides are the main products resulting from the utilization of crustacean wastes. These products are produced and improved merely for human interests, such as for food, industry and health. The utilization of shrimp waste derived products in foods is mainly to increase shelf life in regard to their role as coating and anti-microbial agents. In addition, these products are also developed as agents for the prevention of environmental pollution. The capacity of chitin, chitosan and chitooligosaccharides on heavy metal adsorption has been assessed effectively to increase waste water quality. The most interesting area of shrimp waste utilization observed is for the purpose of promoting good human health. Numerous studies have been done to explore the bioactivities of these products as preventive and therapeutic agents for human health, from gastrointestinal disorder to cancer therapy.

Chitin, chitosan and chitooligosaccharides, with various polymerization degrees, are generally produced by chemical and enzymatic processes. The degree of depolymerization is strongly dependent on the purity and concentration of solvents; the types, purity and concentration of enzymes; the duration and temperature of reactions, etc. Techniques for further purification of these products as well as for sequence determination are now available, but are still quite challenging to exploit. Scaling up purification methods at an economically acceptable cost is another challenge; meaning that, from an economical point of view, it is probably cheaper to produce products that are enriched for a bioactivity, rather than produce pure compounds. Based on this issue, in the future, improvisation, simplification, efficiency and innovation of chitin, chitosan and chitooligosaccharide production, are necessary.

Shrimp waste derived products or their derivates have been applied for the reduction of many degenerative diseases. They have a remarkably wide spectrum of possible bioactivities, including hypochlesterolemic, immune-stimulating effects, tumor growth inhibition, and anti-inflammation, etc. Despite their highly promising future, there is no doubt that these bioactivities need to be validated with further studies of well-defined preparation, as well as by fundamental researches on the molecular mechanism behind the activities.

References

Abd El-Rehim, H.A., N.M. Naeem, El.S.A. Hegazy, El.S.A. Soliman and A.M. El-barbary. 2012. Improvement of antioxidant activity of chitosan by chemical treatment and ionizing radiation. International Journal of Biological Macromolecules. 50: 403–413.

Adeyeye, E.I. and H.O. Adubiaro. 2004. Chemical composition of shell and flesh of three prawn samples from Lagos lagoon. J. Sci. Food. Agric. 84: 411–414.

Adeyeye, E.I., H.O. Adubiaro and O.J. Awodola. 2008. Comparability of chemical composition and functional properties of shell and flesh of *Penaeus notabilis*. Pak. J. Nutr. 7(6): 741–747.

Adour, W., W. Arbia, A. Amrane and N. Mameri. 2008. Combined used of waste material-recovery of chitin from shrimp shell by lactic acid fermentation supplemented with date juice waste or glucose. J. Chem. Technol. and Biotechnol. 83: 1664–1669.

Alasalvar, C. and T. Taylor. 2002. Seafoods—technology, quality and nutraceutical applications. Springer Verlag Berlin, Heidelberg, Germany.

Alemdaroglu, C., Z. Degim, N. Celebi, F. Zor, S. Ozturk and D. Erdogan. 2006. An investigation on burn wound healing in rats with chitosan gel formulation containing epidermal growth factor. Burn. 32: 319–327.

Ali, N.E., N. Hmidet, O. Ghorbel-Bellaj, N. Fahfakh-Zouri, A. Bougatef and M. Nasri. 2011. Solvent-stable digestive alkaline proteinases from striped seabream (*Lithognathus mormyrus*) viscera: Characteristics, application in the deproteinization of shrimp waste, and evaluation in laundry commercial detergent. Appl. Biochem. Biotechnol. 164: 1096–1110.

Aluwihare, L.I., D.J. Repeta, S. Pantoja and C.G. Johnson. 2005. Two chemically distinct pools of organic nitrogen accumulate in the ocean. Science. 308: 1007–1010.

ANTARA. 2010. FAO rates RI as world's fourth biggest shrimp producer. http://www.antaranews.com/en/news/1290415431/fao-rates-ri-as-worlds-fourth-biggest-shrimp producer.

Antonov, S.F., E.V. Kryzhanovskaya, Y.U. Filippov, S.M. Shinkarev and M.A. Frolova. 2008. Study of wound-healing properties of chitosan. Russ. Agric. Sci. 34(6): 426–427.

Aoi, W., Y. Naito, K. Sakuma, M. Kuchide, H. Tokuda, T. Maoka, S. Toyokuni, S. Oka, M. Yasuhara and T. Yoshikawa. 2003. Astaxanthin limits exercise-induced skeletal and cardiac muscle damage in mice. Antioxid Redox Signal. 5: 139–144.

Araki, Y. and E. Ito. 1975. Pathway of chitosan formation in *Mucor rouxii*. Eur. J. Biochem. 55: 71–78.

Arbia, W., L. Adour, A. Amrane and H. Lounici. 2013. Optimization of medium composition for enhanced chitin extraction from *Parapenaeus longirostris* by *Lactobacillus helveticus* using response surface methodology. Food Hydrocoll. 31: 392–403.

Armenta-López, R., I. Guerrerol and S. Huerta. 20002. Astaxanthin Extraction From Shrimp Waste by Lactic Fermentation and Enzymatic Hydrolysis of the Carotenoprotein Complex. J. Food Sci. 67(3): 1002–1006.

Aspan, A. and K. Soderhall. 1991. Purification of prophenoloxidase from crayfish cells, and its activation by an endogenous serine proteinase. Insect Biochem. 21: 363–373.

Aye, K.N. and W.F. Stevens. 2004. Technical note: Improved chitin production by pretreatment of shrimp shells. J. Chem. Technol. Biotechnol. 79: 421–425.

Bajaj, M., J. Winter and Gallert. 2011. Effect of deproteinization and deacetylation conditions on viscosity of chitin and chitosan extracted from *Crangoncrangon* shrimp waste. Biochem. Eng J. 56: 51–62.

Balogun, A.M. and Y. Akegbejo Samson. 1992. Waste yield, proximate and mineral composition of shrimp resources of Nigeria's coastal water. Biores. Technol. 40: 157–161.

Beaney, P., J. Lizardi-Mendoza and M. Healy. 2005. Comparison of chitins produced by chemical and bioprocessing methods. J. Chem. Technol. Biotechno. l80: 145–150.

Benhabiles, M.S., N. Abdia, N. Drouichea, H. Lounicia, A. Pauss, M.F.A. Goosend and N. Mamerib. 2013. Protein recovery by ultrafiltration during isolation of chitin from shrimp shells *Parapenaeus longirostris*. Food Hydrocoll. 32(1): 28–34.

Benjakul, S. and P. Wisitwuttikul. 1994. Improvement of deacetylation of chitin from black tiger thrimp (*Penaeus monodon*) carapace and shell. ASEAN Food J. 9(4): 136–140.

Benjakul, S., W. Visessanguan and M. Tanaka. 2005. Properties of phenoloxidase isolated from the Cephalothorax of kuruma prawn (*Penaeus japonicus*). J. Food Biochem. 29: 470–485.

Bennedsen, M., X. Wang, R. Willen, T. Wadstrom and L.P. Andersen. 1999. Treatment of H. pylori infected mice with antioxidant astaxanthin reduces gastric inflammation, bacterial load and modulates cytokine release by splenocytes. Immunol. Lett. 70: 185–189.

Bokura, H. and S. Kobayashi. 2003. Chitosan decreases total cholesterol in women: a randomized, double-blind, placebo controlled trial. Eur. J. Clin. Nutr. 57: 721–725.

Bosso, C., J. Defaye, A. Domard and A. Gadelle. 1986. The behavior of chitin towards anhydrous hydrogen fluoride. Preparation of β-(1→4)-linked 2-acetamido-2-deoxy-D-glucopyranosyl oligosaccharides. Carbohydr. Res. 156: 57–68.

Brasileiro, O.L., J.M.O. Cavalheiro, J.P. Prado, de S., A.G., dos Anjos and T.T.B. Cavalheiri. 2012. Determination of the chemical composition and functional properties of shrimp waste protein concentrate and lyophilized flour. Ciênc. agrotec. Lavras. 36: 189–194.

Camera, E., A. Mastrofrancesco, C. Fabbri, F. Daubrawa, M. Picardo, H. Sies and W. Stahl. 2009. Astaxanthin, canthaxanthin and beta-carotene differently affect UVA-induced oxidative damage and expression of oxidative stress-responsive enzymes. Exp. Dermatol. 18(3): 222–231.

Cao, W., C. Zhang, P. Hong and H. Ji. 2008. Response surface methodology for autolysis parameters optimization of shrimp head and amino acids released during autolysis. Food Chem. 109: 176–183.

Cardoch, L., J.W. Day, Jr., J.M. Rybczyk and G.P. Kemp. 2000. An economic analysis of using wetlands for treatment of shrimp processing wastewater—A case study in Dulac, LA. Ecol. Econ. 33: 93–101.

Cavalheiro, J.M.O., E.O. de Souza and P.S. Bora. 2007. Utilization of shrimp industry waste in the formulation of tilapia (*Oreochromis niloticus* Linnaeus) feed. Bioresource Technology. 98: 602–606.

Chander, R. and P. Thomas. 1999. Alkaline phosphatase from jawala shrimp (*Acetes indicus*). J. Food Biochem. 25: 91–103.

Charest, D.J., M.O. Balaban, M.R. Marshall and J.A. Cornell. 2001. Astaxanthin extraction from crawfish shells by supercritical CO_2 with ethanol as co solvent. J. Aquat. Food. Prod. Technol. 10(3): 81–96.

Chen, H. and S.P. Meyers. 1982. Extraction of astaxanthin pigment from crawfish waste using a soy oil process. J. Food Sci. 47: 892–900.

Chen, Y.C., L.W. Chung, K.T. Wang, S. Chen and Y. Li. 2002. Antibacterial properties of chitosan in waterborne pathogen. J. Environ. Sci. Health A Tox Hazard Subst. Environ. Eng. 37: 1379–1390.

Cho, Y., R. Shi and R.B. Borgens. 2010. Chitosan produces potent neuroprotection and physiological recovery following traumatic spinal cord injury. J. Exp. Biol. 213: 1513–1520.

Chou, T.C., E. Fu and E.C. Shen. 2003. Chitosan inhibits prostaglandin E2 formation and cyclooxygenase-2 induction in lipopolysaccharide-treated RAW 264.7 macrophages. Biochem. Biophys. Res. Commun. 308(2): 403–407.

Davydova, V.N., V.P. Nagorskaya, V.I. Gorbach, A.A. Kalitnik, A.V. Reunov, T.F. Solov'eva and I.M. Ermak. 2011. Chitosan antiviral activity: Dependence on structure and depolymerization method. Appl. Biochem. Microbiol. 47(1): 103–108.

Dey, S. and K. Dora. 2012. Antioxidative activity of protein hydrolysate produced by alcalase hydrolysis from shrimp waste (*Penaeus monodon* and *Penaeus indicus*). J. Food Technol. 49: 1–9.

Duan, Li, L., Z. Zhuang, W. Wu, S. Hong and J. Zhou. 2012. Improved production of chitin from shrimp waste by fermentation with epiphytic lactic acid bacteria. Carbohydr. Polym. 89: 1283–1288.

FAO. 1990. Proceedings. Shrimp Culture Industry Workshop. Jepara (Indonesia). 25–28 Sep 1989. Jepara (Indonesia). http://www.fao.org/docrep/field/003/AC058E/AC058E00. htm.

FAO. 2012. The State of World Fisheries and Aquaculture 2012. FAO. Rome. Italia.

Fukada, Y., K. Kimura and Y. Ayaki. 1991. Effect of chitosan feeding on intestinal bile acid metabolism in rats. Lipids. 26: 395–399.

Garcia-Carreno, F.L., A.N. del Toro and M. Ezquerra. 1997. Digestive shrimp proteases for evaluation of protein digestibility *in vitro*. I: Effects of protease inhibitors in protein ingredients. J. Mar. Biotechnology. 55: 36–40.

Gollas-Galván, T., J. Hernández-López and F. Vargas-Albores. 1991. Prophenoloxidase from brown shrimp (*Penaeus californiensis*) hemocytes. Comp. Biochem. Physiol. B Biochem. Mol. Biol. 122(1): 77–82.

Goto, S., K. Kogure, K. Abe, Y. Kimata, K. Kitahama, E. Yamashita and H. Terada. 2001. Efficient radical trapping at the surface and inside the phospholipid membrane is responsible for highly potent antiperoxidative activity of the carotenoid astaxanthin. Biochim. Biophys. Acta. 1512: 251–258.

Guillou, M., M. Khalil and L. Adambounou. 1995. Effect of silage preservation of astaxanthin forms and fatty acid profiles of processed shrimp (*Pandalus borealis*) waste. Aquaculture. 130: 351–360.

Guzman, J., I. Saucedo, J. Revilla, R. Navarro and E. Guibal. 2003. Copper sorption by chitosan in the presence of citrate ions: Influence of metal speciation on sorption mechanism and uptake capacities. Int. J. Biol. Macromol. 33: 57–65.

Han, S.S. 2005. Topical formulations of water-soluble chitin as a wound healing assistant— evaluation on open wounds using a rabbit ear model—fibers and polymers. 6(3): 219–223.

Handayani, A.D., Sutrisno, N. Indraswati and S. Ismadji. 2008. Extraction of astaxanthin from giant tiger (*Panaeus monodon*) shrimp waste using palm oil: Studies of extraction kinetics and thermodynamic. Bioresour. Technol. 99: 4414–4419.

Hayashi, K. and M. Ito. 2002. Antidiabetic action of low molecular weight chitosan in genetically obese diabetic KK-Ay. mice. Biol. Pharm. Bull. 25(2): 188–192.

Helander, I.M., E.L. Nurmiaho-Lassila, R. Ahvenainen, J. Rhoades and S. Roller. 2001. Chitosan disrupts the barrier properties of the outer membrane of gram-negative bacteria. Int. J. Food Microbiol. 71: 235–244.

Heu, M.S., J.S. Kim and F. Shahidi. 2003. Components and nutritional quality of shrimp processing by-products. Food Chem. 82(2): 235–242.

Hussein, G., T. Nakagawa, H. Goto, Y. Shimada, K. Matsumoto, U. Sankawa and H. Watanabe. 2007. Astaxanthin ameliorates features of metabolic syndrome in SHR/NDmcr-cp. Life Sci. 80(6): 522–529.

Ibrahim, H.M., M.F. Salama and H.A. El-Banna. 1999. Shrimp's waste: Chemical composition, nutritional value and utilization. Nahrung. 43(6): 418–423.

Islam, M.S., S. Khan and M. Tanaka. 2004. Waste loading in shrimp and fish processing effuents: potential source of hazards to the coastal and nearshore environments. Mar. Pollut. Bull. 49: 103–110.

Iwamoto, T., K. Hosoda, R. Hirano, H. Kurata, A. Matsumoto, W. Miki, M. Kamiyama, H. Itakura, S. Yamamoto and K. Kondo. 2000. Inhibition of low-density lipoprotein oxidation by astaxanthin. J. Atheroscler. Throm. 7(4): 216–222.

Jaffer, S. and J.S. Sampalis. 2007. Efficacy and safety of chitosan HEP-40™ in the management of hypercholesterolemia: A randomized, multicenter, placebo-controlled trial. Alternative Medicine Review. 12(3): 265–273.

Jeon, Y.J. and S.K. Kim. 2000. Continuous production of chitooligosaccharides using a dual reactor system. Proc. Biochem. 35: 623–632.

Jeon, Y.J. and S.K. Kim. 2002. Antitumor activity of chitosan oligosaccharides produced in ultrafiltration membrane reactor system. J. Microbiol. Biotechnol. 12: 503–507.

Jimeno, J., G. Faircloth, J.F. Soussa-Faro, P. Scheuer and K. Rinehart. 2004. New marine derived anticancer therapeutics—A journey from the sea to clinical trials. Mar. Drugs. 2: 14–29.

Johnson, E.A. and M.J. Lewis. 1979. Astaxanthin formation by the yeast *Phafia rhodozyma*. J. Gen. Microbiol. 115: 173–183.

Kannan, A., N.A. Hettiarachchy, M. Marshall, S. Raghavan and H. Kristinsson. 2011. Shrimp shell peptide hydrolysates inhibit human cancer cell proliferation. J. Sci. Food Agric. 91: 1920–1924.

Karakoltsidis, P.A., A. Zotos and S.M. Constantinides. 1995. Composition of commercially important Mediterranean finfish, crustaceans, and molluscs. J. Food Compos. Anal. 8: 258–273.

Katas, H., A. Mohamad and N.M. Zin. 2011. Physicochemical effects of chitosan-tripolyphosphate nanoparticles on antibacterial activity against gram-positive and gram-negative bacteria. J. Med. Sci. 4: 192–197.

Katiyar, D.M. 2011. Evaluation of antidiabetic and hypolipidemic activity of chitooligosaccharides in alloxan-induced diabetes mellitus in mice. International J. Pharm. Bio. Sci. 2(1): 407–416.

Keyhan, N.O. and S. Roseman. 1999. Physiological aspects of chitin catabolism in marine bacteria. Biochim. Biophys. Acta. 1473: 108–122.

Kikkawa, Y., T. Kawada, L. Furukawa and T. Sakuno. 1990. A convenient preparation method of chitooligosaccharides by acid hydrolysis. J. Facul. Agr., Tottori University. 26: 9–17.

Kim, S.H., H.K. No and W. Prinyawiwatkul. 2007. Effect of molecular weight, type of chitosan, and chitosan solution pH on the shelf-life and quality of coated eggs. J. Food Sci. 72(1): S044–8.

Kim, Y.J., Y. Zhao, K.T. Oh, V.N. Nguyen and R.D. Park. 2008. Enzymatic deacetylation of chitin by extracellular chitin deacetylase from a newly screened Mortierella sp. DY-52. J. Microbiol. Biotechnol. 8(4): 759–66.

Kim, M.S., M.J. Sung, S.B. Seo, S.J. Yoo, W.K. Lim and H.M. Kim. 2002. Water-soluble chitosan inhibits the production of pro-inflammatory cytokine in human astrocytoma cells activated by amyloid β peptide and interleukin-1β. Neurosci. Lett. 321: 105–109.

Kim, S.K. and N. Rajapakse. 2005. Enzymatic production and biological activities of chitosan oligosaccharides (COS): A Review. Carbohydr. Polym. 62(4): 357–368.

Kittur, F., A.B.V. Kumar, L.R. Gowda and R.N. Tharanathan. 2003. Chitosanol-ysis by pectinase isozyme of Aspergillus nigera non specific activity. Carbohyd. Polym. 53: 191–196.

Kiyohara, Y., K. Nishiguchi, F. Komada, S. Iwakawa, M. Hirai and K. Okumura. 1993. Cytoprotective effects of epidermal growth factor (EGF) ointment containing nafamostat, a protease inhibitor, on tissue damage at burn site in rats. Biol. Pharm. Bull. 16(11): 1146–9.

Kjartansson, G.T., S. Zyvanovic, K. Kristberg and J. Weiss. 2006. Sonification assisted extraction of chitin North atlantic shrimp (*Pandalus borealis*). J. Agric. Food Chem. 54: 5894–5902.

Khempaka, S., C. Chitsatchapong and W. Molee. 2011. Effect of chitin and protein constituents in shrimp head meal on growth performance, nutrient digestibility, intestinal microbial populations, volatile fatty acids, and ammonia production in broilers. J. Appl. Poult. Res. 20: 1–11.

Kobayashi, T., S. Otsuka and Y. Yugari. 1979. Effect of Chitosan on serum and liver cholesterol levels in cholesterol-fed rats. Nutritional Rep. Int. 19: 327–334.

Kochkina, Z.M. and S.N. Chirkov. 2000. Influence of Chitosan derivatives on the development of phage infection in the Bacillus thuringiensis culture. Microbiology. 69(2): 217–219.

Kristinsson, H.G. and B.A. Rasco. 2000. Fish protein hydrolysates: Production, biochemical and chemical properties. Critical Reviews in Food Science and Nutrition. 43: 43–81.

Kulikov, S.N., S.N. Chirkov, A.V. Il'ina, S.A. Lopatin and V.P. Varlamov. 2006. Effect of the molecular weight of chitosan on its antiviral activity in plants. Appl. Biochem. Microbio. 42(2): 200–203.

Kumar, M.N.V.R. 1999. Chitin and chitosan fibres. Bull. Mater. Sci. 22: 905–915.

Kumar, S.G., M.A. Rahman, S.H. Lee, H.S. Hwang, H.A. Kim and J.W. Yun. 2009. Plasma proteome analysis for anti-obesity and anti-diabetic potentials of chitosan oligosaccharides in ob/ob mice. Proteomics. 9: 2149–2162.

Kuo, J.M. and B.S. Pan. 1992. Occurrence and properties of 12-lipoxygenase in the hemolymph of shrimp (*Penaeus japonicus* Bate). J. Chin. Biochem. Soc. 21: 9–16.

Kurashige, M., E. Okimasu, M. Inoue and K. Utsumi. 1990. Inhibition of oxidative injury of biological membranes by astaxanthin. Physiol. Chem. Phys. Med. NMR. 22: 27–38.

Kurihara, H., H. Koda, S. Asami, Y. Kiso and T. Tanaka. 2002. Contribution of the antioxidative property of astaxanthin to its protective effect on the promotion of cancer metastasis in mice treated with restraint stress. Life Sci. 70(21): 2509–20.

Kurita, K. 2001. Controlled functionalization of the polysaccharide chitin. Prog. Polym. Sci. 26: 1921–1971.

Lamiaa and A.A. Baraka. 2011. Hypolipidemic and antiatherogenic effects of dietary chitosan and wheat bran in high fat-high cholesterol fed rats. Aus. J. Basic. Appl. Sci. 5(10): 30–37.

Latscha, T. 1989. The role of astaxanthin in shrimp pigmentation. Advances in Tropical Aquaculture. 319–325.

Leal, A., P. de Castro, J. de Lima, E. de Souza Correia and R. de Souza Bezerra. 2010. Use of shrimp protein hydrolysate in Nile tilapia (*Oreochromis niloticus*, L.) feeds. Aquacult. Int. 18: 635–646.

Le Boulay, C., A. Van Wormhoudt and D. Sellos. 1996. Cloning and expression of cathepsin like proteinases in the hepatopancreas of shrimp *Penaeus vannamei* during intermolt cycle. J. Comp. Physiol. 166B: 310–318.

Lee, A.A. and N.N. Chuang. 1991. Characterization of different molecular forms of alkaline phosphatase in the hepatopancreas from the shrimp *Penaeus monodon* (Crustacea: Decapoda). Comp. Biochem. Physiol. Biochem. Mol. Biol. 44: 845–850.

Lee, D.X., W.S. Xia and J.L. Zhang. 2008. Enzymatic preparation of chitooligosaccharides by commercial lipase. Food Chem. 111(2): 291–295.

Lee, H.W., Y.S. Park, J.W. Choi, S.Y. Yi and W.S. Shin. 2003a. Antidiabetic effects of chitosan oligosaccharides in neonatal streptozotocin-induced noninsulin-dependent Diabetes Mellitus in rats. Biol. Pharm. Bull. 26(8): 1100–1103.

Lee, S.H., S.K. Roh, K.H. Park and K.R. Yoon. 1999. Effective extraction of astaxanthin pigment from shrimp using proteolytic enzymes. Biotechnol. Bioprocess Eng. 4: 199–204.

Lee, S.J., S.K. Bai, K.S., Lee, S. Namkoong, H.J. Na, K.S. Ha, J.A. Han, S.W. Yim, K. Chang, Kwon, Lee, S.K. and Y.M. Kim. 2003b. Astaxanthin inhibits nitric oxide production and inflammatory gene expression by suppressing Iκβ kinase independent. Mol. Cells. 16: 97–105.

LeHoux, J.G. and F. Grondin. 1993. Some effects of chitosan on liver function in the rat. Endocrinology. 132: 1078–1084.

Li, H., G. Xu, Q. Shang, L. Pan, S. Shefer, A. Batta, J. Bollineni, G.S.T. int, B.T. Keller and G. Salen. 2004. Inhibition of illeal bile acid transport lowers plasma cholesterol levels by inactivating hepatic farnesoid X receptor and stimulating cholesterol 7α-hydroxylase. Metabolism. 53(7): 927–932.

Li, X., W. Wang, G. Wei, G. Wang, W. Zhang and X. Ma. 2010. Immunophilin FK506 loaded in chitosan guide promotes peripheral nerve regeneration. Biotechnol. Lett. 32: 1333–1337.

Lin, C.C., C.I. Li, C.S. Liu, W.Y. Lin, M.M.T. Fuh, S.Y. Yang, C.C. Lee and T.C. Li. 2012. Impact of lifestyle-related factors on all-cause and cause-specific mortality in patients with type 2 diabetes. Diabetes Care. 35: 105–112.

Lin, S.B., Y.C. Lin and H.H. Chen. 2009. Low molecular weight chitosan prepared with the aid of cellulase, lysozyme and chitinase: Characterization and anti-bacterial activity. Food Chem. 116(1): 47–53.

Liu, H., P. Jiravanichpaisal, L. Cerenius, B.L. Lee, I. Soderhall and K. Soderhall. 2007. Phenoloxidase is an important Component of the defense against Aeromonas hydrophila infection in a crustacean, *Pacifastacus leniusculus*. J. Biol. Chem. 282: 33593–33598.

Liu, S., J. Sun, L. Yu, C. Zhang, J. Bi, F. Zhu, M. Qu, C. Jiang and Q. Yang. 2012. Extraction and characterization of chitin from the beetle Holotrichia parallela Motschulsky. Molecules. 17(4): 4604–4611.

Lymer, D., S. Funge-Smith, J. Clausen and W. Miao. 2008. Status and potential of fisheries and aquaculture in Asia and the Pacific 2008. Food and Agriculture Organization of the United Nations Regional Office for Asia and the Pacific. Bangkok.

Maezaki Y., K. Tsuji, Y. Nakagawa, Y. Kawai, M. Akimoto, T. Tsugita, W. Takekawa, A. Terada, H. Hara and T. Mitsuoka. 1993. Hypocholesterolemic effect of chitosan in adult males. Biosci. Biotechnol. Biochem. 57: 1439–1444.

Maeda, Y. and Y. Kimura. 2004. Antitumor effects of various low-molecular-weight chitosans are due to increased natural killer activity of intestinal intraepithelial lymphocytes in sarcoma 180–bearing mice. J. Nutr. 34: 945–50.

Manabe, E., O. Handa, Y. Naito, K. Mizushima, S. Akagiri, S. Adachi, T. Takagi, S. Kokura, T. Maoka and T. Yoshikawa. 2008. Astaxanthin protects mesangial cells from hyperglycemia-induced oxidative signaling. J. Cell Biochem. 103(6): 1925–1937.

Martinou, A., D. Kafetzopoulos and V. Bouriotis. 1995. Chitin deacetylation by enzymatic means: Monitoring of deacetylation processes. Carbohydr Res. 273: 235–242.

Mauldin, M.A. and A.J. Szabo. 1974. Shrimp canning waste treatment study. Environmetal Protection Technology Series. Environmetal Protection Agency. Washinton USA.

Miao, J., X. Zhang, Y. Honga, Y.F. Rao, Li, Xie, J. Woa and M. Li. 2012. Inhibition on hepatitis B virus e-gene expression of 10–23 DNAzyme delivered by novel chitosan oligosaccharide–stearic acid micelles. Carbohydr. Polym. 87: 1342–1347.

Ministry of Marine Affairs and Fisheries. 2012. Export Statistics of Fishery Products. Jakarta.

Miyatake, K., Y. Okamoto, Y. Shigemasa, S. Tokura and S. Minami. 2003. Anti-inflammatory effect of chemically modified chitin. Carbohydrate Polymers. 53: 417–423.

Mònek, J. and H. Bartonová. 2005. Effect of dietary chitin and chitosan on cholesterolemia of rats. Acta Vet. Brno. 74: 491–499.

Montero, M.P., A. Avalos and M. Pérez-Mateos. 2001. Characterization of polyphenoloxidase of prawn (*Penaeus japonicus*). Alternatives to inhibition: Additives and high pressure treatment. Food Chem. 75: 317–324.

Montero, M.P., A. Avalos and M. Pérez-Mateos. 2001. Characterization of polyphenoloxidase of prawn (*Penaeus japonicus*). Alternatives to inhibition: Additives and high pressure treatment. Food Chem. 75: 317–324.

Mortensen, A., L.H. Skibsted, J. Sampson, C. Rice-Evans and S.A. Everett. 1997. Comparative mechanisms and rates of free radical scavenging by carotenoid antioxidants. Febs Lett. 418: 91–97.

Muzzarelli, R.A.A. 2009. Chitin and Chitosan for the repair of wounded skin, nerve, cartilage and bone. Carbohidr. Polymer. 76(2): 167–182.

Muzzarelli, R.A.A., M. Guerrieri, G. Goteri, C. Muzzarelli, T., Armeni, R. Ghiselli and M. Cornilissen. 2005. The biocompatibility of dibutyryl chitin in the context of wound dressings. Biomaterials. 26: 5844–5854.

Naguib, Y.M.A. 2000. Antioxidant activities of astaxanthin and related carotenoids. J. Agric. Food Chem. 48: 1150–1154.

Naito, Y., K. Uchiyama, W. Aoi, G. Hasegawa, N. Nakamura, N. Yoshida, T. Maoka, J. Takahashi and T. Yoshikawa. 2004. Prevention of diabetic nephropathy by treatment with astaxanthin in diabetic db/db mice. Biofactors. 20: 49–59.

Nakagawa, T. and F. Nagayama. 1981. Distribution of cathecol oxidase in crustaceans. Bulletin of the Japanese Society of Science Fisheries. 47: 1645.

Ngoan, L.D., J.E. Linderg, B.B. Ogle and S. Thomke. 2000. Anatomical Proportion and Chemical Amino Acid composition of common shrimp species in central Vietnam. Asian-Aust. J. Anim. Sci. 13(10): 1422–1428.

Nishimura, S.I., H. Kaia, K. Shinada, T. Yoshida, S. Tokura, K. Kurita, H. Nakashima, N. Yamamoto and T. Uryu. 1998. Regioselective syntheses of sulfated polysaccharides: specific anti-HIV-1 activity of novel chitin sulfates. Carbohydr. Res. 306: 427–433.

No, H.K. and S.P. Meyers. 1995. Preparation and characterization of chitin and chitosan: A Review. J. Aquatic Food Prod. Technol. 4(2): 27–52.

No, H.K., S.P. Meyers and K.S. Lee. 1989. Isolation and characterization of chitin from crawfish shell waste. J. Agric. Food Chem. 37(3): 575–579.

Ohgami, K., K. Shiratori, S. Kotake, T. Nishida, N. Mizuki, K. Yazawa and S. Ohno. 2003. Effects of astaxanthin on lipopolysaccharide-induced inflammation *in vitro* and *in vivo*, Invest Ophthalmol. Vis. Sci. 44(6): 2694–2701.

Okada, Y., M. Ishikura and T. Maoka. 2009. Bioavailability of astaxanthin in Haematococcus algal extract: the effects of timing of diet and smoking habits. Biosci. Biotechnol. Biochem. 73(9): 1928–1932.

Olsen, R.L., K. Onerbo and B. Myrnes. 1991. Alkaline phosphatase from the hepatopancreas of shrimp (*Pandalus borealis*): A dimeric enzyme with catalytically active subunits. Comp. Biochem. Physiol. 99B: 755–761.

Omura, Y., M. Shigemoto, T. Akiyama, H. Saimoto, Y. Shigemasa, I. Nakamura and T. Tsuchido. 2002. Re-examination of antimicrobial activity of chitosan having different degrees of acetylation and molecular weights. Advances in Chitin Science. 6: 273–274.

Overcash, M.R. and D. Pal. 1980. Characterization and land application of Seafood industry wastewaters. Water Resources Research Institute. The University of North Carolina. USA.

Pacheco, N., M. Garnica-González, J.Y. Ramírez-Hernández, B. Flores-Albino, M. Gimenob, E. Bárzana and K. Shirai 2009. Effect of temperature on chitin and astaxanthin recoveries from shrimp waste using lactic acid bacteria. Biores. Technol. 100: 2849–2854.

Pérez-Gálvez, A., J.J. Negro-Balmaseda, M.I. Mínguez-Mosquera, Cascajo-Almenara and J. Garrido-Fernández. 2008. Astaxanthin from crayfish (*Procambarus clarkii*) as a pigmentary ingredient in the feed of laying hens. Grasas Y Aceites. 59(2): 47–53.

Pu, J., P.J. Bechtel and S. Sathivel. 2010. Extraction of shrimp astaxanthin with flaxseed oil: Effects on lipid oxidation and astaxanthin degradation rates. Biosys. Eng. 107: 364–371.

Qin, C.Q., Y.M. Du, L. Xiao, Z. Li and X.H. Gao. 2002a. Enzymic preparation of water soluble chitosan and their antitumor activity. Int. J. Biol. Macromol. 31: 111–117.

Qin, C.Q., Y.M. Du, L. Xiao, X.H. Gao, J.L. Zhou and L.L. Liu. 2002b. Effect of molecular weight and structure on antitumor activity of oxidized chitosan. Wuhan Univ. J. Nat. Sci. 7: 231–236.

Raisi, A., S. Azizi, N. Delirezh, B. Heshmatian and K. Amin. 2010. Use of chitosan conduit for bridging small-gap peripheral nerve defect in sciatic nerve transection model of rat. Iran J. Vet. Sur. 5: 89–99.

Randriamahatody, Z., K.S.B. Sylla, H.T.M. Nguyen, C. Donnay-Moreno, L. Razanamparany, N. Bourgougnon and J.P. Bergé 2011. Proteolysis of shrimp by-products (*Peaneusmonodon*) from Madagascar. CyTA—Journal of Food. 9(3): 220–228.

Ravichandran, S., G. Rameshkumar and A. Rosario Prince. 2009. Biochemical composition of shell and flesh of the Indian white shrimp *Penaeus indicus* (H. Milne Edwards 1837). American-Eurasian Journal of Scientific Research. 4(3): 191–194.

Rosa, R. and M.L. Nunes. 2003. Nutritional quality of red shrimp, *Aristeus antennatus* (Risso), pink shrimp, *Parapenaeus longirostris* (Lucas), and Norway lobster, *Nephrops norvegicus* (Linnaeus). J. Sci. Food Agric. 84: 89–94.

Ruttapornvareesakul, Y., M. Ikeda, K. Hara, K. Osatomi, K. Osako, O. Kongpun and Y. Noyaki. 2006. Concentration-dependent suppressive effect of shrimp head protein hydrolysate on dehydration-induced denaturation of lizardfish myofibrils. Biores. Technol. 97: 762–769.

Sachindra, N.M. and N.S. Mahendrakar. 2005a. Process optimization for extraction of carotenoids from shrimp waste with vegetable oils. Biores. Technol. 96: 1195–1200.

Sachindra, N.M., N. Bhaskar and N.S. Mahendrakar. 2005b. Carotenoids in different body components of Indian shrimps. J. Sci. Food Agri. 85: 167–172.

Sachindra, N.M., N. Bhaskar and N.S. Mahendrakar. 2006. Recovery of carotenoids from shrimp waste in organic solvents. Waste Manage. 26: 1092–1098.

Sanchez-Camargo, A.P., M.A.A. Meireles, B.L.F. Lopes and F.A. Cabral. 2011a. Proximate composition and extraction of carotenoids and lipids from Brazilian redspotted shrimp waste (*Farfantepenaeus paulensis*). J. Food Eng. 102: 87–93.

Sanchez-Camargo, A.P., H.A. Martinez-Correab, L.C. Paviania and F.A. Cabrala. 2011b. Supercritical CO$_2$ extraction of lipids and astaxanthin from Brazilian redspotted shrimp waste (*Farfantepenaeus paulensis*). J. of Supercritical Fluids. 56: 164–173.

Savagon, K.A. and A. Sreenivasan. 1978. Activation mechanism of pre-phenolase in lobster and shrimp. Fish Technol. 15(1): 49–55.

Senphan, T. and S. Benjakul. 2012. Compositions and yield of lipids extracted from hepatopancreas of Pacific white shrimp (*Litopenaeus vannamei*) as affected by prior autolysis. Food Chem. 134: 829–835.

Seo, S., J.M. King and W. Prinyawiwatkul. 2007. Simultaneous depolymerization and decolorization of chitosan by ozone treatment. J. Food Science. 72(9): C522–C526.

Seymour, T.A., S.J. Li and T.M. Morrissey. 1996. Characterization of a natural antioxidant from shrimp shell waste. J. Agric. Food Chem. 44: 682–685.

Shahidi, F. and J. Synowiecki. 1991. Isolation and characterization of nutrients and value-added products from snow crab (*Chionoecetes opilio*) and shrimp (*Pandalus borealis*) processing discards. J. Agric. Food Chem. 39(8): 1527–1532.

Shaw, J.E., R.A. Sicree and P.Z. Zimmet. 2010. Global estimates of the prevalence of diabetes for 2010 and 2030. Diabetes Research and Clinical Practices. 87: 4–14.

Shen, H., Z.L. Shen, P.H. Zhang, N.L. Chen, Y.C. Wang, Z.F. Zhang and Y.Q. Jin. 2010. Ciliary neurotrophic factor-coated polylactic–polyglycolic acid chitosan nerve conduit promotes peripheral nerve regeneration in canine tibial nerve defect repair. J. Biomed. Mater. Res. B Appl. Biomater. 95: 161–170.

Simpson, B.K. and N.F. Haard. 1985. The use of proteolytic enzymes to extract carotenoproteins from shrimp wastes. J. Appl. Biochem. 7: 212–222.

Simpson, B.K., M.R. Marshall and W.S. Otwell. 1987. Phenol oxidase from shrimp (Penaeus setiferus): purification and some properties. J. Agric. Food Chem. 35(6): 918–921.

Sindhu, S. and P.M. Sherief. 2011. Extraction characterization antioxidant and anti-inflammatory properties of carotenoids from the shell waste of Arabian red shrimp. The Open Conference Proceeding J. 2: 95–103.

Sosa, M.A.G., F. Fazely, J.A. Koch, S.V. Vercellotti and R.M. Ruprecht. 1991. N-Carboxymethylchitosan-N,O-sulfate as an anti-HIV-1 agent. Biochem. Biophys. Res. Comm. 174(2): 489–496.

Souza, C.P., B.C. Almeida, R.R. and I.N.G. Colwell Rivera. 2011. The importance of chitin in the marine environment. Mar. Biotechnol. 13: 823–830.

Spector, W.G. and D.A.Willoughby. 1963. The inflammatory response. Bacteriological Reviews. 27: 117–149.

Spinelli, J., L. Lehman and D. Wieg. 1974. Composition, processing and utilization of red crab as an aquacultural feed ingredient. J. Fish. Res. Board Can. 31: 1025–1029.

Srijanto, B., I. Paryanto, Masduki, Purwantiningsih. 2006. The effect of different degree of deacetylation of raw material on chitosan depolimerazion. Akta Kimindo. 1: 67–72 (in Indonesian).

Stepnowski, P., G. Olaffson, H. Helgason and B. Jastorff. 2004. Recovery of astaxanthin from seafood waste water utilizing fish scales waste. Chemosphere. 54: 413–417.

Stevens, W.F., P. Cheypratub, S. Haiqing, P. Lertsutthiwong, N.C. How and S. Chandrkrachang. 1998. Alternatives in shrimp biowaste processing. In: T.W. Flegel (ed.). Advances in shrimp biotechnology. National Center for Genetic Engineering and Biotechnology, Bangkok.

Sudarshan, N.R., D.G. Hoover and D. Knorr. 1992. Antibacterial action of chitosan. Food Biotechnol. 6(3): 257–272.

Sugano, M., T. Fujikawa, Y. Hiratsuji, K. Nakashima, N. Fukuda and Y. Hasegawa. 1978. A novel use of chitosan as a hypocholesterolemic agent in rats. Am. J. Clin. Nutr. 33: 787–793.

Tareq, A., M. Alam, S. Raza, T. Sarwar, Z. Fardous, A.Z. Chowdhury and S. Hossain. 2013. Comparative study of antibacterial activity of chitin and chemically treated chitosan prepared from shrimp (Macrobrachium Rosenbergii) Shell Waste. J. Virol. Microbiol. ID 369217.

Teruhisa, K. and K. Tadashi. 1972. The biosynthesis of astaxanthin. VI. The carotenoids in the prawn *Penaeus japonicus* bate (part II). Int. J. Biochem. 3(5): 363–368.

Tipparat, H. and O. Riyaphan. 2008. Effect of deacetylation conditions on antimicrobial activity of chitosans prepared from carapace of black tiger shrimp (*Penaeus monodon*). Songklanakarin J. Sci. Technol. 30, Suppl. 1: 1–9.

Tommeraas, K., K.M. Varum, B.E. Christensen and O. Smidrod. 2001. Preparation and characterization of oligosaccharides produced by nitrous acid depolymerization of chitosans. Carbohydr. Res. 333: 137–144.

Torrissen, O.J., R. Christiansen, G. Struksnæs and R. Estermann. 1995. Astaxanthin deposition in the flesh of Atlantic Salmon, *Salmo salar* L., in relation to dietary astaxanthin concentration and feeding period. Aquacult. Nutr. 1: 77–84.

Tsigos, I., A. Martinou, D. Kafetzopoulos and V. Bouriotis. 2000. Chitin Deacetylases: New, Versatile Tools in Biotechnology. TIBTECH. 18: 305–312.

Ueno, H., H. Yamada, I. Tanaka, N. Kaba, M. Matsuura, M. Okumura, T. Kadosawa and T. Fujinaga. 1999. Accelerating effects of chitosan for healing at early phase of experimental open wound in dogs. Biomaterials. 20: 1407–1414.

Valdez-Peña, A.U., J.D. Espinoza-Pérez, G.C. Sandoval-Fabian, N. Balagurusamy, A. Hernández-Rivera, I.M. De-la garza-Rodríguez and J.C. Contreras-Esquivel. 2010. Screening of industrial enzymes for deproteinization of shrimp head for chitin recovery. Food Sci. Biotechnol. 19: 553–557.

Vane, J. and R. Botting. 1987. Inflammation and the mechanism of action of anti-inflammatory drugs. FASEBJ. 1: 89–96.

Wang, P.H., I.L. Tseng and Hsu S. 2011. Review: Bioengineering Approaches for Guided Peripheral Nerve Regeneration. J. Med. Biol. Eng. 31: 151–160.

Wang, J.J., Z.Q. Chen and W.Q. Lu. 2012. Hypoglycemic effect of astaxanthin from shrimp waste in alloxan-induced diabetic mice. Med. Chem. Res. 21: 2363–2367.

Weska, R.F., J.M. Moura, L.M. Batista, J. Rizzi and L.A.A. Pinto. 2007. Optimization of deacetylation in the production of chitosan from shrimp wastes. J. Food Eng. 80(3): 749–753.

Wild, S., G. Roglic, A. Green and R. Sicree King H. 2004. Global prevalence of diabetes: estimates for the year 2000 and projections for 2030. Diabetes Care. 27: 1047–1053.

Xiao-Fang, L., F. Xiao-Qiang and Y. Sheng. 2010. A mechanism of antibacterial activity of chitosan against gram-negative bacteria. J. Food Sci. 31(13): 148–153.

Xiao-Fang, L., F. Xiao-Qiang, Y. Sheng, W. Ting-Pu and S. Zhong-Xing. 2008. Effects of molecular weight and concentration of chitosan on antifungal activity against *Aspergillus niger*. Iran. Polymr. J. 17(11): 843–852.

Xu, H., Y. Yan and S. Li. 2011. PDLLA/chondroitin sulfate/chitosan/NGF conduits for peripheral nerve regeneration. Biomaterials. 32(20): 4506–4516.

Xu, G., X. Huang, L. Qiu, J. Wu and Y. Hu. 2007. Mechanism study of chitosan on lipid metabolism in hyperlipidemic rats. Asia Pac. J. Clin. Nutr. 6(Suppl. 1): 313–317.

Xing, R., H. Yu, S. Liu, W. Zhang, Q. Zhang, Z. Li and P. Li. 2005. Antioxidative activity of differently regioselective chitosan sulfates *in vitro*. Bioorg. Med. Chem. 13: 1387–1392.

Yamaguchi, K., M. Murakami, H. Nakano, S. Konosu, T. Kokura, H. Yamamoto, M. Kosaka and K. Hata. 1986. Supercritical carbon dioxide extraction of oils from Antarctic krill. J. Agric. Food Chem. 34: 904–907.

Yang, E.J., J.G. Kim, J.Y. Kim, S.C. Kim, N.H. Lee and C.G. Hyun. 2010. Anti-inflammatory effect of chitosan oligosaccharides in RAW 264.7 cells. Cent. Eur. J. Biol. 5(1): 95–102.

Yao, H.T., S.Y. Huang and M.T. Chiang. 2008. A comparative study on hypocholesterolemic effects of high and low molecular weight chitosan in streptozotocin-induced diabetic rats. Food Chem. Toxicol. 46: 1525–1534.

Zamorano, J.P., O. Martínez-Álvarez, P. Montero and M.C. Gómez-Guillén. 2008. Characterisation and tissue distribution of polyphenol oxidase of deepwater pink shrimp (*Parapenaeus longirostris*). Food Chem. 112(1): 104–111.

Zhang, H., Y. Du, X. Yu, M. Mitsutomi and S. Aiba. 1999. Preparation of chitooligosaccharides from chitosan by a complex enzyme. Carbohydr. Res. 320: 257–260.

Zhang, H., Y.J. in, Y. Deng, D. Wang and Y. Zhao. 2012. Production of chitin from shrimp shell powders using *Serratia marcescens* B742 and *Lactobacillus plantarum* ATCC 8014 successive two-step fermentation. Carbohydr. Res. 362: 13–20.

19

Selenium-Health Benefit Values as Seafood Safety Criteria

Nicholas V.C. Ralston,[1,]* Alexander Azenkeng,[1] Carla R. Ralston,[1] J. Lloyd Blackwell III[2] and Laura J. Raymond[1]

1 Introduction

1.1 Seafood safety concerns

Ocean fish are healthy sources of low-fat protein and tend to be rich in essential omega-3 long-chain polyunsaturated fatty acids (n-3 PUFA), selenium (Se), vitamins A and D, iodine, and other nutrients (U.S. Food and Drug Administration [FDA] 2009, Lund 2013). Since these nutrients are essential for normal physiology and are important for optimal fetal neurodevelopment and maternal health, increased ocean fish intake during pregnancy is notably beneficial. However, these benefits can be diminished by potentially deleterious coexposures to methylmercury (MeHg) which accumulates at high levels in species that occupy the upper tiers of the marine food web. Prior studies of highly exposed sentinel populations in New Zealand and the Faroe Islands reported that children with the highest blood MeHg contents were adversely affected (Crump et al. 1998; Grandjean et al. 1997). More recent studies are finding that children are benefitted instead of harmed by maternal consumption of ocean fish (Daniels et al.

[1] Energy & Environmental Research Center, University of North Dakota, 15 North 23rd Street, Grand Forks, ND 58202-9018.
[2] Emeritus Professor of Economics, University of North Dakota, Grand Forks, ND 58202-9018.
* Corresponding author: nralston@undeerc.org

2004; Oken et al. 2005, 2008a,b; Hibbeln et al. 2007; Budtz–Jørgensen et al. 2007; Lederman et al. 2008; Davidson et al. 2011) and that avoiding ocean fish during pregnancy may harm child health (Hibbeln et al. 2007).

Based on the adverse effects of maternal MeHg exposures which were noted in the Faroe Island studies, the FDA and the U.S. Environmental Protection Agency (EPA) generated the 2004 Joint Federal Advisory for Mercury in Fish (U.S. EPA 2004) which advises women who are pregnant or who may become pregnant, nursing mothers, and young children to avoid species of fish that are very high in mercury (shark, swordfish, king mackerel, and tilefish) and 'eat up to 12 ounces (340 grams) a week of a variety of fish and shellfish that are lower in mercury.' However, consumers find the advisory confusing and difficult to understand with apparently conflicting views that leave most consumers unsure whether seafoods are safe and beneficial for them to eat or hazardous due to their MeHg contents. For example, the advisory's message to consumers concludes, 'With a few simple adjustments, you can continue to enjoy these foods in a manner that is healthy and beneficial and reduce your unborn or young child's exposure to the harmful effects of mercury at the same time.' Since this leaves readers with the impression that the Hg present in the fish they are being encouraged to eat is still going to have some harmful effects, many decide their safest alternative is to totally avoid eating seafood.

Unfortunately, consumers are not informed that adverse effects have also been observed among children of mothers who avoided eating seafood during pregnancy. Mothers who choose to avoid seafood during pregnancy fail to realize that they may actually be putting their children at greater risk of harm because they are not receiving the beneficial nutrients obtained from eating seafood (Hibbeln et al. 2007). The Joint FAO/World Health Organization (WHO) Expert Consultation on the Risks and Benefits of Fish Consumption (FAO 2010) concluded that for fish with MeHg concentrations below 0.5 ppm the benefits of maternal fish consumption outweigh the risks for up to seven 100-g (3.5-oz) servings a week.

Consumers are not alone in having difficulty understanding how seafood's health benefits relate to potential risks of MeHg exposures. Physicians, dieticians, regulators, and legislators responsible for protecting and improving public health tend to experience similar levels of confusion. There is a need for an easily understood seafood safety index regarding beneficial seafood choices that are clearly differentiated from those associated with risks. This was the primary motivating factor for developing the selenium-Health Benefit Value (HBV_{Se}) criterion.

2 Seafood Selenium and Mercury

2.1 Seafood as a source of Selenium

Selenium is a nutritionally essential trace element with vitally important functions in brain (Behne et al. 2000) and endocrine tissues (Köhrle et al. 2000). Its significance as a micronutrient is well known, and interest in Se-dependent physiology is steadily increasing. Many metabolic processes and, hence, many clinical conditions and diseases involve disruption of Se physiology (Rayman 2000). Ocean fish are among the richest dietary sources of Se with concentrations in fillets that vary considerably by species, but usually remain relatively constant in each type of fish regardless of its size (Kaneko and Ralston 2007). In the United States, 15 of the top 20 sources of dietary Se are ocean fish (U.S. Department of Agriculture 2012).

Figure 1 shows the sulfur and Se-containing amino acids which occur in animal proteins. The molecular similarities between methionine (Met) and selenomethionine (SeMet) are obvious, as are the similarities among the chalcogen amino acid series: serine (Ser), cysteine (Cys), and selenocysteine (Sec). Although the structural distinctions between Ser, Cys, and Sec simply involve their terminal hydroxyl (OH), thiol (SH), or selenoate (Se⁻) side groups, the genetic and biochemical distinctions between these forms are far more significant. As their pKa's indicate, Ser (pKa ~13) and Cys (pKa =

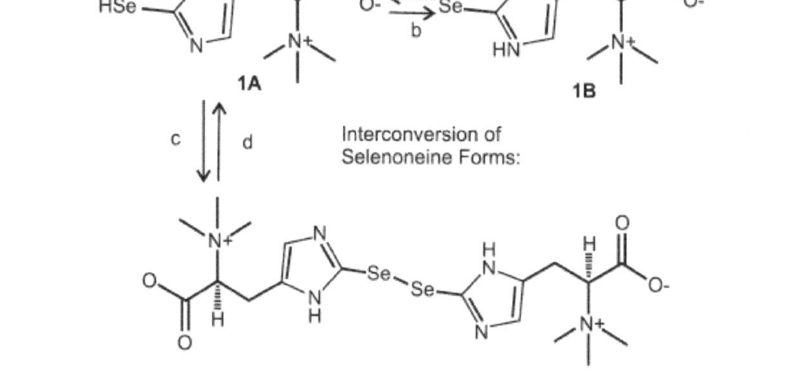

Figure 1. Molecular Forms of Selenomolecules and Related Species.

8.3) are predominantly protonated at physiological pH (7.4), but Sec (pKa = 5.2) is almost exclusively in the highly reactive ionized selenoate form.

The biological importance of Se is primarily dependent on the biochemical functions of Sec, but it is important to understand how these various forms relate to one another. Although SeMet can be present in any protein, the mere presence of Se is not sufficient to define a selenoprotein. This is because SeMet is not biochemically distinguished from Met and does not employ a unique DNA/RNA encoding sequence or use a specific tRNA for its insertion into proteins. Only when SeMet is degraded to release an inorganic Se which can be employed for the *de novo* synthesis of Sec does its Se become involved in the metabolic pathways that are unique to selenoproteins.

Selenocysteine, single letter code U (IUPAC 1999), is the 21st genetically encoded amino acid (Gladyshev 2004). As shown in Fig. 1, Sec is the most nucleophilic member of the chalcogen series of amino acids and is the most powerful intracellular nucleophile. The enzymes that employ Sec are catalytically elite compared to forms where Cys has been substituted by genetic modification. All vertebrates and almost all invertebrates use selenoenzymes to protect their brain tissues against oxidative damage that might otherwise occur as a byproduct of normal cellular respiration (Schweizer et al. 2004a; Chen and Berry 2003). Endocrine tissues also require selenoenzymes to perform these functions as well as perform tissue-specific functions (Köhrle et al. 2000). Therefore, Se is essential to mollusks, crustaceans, and ocean fish (Lobanov et al. 2009) as well as all other vertebrates. Humans express selenoproteins in tightly regulated tissue-specific occurrences and distributions that are similar among all mammals (Hatfield et al. 2006). Fish and mammalian proteomes share the same core selenoenzyme families, but ocean fish also express species-specific selenoproteins which are not present in mammals. Ocean fish possess 30–37 Sec-containing proteins, making their Sec-proteome rank among the largest of all life forms (Lobanov et al. 2009).

The protein-forming amino acids Sec and SeMet are present in ocean fish along with other forms which are not protein-associated. The nonprotein molecular forms of Se that are present in ocean fish are more varied and abundant than those which occur in humans (Yoshida et al. 2011). Recently, selenoneine, a novel organic form of Se (see Fig. 1), was found to be particularly abundant in pelagic fish (Yamashita and Yamashita 2010, Yamashita et al. 2010a,b). This imidazole compound is a Se analogue of ergothioneine, an amino acid made by actinobacteria and filamentous fungi (Fahey 2001) which is well absorbed and accumulates at millimolar levels in brain tissues. In humans, ergothioneine is absorbed from the gut and concentrated in some tissues by specific transporters, but its physiological functions are not yet known (Ey et al. 2007). Selenoneine comprises ~90%

of Se in fish blood and muscle tissues and contributes to their high Se contents in comparison to most terrestrial food products. Selenoneine's imidazole ring has free radical scavenging activity that differs from other sulfur-containing antioxidant molecules such as glutathione, thioredoxin, or lipoic acid (Yamashita and Yamashita 2010). Hypothetically, it may provide antioxidant protection to tissues of fish that experience anoxia during feeding forays into the oxygen-poor depths of the ocean. The Se in fish has been shown to be available for selenoenzyme synthesis (Ralston and Raymond 2010b), but further study with purified selenoneine will be needed for confirmation.

2.2 Selenoenzymes and their functions

Computational analysis of the human genome has identified 25 distinct selenoproteins (Kryukov et al. 2003). As reviewed by Hatfield and Gladyshev (2002), these include the glutathione peroxidase family of enzymes with antioxidant capabilities that prevent and reverse oxidative damage and the thioredoxin reductase family that maintains the cytosolic and mitochondrial lumens in the reducing state required for cellular respiration. These redox control functions are assisted by Selenoprotein M and W. Three Se-dependent deiodinases regulate thyroxin metabolism, and another governs calcium homeostasis. Selenophosphate synthetase is required for *de novo* synthesis of the Sec residue required by all selenoenzymes. The methionine sulfoxide reductase (SelR, or MsrB1) that reverses oxidative damage to proteins is also important. Selenoprotein P (SelP), the only selenoprotein to have more than one Sec residue in its primary sequence, possesses ten Sec residues and is pivotal in Se transport to brain and endocrine tissues. During dietary shortfall, metabolic mechanisms release Se from tissue reservoirs, and SelP is bound and internalized by the ApoER2 receptor expressed on the surfaces of brain and endocrine cells, ensuring these tissues are preferentially supplied. If dietary intakes exceed metabolic needs, multiple excretory pathways (e.g., urine, hair, exhalation) are highly effective in eliminating excess Se from the body.

Selenium is essential for normal physiology, particularly in the brain and neuroendocrine tissues (Schweizer et al. 2004a,b; Chen and Berry 2003; Köhrle et al. 2000). Highly effective homeostatic mechanisms selectively supply these tissues with Se regardless of dietary intakes. Even after prolonged dietary Se deficiencies have caused the Se contents of all other tissues to diminish to 1% of normal, the Se of brain and neuroendocrine tissues will plateau at 60% of normal; however, their selenoenzyme activities will diminish only slightly (Behne et al. 2000). Since brain selenoenzyme activities are maintained at approximately 90% of their normal levels, it appears that the Se retained in brains of rats fed low-Se diets primarily reflects the Sec remaining following degradation of SeMet to support Sec synthesis. No overt signs accompany these slight diminishments in brain selenoenzyme activities that occur during dietary Se deprivation. Because tissues that require Se remain sufficiently supplied, no Se deficiency phenotype occurs. Genetic knockouts of SelP (Hill et al. 2003) or the ApoER2 receptor that binds SelP (Valentine et al. 2008), are accompanied by disabling neurofunctional deficits that can be offset by supplemental dietary Se. The only other treatment that impairs Se distribution and availability in brain and endocrine tissues is MeHg toxicity (Ralston et al. 2007, 2008).

2.3 Methylmercury as an irreversible inhibitor of selenoenzymes

Although the high affinities between sulfur and Hg are well known (e.g., thiol molecules are known as 'mercaptans', meaning mercury capturing), early assumptions that thioenzymes were the critical molecular target of Hg toxicity appear to have been mistaken. Selenium-dependent protective effects which counteract Hg intoxication have been demonstrated in all vertebrates that have been investigated (Cuvin–Aralar and Furness 1991; Chapman and Chan 2000) as well as in invertebrates that express selenoenzymes. Initially it was thought protection occurred because Se trapped Hg. It is now understood that supplemental dietary Se replaces the Se which has actually been sequestered by Hg.

Because MeHg–Cys biochemically resembles Met (i.e., a molecular mimic of Met) and other neutral amino acids that the LAT1 amino acid transporter binds and mobilizes (Simmons–Willis et al. 2002), it is readily distributed between maternal/placental/fetal compartments (Aschner and Clarkson 1989; Bridges and Zalups 2010). Methylmercury readily exchanges binding partners with chemical species of equal or greater affinities. Since the covalent binding affinities between Se and Hg are approximately a million times greater than those between sulfur and Hg (Dyrssen and Wedborg 1991), mass action effects support selective sequestration of Se in association with MeHg. Enzyme kinetics are also a potentiating factor since the Cys thiol of glutathione, thioredoxin, and lipoic acids directly interact with Sec in active sites of selenomolecules (Arnér 2009). Therefore, MeHg bound to the Cys of these substrates enters the enzyme active site in the correct orientation to directly encounter and bind to the far more nucleophilic Se of Sec.

The direct exchange of MeHg from 'suicide substrates' to the enzyme's active site Sec appears to be the molecular mechanism of MeHg toxicity (Carvalho et al. 2008). Since a direct interaction with an enzyme active site via covalent bonding is the defining feature of an irreversible inhibitor, MeHg is, by biochemical definition, a highly selective irreversible inhibitor of selenoenzymes. The MeHg–Sec that forms is eventually degraded to the inorganic mercury selenide (HgSe) that accumulates in lysosomes, especially in highly exposed individuals (Falnoga et al. 2000, 2006; Arai et al. 2004; Huggins et al. 2009; Korbas et al. 2010). This sequestration of Se prevents its participation in Sec synthesis, resulting in a conditioned Se deficiency, meaning that the Se remains present in the cell, but is no longer biologically active or available for redistribution to brain and endocrine tissues. In weanling rats fed low-Se diets, toxic MeHg exposures caused diminishments in brain Se to 43% of normal (Ralston et al. 2008), far less than the ~60% that had otherwise been the lowest brain Se observed.

High maternal Hg^{2+} exposures in mice diminish Se transport across the placenta, reducing the Se supplied to their offspring by more than 50% (Parizek et al. 1971). High maternal MeHg exposures also severely restrict Se transport into fetal brain and decrease selenoenzyme activities to ~30% of normal (Watanabe et al. 1999a,b). Increasing MeHg exposures inevitably sequesters maternal, placental, and fetal Se, preventing or limiting distribution to the developing fetal brain. Since much of the Se present in maternal/placental/fetal compartments becomes bound as HgSe following high MeHg exposures, the actual amounts available for selenoenzyme synthesis in the fetal brain is much less than the total amount present.

Intracellular Se is normally present in substantial excess of MeHg throughout all body tissues, enabling tissue Se redistribution to brain and endocrine tissues when needed. However, high MeHg exposures sequester Se and diminish the amount that is available for selenoenzyme synthesis. Therefore, the previously noted 'protective' effect of supplemental Se in prevention of Hg toxicity actually occurs because dietary Se offsets the Se lost to Hg sequestration (Watanabe et al. 1999a,b), thereby preventing or reducing the interruption of selenoenzyme synthesis and activities in vital tissues.

2.4 Selenium in prevention of mercury toxicity

Mercury and/or MeHg results in oxidative damage, but this is not due to direct mechanisms (Seppänen et al. 2004). Instead, their effects are through inhibition of the selenoenzyme activities that prevent and reverse oxidative damage arising from free radicals and reactive oxygen species formed during normal oxygen metabolism. Selenium's ability to counteract toxic effects of high Hg exposures has been recognized for ~50 years (Parizek and Ostadolova 1967), and subsequent studies have confirmed that dietary Se prevents, ameliorates, and reverses toxic effect of high Hg or MeHg exposures (Iwata et al. 1973; Kosta et al. 1975; Wada et al. 1976; Ohi et al. 1976; Beijer and Jernelov 1978; El–Begearmi et al. 1982; Whanger 1992;

El–Demerdash 2001; Chen et al. 2006; Ralston et al. 2007, 2008; Ralston and Raymond 2010b). In addition, studies have shown that supplemental dietary Se prevents MeHg's dose-dependent inhibition of selenoenzyme activities in the fetal brain (Watanabe et al. 1999a,b; Stringari et al. 2008).

Additional studies demonstrate that seafood Se prevents or ameliorates MeHg toxicity, including studies of Se from yellow fin tuna (Ohi et al. 1976; Ganther et al. 1972), bigeye tuna (Ralston and Raymond 2010b), menhaden (Stillings et al. 1974), swordfish (Freidman et al. 1978; Ralston and Raymond 2010b), rockfish (Ohi et al. 1980). Although MeHg will also bind to sulfur which occurs in high millimolar concentrations while Se is present at low micromolar concentrations, Se does not need to be in molar excess of MeHg in order to prevent loss of selenoenzyme activities. It only needs to be present in amounts sufficient to maintain normal selenoenzyme activities and prevent the oxidative damage to the brain that otherwise accompanies high MeHg exposures. Thus, the Se in ocean fish is readily bioavailable for selenoenzyme synthesis and effective in counteracting MeHg toxicity. It is important to note that the protective effects of Se are much less when slow-release forms of Se such as SeMet are used in diets (e.g., Beyrouty and Chan 2006) instead of forms that are more readily degraded to inorganic selenide, the precursor for Sec synthesis (Ralston and Raymond 2010a). However, consuming foods with Se in molar excess of MeHg will ensure health risks are ameliorated or avoided.

Selenoenzyme activities in the brains of fetal rats are more vulnerable to MeHg exposures than the brains of their dams (Watanabe 1999a,b). In growing rats exposed to amounts of MeHg that were otherwise lethal, supplemental dietary Se preserved brain selenoenzyme activities and completely prevented development of any of the MeHg-dependent neurotoxic consequences that were monitored (Ralston et al. 2008). However, once fetal brain selenoenzyme activities have been abolished by high MeHg exposures, they do not readily recover (Stringari et al. 2008). Since the effects of high MeHg exposures coincide with those expected to occur (Ralston and Raymond 2010a) and depend on dietary Hg:Se molar ratios rather than Hg exposures alone (Ralston et al. 2008), the 'selenoenzyme inhibition and Se sequestration hypothesis' of MeHg toxicity (Salonen et al. 1995; Raymond and Ralston 2004; Carvalho et al. 2008, 2011; Ralston et al. 2008; Ralston 2008; Ralston and Raymond 2010b) are increasingly well supported.

2.5 Effects of methylmercury exposures

Catastrophic poisoning episodes in Japan and Iraq made the world aware of the potential risks associated with MeHg exposures from seafood consumption. In the 1950s and again in the 1960s, thousands of

Japanese were poisoned by extremely high MeHg concentrations that had accumulated in fish from waters that had been polluted by tons of Hg released from chemical factories (Harada 1968; Igata 1993). This was followed in the 1970s by poisoning incidents involving starving Iraqi villagers that had eaten bread made from seed grain impregnated with MeHg as a fungicide. Villagers that ate the contaminated bread did not experience symptoms until months later. This distinctive 'latency effect' is a feature of MeHg poisoning that had been unexplained for decades. However, since the delayed onset of symptoms coincides with the rate of Se sequestration and loss of selenoenzyme activities, the effect now seems self-explanatory. Another characteristic of high MeHg exposures was perhaps even more insidious. It was noted in these incidents that children exposed *in utero* often suffered severe neurological harm, even when their mothers had not suffered any ill effects (Harada 1995; Bakir et al. 1973). The accentuated vulnerability of the fetus to MeHg exposure aroused concern that seafood consumption during pregnancy is potentially harmful to developing children.

To examine this possibility, sentinel studies of populations with high rates of seafood consumption were initiated. Studies of populations that ate seafoods with high MeHg contents such as shark (New Zealand) or pilot whale meats (Faroe Islands) reported that fetal exposures were associated with adverse child outcomes (Crump et al. 1998; Grandjean et al. 1997). But populations with similarly high MeHg exposures due to consumption of ocean fish with lower MeHg contents have not found adverse effects. For example, in the Seychelle Islands, high rates of maternal fish consumption (~12 fish meals per week) resulted in higher average MeHg exposures than observed in the Faroes, but with no adverse effects (Davidson et al. 1998, 2011).

The reason for this discrepancy appears to be related to the Hg:Se molar ratios in top predators such as sharks and pilot whales vs. most types of commercially available ocean fish. The FDA Control Guidance Document for Fish and Fisheries Products has reiterated the 1.0 ppm Hg (~5 μmol MeHg/kg) limit for edible portions of ocean fish (U.S. FDA 2011). Concentrations of MeHg in some commercially harvested apex predator fish species approach or occasionally exceed this limit; however, most commercially available fish contain less than 0.5 ppm (~2.5 μmol MeHg/kg). The high Hg concentrations observed in top predators result in disproportionately high Hg:Se molar ratios in shark (2:1) and pilot whales (5:1), ratios that are far greater than in varieties of ocean fish (~1:4 to 1:20) that are available commercially (see Fig. 2).

Evaluations in the Faroe Islands report adverse effects in children following high maternal MeHg exposures (Grandjean et al. 1997, 1998; Debes et al. 2006). Situated between Scotland and Iceland, the Faroes have

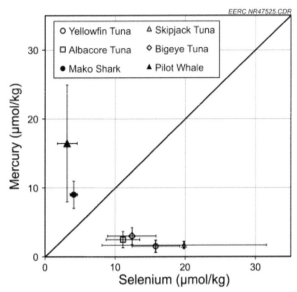

Figure 2. Molar relationships between Hg and Se in various seafood types (1 mg Se/kg = ~12.6 µmol Se/kg and 1 mg Hg/kg = ~5 µmol Hg/kg). Fish data is from Kaneko and Ralston (2007), and pilot whale data are from, Julshamn et al. (1987). The diagonal line defines the 1:1 Hg:Se molar ratio.

a small population of primarily Danish descent and a seafaring tradition that results in seafoods comprising a large portion of their diets. During the period of the first Faroes cohort study (March 1986 to December 1987), mothers consumed ~72 g of cod per day which contained an average of 3.8 µmol Se/kg and 0.3 µmol Hg/kg (Hellou et al. 1992). Cod were a rich source of Se (273.6 µmol Se/day) in their diets, but contributed relatively little MeHg (23.4 µmol MeHg/day). However, the Hg:Se molar ratios in the pilot whale meats they consumed were far different from those of most ocean fish.

Pilot whales are among the largest of oceanic dolphins, only slightly smaller than killer whales. They primarily feed on squid and pelagic fish and occupy the top tier of the ocean food web. During the study period, ~3,600 whales were harvested and distributed among the ~46,000 citizens living in the Faroes at the time of the study. The average Hg and Se contents of pilot whale meat, blubber, liver, and kidney were; 16.4, 3.4, 1,395.9, and 89.7 µmol Hg/kg, and; 1.4, 3.4, 126.7, and 16.5 µmol Se/kg, respectively (Julshamn et al. 1987). However, since much of the Hg present in most whale tissues is in the HgSe form (Huggins et al. 2009), only the biologically available MeHg and Se contents should be considered. The MeHg concentrations in pilot whale meat, blubber, and liver were 7.4, 0.8, and 162.3 µmol MeHg/kg. The amount of MeHg present in the kidneys they ate was not reported. Since inorganic Hg was in molar excess of Se in meat, blubber, and kidney, it can be assumed that the majority of the Se present in these tissues was bound as HgSe. The liver contained ~126 µmol more Se than Hg per kg, so it may have also been a good source of Se, although it was only intermittently consumed. However, since the liver accumulates cadmium at high concentrations (~442 µmol Cd/kg) which also exhibits Se binding behavior, little or no Se may actually have been available. The kidney also contained high cadmium (~674 µmol Cd/kg) in substantial excess of Se.

Pilot whale products were provided as free commodities to all Faroese citizens, and the amount of muscle, blubber, and liver eaten on a daily basis was 9.3, 4.6, and 0.5 g respectively (Anderson et al. 1987); the amounts of kidney consumed were not reported. The frequency and amounts of liver consumed by mothers in the Faroes cohort were not indicated by Grandjean et al. (1992); however, the relative Hg and Se contents of the liver they ate are mentioned. Pilot whale meat, blubber, and liver provided: 68.9, 3.6, and 69.6 µmol MeHg/day, respectively. Thus, consumption of pilot whale products resulted in ~85% of the total MeHg exposure and between 0 and 20% of their seafood Se intakes. Meanwhile cod provided ~15% of the total MeHg exposure and 85% or more of the Se they obtained from seafood.

Increased Hg in maternal hair and umbilical cord blood (see Fig. 3) was directly related to pilot whale meat consumption and the reported adverse outcomes in the children (Grandjean et al. 1997, 1998; Debes et al. 2006). High prenatal MeHg exposures were linked to minor decreased scores in language, attention, and memory of some assessments and, to a lesser extent, sensory and motor functions. As seen in Fig. 3, cord blood Hg in the Faroes children approached a 1:1 stoichiometry with Se.

Cod contributed the bulk of the Se in the Faroese diet, and comparatively small amounts of additional Se would have been provided by other food sources as well. The MeHg in the Faroese diets primarily came from eating pilot whale meat and, therefore, would not necessarily be accompanied by sufficient Se to offset their MeHg intakes. The dietary Se intakes of the Faroese is relatively poor in comparison to the United States, where the RDA of 50 µg Se/day (i.e., ~633 µmol Se/day) is typically met and often exceeded, even by consumers that do not eat Se-rich ocean fish. In cord bloods of individuals with low MeHg exposures, cord blood Se is in 100–1,000-fold molar excess of Hg (see Fig. 3), but as blood Hg increases, Se:Hg molar ratios diminish. Because Se is homeostatically regulated, cord blood Se remained essentially constant at 1 µM in Faroese children but increased significantly (F=26.5; p<0.0001) with increasing blood Hg. This increase may reflect a homeostatic response to accumulation of MeHg–Sec conjugates that render Se biologically unavailable.

Blood Hg levels of 0.2–0.5 mg Hg/kg (~1–2.5 µM Hg) define the threshold for clinical MeHg toxicity (World Health Organization 1976).

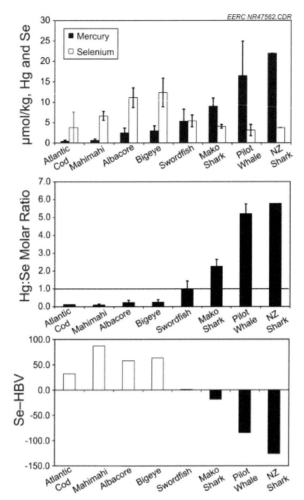

Figure 3. Seafood safety parameters of commonly consumed ocean fish in comparison to shark and pilot whale meats associated with harmful effects to prenatally exposed children.

These amounts correspond with the ~1:1 Se:Hg molar stoichiometry that is likely to compromise selenoenzyme synthesis and activities (Ralston and Raymond 2010a). Since the adverse effects that were associated with cord blood Hg in the Faroes study were proportional to declines in Se:Hg molar ratios in cord blood, their results are consistent with expectations based on the selenoenzyme inhibition theory of MeHg toxicity.

In informal terms, the conventional hypothesis states that maternal MeHg exposures are directly associated with adverse child development outcomes. The selenoenzyme inhibition mechanism of MeHg toxicity suggests the conventional hypothesis should be modified as follows:

Maternal MeHg exposures in excess of Se intakes are directly associated with adverse child development outcomes (Se hypothesis). The largest and most appropriate studies reject the conventional hypothesis but uniformly support the 'selenium hypothesis' (see Table 1).

The hazardously high Hg:Se molar ratios observed in the Faroes study are unlikely to be reproduced in populations that consume ocean fish, but similar or even more serious exposures might occur among freshwater fish consumers exposed to high MeHg contents in low-Se fish in watersheds of regions with poor soil Se availability. Since increasing Hg:Se molar ratios correspond with increasing risks of MeHg toxicity while MeHg exposures alone are an unreliable index of risk (Ralston et al. 2008; Ralston and Raymond 2010a), seafood safety assessments clearly need to consider Se.

The Department of Public and Occupational Health in the Faroe Islands has indicated that pilot whales are not fit for human consumption because of the high concentrations of organochlorines and MeHg in the meat (Uttranrikisradid 2009). The Faroese Food and Veterinary Authority advised the Faroese to no longer eat kidney or liver meats from pilot whales. They also recommend eating no more than one serving of muscle meat per month and for women to refrain from eating blubber if they plan to have children and to completely avoid eating whale meat if they are pregnant, planning to conceive in the next 3 months, or breastfeeding.

Based on the benchmark dose lower limit (BMDL) established in the Faroes (58 ppb Hg in cord blood), recommendations were developed to protect against potentially adverse effects of seafood MeHg exposures.

Table 1. Hypothesis Testing.

Hypothesis 1: Maternal MeHg exposures are directly associated with adverse child development outcomes.

Hypothesis 2: Maternal MeHg exposures in excess of Se intakes are directly associated with adverse child development outcomes.

Location	Outcome[a]	Hypothesis 1	HBV$_{Se}$[b]	Hypothesis 2
Japan	Harm	Retain	−5,000	Retain
Iraq	Harm	Retain	−800	Retain
New Zealand	Harm	Retain	20 to −120	Retain
Faroe Islands	Harm	Retain	3 to −80	Retain
Seychelles	Benefit	Reject	4 to 10	Retain
United Kingdom	Benefit	Reject	3 to 20	Retain
United States	Benefit	Reject	10 to 20	Retain
Denmark	Benefit	Reject	3 to 20	Retain

[a]Maternal seafood (MeHg) exposure effects on child's health outcomes.
[b]Selenium Health Benefit Values of MeHg-containing foods in Japan and Iraq were estimated using best available data. All others reflect the range of HBV$_{Se}$ of commonly consumed seafoods in each nation.

In calculating the MeHg RfD, a composite reflection of a 3-fold factor for pharmacokinetic variability and an additional 3-fold factor reflecting the pharmacodynamic imprecision (that may largely result from omitting consideration tissue Se concentrations), resulted in the adoption of a 10-fold uncertainty factor. Thus, 5.8 ppb Hg is recognized as a fetal blood level expected to be safe in all populations, including sensitive subgroups. Based on these data, the MeHg reference dose (RfD) of 0.1 µg/kg body weight per day was established (National Research Council 2000). The RfD indicates the daily maternal MeHg exposure that is without appreciable risk of adverse effects to their unborn children (Rice et al. 2003).

2.6 Sources of consumer confusion

To most consumers, the term 'seafood' primarily indicates ocean fish but also includes various types of lobster, shrimp, oysters, and clams. For this reason, consumers tend to be surprised when they learn that seafood safety warnings regarding potential risks of MeHg exposures are primarily based on adverse effects associated with eating pilot whale meat. Although there is controversy regarding the safety of consuming certain types of seafood during pregnancy, what have been thought to be conflicting findings of the human studies shown in Table 1 are actually mutually supportive perspectives of the importance of considering both Hg and Se in seafood safety issues. Adverse effects of MeHg exposures only arise in studies that investigate consumers of seafood with high Hg:Se molar ratios. Among consumers of seafood with Se in molar excess of Hg, the net effects have been beneficial. Unfortunately, consumers and most public health professionals remain largely unaware of the importance of Hg:Se molar ratios in seafood safety assessments.

Current maternal seafood consumption warnings are appropriate for protection against MeHg exposures arising from consumption of diets that include pilot whale meat but may not protect or improve the health of populations that consume commercially available fish. This is because it is impossible to properly establish a risk-benefit assessment without evaluating the effects of beneficial nutrients that are also present in seafood. To better serve the needs of seafood consumers, safety criteria should provide reliably accurate indications of risks in relation to the benefits expected to accompany consumption. This type of criteria would provide regulatory agencies, health professionals, and the public with a far more accurate indication of which fish are safe and beneficial to eat and which ought to be avoided by pregnant women.

3 Selenium-Health Benefit Values

3.1 Origin of the Selenium Health Benefit Value

To provide an assessment of the relationship between seafood MeHg exposures and dietary Se intakes which would serve as an index of seafood safety, the selenium-Health Benefit Value (HBV_{Se}) was proposed by Kaneko and Ralston (2007). Because it incorporates the mechanistic relationships between dietary Se intakes in regard to concurrent MeHg exposures, the HBV_{Se} (see Equation 1) provides more reliable criteria for seafood safety than assessments based on MeHg exposures alone (see Fig. 4).

The equation for calculating the HBV_{Se} considers the amounts of Se in relation to MeHg within that seafood. The equation for calculating the HBV_{Se} was structured to reflect the availability of dietary Se for synthesis of selenoenzymes in relation to the risks of MeHg exposures potentially resulting in significant inhibition of selenoenzymes. To predict these mechanistic outcomes, molar concentrations of Se and Hg must be used in the calculation:

$$HBV_{Se} = (Se - Hg)/Se \bullet (Se + Hg) \qquad \text{(Equation 1)}$$

A positive HBV_{Se} indicates Se occurs in molar excess of MeHg in the seafood, while a negative HBV_{Se} indicates MeHg occurs in molar excess of Se. Therefore, consuming seafoods with a positive HBV_{Se} would improve maternal and fetal Se-status, whereas eating seafoods with a negative HBV_{Se} could compromise maternal and fetal Se-status. The magnitudes of the HBV_{Se} reflect the magnitudes of the Se-surplus or deficiency predicted to accompany eating that seafood. Foods that contain far more MeHg than Se have a very negative HBV_{Se}, e.g., pilot whale (HBV_{Se} –82), mako shark (HBV_{Se} –16), and the estimated value for the high Hg shark meat (4.4 ppm) that was consumed in New Zealand (HBV_{Se} –120), but almost all other types of seafood (e.g., skipjack 19.6 ± 11.8, yellowfin tuna 15.6 ± 3.4, bigeye tuna 11.56 ± 3.8, blue marlin 11.56 ± 4.2, pacific cod 2.86 ± 1.2, coho salmon 2.96 ± 0.9) have highly positive HBV_{Se} (Ralston 2008; Kaneko and Ralston 2007; Ralston and Raymond 2010b). Pregnant women and children should limit eating seafoods with a negative HBV_{Se} because of MeHg exposure risks. In contrast, seafoods with a positive HBV_{Se} are a good source of dietary Se and will prevent adverse effects from MeHg exposure.

Children whose mothers had consumed the highly contaminated fish in Minamata, Japan in the 1950's (HBV_{Se} = –5,000) or the MeHg treated seed grain in Iraq during the 1970's (HBV_{Se} = –800) poisoning catastrophes would have had little Se delivered to their developing brain tissues, and even the inadequate amount that was available would have been continually lost to

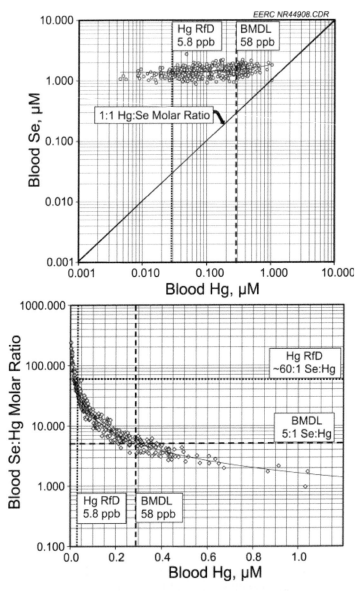

Figure 4. Umbilical cord blood mercury and selenium concentrations from the Faroe Islands study. Dotted lines indicate the BMDL and reference dose on the basis of Hg and Hg:Se. Data were acquired from graphs shown in Choi et al. 2008, using Image J (http://rsbweb. nih.gov/ij/) particle recognition software to identify x and y coordinates and plotted using Microsoft Excel.

sequestration by MeHg present in >50-fold molar excess of Se. The resulting diminishment of selenoenzyme activities in the child's vital but vulnerable brain and endocrine tissues would have been severe.

Extinction of Se availability due to Hg binding did not occur in the highly exposed children of the Faroes, perhaps because they also ate large amounts of Se-rich cod which would have replenished their Se reserves. Hypothetically, intermittent interruptions in selenoenzyme synthesis due to transitory diminishments of Se transport to the developing child's brain could account for the subtle effects that were observed. The same is true of the New Zealand experience with meats of large sharks. The distinct differences in the HBV_{Se} of the seafoods that were consumed in those studies may explain why epidemiological studies of maternal seafood consumption performed in the Seychelle Islands and other populations consuming fish with positive HBV_{Se} have not observed adverse effects. Instead, MeHg exposures from ocean fish consumption were associated with improved cognitive and neurodevelopmental outcomes (Davidson et al. 2011). Since only fish with positive HBV_{Se} were eaten by these populations, disruptions in selenoenzyme synthesis would not have occurred, so the beneficial effects of improved omega-3 and other vital nutrients were observed instead.

3.2 Methodological concerns in MeHg toxicity studies

To evaluate dose-effect outcomes in epidemiological studies which can involve multiple concomitant variables which may enhance or diminish the effects of the primary regressor, careful consideration must be given to model specification in order to limit or avoid potential sources of error. Studies of prenatal MeHg exposures that selectively omit the consideration of variables known to influence neurological outcomes raise methodological concerns.

Multiple regression models are useful tools for estimation of epidemiological models because of the information they yield. Consider a hypothetical model in which neurodevelopmental outcomes (NO) are the regressands (dependent variables), and the regressors (independent variables) are cord blood levels of MeHg, Se, and n-3 PUFA. In this example, we will assume that the model has no difficulties other than in the selection of regressors and is stated as:

$$NO = b_0 + b_1 MeHg + b_2 Se + b_3 (n-3\ PUFA) + e \qquad \text{(Equation 2)}$$

In Equation 2, b_0, b_1, b_2, and b_3 are sample estimates of the corresponding population parameters, and 'e' represents sample errors. Since our example assumes that MeHg, Se, and n-3 PUFAs are the only variables influencing NO, b_0, b_1, b_2, and b_3 are unbiased (Johnston and Dinardo 1997). This means that b_0, b_1, b_2, and b_3 can be used to show the independent influences of each regressor, and the impacts of one-unit changes in blood MeHg, Se, and n-3 PUFA respectively, on NO. Furthermore, by inserting observed values of MeHg, Se, and n-3 PUFA, the expected value of NO can be determined.

Estimated regression models are useful tools for public health issues. For example, if our model tells us that $b_1 = -1.5$, $b_2 = 0.05$, and $b_3 = 0.01$, and we observe blood MeHg=1.5 µM, Se=1.0 µM, and n-3 PUFA =100 µg/mL in a sample from a well-defined population, child NO would be–1.20, an undesirable outcome. If seafood intakes changed following public health advisories and blood MeHg contents diminished to 0.28 µM, while Se intakes improved and increased cord blood Se contents to 2.0 µM, along with blood n-3 PUFA contents increasing to 150 µg/mL, a desirable outcome would be achieved; NO = 1.18.

In this example, nothing has been said about statistical testing of this model, although many researchers appear to believe that only statistically significant regressors can be considered. Also, the possibility of confounding has not yet been mentioned. In the above model, if the estimated coefficient of blood MeHg was influenced by Se and/or n-3 PUFA, the effect of MeHg would be confounded with those of Se and n-3 PUFA. In that case, it would be impossible to determine the independent effects of MeHg and perhaps Se and n-3 PUFA as well.

In regression models, confounding is caused by multicollinearity, which simply means that regressors are linearly related (Wooldridge 2003). In this example, MeHg, Se, and n-3 PUFA were assumed to be the only influences on NO, and multicollinearity was, therefore, assumed not to cause confounding. However, it is important to note that multicollinearity inflates the standard errors of regressors, which can lead to insignificant t-statistics of some regressors. Insignificant regressors are often considered to contribute nothing of value to a model and are summarily dropped. Whether a particular insignificant variable is dropped without further thought will depend on attitudes inherited from statistics training. For example:

"One or several independent variables may be dropped from the model in order to lessen the multicollinearity problem and thereby reduce the standard errors of the estimated regression coefficients of the independent variables remaining in the model. This remedial measure has two important limitations. First, no direct information is obtained about the dropped independent variables. Second, the magnitudes of the regression coefficients for the independent variables remaining in the model are affected by the correlated independent variables not included in the model." (Neter et al. 1985)

Neter et al. clearly understand multicollinearity, but they do not discourage the practice of permanently dropping collinear variables as strongly as they should. Their concerns about inflated standard errors causing insignificant t-statistics was greater than their considerations regarding bias which causes confounding.

It is always important to know whether multicollinearity is present in any model estimated. Diagnostics for this are available (Wooldridge 2003; Greene 2003), but temporarily dropping variables is often a more reliable way to determine whether multicollinearity is present. For example, suppose the t-statistic of b_1, the coefficient of MeHg, is insignificant at the 5% level. If we temporarily drop Se from the model, run the adjusted regression model again, and observe that the standard error of b_1 decreases and the t-statistic increases, there is a linear relation between Se and MeHg. We next need to determine whether both Se and n-3 PUFA are collinear with MeHg, so we drop n-3 PUFA, run a regression with MeHg as the only regressor, and again observe b_1. Additional changes in b_1 and its standard error would indicate that MeHg is linearly related with both Se and n-3 PUFA.

The change in the coefficient of MeHg when Se and/or n-3 PUFA were dropped shows that multicollinearity has caused a bias, and the coefficient of MeHg is now confounded. It is clear that the only way to avoid this confounding of the effects on MeHg is to retain variables which are collinear with MeHg. This clearly leaves the researcher caught between statistically insignificant results and confounded results. It might be argued that things would be fine if the effects of the multicollinearity could be removed, but this is not allowed, because standard errors of the independent, unbiased effects are affected by multicollinearity and any attempt to purge a standard error of the influence of multicollinearity introduces bias to the estimated standard error. On the other hand, the path of blind obedience to statistical significance testing is very likely to lead to confounded coefficients which rob affected coefficients of meaning. Learning to deal with known multicollinearity to achieve unbiased estimates seems preferable.

Science will usually have established the independent variables known or expected to influence the dependent variable in most studies. Indeed, such information should be used to specify variables that should be included. It is also likely that the signs of coefficients of particular regressors and/or plausible ranges of values will be known, so prior scientific knowledge can be used to support inclusion of variables that might otherwise be eliminated. In addition, one can argue that some variables need to be retained because they are collinear with other variables which would otherwise be confounded. It is rarely necessary to surrender to confounding.

4 Conclusions

Seafood is an excellent source of many beneficial nutrients, including Se, which is required for numerous biological activities especially in the brain and endocrine tissues. Methylmercury, also found in fish, is a highly specific irreversible selenoenzyme inhibitor that sequesters tissue Se, resulting in the biologically unavailable HgSe form. Providing adequate dietary Se to offset losses due to MeHg binding negates loss of selenoenzyme activities and alleviates toxicity risks associated with MeHg exposures. Therefore, the molar ratios of Se and MeHg need to be considered from a mechanistically informed perspective in order to perform reliably accurate risk assessments regarding MeHg exposures from seafood consumption.

Risk evaluations based on MeHg exposures alone cannot reliably predict the effects of maternal fish consumption on child development outcomes. For this reason, the more reliably accurate HBV_{Se} criteria should be used to indicate which MeHg containing seafoods should be avoided and which are safe and nutritionally beneficial to consume.

Acknowledgements

The authors thank Kathleen McIntyre for performing the analysis for the studies reported in this article. This research was supported by a grant from the U.S. EPA: EPA–STAR–RD834792–01 (Fish Selenium-Health Benefit Values in Mercury Risk Management) and National Oceanic and Atmospheric Administration Grant NA10NMF4520363 (Selenium Health Initiative Program II) to the University of North Dakota Energy & Environmental Research Center. This article was not reviewed by the funding agencies. Therefore, it does not necessarily reflect their views, and no official endorsements should be inferred.

References

Anderson, A., K. Julshamn, O. Ringdal and J. Morkore. 1987. Trace elements intake in the Faroe Islands II. Intake of mercury and other elements by consumption of pilot whales (*Globicephalus meleanus*). Sci. Total Environ. 65: 63–68.

Arai, T., T. Ikemoto, A. Hokura, Y. Terada, T. Kunito, S. Tanabe and I. Akai. 2004. Chemical forms of mercury and cadmium accumulated in marine mammals and seabirds as determined by XAFS analysis. Environ. Sci. Technol. 38: 6468–6474.

Arnér, E.S.J. 2009. Focus on mammalian thioredoxin reductases—Important selenoproteins with versatile functions. Biochim. Biophys. Acta 1790: 495–526.

Aschner, M. and T.W. Clarkson. 1989. Methyl mercury uptake across bovine brain capillary endothelial cells *in vitro*: The role of amino acids. Pharmacol Toxicol. 64(3): 293–297.

Bakir, F., S.F. Damluji, L. Amin–Zaki, M. Murtadha, A. Khalidi and N.Y. al–Rawi. 1973. Methylmercury poisoning in Iraq. Science 181: 230–241.

Behne, D., H. Pfeifer, D. Rothlein and A. Kyriakopoulos. 2000. Cellular and subcellular distribution of selenium and selenium–containing proteins in the rat. In: A.M. Roussel, A.E. Favier, and R.A. Anderson (eds.). Trace Elements in Man and Animals 10. Kluwer Academic/Plenum Publishers, NY, pp. 29–34.

Beijer, K. and A. Jernelov. 1978. Ecological aspects of mercury–selenium interaction in the marine environment. Environ. Health Perspect. 25: 43–45.

Beyrouty, P. and H.M. Chan. 2006. Co-consumption of selenium and vitamin E altered the reproductive and developmental toxicity of methylmercury in rats. Neurotoxicol. Technol. 28: 49–58.

Bridges, C.C. and R.K. Zalups. 2010. Transport of inorganic mercury and methylmercury in target tissues and organs. J. Toxicol. Environ. Health. 13: 385–410.

Budtz–Jørgensen, E., P. Grandjean and P. Weihe. 2007. Separation of risks and benefits of seafood intake. Environ. Health Perspect. 115: 323–327.

Carvalho, C.M.L., E.–H. Chew, S.I. Hashemy, J. Lu and A. Holmgren. 2008. Inhibition of the human thioredoxin system: A molecular mechanism of mercury toxicity. J. Biol. Chem. 283: 11913–11923.

Carvalho, C.M.L., J. Lu, X. Zhang, E.S.J. Arnér and A. Holmgren. 2011. Effects of selenite and chelating agents on mammalian thioredoxin reductase inhibited by mercury: Implications for treatment of mercury poisoning. FASEB J. 25: 370–381.

Chapman, L. and H.M. Chan. 2000. The influence of nutrition on methyl mercury intoxication. Environ. Health Perspect. 108: 29–56.

Chen, C., L. Qu, J. Zhao, S. Liu, G. Deng, B. Li, P. Zhang and Z. Chai. 2006. Accumulation of mercury, selenium, and their binding proteins in porcine kidney and liver from mercury exposed areas with the investigation of their redox responses. Sci. Total Environ. 366: 627–637.

Chen, J. and M.J. Berry. 2003. Selenium and selenoproteins in the brain and brain diseases. J. Neurochem. 86: 1–12.

Choi, A.L., E. Budtz-Jørgensen, P.J. Jørgensen, U. Steuerwald, F. Debes, P. Weihe and P. Grandjean. 2008. Selenium as a potential protective factor against mercury developmental neurotoxicity. Environ. Res. 107(1): 45–52.

Crump, K.S., T. Kjellstrom, A.M. Shipp, A. Silvers and A. Stewart. 1998. Influence of prenatal mercury exposure upon scholastic and psychological test performance: Benchmark analysis of a New Zealand cohort. Risk Anal. 18: 701–713.

Cuvin-Aralar, M.L. and R.W. Furness. 1991. Mercury and selenium interaction: A review: Ecotoxicol. Environ. Safety 21: 348–364.

Daniels, J.L., M.P. Longnecker, A.S. Rowland and J. Golding. 2004. Fish intake during pregnancy and early cognitive development of offspring. Epidemiology 15: 394–402.

Davidson, P.W., D.A. Cory–Slechta, S.W. Thurston, L.–S. Huang, C.F. Shamlaye, D. Gunzler, G. Watson, E. van Wijngaarden, G. Zareba, J.D. Klein, T.W. Clarkson, J.J. Strain and G.J. Myers. 2011. Fish consumption and prenatal methylmercury exposure: Cognitive and behavioral outcomes in the main cohort at 17 years from the Seychelles child development study. Neurotoxicology 32: 711–717.

Davidson, P.W., G.J. Myers, C. Cox, C. Axtell, C. Shamlaye and J. Sloane–Reeves. 1998. Effects of prenatal and postnatal methylmercury exposure from fish consumption on neurodevelopment: Outcomes at 66 months of age in the Seychelles Child Development Study. J. Am. Med. Assoc. 280: 701–707.

Debes, F., E. Budtz–Jorgensen, P. Weihe, R.F. White and P. Grandjean. 2006. Impact of prenatal methylmercury exposure on neurobehavioral function at age 14 years. Neurotoxicol. Teratol. 28: 536–547.

Dyrssen, D. and M. Wedborg. 1991. The sulfur-mercury(II) system in natural waters. Water, Air, Soil Pollut. 56: 507–519.

El-Begearmi, M.M., H.E. Ganther and M.L. Sunde. 1982. Dietary interaction between methylmercury, selenium, arsenic and sulfur amino acids in Japanese quail. Poultry Sci. 61: 272–9.

El-Demerdash, F.M. 2001. Effects of selenium and mercury on the enzymatic activities and lipid peroxidation in brain, liver, and blood of rats. J. Environ. Sci. Health 36: 489–99.

Ey, J., E. Schömig and D. Tauber. 2007. Dietary sources and antioxidant effects of ergothioneine. J. Agric. Food Chem. 55: 6466–74.

Fahey, R.C. 2001. Novel thiols of prokaryotes. Annu. Rev. Microbiol. 55: 333–56.

Falnoga, I., M. Tusek–Znidaric, M. Horvat and P. Stegnar. 2000. Mercury, selenium, and cadmium in human autopsy samples from Idrija residents and mercury mine workers. Environ. Res. 83: 211–218.

Falnoga, I., M. Tušek–Žnidarič and P. Stegnar. 2006. The influence of long-term mercury exposure on selenium availability in tissues: An evaluation of data. BioMetals 19: 283–294.

FAO Fisheries and Aquaculture Report No. 978. 2010. Report of the Joint FAO/WHO Expert Consultation on the Risks and Benefits of Fish Consumption. www.fao.org/docrep/014/ba0136e/ba0136e00.pdf.

Freidman, M.A., L.R. Eaton and W.H. Carter. 1978. Protective effects of freeze-dried swordfish on methylmercury content. Bull. Environ. Contam. Toxicol. 19: 436–443.

Ganther, H., C. Goudie, M. Sunde, M. Kopeckey, S. Wagner and W. Hoekstra. 1972. Selenium: Relation to decreased toxicity of methylmercury added to diets containing tuna. Science 175: 1122–1124.

Gladyshev, V.N., G.V. Kryukov, D.E. Fomenko and D.L. Hatfield. 2004. Identification of trace element-containing proteins in genomic databases. Annu. Rev. Nutr. 24: 579–596.

Grandjean, P., P. Weihe, R.F. White, F. Debes, S. Araki and K. Murata. 1997. Cognitive deficit in 7-year-old children with prenatal exposure to methylmercury. Neurotoxicol. Teratol. 19: 417–428.

Grandjean, P., P. Weihe, P.J. Jorgenson, T. Clarkson, E. Cernichiari and T. Videro. 1992. Impact of maternal seafood diet on fetal exposure to mercury, selenium, and lead. Arch. Environ. Health 47(3): 185–195.

Grandjean, P., P. Weihe, R.F. White and F. Debes. 1998. Cognitive performance of children prenatally exposed to 'safe' levels of methylmercury. Environ. Res. 77: 165–172.

Greene, W.H. 2003. Econometric Analysis, 5th edition. Upper Saddle River, New Jersey: Prentice Hall.

Harada, M. 1995. Minamata disease: Methylmercury poisoning in Japan caused by environmental pollution. Crit. Rev. Toxicol. 25: 1–24.

Harada, Y. 1968. Congenital (or fetal) Minimata disease, in: Minamata Disease, Study Group of Minamata Disease, eds., Japan: Kumamoto University, pp. 93–117.

Hatfield, D.L. and V. Gladyshev. 2002. How selenium has altered our understanding of the genetic code. Mol. Cell. Biol. 22: 3565–3576.

Hatfield, D.L., M.J. Berry and V.N. Gladyshev. 2006. Selenium: Its molecular biology and role in human health 2nd edition, Springer–Verlag, New York Inc., New York.

Hellou, J., W.G. Warren, J.F. Payne, S. Belkhode and P. Lobel. 1992. Heavy metals and other elements in three tissues of cod, Gadus morhua from the northwest Atlantic. Mar. Poll. Bull. 24: 452–458.

Hibbeln, J.R., J.M. Davis, C. Steer, P. Emmett, I. Rogers and C. Williams. 2007. Maternal seafood consumption in pregnancy and neurodevelopmental outcomes in childhood (ALSPAC study): An observational cohort study. Lancet 369: 578–585.

Hill, K.E., J. Zhou, W.J. McMahan, A.K. Motley, J.F. Atkins, R.F. Gesteland and R.F. Burk. 2003. Deletion of Selenoprotein P alters distribution of selenium in the mouse. J. Biol. Chem. 278: 13640–13646.

Huggins, F., S.A. Raverty, O.S. Nielsen, N. Sharp, J.D. Robertson and N.V.C. Ralston. 2009. An XAFS investigation of mercury and selenium in beluga whale tissues. Env. Bioindicators. 4: 291–302.

Igata, A. 1993. Epidemiological and clinical features of Minamata disease. Environ. Res. 63: 157–169.

IUPAC–IUBMB Joint Commission on Biochemical Nomenclature (JCBN) and Nomenclature Committee of IUBMB. 1999. Eur. J. Biochem. 264: 607–609.

Iwata, H., H. Okamoto and Y. Ohsawa. 1973. Effect of selenium on methylmercury poisoning. Res. Commun. Chem. Path. Pharmacol. 5: 673–680.

Johnston, J. and J. Dinardo. 1997. Econometric Methods, 4th edition McGraw-Hill Companies, Inc., New York, New York.

Julshamn, K., A. Anderson, O. Ringdal and J. Morkore. 1987. Trace elements intake in the Faroe Islands, I. Element levels in edible parts of pilot whales (*Globicephalus meleanus*). Sci. Total Environ. 65: 53–62.

Kaneko, J.J. and N.V.C. Ralston. 2007. Selenium and mercury in pelagic fish in the central north Pacific near Hawaii. Biol. Trace Elem. Res. 119: 242–54.

Köhrle, J., R. Brigelius–Flohé, A. Böck, R. Gärtner, O. Meyer and L. Flohé. 2000. Selenium in biology: Facts and medical perspectives. Biol. Chem. 381: 849–864.

Korbas, M., J.L. O'Donoghue, G.E. Watson, I.J. Pickering, G.J. Singh, G.J. Myers, T.W. Clarkson and G.N. George. 2010. The chemical nature of mercury in human brain following poisoning or environmental exposure. Neuroscience 1: 810–818.

Kosta, L., A.R. Byrne and V. Zelenko. 1975. Correlation between selenium and mercury in man following exposure to inorganic mercury. Nature 254: 238–239.

Kryukov, G.V., S. Castellano, S.V Novoselov, A.V. Lobanov, O. Zehtab, R. Guigo and V.N. Gladyshev. 2003. Science 300: 1439–1443.

Lederman, S.A., R.L. Jones, K.L. Caldwell, V. Rauh, S.E. Sheets and D. Tang. 2008. Relation between cord blood mercury levels and early child development in a World Trade Center cohort. Environ. Health Perspect. 116: 1085–1091.

Lobanov, A.V., D.L. Hatfield and V.N. Gladyshev. 2009. Eukaryotic selenoproteins and selenoproteomes. Biochim. Biophys. Acta 1790: 1424–1428.

Lund, E.K. 2013. Health benefits of seafood: Is it just the fatty acids? Food Chemistry (in press).

National Research Council. 2000. Toxicological Effects of Methylmercury. National Academy Press, Washington, DC.

Neter, J., W. Wasserman and M.H. Kutner. 1985. Applied Linear Statistical Models: Regression, Analysis of Variance, and Experimental Methods. Richard D. Irwin, Inc., Homewood, Illinois.

Ohi, G., S. Nishigaki, H. Seki, Y. Tamura and T. Maki. 1976. Efficacy of selenium in tuna and selenite in modifying methylmercury intoxication. Environ. Res. 12: 49–58.

Ohi, G., S. Nishigaki, H. Seki, Y. Tamura, T. Maki, K. Minowa, Y. Shimamura and I. Mizoguchi. 1980. The protective potency of marine animal meat against the neurotoxicity of methylmercury: its relationship with the organ distribution of mercury and selenium in the rat. Food Cos. Toxicol. 18: 139–145.

Oken, E., M.L. Østerdal, M.W. Gillman, V.K. Knudsen, T.I. Halldorsson and M. Strøm. 2008a. Associations of maternal fish intake during pregnancy and breastfeeding duration with attainment of developmental milestones in early childhood: A study from the Danish National Birth Cohort. Am. J. Clin. Nutr. 88: 789–796.

Oken, E., J.S. Radesky, R.O. Wright, D.C. Bellinger, C.J. Amarasiriwardena and K.P. Kleinman. 2008b. Maternal fish intake during pregnancy, blood mercury levels, and child cognition at age 3 years in a U.S. cohort. Am. J. Epidemiol. 167: 1171–1181.

Oken, E., R.O. Wright, K.P. Kleinman, D. Bellinger, C.J. Amarasiriwardena and H. Hu. 2005. Maternal fish consumption, hair mercury, and infant cognition in a U.S. cohort. Environ. Health Perspect. 113: 1376–1380.

Parizek, J., I. Ostadalove, J. Kalouskove, A. Babicky, L. Pavlik and B. Bibr. 1971. Effect of mercuric compounds on the maternal transmission of selenium in the pregnant and lactating rat. J. Reprod. Fert. 25: 157–170.

Parizek, J. and I. Ostadalova. 1967. The protective effect of small amounts of selenite in sublimate intoxication. Experiential 23(2): 142–143.

Ralston, N.V.C. 2008. Selenium-health benefit values as seafood safety criteria. EcoHealth. 5: 442–455.

Ralston, N.V.C., C.R. Ralston, J.L. Blackwell and L.J. Raymond. 2008. Dietary and tissue selenium in relation to methylmercury toxicity. Neurotoxicology 29: 802–811.

Ralston, N.V.C., J.L. Blackwell and L.J. Raymond. 2007. Importance of molar ratios in selenium-dependent protection against methylmercury toxicity. Biol. Trace Elem. Res. 119: 255–268.

Ralston, N.V.C. and L.J. Raymond. 2010a. Dietary selenium's protective effects against methylmercury toxicity. Toxicology 278: 112–123.

Ralston, N.V.C. and L.J. Raymond. 2010b. Selenium Health Benefit Values as Seafood Safety Criteria. NOAA Final Report. NA08NMF4520492.

Rayman, M. 2000. The importance of selenium to human health. Lancet 356: 233–241.

Raymond, L.J. and N.V.C. Ralston. 2004. Mercury–selenium interactions and health implications. Seychelles Med. Dent. J. 7: 72–75.

Rice, D.C., R. Schoeny and K. Mahaffey. 2003. Methods and rationale for derivation of a reference dose for methylmercury by the U.S. EPA. Risk Anal. 23: 107–115.

Salonen, J.T., K. Seppänen, K. Nyyssonen, H. Korpela, J. Kauhanen, M. Kantola, J. Tuomilehto, H. Esterbauer, F. Tatzber and R. Salonen. 1995. Intake of mercury from fish, lipid peroxidation, and the risk of myocardial infarction and coronary, cardiovascular, and any death in eastern Finnish men. Circulation 91: 645–55.

Schweizer, U., A.U. Bräuer, J. Köhrle, R. Nitsch and N.E. Savaskan. 2004a. Selenium and brain function: A poorly recognized liaison. Brain Res. Rev. 45: 164–178.

Schweizer, U., L. Schomburg and N.E. Savaskan. 2004b. The neurobiology of selenium: Lessons from transgenic mice. J. Nutr. 134: 707–710.

Seppänen, K., P. Soininen, J.T. Salonen, S. Lotjonen and R. Laatikainen. 2004. Does mercury promote lipid peroxidation? An *in vitro* study concerning mercury, copper, and iron in peroxidation of low-density lipoprotein. Biol. Trace Elem. Res. 101: 117–32.

Simmons-Willis, T.A., A.S. Koh, T.W. Clarkson and N. Ballatori. 2002. Transport of a neurotoxicant by molecular mimicry: The methylmercury-L-cysteine complex is a substrate for human L-type large neutral amino acid transporter (LAT) 1 and LAT2. Biochem. J. 367 (Pt 1): 239–46.

Stillings, B.R., H. Lagally, P. Bauersfield and J. Soares. 1974. Effect of cystine, selenium, and fish protein on the toxicity and metabolism of methylmercury in rats. Toxicol. Appl. Pharmacol. 30: 243–254.

Stringari, J., A.K.C. Nunes, J.L. Franco, D. Bohrer, S.C. Garcia, A.L. Dafre, D. Milatovic, D.O. Souza, J.B.T. Rocha, M. Aschner and M. Farina. 2008. Prenatal methylmercury exposure hampers glutathione antioxidant system ontogenesis and causes long-lasting oxidative stress in the mouse brain. Toxicol. Appl. Pharmacol. 227: 147–154.

U.S. Department of Agriculture (USDA). 2012. National Nutrient Database for Standard Reference, Release 25. Nutrient Data Laboratory Home Page, www.ars.usda.gov/ba/bhnrc/ndl.

U.S. Environmental Protection Agency (U.S. EPA). 2001. Water quality criterion for the protection of human health: Methylmercury. http://water.epa.gov/scitech/swguidance/standards/criteria/aqlife/methylmercury/upload/2009_01_15_criteria_methylmercury_mercury–criterion.pdf.

U.S. Environmental Protection Agency (U.S. EPA). 2004. Consumption advice: Joint federal advisory for mercury in fish. Backgrounder for the 2004 FDA/EPA Consumer Advisory: What You Need to Know About Mercury in Fish and Shellfish. http://water.epa.gov/scitech/swguidance/fishshellfish/outreach/factsheet.cfm.

U.S. Food and Drug Administration (U.S. FDA). 2009. Summary of published research on the beneficial effects of fish consumption and omega-3 fatty acids for certain neurodevelopmental and cardiovascular endpoints: Section B – Neurodevelopmental. www.fda.gov/food/foodsafety/productspecificinformation/seafood/foodbornepathogenscontaminants/methylmercury/ucm153054.htm.

U.S. Food and Drug Agency (U.S. FDA). 2011. Fish and Fishery Products Hazards and Controls Guidance, 4th Edition. www.fda.gov/downloads /food/guidancecomplianceregulatory information/guidancedocuments/seafood/ucm251970.pdf.

Uttranrikisradid Ministry of Foreign Affairs. 2009. The consumption of pilot whale meat and blubber in the Faroes.

Valentine, W.M., T.W. Abel, K.E. Hill, L.M. Austin and R.F. Burk. 2008. Neurodegeneration in mice resulting from loss of functional Selenoprotein P or its receptor Apolipoprotein E Receptor 2. J. Neuropathol. Exp. Neurol. 67: 68–77.

Wada, O., N. Yamaguchi and T. Ono. 1976. Inhibitory effect of mercury on kidney glutathione peroxidase and its prevention by selenium. Environ. Res. 12: 75–80.

Watanabe, C., K. Yin, Y. Kasanuma and H. Satoh. 1999a. *In utero* exposure to methylmercury and selenium deficiency converge on the neurobehavioral outcome in mice. Neurotoxicol Teratol 21: 83–88.

Watanabe, C., K. Yoshida,Y. Kasanuma, Y. Kun and H. Satoh. 1999b. *In utero* methylmercury exposure differentially affects the activities of selenoenzymes in the fetal mouse brain. Environ. Res. 80(3): 208–14.

Whanger, P.D. 1992. Selenium in the treatment of heavy metal poisoning and chemical carcinogenesis. J. Trace Elem. Electrolytes 6: 209–221.

Wooldridge, J.M. 2003. Introductory Econometrics, 2nd edition. Thomson Southwestern, Mason, Ohio.

World Health Organization. 1976. International Programme on Chemical Safety Mercury Environmental Health Criteria 1. Geneva.

Yamashita, Y. and M. Yamashita. 2010. Identification of a novel selenium-containing compound, selenoneine, as the predominant chemical form of organic selenium in the blood of bluefin tuna. J. Biol. Chem. 285: 18134–18138.

Yamashita, Y., H. Amlund, T. Suzuki, T. Hara, A.M. Hossain, T. Yabu, K. Touhata and M. Yamashita. 2010a. Selenoneine, total selenium, and total mercury content in the muscle of fishes. Fish. Sci. 77: 679–686.

Yamashita, Y., T. Yabu and M. Yamashita. 2010b. Discovery of the strong antioxidant selenoneine in tuna and selenium redox metabolism. World J. Biol. Chem. 1: 144–150.

Yoshida, S., M. Haratake, T. Fuchigami and M. Nakayama. 2011. Selenium in seafood materials. J. Health Sci. 57: 215–224.

20

Role of Bacteria in Seafood Products

Françoise Leroi

1 Introduction

Fish and fishery products are among the most traded food commodities worldwide, with trade volumes and values expected to carry on rising. The contribution of fisheries to total animal protein intake by humans is significant (at 16.6%) and is probably higher than indicated by official statistics in view of the under-recorded contribution of small-scale and subsistence fisheries. While capture fisheries' production remains stable (90.4 million tons in 2011), aquaculture production keeps on expanding, reaching 63.6 million tons in 2011 (FAO 2012). Aquaculture is set to remain one of the fastest-growing animal food-producing sectors and total production from both capture and aquaculture exceeds that of beef, pork or poultry. Eighty five percent of the 154 million tons of fish and fishery production is used as food for humans and the annual per capita consumption has grown from around 10 kg (live weight equivalent) in the 1990s to 18.8 kg in 2011 (FAO 2012).

Fisheries products are highly perishable. Depending on the region of the world, the discards, losses and wastage at different stages in the food supply chain represent 30 to 50% of the initial catch (fish and seafood harvested). Twenty to 30% of the total losses occur at the distribution level, and in North America and Oceania, 50% of the losses are due to consumer households (FAO 2011).

Laboratoire de Science et Technologie de la Biomasse Marine, Ifremer, Rue de l'Ile d'Yeu, BP 21105, 44311 Nantes Cedex 03, France.
Email: fleroi@ifremer.fr

Microbial activity is one of the major causes responsible for the sensory degradation of most of the fishery and processed seafood, although enzymatic and chemical oxidation of lipids might also contribute to the spoilage of fatty fish. Additionally, the presence of microorganisms which are pathogenic for human may lead to losses and recalls. On the other hand, some naturally occurring bacteria may play a natural positive role in preservation and/or transformation of seafood. This chapter discusses the different roles of bacteria in seafood products.

2 Microbiota of Marine Animals

As in all animals, the muscle of living fish, shellfish and crustaceans is sterile. However, many bacteria are present on the external parts like skin, mucus, gill and shell, as well as in the gastro-intestinal (GI) tract of the animals. The cultivable bacteria of marine finfish varies from 10^2 to 10^5 CFU per cm² of skin, 10^3 to 10^7 CFU per cm² of gills, 10^3 to 10^5 CFU per gram of feces and 10^4 to 10^9 CFU per gram of intestinal content (Abgrall 1988; Shiina et al. 2006; Skrodenyte-Arbaciauskiene 2007). Higher loads (10^{10} to 10^{11} CFU per gram of feces or intestinal content) have been reported when using culture-independent enumeration techniques (Asfie et al. 2003; Sugita et al. 2005; Shiina et al. 2006).

Some authors have hypothesized that the composition of the bacterial community of skin, mucus, gills and intestine is largely determined by the bacteria in the aquatic environment surrounding the fish in the larval stage (Hansen and Olafsen 1999). It therefore varies depending on a number of hydrological parameters such as quality of the water, temperature, deep, salinity, level of dissolved oxygen, etc. However, fish species, age, feeding, stress, etc. may also have an incidence on the microbiota. In a review of recent studies performed with culture-independent methods, Wong and Rawls (2012) emphasized that the intestinal microbiota composition in fish is influenced both by environment and host ecology. Sullam et al. (2012) demonstrated that salinity trophic levels shaped the composition of fish gut microbiota; they also highlighted a correlation between the microbiota composition and host phylogeny. In Atlantic cod, *Gadus morhua*, Wilson et al. (2008) have distinguished between resident flora present all year round and transient flora whose components vary depending on the season. Roeselers et al. (2011) recently suggested that the host gut is a selective environment for bacteria, based on their finding that Zebrafish (*Danio rerio*) from different origins harbor a stable core gut microbiome.

The microbiota of marine fish found in temperate waters consists of Gram-negative psychrotolerant bacteria, whose growth is possible at 0°C but optimal around 25°C. Among these, the majority belong to the subclass γ of proteobacteria: *Pseudomonas, Shewanella, Acinetobacter, Aeromonas, Vibrio, Moraxella, Psychrobacter, Photobacterium*, etc. and to a lesser extent, the CFB (*Cytophaga-Flavobacter-Bacteroides*) group. Nevertheless, Gram-positive bacteria like *Micrococcus, Bacillus, Lactobacillus, Clostridium* or Coryneforms may also be present in variable proportions (Shewan 1971; Hobbs 1983; Mudarris and Austin 1988; Gram and Huss 1996; Gennari et al. 1999; Huber et al. 2004; Wilson et al. 2008). Some genus like *Vibrio, Photobacterium* and *Shewanella* require salt for their growth and so are typical marine bacteria, whereas *Aeromonas* can grow without salt and thus is more frequently isolated from freshwater (Skrodenyte-Arbaciauskiene et al. 2008). In tropical fish, the flora has the same composition overall (Al Harbi and Uddin 2005), but often with a greater proportion of Gram-positive bacteria (*Micrococcus, Bacillus*, Coryneforms) and enterobacteria (Devaraju and Setty 1985; Liston 1992; Huss 1999).

Due to their importance in digestion, nutrition and growth and in disease control in aquaculture, the indigenous microflora of the fish GI tract have been studied much more than those of the skin or mucus (Ringo et al. 1995; Spanggaard et al. 2000). Before the 2000's most of the studies were based on culture techniques, involving growth in laboratory culture medium. However, it has been estimated that cultivability of bacteria from aquatic environment was less than 2% of the GI microbiota of rainbow trout (Huber et al. 2004) and sometimes less than 0.01% of the skin microbiota (Bernadsky and Rosenberg 1992). A new species belonging to the genus *Mycoplasma* has been detected for the first time in fish and found in abundance in the intestine of wild and farmed salmon (Holben et al. 2002). Strict anaerobic bacteria may have been underestimated with classical techniques. Pond et al. (2006) have identified by Restriction Fragment Length Polymorphism (RFLP) and 16S rRNA gene sequence, a strict anaerobe *Clostridium gasigenes* in the GI tract of rainbow trout.

Based on the meta-analysis of 25 bacterial 16S rRNA gene sequence libraries derived from the intestines of different fish species from salt and freshwater, Sullam et al. (2012) showed that the dominant phylum, in term of number of bacterial species, was the Proteobacteria (*Aeromonadales, Enterobacteriales, Vibrionales, Rhizobiales, Desulfovibrionales, Pseudomonadales, Alteromonadales, Rhodobacterales, Neisseriales, Burkholderiales, Legionellales and Sphingomonadales*), with a prevalence (i.e., number of representative species) of 62.5%. This confirmed the results of many authors (for a review, see Nayak (2010)). The other main phyla were Firmicutes (*Clostridiales, Lactobacillales, Bacillales*: 15.2%), Bacteroidetes (*Bacteroidales, Flavobacteriales*: 6.04%) and Actinobacteria (*Actinomycetales and Acidimicrobiales*: 3.70%). In this study, freshwater fish harbored a greater proportion of Aeromonadales and Enterobacteriales while Vibrionales were dominant in seawater fish, and this is in accordance with the salt requirements of the different bacterial species.

Although not in a majority, lactic acid bacteria (LAB) are frequently found in the fish intestine (Yang et al. 2007; Ringo 2008). *Lactobacillus, Leuconostoc, Lactococcus, Vagococcus, Carnobacterium, Streptococcus* and *Weisella* have been isolated from Atlantic salmon, pollock, cod, Arctic char, rainbow trout and other species (for a review, see Leroi (2010)). Yeasts (principally *Metschnikowia zobelii, Trichosporon cutaneum* and *Candida tropicalisare*) have also been found in the GI tract of marine fish (Gatesoupe 2007).

As in fish, the bacterial flora of crustaceans depends on several factors including the species considered, the geographic location and environment, the temperature and salinity of the water, etc. Overall, however, the same species of microorganisms are found in crustaceans and fish. The microbiota of tropical shrimp consists mainly of *Pseudomonas, Vibrio, Acinetobacter, Moraxella, Flavobacterium*, and a high proportion of *Aeromonas* (Vanderzant et al. 1973; Jayaweera and Subasinghe 1988; Jeyasekaran et al. 2006). Significant *Vibrio* spp. including *V. parahaemolyticus* have been reported by Gopal et al. (2005) in shrimp from India. Dominance of coryneform bacteria and *Moraxella* has been found in shrimp from Nicaragua by Benner et al. (2004) followed at a lower level by *Bacillus, Lactobacillus, Micrococcus, Proteus, Acinetobacter, Shewanella* and *Pseudomonas*. These results confirm those of Matches (1982). Chinivasagam et al. (1996) showed the influence of the capture area on the microbiota. Indeed, Gram-positive bacteria dominated in shrimp fished at low depth and *Pseudomonas* on those caught in deep water.

3 Role of Bacteria in Living Marine Animals

The role of the GI microbiota in nutrient metabolism and adsorption by the host, regulation of energy balance, epithelial renewal, angiogenesis and fish immunity has been reviewed by Nayak (2010).

Beside those beneficial effects, some bacteria are pathogenic for fish. If the natural balance between the established bacteria in healthy animal and the transient community is disturbed, some pathogens present in the transient state can cause lethal infection. *Vibrio anguillarum, Vibrio salmonicida* and *Vibrio viscosus* are responsible for classical vibriosis worldwide and cold-water vibriosis in Atlantic salmon and cod. *Aeromonas salmonicida*, present in fresh and seawater, are involved in furunculosis of salmonids, turbot, green flounder. Many other species such as *Pasteurella piscicida, Streptococcus* spp., *Renibacterium salmoninarum, Mycobacterium* spp. and

Piscirickettsia salmonis are pathogenic for fish (Håstein 2013). Although generally recognized as safe, the implication of LAB in fish disease has been reported. *Lactococcus garvieae* is responsible for septicemias, opthalmias and hemorrhages of worldwide marine farmed fish (Eldar et al. 1996). *Carnobacterium maltaromaticum* has been isolated from different diseased fish and its virulence has been established in experimentally infected rainbow trout and striped bass (Baya et al. 1991; Toranzo et al. 1993).

4 Microbiota of Fish Muscle and Role in Spoilage

4.2 Concept of specific spoilage microorganisms

At the time of fish death, collapse of the immune system and destructuration of tissue may favor penetration of microorganisms in the flesh. Additionally, gutting and filleting is a source of contamination of the muscle with endogeneous bacteria present on the surface and in the intestine. Also post-contamination by exogenous bacteria from human, environmental and material origin may occur during different processing steps such as salting, smoking, marinating, slicing, packaging, etc.

After a variable lag phase, bacteria enter into an exponential growth phase and may reach high levels in few days. The lag phase and growth rate vary with the animal species, the capture technique, the bacterial species, the storage temperature, etc. Generally, bacterial growth is more rapid in fish from temperate seawater than in tropical fish, in which lag phase may reach 1–2 weeks, due to the adaptation to chilled storage temperature by mesophilic bacteria present in the muscle (Gram 1995).

Fish muscle contains low carbohydrate concentration (0.2 to 1.5%) and high levels of non-protein low molecular weight nitrogenous compounds such as amino-acids, creatine, taurine, nucleotide, urea and trimethylamine oxide (TMA-O). Combined with a high *post mortem* pH (>6), this favors the growth of Gram-negative bacteria such as *Shewanella*, *Photobacterium*, *Pseudomonas*, *Vibrio* and *Aeromonas*. Conversely, mammalian meat is richer in glycogen, leading to a post rigor acidification <5.5. This prevents growth of *Shewanella* and allows the development of LAB.

Seafood is generally acceptable from a sensory point of view well after the total flora have reached their maximum. Only, few members of the microbial community give rise to the off-flavours associated with seafood spoilage. Those microorganisms are named specific spoilage microorganisms (SSO). In raw fish, the sensory degradation is generally due to one SSO (Hozbor et al. 2006) whereas in processed fish, various species may participate to cause spoilage in a complex manner (Stohr et al. 2001; Joffraud et al. 2006).

Spoilage of fish is characterized by unpleasant taste and odours (sulphurous, ammonia-like, rotten, fruity, etc.), pasty texture, slime, gas in packaging, and sometimes visible colonies on the surface of the flesh. The spoiling potential of bacteria is not easy to determine. Many authors have tried to test it in laboratory model broth or fish juice but results differ from those obtained in fish muscle (Truelstrup Hansen et al. 1995; Stohr et al. 2001). Therefore, it is important to perform the tests in flesh sterilized with a technique that does not modify its composition. Collection of fish fillets in aseptic condition or ionization for contaminated products (i.e., processed products such as salted, smoked, marinated... or crustaceans for which the GI tract in not removed) are methods currently used for this kind of tests (Joffraud et al. 1998; Jorgensen et al. 2000a).

5 Spoilage of Raw Marine Animal

5.2 Raw marine animal stored under air

Among psychrotrophic Gram-negative bacteria, sensitive-pH species such as *Shewanella* spp. are often in a majority, particularly in white fish, demersal fish and crustaceans from temperate seawater. *Shewanella putrefaciens* has been considered the main spoiler of these products because it is able to use TMA-O as a final electron acceptor. TMA-O is thus reduced to TMA which produces the typical ammonia-like fishy off-odour of rotten fish. TMA-O is a typically marine molecule and thus TMA is never incriminated in spoilage of freshwater fish. *S. putrefaciens* is also able to produce H_2S and other sulphurous compounds from the amino-acids cysteine and methionine. Jorgensen et al. (1988) have shown that the number of *S. putrefaciens* is inversely correlated to the remaining shelf-life, and consider that, at approximately 10^8 CFU g^{-1}, the sensory deterioration of the product is no longer acceptable. *S. putrefaciens* consists of a heterogeneous group whose taxonomy has greatly changed over the last few years (Vogel et al. 2005) have shown that the large majority of microorganisms which produce H_2S isolated from Baltic Sea fish chilled for several days should now be identified as *Shewanella baltica*. Furthermore, other strains have been identified as new species: *Shewanella hafniensis, Shewanella morhuae, Shewanella glacialipiscicola* and *Shewanella algidipiscicola* (Satomi et al. 2006, 2007). When fish is stored at higher temperature than in ice, *Aeromonas* can became dominant and also produces TMA.

In marine fish poor in TMA-O and with pH <6, such as pelagic dark meat (tuna, mackerel, garfish), the spoilage is mainly attributed to *Pseudomonas fragi, Pseudomonas fluorescens, Pseudomonas putida* and *Pseudomonas lundensis*). *Pseudomonas* spp. are not capable of anaerobic respiration and do not produce TMA nor H_2S. The typical fruity and rotten off-odours are due to the formation of ammoniac, esters, ketones and aldehydes. The SSO of tropical fish from sea or freshwater is *Pseudomonas*. *S. putrefaciens* has also been isolated from these products but does not appear to play an important part in spoilage. This could be due to the inability of this microorganism to develop in the presence of a large number of *Pseudomonas* (Gram et al. 1990; Gram and Melchiorsen 1996). Urea found in large quantities in selachians can be metabolised into ammonia which also has a strong, unpleasant odour. Deamination of amino acids also leads to the production of ammonia. For different species of fresh fish, the measurement of total volatile basic nitrogen (TVB-N), which includes, among others, the TMA and ammonia, is a quality criterion which is regulated in Europe. The permitted limit of TVB-N varies from 25 to 35 mg-N/100 gram of flesh according to fish species (1022/2008/EC 2008).

Lastly, fatty fish are rich in polyunsaturated fatty acids that can be rapidly oxidised by either chemical chain reactions or lipolysis resulting from autolytic or bacterial enzyme activity respectively.

5.2 Raw marine animal stored under vacuum or modified atmosphere packaging

In order to increase the shelf-life of seafood, fish are often stored under vacuum or modified atmosphere packaging (MAP) containing CO_2 and N_2, and small concentrations of O_2 in some countries. Absence of oxygen in vacuum packed products prevents the growth of *Pseudomonas* but *Shewanella* ssp. can still grow and is the major SSO. In modified atmosphere packaging, CO_2 inhibits the growth of *Shewanella* spp. However, *Photobacterium* spp. are resistant to CO_2 and capable of anaerobic respiration, and therefore produce TMA. Production of H_2S is observed but seems to be strain dependant. *Photobacterium phosphoreum* has long been recognized as the major SSO of cod stored under MAP (Dalgaard 1995). Molecular tools have replaced morphologic and phenotypic characterization and more recently, the new species *Photobacterium iliopiscarium* and *Photobacterium kishitani* have been identified; they also play a role in the degradation of flesh (Ast and Dunlap 2005; Dunlap and Ast 2005; Olofsson et al. 2007). The presence of typical marine bacteria capable of anaerobic respiration with the marine molecule TMA-O explains why the shelf-life extension of seafood with vacuum or MAP is, by far, less important than in meat.

For fish with little TMA-O, such as salmon, and for some tropical fish, LAB and *Brochothrix thermosphacta* can also grow and spoil the product with acidic, sour, rotten egg or butter off-odours. LAB are anaero-aerotolerant bacteria that show a higher resistance to CO_2 and to the lack of O_2 than Gram-negative bacteria (Alfaro et al. 2013). Rudi et al. (2004) reported

C. maltaromaticum, Carobacterium divergens and *B. thermosphacta* as the dominant bacteria in salmon fillets packed in a 60% CO_2 and 40% N_2 atmosphere, while Powell and Tamplin (2012) reported a domination of *Carnobacterium* spp. and *Shewanella* spp. in MAP Australian Atlantic salmon. According to Macé et al. (2012), *Pseudomonas* spp. was dominant at the beginning of the storage of salmon steaks packed under vacuum or 50% CO_2 and 50% N_2 atmosphere. After seven days at 2°C and 8°C, LAB (mainly *Lactococcus piscium*, but also *C. maltaromaticum, C. divergens, Lactobacillus fuchuensis, Vagococcus* sp. and *Leuconostoc gasocomitatum*), *P. phosphoreum/ iliopiscarium* and Enterobacteriaceae (*Serratia proteomaculans* and *Serratia quinivorans*) became dominant. *B. thermosphacta* reached 10^6 CFU/g, which is not sufficient to spoil the product.

In an attempt to identify SSO of MAP salmon, Macé et al. (2013) have inoculated sterile raw salmon cubes with eight bacterial groups/species (*Serratia* spp., *Hafnia alvei, B. thermosphacta, C. maltaromaticum, S. baltica, L. piscium, P. phosphoreum* and a mix of other Enterobacteriaceae containing one strain of *Moellerella* sp., *Morganella* sp. and *Pectobacterium* sp.). After MAP storage for 12 days at 8°C, the greatest spoilers were *C. maltaromaticum, H. alvei* and *P. phosphoreum*, both in pure and mixed cultures. Among the different volatile compounds measured by SPME GC-MS, acetic acid produced by *P. phosphoreum* was clearly correlated to sensory perception.

6 Spoilage of Lightly Preserved Seafood

The lightly preserved fish products (LPFP) refer to lightly processed seafood, such as salting, drying, smoking or marinating, resulting in a final pH >5, and a salt content <6% (water phase) in the final product. Smoked salmon, carpaccio, gravelax and marinated fish are examples of LPFP.

The treatment of LPFP reduces the initial microbial load and lowers the growth rate of most of the microorganisms. However, it is not sufficient for a total elimination of the endogeneous microorganisms. In shrimp, the heat treatment (generally 2 min at 72°C) constitutes a strong barrier (Harrison et al. 2000) but does not destroy all the bacteria, particularly those present in the GI tract. Additionally, products are often recontaminated by the processing equipment and environment.

In cold-smoked salmon (CSS), the initial microbiota depends strongly on the hygiene conditions within the company (Gancnel et al. 1997; Truelstrup Hansen and Huss 1998; Leroi et al. 2001) and is often dominated by Gram-negative bacteria typical of fresh fish such as *Shewanella, Photobacterium* and *Vibrio* (Leroi et al. 1998; Paludan Müller et al. 1998; Gonzalez-Rodriguez et al. 2002; Olofsson et al. 2007). During storage, Gram-positive bacteria, particularly LAB, become predominant. In LPFP, LAB can easily reach 10^{7-8} CFU g^{-1} and such amounts have been found in CSS (Leroi et al.

1998, 2000), smoked trout (Lyhs et al. 1998), smoked herring (Gancel et al. 1997), salted lumpfish roe (Basby et al. 1998), cooked cold-water shrimp (Dalgaard et al. 2003) and tropical shrimp (Mejlholm et al. 2005; Jaffrès et al. 2009). In contrast, in gravad salmon, LAB are present but not predominant (Lyhs et al. 2001b). A more detailed taxonomic study revealed that *C. maltaromaticum, Lactobacillus curvatus* and *Lactobacillus sakei* are often in a majority. Carnobacteria are microorganisms resistant to freezing that grow very well at refrigerated temperatures, in all packaging conditions and in the presence of many preservatives (Leroi et al. 2000; Laursen et al. 2005; Alfaro et al. 2013), explaining why this genus is very often found in refrigerated meat or fish products. Other species like *C. divergens, Lactobacillus farciminis, Lactobacillus alimentarius, Lactobacillus plantarum, Lactobacillus homohiochii, Lactobacillus delbrueckii, Lactobacillus casei, Lactobacillus coryniformis, Leuconostoc mesenteroides, Enterococcus faecalis, Weisella kandleri, Vagococcus fluvialis* or *Vagococcus penaei* are also recorded (for a review, see Leroi (2010)). Depending on the products and factories, Enterobacteriaceae (*Serratia liquefaciens, Hafnia alvei*), *P. phosphoreum* and *B. thermosphacta*) may also be present at high levels (Truelstrup Hansen and Huss 1998; Cardinal et al. 2004; Gonzales-Fandos et al. 2004; Rachman et al. 2004; Jaffrès et al. 2009).

The role of LAB in spoilage of LPFP depends on species but is also strain dependant. Stohr et al. (2001) showed that *L. sakei* was the main SSO in CSS whereas *L. alimentarius* was neutral. *L. sakei* generally produces sulphurous and acidic odours (Nilsson et al. 1999; Stohr et al. 2001), associated with the production of H$_2$S, acetic acid and ethyl and *n*-propyl acetate (Joffraud et al. 2001), but some *L. sakei* strains do not affect the organoleptic quality of this product (Weiss and Hammes 2006). *L. alimentarius* which does not spoil CSS has been identified as the bacterium responsible for the sensory deterioration of marinated herring (Lyhs et al. 2001a). The status of Carnobacteria as an SSO of LPFP is unclear (Laursen et al. 2005; Leisner et al. 2007). Many studies have showed that the inoculation of CSS by various strains of *C. maltaromaticum* and *C. divergens* leads to few or no changes in organoleptic quality (Leroi et al. 1996; Paludan Müller et al. 1998; Duffes et al. 1999; Nilsson et al. 1999; Brillet et al. 2005). When the carnobacteria reached a high enough level, flavours of butter and of plastic were detected, associated with the production of 2,3-butanedione (diacetyl) and 2,3-pentanedione (Joffraud et al. 2001; Stohr et al. 2001) but they were not sufficient for a trained panel to reject the product (Brillet et al. 2005). However, Jaffrès et al. (2011) have shown that *C. maltaromaticum* was one of the major SSO of MAP tropical shrimp. When inoculated in sterile shrimp, this species released cheese, feet, sour, fermented and butter odours and produced 3-methyl-1-butanal, 2-methyl-1-butanal, 2-methyl-1-butanol, ethyl acetate, 2-methyl-1-propanal, 2,3-butanedione, 3-methyl-2-butanone,

3-methyl-1-butene, thiocarbamide, cyclopentanol and acetaldehyde. In this study, *C. divergens* also spoiled the product, although off-odours were less pronounced. *Carnobacterium alterfunditum* did not modify the sensory quality of shrimp in a significant manner. Two strains of *C. maltaromaticum* and *divergens* inoculated into Arctic shrimp generated strong chlorine, malt, nut and sour odours and the samples were judged unfit for consumption (Laursen et al. 2006). Ammonia and numerous alcohols, aldehydes and ketones were produced. Nevertheless, here again, there was variability depending on the strain.

S. *liquefaciens* is also recognized as a major SSO of CSS, with production of amines, cheese, acid or rubber odours, associated with the molecules TMA, dimethyldisulphur, 2,3 butanediol and 2-pentanol (Joffraud et al. 2001; Stohr et al. 2001). This species is also responsible for spoilage of tropical shrimp, releasing strong cabbage, gas, garlic, amine and urine odours (Jaffrès et al. 2011).

In CSS, *B. thermosphacta* produces typical odours of blue cheese and plastic, closely correlated with 2-heptanone and 2-hexanone. Nevertheless, it is quite rare for these bacteria to reach sufficiently levels in naturally contaminated products to be the sole explanation for sensory rejection. In Artic and tropical shrimp (Laursen et al. 2006; Fall et al. 2010a; Jaffrès et al. 2011; Fall et al. 2012), have clearly identified *B. thermosphacta* as a major SSO. Butter-like, cheese/feet and sour/fermented odours were detected in the products, associated to 3-methyl-1-butanal, 2,3-butanedione, 2-methyl-1-butanal, 2-methyl-1-butanol, acetaldehyde, 2-methyl-1-propanal and ethyl acetate.

P. *phosphoreum* seems to play a more moderate role in the deterioration of LPFP than in fresh fish. Sterile CSS cubes inoculated with different strains of *P. phosphoreum* have been judged as lightly spoiled, with weak odours of acid, amine and feet (Joffraud et al. 2001). Jorgensen et al. (2000b) give much greater weight to the spoiling action of this species. A good correlation between the sensory quality of CSS and the production of tyramine and histamine was found. As *P. phosphoreum* was the only histamine producer present in CSS, they concluded about its SSO status.

S. *putrefaciens,* the most common spoiling bacteria in fresh fish, and *Vibrio* sp. have never been implicated in spoilage of CSS, even when inoculated at high concentrations.

Spoilage of LPFP is probably much more complex than of fresh fish and the interaction between the different species has to be taken into consideration. Joffraud et al. (2006) have shown that the spoilage observed with *L. sakei* was weakened in the presence of *S. liquefaciens* even though the latter also had a spoiling effect in monoculture. The association of *C. maltaromaticum* and *Vibrio* spp. or *B. thermosphacta* produced strong off-odours due to *de novo* synthesis of TVBN (Brillet et al. 2005; Joffraud et al. 2006) although none of the species was considered as an SSO in pure culture. Similarly, Laursen et al. (2006) showed that the unpleasant odours generated in cooked shrimp by an association of *Carnobacterium* spp. and *B. thermosphacta* were different from those due to these two bacteria in pure culture. In raw salmon steak, association of species did not significantly modify the sensory perception, and the dominant species determined the spoilage characteristics (Macé et al. 2013).

7 Microbiology and Safety of Fish Muscle

Ten to twenty percent of foodborne outbreaks are attributed to the consumption of seafood. This figure obviously varies with the quality of the monitoring plan implemented in each country, the level of consumption and dietary habits. Finfish are responsible for more outbreaks than shellfish; however, with a less significant number of cases per outbreak (Huss et al. 2000). In fish, most diseases are caused by bacteria, whereas viruses are responsible for 50 percent of outbreaks linked to the consumption of bivalve molluscs, which are filter feeding animals (Lee and Rangdale 2008).

Pathogenic bacteria naturally present in the marine environment may contaminate the animals and be responsible for specific diseases due to the consumption of fish and shellfish. This is the case for some *Vibrio* spp., *L. monocytogenes*, *Clostridium botulinum* and *Aeromonas hydrophila*, as well as histamine-producing bacteria. Normally, these bacteria are present at too low a level to cause the disease. In addition, proper cooking can eliminate the microorganisms or eliminate their toxins (except histamine). Therefore, the risk focuses on products in which the growth of these bacteria is possible during the storage period and which are consumed raw or undercooked.

Just like other foods, seafood may also be post-contaminated during the process of transformation, by bacteria such as *Staphylococcus aureus*, *Salmonella* spp., *Shigella* spp., *Clostridium perfringens*, *Bacillus cereus*, *Yersinia enterocolitica*, *Listeria monocytogenes* and *Escherichia coli*. Some of these microorganisms, including salmonella and viruses, can also be found in coastal and estuarine waters contaminated by human activities or in aquaculture ponds if the quality of the feed is not controlled.

8 Histamine Producing Bacteria

Histamine is responsible for 51% of related seafood illness in France and in Europe and for 20% in United States (Delmas et al. 2010; Helwigh et al. 2010; CDC 2011). Histamine fish poisoning (HFP) is related to fish species that contain high levels of histidine, such as tuna (84% of all cases), swordfish, mackerel, bonito, sardine, herring or anchovies. HFP is rarely lethal but causes serious problems soon after ingestion, with allergic symptoms (rash, hives, headaches, diarrhea, vomiting) that may lead to hospitalization. Histamine is produced after the death of fish by decarboxylation of free histidine under the action of the bacterial enzyme histidine decarboxylase. Mesophilic enterobacteria like *Morganella morganii, Hafnia alvei, Raoultella planticola* or *Enterobacter aerogenes* can produce high levels of histamine (>5,000 mg/kg) (Björnsdóttir-Butler et al. 2010), especially when fish is stored at abuse temperature, for instance during storage on the vessels or during the thawing step before processing (for a review, see Hungerford (2010)). Histamine content is regulated in Clupeid and Scombrid fish and in other histidine-rich fish species (<100 mg/kg in Europe and 200 mg/kg in products which have undergone enzymatic maturation in brine, <50 mg/kg in USA). Exceeding the concentration leads to a systematic rejection of the batch, regardless of the further processing, because histamine is thermostable. Prevention consists of the application of good hygienic practices and strict compliance with the cold chain. However, psychrotolerant bacteria as *P. phosphoreum* and the recently discovered species, *Morganella psychrotolerans,* have been found to produce histamine at low temperatures (0–2°C and above). Both species have recently been implicated in cases of HFP (Emborg and Dalgaard 2006; Emborg et al. 2006). Development of methods to prevent the formation of histamine in fish remains an important issue for the industry. In fish sauces, histamine can reach very high concentrations. The main producers are *Bacillus* spp. (Tsai et al. 2006; Lin et al. 2012) and LAB such as *Tetragenococcus halophilus, Enterococcus faecium* or *Lactobacillus* spp. (Dapkevicius et al. 2000; Satomi et al. 2008).

8.1 Listeria monocytogenes

L. monocytogenes is the pathogenic bacterium responsible for listeriosis, which is a food-borne disease. Listeriosis is generally associated with a high mortality rate (20–40%) and is regarded as the most fatal food-borne infection (Feldhusen 2000; Rocourt et al. 2000). *L. monocytogenes* is a major pathogen in LPFP eaten raw, like carpaccio, cold-smoked or lightly marinated fish and cooked shrimp packed in a protective atmosphere (Mejlholm et al. 2012). The prevalence of *L. monocytogenes* in these products is high, varying from 2 to 60% depending on the studies (Jorgensen and Huss 1998; Valdimarsson et al. 1998; Nakamura et al. 2004; Gudmundsdóttir et al. 2005; Hu et al. 2006; Beaufort et al. 2007). *L. monocytogenes* is not destroyed during the various stages of the transformation process. Although the initial contamination level is often inferior to 1 CFU/g, *L. monocytogenes* can grow in LPFP and sometimes pass the European tolerated limit of 100 CFU/g (1441/2007) at the end of the shelf-life. Despite these figures, relatively few cases of listeriosis linked to the consumption of seafood have been reported (10 cases with shrimp, 5 with smoked trout, 9 with marinated trout, 6 with smoked mussels, 2 with imitation of crabmeat, and 1 with fish) (Rocourt et al. 2000; Warriner and Namvar 2009). However, *L. monocytogenes* is a pathogen under high surveillance in seafood, as in all ready-to-eat food.

8.2. Vibrio spp.

Of the 2,500 reported cases of illnesses due to bivalve molluscs, 50% were due to *Vibrio*, with 95 deaths during the period 1984–1993 in the US (Wittman and Flick 1995). The genus *Vibrio* includes a multitude of species among which some are pathogenic for humans. The main pathogenic species are *Vibrio parahaemolyticus, Vibrio vulnificus* and *Vibrio cholerae*. These mesophilic and halophilic bacteria (except *V. cholerae*) are found in tropical and temperate waters at the end of summer. They may represent a risk in bivalves that concentrate the particles during their filter-feeding. They can also be naturally present in fish and crustaceans and their presence is often reported in farmed tropical shrimp (Gopal et al. 2005). The growth rate of *Vibrio* spp. is very high at temperatures above 8°C. The infectious dose is in the range of 10^6 CFU/g. Less than 5% of the strains of *V. parahaemolyticus* are pathogenic with the production of hemolysin, whereas most strains of *V. vulnificus* cause septicemia, fatal in 50% of cases. Some strains of *V. cholerae* O1 and O139 serotypes are responsible for cholera. Although the primary infection is related to the ingestion of contaminated products, the epidemic can spread rapidly through feces. In Peru in 1991, 400,000 cases of cholera were reported, including 4,000 deaths, due to the consumption of pickled fish (ceviche).

8.3. Clostridium botulinum

C. botulinum may be present in marine sediments and contaminated animals. The types B, E and F of non-proteolytic psychrotolerant *C. botulinum* are the most frequent and they can be found in cold or temperate water marine animals. *C. botulinum* is an obligate anaerobe, now well controlled in canned seafood. Most cases of botulism identified are related to slightly processed products (smoked, salted, fermented), those which are homemade and kept in poor conditions (Korkeala et al. 1998). Storage at 5°C, combined with a salt content of 3–5% in aqueous phase, is used to prevent the growth and thus the production of toxin. In the case of products with less salt, it is advisable to market products as frozen, or to avoid anaerobiosis storage by using oxygen-permeable films or addition of O_2 in the gas mixture for products stored in a protective atmosphere.

9 Protective Bacteria in Fish Muscle

Among mixed population of microorganisms, different types of interaction have been described and among them, antagonism has currently been observed. In 1962, Jameson (1962) observed the suppression of growth of *Salmonella* when total flora had reached its maximum. The so-called 'Jameson effect' has also been reported by Ross and McMeekin (1991) on *S. aureus* and Gimenez and Dalgaard (2004) for *L. monocytogenes*.

Since the nineties, many studies have been conducted to select microorganisms with antagonistic properties that could be added to seafood to improve their safety. This technology is termed biopreservation. *L. monocytogenes* is the main target strain studied but more recently, some authors have extended this concept to prevent growth of SSO and improve sensory quality of seafood (for a review on biopreservation in fish products, see Leroi (2010), Pilet and Leroi 2011).

LAB are the main candidates for biopreservation as they are generally recognized as safe by US FDA and they often possess antimicrobial activity. However, the selection of protective LAB for an application in seafood remains a challenge. Indeed, the protective strain must be able to grow in the seafood matrix which is poor in sugar. Their metabolic activities should not change the initial characteristics of the product (such as in fermented products) and not induce spoilage that could lead to sensory rejection.

As shown in this chapter, LAB can easily grow and be competitive in LPFP, whereas their implantation in fresh fish is more difficult, explaining why most of the tests have been conducted in LPFP.

9.1 Inhibition of L. monocytogenes

LAB from the *Carnobacterium* genus have been widely studied in CSS as they are part of the natural ecosystem and they are non aciduric. Different strains of *C. divergens* and *C. piscicola* have shown promising inhibition potential, many often due to production of bacteriocins (Nilsson et al. 1999; Brillet et al. 2004; Tahiri et al. 2009). Most of the times, the inoculation of CSS with 10^3–10^7 CFU/g of the selected strain reduces the final level of *L. monocytogenes* by 1.5 to 4 Log. *C. divergens* V41 is able to maintain *L. monocytogenes* at its initial level (<100 CFU/g) over the storage period (four weeks at 4 and 8°C) (Brillet et al. 2004). The inhibition is due to divercin V41 (Richard et al. 2003). This strain does not produce any adverse odour and flavour (Brillet et al. 2005). An Italian company (SACCO, Cadorago) commercializes strains of *Carnobacterium* spp. (Lyoflora FP18) producing bacteriocins for an antilisterial application in CSS. Other promising results have been obtained with strains that do not produce bacteriocins (Yamazaki et al. 2003; Nilsson et al. 2005). Apart *Carnobacterium* spp., some isolates of *L. plantarum*, *L. casei*, *L. sakei*, and *E. faecium* also exhibit antilisterial properties in CSS but have been tested on *Listeria innocua* only (Katla et al. 2001; Vescovo et al. 2006; Weiss and Hammes 2006; Tomé et al. 2008).

10 Prevention of Histamine Formation

Most of the published studies focus on the use of strains which are able to degrade histamine by production of amino-oxidase. *Staphylococcus xylosus* reduced the concentration of histamine by 38% in phosphate buffer and by 16% in salted and fermented herring (Mah and Hwang 2009). In fermented sauce, strains of *Staphylococcus carnosus* and *Bacillus amyloliquefasciens* allowed a histamine reduction of 28to 15% respectively (Zaman et al. 2011). Other authors have obtained significant reduction of histamine in fish products but it is unclear if it is due to consumption of histamine or inhibition of production. Salted anchovy, in which 580 mk/kg of histamine was recorded after 15 days, was free of histamine when inoculated by the extremely halophilic archaea strains, *Halobacterium salinarium* and *Haloarcula marismortui* (Aponte et al. 2010). Mixed starters of *L. plantarum*, *S. xylosus*, *Pediococcus pentosaceus and L. casei* prevented the formation of histamine in fermented silver carp sausages by 90% (Hu et al. 2007).

Another option is to limit the development of histamine-producing bacteria. *L. sakei* LHIS2885 is able to both, inhibit growth of *M. morganni* and *M. psychrotolerans* in cooked tuna stored under vacuum-packaging and reduce histamine concentration, which remains under the European limit of 100 mg/kg (Podeur et al. 2013). The company Biocéane (Nantes, France) commercializes the ferment LLO (*L. lactis*) which allows a 10-fold reduction of histamine naturally produced in raw tuna loins and imported from Ecuador under vacuum-packaging.

11 Improvement of Sensory Quality and Shelf-life Extension

The use of protective culture to extend shelf-life of seafood has not been studied extensively. It is much more complex than pathogens, as spoilage is the result of the activity of various SSO, so the inhibitory spectrum of the protective culture much be very large. Additionally, spoilage of a product may vary according to the initial contamination of fish, hygienic conditions in the factory, etc. So it is very important to assess the effect of a protective culture in different batches of naturally contaminated products.

The best indicator to assess the effect of a strain is the sensory analysis. Unfortunately, some studies only provide data on the effect on total flora or TMA and TVBN, which is not sufficient to clearly estimate the potential of a strain. As an example, El Bassi et al. (2009) have selected 2 strains of *L. plantarum* and *L. pentosus,* originally isolated from seafood, able to prevent growth of psychrotrophic SSO and pathogenic bacteria in a liquid medium. Their addition in vacuum-packed minced sea bass (*Dicentrarchus labrax*) limited the development of total coliform and slightly reduced the production of TMA and TVBN, but no tests were performed to evaluate the sensory quality of the bioprotected products.

To our knowledge, most of the studies have concentrated on different strains of *L. gelidum* (EU2247, EU2262) and *L. piscium* (EU2229, EU2241) isolated from vacuum-packed fresh salmon steaks by Matamoros et al. (2009a). These strains, inoculated at 10^{5-6} CFU/g, were able to delay the sensory shelf-life of two batches of cooked, peeled shrimp (*Penaeus vanamei*) stored under MAP at 8°C. The controls were highly spoiled after 28 days, whereas the bioprotected shrimp were still considered as fresh by a trained sensory panel. This effect was not correlated with the inhibition of the microbiological indicators classically measured for this product (total psychrotrophic flora, LAB, Enterobacteriaceae) (Matamoros et al. 2009b). Thereafter, Fall et al. (Fall et al. 2010a; Fall et al. 2012) showed that improved sensory quality could be linked to the specific inhibition of *B. thermosphacta*. The mechanism involved in this inhibition does not either involve bacteriocin nor lactic acid (Fall et al. 2010b). These strains have also been tested in CSS, and the effect was strain dependant, one strain of *L. piscium* (EU2241) being able to delay the production of off-odours by 2 weeks. Trials to improve the shelf-life of two batches of fresh salmon fillets packaged under MAP were performed using the strains of *L. piscium* and *L. gelidum* as well as a commercial ferment (LLO, Biocéane, Nantes, France). The best results were obtained with the strain *L. gelidum* EU2247, and, in a lesser extent, LLO. After nine days of storage at 2 and 8°C, the uninoculated control was totally spoiled whereas the biopreserved products had kept their freshness (Brillet-Viel et al. 2011).

Other tests have been performed with strains of *Bifidobacterium bifidum* in plaice fillets (Altieri et al. 2005) or *E. faecium* in gutted sardines and tiger shrimp (Paari et al. 2012) with a small beneficial effect.

Conclusion

The microbial ecosystem of marine environments and animals is very specific. The composition of the flesh, poor in carbohydrates and rich in nitrogenous compounds, differs from terrestrial mammalian. This contributes to the selection of a core microbiota, generally composed of Gram-negative psychrotrophic bacteria (*Photobacterium, Shewanella, Vibrio, Pseudomonas*...). The qualitative and quantitative evolution of the microbiota varies with fish species, process and storage conditions. Among the different groups of microorganisms, some are responsible for spoilage and lead to depreciation and loss of products. This is particularly the case for *Shewanella* and *Photobacterium*, which are able to metabolize the marine molecule TMA-O into TMA, giving the specific off-odour of rotten fish. In LPFP, the different processing steps associated with storage under vacuum or MAP tend to select Gram-positive bacteria, and the microbiota is comparable to lightly processed meat products, with dominance of LAB and *B. thermosphacta*. Some LAB species play a significant role in degradation of LPFP, but the spoiling process is very complex because it is strain dependant. Moreover, bacterial interactions between the various species give rise to variable odours and flavours. Sometimes, LAB present at high level in flesh may be totally neutral and others may have interesting antimicrobial properties. In case of poor hygienic conditions, bad manufacturing practices or rupture of cold-chain storage, some pathogens from marine origins (histamine-producing bacteria, *Vibrio, Aeromonas, Clostridium*, etc.) and recontamination pathogens may develop in the flesh and be responsible of serious outbreaks. LAB naturally present or intentionally added may constitute an additional hurdle to limit this growth and secure the products.

References

1022/2008/EC, C.R.E.N. 2008. Commission Regulation 1022/2008/EC of 17 October 2008 amending Regulation (EC) No 2074/2005 as regards the volatile basic nitrogen (TVB-N) limits. pp. 18–20. *In*: O J E U.

Abgrall, B. 1988. Poissons et autres produits de la mer. pp. 251–264. *In*. Microbiologie alimentaire. Lavoisier, Paris.

Al Harbi, A.H. and N. Uddin. 2005. Bacterial diversity of tilapia (*Oreochromis niloticus*) cultured in brackish water in Saudi Arabia. Aquaculture. 250: 566–572.

Alfaro, B., I. Hernandez, Y. Le Marc and C. Pin. 2013. Modelling the effect of the temperature and carbon dioxide on the growth of spoilage bacteria in packed fish products. Food Control. 29: 429–437.

Altieri, C., B. Speranza, M.A. Del Nobile and M. Sinigaglia. 2005. Suitability of bifidobacteria and thymol as biopreservatives in extending the shelf life of fresh packed plaice fillets. J. Appl. Microbiol. 99: 1294–1302.

Aponte, M., G. Blaiotta, N. Francesca and G. Moschetti. 2010. Could halophilic archaea improve the traditional salted anchovies' (*Engraulis encrasicholus* L.) safety and quality? Lett. Appl. Microbiol. 51: 697–703.

Asfie, M., T. Yoshijima and H. Sugita. 2003. Characterization of the goldfish fecal microflora by the fluorescent *in situ* hybridization method. Fish. Sci. 69: 21–26.

Ast, J.C. and P.V. Dunlap. 2005. Phylogenetic resolution and habitat specificity of members of the *Photobacterium phosphoreum* species group. Environ. Microbiol. 7: 1641–1654.

Basby, M., V.F. Jeppesen and H.H. Huss. 1998. Characterization of the microflora of lightly salted lumfish (*Cyclopterus lumpus*) roe stored at 5°C. J. Aquat. Food Prod. Technol. 7: 35–51.

Baya, A.M., A.E. Toranzo, B. Lupiani, T. Li, B.S. Roberson and F.M. Hetrick. 1991. Biochemical and serological characterization of *Carnobacterium* spp. isolated from farmed and natural populations of striped bass and catfish. Appl. Environ. Microbiol. 57: 3114–3120.

Beaufort, A., S. Rudelle, N. Gnanou-Besse, M.T. Toquin, A. Kerouanton, H. Bergis, G. Salvat and M. Cornu. 2007. Prevalence and growth of *Listeria monocytogenes* in naturally contaminated cold-smoked salmon. Lett. Appl. Microbiol. 44: 406–411.

Benner, R.A., W.F. Staruszkiewicz and W.S. Otwell. 2004. Putrescine, cadaverine, and indole production by bacteria isolated from wild and aquacultured Penaeid shrimp stored at 0, 12, 24, and 36°C. J. Food Prot. 67: 124–133.

Bernadsky, G. and E. Rosenberg. 1992. Drag-reducing properties of bacteria from the skin mucus of the cornetfish (*Fistularia commersonii*). Microb. Ecol. 24: 63–74.

Björnsdóttir-Butler, K., G.E. Bolton, L.A. Jaykus, P.D. McClellan-Green and D.P. Green. 2010. Development of molecular-based methods for determination of high histamine producing bacteria in fish. Int. J. Food Microbiol. 139: 161–167.

Brillet-Viel, A., M.F. Pilet, F. Chevalier, M. Cardinal, J. Cornet, X. Dousset, J.J. Joffraud and F. Leroi. 2011. Biopreservation, a new hurdle technology to improve safety and quality of seafood products CIGR Section VI International Symposium. Towards a Sustainable Food Chain. Food Process, Bioprocessing and Food Quality Management. Nantes, 18–20 April.

Brillet, A., M.F. Pilet, H. Prévost, A. Bouttefroy and F. Leroi. 2004. Biodiversity of *Listeria monocytogenes* sensitivity to bacteriocin-producing *Carnobacterium* strains and application in sterile cold-smoked salmon. J. Appl. Bacteriol. 97: 1029–1037.

Brillet, A., M.F. Pilet, H. Prévost, M. Cardinal and F. Leroi. 2005. Effect of inoculation of *Carnobacterium divergens* V41, a biopreservative strain against *Listeria monocytogenes* risk, on the microbiological and sensory quality of cold-smoked salmon. Int. J. Food Microbiol. 104: 309–324.

Cardinal, M., H. Gunnlaugsdottir, M. Bjoernevik, A. Ouisse, J.L. Vallet and F. Leroi. 2004. Sensory characteristics of cold-smoked Atlantic salmon (*Salmo salar*) from European markets and relationships with chemical, physical and microbiological measurements. Food Res. Int. 37: 181–193.

CDC. 2011. Center for disease control and prevention. Outbreak Surveillance Data. http://www.cdc.gov. Access March 2013.

Chinivasagam, H.N., H.A. Bremner, S.J. Thrower and S.M. Nottingham. 1996. Spoilage pattern of five species of Australian prawns: Deterioration is influenced by environment of capture and mode of storage. J. Aquat. Food Prod. Technol. 5: 25–30.

Dalgaard, P. 1995. Qualitative and quantitative characterization of spoilage bacteria from packed fish. Int. J. Food Microbiol. 26: 319–333.

Dalgaard, P., M. Vancanneyt, N. Euras Vilalta, J. Swings, P. Fruekilde and J.J. Leisner. 2003. Identification of lactic acid bacteria from spoilage associations of cooked and brined shrimps stored under modified atmosphere between 0°C and 25°C. J. Appl. Microbiol. 94: 80–89.

Dapkevicius, M., M.J.R. Nout, F.M. Rombouts, J.H. Houben and W. Wymenga. 2000. Biogenic amine formation and degradation by potential fish silage starter microorganisms. Int. J. Food Microbiol. 57: 107–114.

Delmas, G., N. Jourdan da Silva, N. Pihier, F.X. Weill, V. Vaillant and H. de Valk. 2010. Les toxi-infections alimentaires collectives en France entre 2006 et 2008. INVS. http://opac.invs.sante.fr/doc_num.php?explnum_id=265. Access Mai 2013.

Devaraju, A.N. and T.M.R. Setty. 1985. Comparative study of fish bacteria from tropical and cold temperature marine waters. FAO Fishery Report. 97–107.

Duffes, F., C. Corre, F. Leroi, X. Dousset and P. Boyaval. 1999. Inhibition of *Listeria monocytogenes* by *in situ* produced and semi purified bacteriocins of *Carnobacterium* spp. on vacuum-packed, refrigerated cold-smoked salmon. J. Food Prot. 62: 1394–1403.

Dunlap, P.V. and J.C. Ast. 2005. Genomic and phylogenetic characterization of luminous bacteria symbiotic with the deep-sea fish *Chlorophthalmus albatrossis* (Aulopiformes: *Chlorophthalmidae*). Appl. Environ. Microbiol. 71: 930–939.

El Bassi, L., M. Hassouna, N. Shinzato and T. Matsui. 2009. Biopreservation of refrigerated and vacuum-packed *Dicentrarchus labrax* by lactic acid bacteria. J. Food Sci. 74: M335–M339.

Eldar, A., C. Ghittino, L. Asanta, E. Bozzetta, M. Goria, M. Prearo and H. Bercovier. 1996. *Enterococcus seriolicida* is a junior synonym of *Lactococcus garvieae*, a causative agent of septicemia and meningoencephalitis in fish. Curr. Microbiol. 32: 85–88.

Emborg, J. and P. Dalgaard. 2006. Formation of histamine and biogenic amines in cold-smoked tuna: An investigation of psychrotolerant bacteria from samples implicated in cases of histamine fish poisoning. J. Food Prot. 69: 897–906.

Emborg, J., P. Dalgaard and P. Ahrens. 2006. *Morganella psychrotolerans* sp. nov., a histamine-producing bacterium isolated from various seafoods. Int. J. Syst. Evol. Microbiol. 56: 2473–2479.

Fall, P.A., F. Leroi, M. Cardinal, F. Chevalier and M.F. Pilet. 2010a. Inhibition of *Brochothrix thermosphacta* and sensory improvement of tropical peeled cooked shrimp by *Lactococcus piscium* CNCM I-4031. Lett. Appl. Microbiol. 50: 357–361.

Fall, P.A., F. Leroi, F. Chevalier, C. Guerin and M.F. Pilet. 2010b. Protective effect of a non-bacteriocinogenic *Lactococcus piscium* CNCM I-4031 strain against *Listeria monocytogenes* in sterilized tropical cooked peeled shrimp. J. Aquat. Food Prod. Technol. 19: 84–92.

Fall, P.A., M.F. Pilet, F. Leduc, M. Cardinal, G. Duflos, C. Guérin, J.J. Joffraud and F. Leroi. 2012. Sensory and physicochemical evolution of tropical cooked peeled shrimp inoculated by *Brochothrix thermosphacta* and *Lactococcus piscium* CNCM I-4031 during storage at 8ºC. Int. J. Food Microbiol. 152: 82–90.

FAO (Food and Agriculture Organization). 2011. Global food losses and food waste. Extent, causes and prevention. Düsseldorf, Germany.

FAO (Food and Agriculture Organization). 2012. The state of world fisheries and aquaculture 2012. http://www.fao.org/docrep/016/i2727e/i2727e.pdf. Access February 2013. Rome, Italy.

Feldhusen, F. 2000. The role of seafood in bacterial foodborne diseases. Method. Microbiol. 2: 1651–1660.

Gancel, F., F. Dzierszinski and R. Tailliez. 1997. Identification and characterization of *Lactobacillus* species isolated from fillets of vacuum-packed smocked and salted herring (*Clupea harengus*). J. Appl. Microbiol. 82: 722–728.

Gatesoupe, F.J. 2007. Live yeasts in the gut: Natural occurrence, dietary introduction, and their effects on fish health and development. Aquaculture. 267: 20–30.

Gennari, M., S. Tomaselli and V. Cotrona. 1999. The microflora of fresh and spoiled sardines (*Sardina pilchardus*) caught in Adriatic (Mediterranean) Sea and stored in ice. Food Microbiol. 16: 15–28.

Gimenez, B. and P. Dalgaard. 2004. Modelling and predicting the simultaneous growth of *Listeria monocytogenes* and spoilage micro-organisms in cold-smoked salmon. J. Appl. Microbiol. 96: 96–109.

Gonzales-Fandos, E., M.C. Garcia-Linares, A. Villarino-Rodriguez, M.T. Garcia-Arias and M.C. Garcia-Fernandez. 2004. Evaluation of the microbiological safety and sensory quality of rainbow trout (*Oncorhynchus mykiss*) processed by the sous vide method. Food Microbiol. 21: 193–201.

Gonzalez-Rodriguez, M.N., J.J. Sanz, J.A. Santos, A. Otero and M.L. Garcia-Lopez. 2002. Numbers and types of microorganisms in vacuum-packed cold-smoked freshwater fish at the retail level. Int. J. Food Microbiol. 77: 161–168.

Gopal, S., S.K. Otta and S. Kumar. 2005. The occurrence of *Vibrio* species in tropical shrimp culture environments; implications for food safety. Int. J. Food Microbiol. 102: 151–159.

Gram, L., C. Wedell-Neegaard and H.H. Huss. 1990. The bacteriology of fresh and spoiling Lake Victoria Nile perch (*Lates niloticus*). Int. J. Food Microbiol. 10: 303–316.

Gram, L. 1995. Quality and quality changes in fresh fish. FAO Fish. Tech. Pap.

Gram, L. and J. Melchiorsen. 1996. Interaction between fish spoilage bacteria *Pseudomonas* sp. and *Shewanella putrefaciens* in fish extracts and on fish tissue. J. Appl. Bacteriol. 80: 589–595.

Gram, L. and H.H. Huss. 1996. Microbiological spoilage of fish and fish products. Int. J. Food Microbiol. 33: 121–137.

Gudmundsdóttir, S., B. Gudbjörnsdottir, H. Lauzon, H. Einarsson, K.G. Kristinsson and M. Kristjansson. 2005. Tracing *Listeria monocytogenes* isolates from cold smoked salmon and its processing environment in Iceland using pulsed-field gel electrophoresis. Int. J. Food Microbiol. 101: 41–51.

Hansen, G.H. and J.A. Olafsen. 1999. Bacterial interactions in early life stages of marine cold water fish. Microb. Ecol. 38: 1–26.

Harrison, W.A., A.C. Peters and L.M. Fielding. 2000. Growth of *Listeria monocytogenes* and *Yersinia enterolitica* colonies under modified atmospheres at 4 and 8°C using a model food system. J. Appl. Microbiol. 88: 38–43.

Håstein, T. 2013. Surveillance and control of marine fish diseases. http://www.oie.int/doc/ged/d5633.pdf. Access February 2013.

Helwigh, B., H. Korsgaard, A.J. Gronlund, A.H. Sorensen, A. Nygaard Jensen, J. Boel and B. Borck Hog. 2010. Microbiological contaminants in food in the European Union in 2004–2009. EFSA. http://www.efsa.europa.eu/fr/supporting/pub/249e.htm. Access May 2013.

Hobbs, G. 1983. Microbial spoilage of fish. pp. 217–229. *In*: Food Microbiology Advances and Prospects. Academic Press (ed.). London.

Holben, W. E., P. Williams, M. Saarinen, L.K. Särkilahti and J.H.A. Apajalahti. 2002. Phylogenetic analysis of intestinal microflora indicates a novel mycoplasma phylotype in farmed and wild salmon. Microb. Ecol. 44: 175–185.

Hozbor, M.C., A.I. Saiz, M.I. Yeannes and R. Fritz. 2006. Microbiological changes and its correlation with quality indices during aerobic iced storage of sea salmon (*Pseudopercis samifasciata*). LWT Food Sci. Tech. 39: 99–104.

Hu, Y., K. Gall, A. Ho, R. Ivanek, Y.T. Grohn and M. Wiedmann. 2006. Daily variability of *Listeria* contamination patterns in a cold-smoked salmon processing operation. J. Food Prot. 69: 2123–2133.

Hu, Y., W. Xia and X. Liu. 2007. Changes in biogenic amines in fermented silver carp sausages inoculated with mixed starter cultures. Food Chem. 104: 188–195.

Huber, I., B. Spanggaard, K.F. Appel, L. Rossen, T. Nielsen and L. Gram. 2004. Phylogenetic analysis and *in situ* identification of the intestinal microbial community of rainbow trout (*Oncorhynchus mykiss*, Walbaum). J. Appl. Microbiol. 96: 117–132.

Hungerford, J.M. 2010. Scombroid poisoning: A review. Toxicon. 56: 231–243.

Huss, H.H. 1999. La qualité et son évolution dans le poisson frais. Changements *post-mortem* dans le poisson. FAO Document technique sur les pêches. 348: 1–26.

Huss, H.H., A. Reilly and P.K. Ben Embarek. 2000. Prevention and control of hazards in seafood. Food Control. 11: 149–156.

Jaffrès, E., D. Sohier, F. Leroi, M.F. Pilet, H. Prévost, J.J. Joffraud and X. Dousset. 2009. Study of the bacterial ecosystem in tropical cooked and peeled shrimps using a polyphasic approach. Int. J. Food Microbiol. 131: 20–29.

Jaffrès, E., V. Lalanne, S. Macé, J. Cornet, M. Cardinal, T. Sérot, X. Dousset and J.J. Joffraud. 2011. Sensory characteristics of spoilage and volatile compounds associated with bacteria isolated from cooked and peeled tropical shrimps using SPME-GC-MS analysis. Int. J. Food Microbiol. 147: 195–202.

Jameson, J.E. 1962. A discussion of dynamics of *Salmonella* enrichment. J. Hyg. (Lond). 60: 193–207.

Jayaweera, V. and S. Subasinghe. 1988. Some chemical and microbiological changes during chilled storage of prawns (*Penaeus indicus*). FAO Fish. Rep. 401: 19–22.

Jeyasekaran, G., P. Ganesan, R. Anandaraj, R. Jeya Shakila and D. Sukumar. 2006. Quantitative and qualitative studies on the bacteriological quality of Indian white shrimp (*Penaeus indicus*) stored in dry ice. Food Microbiol. 23: 526–533.

Joffraud, J.J., F. Leroi and F. Chevalier. 1998. Development of a sterile cold-smoked fish model. J. Appl. Microbiol. 85: 991–998.

Joffraud, J.J., F. Leroi, C. Roy and J.L. Berdagué. 2001. Characterization of volatile compounds produced by bacteria isolated from the spoilage flora of cold-smoked salmon. Int. J. Food Microbiol. 66: 175–184.

Joffraud, J.J., M. Cardinal, J. Cornet, J.S. Chasles, S. Léon, F. Gigout and F. Leroi. 2006. Effect of bacterial interactions on the spoilage of cold-smoked salmon. Int. J. Food Microbiol. 112: 51–61.

Jorgensen, B.R., D.M. Gibson and H.H. Huss. 1988. Microbiological quality and shelf life prediction of chilled fish. Int. J. Food Microbiol. 6: 295–307.

Jorgensen, L.V. and H.H. Huss. 1998. Prevalence and growth of *Listeria monocytogenes* in naturally contaminated seafood. Int. J. Food Microbiol. 42: 127–131.

Jorgensen, L.V., H.H. Huss and P. Dalgaard. 2000a. The effect of biogenic amine production by single bacterial cultures and metabolisis on cold-smoked salmon. J. Appl. Microbiol. 89: 920–934.

Jorgensen, L.V., P. Dalgaard and H.H. Huss. 2000b. Multiple compound quality index for cold-smoked salmon (*Salmo salar*) developed by multivariate regression of biogenic amines and pH. J. Agric. Food Chem. 48: 2448–2453.

Katla, T., T. Moretro, I.M. Aasen, A. Holck, L. Axelsson and K. Naterstad. 2001. Inhibition of *Listeria monocytogenes* in cold smoked salmon by addition of sakacin P and/or live *Lactobacillus sakei* cultures. Food Microbiol. 18: 431–439.

Korkeala, H., G. Stengel, E. Hyytiä, B. Vogelsang, A. Bohl, H. Wihlman, P. Pakkala and S. Hielm. 1998. Type E botulism associated with vacuum-packaged hot-smoked whitefish. Int. J. Food Microbiol. 43: 1–5.

Laursen, B.G., L. Bay, I. Cleenwerck, M. Vancanneyt, J. Swings, P. Dalgaard and J.J. Leisner. 2005. *Carnobacterium divergens* and *Carnobacterium maltaromicum* as spoilers or protective cultures in meat and seafood: phenotypic and genotypic characterization. Syst. Appl. Microbiol. 28: 151–164.

Laursen, B.G., J.J. Leisner and P. Dalgaard. 2006. *Carnobacterium species*: Effect of metabolic activity and interaction with *Brochothrix thermosphacta* on sensory characteristics of modified atmosphere packed shrimp. J. Agric. Food Chem. 54: 3604–3611.

Lee, R.J. and R.E. Rangdale. 2008. Bacterial pathogens in seafood. pp. 247–291. *In*: T. Borresen (ed.). Improving seafood products for the consumer. Woodhead Publishing Limited, Cambridge.

Leisner, J.J., B.G. Laursen, H. Prevost, D. Drider and P. Dalgaard. 2007. *Carnobacterium*: Positive and negative effects in the environment and in foods. FEMS Microbiol. Rev. 31: 592–613.

Leroi, F., N. Arbey, J.J. Joffraud and F. Chevalier. 1996. Effect of inoculation with lactic acid bacteria on extending the shelf-life of vacuum-packed cold-smoked salmon. Int. J. Food Sci. Technol. 31: 497–504.

Leroi, F., J.J. Joffraud, F. Chevalier and M. Cardinal. 1998. Study of the microbial ecology of cold smoked salmon during storage at 8°C. Int. J. Food Microbiol. 39: 111–121.

Leroi, F., J.J. Joffraud and F. Chevalier. 2000. Effect of salt and smoke on the microbiological quality of cold smoked salmon during storage at 5°C as estimated by the factorial design method. J. Food Prot. 63: 502–508.

Leroi, F., J.J. Joffraud, F. Chevalier and M. Cardinal. 2001. Research of quality indices for cold-smoked salmon using a stepwise multiple regression of microbiological counts and physico-chemical parameters. J. Appl. Microbiol. 90: 578–587.

Leroi, F. 2010. Occurrence and role of lactic acid bacteria in seafood products. Food Microbiol. 27: 698–709.

Lin, C.S., F.L. Liu, Y.C. Lee, C.C. Hwang and Y.H. Tsai. 2012. Histamine contents of salted seafood products in Taiwan and isolation of halotolerant histamine-forming bacteria. Food Chem. 131: 574–579.

Liston, J. 1992. Bacterial spoilage of seafood. Quality assurance in the fish industry. Proceedings of an international conference, Copenhagen, Denmark, August 1992. Elsevier, Amsterdam. 93–105.

Lyhs, U., J. Björkroth, E. Hyytiä and H. Korkeala. 1998. The spoilage flora of vacuum-packaged, sodium nitrite or potassium nitrate treated, cold smoked rainbow trout stored at 4°C ou 8°C. Int. J. Food Microbiol. 45: 135–142.

Lyhs, U., H. Korkeala, P. Vandamme and J. Björkroth. 2001a. *Lactobacillus alimentarius*: A specific spoilage organism in marinated herring. Int. J. Food Microbiol. 64: 355–360.

Lyhs, U., J. Lahtinen, M. Fredriksson Ahomaa, E. Hyytiä-Trees, K. Elfing and H. Korkeala. 2001b. Microbiological quality and shelf life of vacuum-packaged gravad rainbow trout stored at 3 and 8°C. Int. J. Food Microbiol. 70: 221–230.

Macé, S., J. Cornet, F. Chevalier, M. Cardinal, M.F. Pilet, X. Dousset and J.J. Joffraud. 2012. Characterization of the spoilage microbiota in raw salmon (*Salmo salar*) steaks stored under vacuum or modified atmosphere packaging combining conventional methods and PCR-TTGE. Food Microbiol. 30: 164–172.

Macé, S., J.J. Joffraud, M. Cardinal, M. Malcheva, J. Cornet, V. Lalanne, F. Chevalier, T. Sérot, M.F. Pilet and X. Dousset. 2013. Evaluation of the spoilage potential of bacteria isolated from spoiled raw salmon (*Salmo salar*) fillets stored under modified atmosphere packaging. Int. J. Food Microbiol. 160: 227–238.

Mah, J.H. and H.J. Hwang. 2009. Inhibition of biogenic amine formation in a salted and fermented anchovy by *Staphylococcus xylosus* as a protective culture. Food Control. 20: 796–801.

Matamoros, S., M.F. Pilet, F. Gigout, H. Prévost and F. Leroi. 2009a. Selection and evaluation of seafood-borne psychrotrophic lactic acid bacteria as inhibitors of pathogenic and spoilage bacteria. Food Microbiol. 26: 638–644.

Matamoros, S., F. Leroi, M. Cardinal, F. Gigout, F. Kasbi Chadli, J. Cornet, F. Prevost and M.F. Pilet. 2009b. Psychrotrophic lactic acid bacteria used to improve the safety and quality of vacuum-packaged cooked and peeled tropical shrimp and cold-smoked salmon. J. Food Prot. 72: 365–374.

Matches, J.R. 1982. Effects of temperature on the decomposition of Pacific Coast shrimp (*Pandalus jordani*). J. Food Sci. 47: 1044–1047.

Mejlholm, O., N. Boknaes and P. Dalgaard. 2005. Shelf life and safety aspects of chilled cooked and peeled shrimps (*Pandalus borealis*) in modified atmosphere packaging. J. Appl. Microbiol. 99: 66–76.

Mejlholm, O., T.D. Devitt and P. Dalgaard. 2012. Effect of brine marination on survival and growth of spoilage and pathogenic bacteria during processing and subsequent storage of ready-to-eat shrimp (*Pandalus borealis*). Int. J. Food Microbiol. 157: 16–27.

Mudarris, M. and B. Austin. 1988. Quantitative and qualitative studies of the bacterial microflora of turbot, *Scophthalmus maximus* L., gills. J. Fish Biol. 32: 223–229.

Nakamura, H., M. Hatanaka, K. Ochi, M. Nagao, J. Ogasawara, A. Hase, T. Kitase, K. Haruki and Y. Nishikawa. 2004. *Listeria monocytogenes* isolated from cold-smoked fish products in Osaka city, Japan. Int. J. Food Microbiol. 94: 323–328.

Nayak, S.K. 2010. Role of gastrointestinal microbiota in fish. Aquacult. Res. 41: 1553–1573.

Nilsson, L., L. Gram and H.H. Huss. 1999. Growth control of *Listeria monocytogenes* on cold-smoked salmon using a competitive lactic acid bacteria flora. J. Food Prot. 62: 336–342.

Nilsson, L., T.B. Hansen, P. Garrido, C. Buchrieser, P. Glaser, S. Knochel, L. Gram and A. Gravesen. 2005. Growth inhibition of *Listeria monocytogenes* by a non bacteriocinogenic *Carnobacterium piscicola*. J. Appl. Microbiol. 98: 172–183.

Olofsson, T.C., S. Ahrné and G. Molin. 2007. The bacterial flora of vacuum-packed cold-smoked salmon stored at 7°C, identified by direct 16S rRNA gene analysis and pure culture technique. J. Appl. Microbiol. 103: 109–119.

Paari, A., P. Kanmani, R. Satishkumar, N. Yuvaraj, V. Pattukumar and V. Arul. 2012. Potential Function of a Novel Protective Culture *Enterococcus faecium*-MC13 Isolated From the gut of Mughil cephalus: Safety assessment and its custom as biopreservative. Food Biotechnol. 26: 180–197.

Paludan.Müller, C., P. Dalgaard, H.H. Huss and L. Gram. 1998. Evaluation of the role of *Carnobacterium piscicola* in spoilage of vacuum and modified atmosphere-packed-smoked salmon stored at 5°C. Int. J. Food Microbiol. 39: 155–166.

Pilet, M.F. and F. Leroi. 2011. Applications of protective cultures, bacteriocins and bacteriophages in fresh seafood and seafood products. pp. 324–347. *In*: C. Lacroix (ed.). Protective cultures, antimicrobial metabolites and bacteriophages for food and beverage biopreservation. Woodhead Publishing Limited, Oxford, Cambridge, Philadelphia, New Delhi.

Podeur, G., F. Leroi, M.F. Pilet and H. Prévost. 2013. Utilisation de *Lactobacillus sakei* pour la biopréservation des produits de la mer. Patent demand to INPI n°1353586, April 2013.

Pond, M.P., D.M. Stone and D.J. Alderman. 2006. Comparison of conventional and molecular techniques to investigate the intestinal microflora of rainbow trout (*Oncorhynchus mykiss*). Aquaculture. 261: 194–203.

Powell, S.M. and M.L. Tamplin. 2012. Microbial communities on Australian modified atmosphere packaged Atlantic salmon. Food Microbiol. 30: 226–232.

Rachman, C., A. Fourrier, A. Sy, M.F. De La Cochetiere, H. Prevost and X. Dousset. 2004. Monitoring of bacterial evolution and molecular identification of lactic acid bacteria in smoked salmon during storage. Le Lait. 84: 145–154.

Richard, C., A. Brillet, M.F. Pilet, H. Prévost and D. Drider. 2003. Evidence on inhibition of *Listeria monocytogenes* by divercin V41 action. Lett. Appl. Microbiol. 36: 288–292.

Ringo, E., E. Strom and J.A. Tabachek. 1995. Intestinal microflora of salmonids: A review. Aquacult. Res. 26: 773–789.

Ringo, E. 2008. The ability of *carnobacteria* isolated from fish intestine to inhibit growth of fish pathogenic bacteria: a screening study. Aquacult. Res. 39: 171–180.

Rocourt, J., C. Jacquet and A. Reilly. 2000. Epidemiology of human listeriosis and seafoods. Int. J. Food Microbiol. 62: 197–209.

Roeselers, G., E.K. Mittge, W.Z. Stephens, D.M. Parichy, C.M. Cavanaugh, K. Guillemin and J.F. Rawls. 2011. Evidence for a core gut microbiota in the zebrafish. ISME J. 5: 1595–1608.

Ross, T. and T.A. McMeekin. 1991. Predictive microbiology—Application of a square root model. Food Aust. 43: 202–207.

Rudi, K., T. Maugesten, S.E. Hannevik and H. Nissen. 2004. Explorative multivariate analyses of 16S rRNA gene data from microbial communities in modified-atmosphere-packed salmon and coalfish. Appl. Environ. Microbiol. 70: 5010–5018.

Satomi, M., B.F. Vogel, L. Gram and K. Venkateswaran. 2006. *Shewanella hafniensis* sp. nov. and *Shewanella morhuae* sp. nov., isolated from marine fish of the Baltic Sea. Int. J. Syst. Evol. Microbiol. 56: 243–249.

Satomi, M., B.F. Vogel, K. Venkateswaran and L. Gram. 2007. Description of *Shewanella glacialipiscicola* sp. nov. and *Shewanella algidipiscicola* sp. nov., isolated from marine fish of the Danish Baltic Sea, and proposal that *Shewanella affinis* is a later heterotypic synonym of *Shewanella colwelliana*. Int. J. Syst. Evol. Microbiol. 57: 347–352.

Satomi, M., M. Furushita, H. Oikawa, M. Yoshikawa-Takahashi and Y. Yano. 2008. Analysis of a 30 kbp plasmid encoding histidine decarboxylase gene in *Tetragenococcus halophilus* isolated from fish sauce. Int. J. Food Microbiol. 126: 202–209.

Shewan, J.M. 1971. The microbiology of fish and fishery products—A progress report. J. Appl. Bacteriol. 34: 299–315.

Shiina, A., S. Itoi, S. Washio and H. Sugita. 2006. Molecular identification of intestinal microflora in Takifugu niphobles. Comparative Biochemistry and Physiology Part D: Genomics and Proteomics. 1: 128–132.

Skrodenyte-Arbaciauskiene, V. 2007. Enzymatic activity of intestinal bacteria in roach *Rutilus rutilus* L. Fish. Sci. 73: 964–966.

Skrodenyte-Arbaciauskiene, V., A. Sruoga, D. Butkauskas and K. Skrupskelis. 2008. Phylogenetic analysis of intestinal bacteria of freshwater salmon *Salmo salar* and sea trout *Salmo trutta trutta* and diet. Fish. Sci. 74: 1307–1314.

Spanggaard, B., I. Huber, J. Nielsen, T. Nielsen, K.F. Appel and L. Gram. 2000. The microflora of rainbow trout intestine: a comparison of traditional and molecular identification. Aquaculture. 182: 1–15.

Stohr, V., J.J. Joffraud, M. Cardinal and F. Leroi. 2001. Spoilage potential and sensory profile associated with bacteria isolated from cold-smoked salmon. Food Res. Int. 34: 797–806.

Sugita, H., M. Kurosaki, T. Okamura, S. Yamamoto and C. Tsuchiya. 2005. The culturability of intestinal bacteria of Japanese coastal fish. Fish. Sci. 71: 956–958.

Sullam, K.E., S.D. Essinger, C.A. Lozupone, M.P.O'Connor, G.L. Rosen, R.O.B. Knight, S.S. Kilham and J.A. Russell. 2012. Environmental and ecological factors that shape the gut bacterial communities of fish: a meta-analysis. Mol. Ecol. 21: 3363–3378.

Tahiri, I., M. Desbiens, C. Lacroix, E. Kheadr and I. Fliss. 2009. Growth of *Carnobacterium divergens* M35 and production of Divergicin M35 in snow crab by-product, a natural-grade medium. LWT Food Sci. Tech. 42: 624–632.

Tomé, E., P.A. Gibbs and P.C. Teixeira. 2008. Growth control of *Listeria innocua* 2030c on vacuum-packaged cold-smoked salmon by lactic acid bacteria. Int. J. Food Microbiol. 121: 285–294.

Toranzo, A.E., B. Novoa, A.M. Baya, F.M. Hetrick, J.L. Barja and A. Figueras. 1993. Histopathology of rainbow trout, *Onchorhynchus mykiss* (Walbaum), and striped bass, *Morone saxatilis* (Walbaum), experimentally infected with *Carnobacterium piscicola*. J. Fish Dis. 16: 261–267.

Truelstrup Hansen, L., T. Gill and H.H. Huss. 1995. Effects of salt and storage temperature on chemical, microbiological and sensory changes in cold-smoked salmon. Food Res. Int. 28: 123–130.

Truelstrup Hansen, L. and H.H. Huss. 1998. Comparison of the microflora isolated from spoiled cold-smoked salmon from three smokehouses. Food Res. Int. 31: 703–711.

Tsai, Y.H., C.Y. Lin, L.T. Chien, T.M. Lee, C.I. Wei and D.F. Hwang. 2006. Histamine contents of fermented fish products in Taiwan and isolation of histamine-forming bacteria. Food Chem. 98: 64–70.

Valdimarsson, G., H. Einarsson, B. Gudbjörnsdottir and H. Magnusson. 1998. Microbiological quality of Icelandic cooked-peeled shrimp (*Pandalus borealis*). Int. J. Food Microbiol. 45: 157–161.

Vanderzant, C., B.F. Cobb, C.A. Thompson and J.C. Parker. 1973. Microbial flora, chemical characteristics and shelf life of four species of pond reared shrimp. J. Milk Food Technol. 36: 443–449.

Vescovo, M., G. Scolari and C. Zacconi. 2006. Inhibition of *Listeria innocua* growth by antimirobial-producing lactic acid cultures in vacuum-packed cold-smoked salmon. Food Microbiol. 23: 689–693.

Vogel, B.F., K. Venkateswaran, M. Satomi and L. Gram. 2005. Identification of *Shewanella baltica* as the most important H$_2$S-producing species during iced storage of Danish marine fish. Appl. Environ. Microbiol. 71: 6689–6697.

Warriner, K. and A. Namvar. 2009. What is the hysteria with *Listeria*? Trends Food Sci. Technol. 20: 245–254.

Weiss, A. and W.P. Hammes. 2006. Lactic acid bacteria as protective cultures against *Listeria* spp. on cold-smoked salmon. Eur. Food Res. Technol. 222: 343–346.

Wilson, B., B.S. Danilowicz and W.G. Meijer. 2008. The diversity of bacterial communities associated with Atlantic cod *Gadus morhua*. Microb. Ecol. 55: 425–434.

Wittman, R.J. and G.J. Flick. 1995. Microbial contamination of shellfish: prevalence, risk to human health, and control strategies. Annu. Rev. Public Health. 16: 123–140.

Wong, S. and J.F. Rawls. 2012. Intestinal microbiota composition in fishes is influenced by host ecology and environment. Mol. Ecol. 21: 3100–3102.

Yamazaki, K., M. Suzuky, Y. Kawai, N. Inoue and T.J. Montville. 2003. Inhibition of *Listeria monocytogenes* in cold-smoked salmon by *Carnobacterium piscicola* CS526 isolated from frozen surimi. J. Food Prot. 66: 1420–1425.

Yang, G.M., B.L. Bao, E. Peatman, H.R. Li, L.B. Huang and D.M. Ren. 2007. Analysis of the composition of the bacterial community in puffer fish Takifugu obscurus. Aquaculture. 262: 183–191.

Zaman, M.Z., F. Abu Bakar, S. Jinap and J. Bakar. 2011. Novel starter cultures to inhibit biogenic amines accumulation during fish sauce fermentation. Int. J. Food Microbiol. 145: 84–91.

21

Health Risks Associated with Seafood

Samanta S. Khora

1 Introduction

As the health benefits associated with seafood continue to expand, fears of seafood contaminants sometimes overshadow the good associated with eating more fish and shellfish from seas. The occurrence of episodes of human poisoning resulting from ingestion of toxic seafood in several areas of the world (Hallegraeff 2004) has certainly called for more attention. Although the safety of seafood has increased globally in recent decades there are still a number of environmental chemical contaminants, naturally occurring marine toxins and microbiological hazards that are present in seafood. The estimate of the probability and severity of the hazard to populations caused by consumption of foods is called risk. Seafood-borne illness can be broadly divided into intoxications and infections. In the first case, the causative agent is a toxic compound that contaminates the seafood or is produced by a biological agent in the marine product. In the case of infections, the causative agent (bacteria, viruses, or parasites) must be ingested alive resulting in its invasion of the intestinal mucous membrane or other organs to produce endotoxins (Venkitanarayanan and Doyle 2002). The severity of these risks depends upon the nature of contamination, and

Senior Professor and Division Chair, Medical Biotechnology Division, Program Chair, M.Sc. (Integrated) Biotechnology, School of Biosciences and Technology, VIT University, Vellore - 632014, Tamilnadu, India.
Email: sskhora@vit.ac.in

may range from mild diarrhoea to death. Fish, shellfish, and other marine organisms are responsible for at least one in six food poisoning outbreaks with a known etiology in the United States (Richards 1986; Lipp and Rose 1997; Anonymous 2002). Risks associated with various types of seafood over the period 1970–84 have been found that "fish" was most frequently involved followed by bivalve molluscan shellfish and crustaceans (Bryan 1980, 1987). During 1971–1990, seafood caused 32% of total food poisoning outbreaks in Korea and 22% in Japan (Lee et al. 1996). Seafood is also responsible for diseases in other parts of the world.

Ciguatera is a highly regionalized and intense risk for inhabitants and visitors consuming certain reef-associated toxin-containing finfish. Scombroid poisoning is widely distributed geographically but is specifically associated with consumption of certain fish species, particularly tuna, mackerel, mahimahi (dolphin), and bluefish. Puffers are the principal fish group causing potential threat to life and health. Consumption of seafood contaminated by algal toxins results in various seafood poisoning syndromes: Paralytic Shellfish Poisoning (PSP), Neurotoxic Shellfish Poisoning (NSP), Amnesic Shellfish Poisoning (ASP), Diarrheic Shellfish Poisoning (DSP), Ciguatera Fish Poisoning (CFP) and Azaspiracid Shellfish Poisoning (AZP). Consumption of aquatic food is the major route of human exposure to methylmercury (MeHg). There has been a fear regarding the presence of higher levels of mercury in some marine fish such as swordfish, shark, king mackerel, and tilefish. Persistent organic pollutant (POPs), including dioxins and Polychlorinated Biphenyls (PCBs) can be found in seafood too. In contrast to heavy metal contaminants and POPs, the number of reported illness from seafood-borne microbial contaminants has remained steady over the past several decades. Exposure to vibrio ad norovirus infection is still a concern. In addition, seafood allergies, distinguished as immunological reactions rather than the inability to digest and appear to be more prevalent, but they are difficult to diagnose and document. Seafood-borne illnesses on ingesting fish and shellfish is a concern especially when traveling to new destinations. Strategies for minimizing the risk of seafood-borne illness are, to some extent, hazard-specific, but overall include avoiding types of seafood identified as being more likely to contain certain contaminants, and following general food safety guidelines, which include proper cooking. To limit the scope of this chapter, it focuses on the health risks of natural marine toxins, environmental chemical contaminants and microbiological pathogenic risks associated with seafood.

2 Historical Perspective

Poisoning caused by the ingestion of toxin-containing seafood has been part of mankind's existence and concern for centuries. Records are dated

2800 BC against eating of Puffer fish from China (Kao 1966). References to fish poisoning are found in Homer's Odyssey (800 BC) and were noted during the time of Alexander the Great (356–323 BC), when his soldiers were forbidden to eat fish to avoid the accompanying maladies and malaise that could have threatened his conquests (Halstead 1965).

Puffer Fish Poisoning (PFP) by tetrodotoxin (TTX) is a deadly neurotoxin which was named by Toshizumi Tahara in the late 1880s after the order of Tetraodontiformes (puffer fish), where it was first observed. CFP has been known since 16th century and appears to be the most commonly occurring risks associated with Seafood (Hughes et al. 1977). While the name has its origin in the West Indies, the phenomenon was observed and recorded in the Indian and Pacific Oceans as early as in the sixteenth century (Bagnis 1981). The recorded outbreaks of ciguatera, both in the New Hebrides aboard the vessel of the Portuguese explorer Pedro Fernandez de Queiros in 1606 and in 1774 aboard Captain Cook's Resolution, were probably two of the earlier vivid accounts of the illness (Banner 1976; Bagnis 1981). In 1866 in Cuba, Mr. Poey, reported intoxication due to consumption of a gastropod (*Livona pica*) which was known locally as "cigua". In this way the name ciguatera was introduced. A fatal case occurred in Kaua'i in 1978 from the consumption of the Marquesan sardine (*Sardinella marquesensis*) representing sardine fish poisoning or clupeitoxicsm (Melton et al. 1984).

Hallucination Fish Poisoning became a recreational drug during the Roman Empire, and are called "the fish that make dreams" in Arabic. In 2006, two men who ate fish, apparently the *Sarpa salpa* caught in the Mediterranean were affected by ichthyoallyeinotoxism and experienced hallucinations lasting for several days. During the 1920s the name "Haff disease" was given to a disorder which affected approximately 1000 people around Köningsberg Haff, a brackish water bay in the Baltic Sea.

Many proposed theories were there on the occurrence of poisonous shellfish during 18th and 19th Centuries (Sommer et al. 1937; Sommer and Meyer 1938). "Tainted shellfish" related illness was documented in 1880 with dead birds and fish kills were also noted during this same period in Tampa (Morgan et al. 2009). The first PSP event was reported in 1927 near San Francisco, USA, which resulted in 102 people falling ill and six deaths (Sommer and Meyer 1937; Schantz 1984). Until the 1970s PSP toxins were only detected in European, North American and Japanese waters. It was not until the 1950s that the relationship between red-tide events and NSP became better understood and researched (Ishida et al. 2004; Van Egmond et al. 2004). DSP is relatively recent and new type of shellfish poisoning was first discovered in Japan by Yasumoto in 1978. The first documented human intoxication caused by DSP toxins was in The Netherlands in 1961. The first reported case of ASP was in 1987 on Prince Edward Island, Canada where 3 people died and over 100 were sickened (CIMWI 2006). It was chemically identified after its isolation in 1958 from the seaweed Chondria armata, found off the coast of Japan. In 1995, the first AZP was reported when at least eight people got ill in the Netherlands after consumption of mussels imported from Ireland but later becoming a continuing problem in Europe (Statake et al. 1998) is a newly identified marine toxin disease.

The only documented account of mercury poisoning involving seafoods occurred in people living around Minamata Bay in Japan during the 1950s. PCBs were legally widely discharged into rivers, streams, and open landfills between 1940 and the early 1970s. In the late 1960s and 1970s, concerns over the health effects of PCBs were raised due to poisoning events resulting from the consumption of PCB-contaminated food in Japan (Masuda et al. 1996) and Taiwan (Ross 2004). Due to these events and the persistence of PCBs in the environment, these chemicals were banned in the United States in 1979. In 1976, the Toxic Substance Control Act (TSCA) was passed, calling for a ban on the manufacture, processing, distribution, and use of PCBs in all products.

Viral disease transmission to human via consumption of seafood has been known since the 1950's (Roos 1956), and human enteric viruses appear to be the major cause of shellfish-associated disease.

Based on above records Environmental Defence Fund (EDF) on seafood and health of USA launched first seafood selector consumer guide in 2001; Added consumption advice to seafood selector in 2004 and Supplemental Gulf seafood safety testing in 2011.

3 Seafood and Human Health

Seafood has become an important item and one of the most highly traded commodities in the world. More than 3.5 billion people depend on the oceans for their primary source of food. Seafood consists of mainly finfish and shellfish include crustaceans and mollusks. The crustaceans comprises crayfish, crab, shrimp, and lobster, whereas the mollusks could be bivalves such as mussel, oyster, scallop; univalve creatures such as abalone, snail, and conch; and cephalopods (squid, cuttlefish, and octopus).

Seafood is a nutrient-rich food that is a good source of high quality proteins, vitamins, minerals and are low in total fat, saturated fat and sodium. Marine fish are also known to contain other nutrients important to maintaining good health (selenium, potassium, phosphorus, iodine, magnesium) and vitamins (biotin, niacin, vitamin B6, vitamin B12) (Parker 2001). Some seafood species are rich in minerals such as calcium (salmon and sardine), iron and copper (oysters) and iodine (marine fish and shellfish).

The most remarkable health-beneficial nutrients of seafood are the poly-unsaturated fatty acids (PUFAs). Seafood is the primary source of health promoting long chained omega 3 fatty acids (EPA and DHA). As early as

1940s, evidence suggested that the deficient intake of some fatty acids would lead to the coronary heart disease (Lee and Lip 2003). Researchers at first found that the occurrence of cardiovascular disease was relatively lower in those populations with higher fish and seafood consumption. Evidence suggests that omega-3's provide benefits to the developing infant including full term pregnancy, greater birth weight, better visual acuity, and improved cognitive development. For the general population and for those at risk for coronary heart disease, there is strong evidence to suggest that the risk of heart disease is reduced when marine fish is included in the diet. Scientific studies continue to explore the relationship between the unique type of fat found in seafood, the omega-3 fatty acids DHA and EPA, in the prevention or mitigation of common chronic diseases.

Marine fish represent 17 of the top 25 food sources of selenium in the American diet according to the U.S. Department of Agriculture. Selenium has essential functions as a component of the antioxidant system vital to protecting the brain and other sensitive organs. There is also strong evidence to suggest that selenium boosts the immune system, has anti-inflammatory effects, boosts the beneficial antioxidant effects of Vitamin E and even detoxifies mercury. Observational studies continue to support the conclusion that seafood is good for our health. The American Heart Association recommends eating at least 2 fish meals per week for heart health. People in Japan eat nearly 10 times more seafood than the average American and have longer life expectancy, and lower rates of heart disease and some cancers.

4 The Risks to Human Health

Most seafood contains detectable levels of contaminants because these are part of the environment and food chain, so it is not always risk-free. Risks from several potential hazards (Table 1) that have been found in seafood, including microbiological pathogens, marine toxins, environmental chemicals pollutants, and heavy metals (Yasumoto and Murata 1993; Plessi et al. 2001; Storelli et al. 2003; Iwamoto et al. 2010). These stand serious concern to human health particularly with respect to rapidly growing international seafood trade.

Naturally, toxic fish and shellfish are not distinguishable from nontoxic ones by sensory inspection, and the toxins are not destroyed by normal cooking or processing. Except for Scombroid Fish Poisoning (SFP), natural intoxications are both highly regional and species associated, and toxins are present in the fish or shellfish at the time of capture. Incidents of illness due to naturally occurring seafood toxins reported to CDC in 1978–1987 were limited to ciguatera, SFP, PSP, NSP. Other intoxications, including Puffer Fish Poisoning (PFP), were reported earlier; DSP and ASP are possible risks that should be anticipated. Scombroid poisoning is due to histamine

Table 1. Risks from several potential seafood hazards.*

Hazard Type	Description
Naturally occurring toxins	Toxins such as biotoxins, cyanogenic glycosides, and mycotoxins such as aflatoxin and ochratoxin A
Organic contaminants	Compounds that accumulate in the environment and the human body. Examples are dioxins and Polychlorinated biphenyls (PCBs). Exposure may result in a wide variety of adverse effects in humans
Heavy metals	Include lead, mercury, and cadmium. These cause neurological damages, particularly in infants and children
Unconventional agents	Examples are agents responsible for bovine spongiform encephalopathy (BSE or "mad cow disease")

*Source: Adapted from Venugopal 2009.

produced by bacteria multiplying on fish that are mishandled after capture; illnesses are widely reported. Toxins produced in marine algae are a major problem for the shellfish industry and for individuals who collect molluscs to eat along the coast. Phycotoxins (odourless and tasteless) accumulate in the digestive glands of shellfish without causing any toxic effect on it. However, when human consume a sufficient amount of contaminated seafood, intoxication occurs. Until now, six different types of poisoning have been identified, each one being responsible for different symptoms: PSP, NSP, DSP, ASP, AZP and CFP.

Of greatest concern is MeHg, a heavy metal readily absorbed and potentially toxic. The concern about MeHg is 3-fold: it accumulates through the food chain and is most concentrated in large predator and long-lived fish; seafood is the major source of mercury in humans; and mercury is potentially toxic to the developing nervous system of the fetus and infant. Since the developing fetus is at the greatest risk from exposure to MeHg, seafood consumption advice has been developed for and directed to pregnant women rather than the general population. PCBs and dioxins can bioaccumulate and have been characterized by the EPA as likely human carcinogens. In addition to the potential carcinogenic effects of PCBs and dioxins, noncancer effects, including changes in hormone levels and fetal development, have been observed at levels of about 10 times above the normal background exposure (EPA 2010). The most toxic dioxin congener, 2,3,7,8-tetrachlorodibenzo-p-dioxin (TCDD), was classified as a known human carcinogen in 1997 and accounts for about 10 percentage of the total background dioxin risk (EPA 2010).

The greatest risk to human health is from pathogens in seafood, negative health effects from seafood pathogens. Reported numbers of illnesses from seafoodborne microbes have remained steady over the past several decades. Exposure to *Vibrio*, a bacterium that contaminates raw oysters and causes illness, and norovirus infection is still a concern, however, as is consumption of raw molluscan shellfish. Steps to take to minimize the risk of seafood borne microbial illnesses include avoiding types of seafood identified as being more likely to be contaminated, and following general food safety guidelines, e.g., proper cooking.

5 Types of Hazards from Seafood

With increasing interests in marine products and rising international trade of the seafood commodity, there has been rapid rise in hazards related to marine products. The potential risks associated with chronic exposure to particular seafoodborne contaminants and risks associated with certain more acute seafoodborne hazards. The amount of a given contaminant or hazardous microbe in seafood depends on the type, size, geographic source, and age and diet of the fish or shellfish. There are a number of contaminants that may be associated with seafood, including chemicals, metals, and other substances as well as potentially harmful microbes.

These hazards in seafood broadly be *Chemical hazards*: organic, inorganic heavy metals, metals, natural toxins, toxins formed during food processing; *Biological hazards*: pathogenic microorganisms (bacteria, viruses, parasites, etc.) and *Allergens*. Depending on the target population segments, the various hazards associated with seafood could be broadly grouped into three (1) those which can cause illness in healthy adults (microbial pathogens such as *Clostridium botulinum*, *C. perfringens*, *Salmonella typhimurium*, *Shigella*, *Vibrio cholerae*, *V. parahaemolyticus*, hepatitis A virus, Norwalk-like viruses), and biotoxins from algae and other sources and chemical contaminant such as heavy metals, and polychlorinated biphenyls (PCBs); (2) those not capable of causing illness in healthy adults, but are dangerous to susceptible people such as immuno-compromised individuals, children, elderly people, and pregnant women (pathogens such as *Listeria monocytogenes* and *V. vulnificus*); and (3) those micro-organisms having uncertain pathogenicity (*Aeromonas hydrophila* and *Plesiomonas shigelloides*) (Mcentire 2004). Table 2 gives a general ranking of hazards with respect to seafood safety. Microbial hazard is the most important concern. However, possible loss in nutritional value as a result of different processing methods is a concern that is probably equally important to microbiological hazards. Concerns such as presence of toxins, environmental contamination, pollutants, and food additives are relatively of less importance (Ashwell 1990). Table 3 presents seafood hazard categories in order of decreasing risks. The various types of hazards with respect to marine products are discussed followed by possible control measures to contain these hazards.

Table 2. Ranking of food safety hazards.

Ranking	Hazard	Relative Risk
1	Microbial content	100,000
2	Pollutant chemicals	100
3	Natural toxins	100
4	Pesticide residue	1
5	Food additive	1

Source: Adapted from Ashwell, M.1990. J. Royal College Phys. 24: 23.

Table 3. Seafood hazard categories in order of decreasing risks.

Category	Description	Example
1	Those consumed raw without any cooking	Mollusks including fresh and frozen mussels, clam, oysters, and raw fish such as *sushi/sashimi*
2	Nonheat processed raw foods often consumed with additional cooking	Fresh/frozen fish and crustacean
3	Lightly preserved fish products (with <6% salt in water phase, pH >5.0)	Salted marinated, fermented, cold smoked fish
4	Semi preserved fish (salt >6%) or pH <5.0 with added preservatives	Salted, marinated fish, fermented fish, caviar
5	Mildly heat-processed (pasteurized, cooked, hot smoked) fish products	Precooked, breaded fillets
6	Heat processed (sterilized, packed in sealed containers)	Canned, retort-pouch packaged items

Source: Adapted from Huss, H.H., A. Reilly and P.K. Ben Embarek. 2000. Food Control. 11: 149; Ashwell, M. 1990. J. Royal College Phys. 24: 23.

5.1 Risks of natural marine toxins in seafood

Marine toxins produce neurological, gastrointestinal, and cardiovascular syndromes, some of which result in high mortality and long-term morbidity (Jeremy and John 2005; Khora 2004). Substantial increases in seafood consumption in recent years, together with globalization of the seafood trade, have increased potential exposure to these agents. It should be noted that marine biotoxins are naturally produced compounds. The possible presence of natural toxins in fish and shellfish has been known for a long time. Most of these toxins are produced by species of naturally occurring marine dinoflagellates (phytoplankton). The term phycotoxin indicates natural metabolites produced by unicellular microalgae (protists). Most phycotoxins are produced by dinoflagellates although cyanobacteria have also been reported to produce saxitoxin (STX), and domoic acid (DA) is produced by diatoms. Some of the toxins have initially been identified in associated organisms, e.g., okadaic acid (OA), which was initially identified in the sponge *Halichondria okadaii* (Tachibana et al. 1981), or palytoxin (PLTX) in the soft coral *Palythoa toxica* (Moore and Scheuer 1971).

Through accumulation in the food chain, these toxins may concentrate in a variety of marine organisms, including filter-feeding bivalves, burrowing and grazing organisms (tunicates and gastropods) as well as herbivorous and predatory fish. Marine animals such as oysters, crustacea and different types of fish may eat the toxic dinoflagellates storing the toxins. The toxins are accumulated in the digestive gland of the shellfish (hepatopancreas) and do not affect the shellfish themselves. These toxins have been responsible for incidents of wide-scale death of sea-life and are increasingly responsible for human intoxication. They often accumulate in shellfish or fish, and when these are eaten by humans they cause diseases. People are exposed principally to the toxins produced by harmful dinoflagellates through the consumption of contaminated seafood products. During dinoflagellate blooms humans eating seafood from infested areas can be poisoned largely. This can lead to serious poisoning. These intoxications (poisonings) are caused by the ingestion of preformed toxins and are not transmitted from person to person. These are marine toxins, non-proteinaceous compounds of low molecular weight, which widely differ in the chemical structures, physical properties and mechanisms of action, producing different effects on contaminated marine fish and shellfish consumers. The most important marine phycotoxins are shellfish toxins and finfish-ciguatoxins risks have been dealt latter.

The seafood poisonings will be limited to poisonings resulting from ingestion of the seafood containing naturally-occurring toxins that cause different syndromes. Some classifications have been proposed, the most accepted being that based on the syndromes caused by the involved toxins. Natural toxin risks are highly regional or species associated. In recent years the impact of Seafood toxins on the utilization of seafood resources is widespread and well recognized throughout the world but largely in various areas of the tropics where an abundant, accessive seafood resources resides.

Some dinoflagellates produce toxins that become concentrated in the bodies of organisms higher in the food chain, such as fish and shellfish. Based on contaminants of toxic dinoflagellates in seafood there are two main types of human poisoning. The terms "fish" and "shellfish" are associated with these illnesses because the toxins concentrate in the fish and shellfish that ingest the harmful dinoflagellates poisonings occur worldwide. However, some phycotoxins do not fit in such classification, and other groups have been included according to the name of the first identified toxin of the group (azaspiracid) or the name of the contaminated fish ("cigua" and puffer fish).

5.1.1 Risks associated with marine finfish

Eventhough marine fish is an integral component of a balanced diet, providing a healthy source of dietary protein and nutrients such as LCn3PUFAs, as with any food there is always the risk of getting food poisoning from eating some marine fish. Most fish poisonings occur from eating fish that normally are considered to be safe to eat. However, fish can become poisonous at different times of the year because of their consumption of poisonous algae and plankton (red tide) that occur in certain locations. Fish poisoning in humans by naturally occurring marine toxins can be a serious medical issue. There are many types of human intoxications in which you can get food poisoning from consuming fish. Fish are implicated in 25 percent of food-borne disease outbreaks in the United States; 86 percent of them due to biotoxins, mostly ciguatera (Olsen et al. 2000). Contamination of fish with natural toxins from the harvest area can cause consumer illness. Most of these toxins are produced by species of naturally occurring marine algae (phytoplankton). They accumulate in fish when they feed on the algae or on other fish that have fed on the algae. There are also a few natural toxins which are naturally occurring in certain species of fish create toxicological risks (Khora 1986). Some toxicological risks created by certain species of marine fish (Table 4) are classified as:

Table 4. Marine finfish poisoning syndromes.*

Poisoning	Causative toxin(s)	Commonly implicating fish	Range
Scombroid Fish Poisoning (SFP)	Scombrotoxin (Histamine)	Most commonly in tuna Mahi mahi, Bluefish, Mackerel, Sardines, Anchovies	Worldwide
Ciguatera Fish Poisoning (CFP)	Ciguatoxin (CTX), Mitotoxin (MTX)	Groupers, barracudas, Cubera snapper, Horse-eye jack, King mackerel, triggerfish, Hogfish	Tropical and Subtropical
Puffer Fish Poisoning (PFP)	Tetrodotoxin (TTX)	Puffers of order Tetraodontiformes	Cosmopolitan
Sardine poisoning (clupeotoxism)	Palytoxin (PLTX) (?)	clupeiformes, which include sardines, herrings, and anchovies.	*Tropical to temperate oceans*
Hallucinogenic fish poisoning	Indole compounds (?)	Fish species many of which are implicated in ciguatera Fish Poisoning	Mediterranean Sea and the Indian and Pacific Oceans
Haff disease	One or more toxins (?)	Certain carp-like fish (ciprinoids)	Baltic Sea, Sweden, the USSR (various lakes) and the USA

*Adapted from: Sandifer P., C. Sotka, D. Garrison and V. Fay (2007); and Van Dolah (2000).

5.1.1.1 Ciguatera Fish Poisoning (CFP)

CFP is also called Ciguatera is the most common illness caused by consumption of finfish. Ciguatera is caused by eating contaminated tropical reef fish. It is most commonly associated with larger reef-dwelling fish. As the concentration of toxin is highest in the viscera, the consumption of whole, ungutted fish generally has the most severe consequences. Ciguatoxic fish are not recognisable as such by means of external features. Contaminated fish smell and taste normal. Frying, boiling, deep-freezing or smoking the fish do not decontaminate it.

Fish species most commonly implicated. Fish that are most likely to cause ciguatera poisoning are carnivorous reef fish, including barracuda, grouper, moray eel, amberjack, sea bass, or sturgeon. Omnivorous and herbivorous fish such as parrot fish, surgeonfish, and red snapper can also be a risk. More than 400 fish species have been described which may contain ciguatoxin. They belong to the following families: Murenidae (Moray eels), Sphyraenidae (Barracudas), Lutjanida (Snappers), Serranidae (Groupers), Carangidae (Jacks), Acanthuridae (Surgeon fish), Balistidae (Trigger fish), Scaridae (Parrot fish) and to a lesser extent: Belonidae (Needlefish), Holocentridae (Soldierfish and Squirrelfish), Labridae (Wrasses), Mugilidae (Mullets), Mullidae (Surmullets, Goatfish) and Scombridae (Mackerels and their allies) (Louhija 2002).

Outbreaks and prevalence. There are probably some 10,000–50,000 cases each year, but estimates show wide variation. The last estimate on annual global incidence of ciguatera poisoning range from 25,000 to 500,000. Ciguatera outbreaks are difficult to predict, and cases of poisoning often occur in the form of an epidemic. The average incidence in endemic regions varies from 5–50 cases per 100,000 inhabitants per year, but in some years this can reach as high as 500/100,000 in the South Pacific. In the United States, 5–70 cases per 10,000 persons are estimated to occur yearly in ciguatera-endemic states and territories (Gessner and Mclaughlin 2008). The largest and most damaging outbreak occurred in Madagascar in 1994 when 500 people were poisoned and 98 died following consumption of shark (*Carcharhinus* sp.) (Habermehl et al. 1994). More than 50,000 cases of ciguatera poisoning occur globally every year. In the Caribbean, Ruff and Lewis (1994) report rates of 30 cases/10,000 population/annum (Guadeloupe) and 73 cases/10,000 population/annum (US Virgin Islands). In the South Pacific, rates are around 100 cases/10,000 population/annum (Kiribati) and 300 cases/10,000 population/annum (Tuvalu). The incidence in travelers to highly endemic areas has been estimated as high as 3 percent. While it is likely that a large proportion of cases go unreported, CFP rates in some regions are still high.

Range. CFP is widespread in tropical and subtropical waters, usually between the latitudes of 35°N and 35°S; it is particularly common in the Pacific and Indian Oceans and the Caribbean Sea. In the Pacific, the severity and incidence of poisoning increase from west to east. It is endemic in the Caribbean and in subtropical Indo-Pacific regions. In countries that import reef fish and/or have reef systems, such as the United States, Australia and Canada, CFP is a major cause of seafoodborne illness.

Causative toxins. The main casuative marine dinoflagellate is *Gambierdiscus toxicus* which originally produces maitotoxins (MTXs), the lipophilic precursors of ciguatoxin (CTX) (Yasumoto et al. 1977). Three forms of MTX, MTX-1, MTX-2 and MTX-3 have been identified from *G. toxicus* (Holmes and Lewis 1992). MTX has been proved to be the most potent toxin identified on a weight basis: the LD_{50} of MTX in mice is less than 0.2 µg/Kg (intraperitoneally) and it is at least 5-fold more toxic than tetrodotoxin. These precursors are biotransformed to CTXs by herbivorous fishes and invertebrates grazing on *G. toxicus* and then accumulated in higher trophic levels (Legrand 1998). The toxins form a family of very closely related structures with a molecular weight of 941–1117 Dalton. There are a number of variants, depending on whether certain chemical groups (-H, $-CH_3$, etc.) are present or not. Scaritoxin was isolated in 1976 in Tahiti from a parrotfish (*Scarus gibbus*). It is a metabolite of CTX. but dinoflagellates are also said to be able to produce the poison in *in vitro* culture. The CTXs are a family of heat-stable, lipid-soluble, highly oxygenated, cyclic polyether molecules with a structural framework reminiscent of the brevetoxins (Scheuer et al. 1967; Tachibana et al. 1987; Murata et al. 1990), and more than 20 toxins may be involved in CFP (Lewis et al. 1998).

Action mechanism. CTX and the closely-related MTX both cause symptoms by interfering with ion channels on cell membranes. CTX opens sodium channels and MTX opens calcium channels, disrupting the signaling between nerves and muscles.

Lethal dose. In mice, ciguatoxin is lethal at 0.45 µg/kg ip, and maitotoxin at a dose of 0.15 µg/kg ip. Oral intake of as little as 0.1 µg ciguatoxin can cause illness in the human adult. Pathogenic dose for humans 23–230 µg.

Regulatory tolerances. By FDA is 0.01 ppb P-CTX-1 equivalents for Pacific ciguatoxin and 0.1 ppb C-CTX-1 equivalent for Caribbean ciguatoxin. In most of the countries, included those of the European Union, USA and Australia, a directive has been imposed, forbidding the sale of some fishery products known to be potentially toxic with CFP toxins (Legrand 1998). Nevertheless, a "safe" level of 0.01 g/kg of fish flesh, a real challenge for the analyst, has been proposed (Tachibana et al. 1987).

Incubation period. Ciguatoxin usually causes symptoms within a few minutes to 30 hours after eating contaminated fish, and occasionally it may take up to 6 hours.

Signs and symptoms. Since more than 20 toxins are involed they produce more than 175 ciguateric symptoms, classified into four categories: gastrointestinal, neurological, cardiovascular and general symptoms (Guzman-Perez and Park 2000; Terao 2000). It should be emphasized that the symptoms of ciguatera vary in different oceans: in the Pacific Ocean neurological symptoms predominate, while in the Caribbean Sea the gastrointestinal symptoms dominate due to the difference in toxin composition. The onset of GI and neurologic symptoms after the consumption of fish are the hallmarks of ciguatera poisoning. Symptoms are usually evident within 2–6 hours after ingestion and usually resolve within 24 hours, although a late-presenting and extended course is not uncommon. GI symptoms generally consist of diaphoresis, abdominal cramps, nausea, vomiting, profuse watery diarrhea, and dysuria. GI symptoms are usually reported to occur prior to neurological symptoms. Interesting, however, is that the predominance of GI or neurologic symptoms seems to vary according to regions, with GI-predominant illness seen in the Caribbean, while neurologic symptoms predominate in the Indo-Pacific region.

Neurologic symptoms tend to occur later (up to 72 h) and may persist for months. These are predominantly paresthesias, but myriad other sometimes bizarre neurologic symptoms may also be observed, including pruritus; the sensation of loose painful teeth; tingling in the lips, tongue, throat, and perioral tissues; metallic taste; reversal of temperature sensation; and the sensation of heat in the superficial tissues of the extremities with concomitant sensation of cold in the deeper tissues. Further neurologic symptoms can include vertigo, ataxia, visual changes, and seizures. In more severe poisonings, bradycardia with hypotension and cardiovascular collapse may occur.

Common nonspecific symptoms include nausea, vomiting, diarrhea, cramps, excessive sweating, headache, and muscle aches. The sensation of burning or "pins-and-needles," weakness, itching, and dizziness can occur. Patients may experience reversal of temperature sensation in their mouth (hot surfaces feeling cold and cold, hot), unusual taste sensations, nightmares, or hallucinations. Symptoms of exposure include eye and respiratory irritation, headache, and gastrointestinal complaints; skin irritation; and difficulties with learning and memory (Lewis et al. 2000).

Morbidity and mortality. CFP is rarely fatal. The overall death rate from ciguatera poisoning is approximately 0.1% but varies according to the toxin dose and availability of medical care to deal with complications. Death from CFP is rare (<1% worldwide).

Treatment. Treatment is primarily supportive. Intravenous mannitol was thought to be the most promising of the pharmacotherapy treatments; however, it has experienced a relative decline in acceptance after a randomized, double-blind trial in 2002 failed to confirm its efficacy (Lewis et al. 2000; Stewart et al. 2010; Isbister and Kiernan 2005). However, with recent case reports supporting its use and a better understanding of ciguatera poisoning, many experts in the field believe the use of mannitol for the treatment of acute CFP arguably deserves revisiting (Lewis et al. 2000; Schnorf et al. 2002).

5.1.1.2 Scombroid Fish Poisoning (SFP)

SFP also known as histamine fish poisoning (HFP) is another cause of illness from particular species of finfish. This is the second most common finfish poisoning reported in the world after Ciguatera. Traditionally, HFP has been associated with consumption of scombroid fish from the families Scombridae and Scomberosocidae (mackerels, tunas and kingfish). More recently, non-scombroid fish have also caused identical symptoms and so "Scombroid poisoning" may not be the best description—hence the use of HFP to describe the symptoms. HFP is caused by the ingestion of foods that contain high levels of histamine and possibly other amines and compounds. The biogenic amines are produced in fish tissues by bacteria in thefamily Enterobacteriaceae, e.g., *Morganella*, *Klebsiella* and *Hafnia*. The bacteria produce decarboxylases. Once histidine decarboxylase has been produced, it may continue to produce histamine, even though bacterial growth has been prevented by chilling to 4°C. Ababouch et al. (1991) showed that histamine production can increase even in ice storage. Lehane and Olley (1999) and Clifford et al. (1991) both consider compounds other than histamine are involved. Neither cooking, canning, nor freezing reduces the toxic effect (Shalaby 1996; FDA 1999).

Fish species most commonly implicated. The term "Scombroid" gets its name from the Scombridae fish family. Species in the families Scombridae and Scomberosocidae that have been implicated in outbreaks of HFP include: mackerel (*Scomber* spp.), tuna (*Thunnus* spp.), saury (*Cololabis saira*) and bonito (*Sarda* spp.). Non-scombroid fish include: mahi-mahi (*Coryphaena* spp.), sardines (*Sardinella* spp.), pilchards (*Sardina pilchardus*), marlin (*Makaira* spp.), bluefish (*Pomatomus* spp.), sockeye salmon (*Oncorhynchus nerka*), yellowtail (*Seriola lalandii*) and Australian salmon (*Arripis trutta*). Fish of concern included representatives of 19 families, 71 genera, and more than 111 individual species. Enteric bacteria have been found to be the most important histamine forming bacteria (HFB) in fish. *Morganella morganii*, *Klebsiella pneumoniae*, *Proteus vulgaris* and *Hafnia alvei* are known to originate from fish implicated incidents of HFP (Frank 1985; Lehane and Olley 2000; Huss et al. 2003).

Outbreaks and prevalence. SFP occurs throughout the world and is perhaps the most common form of toxicity caused by the ingestion of fish. Japan, the United States and the United Kingdom are the countries with the highest number of reported incidents, although this possibly reflects better reporting systems. Frequent incidents have been reported elsewhere in Europe, Asia, Africa, Canada, New Zealand and Australia (Ababouch et al. 1991; Lehane and Olley 2000). The estimated number of unreported cases is high, as many cases are not recognised as scombrotoxism, or the poisoning is mild and a doctor is not consulted. In the United Staes Scombroid poisoning accounts for about 5% of food-borne poisonings and even for 38% of all seafood associated outbreaks. Especially in Hawaii Scombrotoxism is a common cause for seafood poisoning and the risk seems to be highest after ingestion of recreational catches. In other countries with high outbreak rates like Denmark, France, Finland or New Zealand the numbers range from 2 to 5 outbreaks/year/million people. In the USA between 1973 and 1986, 178 outbreaks were reported with 1,096 people involved. Most of the cases of poisoning in these outbreaks were traced to *Coryphaena hippurus* (Mahi mahi), followed by tuna and the Bluefish *Pomatomus saltatrix* (MMWR 1989).

Range. Although HFP cases or outbreaks have been reported worldwide (Kanki et al. 2004; Tsai et al. 2007), many incidents have been claimed to be left unreported (Mah et al. 2009; Rabie et al. 2009).

Causative toxins. There is compelling evidence that histamine is a significant causative agent of SFP. Examples of the most convincing evidence include high levels of histamine in most incriminated fish, elevated blood or urine histamine in poisoned patients, and effectiveness of antihistamine drugs to reduce the symptoms. However, oral administration of pure histamine at the same dose found in spoiled fish does not elicit the same toxicological effects seen in SFP (Taylor 1986). Some studies suggest that there are histamine potentiators in spoiled fish that contribute to the histamine-related SFP. By competitively inhibiting histamine detoxification enzymes DAO and HMT, histamine potentiators can decrease the threshold dose of histamine needed to provoke an adverse reaction in humans (Al Bulushi et al. 2009; Bjeldanes et al. 1978; Taylor and Lieber 1979; Taylor 1986). Cadaverine and putrescine have been implicated as possible histamine potentiators based on both *in vivo* and *in vitro* animal studies (Bjeldanes et al. 1978; Lyons et al. 1983; Mongar 1957). Lehane and Olley 2000 speculate that urocanic acid may be the missing factor "scombroid toxin" in histamine fish poisoning. The conversion of histidine to histamine occurs at temperature greater than 15°C (temperature well above proper refrigeration), and even proper cooking is not a remedy for improper storage, as histamine is heat stable.

Action mechanism. High levels of histamine are often found in seafood that has caused the reaction. Histidine is converted to histamine by bacterial overgrowth in fish that has been improperly stored (>20°C) after capture. Histamine and other scombrotoxins are resistant to cooking, smoking, canning, or freezing.

Lethal and infectious dose. The threshold toxic dose for histamine is not precisely known and scombroid poisoning has occurred at histamine levels as low as 50 mg/kg. In most cases, histamine levels in illness-causing fish have been above 200 ppm, often above 500 ppm. A hazardous level of histamine for human health has been suggested as 500 mg/kg although low levels as 50 mg/kg (50 ppm) have been reported in histamine poisoning (FDA 2001; Huss et al. 2003).

Regulatory tolerances. Specify 50 mg/100 g as the toxicity level, and 5 mg/100 g as the defect action level because histamine is not uniformly distributed in fish that has undergone temperature abuse. Therefore, if 5 mg/100 g is found in one section, there is a possibility that other units may exceed 50 mg/100 g (FDA 2001a). In the United Kingdom, guidelines for histamine levels in fish (Scoging 1998) are: Safe (<10 mg/100 g); Potentially toxic (10–50 mg/100 g); Probably toxic (50–100 mg/100 g) and Toxic (>100 mg/100 g).

Incubation period. Symptoms of SFP resemble an acute allergic reaction and usually appear 10–60 minutes after eating contaminated fish. Symptoms are typical allergic reactions caused by histamine—often within a few minutes or immediate to 2–8 hours of consuming the affected food item.

Signs and symptoms. It is widely believed that all humans are susceptible to scombroid poisoning though symptoms can be severe for the elderly (FDA 1999). A constellation of symptoms is seen in scombroid poisoning. These can include skin flushing, throbbing headache, peppery taste, oral numbness, abdominal cramps, nausea, diarrhea, palpitations, and anxiety. Generally, they are self-limited of 8 hours duration. Physical signs may include a diffuse blanching erythema, tachycardia, wheezing, and hypotension or hypertension. More severe cardiovascular manifestations attributed to scombroid poisoning have been reported but are rare.

Morbidity and mortality. Mortality has become rare. Patients with scomorbid illnesses such as coronary artery disease risk acute coronary syndromes caused by the tachycardia and hypotension associated with severe cases of scombroid poisoning.

Treatment. Antihistamine therapy works relatively quickly (usually less than eight hours) and for those taking medications such as isoniazid, a potent histaminase inhibitor (Morinaga et al. 1997). Scombroid poisoning usually responds well to antihistamines (H_1-receptor blockers, although H_2-receptor blockers may also be of benefit).

5.1.1.3 Puffer Fish Poisoning (PFP)

PFP or tetrodotoxin (TTX) poisoning results from ingestion of the flesh of certain species of fish belonging to the Tetraodontidae (Halstead 1967; Khora 1990, 1994). Tetrodotoxin poisoning is caused by the consumption of TTX, found most commonly in the liver, intestines, and skin of puffer fish. TTX has also been found in other fish and non-fish species including parrotfish, porcupine fish, ocean sunfish, newts and salamanders, frogs, blue-ringed octopus, starfish, and xanthid crabs. Puffers are the principal fish group causes potential threat to life and health. Puffer fish is therefore considered to be a very high risk food. The meat of some species is a delicacy in both Japan (as *fugu*) and Korea (as *bok-uh*) but the problem is that the skin and certain organs of many puffer fish are very poisonous to humans. For example, in 2007 several serious illnesses resulted from the illegal importation of toxic pufferfish that had been mislabeled as monkfish to circumvent U.S. import restrictions for this product (Cohen et al. 2009).

Fish species most commonly implicated. PFP implicated to puffer fish belong to the order Tetraodontiforme. There are more than 189 species of Puffers are available worldwide but true puffers are about 120 species. Although most species live in inshore and estuarine waters, 29 species spend their entire life in freshwater. In different parts of the world, they have different names such as Fugu, Blowfish, Swellfish, Baloonfish, Toadfish, Globefish, Porcupinefish, Sunfish, etc. Most deaths from puffers happen when untrained people catch and prepare the fish.

Outbreaks and prevalence. Many fatalities have occurred and still occurring along the coasts of China for thousands of years and in Japan for hundreds. Japan accounts for the most poisonings worldwide (30–50/y), owing to its consumption of puffer fish, known as fugu, which is seen as a delicacy. Between 1886 and 1963, 10,745 cases were recorded in Japan, with a mortality rate of almost 60% (Halstead 1967). From 1974 through 1983 there were 646 reported cases of fugu (pufferfish) poisoning in Japan, with 179 fatalities. Recent data from Japan indicate a much lower mortality rate of barely 7% in a total of 488 patients in the years 1987–1996 (Yoshikawa-Ebesu et al. 2001). Statistics show that there were 20 to 44 incidents of *fugu* poisoning per year between 1996 and 2006 in all of Japan and up to six incidents per year led to death. In November 2011, a two-Michelin star chef was suspended from his post at "Fugu Fukuji" restaurant in Tokyo. The chef served fugu liver to a female customer who subsequently required hospital treatment for mild symptoms of tetrodotoxin paralysis, but made a full recovery. Estimates as high as 200 cases per year with mortality approaching 50% have been reported. Deaths have also been reported in Hong Kong, Malaysia, Indonesia, Singapore, Sri Lanka, Pakistan, Madagascar and Australia. Some prominent cases like—on August 23, 2007, a doctor in Thailand reported that unscrupulous fish sellers sold puffer meat disguised as salmon, which resulted in the deaths of fifteen people over three years. In March 2008, a fisherman in the Philippines died and members of his family became ill from pufferfish. The previous year, four people in the same town died and five others fell ill after eating the same variety of pufferfish. Multiple outbreak of Puffer poisoning is sporadically encountered in Bangladesh throughout the year. In one study reported that death rate is approximately 15% in total and 3% annual. Pufferfish have been incriminated for death of people living in all coastal states of India too. Several reported cases of poisoning, including fatalities involved from transported/imported puffers from the Atlantic Ocean, Gulf of Mexico and Gulf of California. Only a few cases have been reported in the United States, and outbreaks in countries outside the Indo-Pacific area are not common.

Range. PFP are likely to be encountered wherever these fishes occur throughout their geographical range but reports of these illness have been mainly limited to Southeast Asian countries include Japan, Taiwan, China, Hong Kog, Thailand, Korea, Singapore, Malaysia, India, Australia, Kiribati, Papua New Guinea and Fiji. Illegal importation continues in response to consumer demand in USA has resulted in multiple poisonings (CDC 1996; Cohen et al. 2009). In other regions tetrodotoxin poisoning is only rarely a cause of fish poisoning.

Causative toxins. The greatest threat to the public health from puffers is the TTX, constitutes a specific chemical hazard deserving appropriate attention by WHO. TTX molecule has an unusual tricyclic structure and consists of a positively charged guanidinium group, which gives the name to this class of neurotoxins q.v., guanidinium toxins and a pyrimidine ring with additional fused ring systems. In puffer fish, tetrodotoxin (TTX) exists as the major toxin with chemically equilibrium analogs (4-epiTTX, 4,9-anhydroTTX) and chemically non-equilibrium analogs (deoxy analogs, 11-oxoTTX, 4-S-cysteinylTTX). The toxin is named after the order Tetraodontiformese since many of these fish often carry the toxin. TTX is one of the most potent nonprotein poisons found in nature. The toxicity of poisonous puffers fluctuates greatly (Halstead 1988). In the puffer fish, TTX concentration and distribution depend on the species, although generally the ovary, liver, and skin usually contain highest amount.

Action mechanism. TTX when introduced to the body will bind to the sodium ion channels in the nerve cells in that specific area and will spread. When it binds to the channels it prevents the uptake of Na^+ ions so it can't depolarize. Without the cell be depolarized no signal can be sent ultimately the victim can't do anything in that specific section causing paralysis and death.

Lethal dose. As estimated by experts, the lethal dose of TTX in human is around 1 mg to 2 mg and the minimum dose necessary to cause symptoms of poisoning is 0.2 mg. Reported cases from the Centers for Disease Control and Prevention (CDC 1996) documented toxicity with ingestion of as little as 1.4 ounces of puffer fish.

Regulatory tolerances. "Detention Without Physical Examination of Puffer Fish" Import Alert 16–20 by FDA. Puffers poisoning highlights the need for continued stringent regulations of puffers importation by US Food and Drug Administration. There is a European Commission directive that states that fish species of the families Tetraodontidae, Molidae, Diodontidae or Canthigasteridae (puffer, porcupine and toby fish) may not be placed on the market. Since 1958, fugu chefs must also earn a license to prepare and sell fugu to the public. Selling or serving the liver (the most toxic part) is illegal in Japan. Puffers were banned in Thailand since 2002.

Incubation period. In human beings, the onset of signs and symptoms of intoxication usually occurs from 10–45 minutes after ingestion, but may be delayed by 3 hours or more.

Signs and symptoms. The first symptoms include numbness of the lips and tongue, spreading to the face and extremities and may be followed by sensations of lightness, floating, or numbness. Nausea, vomiting, diarrhea, and epigastric pain may also be present. Later, respiratory symptoms become prominent with dyspnoea, shallow, rapid respiration and the use of auxiliary muscles. Cyanosis and hypotension follow and convulsions and cardiac arrhythmia may occur. Although victim can think clearly, cannot speak or move and soon cannot breathe as it prevents the lung working eventually dies from asphyxiation". In most instances, the victims retain consciousness until shortly before death, which usually takes place within the first 6 hours. Acute poisoning is an important clinical emergency and contributor to morbidity and mortality.

Morbidity and mortality. The puffer fish poison produces a rapid and violent death. Victims die from suffocation as diaphragm muscles are paralyzed. Most of the victims die after 4 to 24 hours. Paralysis, convulsions, mental impairment, and cardiac arrhythmia cause death in up to 60% of cases.

Treatment. There are currently no known antidotes or antitoxins to TTX. The treatment of symptoms is supportive in nature for relieving the symptoms. Administration of fluids and anti-cholinesterases can help the patients.

5.1.1.4 Sardine poisoning (Clupeotoxism)

Sardine poisoning or clupeotoxism is rare and resembles ciguatera, but is very rapid in its action. Some sardines contain toxins (Clupeotoxin) which can be poisonous to humans if eaten. The concentration of poison appears to be highest in the viscera. The toxin appears to be present in higher concentrations in summer and is believed to be possible linked to the consumption of toxic food in its food web. Preparing the fish by boiling, salting or drying does not decontaminate them.

Fish species most commonly implicated. Clupeotoxicity is most likely to occur during the warmest months caused by the ingestion of clupeiformes, which include sardines, herrings, and anchovies. The size and age of the sardines does not appear to be related to the toxicity. Individual toxicity can be very variable even within the same school of fish. Since clupeid fish are primarily plankton feeders, it is likely that some of them ingest highly toxic dinoflagellates, such as *Ostreopsis siamensis*. Clupeotoxicity is suspected to also be associatedwith Bonefishes (*Albulidae*), Slickheads (*Alepocephalidae*), Ladyfishes (*Elopidae*) or *Pterothrissidae*. Herrings, sardines and anchovies can also be ciguatoxic.

Outbreaks and prevalence. According to a study in England, the most significant cause of poisoning there after mackerels and tuna are sardines (Bartholomew et al. 1987). A fatal case occurred in Kaua'i in 1978 from the consumption of the Marquesan sardine (*Sardinella marquesensis*). This species has been replaced in abundance in the Hawaiian Islands by another import, the Goldspot Sardine (*Herklotsichthys quadrimaculatus*). Recently, 14 people died from eating poisonous sardines and other 120 people were serious ill after consumption of small quantities of a sardine (*H. quadrimaculatus*) in Madagaskar palytoxin was found in the tissue of the fish (Onuma et al. 1999). They are caused by fish caught close to the coast.

Range. Poisoning occur practically only in tropical and subtropical island regions, possibly also isolated cases in the Mediterranean. Sporadic cases of poisoning due to clupeotoxic fish are known in island regions of the tropical Atlantic and Pacific as well as in the Caribbean.

Causative toxins. While the absolute identity of the toxin is unknown, experts in marine toxins believe it to be a PLTX (Onuma et al. 1999; Randall 2005). It is suggested that PLTX is the cause of clupeotoxism. Heat does not destroy the toxin and there is still uncertainty as to the origin of the toxin.

Action mechanism. Data deficient.

Lethal dose. Data deficient.

Regulatory tolerances. Data deficient.

Incubation period. The onset of symptoms is very rapid, usually within 15 minutes. Death may occur within hours (Randall 2005).

Signs and symptoms. The most distinctive and immediate symptom is a sharp metallic or bitter taste accompanied by nausea and tingling of tongue and lips. This is soon followed by GI symptoms that include vomiting, abdominal pain, and severe diarrhea, which may be accompanied by tachycardia, clammy skin, hypotension, and other signs of impending vascular collapse. Neurological symptoms reported include nervousness, dilated pupils, severe headache, numbness, tingling, hyper salivation, dyspnea, progressive muscular paralysis, convulsions, coma, and death.

Morbidity and mortality. While sporadic and very rare, significant mortality is reported. Clupeotoxism is said to have a high mortality rate and death may occur in less than 15 minutes (Halstead 2001a).

Treatment. No antidote is available, and treatment is supportive care only. Decontamination—induced vomiting; hydration; symptomatic therapy.

5.1.1.5 Hallucinogenic fish poisoning

Hallucinogenic fish poisoning, or ichthyoallyeinotoxism, is sporadic, uncommon and absolutely unpredictable form of seafood poisoning that can occur with the ingestion of a number of fish species.

Fish species most commonly implicated. Several species from the *Kyphosus,* including *Kyphosus fuscus*, *K. cinerascens* and *K. vaigiensis* are most commonly claimed to be capable of producing this kind of toxicity. This has given rise to the collective common name "dream fish" for ichthyoallyeinotoxic fish. The effects of eating ichthyoallyeinotoxic fishes are reputed to be similar in some aspects to LSD. Experiences may include vivid auditory and visual hallucinations. *Sarpa salpa*, a species of bream, can induce LSD-like hallucinations if it is eaten. It is unclear whether the toxins are produced by the fish themselves or by marine algae in their diet, but a dietary origin may be more likely.

Outbreaks and prevalence. Sarpa salpa is widely distributed coastal fish (Froese et al. 2009) became a recreational drug during the Roman Empire, and are called "the fish that make dreams" in Arabic. In 2006, two men who ate fish, apparently the *Sarpa salpa* caught in the Mediterranean were affected by ichthyoallyeinotoxism and experienced hallucinations lasting for several days (de Haro and Pommier 2006; Clarke 2006). Other hallucinogenic fish are *Siganus spinus* (Froese et al. 2009) called "the fish that inebriates" in Reunion Island, and *Mulloides flavolineatus* called "the chief of ghosts" in Hawaii (Thomas et al. 1997). Cases of poisoning in Hawaii occur only in the months of June, July and August.

Range. It is mainly found in reef fish from the Mediterranean Sea and the Indian and Pacific Oceans. Outbreaks are known to occur in Hawaii and on the Fiji Islands and in several parts of the tropics.

Causative toxins. The exact toxin is unknown, although indole compounds formed by macroalgae have been implicated (Helferich and Banner 1960; de Haro and Pommier 2006).

Action mechanism. Data deficient.

Lethal dose. Data deficient.

Regulatory tolerances. Data deficient.

Incubation period. Ichthyoallyeinotoxism occur within a few minutes to 2 hours after ingestion of toxic fish.

Signs and symptoms. The first symptoms usually seen are a loss of balance and coordination and generalized malaise primarily affect the central nervous system (de Haro and Pommier 2006). GI symptoms are generally mild and include nausea, abdominal pain, and diarrhea. Within a few hours, specific signs of poisoning occur including delirium, visual and/or auditory hallucinations (often involving animals), depression, feeling of impending death with reactive tachycardia and hyperventilation, and disturbed behavior. If they are able to sleep, patients classically report terrifying nightmares (Helferich and Banner 1960). While hallucinogenic fish poisoning does share many similarities with ciguatera, it has pronounced CNS involvement, whereas ciguatera features peripheral nervous system involvement. They last for about 24 hours and then disappear entirely (Halstead 2001a): dizziness, loss of equilibrium, lack of motor coordination, hallucinations and mental depression. Sensation of tight constriction around the chest. Terror, nightmares. itching, burning of the throat, muscular weakness, rarely abdominal distress. Symptoms generally resolve within 24–36 hours, but weakness may persist for several days.

Morbidity and mortality. There have been no known fatalities. In comparison to the more severe forms of poisoning, such as CFP or PFP, this form is classed as mild.

Treatment. No specific treatment or antidote treatment is available. Appropriate management of transient behavioral disturbances (e.g., using benzodiazepine or neuroleptics) is important to prevent self-inflicted or other injury. Symptomatic treatment for GI manifestations can enhance patient comfort.

5.1.1.6 Haff disease

Haff disease is a syndrome characterised by rhabdomyolysis, which results from eating certain carp-like fish (Ciprinoids).

Fish species most commonly implicated. Results from eating certain carp-like fish (Ciprinoids).

Outbreaks and prevalence. During the 1920s the name "Haff disease" was given to a disorder which affected approximately 1000 people around Köningsberg Haff, a brackish water bay in the Baltic Sea. Similar cases occurred later in Sweden, the USSR (various lakes) and the USA. It was assumed that eating toxic burbot or eelpout (*Lota lota*), a species of fish, was responsible. In 1997, 6 cases were identified in California and Missouri. The symptoms began after eating *Ictiobus cyprinellus*, known as buffalofish. This is a benthic species (it feeds on the river bottom) and is found in the Mississippi and its tributaries.

Causative toxins. The disorder is caused by one or more toxins, the structure of which has not to date been clarified. Possibly it is a toxin originating from cyanobacteria. The toxin is heat-stable and is therefore not destroyed by boiling or baking.

Action mechanism. Data deficient.

Lethal dose. Data deficient.

Regulatory tolerances. Data deficient.

Incubation period. The onset is acute, on average 8 hours after eating the fish. Sometimes the incubation period is longer, up to 18 hours.

Signs and symptoms. There is pronounced muscular pain and muscle stiffness. Due to muscle necrosis large amounts of myoglobin pass into the blood stream. The urine is stained brown by the myoglobin. This is sometimes confused with haematuria. The concentration of muscle enzymes in the peripheral blood increases greatly, initially the CK, but also LDH. Hyperkalaemia can be expected. The symptoms generally last 2–3 days. Tachypnoea, tachycardia, hypertension and hypothermia may occur. Renal insufficiency is common. Residual muscle weakness may persist after the acute episode.

Mortality. is approximately 1%.

Treatment. is symptomatic and based on administration of sufficient fluid to prevent myoglobin nephrotoxicity. It is not clear whether administration of mannitol IV has a favourable effect on the course.

5.1.2 Risks associated with marine shellfish

Toxic illness caused by shellfish has been recognized for several hundred years. Several forms of shellfish poisoning may occur after ingesting filter-feeding bivalve mollusks such as mussels, oysters, clams, scallops, and cockles or crustaceans such as crabs and lobsters that accumulate potent marine toxins produced by single-celled microscopic marine algae (dinoflagellates or diatoms) associated with algal blooms or "red tide" (Scully 1990). Although there are thousands of species of microalgae that form the base of the food chain, fewer than 60 species are toxic or harmful. These toxic species may cause killing of fish and shellfish, mortality among seabirds and marine mammals, and human illnesses and death. Algal toxins have resulted in more than 500,000 incidents per year, with an overall mortality rate of 1.5% on a global basis (Wang 2008). In the United States, harmful algal blooms now threaten virtually every coastal state, and the number of toxic species is increasing. Thus, toxins are accumulated actively by shellfish and concentrated in the hepatopancreas (HP) of bivalves, their digestive organ. The factors influencing this accumulation are studied intensively. Although some toxins are accumulated very regularly by specific shellfish species in some areas. Contaminated shellfish may be found in temperate and tropical waters, typically during or after dinoflagellate blooms or "red-tides". High concentrations of marine biotoxins in these animals can cause illness amongst people who eat contaminated shellfish (Table 5). Shellfish poisoning usually occurs in the form of epidemics in these regions, but outbreaks in areas remote from the coast, where poisonous shellfish have been imported are also possible. A range of marine biotoxins selected for their involvement in poisoning events or their bioactivity observed in laboratory animals in combination with their repeated occurrence in shellfish. Numerous varieties of shellfish should not be eaten at all. Illnesses that result from marine biotoxins are not related to the manner in which food is handled or prepared, but the exposure that shellfish has to the species of toxigenic dinoflagellates while living in the marine environments. An easy classification of shellfish poisonings were based on the distinct clinical syndromes experienced by humans following consumption of contaminated shellfish (Table 6) are as follows:

Table 5. Marine dinoflagellate-biotoxins and the associated human poisonings (Taylor et al. 2003; Wang 2008).

Humans poisoning	Causative dinoflagellates (examples)	Usual transvector (s)	Distribution	Main toxins	Action target
Paralytic shellfish poisoning (PSP)	*Alexandrium acatenella, A. andeonii, A. catenella, A. cohorticula, A. fundyense, A. fraterculus, A. leei, A. minutum, A. monilatum, A. tamarense, A. ostenfeldii, A. pseudogonyaulax, A. tamiyavanichii Gymnodinium catenatum, Lingulodinium polyedrum, Pyrodiniumbahamense var. compressum, Cochlodinium catenatum, C. polykrikoides*	Clams, mussels, oysters, cockles, gastropods, scallops, whelks, lobsters, copepods, crabs, fish	Temperate areas worldwide Cosmopolitan (Northwest, West, Northeast, Florida)	Saxitoxins (STXs) Gonyautoxins	Voltage-gated sodium channel 1
Diarrhetic shellfish poisoning (DSP)	*Dinophysis acuta, D. acuminata, D. caudata, D. fortii, D. norvegica, D. mitra, D. rotundata D. sacculus, D. fortii, D. miles D. norvegica, D. tripos, Prorocentrum arenarium, P. balticum, P. belizeanum, P. concacum, P. faustiae, P. hoffmannianum, P. lima, P. maculosum, P. mexicanum, P. micans, P. minimum, P. ruetzlerianum, P. arenarium, P. belizeanum, P. cassubicum, P. concacum, P. faustiae, P. hoffmannianum, P. maculosum P. reticulatum, Coolia sp. Protoperidium oceanicum, P. pellucidum, Phalacroma rotundatum*	Mussels, scallops, clams, Gastropods	Europe, Japan, Cold and warm-temperate Atlantic, Pacific, Indo-Pacific (Canada, Northeast?)	Okadaic acid, dinophysis toxins (DTXs), yessotoxins (YTXs) and pectenotoxins (PTXs)	Inhibitors of protein phosphatases 1A and 2A. They are possibly carcinogenic
Ciguatera fish poisoning (CFP)	*Gambierdiscus toxicus, Prorocentrummicans, P. lima, P. concavum, P. hoffmannianum, P. mexicanum, P. rhathytum, Gymnodinium sangineum, Gonyaulax polyedra, G. polygramma Ostreopsisheptagona, O. lenticularis, O. mascarenensis, O. ovata, O. siamensis Amphidinium sp., Coolia monotis*	Fish, snail, shrimps, crabs	Tropical coral reefs (Southeast, Hawaii, Puerto Rico)	Ciguatoxins (CTXs), maitotoxins (MTXs), palytoxin, gambierol	Voltage-gated sodium channel 5 Voltage-gated calcium channel

Table 5. contd....

Table 5. contd.

Humans poisoning	Causative dinoflagellates (examples)	Usual transvector (s)	Distribution	Main toxins	Action target
Neurotoxic shellfish poisoning (NSP)	*Karenia brevis, K. papilonacea K. sellformis, K. bicuneiformis K. Concordia, Procentrum borbonicum? Gymnodinium breve, G. catenatum, G. mikimotoi, G. pulchellum, G. veneficum Gyrodinium galatheanum*	Oyster, clams, mussels, cockles, whelks	Gulf of Mexico, southern U.S. coast, New Zealand Subtropical/ warm temperate Gulf Coast, eastern Florida, North Carolina	Brevetoxins (PbTxs)	Voltage-gated sodium channel 5
Azaspiracid shellfish poisoning (AZP)	*Protoperidinium crassipes*	Mussels, oysters	Europe, Ireland, UK, Norway, France, Portugal, Northern Africa (Morocco), South America (Chile) and the USA	Azaspiracids (AZAs)	Voltage-gated calcium channel
Palytoxin poisoning	*Ostreopsis siamensis*	Crabs, Sea urchin	Brazil, Mediterranean Sea (Italy, Spain, Greece and France)	Palytoxin (PTXs)	Na$^+$-K$^+$ ATPase
Yessotoxin poisoning	*Protoceratium reticulatum, Lingulodinium polyedrum and Gonyaulax spinifera*	Scallops, Mussels	Italy, Norway and Portugal	Yessotoxins (YTXs)	Voltage-gated calcium/sodium channel?

Table 6. Summary of clinical entities of human poisoning by marine dinoflagellate.

Entities	CFP	PSP	NSP	DSP	ASP	PLTXs	YTXs
Incubation period	24 h	30′–3 h	5′–3 h	30′–2 h	15′–38 h	In minutes to 2–4 days	Unknown
Early Symptoms	Abdominal cramp, Nausea, Vomiting Profuse diarrhea Dysuria	Nausea, Vomiting Tingling mouth, lips, throat. Floating feeling	Nausea, Vomiting Diarrhoea Abdominal pain	Nausea, Vomiting Diarrhoea Abdominal pain	Nausea, Vomiting Abdominal pain	Muscle cramps include fever inaction	Data deficient
Mild	Paresthesias Pruritus Pain ful teeth Vertigo Ataxia Visual change Seizures	Paresthesia ++ Muscular weakness Ataxia Headache	Paresthesia Vertigo Ataxia Headache	Severe diarrhoea, Dehydration	Diarrhoea Headache Memory problems Mutism	Haemolysis, rhabdomyolysys, ataxia, drowsiness	Data deficient
Severe	Bradycardia with Hypotension and cardiovascular collapse	Dysphagia Dysarthria Diplopia Paralysis	Bradycardia Convulsions Mydriasis No paralysis	Shock	Hemiparesis Ophthalmoplegia Convulsions Hypotension Cardiac arrhythmias	Weakness of limbs followed by death	Data deficient
Foremost	GI and neurologic	Paralysis	Paresthesia	Severe diarrhea	Progressing paralysis	Respiratory failure	Data deficient
Duration	1–4 weeks	2–5 days	2–3 days	3 days	1–100 days Sometimes permanent memory problems	1–2 weeks	Data deficient
Mortality	0.1%	6% average	0%	0%	4%	0.1%?	Data deficient
Treatment	Supportive care and treatment with mannitol	Supportive care and mechanical ventilation	Supportive care	Supportive care	Supportive care	Supportive care	Supportive care

5.1.2.1 Paralytic Shellfish Poisoning (PSP)

PSP is caused by the consumption of shellfish that have ingested dinoflagellates that produce toxins. High concentrations of these toxins occur primarily during periods of algae blooms, known as "red-tides". PSP is the first hazard found to be related "red-tide" phenomenan in coastal waters due to massive dinoflagellates blooms. PSP is an important global hazard have been known for a longtime and more recent reports on the occurrence are in Canada, U.S.A. West Europe, German Federal Republic, Switzerland, Norway, UK, Mexico, South Africa, Venezuala, Japan, Philippine, Thailand, Malaysia, Indonesia, India and some more. Intoxication after consumption of shellfish is a syndrome that has been known for centuries, the most common being PSP. PSP is caused by a group of toxins (saxitoxins and derivatives) produced by dinoflagellates of the genera *Alexandrium, Gymnodinium* and *Pyrodinium*.

Shellfish species most commonly implicated. The shellfish most associated with saxitoxin (PSP) include bivalve molluscs (oysters, clams, mussels), non-bivalve shellfish (whelks, moon snails and dogwinkles) or tomalley of crustaceans (crabs, lobster). Filter-feeding bivalve shellfish (oysters, mussels, clams) which feed on dinoflagellates—*Alexandrium (Gonyaulax) tamarense, A. catenella, P. bahamense, G. catenatum* and *Cochlodinium catenatum*.

Outbreaks and prevalence. PSP is not only the most common form of shellfish poisoning, but it is also the deadliest, with a mortality rate of 6% worldwide (higher in developing countries) (Isbister and Kiernan 2005; Meier and White 1995; Lehane 2001). PSP usually occurs in outbreaks and is observed most commonly in recreational diggers. There was a clear correlation between the number of red tide outbreaks per year in Tolo Harbour between 1976 and 1986 and population growth in Hong Kong. The number of outbreaks increased steadily from 0 to 18 per year. In the same time the human population increased from <1 million to nearly 5 million (Hallegraeff 2006). Clinically documented PSP outbreaks with several dozen to around 200 patients have been described in western Europe (McCollum 1968; Zwahlen et al. 1977), Taiwan (Cheng et al. 1991) and Guatemala (Rodrigue et al. 1990). Cases have also been described in Massachusetts and Alaska (MMWR 1991; Gessner and Schloss 1996). Between 1927 and 1985, 505 PSP cases were recorded in California, of which 32 ended fatally (MMWR 1983). In contrast, in 10 PSP epidemics in the USA between 1971 and 1977, there were no recorded fatalities (Hughes 1979). Since 1927, a total of 500 cases of PSP and 30 deaths have been reported in California. The 2009 Annual Report of the American Association of Poison Control Centers' National Poison Data System documented 136 single exposures to paralytic shellfish but no deaths.

Range. Until 1970, cases of PSP had only been reported in the Northern hemisphere, but by 1990, PSP had spread to southern Africa, Australia, New Zealand, India, Thailand, Brunei, Sabah, the Philippines and Papua New Guinea. Since 1990, PSP has continued to spread.

Causative toxins. Saxitoxin (STX) takes its name from the Alaskan butter clam *Saxidomus giganteus*. PSP is the result of ingestion of STX, a purine alkaloid. Within the STX group around 30 different analogues have been detected [45]. Many derivatives of STXs are known as gonyautoxins. The name refers to *Gonyaulax*, the former name of *Alexandrium* dinoflagellates. The basic chemical stucture of these gonyautoxins is identical, but they are distinguished by chemical side-chains such as: -H, $-OSO_3$, $-CONH_2$, $-CONHSO_3$). The toxins are heat-stable and water-soluble. STX blocks sodium channels, which leads to paralysis. 2/116 died in an epidemic in Taiwan (Cheng et al. 1991 In September 1997, an outbreak of PSP was reported from Vizhinjam, Kerala, resulting in the death of seven persons and hospitalization of over 500 following consumption of mussel, *Perna indica*. Recently, in September 2004, an unusual nauseating smell emanating from the coastal waters was recorded from Kollam to Vizhinjam in the southwest coast of India. More than 200 persons, especially children, complained of nausea and breathlessness for short duration due to the smell which coincided with massive fish kills.

Action mechanism. The pharmacological action of the PSP toxins strongly resembles that of TTX. STX and several other PSP toxins block the voltage-gated sodium channel with great potency, thus slowing or abolishing the propagation of the action potential. However, they leave the potassium channel unaffected.

Lethal dose for humans is 0.1 to 1 mg. Consequently the toxin is extremely powerful (as toxic as tetrodotoxin). It is even regulated under the Chemical Weapons Convention. It is estimated that 0.5–1 mg can be fatal to humans (Clark et al. 1999). There is evidence that children are more susceptible to STX than adults (Rodrigue et al. 1990).

Regulatory tolerances. By FDA, 0.8 ppm STX equivalent (80 ug/100g) in all fish. The European Union directive 91/492/EEC permits a maximum level of PSP toxins of 80 g STX eq/100 g shellfish flesh (EC 1991). This maximum permissible level (referred to PSP toxins in general or only to STX, in a variety of molluscs or specifically in bivalves) is also followed by countries such as Canada, USA, Chile, Guatemala, Venezuela, Morocco, Singapore, Australia and New Zealand (Fernandez 1998; Shumway et al. 1995; Sim and Wilson 1997). Mexico and The Philippines lower this value to 30 and 40 g STX eq/100 g shellfish, respectively, due to the frequently reported PSP outbreaks (Aune 2001). In other countries, such as Argentina,

Uruguay, Panama, China, Hong Kong, Japan and The Republic of Korea, a tolerance limit of 400 MU/100 g has been set (Fernandez 1998; Shumway et al. 1995).

Incubation period. Symptoms may occur within a few minutes and up to 10 hours after ingestion. Usually within 2 hours.

Signs and symptoms. Clinical illness is characterized by neurological symptoms such as paresthesia and/or paralysis involving the mouth and extremities, which may be accompanied by gastrointestinal symptoms (nausea, vomiting, diarrhea, and abdominal pain) within 12 hours of ingestion of contaminated shellfish. A floating sensation often is described. Dysphonia, ataxia, weakness, and paralysis of skeletal muscles (leading to respiratory failure) can occur within 2–12 hours in severe poisoning and may persist for as long as 72 hours to a week. If a patient survives the first 12–18 h, the prognosis is good (Eastaugh and Shepherd 1989). Muscle weakness can persist for days to weeks. Severe cases involving ataxia, muscle paralysis, and respiratory arrest may result in death.

Morbidity and mortality. Based on mortality figures from recent outbreaks, children appear to be more sensitive to the STXs of PSP than adults. The case-fatality ratio averages 6%. In an epidemic in Guatemala, the mortality rate was 26/187, whereby children <6 years had a mortality rate of 50% and adults >18 years of 7% (Rodrigue et al. 1990).

Treatment. No specific treatment exists. Treatment is symptomatic and supportive. Severe cases of PSP may require mechanical ventilation. Endotracheal intubation and artificial respiration (may be necessary for a period of several days). The use of monoclonal neutralising antibodies has been proposed

5.1.2.2 Neurotoxic Shellfish Poisoning (NSP)

NSP is the least common of the shellfish poisonings. The illness is self-limiting and resolves within several days with no lasting effects. People can become ill after eating affected shellfish, but more commonly, respiratory distress occurs in compromised individuals, such as older citizens and asthmatics, who have inhaled the toxin while in or near the water. It was not until the 1950s that the relationship between red tide events and NSP became better understood and researched (McFarren 1965; Viviani 1992). NSP usually presents as gastroenteritis accompanied by minor neurologic symptoms, resembling mild CFP or mild PSP. Inhalation of aerosolized toxin in the sea spray associated with a red-tide may cause an acute respiratory illness, rhinorrhea, and bronchoconstriction.

Shellfish species most commonly implicated. The causative agents are brevetoxins, which are produced by dinoflagellates of the genus *Karenia*, especially *Karenia brevis* ingested by filter-feeding bivalve shellfish (oysters, mussels, clams) that concentrate the toxin and are subsequently consumed by predators, including humans. Brevetoxins may be present in molluscs (oysters, mussels) during an algal bloom, but are not present in fish, crabs or snails. *Gymnodinium breve* (formerly *Ptychodiscus brevis*) is found in the Caribbean and the Gulf of Mexico. This kind of aerosol is facilitated by the fact that *Gymnodinium* is a very fragile organism which easily breaks in the surf, releasing the endotoxins.

Outbreaks and prevalence. Scattered reports of cases have occurred in Florida with 2 cases in 1995, 3 in 1996, 2 in 2001, and 4 in 2005 (Terzagian 2006). Many reports of NSP involve a single case or small case series (Sakamoto 1987; Ahmed 1991) with few large outbreaks recorded. The largest documented outbreak of NSP occurred in New Zealand in 1992–1993 with over 180 cases reported over a period of several weeks. Green mussel, cockles and oysters were implicated in the New Zealand clusters (Ishida 1995, 1996; Morohashi 1995, 1999; Mackenzie 1995). The largest and best documented outbreak in the United States occurred in North Carolina (Morris 1991). It began in October 1987 when a *K. brevis* bloom became entrained in the Gulf Stream off eastern Florida and was transported up the eastern seaboard (Fowler 1989; Morris 1991). An NSP outbreak with 48 documented cases in North Carolina was reported in 1987 (Morris et al. 1991).

Range. It causes health problems around the Gulf of Mexico. Scattered reports of cases have occurred in Florida (Terzagian 2006). Many reports of NSP involve a single case or small case series (Sakamoto 1987; Ahmed 1991) with few large outbreaks recorded. The largest documented outbreak of NSP occurred in New Zealand in 1992–1993 with over 180 cases reported over a period of several weeks.

Causative toxins. This dinoflagellate produces at least two brevetoxins (BTXs). Although brevetoxins were implicated, more than one group of marine toxins and more than one algal species appear to have been involved (Todd 2002). The FDA action level for BTXs in shellfish is 20 MU/100 grams of shellfish tissue. These are fat-soluble complex molecules (polyketides). Like many marine toxins, the brevetoxins are tasteless, odorless, and heat stable. After being inhaled as aerosol they cause bronchial spasms. This may be manifested as an "asthma" crisis, rhinitis, sneezing, cough or burning eyes after walking on the beach while a strong breeze which splashes up water (with the toxin). A no observed adverse effect level (NOEL) for brevetoxins in humans has not yet been established. Although toxicity occurs in the nanomolar concentration range (Baden 1989; Toyofuku 2006).

Action mechanism. BTXs act by disrupting the flow of Na+ ions in nerve cells. They are similar to CTXs in that they are sodium channel openers that cause neuroexcitatory effects. They bind to sites near the voltage gated sodium channels, allowing an unchecked flow of Na+ ions into or out of the cell. This disruption of ion flow within nerve cells is responsible for the neurological effects associated with NSP. Incidentally, BTXs have nearly the opposite effect as STXs, which bind to a different site and effectively block Na+ ions from passing through the sodium channel (NIEHS 2000).

Lethal dose. Pathogenic dose for humans is in the order of 42–72 mouse units. In human cases of NSP, the BTX concentrations present in contaminated clams have been reported to be 30–18 ug (78–120 ug/mg).

Regulatory tolerances. A regulatory level of 80 g brevetoxin/100 g shellfish flesh, equivalent to 20 MU/100 g shellfish flesh, and the mouse bioassay as analytical method, have been adopted by USA and New Zealand as guidelines for the regulation of NSP toxins (Poli 2000; Terzagian 2006).

Incubation period. The time to onset is anywhere from 15 minutes to 12 hours (mean time of 3 h).

Signs and symptoms. NSP produces an intoxication syndrome nearly identical to that of ciguatera in which gastrointestinal and neurological symptoms predominate. In addition, formation of toxic aerosols by wave action can produce respiratory asthma-like symptoms. This results in neurological complaints such as paresthesia, temperature reversal, and ataxia, as well as GI symptoms such as nausea, abdominal pain, and diarrhea. Symptoms are generally mild and self-limited. If the toxins are absorbed in the intestine, nausea and vomiting, abdominal pain and diarrhoea occur. There then follows paresthesia around the mouth, which extends further to the throat, trunk and limbs. Ataxia, mydriasis, vertigo, breathing difficulties, headache and bradycardia may follow.

Morbidity and mortality. No deaths have been reported and the syndrome is less severe than ciguatera, but nevertheless debilitating. To date, no deaths have been reported for NSP.

Treatment. The diagnosis is clinical. There is no antidote. Treatment is supportive care. Rehydration in the case of inhalation: treatment of asthma attacks.

5.1.2.3 Diarrheal Shellfish Poisoning (DSP)

DSP is a mild poisoning. DSP produces gastrointestinal symptoms, usually beginning within 30 minutes to a few hours after consumption of toxic shellfish (Yasumoto and Murata 1990). The illness, which is not fatal, is characterized by incapacitating diarrhea, nausea, vomiting, abdominal cramps, and chills. Symptoms are generally mild and consist of nausea, vomiting, diarrhea, and abdominal pain.

Shellfish species most commonly implicated. Molluscan shellfish feeding on algae (*Dinophysis* and *Prorocentrum* spp.). Commonly affected filter-feeding bivalve shellfish (oysters, mussels, clams, scallops) that concentrate the toxin from dinoflagellates *P. lima* and species of the genus *Dinophysis*, and are subsequently consumed.

Outbreaks and prevalence. Thousands of cases of gastrointestinal disorders caused by DSP have been reported in Europe, Japan, South East Asia, North- and South-America (Sechet et al. 1990). It causes health problems along the Atlantic coast of Canada. Sixty two clinical cases of Diarrhetic Shellfish Poisoning (DSP) were reported in British Columbia linked to the consumption of mussels between July 28th and August 6th, 2011.

Range. Now-a-days, high levels of Okadaic acid (OA) group toxins are repeatedly reported in shellfish or algae along the coasts of Europe (UK, Ireland, Denmark, Sweden, Norway, France, Spain, Italy, Portugal, the Netherlands and Belgium), Canada, South America (Chile), Japan, Australia and Africa (Morocco) (Garcia et al. 2005; Scoging and Bahl 1998; Elgarch et al. 2008). It causes health problems along the Atlantic coast of Canada. These dinoflagellates are widespread, which means that this illness could also occur in any other parts of the world.

Causative toxins. The main causative agent appears to be OA. The substance takes its name from the marine sponge *Halichondria okadai*. OA has several derivatives (polyketides). They are known as dinophysitoxins (DTXs) and pectenotoxins (PTXs). Severe diarrhoea results from acute intoxication. Within Europe the permitted level for the total amount of OA, DTXs and PTXs in shellfish has been set at 160 µg OA-equivalents/kg shellfish. In 2008, the EFSA panel concluded in their opinion on OA and analogues that OA and DTXs should not exceed 45 µg OA-equivalents/kg shellfish in order to not exceed the ARfD. For PTXs, a separate EFSA opinion has been prepared (Alexander et al. 2009).

Action mechanism. The mechanisms by which okadaic acid causes diarrhea are not well understood but it is generally believed that it is a potent inhibitor of protein phosphatases. This probably causes diarrhea by stimulating the phosphorylation that controls sodium secretion by intestinal cells (Yasumoto et al. 1993). OA is also known to affect the flow of Ca^{2+} ions across cell membranes and some evidence shows that it promotes cancerous tumor growth in mice (Yasumoto et al. 1993). Potent inhibitors of protein phosphatases type PP1 and PP2A, causing protein hyperphosphorylation and tumorigenesis.

Lethal dose. 35–40 µg.

Regulatory tolerances. By FDA 0.2 ppm okadaic acid plus 35-methyl okadaic acid (DXT 1) in all fish [0.60 ug/100g.]. Within Europe the permitted level for the total amount of OA, DTXs and PTXs in shellfish has been set at 160 µg OA-equivalents/kg shellfish. The European Commission agreed in 2002 to set a tolerable level for OA, DTXs and PTXs present at the same time in edible tissues of 160 g OA eq/kg shellfish (equivalent to 40 MU/kg), and a maximum level of YTXs of 1 mg YTX eq/kg shellfish. Australia, New Zealand, Canada, Japan and The Republic of Korea recommend an amount of DSP toxins not exceeding 200 g OA eq/kg shellfish, equivalent to 5 MU/100 g (Manerio et al. 2008; Scoging and Bahl 1998).

Incubation period. Time to onset is within 30 minutes to a few hours, with complete recovery within 3 days.

Signs and symptoms. The illness, which is not fatal, is characterized by incapacitating diarrhea, nausea, vomiting, abdominal cramps, and chills. Symptoms are generally mild and consist of nausea, vomiting, diarrhea, and abdominal pain. Severe diarrhoea results from acute intoxication. In contrast to the other forms of shellfish poisoning, no neurological signs or symptoms occur. Symptoms are gastrointestinal disorder (diarrhoea, vomiting, abdominal pain) and victims recover within 3–4 days with or without treatment.

Morbidity and mortality. Self-limiting illness that resolves within several days with no lasting effects. To date, no deaths have been reported for DSP. No deaths have been reported.

Treatment. It is symptomatic and supportive, with particular attention to fluid replacement by rehydration.

5.1.2.4 Amnestic Shellfish Poisoning (ASP)

Because short term memory loss is permanent, the name ASP was given to the problem (CIMWI 2006). ASP can be a life-threatening syndrome that is characterized by both gastrointestinal and that may be accompanied by headache, confusion, and permanent short-term memory loss. ASP is caused by the consumption of shellfish containing domoic acid, a toxin produced by algae known as *Pseudonitzschia* species. ASP is the only shellfish poison produced by a diatom. All humans are susceptible to this shellfish poisoning. The elderly are particularly predisposed to serious neurological deficits. All fatalities to date have involved elderly persons.

Shellfish species most commonly implicated. After eating toxin concentrated mussels (blue and red horse), clams (hard and soft shell), oysters, scallops, lobster and crabs people experience ASP is caused by marine red algae of the genus *Chondria* and diatoms of the *Pseudonitzschia* f. *multiseries*. DA is heat stable and cooking does not destroy the toxin, although normal home cooking processes, such as boiling and steaming, could reduce the amount of DA in shellfish meat due to partial leaching of the toxin into the cooking fluids. In scallops, redistribution of DA from the hepatopancreas into the other tissues could occur. For other types of shellfish it is unlikely that processing would have a major effect on the DA concentration in shellfish meat.

Outbreaks and prevalence. ASP is a potentially serious poisoning, although only one human outbreak has ever been reported (Perl et al. 1990). Outbreaks have so far been confined to Canada and the USA, although the responsible algae has been found in many other areas. The first reported case of ASP was in 1987 on Prince Edward Island, Canada where 3 people died and over 100 were sickened (CIMWI 2006).

Range. ASP was first identified in late 1987 in Canada and later was also observed in Washington State, Oregon, and along the coast of Texas. Most of these toxigenic species are an important component of phytoplankton and have been recorded in the Southern Cone of South America (Argentina, Brasil, Chile and Uruguay).

Causative toxins. ASP is caused by domoic acid, a neurotoxic tricarboxylic amino acid structurally related to glutamic acid. Although several isomers of DA (diastereoisomer epi-domoic acid (epi-DA) and isodomoic acids (iso-DAs)) have been identified data on the occurrence only of DA and epi-DA (expressed as sum DA) have been reported. Consumption of a 400 g portion of shellfish meat containing DA and epi-DA at the current EU limit of 20 mg DA/kg shellfish meat would result in a dietary exposure of 8 mg DA (equivalent to about 130 µg DA/kg b.w. for a 60 kg adult). This is about four times higher than the acute reference dose (ARfD) of 30 µg DA/kg b.w. (equivalent to 1.8 mg DA per portion for a 60 kg adult) and is considered to constitute a potential health risk. Based on current consumption and occurrence data there is a chance of about 1 per cent of exceeding the ARfD of 30 µg DA/kg b.w. when consuming shellfish currently available on the European market. In 1998, domoic acid (DA) toxicosis was first documented in marine mammals, when more than 400 California sealions (*Zalophus californianus*) were determined to have been exposed to DA through contaminated prey that was linked to a bloom of toxin-producing diatoms. Evidence with laboratory animals show that DA can cause epilepsy, may affect brain development, and may have synergistic effects with some pollutants.

Action mechanism. Gastroenteritis. Acts as agonist to glutamate receptor, which conducts Na^+ channels inducing depolarization which in turns increases Ca^{++} permeability and ultimately leads to cell death.

Lethal dose. If the concentration of domoic acid is more than 20 ppm, the seafood is unsuitable for human consumption (Isbister and Kiernan 2005). In the Canadian 1987 outbreak, human toxicity occurred at 1–5 mg/kg (Todd 1993).

Regulatory tolerances. By FDA 20 ppm domoic acid in all fish; 30 ppm domoic acid in viscera of Dungeness crab. The European Union establishes an upper safe limit of 20 mg/kg for the total ASP toxins content in the edible parts of molluscs. This general guideline is appliedalso by Canada, USA and New Zealand.

Incubation period. Time between ingestion and onset of symptoms (all symptoms): 15 min–38 h (median 5.5 h) (Perl et al. 1990). ASP: Symptoms usually occur 30 minutes to 6 hours after ingestion. GI symptoms usually develop within 24 hours, neurological symptoms within 48 hours.

Signs and symptoms. Clinical illness is characterized by rapid onset of gastrointestinal symptoms such as nausea, vomiting, abdominal cramps and diarrhea within 30 minutes to six hours of ingestion of contaminated shellfish, followed in some cases by neurological manifestations such as headache, confusion, loss of memory, seizures and coma. Severe persistent central nervous disorders distinguish ASP from other forms of shellfish poisoning with neurological disturbances. In severe cases seizures followed by coma and death may occur. The short-term memory loss seems to be permanent in surviving victims. Some persons develop permanent neurological deficits, especially dementia.

Morbidity and mortality. To date, all the reported deaths from ASP have been in elderly persons who had more severe neurologic symptoms. 18% of patients were hospitalised for 4–101 days (median 37.5 days). The reason for hospitalisation was almost exclusively neurological disorders (Perl et al. 1990). Persistence of neurological disorders: chronic residual memory deficit, motor neuronopathy and axonopathy (Teitelbaum et al. 1990). In severe cases, seizures, paralysis, and death may occur. The mortality rate in the only known outbreak of ASP was 3%.

Treatment. Treatment is supportive care, with neurological follow-up recommended. Endotracheal intubation due to bronchial hypersecretion

(Perl et al. 1990), possibly artificial respiration. There are efforts to develop antagonists (Glavin et al. 1990).

5.1.2.5 Azaspiracid Shellfish Poisoning (AZP)

AZP is one of the more recently discovered seafood poisonings. It was identified following cases of severe GI illness from the consumption of contaminated mussels from Ireland, and now contamination has been confirmed throughout the western coastline of Europe. The dinoflagellate *Azadinium spinosum* is smaller (12–16 µm) than any of the other toxin-producing dinoflagellates has been identified as a species producer of azaspiracids (AZAs), marine toxins reported to cause human poisoning etiologically similar to DSP.

Shellfish species most commonly implicated. AZAs are accumulated by molluscs via food web which, in turn, may cause human poisoning after consumption of contaminated shellfish. The implicated toxins, azaspiracids, accumulate in bivalve mollusks that feed on toxic microalgae. Recently, it was discovered that the AZAs are actually produced by a minute dinoflagellate (Tillmann et al. 2009; Krock 2009).

Outbreaks and prevalence. AZP first reported from the Netherlands but later becoming a continuing problem in Europe (Statake et al. 1998), is a newly identified marine toxin disease. Nowadays about one dozen derivatives (AZA2 to 11) of azaspiracid (AZA1) have been identified and characterized from *P. crassipes* and contaminated shellfish (Oufji et al. 1999; James et al. 2002; James et al. 2003). Incidents of human intoxications throughout Europe, following the consumption of mussels have been attributed to AZP. Within a few years of its discovery, cases of human illness from several European countries were attributed to this toxin, and azaspiracid-producing dinoflagellates were isolated in extensive regions of northern European waters (James et al. 2003). In 1993, a total of 188 persons were admitted to a hospital in Madagascar after consuming a shark with symptoms of burning perioral pain, parasthesias, ataxia, cranial nerve palsies, coma, convulsions, and respiratory distress. Fifty patients (27%) died. On the basis of clinical findings, 1 report diagnosed the illness as ciguatera, attributing the high mortality rate to very high toxin levels in the shark (Habermehl et al. 1994).

Range. Since then several AZP outbreaks have occurred in Ireland and by now AZAs have been detected in Ireland, UK, Norway, France, Portugal, Northern Africa (Morocco), South America (Chile) and the USA (Elgarch et al 2008; James et al. 2004; Torgersen et al. 2008; Klontz et al. 2009). Although first discovered in Ireland, the search for the causative toxins, named azaspiracids, in other European countries has now led to the first discovery of these toxins in shellfish from France and Spain.

Causative toxins. Until now, 24 different AZAs have been described, with azaspiracid-1 (AZA1), -2 (AZA2), -3 (AZA3) as the predominant ones (Rehmann et al. 2008). Toxicological studies have indicated that azaspiracids can induce widespread organ damage in mice and that they are probably more dangerous than previously known classes of shellfish toxins (James et al. 2004; Furey et al. 2010).

Action mechanism. The action mechanism of AZAs is unknown at present. Necrosis in the lamina propria of the small intestine and lymphoid tissues. Some studies indicate that AZAs might have different targets, since AZA1 and AZA2 increase $[Ca^{2+}]_i$ by activation of Ca^{2+}-release from internal stores and Ca^{2+}-influx, while AZA3 induces only Ca^{2+}-influx. AZA5 does not modify intracellular Ca^{2+} homeostasis. Recent investigation of the effect of AZA4 on cytosolic calcium concentration $[Ca^{2+}]_i$ in fresh human lymphocytes demonstrated that AZA4 inhibits store-operated Ca^{2+} channels (SOC channels) and Ca^{2+} influx and that this process is reversible. It was postulated that AZA4 inhibits SOC channels by direct interaction with the channel pore, with another region of channel protein or with a closely associated regulatory protein and it was also found that AZA4 acts through another type of Ca^{2+} channel, probably some nonselective cation channel usually activated by MTX (Furey et al. 2010). AZA groups are novel inhibitors of Ca^{2+} channels, SOC and non-SOC channels.

Lethal dose. EFSA reviewed all available toxicity data and suggested that a safe level of AZA toxins in shellfish is below the ARfD of 30 μg AZA-1 equivalents /kg shellfish.

Regulatory tolerances. By FDA is 160 μg azaspiracid equivalents/kg. Although the European Commission decided in 2002 to set the maximum permitted level of AZP toxins in bivalve molluscs, echinoderms, tunicates and marine gastropods, to 160 g/kg; however, the lack of information about these AZP toxins seems to recommend a review of this limit (James et al. 2003). In fact, this limit was set based on the fact that no lower limits can be detected by the mouse bioassay, proposed as the preferred method of analysis. However, a tolerable limit of 80 g/kg seems to be more adequate, since it ensures no appreciable risk for human health. According to current EU legislation the total amount of AZAs should not exceed160 μg/kg AZA1-equivalents (Regel 2004).

Incubation period. In 15 minutes to 38 hours.

Signs and symptoms. The symptoms of AZP include nausea, vomiting, severe diarrhea and stomach cramps. Neurotoxic symptoms were also observed (Furey et al. 2010; Alfonso et al. 2005; Oufji et al. 2009). Symptoms closely resemble those associated with DSP.

Morbidity and mortality. To date, no deaths and no fatalities have ever been observed.

Treatment. It is symptomatic and supportive.

5.1.2.6 Palytoxin (PLTX) poisoning

PLTX is a large, water soluble polyalcohol first isolated from the soft coral *Palythoa toxica*, subsequently found in a variety of marine organisms ranging from dinoflagellates to fish and implicated in seafood poisoning with potential danger to public health. PLTX is a polyhydroxylated compound that shows remarkable biological activity at an extremely low concentration. This toxin was first isolated from the soft coral *P. toxica* and subsequently from many other organisms such as seaweeds and shellfish. PLTX, a polyether marine toxin originally isolated from the zooxanthid *Palythoa toxica*, is one of the most toxic non-protein substances know. Fatal poisonings have been linked to ingestion of PLTX-contaminated seafood, and effects in humans have been associated with dermal and inhalational exposure to PLTX containing organisms and water.

Shellfish species most commonly implicated. Cases of death resulting from PLTX have been reported to be due to consumption of contaminated crabs in the Philippines (Faimali 2012) sea urchins in Brazil (Alcala et al. 1998) and fish in Japan (Granéli et al. 2002; Fukui et al. 1987; Onuma et al. 1999). PLTX has become of worldwide concern due to its potential impact on animals including humans.

Outbreaks and prevalence. Results from 2008 and 2009 showed that there is a real danger of human poisoning, as these demonstrated bioaccumulation of the PLTX group (PLTX and ovatoxin-a) in both filter-feeding bivalve molluscs (mussels) and herbivorous echinoderms (sea urchins) (De la Rosa et al. 2001). So far in temperate areas, *O. ovate* blooms were reported to cause intoxications of humans by inhalation and irritations by contact (Amzil 2012).

Range. Caribbean Sea, Mediterranean Sea, Pacific Ocean and Indian Ocean.

Causative toxins. Recently, PLTX was also found in a benthic dinoflagellate, *Ostrepsis siamensis*, which caused blooms along the coast of Europe (Taniyama et al. 2003; Penna et al. 2005; Gallitelli et al. 2005; Sansoni et al. 2003; Ciminiello et al. 2006; Riobó et al. 2006), extensive death of edible mollusks and echinoderms (Gallitelli et al. 2005; Sansoni et al. 2003) and human illnesses (Penna et al. 2005; Gallitelli et al. 2005). PLTX is a large, very complex molecule with both lipophilic and hydrophilic regions, and has the longest chain of continuous carbon atoms in any known natural product. Recently several analogues, ostreocin-D (42-hydroxy-3, 26-didemethyl-9, 44-dideoxypalytoxin) and mascarenotoxins were identified in *O. siamensis*.

Action mechanism. Over the past few decades much effort has been devoted to define the action mechanisms of PLTXs; however these have not been identified. Pharmacological and electrophysiological studies have demonstrated that PLTXs act as a haemolysin and alter the function of excitable cells. PLTX selectively binds to the Na^+, K^+-ATPase with a Kd of 20 pM and transforms the pump into a channel permeable to monovalent cations with a single-channel conductance of 10 pS (Uemura 1991; Habermann 1989; Kim et al. 1985). Presently, three primary sites of action of PLTXs have been postulated: PLTX first opens a small conductance, non-selective cationic channel which results in membrane depolarization, K^+ efflux and Na^+ influx. Subsequently, the membrane depolarization may open voltage dependent Ca^{2+} channels in synaptic nerve terminals, cardiac cells and smooth muscle cells, while Na^+ influx may load cells with Na^+ and favor Ca^{2+} uptake by the Na^+/Ca^{2+} exchanger in synaptic terminals, cardiac cells and vascular smooth muscle cells. Then the increase of [Ca^{2+}] stimulates the release of neurotransmitters by nerve terminals, of histamine by mast cells and of vasoactive factors by vascular endothelial cells as a signal. It also induces contractions of striated and smooth muscle cells. Additional effects of a rise in [Ca^{2+}] may be activation of phospholipase C (Hirsh and Wu 1997) and phospholipase A2 (Habermann and Laux 1986). There are reports that PLTX opens an H^+ conductive pathway which results in activation of the Na^+/H^+ exchanger (Levine and Fujiki 1985; Frelin et al. 1991). Other investigators suggest that PLTX raises [Ca^{2+}] independently of the activity of voltage dependent Ca^{2+} channels and Na^+/Ca^+ exchange (Yoshizumi et al. 1991). The last two actions might act as the opening of H^+ specific and Ca^{2+} specific channels. Overall, PTX might possess more than one site of action in excitable cells and act as an agonist for low conductance channels conducting Na^+/K^+, Ca^{2+} and H^+ ions.

Lethal dose. PLTX is regarded as one of the most potent toxins so far known (Yoshizumi et al. 2007). The LD_{50} 24 hours after intravenous injection vary from 0.025 µg/kg in rabbits and about the same in dogs to 0.45 µg/kg in mice, with monkeys, rats and guinea pigs around 0.9 µg/kg. By extrapolation, a toxic dose in humans would range between 2.3 and 31.5 µg (Moore et al. 1982).

Regulatory tolerances. FDA makes no recommendations in this guidance and has no specific expectations with regard to controls for PLTX.

Incubation period. In minutes to 2–4 days.

Signs and symptoms. Include fever inaction, ataxia, drowsiness, and weakness of limbs followed by death.

Morbidity and mortality. Onset of symptoms rapid, with death occurring within minutes.

Treatment. Treatment is symptomatic and supportive therapy probably as for a coronary spasm.

5.1.2.7 Yessotoxins (YTXs) poisoning

YTXs are a group of structurally related polyether toxic compound originally isolated in Japan from the digestive gland of the scallop *Patinopecten yessoensis*, and is produced by phytoplanktonic microalgae of dinoflagellates *Protoceratium reticulatum*, *Lingulodinium polyedrum* and *Gonyaulax spinifera* (Ito et al. 2000).

Shellfish species most commonly implicated. Scallops and from mussels of different origin (Murata et al. 1987; Draisci et al. 1999).

Outbreaks and prevalence. YTX and it analogues, which are disulphated polyether compounds are of increasing occurrence in seafood and are a worldwide concern due to their potential risk to human health.

Range. YTX was originally isolated from the scallop *P. yessoensis*, collected at Mutsu Bay, Japan (Beatriz et al. 2008). Since then, YTXs have been found in Europe, South America and New Zealand, and become a worldwide concern due to its potential risk to human health.

Causative toxins. Three dinoflagellate species, *P. reticulatum*, *L. polyedrum* and *G. spinifera* produce YTXs (Murata et al. 1987; Amzil et al. 2008). YTX and its derivatives, 45-hydroxy YTX (45-OH-YTX), 45,46,47-trinor YTX, homo YTX, and 45-hydroxyhomo YTX (Rhodes et al. 2006; Satake et al. 1996) are disulfated polyether lipophilic toxins (Murata et al. 1987). Recently several new YTX analogues: carboxyyessotoxin (with a COOH group on the C_{44} of YTX instead of a double bond); carboxyhomoyessotoxin (with a COOH group on the C_{44} of homoYTX instead of a double bond); 42,43,44,45,46,47,55-heptanor-41-oxo YTX and 42,43,44,45,46,47,55-heptanor-41-oxohomo YTX in Adriatic mussels (*M. galloprovincialis*) have been identified in dinoflagellates (Satake et al. 1997; Ciminiello et al. 2001). Originally, YTXs were classified among the toxins responsible for DSP, mainly because they appear and are extracted together with the DSP toxins, OA and the dinophysistoxins (DTXs) (Murata et al. 1987). However, YTXs are proved to be not diarrheogenic compared to OA and its derivatives, the DTXs, which cause intestinal fluid accumulation or inhibition of protein phosphatase 2A.

Action mechanism. Recently it was demonstrated that YTX is a potent neurotoxin to neuronal cells. However, the action site and the mechanism are unknown (Paz et al. 2008). YTX was observed to induce a two-fold increase in cytosolic calcium in cerebellar neurons that was prevented by the voltage-sensitive calcium channel antagonists nifedipine and verapamil. These results suggest YTX might interact with calcium channels and/or sodium channels directly. Previous studies also showed that YTX activated nifedipine-sensitive calcium channels in human lymphocytes (Perez-Gomez et al. 2006), and YTX was postulated to activate non-capacitative calcium entry and inhibit capacitive calcium entry by emptying of internal calcium stores.

Lethal dose. Toxicological studies indicated that acute oral administration at doses up to 10 mg/kg YTX or repeated (seven days) oral exposure to high (2 mg/kg/day) doses of the toxin caused no mortality nor strong signs of toxicity in mice (Ciminiello et al. 2002; Aune et al. 2002; Tubaro et al. 2004). YTX caused motor discoordination in the mouse before death due to cerebellar cortical alterations (Tubaro et al. 2004; Tubaro et al. 2003). Histopathological study revealed that YTX provoked alterations in the Purkinje cells of the cerebellum, including cytological damage to the neuronal cell body and change in the neurotubule and neurofilament immunoreactivity (Tubaro et al. 2003). Assumed that health risks are less than those of other DSP toxins.

Regulatory tolerances. 1 mg yessotoxin equivalent/kg implemented a maximum permitted level (MPL) of 1 mg YTX equivalents/kg shellfish intended for human consumption (Directive 2002/225/EC) (Franchini et al. 2004).

Incubation period. Data deficient.

Signs and symptoms. Symptoms of intoxication produced by YTX in humans are relatively unknown due to the fact that no human intoxication has been reported to date. However, it seems clear that YTX does not produce diarrhea in humans. The scarcity of toxicological studies on YTX and its analogues, which are necessary to assess its human health risks, have been hampered until now, by the limited availability of the toxin.

Morbidity and mortality. Data deficient.

Treatment. Supportive care.

5.2 Risks of environmental chemical contaminants

The ocean dumping of hundreds of millions tons of waste material from industrial processing, sludge from sewage treatment plants, draining into the sea of chemicals used in agriculture and raw untreated sewage from large urban populations and industries all participate in contaminating the coastal marine environments. Chemical contamination of the environment are nearly all man-made. The chemicals find their way into fish and other aquatic organisms. In the process of concentrating fish proteins as a food source, a variety of protein-bound, non–water-soluble, or non–alcohol-soluble toxic compounds may be preserved. These include organic mercurials, hydrocarbons, dioxins, polychlorinated dibenzofurans, chlorinated pesticides, and heavy metals as antimony, arsenic, cadmium, chromium, cobalt, lead, phosphorus, mercury, nickel, and zinc (Ayotte et al 1997; Connell et al. 1998; Svensson et al. 1991).

Increasing amounts of chemicals may be found in predatory species as a result of biomagnification or of bioaccumulation, when increasing concentrations of chemicals in the body tissues accumulated over the life span of the individual. The chemical contaminants in seafood is therefore highly dependent on geographic location, species and fish size, feeding patterns, solubility of chemicals and their persistence in the environment. These contaminants include inorganic compounds such as methylmercury (MeHg) and other metals, as well as persistent organic pollutants (POPs) such as dioxins and polychlorinated biphenyls (PCBs). The potential health risks associated with the presence of chemical contaminants, both those occurring naturally and those resulting from human activities, in seafood. The potential risks include contaminants from chemicals, heavy metals, chemicals, marine toxins, and inorganic compounds (Hites et al. 2004a; Nesheim and Yaktine 2007; Iwamoto et al. 2010).

Health risks resulting from chronic exposure to these chemicals may also be higher for infants and young children. During pregnancy and lactation, mothers can pass DDTs and PCBs on to their infants. These chemicals can then affect overall growth and development, and brain development and function.

5.2.1 Inorganic compounds/metals contaminants

The risks of inorganic compounds such as MeHg and other metals, PCBs and dioxins and Dixins Like compounds (DLC) involved in seafood consumption. Of these, MeHg is the contaminant that has elicited the most concern among consumers.

5.2.1.1 Methylmercury (MeHg) poisoning

Mercury poisoning is the common phrase for *mercuralism*, a condition caused by the body absorbing mercury. There are two forms of mercury in the environment, namely inorganic and organic mercury (MeHg) (US EPA 2001). Nearly all the human exposure to MeHg occurs via fish consumption. Therefore individuals who regularly consume large amounts of fish, particularly those fish with high mercury levels could be exposed to high levels of mercury. MeHg is toxic to the nervous system, especially the developing brain (Inskip and Piotrowski 1985; Bernier 1995; Mahaffey 1999; Järup 2003; Elhamri et al. 2007). MeHg is highly toxic and causes severe effects following incidents in Iraq (grain) and Japan (seafood). Because of the health risks for babies associated with MeHg exposure, the United States FDA and Environmental Protection Agency (EPA) have recommended that women of childbearing age, pregnant women, and breastfeeding mothers limit their intake of fish and avoid all consumption of shark, swordfish, kingmackerel, or tilefish. Some evidence also suggests that mercury can increase risks for heart disease and neurological problems in adulthood.

Bioaccumulation of MeHg in seafood. Inorganic mercury gets transformed into organic MeHg through the process of methylation by bacteria, and get absorbed by aqatic organisms. After consumption by fish and shellfish, MeHg is accumulated in bodies of these species (Hughner et al. 2008). Then, the MeHg concentrations are gradually magnified through the food chain. Predatory fish or mammals (whales) at the top of the food web have the largest amounts. Hence, the higher MeHg contents are found in the longer-living, larger, and predatory species on higher levels of the food chain, e.g., swordfish, marlin and shark. From the perspective of human health, negative effects of mercury are trivial compared with MeHg, as the latter is more easily absorbed into human blood stream. The half-life of MeHg is about 70 days in humans.

Adverse health effects from high-level exposure. People who were exposed to high mercury levels when they were adults underwent sensory and motor impairment. An investigation of Faroe Islands marine food consumption and contaminants exposure found that prenatal mercury and PCB exposure led to children's neurobehavioral dysfunctions (Weihe et al. 1996). In addition, MeHg could cross the placenta and blood-brain barrier, and become toxic to nervous system of fetus (US EPA 2001). It was reported high MeHg exposures lead to mental retardation and cerebral palsy (Myers et al. 2003). One possible explanation to the mercury exposure and cardiovascular diseases risk is due to the effects of mercury in oxidative stress propagation (Virtanen et al. 2007). More than neurobehavioral deficits, with high dietary mercury exposures, symptoms as cytogenetic damage, immune alterations, and cardiovascular toxicity were reported in several Amazon countries (Passos et al. 2003).

Adverse health effects from low-level exposure. Recently, it has been suggested that low-dose exposure of the foetus to MeHg may lead to impaired performance, which appears when the individual reaches early childhood. According to Kjellstrom et al. 1989; Mahaffey 2004; Myers et al. 2003, young children exposed as foetuses perform badly in tests that measure attention, language, memory and fine-motor function (called neurobiological tests). There is also evidence that exposure to MeHg can affect the cardiovascular system (blood pressure regulation, variable heart rate and heart disease). Exposure during the first trimester (three months) of pregnancy appears to be the critical period. The study of Oken et al. (2005) also showed mercury exposure contributes to negative cognitional effects. By applying an IQ index, an estimate of infant neurodevelopment, it was found that maternal mercury had negative influence on IQ points. The IQ points reduce –0.18 with one more part per million of hair mercury of the mother (Axelrad et al. 2007). So, MeHg is viewed as one strong neurotoxicant, particularly on fetuses (Díez 2009).

Regulatory tolerances. A MeHg intake of 3.3 ug/kg body weight per week may be used as a guideline to protect against non-developmental adverse effects.

5.2.1.2 Other metals risks

Metal contaminants other than mercury, including lead, manganese, chromium, cadmium, and arsenic may be present in seafood, although on a population basis, seafood consumption does not appear to be a major route of exposure to these metals. In analyses of farmed Atlantic and wild salmon, Foran et al. (2004) found that for none of nine metals measured did the levels exceed federal standards. For three of the metals measured (cobalt, copper, and cadmium), levels were significantly higher in wild than farmed salmon. Burger and Gochfeld (2005) measured the levels of seven metals (arsenic, cadmium, chromium, lead, manganese, mercury, selenium) in fish obtained exceeded health-based standards from New Jersey markets. Kong et al. (2005) found levels of lead and chromium in farmed tilapia from China that exceeded local guidelines. However, coexposure to selenium may diminish the toxic effects of some forms of mercury and other heavy metals, including cadmium and silver (Whanger 1985). Selenium was first reported by Parizek and Ostadalova (1967) to counteract acute mercuric chloride toxicity. Later, Ganther et al. (1972) showed the mitigating effect of sodium selenite on the toxicity of MeHg. The overall public health risk for environmental contamination is concerning; however, the true risks of exposure is unknown.

5.2.2 Persistent Organic Pollutants (POPs)

POPs are defined as organic chemicals that remain intact in the environment for long periods, become widely distributed geographically, bioaccumulate up the food chain by amassing in fatty tissues of animals, and are toxic to humans, wildlife, and the environment (Bidleman and Harner 2000; IOM 2003; UNEP Global Environmental Facility 2003; Robson and Hamilton 2005). Many POPs are chlorinated compounds, but brominated and fluorinated compounds also exist (e.g., brominated flame retardants and Freon) and may have a detrimental impact on the environment.

Owing to their toxic characteristics, they can pose a threat to humans and the environment. Because of their lipophilic character, POPs are absorbed and transported to fatty tissues in fish and marine mammals. Uptake of POPs can occur through exposure from sediments in water or via consumption of smaller fish by predatory species (Geyer et al. 2000). The POPs to which seafood consumers are most likely exposed are the dioxins, DLCs, and PCBs. More than mercury, organic contaminants such as PCBs and dioxins are potential risks to seafood consumers (Ahmed et al. 1993). The biological activity of dioxins, DLCs, and PCBs varies due to differences in toxicity and half-life of the various congeners. Variations in toxicity among congeners are related to a number of factors, including binding interaction at the cellular level with the arylhy-drocarbon receptor (AhR) and variability in pharmacokinetics *in vivo*.

Levels of POPs in seafood. PCBs and dioxins exist in natural water bodies at low levels. However, with less degradable and lipophilic attributes, PCBs and dioxins become bioaccumulated through food chains, and are at much higher levels in the bodies of large, predatory fishes and other seafood species, and then in the body of human (Tryphonas 1995; Kris-Etherton et al. 2002). Hites et al. (2004a) found that, because of their higher fat levels, some farmed salmon contain significantly higher concentrations of certain organochlorine contaminants, including PCBs, than wild-caught salmon. In addition, PCB concentrations in samples of commercial salmon feed purchased in Europe were higher than those in samples purchased in North and South America, suggesting that regional differences in the composition of feed contribute to regional differences in the PCB concentrations in farmed salmon. The mean wet weight concentration of PCBs in farmed salmon was 50 ng/g or below (Hites et al. 2004a), regardless of source, and thus below the FDA action level of 2 ppm for PCBs in food. Using the US EPA risk assessment for PCB and cancer risk, Hites et al. (2004a) concluded that, given the PCB levels in the fish samples, a consumer's risk will not be increased if consumption is limited to no more than 1 meal per month of farmed salmon. Similar to metals, particularly arsenic and cadmium, PCBs and dioxins were found of higher levels in large crustaceans than fish (Guldner et al. 2007).

5.2.2.1 Dioxins and Dioxin-like Compounds (DLCs)

Dioxins and DLCs are unintentional by-products of combustion of organic material from industries. Sources of dioxins include herbicides (2,4,5-T), wood pre-servatives, diesel and gasoline fuel combustion, and industrial combustionand backyard barrel burning. Currently, new dioxin releases into the environment are mostly from backyard and agricultural burning (IOM 2003). However, since dioxins are persistent compounds, they can be expected to remain in the environment andthe food supply for many years to come (IOM 2003). The reference compound for the Toxic Equivalency Factors (TEFs) is the dioxin compound 2,3,7,8-tetrachlorodibenzo-*p*-dioxin (TCDD). The NRC committee on EPA's Exposure and Human Health Reassessment of TCDD and Related Compounds (NRC 2006) noted that the classification of DLCs as "carcinogenic to humans" vs. "likely to be carcinogenic to humans" is dependent on "the definition and interpretation of the specific criteria used for classification, with the explicit recognition that the true weight of evidence lies on a continuum with no bright line that easily distinguishes between these two categories."

Bioaccumulation of Dioxins in seafood. Exposure to dioxins and DLCs occurs when fish consume aquatic invertebrates that come in direct contact with dioxin particles that settle in sediment; through direct absorption through the gills; or by eating contaminated sediment, insects, and smaller fish (Evans 1991). Because of their lipophilic character, dioxins and DLCs are distributed to fatty tissues in fish, including the liver and gonads. Muscle tissue is less contaminated, depending on the fat content of the muscle, which is likely to be greater in the older, larger, and oily fish.

Adverse health effects. Adverse health effects associated with exposure to dioxins have been identified in populations exposed through unintended industrial releases. One of the largest population exposures to TCDD occurred from an unintended industrial release in Seveso, Italy. Those who were exposed to the highest doses, primarily children, exhibited chloracne (Mocarelli et al. 1999), a severe skin disease with acne-like lesions that occur mainly on the face and upper body. Other adverse health outcomes included an increased risk for cancer. When compared to the nonexposed general population, the exposed population did not show an increased overall cancer mortality, but did have a significant excess mortality risk for esophageal cancer in males and bone cancer in females among those who were exposed to the lowest doses (Bertazzi et al. 1997). The US EPA (2000a) considers TCDD as a non-genotoxic carcinogen and a potent tumor promoter. Dioxins have shown similar health effects as PCBs in animal experiments. Moreover, some studies suggest that there are threats of liver damage and cancer to human with exposures of dioxins (US EPA 1999b).

Some studies indicated developmental human dioxins exposure increases heart disease risks in later life. Animal research also confirmed adult dioxin exposure can lead to hypertension and cardiovascular disease (Kopf and Walker 2009).

Regulatory tolerances. WHO recommends a tolerable daily intake of DLCs and PCBs of 1–4 pg/TEQ/kg/day (IOM 2003). The US EPA has estimated 0.001 pg/kg/day of TCDD as the level associated with a 1 in 1 million excess risk for human health effects from exposure to DLCs and PCBs (IOM 2003).

5.2.2.2 Polychlorinated Biphenyls (PCBs)

PCBs are also long-lived chlorinated aromatic compounds. Production of PCBs began in 1929, and the compounds were used as coolants and lubricants in transformers and other electrical equipment. Because of their noncombustible insulating characteristics, PCBs were used to reduce the flammability of materials used in schools, hospitals, factories, and office buildings. A variety of commercial products, including paints, plastics, newsprint, fluorescent light ballasts, and caulking materials contained PCBs until production was banned in the 1970s. They include over 200 chemical compounds in the form of oily fluids to heavier grease or waxy substances. Local sources of PCBs may be more important than local sources of dioxins and DLCs for contamination of aquatic organisms. The TSCA was based on three concerns: first, PCBs persist in the environment and resist biodegradation; second, a population-wide incident of human poisoning in Japan in 1968 was attributed to introduction of PCB-contaminated oil into a community; and third, in 1975 the CDC reported that, in rat experiments, oral gavage with Aroclor 1260 (a mixture of PCBs) caused liver cancer (Kimbrough et al. 1975). A significant correlation has been observed between blood PCB levels and the quantity of fish consumed by humans (Humpfrey 1988; Jacobson et al. 1990; Smith and Gangolli 2002). The levels of dioxins, DLCs, and PCBs in seafood are generally greater than those in meat (IOM 2003). The mean wet weight concentration of PCBs in farmed salmon was 50 ng/g or below (Hites et al. 2004a), regardless of source, and thus below the FDA action level of 2 ppm for PCBs in food. Using the US EPA risk assessment for PCB and cancer risk, Hites et al. (2004a) concluded that, given the PCB levels in the fish samples, a consumer's risk will not be increased if consumption is limited to no more than 1 meal per month of farmed salmon. Consuming farmed salmon in amounts that provides 1 g/day of EPA/DHA would produce a cumulative cancer risk that is 24 times the acceptable cancer risk level. For wild-caught salmon, the cumulative cancer risk would be eight times the acceptable level. Both farmed and wild-caught salmon could be consumed in amounts that provide at least 1 g/day of EPA/DHA per unit of noncarcinogenic risk (Foran et al. 2005).

Bioaccumulation of PCBs. A significant correlation has been observed between blood PCB levels and the quantity of fish consumed by humans (Humpfrey 1988; Jacobson et al. 1990; Smith and Gangolli 2002). Bioaccumulation of dioxins and PCBs in the fatty tissues of food animals contributes to human body burdens through ingestion of animal fats in foods such as meat and full-fat dairy products. These foods are the largest contributors of dioxins and DLCs from the US food supply.

Adverse health effects. Experimental studies have shown the adverse health effects of PCBs, including toxicity to liver, blood, skin, gastrointestinal system, endocrine system, immune system, nervous system, and reproductive system (UK SACN 2004; Costa 2007). An extensive experimental literature on rodent and nonhuman primate models demonstrates that prenatal exposure to PCBs can interfere with neurodevelopment (Rice 2000; Faroon et al. 2001; Bowers et al. 2004; Nguon et al. 2005). This literature is complemented by numerous prospective epidemiological studies of children conducted in Michigan, North Carolina, Oswego, NY; Germany, Faroe Islands, and the Netherlands (Schantz et al. 2003). The cohorts were often chosen to include children born to women who consumed fish from waters known to be contaminated with PCBs. The results of these epidemiological studies are generally congruent with those using animal models, although, as in most areas of observational research in humans, results are not always consistent across studies or consistent over time in a particular study. PCBs and dioxins are toxic to the human immune system, particularly to the fetus and newborn (Tryphonas 1998). The "Yusho" disease in Japan and "Yu-Cheng" disease reported in Taiwan are the typical cases of the effect of PCBs. Higher concentrations of PCBs and dioxin-like compounds are found in Inuit people in the Arctic because of their traditional diet, which includes large quantities of sea mammal fat. Data suggest that there may be an elevated risk of multiple myeloma in groups with high consumption of dioxin-contaminated fish from the Baltic Sea and Alaska.

Regulatory tolerances. WHO has recommended a Tolerable Daily Intake (TDI) of 1–4 pg/kg body weight per day for TCDD, and the TDI is applied to mixtures of dioxins and PCBs (IOM 2003). Based on its estimate of cancer potency for DLCs, the US EPA concludes that intakes should not exceed 1–4 pg TEQ/kg/day in the general population (IOM 2003).

5.2.2.3 Polybrominated Diphenyl Ethers (PBDEs)

PBDEs are synthetic compounds that are added to a variety of materials to increase their fire resistance. These are structurally similar to PCBs, and can exist, theoretically, as 209 distinct isomers. The patterns of use of PBDEs are changing rapidly. These are released into the environment as emissions from facilities manufacturing them and as a result of degradation, recycling, or disposal of products that contain them. Although the concentrations of PBDEs have been found to vary widely across countries, market basket surveys in developed countries, total diet studies, duplicate diet studies, and commodity-specific surveys have repeatedly shown that, within a region, fish and shellfish tend to have PBDE concentrations that are greater than those found in dairy products, eggs, fats, and oils, and other meat products are important sources of exposure to PBDEs. In terms of total intake of PBDEs, fish and shellfish are the major contributors in Europe and Japan, while meats and poultry are the major contributors in the United States and Canada (FAO/WHO JECFA 2005).

Bioaccumulation of PBDEs. PBDE levels in aquatic wildlife have increased rapidly in recent decades (Ikonomou et al. 2002; Law et al. 2003), with doubling times of between 1.6 years and 6.0 years (Lunder and Sharp 2003; Rayne et al. 2003; Hites et al. 2004a). The concentrations of PBDEs in biological tissues collected in North America are at least 10 times greater than those collected in Europe or Japan (Peele 2004). The PBDE concentration tends to be greater in fish at higher trophic levels, i.e., predatory fish (Rice 2005). In a market basket survey conducted in Dallas, Texas (Schecter et al. 2004), the highest levels of total PBDEs were found in samples of salmon, catfish, and shark. It is notable that the congener pattern was highly variable across samples, even within types (e.g., catfish). Similar findings were reported in a market basket survey of foods conducted in California (Luksemburg et al. 2004), in which the highest PBDE levels were found in swordfish, Alaskan halibut, and Atlantic salmon. PBDE levels were 15 times greater in Pacific farm-raised salmon than in Pacific wild salmon (Easton et al. 2002). PBDE levels are higher in salmon farmed in the United States and Europe than in Chile (Hites et al. 2004a).

Adverse health effects. Experimental animal studies indicate that PBDEs affect the nervous (Viberg et al. 2003), endocrine (Stocker et al. 2004), and immune systems (Fowles et al. 1994), and that the potency of PBDEs might be comparable to that of PCBs, although considerable uncertainty remains (Kodavanti and Ward 2005). In light of the fact that *in vitro* studies with purified PBDE congeners do not show AhR activation. It is possible that the presence of trace amounts of DLCs have confounded these assessments of PBDE toxicity (FAO/WHO JECFA 2006). In various species of fish collected in 1998–1999 from the mid-Atlantic region of the United States (Virginia), sum PBDEs ranged from less than 5 µg/kg (detection limit) to the highest recorded value for PBDEs in edible fish tissue of 47.9 mg/kg (lipid based) (Hale et al. 2001b). PBDE 47 accounted for 40–70% of the sum PBDEs and, in certain fish, exceeded the concentration of PCB 153 and p,p'-DDE. Farmed salmon available on Scottish and European markets had lower concentrations of total PBDEs (1.1–85.2 ng/g lipid) but, as with the wild fish, PBDE congeners 47, 99 and 100 represented on average 77% of the total (Gilron et al. 2007). PBDE levels in aquatic wildlife have increased rapidly in recent decades (Ikonomou et al. 2002; Law et al. 2003), with doubling times of between 1.6 years and 6.0 years (Lunder and Sharp 2003; Rayne et al. 2003; Hites et al. 2004a). Of PBDEs detected in pilot whale blubber samples, congeners 47 and 99 accounted for on average over 70% of the total, with the former responsible for approximately 53% (Lindström et al. 1999).

Regulatory tolerances. Swiss tolerance value of 2 pg/g is based on the European Council regulation (EC) No 2375/2001 from 2001.

5.2.2.4 Pesticides such as Dichloro-Diphenyl-Trichloroethane (DDT)

The chlorinated hydrocarbon pesticides breakdown products, DDD and DDE, persist in the environment, which is why DDT still poses a public health risk more than 45 years after it was banned. Its metabolites are detected in food samples in every country in the world, and absorption through digestive tract is likely the greatest source of exposure for the general population. Consumption of fish and marine mammals in the Artic appears to be a significant dietary contributor to human exposure to DDT (Laden et al. 1999). DDT is still being used in some countries such as Bangladesh, India and Mexico.

5.2.3 Risks of chemotherapeutants

Most aquaculture/mariculture operations depend on the use of various chemotherapeutants to control infectious diseases (FAO/NACA/WHO Study Group 1999; FDA 2001a). Aquaculture/mariculture initially relied upon the same antimicrobials employed for production of beef and poultry and other land-based farming. The resultant seafood safety concerns, as for land-based agriculture, include possible toxic residue in the edible portions, contributions to potential antibiotic-resistant diseases (for both organisms/ animals and consumers), and concomitant issues involving environmental contamination. Toxic heavy metals, i.e., antimony, arsenic, cadmium, lead, mercury, selenium, sulfites (used in shrimp processing) are contaminants. Although the volume of chemotherapeutants used in mariculture/aquaculture is far less than for other medical practices and agricultural production, international mariculture/aquacultural use with less scrutiny may increase. Product seizures due to the presence of chemotherapeutants in some imported farm-raised seafood have occurred (Allshouse 2003).

Compounds of concern have included chloramphenicol, nitrofurans, fluoroquinolone, malachite green, and others (Table 7). All of these antimicrobial/antifungal agents have been used at some time for aquaculturalproduction in the United States, prior to the implemention of restrictions by federal agencies.

Table 7. Antimicrobial/Antifungal agents used at some time for aquaculture production in the United States.

Illegal Antibiotic or Chemotherapeutant	Action Level Based on Detection Limit
Chloramphenicol	0.3 ppb
Nitrofurans	1.0 ppb
Malachite green	1.0 ppb
Fluoroquinolones	5.0 ppb
Quinolones (Oxolinic Acid, Flumequine)	10.0 ppb (oxolinic acid) and 20.0 ppb
Ivermectin	10.0 ppb
Oxytetracycline	2.0 ppm

Notes: ppb = parts per billion; ppm = parts per million.
Source: Food and Drug Administration 2006.

5.3 Microbiological risks associated with seafood

Risks associated with more acute seafoodborne illness are microbiological hazards. Seafoodborne pathogens represent a serious public health concern. Seafoods, like any food item, have the potential to cause disease from bacterial, viral, and parasitic microorganisms under certain circumstances. At low levels, pathogenic microorganisms cause no problems. At illness thresholds, however, they can make people ill and cause death. Most bacterial and viral food poisonings appear within 8 hours of ingesting food. The signs and symptoms of poisoning include nausea, vomiting, diarrhea, muscle aches, and low-grade fever. Food poisoning can also occur from ingestion of parasites. These agents are acquired from three sources: (1) mainly fecal pollution of the aquatic environment, (2) the natural aquatic environment, and (3) industry, retail, restaurant, or home processing and preparation. With the exception of foods consumed raw, however, the reported incidences of seafood-related microbial diseases are low.

Shellfish, particularly bivalve mollusks, contaminated with bacteria or viruses are implicated more than any other marine animal in seafood-related human illness (Potasman et al. 2002). As filter feeders, bivalve mollusks concentrate bacteria and viruses. Oysters and other filter feeding shellfish can concentrate *Vibrio* bacteria that are naturally present in sea water, or other microbes such as norovirus that are present in human sewage dumped into the sea. Standard depuration in purified (with ultraviolet light or ozone) water for 48 to 72 hours may not significantly reduce these contaminants, or effectively remove viruses (Jones et al. 1991; Schwab et al. 1998). Viruses and naturally occurring bacteria that cause disease and death are of great concerns because they are so common. When seafood or any other food is eaten raw or partially cooked, the risk of illness is significantly increased.

5.3.1 Risks of pathogenic bacteria in seafood

Seafood may be a vehicle for many bacterial pathogens from various sources. Pathogenic bacteria can be introduced into marine waters mainly by commercial shipping activity through fouling, oil fall, discharge of sewage, and sediment, although visible light and biotic components of seawater were the important inactivating factors of marine pathogens. Shellfish, especially the filter-feeding bivalve mollusks (oysters, scallops, mussels, clams and cockles) can accumulate pathogenic bacteria in the alimentary tract. Since the alimentary tract of these bivalves forms the major edible portion for humans, these mollusks can pose potent hazards to humans (Table 8), unless care is taken to thoroughly clean them before consumption. The risk is severe due to the fact that many of these pathogens can survive chill temperatures. Consumption of seafood contaminated with fecal organisms continues to pose large-scale health threat. The degree of processing of seafood has implications for the associated risks: botulism, listeriosis and cholera are more frequently associated with consumption of smoked, fermented, salted and pickled products, which are usually ingested without further processing (FAO1994; Jay 2000).

Broadly the pathogenic bacteria associated with seafoodborne diseases may conveniently be divided into:

5.3.1.1 Indigenous bacteria

The bacteria are common naturally and widely distributed in the aquatic environment, such as *Clostridium botulinum, Vibrio* spp., *Aeromonas hydrophila, Plesiomonas shigelloides* and *Listeria monocytogenes* in various parts of the world. The more psychrotrophic *(C. botulinum* and *L. monocytogenes)* are common in Arctic and colder climates, while the more mesophilic types

Table 8. Seafood-borne illness associated with Bacterial Pathogens.

Pathogenic bacteria	Seafood vector	Minimal Dose for infection (cfu/g)[a]	Clinical presentation
Vibrio paramolyticus	Crustaceans, fish	105–106	Diarrhea, nausea, vomiting
V. cholera	Shellfish, fish	102–106	Abdominal pain, vomiting, and profuse watery diarrhea, may lead to severe dehydration andpossibly death, unless fluid and salt are replaced
C. botulinum type E	Fish, shellfish, smoked	0.1–1 mg toxin	Paralysis, diarrhea, death
C. perfringens	Sporadic incidences	105–108	Diarrhea, seldom lethal
A. hydrophila	Shellfish	105–106	Vomiting, diarrhea
Listeria monotycogenes	Raw seafood, smoked, salted	102	Diarrhea, vomiting, nausea
Bacillus cereus	Seafood, squid, prawn	106–109	Diarrhea, nausea, vomiting
Salmonella spp.	Shrimp, mollusks, fish	102	Fever, headache, nausea, vomiting, abdominal pain, and diarrhea
Shigella	Fish, mollusks	101–102	Severe diarrhea, cramps, vomiting
Y. enterocolitica	Fish/shellfish	107–109	Diarrhea, vomiting, fever
E. coli	Fish/shellfish	101–109, depends on strain	Diarrhea, fever
S. aureus	Contamination from infected persons	105–106 or 0.14–0.19 mgenterototoxin	Diarrhea, cramps, vomiting

a Colony forming units per gram.
Source: Adapted from Venugopal 2009.

(*V. cholerae, V. parahaemolyticus*) are representing part of the natural flora on fish from coastal and estuarine environments of temperate or warm tropical zones.

5.3.1.1.1 *Clostridium botulinum*

The organism is widely distributed in ocean sediments, and hence contaminates fish *C. botulinum* is an anaerobic pathogen, responsible for foodborne botulism, a potentially lethal paralytic disease caused by ingesting preformed botulinum neurotoxin released by the bacterium. *C. botulinum* type E is the most common clostridia in fishery products. Human botulism is a serious but relatively rare disease. Incubation of 2 to 8 days most usually 12 to 36 hours. Symptoms may include nausea and vomiting followed by a number of neurological signs and symptoms: visual impairment (blurred or double vision), loss of normal mouth and throat functions, weakness or total paralysis, respiratory failure. Duration of botulism is long term and may lead to death.

5.3.1.1.2 *Vibrio* spp.

Most vibrios are of marine origin and they require Na+ for growth. *Vibrio* spp. are frequently isolated from seafood. Some *Vibrio* spp. are both human and fish pathogens. Currently, more than 10 *Vibrio* spp. are known to be involved in human infections acquired by consumption of contaminated foods and water. The pathogenic species are mostly mesophilic, i.e., generally occurring (ubiquitous) in tropical waters and in highest numbers in temperate waters during late summer or early fall. Only two cholera serotypes, *V. cholera* 01 and 0139, have been shown to cause the disease. The 01 serotype occurs in two biovars: the classic and the El tor. The classical biovar, serovar 01 is today restricted to parts of Asia (Bangladesh), and most cholera is caused by the E1 tor biovar. Marinated raw fish, crabs, and undercooked seafood or shellfish have been implicated as vehicles for the transmission of cholera. The hazards associated with consumption of raw fishery products, particularly farmed finfish and crustaceans harboring *V. cholerae* and *V. parahaemolyticus* have been major causes of gastroenteritis in Japan. In 1997 and 1998 there were large outbreaks of food poisoning from consumption of oysters in North America in which *V. parahaemolyticus* was the cause. Its incubation period is 2 to 48 hours, usually 12 to 18 hours but duration of poisoning 1 to 7 days.

The diseases associated with *Vibrio* sp. are characterized by gastro-enteritic symptoms varying from mild diarrhea to the classical cholera, with profuse watery diarrhea. One exception is infections with *V. vulnificus*, which are primarily characterized by septicaemias.

5.3.1.1.3 *Aeromonas* sp.

This organism may also be readelly l isolated from meat, fish and seafood, ice-cream and many other foods as reviewed by KnΦchel (1989). Indeed the organism has been identified as the main spoilage organism of raw meat (Dainty et al. 1983), raw salmon packed in vacuum or modified atmospheres, and fish from warm, tropical waters (Gram et al. 1990; Gorczyca and Pek Poh Len 1985). In recent years the motile *Aeromonas hydrophila* has received increasing attention as a possible agent of foodborne diarrheal disease. Species of *Aeromonas* produce a wide range of toxins such as cytotoxic enterotoxin, hemolysins and a tetrodotoxin-like sodium channel inhibitor (Varnam and Evans 1991).

5.3.1.1.4 *Listeria monocytogenes*

Recently it has been identified as the causative agent of listeriosis in humans. *Listeria* has been isolated from seafood (Lennon et al. 1984), beef and poultry (Peters 1989). Seafood that have tested positive for *Listeria* include: raw fish (NFI 1989), cooked crabs (Anonymous 1987), raw and cooked shrimp (Anonymous 1987), raw lobster, surimi and smoked fish (NFI 1989). A recent epidemic of perinatal listeriosis in New Zealand was loosely linked to the consumption of shellfish and raw fish, but a definitive connection to seafood could not be drawn (Lennon et al. 1984).

5.3.1.2 Non-indigenous bacteria

The second group consists of mesophilic bacteria that contaminate the aquatic environment from human or animal reservoirs such as *Salmonella* spp., *Escherichia coli*, *Shigella* spp. or *Staphylococcus aureus*. The United States General Accounting Office (1996a) targeted four bacteria as of major concern in foodborne diseases in the United States: *E. coli* O157:H7, *Salmonella enteriditis*, *Listeria monocytogenes* and *Campylobacter jejeuni*.

5.3.1.2.1 *Salmonella* spp.

Salmonella are members of the family Enterobacteriaceae and occur in more than 2000 serovars. These mesophilic organisms are distributed geographically all over the world, but principally occurring in the gut of man and animals and in environments polluted with human or animal excreta. There are reports on 7 outbreaks of seafoodborne salmonellosis in USA in the period 1978–1987 (Ahmed 1991). Three of these outbreaks were due to contaminated shellfish including 2 outbreaks after consumption of raw oysters harvested from sewage-polluted waters. Salmonella can be transferred to shellfish by sewage pollution of coastal waters. Its incubation period is 6 to 72 hours, more commonly 12 to 36 hours but duration for 1 to 7 days. Salmonella infections cause nausea, vomiting, abdominal cramps, and fever. The 10% mortality rate of Salmonella typhi and Salmonella paratyphi is high compared to other Salmonella species (FDA/CFSAN 2003b).

5.3.1.2.2 *Shigella* spp.

Shigella were discovered over 100 years ago by a Japanese scientist named Shiga, for whom they are named. This bacterium causes symptoms very similar to Salmonella. Shigella is found only in the human intestinal tract and is not a result of contamination by animal species. There are several different kinds of Shigella bacteria: *Shigella sonnei*, also known as "Group D" Shigella, accounts for over two-thirds of shigellosis in the United

States. Shigella flexneri, or "group B" Shigella, accounts for almost all the rest. One type found in the developing world, Shigella dysenteriae type 1, can cause deadly epidemics. An estimated 20% of the total number of cases of shigellosis involve food as the vehicle of transmission (Mead et al. 1999). The principal foods involved in outbreaks include a variety of salads and seafoods (Morris 1986) contaminated during handling by infected workers (Bryan 1978).

5.3.1.2.3 *Escherichia coli*

This bacterium is one of the fecal coliforms. Most types of *E. coli* are essential inhabitants of the human intestinal tract and are needed for proper digestion and processing of foods. Its incubation period is 10 to 72 hours more commonly by 12 to 24 hours but the duration may be as long as 5 days. Pathogenic forms can cause abdominal cramps, diarrhea, fever, nausea and vomiting. Death may occur among the very young, the elderly, or immuno-compromised individuals.

5.3.1.2.4 *Staphylococcus aureus*

In addition to disease caused by direct infection, some seafoodborne diseases are caused by the presence of a toxin in the food that was produced by a microbe in the food.The bacterium *Staphylococcus aureus* can grow in some foods and produce a toxin that causes intense vomiting. Its incubation period is 1 to 6 hours but duration of toxicity for 6 to 24 hours. Staphylococcal food intoxication is estimated to cause 185,000 cases of foodborne illness annually (Mead et al. 1999).

5.3.2 Viral risks in seafood

Viral disease transmission to human via consumption of seafood has been known since the 1950's (Roos 1956), and human enteric viruses appear to be the major cause of shellfish-associated disease. Man is the only known reservoir for the major viruses causing foodborne diseases: calicivirus and hepatitis A virus (Svensson 2000). Bivalve molluscs have consistently been proven to be an effective vehicle for the transmission of viral diseases (Lees 2000). The high risk results from two facts: shellfish are filter feeders and many of them are consumed whole and raw or only lightly cooked. The risk is increased because many species are cultivated in near-coastal waters, where contamination with human sewage—which may contain high levels of viral particles may easily occur. Viruses have been isolated from hard clams, oysters, mussels, soft clams, crabs, cockles, lobster and conch. In filter-feeding mollusks, the viruses can become concentrated at a level higher than the surrounding water. Mobile shellfish, such as crabs and lobsters, also present a problem since they can accumulate viruses in polluted waters and move to cleaner areas and act as vectors of viral disease. Ingestion of pathogenic viruses can cause polio, gastroenteritis and hepatitis (Lees 2000; Svensson 2000).

Shellfishborne gastroenteritis was linked to viruses for the first time in the United Kingdom in 1976–77 when cooked cockles were epidemiologically linked to 33 incidents affecting almost 800 persons (Appleton and Pereira 1977). A large outbreak involved over 2000 persons in Australia (Murphy et al. 1979), and other outbreaks have been registered in Japan, Canada, the United Kingdom and the Scandinavian countries (Lees 2000). Presently there are more than 100 known enteric viruses which are excreted in human faeces and find their way into domestic sewage. Caliciviruses, such as Norovirus (Norwalk virus) were recognized as the major cause of seafood-associated gastro-enteritis. Over 80 percent of the outbreaks of non-bacterial gastro-enteritis in the United States and Europe are currently attributed to caliciviruses (Svensson 2000). The largest known outbreak of hepatitis A occurred in Shanghai, China, in 1988, where almost 300,000 cases were traced to the consumption of clams harvested from a sewage-polluted area (Halliday et al. 1991; Tang et al. 1991).

5.3.2.1 Hepatitis-type A (HAV)

HAV is 27 nm in diameter and has single-stranded RNA (Gerba et al. 1985). The first outbreak of seafood-borne (oysters) HAV occurred in Sweden in 1955 (Lindberg-Braman 1956). Also, outbreaks of HAV have been associated with oysters harvested from certified grounds (Mackowiak et al. 1976; Portnoy et al. 1975). Both raw and steamed hard clams (Feingold 1973), oysters (Mackowiak et al. 1976; Portnoy et al. 1975), mussels (Dienstag et al. 1976) and soft clams (Gerba and Goyal 1978), have been implicated in outbreaks of HAV. Symptoms of HAV infection usually begin within 4 weeks (mostly 2–6 weeks) of exposure to the virus. The initial symptoms are usually weakness, fever, malaise and abdominal epigastric pain. As the illness progresses, the individual usually becomes jaundice, and may have dark urine. The severity of the illness ranges from very mild (young children are often asymptomatic), to severe, requiring hospitalization. The fatality rate is low (<0.1%), and deaths primarily occur among the elderly and individuals with underlying diseases (Anonymous 1989; Bryan 1986; Feingold 1973). Residence of coastal states have a higher incidence of infection than inland states (Goyal et al. 1979; Goyal 1984). The CDC reported 4 outbreaks of HAV traced to seafood consumption between 1977 and 1981 (USFDA 1984).

5.3.2.2 Norovirus

Norovirus (Norwalk virus) was first recognized as a pathogen during an outbreak of gastroenteritis in Norwalk, Ohio in 1968 (Gerba et al. 1985). It is now considered a major cause of non-bacterial gastroenteritis. From 1976 to 1980, the CDC reported that 42% of the outbreaks of non-bacterial gastroenteritis were caused by Norovirus (Gerba et al. 1985). Illness from norwalk virus has been associated with eating clams (both raw and steamed) (Morse et al. 1986; Porter et al. 1987), oysters (Gunn et al. 1982; Eyles et al. 1981) and cockles (Gunn et al. 1981). The main reservoir for this virus is man. It causes nausea, vomiting, diarrhea, abdominal cramps and occasionally fever in humans. Symptoms of gastroenteritis usually begin within 40 hours (mostly 12–72 hours) of consuming contaminated food. Gastroenteritis coused by norovirus is a self-limiting illness which usually persists <48 hours, but can last a long as 1 week (Grohmann et al. 1981; Gunn et al. 1982; Bryan 1986; Morse et al. 1986; Porter et al. 1987).

5.3.2.3 Poliovirus

Some of the more frequently recovered viruses from shellfish are the polioviruses because of the common practice of immunizing American children against polio (Larkin and Hunt 1982). The vaccine consists of live attenuated viruses that replicate in the intestine but produce few or no clinicval symptoms. Children who have been immunized excrete viruses (from 1000 to 1,000,000 viruses/gram feces) for several days after the vaccine is administered. An examination of 20% of the polioviruses isolated from the Texas Gulf showed that all were of vaccinal origin. Since the viruses in the vaccine are modified, they present no health hazard if consumed by humans.

5.3.3 Risks of parasites in seafood

Foodborne parasitic diseases affect millions of people world-wide. Protozoa, platyhelminths and nematodes can also cause foodborne diseases so health concern. Very often sea-snails or crustaceans are involved as first intermediate host and marine fish as second intermediate host, while the sexually mature parasite is found in mammals as the final host. The presence of parasites in fish is very common. Most are rare and involve only slight to moderate injury but some pose serious potential health risk. The most important are listed in Table 9.

There are 50 species of helminths found worldwide in fishes, crabs, crayfishes, and bivalves that can cause human infections (Huss et al. 2003). With increasing consumption of raw seafood such as sushi and sashimi, the

Table 9. Pathogenic parasites transmitted by marine fish and shellfish (WHO 1995).

Parasite	Occurrence	Fish and shellfish
Nematodes or round-worms		
Anisakis simplex	North Atlantic	Herring
Pseudoterranova dicipiens	North Atlantic	Cod
Cestodes or tape-worms		
D. pacificum	Peru, Chile, Japan	seawater fish
Trematodes or flukes		
Metagonimus yokagawai	Far East	
Heterophyes sp.	Middle East, Far East	snails, freshwater fish brackish water fish
Paragonimus sp.	Asia, America, Africa	snails, crustaceans, fishes

number of documented human infections is increasing. Hawai'i consumers eat seafood at nearly 3 times the US national average rate, with a long tradition and high level of raw fish consumption (Kaneko and Medina 2009; FDA 2001) reported several TFPs implicated in human infection, which are ceviche (fish and spices marinated in lime juice), lomi lomi (salmon marinated in lemon juice, onion and tomato), poisson cru (fish marinated in citrus juice, onion, tomato and coconut milk), green herring (lightly brined herring) and coldsmoked fish. Among parasites, the nematodes (round-worms) *Anisakis* spp., *Pseudoterranova* spp., *Eustrongylides* spp. and *Gnathostoma* spp.; cestodes (tape-worms) *Diphyllobothrium* spp., and trematodes (flukes) *Chlonorchis sinensis, Opisthorchis* spp., *Heterophyes* spp., *Metagonimus* spp., *Nanophyetes salminicola* and *Paragonimus* spp., are of the most concern in seafood (Rim 1998; FDA 2001; Murrell 2002; Huss et al. 2003). Some traditional fish products (TFPs) such as fermented and marinated do not include cooking step to kill these parasites. Therefore, they are very likely to cause disease and must be regarded as a significant hazard (Köse 2010).

5.3.3.1 Nematodes (round-worms)

Nematodes are common and found in marine fish all over the world. The anisakis nematodes Anisakis simplex (herring/whaleworm) and Pseudoterranova (formerly Phocanema) decipiens (cod/sealworm) are typical round worms, 1–6 cm long and both parasites have several intermediate hosts. If these live worms are ingested by humans they may penetrate into the wall of the gastrointestinal tract and cause an acute inflammation ("herring worm disease"). Anisakis simplex which is usually associated with herring, cod, mackerel and salmon is considered to be the most pathogenic, while P. decipiens, which is associated with cod, halibut and flatfish family or Pacific red snapper is frequently reported to cause symptoms (Murrell 2002). More than 95% of cases manifest as acute gastric anisakiasis, in which severe epigastric pain is experienced shortly after ingestion of fish carrying parasite larvae incorporated in the fish muscle (Murrell 2002). Huss et al. (2003) reported that 65–100% of wild salmon samples originated from Washington, Atlantic and Japan, 86–88% of herring samples from Mediterranean sea and Pacific Ocean, and 84% of cod samples obtained from Pacific Ocean contained A. simplex. Anisakiasis is associated with the consumption of raw or undercooked fish. Most documented cases of anisakiasis have occurred in areas where raw fish is commonly eaten, in Japan, the Netherlands and Western U.S. (Hawaii, Alaska, and California).

5.3.3.2 Cestodes (tape-worms)

Cestodes are flatworms which have an anterior attachment structure, called the scolex, and body segments, called proglottides. Cestodes found in infected fish range from a few millimeters to several centimeters in length (Olson 1986). *D. latum* is a cestode, or tapeworm, which parasitizes a variety of piscivorous mammals of the upper northern latitudes (Olson 1986). In the United States, the consumption of raw fish (sushi) has led to more frequent recognition of infestation with D. latum. Most Diphyllobothrium tapeworms infect freshwater fish, however, anadromous salmonid fish can also carry the parasite (Schantz 1989; Olson 1986). Salmon appears to be a popular culprit (Hutchinson et al. 1997; Curtis and Bylund 1991; Torres et al. 1993). Diphyllobothriasis is also reported from eating raw flesh of redlip mullet (Chung et al. 1997). The fish tapeworm has a complex life cycle, in which a gravid egg released into freshwater releases a ciliated coracidium, which is eaten by a crustacean intermediate host. The coracidium penetrates the intestinal wall of the crustacean and then develops into a procercoid larva.

5.3.3.3 Trematodes (flukes)

Paragonimiasis is caused by a lung fluke *Paragonimus westermani* is predominant in Asia, but is also found in Africa and South and Central America, while *P. kellicotti* is more frequent in North and Central America. Paragonimiasis is acquired by ingesting crustaceans infected with metacercariae which hatch and bore their way as young flukes through the walls of the duodenum and then move to the lungs, where they become enclosed in connective tissue cysts.

Clonorchiasis caused by Chinese liver fluke *Clonorchis (Opisthorchis) sinensis* causes oriental biliary cirrhosis. *Clonorchiasis* is due to the ingestion of fish containing the metacercariae: the cyst wall dissolves in the intestine, the young flukes emerge and migrate through the body to the bile ducts of the liver. *C. sinensis* also causes cirrhosis and liver cancer. *Clonorchiasis sinensis* is widespread throughout Asia and the former Soviet Union.

Opisthorchiasis is caused by *Opisthorchis felineus* and *O. viverrini*. The infections are mostly contracted from the ingestion of raw or improperly cooked crabs or fish. *O. felineus* is generally found in Eastern Europe, Poland, Germany, and Siberia (Murrell 2002).

Fishborne endemic trematode infections

World-wide about 700 million people are considered to be at risk of contracting foodborne trematode infections (FBT) and 40 to 50 million are believed to be infected by one or more trematode parasites (WHO 1995). In general, parasitic infections are concentrated in certain ethnic groups that favor the consumption of raw or partially cooked seafoods. FDA (2001) reported that parasites (in the larval stage) consumed in uncooked, or undercooked, unfrozen seafood can present a human health hazard. The method of capture, handling and storage of the catch can directly affect the quality of the seafood with regard to the presence and numbers of parasites. Most of these infections occur in Asia as a direct result of the consumption of raw or improperly cooked fish and crustaceans containing a viable and infective stage of the parasite. Most of FBT infections occur in areas where there is poverty, pollution and increasing population growth. The use of water containing human and animal faeces to fertilize plants and to feed fish, allows the life cycle of these parasites to be completed and perpetuates the infection.

Nowadays, increased access of the ethnic communities in the EU to their traditional food items as well as the fact that many of those habits have become fashionable (Chinese food, sushi and sashimi) have increased the risk of FBT in Europe. Table 10 shows where the diseases are endemic, the estimated number of infected people, and the sources of infection.

5.3.4 Seafood allergens

Seafood allergies have been estimated at less than one tenth of one percent. A seafood allergy involves an immunologic reaction following exposure to a seafood (Anderson 1986; Adverse Reactions to Food Committee 2003). True food allergies are abnormal responses of the body's immune system to certain foods or food ingredients. Most allergies are specific to a certain specie or type of seafood. For instance, someone who is allergic to finfish, may not be allergic to shellfish and vice versa. The immune system produces antibodies to help fight the "invasion". Antibodies bind with other cells and this complex releases histamines. Histamines are responsible for allergy

Table 10. Location and estimates for some endemic fish-borne trematode infections (FBT) (WHO 1995).

Disease	Caused by	Endemic infected	Estimated infected	Acquired from
Clonorchiasis	*Clonorchis sinensis*	China, Republic of Korea, Japan, Hong Kong, Viet Nam, Russian Federation	7 million (more than half in China)	113 fish ssp., mainly Cypriniade
Opisthorchiasis	*Opisthorchis felineus*	Kazakhstan, Russian Federation, Ukraine, Thailand,	10 million 7 million in Thailand alone	Cyprinidae
	O. viverrini	Lao People's Democratic Republic		
Paragonimiasis	*Paragonimus westermani*	At least 20 countries. In Asia endemic in regions of China, Republic of Korea Lao People's	22 million	Crabs and crayfish
	Paragonimus spp.	Democratic Republic Philippines Thailand		

symptoms which can range from mild to severe. Symptoms can affect the gastrointestinal tract (nausea, stomach cramps, vomiting, diarrhea), skin (localized itching, hives or rashes, swelling of the lips), and/or respiratory system (breathing problems). People who already have asthma may be more likely to have food allergies. Although allergic reactions are usually mild, some individuals can experience severe symptoms, like anaphylactic shock, which can cause death.

Seafood allergies in the United States remain among the most common food-induced allergies (Taylor and Bush 1988; O'Neil et al. 1993; Hefle 1996). But actual occurrence of seafood allergies is estimated to affect less than 2 percent of the US population (Hefle 1996). Of this group, 280,000 to 500,000 consumers may be at risk for developing allergic reactions to seafood (Lehrer 1993; O'Neil and Lehrer 1995). Since exposure is the mediating factor, occurrence tends to be more prevalent near coastal regions and will likely increase as per capita seafood consumption increases (Lehrer 1993; O'Neil et al. 1993; O'Neil and Lehrer 1995). Exposure can involve ingestion, inhalation (of vapors), or product handling for consumption or occupation. Likewise, potential exposure can behidden as the presence of the particular seafood item may not be obvious or expected due to an unidentified ingredient or misidentified ingredients (fish-based surimi used in a "crab" salad). It can also result from cross-contamination of nonallergenic foods from handling either with the same improperly cleaned utensils or through subsequent cooking in the same containers or cooking media (frying oil or boiling water) as seafood (O'Neil and Lehrer 1995; Hefle 1996).

Hypersensitivity reactions secondary to occupational exposure to fish and shellfish in the seafood processing industry are increasingly recognized. Occupational reactions have been reported in a variety of seafood workers, including fishermen, seafood processing workers, canners, restaurant cooks, delivery persons, and other workers associated with the seafood industry (Barton et al. 1995; Bossart et al. 1998; Goldmann 1985). Occupational seafood allergy can manifest as rhinitis, conjunctivitis, asthma, urticaria, contact dermatitis, or OAS (Ahasan et al. 2004; Chan and Wang 1993) Studies performed on snow crab workers demonstrated a 33% incidence of asthma, 24% incidence of skin rash, and 18% rate of rhinitis or conjunctivitis related to inhalational exposure or skin contact with snow crab meat or by-products (Barton et al. 1995).

Clinical manifestations of fish and shellfish allergies are similar to other IgE-mediated food allergy reactions, ranging from mild urticaria to life-threatening anaphylaxis. In the U.S. telephone survey, 55% of finfish reactions and 40% of shellfish reactions were severe enough that evaluation by a physician was sought (Greenwood et al. 1998). IgE-mediated reactions are generally rapid in onset, with allergic symptoms developing within minutes to an hour of exposure and most reactions occurring within 30 minutes (Boisier et al. 1994; Centers for Disease Control 1997). However, delayed onset of symptoms may occur (3 to 24 hours after exposure) and have been noted with, among other seafood, dogfish (Endean et al. 1993; Gessner and Middaugh 1995), cuttlefish (Gras-Rouzet et al. 1996), abalone (DePaola et al. 2010), and limpets (Endean et al. 1993). In 25% to 30% of cases, a biphasic reaction occurs whereby the patient will appear to recover and then experience a late-phase reaction with a recrudescence of symptoms after an asymptomatic period of 1 to 72 hours (Glazious and Legrand 1994).

6 Risks Reduction Measures

Risk management and communication are often performed by public health organizations including risk evaluation, option assessment, option implementation, monitoring and review, and dissemination of information (FAO/WHO 1997). Extensive environmental monitoring and seasonal quarantine of a harvest are employed to reduce the risk of exposure. The conditions of the seawater can be monitored by taking water samples and via satellite pictures. All countries have their own guidelines for acceptable toxin levels. If these levels are exceeded the government will close commercial mussel and oyster banks, forbid the sale of certain seafood and advise against the use of it. For STX, for example, the limit is set at more than 500 cells of *Pyrodinium bahamense* per litre of sea water or more than 40 µg of STX per 100 gram of mollusc. For brevetoxin there is a guideline that only total absence of the toxin can be accepted. A guideline such as this leads to practical problems. The most frequently used monitoring technique is that of the mouse bioassay. For STX one mouse unit corresponds to 0.18 µg of STX. 1MU kills a 20 gram mouse within 15 minutes if the toxin is administered into the peritoneum. An *in vitro* toxicity test via tissue cell cultures will possibly form a good alternative.

Current contaminant monitoring and surveillance programs provide an inadequate representation of the presence of contaminants in edible portions of domestic and imported seafood, resulting in serious difficulties in assessing both risks and specific opportunities for control. Due to the unevenness of contamination among species and geographic sources, it is feasible to narrowly target control efforts and still achieve meaningful reductions in exposure. The data base for evaluating the safety of certain chemicals that find their way into seafood via aquaculture and processing is too weak to support a conclusion that these products are being effectively controlled. Recently, the FDA has expanded its regulatory program and implemented a new system for seafood safety control called Hazard Analysis Critical Control Point (HACCP). Consumption of some types of contaminated seafood poses enough risk that efforts toward evaluation, education and control of that risk must be improved. Public education on specific chemical contaminant hazards should be expanded by government agencies and the health professions. For specific contaminants in particular species from high-risk domestic or foreign geographic areas, government agencies should consider the option of mandatory labeling. Shellfish poisoning can be prevented by avoiding potentially contaminated bivalve mollusks. This is particularly during or shortly after "red-tides." The health risk caused by the possible accumulation of marine toxins in shellfish when sporadic algal blooms appear in imposes the monitoring of possibly contaminated fishery products and of the marine environment. Increased environmental monitoring should be initiated at the state level as part of an overall federal exposure management system. Some countries have implemented shellfish and seawater screening programmes, according to the established regulations, as primary preventive tools (Table 11).

Marine toxins are heat stable and largely unaffected by cooking or freezing. In some cases, appropriate methods of food processing and thorough cooking can be employed to destroy or reduce the level of toxin. In other cases where the toxin cannot be reduced or removed, intake should be limited. Removal of gonads, skin, and parts of certain fish eliminates toxins concentrated in these tissues. In general, whether a substance poses harm depends on its concentration, amount of intake and the health status of individual since the body can detoxify low levels of many potentially

Table 11. Monitoring programmes to protect public health and environment from marine toxins contamination.*

Countries	Monitoring programmes
Argentina	Monitoring programme for mussel toxicity
Australia	Monitoring of mussels and algae
Brazil	Proposal of a pilot monitoring initiative, as a first step towards a national monitoring programme
Canada	Monitoring programme for *Alexandrium* spp., *Dynophysis* spp., *Prorocentrum* spp. and *Pseudo-nitzschia pungens*
Chile	Two different national toxicity monitoring programmes, established by The National Health Service (using bioassays) and by the Fisheries Research Institute in conjunction with universities
Denmark	Assessment of shellfish and monitoring of *Pseudo-nitzschia pungens* and *Gymnodinium* spp.
India	Ministry of Earth Sciences commenced a national programme on monitoring HABs since 1998
Ireland	Biotoxin monitoring programme based on weekly shellfish testing using DSP mouse bioassay, LC-MS for OA, DTX-2 and azaspiracids, and LC for DA, as well as phytoplankton analysis
Italy	Monitoring of NSP-producing algae
Japan	Monitoring programme for both plankton and shellfish and for the detection of *Dinophysis* spp. carried out by researchers from Prefectural Fisheries Experimental Stations
Malaysia	Two shellfish toxicity monitoring programmes, established by the Malaysian Department of Fisheries and by the Fisheries Research Institute of Penang
New Zealand	Biotoxin monitoring programme combining regular shellfish testing and phytoplankton monitoring
Korea	Bi-weekly assessment of plankton run by The National Fisheries Research and Development Institute, for the detection of ASP, PSP and DSP toxins, combined with monitoring of *Prorocentrum* spp.
United Kingdom	Programme co-coordinated by the Ministry of Agriculture, Fisheries and Food, based on testing bivalve mollusks (and some crustaceans) for the detection of PSP and DSP toxins by the mouse bioassay
USA	Sampling programme for DA conducted by The Department of Marine Resources and control programme for NSP-contaminated shellfish carried out by The Florida Department of Natural Resources
Uruguay	National monitoring programme for mussel toxicity and toxic phytoplankton

*Source: Adated from Campas' et al. 2007.

dangerous substances. On the basis of the COT opinion, the FSA has advised that pregnant women, women intending to become pregnant and children under 16 should avoid eating shark, marlin and swordfish (US FDA and US EPA 2004a). Likewise, Scientific Advisory Committee on Nutrition of United Kingdom (UK SACN) suggested pregnant and nursing women need to avoid some species, such as the oily or large predatory fish with high mercury, PCBs and dioxin concentrations. However, other population groups (e.g., men, women who are not of childbearing years) can consume these species within guideline values (UK SACN 2004). Thorough cooking of seafood products would virtually eliminate all microbial and parasitic pathogens. Individuals who choose to eat raw seafood should be educated about the potential risks involved and how to avoid or mitigate them. In particular, immunocompromised individuals and those with defective liver function should be warned never to eat raw shellfish. The FDA has also advised consumers to avoid eating raw oysters harvested in the Pacific Northwest as a result of increased reports of illnesses associated with the naturally occurring bacteria *Vibrio parahaemolyticus* known to cause gastrointestinal illness. Consumers were advised to cook oysters before consumption to reduce the risk of infection from bacteria that may be found in raw oysters.

Obtaining a history of seafood consumption, and possibly a history of illness in people exposed to shared seafood, may be critical in determining a diagnosis. Diagnostic tests for any of the marine toxins are not available in health care clinics, nor does an antidote exist for any toxin except histamine. Accordingly, diagnosis is based on clinical presentation and a history of seafood consumption in the preceding 24 h, and treatment is essentially supportive. For patients affected by marine toxins that may cause rapid death, a high index of suspicion and placement in an intensive care unit are essential. Suspected cases should be reported immediately to local or state health departments, because rapid investigation may result in the identification of the contaminated food and the prevention of additional illnesses. Finfish or shellfish sold as bait should not be eaten. Bait products do not need to meet the same food safety regulations as seafood for human consumption.

Seafood consumers can take the following precautions to prevent the occurrence of CFP: Avoid or limit consumption of the reef fish listed above, particularly when the fish weighs 6 lb or more; Never eat high-risk fish such as barracuda or moray eel, sturgeon, king mackerel, snapper, amberjack, and grouper; As the higher concentrate of the toxins will be present in the fish's internal organs, so no one should ever consume those particular parts, i.e., liver, intestines, roe, and head. As ciguatera toxins do not affect the texture, taste, or smell of fish, and they are not destroyed by gastric acid, cooking, smoking, freezing, canning, salting, or pickling. Commercial kits (if available) can be used to check if the fish is safe to eat. In order to avoid scombroid poisoning it is important to be conscious of the care the fish has received after being caught. Do not eat any fish but has not received proper refrigeration immediately after being caught. When eating fish like sardines, tuna, mahi-mahi, anchovies or mackerel, use extreme caution. If something doesn't smell right and doesn't look right, chances are you should avoid it. It is just important that you understand the different ways in which poisoning can occur, so that you can avoid food poisoning caused by fish. Fish contaminated with histamine may have a peppery, sharp, salty, or bubbly taste but may also look, smell, and taste normal. The key to prevention is to make sure that the fish is promptly chilled (below 38°F) after capture. Cooking, smoking, canning, or freezing will not destroy histamine in contaminated fish. PFP is now a common form of food poisoning throughout coastal countries, but its diagnosis and management are still unclear. The history of blowfish is the history of prohibition by authorities in Japan. Puffers were banned in Thailand since 2002. PFP highlights the need for continued stringent regulations of puffers importation by U.S. Food and Drug Administration.

Control systems for microbiological hazards must include inspection techniques, preferably HACCP based, that specifically test for the hazard itself or for some condition that enhances or reduces hazard. Adequate and proper treatment and disposal of sewage must be implemented to avoid contamination of harvest areas by human enteric pathogens.

Means must be investigated and implemented to eliminate, or at least reduce, levels of potentially pathogenic *Vibrio* species in raw shellfish. Consideration should be given to monitoring *Vibrio* counts in molluscan shellfish during warm months. Valid microbiological guidelines, established with an appropriate epidemiologic data base, are needed for seafood products. Seafood-related parasitic infections are less common than bacterial and viral infections, with anisakids and cestodes having the greatest public health significance. All of the helminths mentioned above are associated with social-cultural and behavioural factors, in particular the consumption of raw or undercooked seafood. Measures can be taken during harvesting, processing or post-processing (e.g., by the consumer) to mitigate the risks of infection. The seafood industry and government authorities can apply various programmes to reduce these risks, including good manufacturing practices (GMPs) andHACCP systems. The extent of processing—including heading and gutting, candling and trimming—and the type of product derived (fresh, frozen, salted or pickled) can all contribute to the control of the risks posed by helminths. The most effective means of killing the parasites are either freezing or heat inactivation Consumers should be advised to cook seafood sufficiently to destroy parasites and bacterial contaminants before consumption. Seafoodborne infections by human enteric viruses in raw and improperly cooked molluscan shellfish could be decreased significantly by the development of valid growing water indicator(s) and of direct detection methodologies for enteric viruses. Most viruses (excluding Hepatitis A) are inactivated when the internal temperature of the mollusk reaches 140°F, which requires 4 to 6 minutes of steaming (Koff and Sear 1967; Giusti and Gaeta 1981). A common cooking practice is to steam mollusks only until the shell opens. It has been demonstrated that shells open after only about 1 minute of steaming, which is not sufficient time to inactivate all of the viruses.

The Opinion of the Scientific Steering Committee on Antimicrobial Resistance-28 May 1999 (European Commission 1999) indicates very clearly the need to reduce the overall use of antimicrobials. This implies elimination of unnecessary and improper use, more precise diagnoses of infectious agents, and monitoring of antimicrobial resistance. In order to design sensible disinfection procedures, Langsrud and Sundheim (1997) suggested the convenience of alternating the use of QACs with chlorine, phenolics, and alkylaminoacetate, to avoid the build-up of resistant strains.

The allergy is managed by avoiding the offending food. There is no cure for food allergies. Consumer awareness and labeling remain the most effective measures to prevent exposure to seafood that could elicit a food sensitivity response. Commercial practices for dual processing or preparation of other foods in facilities or with utensils used for seafood must avoid potential cross-contamination that could result in unanticipated exposures. Requirements to identify seafood or any use of seafood ingredients, as well as certain food additives, have been emphasized by the HACCP mandate (FDA 2001a) requiring appropriate hazard analysis to identify any potential food sensitivity risks controlled through proper cleaning, product segregation, or product identification in order to prevent a potential hazard. However, if you know for certain you are allergic to a certain type or specie of seafood, read ingredient listings on food labels and question restaurant staff carefully before eating food you suspect contains a potential allergen. Additional study of potential chemical contamination risks associated with both domestic and imported aquaculture products is required. Overall, several chemical contaminants in some species of aquatic organisms in particular locations have the potential to pose hazards to public health that are great enough to warrant additional efforts of control.

7 Discussions and Conclusion

While benefits in consuming seafood have been supported by numerous studies, seafood consumption may contain some risks. It is worth emphasizing that because the food supply is dynamic, benefit-risk analyses are not static (Willett 2006). Our slow progress towards a better understanding of the nature and strong evidences of an increase in magnitude, duration and geographical distribution of seafoodborne illness caused by seafood (mainly Fish and Shellfish) continue to present serious economic, environmental, and public health risks. PFP and PSP can be rapidly fatal, and antidotes do not exist. Fatality rates from PSP, the most severe of the all shellfish syndromes, ranges from 1–12% in isolated outbreaks. PFP is now a common form of food poisoning throughout coastal countries, but its diagnosis and management are still unclear. However, the incidence of PFP is decreasing, presumably because of heightened awareness and proper preparation of fish.

Nearly 20 percent of foodborne disease outbreaks in the United States may result from seafood consumption, with as many as half of those that is the result of naturally occurring algal toxins. There are monitoring routines in place to protect the public from marine algal toxins, but too much is still unknown about the basic biology of toxin production, analytical methods, and the impacts of toxins on health. Natural seafood toxinsmainly ciguatera and scombroid poisoning and, to a lesser extent, PSP were responsible for 62.5% of all seafoodborne outbreaks of illness, but constituted only 28% of all reported cases. The frequency, intensity and duration of spontaneous mass increases in microalgae appear to be on the increase, and toxic algae are advancing into new regions. This is chiefly blamed on increasing pollution and eutrophication of the water. The spread of algal blooms can be ascribed to the dispersal of resting forms of algae (resting cysts) in the ballast water of ships and infected mussel stocks in aquacultures. The extent to which it might affect the toxicity of seafood-borne contaminants is largely unknown. Routine clinical diagnostic tests are not available for these toxins; diagnosis is based on clinical presentation and a history of eating seafood in the preceding 24 h. There is no antidote for any of the marine toxins, and supportive care is the mainstay of treatment. In particular, paralytic shellfish poisoning and puffer fish poisoning can cause death within hours after consuming the toxins and may require immediate intensive care. Rapid notification of public health authorities is essential, because timely investigation may identify the source of contaminated seafood and prevent additional illnesses.

The long-term health risks are mainly from exposures to heavy metals, particularly mercury in certain species, and organic contaminants, such as PCBs and dioxins. The risks of mercury exposure include damage to brain and neurologic development, and negative impacts to the human immune system and cardiovascular health (Price 1992). The danger of Hg poisoning may seem like a good reason to refrain from consuming seafood, the benefits of eating may outweigh many of the risks (Pandey et al. 2012). However, the FDA's advice to pregnant and nursing women is to avoid eating shark, swordfish, tilefish and king mackerel. Potential adverse effects of PCBs are threats to liver damage and risk of cancer, damaging human immune system, and create neuro-toxic effects. However, other food products have higher levels of PCBs and dioxins than seafood and yet are below the U.S. FDA threshold level. Some examples of risks that may be significant include reproductive effects from PCBs and methylmercury; carcinogenesis from selected congeners of PCBs, dioxins and possibly, parkinsonism in the elderly from long-term mercury poisoning. Several other metallic and pesticide residues also warrant attention. Present quantitative risk assessment procedures used by government agencies should be improved and extended to non cancer effects.

The microbiological risk associated with seafood other than raw molluscan shellfish is much lower and appears to result from recontamination or cross-contamination of cooked with raw products, or due to contamination during preparation followed by time/temperature abuse (e.g., holding at warm temperature long enough for microbial growth or toxin production to occur). This occurs mainly at the food service (post processing) level, which is common to all foods and not specific for seafood products. Because of the high risks associated with raw molluscan shellfish, the importation of live shellfish for raw consumption should not be permitted.

World Public Health authorities must assume responsibility to take proper consideration of the problem, establish a proper diagnosis and universal therapy program, and develop specific assay kits for red tide toxins and in their disperse geographic localization, and chemical test capable of detecting all toxins in any suspected food source. While a majority of research has shown that seafood consumption greatly outweighs the risks, it is important to keep in mind that this field of science is just at the incipient stages of determining how to accurately assess the everyday choices we make in our diet and how these ultimately affect our lives.

Acknowledgements

The author is grateful to the authorities of VIT University, Vellore-632014, Tamil Nadu for the facilities and support. He also thanks one of his Research Scholars (Ms Niharika Mandal) for her assistance with preparation of the references.

References

Ababouch, L., M.E. Afilal, H. Benabdeljelil and F.F. Busta. 1991. Quantitative changes in bacteria, amino acids and biogenic amines in sardines (Sardina pilchardus) stored at ambient temperature (25–28°C) and in ice. Int. J. Food Sci. Technol. 26: 297–306.

Adverse Reactions to Food Committee. 2003. Academy Practice Paper: Current Approach to the Diagnosis and Management of Adverse Reactions to Foods. Milwaukee, WI: American Academy of Allergy, Asthma, and Immunology.

Ahasan, H.A., A.A. Mamun, S.R. Karim et al. 2004. Paralytic complications of puffer fish (tetrodotoxin) poisoning. Singapore Med. J. 45: 73.

Ahmed, F.E. 1991. Naturally Occurring Seafood Toxins. J. Toxicol. Toxin Reviews. 10: 263–287.

Ahmed, F.E., D. Hattis, R.E. Wolke et al. 1993. Risk assessment and management of chemical contaminants in fishery products consumed in the USA. Journal of Applied Toxicology. 13: 395–410.

Ahmed, F.E. (ed.). 1991. Seafood Safety. National Academy Press, Washington D.C., USA.

Al Bulushi, I., S. Poole, H.C. Deeth and G.A. Dykes. 2009. Biogenic Amines in Fish: Roles in Intoxication, Spoilage, and Nitrosamine Formation—A Review. Critical Reviews in Food Science and Nutrition. 49 (4): 369–377.

Alcala, A.C., L.C. Alcala, J.S. Garth, D. Yasumura and T. Yasumoto. 1998. Human fatality due to ingestion of the crab Demania reynaudii contained a palytoxin-like toxin. Toxicon. 26: 105–107.

Alexander, J., D. Benford, A. Cockburn, J.-P. Cradevi, E. Dogliotti, A.D. Domenico, M.L. Fernandez-Cruz, J. Fink-Gremmels, P. Furst, C. Galli, P. Grandjean, J. Gzyl, G. Heinemeyer, N. Johansson, A. Mutti, J. Schlatter, R. van Leeuwen, C. van Peteghem and P. Verger. 2009. Marine biotoxins in shellfish—pectenotoxin group. EFSA J. 1109: 1–47.

Alfonso, A., Y. Roman, M.R. Vieytes, K. Ofuji, M. Statake, T. Yasumoto and L.M. Botana. 2005. Azaspiracid-4 inhibits Ca^{2+} entry by stored operated channels in human T lymphocytes. Biochem. Pharmacol. 69: 1627–1636.

Allshouse, J., J.C. Buzby, D. Harvey and D. Zorn. 2003. International trade and seafood safety. *In:* J.C. Buzby (ed.). International Trade and Food Safety: Economic Theory and Case Studies. Agricultural Economic Report. Washington DC, Economic Research.

Amzil, Z., M. Sibat, F. Royer and V. Savar. 2008. First report on azaspiracid and yessotoxin groups detection in French shellfish. Toxicon. 52: 39–48.

Amzil, Z. 2012. Ovatoxin-a and palytoxin accumulation in seafood in relation to Ostreopsis cf. ovata blooms on the French Mediterranean coast. Marine Drugs. 10: 477–496.

Anderson, J.A. 1986. The establishment of common language concerning adverse reactions to food and food additives. Journal of Allergy and Clinical Immunology. 78: 140–143.

Anonymous. 1987. FDA checking imported, domestic shrimp, crabmeat for Listeria. Food Chem. News. August 17, pp 7–8.

Anonymous. 1989. Epidemiology of hepatitis A in North Carolina in 1988. Epi Notes, Report No. 89-1. Epidemiology Section, Division of Health Services, North Carolina.

Anonymous. 2002. Seafood is leading cause of food-borne illness outbreaks. Dairy Food Environ. Sanit. 22: 38.

Appleton, H. and M.S. Pereira. 1977. A possible virus actiology in outbreaks of food poisoning from cocles. Lancet. 1: 780–781.

Ashwell, M. 1990. How safe is our food—a report of the British nutrition foundation's 11th annual conference. J. R. Coll. Phys. 24: 23.

Aune, T. 2001. Mycotoxins and Phycotoxins in Perspective at the Turn of the Millennium. *In:* W.J. De Koe, R.A. Samson, H.P. van Egmond, J. Gilbert and M. Sabino (eds.). Proceedings of the X International IUPAC Symposium on Mycotoxins and Phycotoxins, Ponsen & Looyen, Wageningen.

Aune, T., R. Sorby, T. Yasumoto, H. Ramstad and T. Landsverk. 2002. Comparison of oral and intraperitoneal toxicity of yessotoxin towards mice. Toxicon. 40: 77–82.

Axelrad, D.A., D.C. Bellinger, L.M. Ryan et al. 2007. Dose-response relationship of prenatal mercury exposure and IQ: an integrative analysis of epidemiologic data. Environmental Health Perspectives. 115: 609–615.

Ayotte, P., E. Dewailly, J.J. Ryan et al. 1997. PCBs and dioxin-like compounds in plasma of adult Inuit living in Nunavik (Arctic Quebec). Chemosphere. 34: 1459.

Babnis, R. 1981. Oceanol. Acta. 4: 375–87.

Baden, D.G. 1989. Brevetoxins: Unique Polyether Dinoflagellate Toxins. FASEB J. 3: 1807–1817.

Banner, A.H. 1976. Vol. III, Chap. 6, pp. 177–213. *In:* O.A. Jones and R. Endean (eds.). Biology and Geology of Coral Reefs. Academic Press, New York.

Bartholomew, B.A., P.R. Berry, J.C. Rodhouse, R.J. Gilbert and C.K. Murray. 1987. Scombrotoxic fish poisoning in Britain: features of over 250 suspected incidents from 1976 to 1986. Epidemiol Infect. 99(3): 775–782.

Barton, E.D., P. Tanner, S.G. Turchen et al. 1995. Ciguatera fish poisoning: A southern California epidemic. West J. Med. 163: 31.

Beatriz, P., H. Antonio, N. Manuel, R. Pilar, M.F. José and J.F. José. 2008. Yessotoxins, a Group of Marine Polyether Toxins: an Overview. Mar. Drugs. 6: 73–102.

Bernier, J., P. Brousseau and K. Krzystyniak et al. 1995. Immunotoxicity of heavy metals in relation to Great Lakes. Environmental Health Perspectives. 103(S9): 23–34.

Bertazzi, P.A., C. Zocchetti, S. Guercilena, D. Consonni, A. Tironi, M.T. Landi and A.C. Pesatori. 1997. Dioxin exposure and cancer risk. A 15-year mortality study after the "Seveso accident." Epidemiology. 8(6): 646–652.

Bidleman, T.F. and T. Harner. 2000. Chapter 10: Sorption to aerosols. pp. 233–260. *In:* R.S. Boethling and D. Mackay (eds.). Handbook of Property Estimation Methods for Chemicals: Environmental and Health Sciences. Lewis Publishers, Boca Raton, Florida.

Bjeldanes, L.F., D.E. Schutz and M.M. Morris. 1978. On the aetiology of scombroid poisoning: cadaverine potentiation of histamine toxicity in the guinea pig. Food Cosmetic Toxicol. 16: 157.

Boisier, P., G. Ranaivoson, N. Rasolofonirira et al. 1994. Fatal ichthyosarcotoxism after eating shark meat: Implications of two new marine toxins. Arch. Inst. Pasteur. Madagascar. 61: 81.

Bossart, G.D., D.G. Baden, R.Y. Ewing et al. 1998. Brevetoxicosis in manatees (Trichechus manatus latirostris) from the 1996 epizootic: Gross, histologic, and immunohistochemical features. Toxicol. Pathol. 26: 276.

Bowers, W.J., J.S. Nakai, I. Chu, M.G. Wade, D. Moir, A. Yagminas, S. Gill, O. Pulido and R. Meuller. 2004. Early developmental neurotoxicity of a PCB/organochlorine mixture in rodents after gestational and lactational exposure. Toxicological Sciences. 77(1): 51–62.

Bryan, F.L. 1978. Factors that contribute to outbreaks of foodborne disease. J. Food Prot. 41: 816.

Bryan, F.L. 1986. Seafood-transmitted infections and intoxications in recent years. pp. 319–337. *In:* D.E. Kramer and J. Liston (eds). Seafood Quality Determination. Elsevier Science Publishers, Amsterdam.

Burger, J. and M. Gochfeld. 2005. Heavy metals in commercial fish in New Jersey. Environmental Research. 99(3): 403–412.

Campas, M., B. Prieto Simon and J. Loius Marty. 2007. Biosensors to detect marine toxins: Assessing seafood safety. Talanta. 72: 884–895.

CDC (Centers for Disease Control and Prevention). 1996. Tetrodotoxin poisoning associated with eating puffer fish transported from Japan—California. 275(21): 1631.

CDC (Centers for Disease Control and Prevention). 1989. Epidemiologic notes and reports: scombroid fish poisoning—Illinois, South Carolina. MMWR 38: 140–2,147.

Chan, T.Y.K. and A.Y.M. Wang. 1993. Life-threatening bradycardia and hypotension in a patient with ciguatera fish poisoning. Trans. R. Soc. Trop. Med. Hyg. 87: 71.

Cheng, H.S., S.O. Chua, J.S. Hung and K.K. Yip. 1991. Creatine kinase MB elevation in paralytic shellfish poisoning. Chest. 99: 1032–1033.

Chung, P.R., W.M. Sohn, Y. Jung et al. 1997. Five human cases of Diphyllobothrium latum infection through eating raw flesh of redlip mullet, Liza haematocheila. Korean J. Parasitol. 35: 283.

Ciminiello, P., C. Dell'Aversano, E. Fattorusso, M. Forino, G.S. Magno, L. Tartaglione, C. Grillo and N. Melchiorre. 2006. The Genoa 2005 outbreak: Determination of putative palytoxin in Mediterranean Ostreopsis ovata by a new liquid chromatography tandem mass spectrometry method. Anal. Chem. 78: 6153–6159.

Ciminiello, P., C. Dell'Aversano, E. Fattorusso, M. Forino, S. Magno and R. Poletti. 2002. Direct detection of yessotoxin and its analogues by liquid chromatography coupled with electrospray ion trap mass spectrometry. J. Chromatogr. A. 968: 61–69.

Ciminiello, P., E. Fattorusso, M. Forino and R. Poletti. 2001. 42,43,44,45,46,47,55-Heptanor-41-oxohomoyessotoxin, a new biotoxin from mussels of the northern Adriatic Sea. Chem. Res. Toxicol. 14: 596–599.

CIMWI Channel Islands Marine and Wildlife Institute. 2006. Domoic Acid Information and History.

Clark, R.F., S.R. Williams, S.P. Nordt and A.S. Manoguerra. 1999. A review of selected seafood poisonings. Undersea Hyperb. Med. 26(3): 175–84.

Clarke, M. 2006-04-19. Men hallucinate after eating fish. Practical Fishkeeping.

Clifford, M.N., R. Walker, J. Wright, P. Iiomah, R. Hardy, C.K. Murray and K.D. Rainsford. 1991. Scombroid-fish poisoning. New Engl. J. Med. 325: 515–516.

Cohen, N.J., J.R. Deeds, E.S. Wong, R.H. Hanner, H.F. Yancy and K.D. White. 2009. Public health response to puffer fish (Tetrodotoxin) poisoning from mislabeled product. J. Food Prot. 72(4): 810–7.

Cohen, P., C.F.B. Holms and Y. Tsukitani. 1990. Okadaic acid: A new probe for the study of cellular regulation. Trends Biochem. Sci. 98.

Connell, D.W., R.S. Wu, B.J. Richardson et al. 1998. Fate and risk evaluation of persistent organic contaminants and related compounds in Victoria Harbor, Hong Kong. Chemosphere. 36: 2019.

Costa, L.G. 2007. Contaminants in fish: risk-benefit considerations, Arhiv za higijenu rada i toksikologiju. 58: 367–374.

Curtis, M.A. and G. Bylund. 1991. Diphyllobothriasis: Fish tapeworm disease in the circumpolar north. Arctic Med. Res. 50: 18.

Dainty, R.H., B.G. Shaw and T.A. Roberts. 1983. Microbial and chemical changes in chill—stored red meats. pp. 151–178. In: T.A. Roberts and F.A. Skinner (eds.). Food Microbiology. Advances and Prospects. Academic Press.

De Haro, L.P. Pommier. 2006. Hallucinatory fish poisoning (ichthyoallyeinotoxism): two case reports from the Western Mediterranean and literature review. Clin. Toxicol. (Phila). 44(2): 185–8.

De la Rosa, L.A., A. Alfonso, M.R. Vieytes and L.M. Botana. 2001. Modulation of cytosolic calcium levels of human lymphocytes by yessotoxin, a novel marine phycotoxin. Biochem. Pharmacol. 61: 827–833.

DePaola, A., J.L. Jones, J. Woods et al. 2010. Bacterial and viral pathogens in live oysters: 2007 United States market survey. Appl. Environ. Microbiol. 76: 2754.

Dienstag, J.L., I.D. Gust, C.R. Lucas, D.C. Wong and R.H. Purcell. 1976. Mussel-associated viral hepatitis, type A: serological confirmation. Lancet. i: 561–564.

Díez, S. 2009. Human health effects of methylmercury exposure. Reviews of Environmental Contamination and Toxicology. 198: 111–132.

Draisci, R., E. Ferretti, L. Palleschi, C. Marchiafava, R. Poletti, A. Milandri, A. Ceredi and M. Pompei. 1999. High levels of yessotoxin in mussels and presence of yessotoxin and homoyessotoxin in dinoflagellates of the Adriatic Sea. Toxicon. 37: 1187–1193.

Eastaugh, J. and S. Shepherd. 1989. Infectious and Toxic Syndromes from Fish and Shellfish Consumption. Arch. Intern. Med. 149: 1735–1740.

Easton, M.D., D. Luszniak and G.E. Von der. 2002. Preliminary examination of contaminant loadings in farmed salmon, wild salmon and commercial salmon feed. Chemosphere. 46(7): 1053–1074.

Ebe, T., M. Matsumura, T. Mori et al. 1990. Eight cases of diphyllobothriasis. Kansenshogaku Zasshi. 64: 328.

EC (European Commission). 1999. Opinion of the Scientific Steering Committee on Antimicrobial Resistance (28 May 1999).

EC (European Commission). 1991. Off. J. Eur. Communities L 268/1–14.

Elgarch, A., P. Vale, S. Rifai and A. Fassouane. 2008. Detection of diarrheic shellfish poisoning and azaspiracid toxins in Moroccan mussels: Comparison of the LC-MS method with the commercial immunoassay kit. Mar. Drugs. 6: 587–594.

Elhamri, H., L. Idrissi, M. Coquery et al. 2007. Hair mercury levels in relation to fish consumption in a community of the Moroccan Mediterranean coast. Food Additives and Contaminants. 24: 1236–1246.

Endean, R., J.K. Griffith and J.J. Robins. 1993. Variation in the toxins present in ciguateric narrow-barred Spanish mackerel, Scomberomorus commersoni. Toxicon. 31: 723.

Evans, M.S., G.E Noguchi and C.P. Rice. 1991. The biomagnification of polychlorinated biphenyls, toxaphene, and DDT compounds in a Lake Michigan offshore food web. Archives of Environmental Contamination and Toxicology. 20(1): 87–93.

Eyles, M.J., G.R. Davey and E.J. Hunmtley. 1981. Demonstration of viral contamination of oysters responsible for an outbreak of viral gastroenteritis. Journal of Food Protection. 44(4): 294–296.

Faimali, M. 2012. Toxic effects of harmful benthic dinoflagellate Ostreopsis ovata on invertebrate and vertebrate marine organisms. Marine Environmental Research. 76: 97–107.

FAO/NACA/WHO Study Group. 1999. Food Safety Issues Associated with Products from Aquaculture. Technical Report Series, No. 883. Geneva, Switzerland: World Health Organization.

FAO/WHO. 1997. Report from a joint consultation: risk management and food safety.

FAO/WHO JECFA. 2005. Summary and Conclusions, Meeting, Rome, Italy.

FAO/WHO JECFA. 2006. Evaluation of Certain Food Contaminants, Sixty-fourth report of the Joint.

FAO/WHO. Expert Committee on Food Additives. WHO Technical Report Series, No. 930. Geneva, Switzerland.

Faroon, O., D. Jones and C. de Rosa. 2001. Effects of polychlorinated biphenyls on the nervous system. Toxicology and Industrial Health. 16: 305–320.

FDA. 2004a. Backgrounder for the 2004 FDA/EPA consumer advisory: what you need to know about mercury in fish and shellfish.

FDA. 1999. Food Code—Chapter 3: Food, U.S. Department of Health and Human Services, Public Health Service.

FDA. 2001. Fish and Fisheries Products Hazards and Controls Guidance. 3rd Edition. Center for Food Safety and Applied Nutrition, Washington, DC, USA.

FDA. 2001a. Fish and Fishery Products Hazards Controls Guidance: Third Edition. Center for Food Safety and Applied Nutrition. Washington DC.

Feingold, A.O. 1973. Hepatitis from eating steamed clams. Journal of the American Medical Association. 225(5): 526–527.

Fernandez, M.L. 1998. pp. 503–516. *In:* M. Miraglia, H.P. Van Egmond, C. Brera and J. Gilbert (eds.). Mycotoxins and Phycotoxins: Developments in Chemistry. Toxicology and Food Safety, Alaken Inc., Fort Collins, CO.

Foran, J.A., D.H. Good, D.O. Carpenter et al. 2005. Quantitative Analysis of the Benefits and Risks of Consuming Farmed and Wild Salmon. Journal of Nutrition. 135: 2639–2643.

Foran, J.A., R.A. Hites, D.O. Carpenter, M.C. Hamilton, A. Mathews-Amos and S.J. Schwager. 2004. A survey of metals in tissues of farmed Atlantic and wild Pacific salmon. Environmental Toxicology and Chemistry. 23(9): 2108–2110.

Fowler, P.K. 1989. Impacts of the 1987–88 North Carolina Red Tide. J. Shellfish Res. 8: 440.

Fowles, J.R., A. Fairbrother, L. Baecher-Steppan and N.I. Kerkvliet. 1994. Immunologic and endocrine effects of the flame-retardant pentabromodiphenyl ether (DE-71) in C57BL/6J mice. Toxicology. 86 (1–2): 49–61.

Franchini, A., E. Marchesini, R. Poletti and E. Ottaviani. 2004. Acute toxic effect of the algal yessotoxin on Purkinje cells from the cerebellum of Swiss CDq mice. Toxicon. 43: 347–352.

Frank, H.A. 1985. Histamine-forming bacteria in tuna and other marine fish. *In:* B.S. Pan and D. James (eds.). Histamine in marine products: production by bacteria, measurement and prediction of formation, FAO Fish Technical Paper, Rome. 252: 2–3.

Frelin, C., P. Vigne and J.P. Breittmayer. 1991. Mechanism of the cardiotoxic action of palytoxin. Mol. Pharmacol. 38: 904–909.

Froese, R. and D. Pauly (eds.). 2009. Sarpa salpa in FishBase. October 2009 version.

Fukui, M., M. Murata, A. Inoue, M. Gawel and T. Yasumoto. 1987. Occurrence of palytoxin in the Trigger fish Melichtys vidua. Toxicon. 25: 1121–1124.

Furey, A., S. O'Doherty, K. O'Callaghan, M. Lehane and K.J. James. 2010. Azaspiracid poisoning (AZP) toxins in shellfish: toxicological and health considerations. Toxicon. 56: 173–90.

Gallitelli, M., N. Ungaro, L.M. Addante, N. Gentiloni and C. Sabbà. 2005. Respiratory illness as a reaction to tropical algal blooms occurring in a temperate climate. JAMA. 293: 2599–2600.

Ganther, H.E., C. Goudie, M.L. Sunde, M.J. Kopecky, P. Wagner, S.-H. Oh and W.G. Hoekstra. 1972. Selenium: Relation to decreased toxicity of methylmercury added to diets containing tuna. Science. 175(26): 1122–1124.

Garcia, C., D. Truan, M. Lagos, J.P. Santelices, J.C. Diaz and N. Lagos. 2005. Metabolic transformation of dinophysistoxin-3 into dinophysistoxin-1 causes human intoxication by consumption of O-acyl-derivatives dinophysistoxins contaminated shellfish. J. Toxicol. Sci. 30: 287–296.

Gerba, C.P. and S.M. Goyal. 1978. Detection and occurance of enteric viruses in shellfish: A review. J. Food Protect. 41: 742.

Gerba, C.P., J.B. Rose and S.N. Singh. 1985. Waterborne gastroenteritis and viral hepatitis. Critical Reviews of Environmental Control. 15(3): 213–236.

Gessner, B.D. and J.P. Middaugh. 1995. Paralytic shellfish poisoning in Alaska: A 20-year retrospective analysis. Am. J. Epidemiol. 141: 766.

Gessner, B. and J. Mclaughlin. 2008. Epidemiologic impact of toxic episodes: neurotoxic toxins. pp. 77–104. *In:* L.M. Botana (ed.). Seafood and freshwater toxins pharmacology, physiology, and detection. CRC Press, Boca Raton, Florida.

Gessner, B.D. and M. Schloss. 1996. A population-based study of paralytic shell fish poisoning in Alaska. Alaska Med. 38(2): 54–58/68.

Geyer, H.J., G.G. Rimkus, I. Scheunert, A. Kaune, K.-W. Schramm, A. Kettrup, M. Zeeman, D.C.G. Muir, L.G. Hansen and D. Mackay. 2000. Bioaccumulation and occurrence of endocrinedisrupting chemicals (EDC), persistent organic pollutants (POPs), and other organic compounds in fish and other organisms including humans. pp. 1–166. *In:* O. Hutzinger and B. Beek (eds.). Bioaccumulation, New Aspects and Developments. The Handbook of Environmental Chemistry, Part J. Springer Verlag, Berlin, Germany.

Gilron, G., J. Archbold, S. Goldacker and J. Downie. 2007. Issues Related to Chemical Analysis, Data Reporting, and Use: Implications for Human Health Risk Assessment of PCBs and PBDEs in Fish Tissue. Human and Ecological Risk Assessment. An International Journal. 13(4): 773–791.

Giusti, G. and G.B. Gaeta. 1981. Doctors in the kitchen: experiments with cooking bivalve mollusks. New England Journal of Medicine. 304(22): 1371–1372.

Glavin, G.B., R. Bose and C. Pinsky. 1990. Infections and toxic syndromes from fish and shellfish consumption. Arch. Intern. Med. 150: 2425.

Glazious, P. and A.M. Legrand. 1994. The epidemiology of ciguatera fish poisoning. Toxicon. 32: 863.

Goldmann, D.R. 1985. Hold the sushi. JAMA. 253: 2495.

Gorczyca, E. and L. Pek Poh. 1985. Mesophilic spoilage of bay trout (Arripis trutta), bream (Acanthropagrus butcheri) and mullet (Aldrichetta forsteri). In Spoilage of tropical fish and product development. Proceedings of a symposium held in conjunction with the Sixth Session of The Indo-PacificFishery Commission Working Party on Fish Technology and Marketing. A. Reilly (ed.). Royal Melbourne Institute of Technology, Melbourne, Australia 23–26 October, 1984. FAO Fish Rep. (317) Suppl. 123–132.

Goyal, S.M. 1984. Viral pollution of the marine environment. Critical Reviews of Environmental Control. 14(1): 1–32.

Goyal, S.M., C.P. Gerba and J.L. Melnick. 1979. Human enteroviruses in oysters and their overlying waters. Applied and Environmental Microbiology. 37(3): 572–581.

Gram, L., C. Wedell-Neergaard and H.H. Huss. 1990. The bacteriology of fresh and spoiling Lake Victorian Nile perch (Lates niloticus). Int. J. Food Microbiol. 10: 303–316.

Granéli, E., C.E.L. Ferreira, T. Yasumoto, E. Rodrigues and M.H.B. Neves. 2002. Book of Abstracts of Xth International Conference on Harmful Algae. Sea urchins poisoning by the benthic dinoflagellate Ostreopsis ovata on the Brazilian coast.

Gras-Rouzet, S., P.Y. Donnio, F. Juguet et al. 1996. First European case of gastroenteritis and bacteremia due to Vibrio hollisae. Eur. J. Clin. Microbiol. Infect. Dis. 15: 864.

Greenwood, M., G. Winnard and B. Bagot. 1998. An outbreak of Salmonella enteritis phage type 19 infection associated with cockles. Commun Dis. Public Health. 1: 35.

Grohmann, G.S., A.M. Murphy, P.J. Christopher, E. Auty and H.B. Greenberg. 1981. Norwalk virus gastroenteritis in volunteers consuming depurated oysters. Australian Journal of Experimental Biology and Medical Science. 59: 219–228.

Guldner, L., C. Monfort, F. Rouget et al. 2007. Maternal fish and shellfish intake and pregnancy outcomes: A prospective cohort study in Brittany, France. Environmental Health. 6: 33.

Gunn, R.A., H.T. Janowski, S. Lieb, E.C. Prather and H.B. Greenberg. 1982. Norwalk virus gastroenteritis frollowing raw oyster consumption. American Jouranl of Epidemiology. 115(3): 348–351.

Guzman-Perez, S.E. and D.L. Park. 2000. Ciguatera toxins: Chemistry and diction. pp. 401–418. *In:* L. Botana (ed.). Seafood and Freshwater Toxins: Pharmacology, Physiology and Detection. Marcel Dekker, New York.

Habermann, E. 1989. Palytoxin acts through the Na+, K+-ATPase. Toxicon. 27: 1171–1187.

Habermann, E. and M. Laux. 1986. Depolarization increases inositol phosphate production in a particulate preparation from rat brain. Naunyn-Schmiederberg's Arch. Pharmac. 334: 1–15.

Habermehl, G.G., H.C. Krebs, P. Rasoanavio and A. Ramialihorisoa. 1994. Severe ciguatera poisoning in Madagascar: a case report. Toxicon. 32: 1539–1542.

Hale, R.C., M.J. La Guardia, E.P. Harvey, T.M. Mainor, W.H. Duff and M.O. Gaylor. 2001b. Polybrominated diphenyl ether flame retardants in Virginia freshwater fishes (USA). Environmental Science and Technology. 35: 4585–4591.

Hallegraef, G.M. 2006. The red tide. A world of Science. 4(3): 2–7.

Hallegraeff, G.M. 2004. Harmful algal blooms: a global overview. *In:* G.M. Hallegraeff, D.M. Anderson and A.D. Cembella (eds.). Manual on harmful marine microalgae. UNESCO, Paris.

Halliday, M.L., L.Y. Kang, T.K. Zhou, M.D. Hu, Q.C. Pan, T.Y. Fu, Y.S. Huang and S.L. Hu. 1991. An epidemic of hepatitis A attributable to the ingestion of raw clams in Shanghai, China. J. Infect. Dis. 164: 852–859.

Halstead, B.W. 2001a. Fishtoxins. pp. 23–50. *In:* Y.H. Hui, D. Kitts and P.S. Stanfield (eds.). Foodborne diseas handbook. 2nd ed., Vol. IV: Seafood and environmental Toxins, Marcel Dekker, New York.

Halstead, B.W. 1967. Poisonous and venomous marine animals of the world (Vol 2). United States Government Printing Office, Washington, DC.

Halstead, B.W. 1965. Poisonous and Venomous marine Animals of the World. Darwin Princeton, N. J.

Hefle, S.L. 1996. The chemistry and biology of food allergens. Food Technology. 50(3): 86–92.

Helfrich, P. and A.H. Banner. 1960. Hallucinatory mullet poisoning: a preliminary report. J. Trop. Med. Hyg. 63: 86–9.

Hirsh, J.K. and C.H. Wu. 1997. Palytoxin-induced single-channel currents from the sodium pump synthesized by *in vitro* expression. Toxicon. 35: 169–176.

Hites, R.A., J.A. Foran, D.O. Carpenter et al. 2004a. Global Assessment of Organic Contaminants in Farmed Salmon. Science. 303: 226–229.

Holmes, M.J. and R.J. Lewis. 1992. Purification characterization of large and small maitotoxins from cultured Gambierdiscus toxicus. Nat Toxins. 2: 64–72.

Hughes, T., G.H. Denton and M.G. Grosswald. 1977. Was there a late-Wurm Arctic ice sheet? Nature. 226: 596–602.

Hughner, R.S., J.K. Maher and N.M. Childs. 2008. Review of food policy and consumer issues of mercury in fish. Journal of the American College of Nutrition. 27: 185–194.

Humphrey, H. 1988. Human exposure to persistent aquatic contaminants. pp. 227–238. *In:* N.W. Schmidtke (ed.). Toxic Contamination in Large Lakes. MI: Lewis Publishers, Chelsea.

Huss, H.H. 1994. Assurance of seafood quality. FAO Fisheries Technical Paper No. 34. Rome, 169 pp.

Huss, H.H., L. Ababouch and L. Gram. 2003. Assessment and Management of Seafood Safety and Quality, FAO Fisheries Technical Paper 444, Rome, 230 pp.

Hutchinson, J.W., J.W. Bass, D.M. Demers et al. 1997. Diphyllobothriasis after eating raw salmon. Hawaii Med. J. 56: 176.

Ikonomou, M.G., S. Rayne and R.F. Addison. 2002. Exponential increases of the brominated flame retardants, polybrominated diphenyl ethers, in the Canadian Arctic from 1981 to 2000. Environmental Science and Technology. 36(9): 1886–1892.

Inskip, M.J. and J.K. Piotrowski. 1985. Review of the health effects of methyl mercury. Journal of Applied Toxicology. 5: 113–13322.

IOM. 2003. Dioxins and Dioxin-Like Compounds in the Food Supply: Strategies to Decrease Exposure. The National Academies Press, Washington, DC.

Isbister, G.K. and M.C. Kiernan. 2005. Neurotoxic marine poisoning. Lancet Neurol. 4(4): 219–28.

Ishida, H. 1995. Brevetoxin B1, a New Polyether Marine Toxin from the New Zealand Shellfish, Austrovenus Stutchburyi. Tetrahedron Lett. 36: 725–728.

Ishida, H. 1996. Study on Nuerotoxic Shellfish Poisoning Involving the Oyster, Crassostrea Gigas, in New Zealand. Toxicon. 34: 1050–1053.

Ishida, H., A. Nozawa, H. Nukaya and K. Tsuji. 2004. Comparative concentrations of brevetoxins PbTx-2, PbTx-3, BTX-B1 and BTX-B5 in cockle, Austrovenus stutchburyi, greenshell mussel, Perna canaliculus, and Pacific oyster, Crassostrea gigas, involved neurotoxic shellfish poisoning in New Zealand. Toxicon. 43: 779–789.

Ito, E., M. Statake, K. Ofuji, N. Kurita, T. McMahon and K. James. 2000. Multiple organ damage caused by a new toxin azaspiracid, isolated from mussels produced in Ireland. Toxicon. 38: 917–930.

Iwamoto, M., T. Ayers, B.E. Mahon and D.L. Swerdlow. 2010. Epidemiology of seafood-associated infections in the United States. Clin. Microbiol. Rev. 23(2): 399–411.

Jacob de, B., T.E. van der Zande, H. Pieters, F. Ariese, C.A. Schipper, T. van Brummelen and A.D. Vethaak. 2001. Organic contaminants and trace metals in flounder liver and sediment from the Amsterdam and Rotterdam harbours and off the Dutch coast. J. Environ. Monit. 3: 386–393.

Jacobson, J.L., S.W. Jacobson and H.E. Humphrey. 1990. Effects of *in utero* exposure to polychlorinated biphenyls and related contaminants on cognitive functioning in young children. Journal of Pediatrics. 116(1): 38–45.

James, K., M.D. Sierra, M. Lehane, A. Brana Magdalena and A. Furey. 2003. Detection of five newly hydroxyl analogues of azaspiracids in shellfish using multiple tandem mass spectrometry. Toxicon. 41: 277–283.

James, K., M. Lehane, C. Moroney, P. Fernandez-Puente, M. Statake and T. Yasumoto. 2002. Azaspiracid shellfish poisoning: unusual toxin dynamics in shellfish and the increased risk of acute human intoxications. Food Addit. Contam. 19: 555–561.

James, K.J., M.J.F. Saez, A. Furey and M. Lehane. 2004. Azaspiracid poisoning, the food-borne illness associated with shellfish consumption. Food Addit. Contam. 21(9): 879–892.

James, K.J., C. Moroney, C. Roden et al. 2003. Ubiquitous 'benign' alga emerges as the cause of shellfish contamination responsible for the human toxic syndrome, azaspiracid poisoning. Toxicon. 41: 145–151.

Järup, L. 2003. Hazards of heavy metal contamination. British Medical Bulletin. 68: 167–182.

Jay, J.M. 2000. Modern food microbiology. 6th edition. Aspen Publishers, Gaithersburg, Maryland, USA.

Jenkins, D., J. Sievenpiper, and D. Pauly et al. 2009. Are dietary recommendations for the use of fish oils sustainable? Canadian Medical Association Journal. 180: 633–637.

Jones, S.H., T.L. Howell and K.R. O'Neill. 1991. Differential elimination of indicator bacteria and pathogenic *Vibrio* spp. from Easter oysters (Crassostrea virginica Gmelin, 1791) in a commercial controlled purification facility in Maine. J. Shellfish Res. 10: 105.

Kaneko, J.J. and L.B. Medina. 2009. Risk of parasitic worm infection from eating raw fish in Hawai'i: a physician's survey. Hawaii Med. J. 68(9): 227–9.

Kanki, M., T. Yoda, M. Ishibashi and T. Tsukamoto. 2004. Photobacterium phosphoreum caused a histamine fish poisoning incident. International Journal of Food Microbiology. 92: 79–87.

Kao, C.Y. 1966. Tetrodotoxin, saxitoxin and their significance in the study of excitation phenomena. Pharmacol Rev. 18: 997–1049.

Khora, S.S. 2004. How safe is our food? Science Reporter. 10–43.

Khora, S.S. 1994. Puffer fish toxins. Fish Technology (Special Issue). 171–176.

Khora, S.S. 1990. Toxicity studies on puffer fish from tropical waters, D.Ag. Thesis, Tohoku University: Sendai, Japan, 130 pp.

Khora, S.S. 1986. A systematic review of poisonous and venomous marine fishes of India. Ph.D. Thesis. Berhampur University, Berhampur, India, 623 pp.

Kim, S.Y., K.A. Marx and C.H. Wu. 1995. Involvement of the Na+, K+-ATPase in the introduction of ion channels by palytoxin. Naunyn-Schmiedeberg's Arch. Pharmacol. 351: 542–554.

Kimbrough, R.D., R.A. Squire, R.E. Linder, J.D. Strandberg, R.J. Montalli and V.W. Burse. 1975. Induction of liver tumor in Sherman strain female rats by polychlorinated biphenyl aroclor 1260. Journal of the National Cancer Institute. 55(6): 1453–1459.

Kjellstrom, T., S. Kennedy, S. Wallis, A. Stewart, L. Friberg, B. Lind, T. Wutherspoon and C. Mantell. 1989. Physical and Mental De elopment of Children with Prenatal Exposure to Mercury from Fish. Stage II: Inter iews and Psychological Tests at Age Solna, Sweden: National Swedish Environmental Protection Board.

Klontz, K.C., A. Abraham, S.M. Plakas and R.W. Dickey. 2009. Mussel-associated azaspiracid intoxication in the United States. Ann. Intern. Med. 150: 361.

KnΦchel, S. 1989. *Aeromonas* spp.—Ecology and significance in food and water hygiene. Ph.D. Thesis. The Royal Veterinary and Agricultural University, Copenhagen, Denmark.

Kodavanti, P.R.S. and T.R. Ward. 2005. Differential effects of commercial polybrominated diphenyl ether and polychlorinated biphenyl mixtures on intracellular signalling in rat brain *in vitro*. Toxicol. Sci. 85: 952–962.

Koff, R.S. and H.S. Sear. 1967. Internal temperature of steamed clams. New England Journal of Medicine. 276: 737–739.

Kong, K.Y., K.C. Cheung, C.K. Wong and M.H. Wong. 2005. Residues of DDTs, PAHs and some heavy metals in fish (Tilapia) collected from Hong Kong and mainland China. Journal of Enironmental Science and Health. Part A: Toxic/Hazardous Substances and Environmental Engineering. 40(11): 2105–2115.

Kopf, P.G. and M.K. Walker. 2009. Overview of developmental heart defects by dioxins, PCBs, and pesticides. Journal of Environmental Science and Health. 27: 276–285.

Köse, S. 2010. Evaluation of Seafood Safety Health Hazards for Traditional Fish Products: Preventive Measures and Monitoring Issues. Turkish Journal of Fisheries and Aquatic Sciences. 10: 139–160.

Kris-Etherton, P.M., W.S. Harris and L.J. Appel. 2002. Fish consumption, fish oil, omega-3 fatty acids, and cardiovascular disease. Circulation. 106: 2747–2757.

Krock, B., U. Tillmann, U. John and A.D. Cembella. 2009. Characterization of azaspiracids in plankton size-fractions and isolation of an azaspiracid-producing dinoflagellate from the North Sea. Harmful Algae. 8: 254–263.

Laden, F., L.M. Neas, D. Spiegelman et al. 1999. Predictors of plasma concentrations of DDE and PCBs in a group of U.S. women. Environ. Health Perspect. 107: 75–81.

Langsrud, S. and G. Sundheim. 1997. Factors contributing to the survival of poultry associated *Pseudomonas* spp. exposed to a quaternary ammonium compound. J. Appl. Microbiol. 82: 705–712.

Larkin, E.P. and D.A. Huint. 1982. Bivalve mollusks: control of microbial contaminants. Bioscience. 32(3): 193–197.

Law, R.J., M. Alaee, C.R. Allchin, J.P. Boon, M. Lebeuf, P. Lepom and G.A. Stern. 2003. Levels and trends of polybrominated diphenylethers and other brominated flame retardants in wildlife. Environment International. 29(6): 757–779.

Lee, K.W. and G.Y. Lip. 2003. The role of omega-3 fatty acids in the secondary prevention of cardiovascular disease. Quarterly Journal of Medicine. 96: 465–480.

Lee, W.C. et al. 1996. An epidemiological study of food poisoning in Korea and Japan. Int. J. Food Microbiol. 29: 1141.

Lees, D. 2000. Viruses and bivalve shellfish. Int. J. Food Microbiol. 59: 81–116.

Legrand, A.M. 1998. Ciguatera toxins: origin, transfer through the food chain and toxicity to humans. pp. 39–43. *In:* B. Reguera, J. Blanco, M. Fernandez and T. Wyatt (eds.). Harmful Algae, Proceedings of the VIII International Conference on Harmful Algae. Xunta de Galicia and IOC of UNESCO; Vigo, Spain.

Lehane, L. and J. Olley. 2000. Histamine (Scombroid) Fish Poisoning. A review in a risk-assessment framework. National Office of Animal and Plant Health Canberra 1999. Revised 2000. Agricultural, Fisheries and Forestry of Australia.

Lehrer, S.B. 1993. Seafood allergy. Introduction. Clinical Reviews in Allergy. 11(2): 155–157.

Lennon, D., B. Lewis, C. Mantell, D. Becroft, B. Dove, K. Farmer, S. Tonkin, N. Yeats, R. Stamp and K. Mickleson. 1984. Epidemic perinatal listeriosis. Pediatr. Infect. Dis. 3: 30–34.

Levine, L. and H. Fujiki. 1985. Stimulation of arachidonic acid metabolism by different types of tumor promoters. Carcinogenesis. 6: 1631–1635.

Lewis, R.J., J. Molgo and D.J. Adams. Pharmacology of toxins involved in ciguatera and related fish poisonings. pp. 419–447. *In:* L. Botana (ed.). Seafood and Freshwater Toxins: Pharmacology, Physiology and Detection. Marcel Dekker, New York.

Lewis, R.J., J.P. Vernoux and I.M. Brereton. 1998. Structure of Caribbean ciguatoxin isolated from Caranx latus. J. Am. Chem. Soc. 120: 5914–5920.

Lindberg-Braman, A.M. 1956. Clinical observations on the so-called oyster hepatitis. American Jouranl of Public Heqalth. 53: 1003–10011.

Lindström, G., H. Wingfors, M. Dam and B. van Bavel. 1999. Identification of 19 polybrominated diphenyl ethers (PBDEs) in long-finned pilot whale (Globicephala melas) from the Atlantic. Arch Environ Contam Toxicol. 36: 355–363.

Lipp, E.K.P. and J.B. Rose. 1997. The role of seafood in food borne diseases in the United States of America. Rev. Sci. Tech. Off. Int. Epiz. 16: 620.

Louhija, A. 2002. A pilot study for the detection of acute ciguatera intoxication in human blood. J. Toxicol. Clin. Toxicol. 40(1): 49–57.

Luksemburg, W., R. Wenning, M. Maier, A. Patterson and S. Braithwaite. 2004. Polybrominated diphenyl ethers (PBDE) and polychlorinated dibenzo-p-dioxins (PCDD/F) and biphenyls (PCB) in fish, beef, and fowl purchased in food markets in northern California U.S.A. Organohalogen Compounds. 66: 3982–3987.

Lunder, S. and R. Sharp. 2003. Tainted Catch: Toxic Fire Retardants Are Building Up Rapidly in San Francisco Bay Fish—And People. Environmental Working Group. Washington DC.

Lyons, D.E., J.T. Beery, S.A. Lyons et al. 1983. Cadaverine and aminoguanidine potentiate the uptake of histamine *in vitro* in perfused intestinal segments of rats. Toxicol. Appl. Pharmacol. 70: 445–58.

Mackenzie, L.L. 1995. A Gymnodinium bloom and the contamination of shellfish with lipid soluble toxins in New Zealand, Jan–April 1993. pp. 795–800. *In:* P. Lassus (ed). Harmful Marine Algal Blooms. Lavoisier Intercept LTD, Paris, France.

Mackowiak, P.A., C.T. Caraway and B.L. Portnoy. 1976. Oyster-associated hepatitis: lessons from the Louisiana experience. American Journal of Epidemiology. 103(2): 181–191.

Mah, J.H., Y.J. Kim and H.J. Hwang. 2009. Inhibitory effects of garlic and other spices on biogenic amine production in Myeolchi-jeot, Korean salted and fermented anchovy product. Food Control. 20: 449–454.

Mahaffey, K.R. 1999. Methylmercury: a new look at the risks, Public Health Reports. 114: 396–399.

Mahaffey, K.R., R.P. Clickner and C.C. Bodurow. 2004. Blood organic mercury and dietary mercury intake: National Health and Nutrition Examination Survey, 1999 and 2000. Environmental Health Perspectives. 112(5): 562–570.

Manerio, E., V.L. Rodas, E. Costas and J.M. Hernandez. 2008. Shellfish consumption: A major risk factor for colorectal cancer. Med. Hypotheses. 70: 409–412.

Martinez, A.J. and G.S. Visvesvara. 1997. Free-living, amphizoic and opportunistic amebas. Brain Pathol . 7: 583.

Masuda, Y., A.J. Schecter and O. Päpke. 1996. Concentration of PCBs, PCDFs and PCDDs in the blood of Yusho patients and their toxic equivalent contributions. Organohalogen Compounds. 30: 146–9.

McCollum, J.P.K. 1968. An epidemic of mussel poisoning in North-East England. Lancet 2: 767–770.

Mcentire, J.C. 2004. IFT issues update on food borne pathogens. Food Technol. 58(7): 20.

McFarren, E.F. 1965. The Occurrence of Ciguatera-like Poison in Oysters, Clams and Gymnodinium BreveCultures. Toxicon. 3: 111–123.

Mead, P.S., L. Slutsker, V. Dietz, L.F. McCaig, J.S. Bresee, C. Shapiro, P.M. Griffine and R.V. Tauxe. 1999. Food-related illness and death in the United States. Emerging Infectious Diseases. 5(5): 607–625.

Melton, R.J., J.E. Randall, N. Fusetani et al. 1984. Fatal sardine poisoning: A fatal case of fish poisoning in Hawaii associated with the Marquesan sardine. Hawaii Med. J. 43: 114.

MMWR Morb Mortal Wkly Rep. 1991. Paralytic shellfish poisoning-Massachusetts and Alaska, 1990. 40: 157–161.

MMWR Morb Mortal Wkly Rep. 1996. Centers for Disease Control and Prevention. Tetrodotoxin poisoning associated with eating puffer fish transported from Japan—California. 45: 389–391.

MMWR Morb Mortal Wkly Rep. 1997. Centers for Disease Control. Viral gastroenteritis associated with eating oysters—Louisiana, December 1996–January 1997. 46: 1109.

Mocarelli, P., P.M. Gerthoux, P. Brambilla, A. Marocchi, C. Beretta, M. Bertona, M. Cazzaniga. L. Colombo, C. Crespi, E. Ferrari, G. Limonta, C. Sarto and P.L. Signorini Tramacere. 1999. Dioxin health effects on humans twenty years after Seseso: A summary. pp. 41–52. *In:* A. Ballarin-Denti, P.A. Bertazzi, S. Facchetti and P. Mocarelli (eds.). Chemistry, Man, and En ironment: The Seveso Accident 0 Years On: Monitoring. Epidemiology and Remediation.

Mongar, J.L. 1957. Effect of chain length of aliphatic amines on histamine potentiation and release. Br. J. Pharmacol. Chemother. 12(2): 140–148.

Moore, R.E., G. Bartolini, J. Barchi, A.A. Bothmer-By, J. Dadok and J. Ford. 1982. Absolute stereochemistry of palytoxin. J. Am. Chem. Soc. 104: 3776–3779.

Moore, R.E. and P.J. Scheuer. 1971. Palytoxin: a new marine toxin from a coelenterate. Science. 172: 495–498.

Morgan, K.L., S.L. Larkin and C.M. Adams. 2009. Firm-level economic effects of HABS: A tool for business loss assessment. Harmful Algae. 8: 212–218.

Morinaga, S. et al. 1997. Histamine poisoning after ingestion of spoiled raw tuna in a patient taking isoniazid. Intern Med. 36(3): 198–200.

Morohashi, A. 1995. Brevetoxin B3, a New Brevetoxin Analog Isolated from the Greenshell Mussel Perna Canaliculus Involved in Neurotoxic Shellfish Poisoning in New Zealand. Tetrahedron Lett. 36: 8895–8998.

Morohashi, A. 1999. Brevetoxin B4 Isolated from Greenshell Mussel, Perna Canaliculus, the Major Toxin Involved in Neurotoxic Shellfish Poisoning in New Zealand. Nat. Toxins. 7: 45–48.

Morris, G.K. 1986. Shigella. *In:* D.O. Cliver and B.A. Cochrane (eds.). Progress in Food Safety. FoodRes. Inst., Univ. of Wisconsin-Madison.

Morris, P.D. 1991. Clinical and Epidemiological Features of Neurotoxic Shellfish Poisoning in North Carolina. Am. J. Pub. Health. 81: 471–474.

Morris, P.D., D.S. Campbell, T.J. Taylor et al. 1991. Clinical and epidemiological features of neurotoxic shellfish poisoning in North Carolina. Am. J. Public Health. 81: 471.

Morse, D.L., J.J. Guzewich, J.P. Hanrahan, R. Stricof, M. Shayegani, R. Deibel, J.C. Grabau, N.A. Nowak, J.E. Herrmann, G. Cukor and N.R. Blacklow. 1986. Widespread outbreaks of clam- and oyster-associated gastroenteritis: role of norwalk virus. New England Jouranl of Medicine. 314(11): 678–681.

Murata, M., A.M. Legrand, Y. Ishibashi and T. Yasumoto. 1990. Structures and configurations of ciguatoxin from the moray eel Gymnothorax javanicus and its likely precursor from the dinoflagellate Gambierdiscus toxicus. J. Am. Chem. Soc. 112: 4380–4386.

Murata, M., M. Kumagai, J.S. Lee and T. Yasumoto. 1987. Isolation and structure of yessotoxin, a novel polyether compound implicated in diarrheic shellfish poisoning. Tetrahedron Lett. 28: 5869–5872.

Murphy, A.M., G.S. Grohmann, P.J. Christopher, W.A. Lopez, G.R. Davey and R.H. Millsom. 1979. An Australia-wide outbreak of gastroenteritis from oysters caused by Norwalk virus. Med. J. Aust. 2: 329–333.

Murrell, K.D. 2002. Fishborne zoonotic parasites: epidemiology, detection and elimination. Lactic acid bacteria in fish preservation. pp. 114–141. *In:* H.A. Bremner (ed.). Safety and quality issues in fish processing. Woodhead Publishing Ltd. CRC pres, New York.

Myers, G.J., P.W. Davidson, C. Cox, C.F. Shamlaye, D. Palumbo, E. Cernichiari, J. Sloane-Reeves, G.E. Wilding, J. Kost, L.S. Huang and T.W. Clarkson. 2003. Prenatal methylmercury exposure from ocean fish consumption in the Seychelles child development study. Lancet. 361(9370): 1686–1692.

Nesheim, M.C. and A.L. Yaktine (eds.). 2007. Seafood Choices: Balancing Benefits and Risks. National Academy of Sciences, Committee on Nutrient Relationships in Seafood: Selections to Balance Benefits and Risks Food and Nutrition Board, M.C. National Academies Press, Washington DC.

NFI (National Fisheries Institute). 1989. Some considerations for control of Listeria. Washington, DC.

Nguon, K., M.G. Baxter and E.M. Sajdel-Sulkowska. 2005. Perinatal exposure to polychlorinated biphenyls differentially affects cerebellar development and motor functions in male and female rat neonates. Cerebellum. 4(2): 112–122.

NIEHS (National Institute of Environmental Health Sciences). 2000. Marine and Freshwater Biomedical Sciences Center Marine Toxins and Human Health Internet.

NRC. 2006. Health Risks from Dioxin and Related Compounds: Evaluation of the EPA Reassessment. The National Academies Press, Washington, DC.

O'Neil, C.E. and S.B. Lehrer. 1995. Seafood allergy and allergens: A review. Food Technology. 49(10): 103–116.

O'Neil, C., A.A. Helbling and S.B. Lehrer. 1993. Allergic reactions to fish. Clinical Reviews in Allergy. 11(2): 183–200.

Oken, E., R.O. Wright, K.P. Kleinman, D. Bellinger, H. Hu, J.W. Rich-Edwards and M.W. Gillman. 2005. Maternal fish consumption, hair mercury, and infant cognition in a US cohort. Environmental Health Perspectives. 113(10): 1376–1380.

Olsen, S.J., L.C. MacKinon, J.S. Goulding, N.H. Bean and L. Slutsker. 2000. Surveillance for foodborne disease outbreaks—United States, 1993–1997. Morbidity and Mortality Weekly Report 49(Surveillance Summary 1): 1–51.

Olson, R.E. 1986. Marine fish parasites of public health importance. pp. 339–355. *In:* D.E. Kramer and J. Liston (eds.). Seafood Quality Determination. Elsevier Science Publishers, Amsterdam, the Netherlands.

Onuma, Y., M. Satake, T. Ukena, J. Roux, S. Chanteau, N. Rasolofonirina, N. Ratsimaloto, H. Naoki and T. Yasumoto. 1999. Identification of putative palytoxin as the cause of clupeotoxism. Toxicon. 37(1): 55–65.

Oufji, K., M. Statake, T. McMahon, J. Silker, K.J. James and H. Naoki. 1999. Two analogs of azaspiracid isolated from mussels, Mytilus edulis, involved in human intoxication in Ireland. Nat Toxins. 7: 99–102.

Pandey, G., S. Madhuri and A.B. Shrivastav. 2012. Contamination of Mercury in fish and its toxicity to both fish and humans: An overview. International Research Journal of Pharmacy. 3(11): 44–47.

Parizek, J. and I. Ostadalova. 1967. The protective effect of small amounts of selenite in sublimate intoxication. Experientia 23(2): 142–143.

Parker, R. 2001. Introduction to Food Science. Delmar Cengage Learning.

Passos, C.J., D. Mergler, E. Gaspar, S. Morais, M. Lucotte, F. Larribe, R. Davidson and S. de Grosbois. 2003. Eating tropical fruit reduces mercury exposure from fish consumption in the Brazilian Amazon. Environmental Research. 93(2): 123–130.

Paz, B., A.H. Daranas, M. Norte, P. Riobó, J.M. Franco and J.J. Fernández. 2008. Yessotoxins, a group of marine polyether toxins: an overview. Mar. Drugs. 6: 73–102.

Peele, C. 2004. Washington State Polybrominated Diphenyl Ether (PBDE) Chemical Action Plan: Interim Plan. Olympia, WA: Washington State Department of Ecology, Department of Health.

Penna, A.M. Vila, S. Fraga, M.G. Giacobbe, F. Andreoni, P. Riobó and C. Veronesi. 2005. Characterization of Ostreopsis and Coolia (Dinophyceae) isolates in the western Mediterranean Sea based on morphology, toxicity, and internal transcribed spacer 5.8S rDNA sequences. J. Phycol. 41: 212–225.

Perez-Gomez, A., A. Ferrero-Gutierrez, A. Novelli, J.M. Franco, B. Paz and M.T. Fernandez-Sanchze. 2006. Potent neurotoxic action of the shellfish biotoxin yessotoxin on cultured cerebellar neurons. Toxicol. Sci. 90: 168–177.

Perl, T.M., L. Bedard, T. Kosatsky, J.C. Hockin, E.C. Todd and RS. Remis. 1990. An outbreak of toxic encephalopathy caused by eating mussels contaminated with domoic acid. N. Engl. J. Med. 322(25): 1775–80.

Peters, J.B. 1989. Listeria monocytogenes: a bacterium of increasing concern. Washington Sea Grant, Seafood Processing Series. Seattle, Washington.

Plessi, M., D. Bertelli and A. Monzani. 2001. Mercury and selenium content in selected seafood. J. Food Comp. Anal. 14(5): 461–7.

Poli, M.A. 2000. Neurotoxic Shellfish Poisoning and Brevetoxin Metabolites: A Case Study from Florida. Toxicon. 38: 981–993.

Porter, J. and W.P. Sarkin. 1987. Outbreaks of clam-associated gastroenteritis in New Jersey: 1983–1984. New Jersey Medicine. 84(9): 649–651.

Portnoy, B.L., P.A. Mackowiak, C.T. Carawayy, J.A. Walker, T.W. McKinley and C.A. Klein. 1975. Oyster-associated hepatitis failure of shellfish certification programs to prevent outbreaks. Journal of the American Medical Association. 233(10): 1065–1068.

Potasman, I., A. Paz and M. Odeh. 2002. Infectious outbreaks associated with bivalve shellfish consumption: A worldwide perspective. Clin. Infect. Dis. 35: 921.

Price, R.J. 1992. Residue concerns in seafoods. Dairy, food and Environmental Sanitation. 12: 139–143.

Rabie, M., R.L. Simon-Sarkadi, H. Siliha, S. El-seedy and A.A. El Badawy. 2009. Changes in free amino acids and biogenic amines of Egyptian salted-fermented fish (Feseekh) during ripening and storage. Food Chemistry. 115(2): 635–638.

Randall, J. 2005. Review of Clupeotoxism, an Often Fatal Illness from the Consumption of Clupeoid Fishes. Pacific Science. 59(1): 73–77.

Rayne, S., M. G. Ikonomou and B. Antcliffe. 2003. Rapidly increasing polybrominated diphenyl ether concentrations in the Columbia River system from 1992 to 2000. Environmental Science and Toxicology. 37(13): 2847–2854.

Regel, E. 2004. Commission Directive 2004/853/EC Specific hygiene rules for food of animal origin. Off. J. Eur. Commun. 22–82.

Rehmann, N., P. Hess and M.A. Quilliam. 2008. Discovery of new analogs of the marine biotoxin azaspiracid in blue mussels (*Mytilus edulis*) by ultra-performance liquid chromatography/ tandem mass spectrometry. Rapid Commun. Mass Spectrom. 22: 549–558.

Rhodes, L., P. McNabb, M. Salas, L. Briggs, V. Beuzenberg and M. Gladstone. 2006. Yessotoxin production by Gonyaulax spinifera. Harmful Algae. 5: 148–155.

Rice, D.C. 2000. Identification of functional domains affected by developmental exposure to methylmercury: Faroe Islands and related studies. Neurotoxicology. 21(6): 1039–1044.

Rice, D.C. 2005. Brominated Flame Retardants: A Report to the Joint Standing Committee on Natural Resources and Maine Legislature. Prepared by the Bureau of Health and Department of Environmental Protection.

Richards, G. 1986. Shellfish associated enteric virus illness in the United States, 1934–1984. Estuaries. 10: 84.

Rim, H.J. 1998. Field investigations on epidemiology and control of fish-borne parasites in Korea. International Journal of Food Science and Technology. 33: 157–168.

Riobó, P., B. Paz and J.M. Franco. 2006. Analysis of palytox-in- like in Ostreopsis cultures by liquid chromatography with precolumn derivatization and fluorescence detection. Anal. Chim. Acta. 566: 217–223.

Rippey, S.R. 1994. Infectious diseases associated with molluscan shellfish consumption. Clin Microbiol Rev. 7: 419.

Robson, M.G. and G.C. Hamilton. 2005. Pest control and pesticides. pp. 544–580. *In:* H. Frumkin (ed.). Environmental Health: From Global to Local. John Wiley and Sons, San Fransisco. Inc.

Rodrigue, D.C., R.A. Etzel, S. Hall, E. de Porras, O.H. Velasquez, R.V. Tauxe, E.M. Kilbourne and P.A. Blake. 1990. Lethal paralytic shellfish poisoning in Guatemala. Am. J. Trop. Med. Hyg. 42(3): 267–271.

Roos, R. 1956. Heapatitis epidemic conveyed by oysters. Svenska Läkartidningen 53: 989.

Ross, G. 2004. The public health implications of polychlorinated biphenyls (PCBs) in the environment. Ecotoxicol Environ Safety. 59(3): 275–91.

Ruff, T.A. and R.J. Lewis. 1994. Clinical aspects of ciguatera: an overview. Mem. Queensl. Mus. 34: 609–619.

Sakamoto, Y. 1987. Shellfish and Fish Poisoning Related to the Toxic Dinoflagellates. S. Med. J. 80: 866–872.

Sansoni, G., B. Borghini, G. Camici, M. Casotti, P. Righini and C. Rustighi. 2003. Fioriture algali di Ostreopsis Ovata (Gonyaulacales: Dinophyceae): Unproblema emergente. Biol Ambientale. 17: 17–23.

Satake, M., K. Terasawa, Y. Kadowaki and T. Yasumoto. 1996. Relative configuration of yessotoxin and isolation of two new analogs from toxic scallops. Tetrahedron Lett. 37: 5955–5958.

Satake, M., R. Viviani and T. Yasumoto. 1997. Yessotoxin in mussels of the northern Adriatic Sea. Toxicon. 35: 177–183.

Schantz, S.L., J.J. Widholm and D.C. Rice. 2003. Effects of PCB exposure on neuropsychological function in children. Environmental Health Perspectives. 111(3): 1–27.

Schantz, P.M. 1989. The dangers of eating raw fish. N. Engl. J. Med. 320: 1143–1145.

Schantz, E.J. 1984. Historical perspective on paralytic shellfish poison. *In:* E.P. Ragelis (ed.). Seafood toxins. American Chemical Society, Washington DC.

Schecter, A., O. Päpke, K.-C. Tung, D. Staskal and L. Birnbaum. 2004. Polybrominated diphenyl ethers contamination of United States food. Environmental Science and Technology. 38(20): 5306–5311.

Scheuer, P.J., W. Takahashi, J. Tsutsumi and T. Yoshida. 1967. Ciguatoxin: isolation and chemical nature. Science. 155(58): 1267–1268.

Schnorf, H., M. Taurarii and T. Cundy. 2002. Ciguatera fish poisoning: a double-blind randomized trial of mannitol therapy. Neurology. 58(6): 873–80.

Schwab, K.J., F.H. Neill, M.K. Estes et al. 1998. Distribution of Norwalk virus within shellfish following bioaccumulation and subsequent depuration by detection using RT-PCR. J. Food Prot. 61: 1674.

Scoging, A. and M. Bahl. 1998. Diarrhetic shellfish poisoning in the UK. Lancet. 352: 117.

Scully, R.E. 1990. Case records of the Massachusetts General Hospital. N. Engl. J. Med. 323: 467.

Sechet, V., P. Safran, P. Hovgaard and T. Yasumoto. 1990. Causative species of diarrhetic shellfish poisoning (DSP) in Norway. Mar. Biol. 105: 269–274.

Shalaby, A.R. 1996. Significance of biogenic amines to food safety and human health. Food Research International. 29(7): 675–690.

Shumway, S.E., H.P. van Egmond, J.W. Hurst and L.L. Bean. 1995. pp. 433–461. *In:* G.M. Hallegraeff, D.M. Anderson, A.D. Cembella and H.O. Enevoldsen (eds.). Manual on harmful marine microalgae, Manuals and Guides No. 33, UNESCO.

Sim, J. and N. Wilson. 1997. N.Z. Public Health Rep. 4: 9.

Smith, A.G. and S.D. Gangolli. 2002. Organochlorine and chemicals in seafood: Occurrence and health concerns. Food and Chemical Toxicology. 40: 767–779.

Sobel, J. and J. Painter. 2005. Illnesses Caused by Marine Toxins. Clin. Infect. Dis. 41(9): 1290–1296.

Sommer, H. and K.F. Meyer. 1937. Paralytic shellfish poisoning. Arch. Pathol. 24: 560–598.

Sommer, H., W.F. Whedon, C.A. Kofoid and R. Stohler. 1937. Arch. Pathol. 24: 537.

Statake, M., K. Ofuji, H. Naoki, K.J. James, A. Furey and T. McMahon. 1998. Azaspiracid, a new marine toxin having unique spiro ring assembles, isolated from Irish mussels, Mytilus edulis. J. Am. Chem. Soc. 120: 9967–9968.

Stewart, I., R.J. Lewis, G.K. Eaglesham, G.C. Graham, S. Poole and S.B. Craig. 2010. Emerging tropical diseases in Australia. Part 2. Ciguatera fish poisoning. Ann. Trop. Med. Parasitol. 104(7): 557–71.

Stoker, T.E., S.C. Laws, K.M. Crofton, J.M. Hedge, Ferrell and R.L. Cooper. 2004. Assessment of DE-71, a commercial polybrominated diphenyl ether (PBDE) mixture, in the EDSP male and female pubertal protocols. Toxicological Sciences. 78(1): 144–155.

Storelli, M.M. 2008. Potential human health risks from metals (Hg, Cd, and Pb) and polychlorinated biphenyls (PCBs) via seafood consumption: estimation of target hazard quotients (THQs) and toxic equivalents (TEQs). Food Chem. Toxicol. 46(8): 2782–88.

Storelli, M.M., R. Giacominelli-Stuffler, A. Storelli and G.O. Marcotrigiano. 2003. Polychlorinated biphenyls in seafood: contamination levels and human dietary exposure. Food Chem. 82(3): 491–6.

Svensson, L. 2000. Diagnosis of food-borne viral infections in patients. Int. J. Food Microbiol. 59: 117–126.

Svensson, B.G., A. Nilsson, M. Hansson et al. 1991. Exposure to dioxins and dibenzofurans through the consumption of fish. N. Engl. J. Med. 324: 8.

Tachibana, K., M. Nukina, Y.D. John and P.J. Scheuer. 1987. Recent developments in the molecular structure of ciguatoxin. Biol. Bull. 172: 122–127.

Tachibana, K., P.J. Scheuer, Y. Tsukitani, H. Kikuchi, D. Van Engen, J. Clardy, Y. Gopichand and J. Schmitz. 1981. Okadaic acid, a cytotoxic polyether from two marine sponges of the genus Halichondria. J. Am. Chem. Soc. 103: 2469–2471.

Tang, Y.W., J.X. Wang, Z.Y. Xu, Y.F. Guo, W.H. Qian and J.X. Xu. 1991. A serologically confirmed, case-control study, of a large outbreak of hepatitis A in China, associated with consumption of clams. Epidemiol. Infect. 107: 651–657.

Taniyama, S., O. Arakawa, M. Terada, S. Nishi, T. Takatani, Y. Mahmud and T. Noguchi. 2003. *Ostreopsis* sp., a possible origin of palytoxin (PTX) in parrotfish Scarus ovifrons. Toxicon. 42: 29–33.

Taylor, F.J.R., Y. Fukuyo, J. Larsen and G.M. Hallegraeff. 2003. Taxonomy of harmful dinoflagellates. pp. 389–432. *In:* G.M. Hallegraeff, D.M. Anderson and A.D. Cembella (eds.). Manual on harmful marine microalgae. UNESCO Publishing, Paris.

Taylor, S.L. and E.R. Lieber. 1979. *In vitro* inhibition of rat intestinal histaminemetabolizing enzymes. Food Cosmet. Toxicol. 17(3): 237–240.

Taylor, S.L. and R.K. Bush. 1988. Allergy by ingestion of seafood. pp. 149–183. *In:* A.T. Tu (ed.). Handbook of Natural Toxins, Marine Toxins and Venoms. Marcel Dekker. New York.

Taylor, S.L. 1986. Histamine poisoning: toxicology and clinical aspects. CRC Critical Reviews in Toxicology. 17(2): 91–128.

Teitelbaum, J.S., R.J. Zatorre, S. Carpenter, D. Gendron, A.C. Evans, A. Gjedde and N.R. Cashman. 1990. Neurologic sequelae of domoic acid intoxication due to the ingestion of contaminated mussels. N. Engl. J. Med. 322: 1781–1787.

Terao, K. 2000. Ciguatera toxins: toxicology. pp. 449–472. *In:* L. Botana (ed.). Seafood and Freshwater Toxins: Pharmacology, Physiology and Detection. Marcel Dekker, New York.

Terzagian, R. 2006. Five Cluster of Neurotoxic Shellfish Poisoning (NSP) in Lee County, July 2006. Florida Department of Health Epi Updates. The Netherlands. 515–526.

Thomas, Craig, M.D. and S. Scott. 1997. All Stings Considered: First Aid and Medical Treatment of Hawai'i's Marine Injuries. University of Hawai'i Press, Hawaii, 120 pp.

Tillmann, U., M. Elbrachter, B. Krock, U. John and A. Cembella. 2009. Azadinium spinosum gen. et sp. nov (Dinophyceae) identified as a primary producer of azaspiracid toxins. Eur. J. Phycol. 44: 63–79.

Todd, Ewen. 1993. Domoic Acid and Amnesic Shellfish Poisoning—a review. Journal of Food Protection. 56: 69–83.

Todd, K. 2002. A Review of NSP Monitoring in New Zealand in Support of a New Programme. Cawthron Institute; Nelson, New Zealand. Cawthron Report No. 660.

Torgersen, T., N.B. Bremnes, T. Rundberget and T. Aune. 2008. Structural confirmation and occurrence of azaspiracids in Scandinavian brown crabs (Cancer pagurus). Toxicon. 51: 93–101.

Torres, P., R. Franjola, J.C. Weitz et al. 1993. New records of human diphyllobothriasis in Chile (1981–1992), with a case of multiple Diphyllobothrium latum infection. Boletin Chileno de Parasitologia. 48: 39.

Toyofuku, H. 2006. Joint FAO/WHO/IOC Activities to Provide Scientific Advice on Marine Biotoxins (research report) Marine Poll. Bull. 52: 1735–1745.

Tryphonas, H. 1995. Immunotoxicity of PCBs (Aroclors) in relation to Great Lakes. Environmental Health Perspectives. 103 (S9): 35–46.

Tryphonas, H. 1989. The impact of PCBs and dioxins on children's health: immunological considerations. Canadian Journal of Public Health. 89(1): S49–S52, S54–S57.

Tsai, Y.H., H.F. Kung, H.C. Chen, S.C. Chang, H.H. Hsu and C.I. Wei. 2007. Determination of histamine and histamine-forming bacteria in dried milkfish (Chanos chanos) implicated in a food-borne poisoning. Food Chemistry. 205: 1289–1296.

Tubaro, A., S. Sosa, G. Altinier, M.R. Soranzo, M. Satake, R. Della Loggia and T. Yasumoto. 2004. Short-term toxicity of homoyessotoxins, yessotoxin and okadaic acid in mice. Toxicon. 43: 439–445.

Tubaro, A., S. Sosa, M. Carbonatto, G. Altinier, F. Vita, M. Melato, M. Satake and T. Yasumoto. 2003. Oral and intraperitoneal acute toxicity studies of yessotoxin and homoyessotoxins in mice. Toxicon. 41: 783–792.

Uemura, D. 1991. Bioactive polyethers. Bioorganic Marine Chemistry. P.J. Scheuer. Springer-Verlag Berlin Heidelberg.

UK SACN 2004. Advice on fish consumption: benefits & risks, 2004, for the Food Standards Agency and the Department of Health United Kingdom.

UNEP Global Environmental Facility. 2003. Regionally Based Assessment of Persistent Toxic Substances. Global Report. Geneva, Switzerland.

United States General Accounting Office. 1996a. Report to Congressional Committees: Food Safety. Information on Food-borne Illnesses. 31 pp.

US EPA 2010. Persistent bioaccumulative and toxic chemical program. Dioxins and furans.

US EPA. 2001. Mercury Update: Impact on Fish Advisories. EPA-823-F-01-011.

US EPA. 2001. Water Quality Criterion for the Protection of Human Health: Methylmercury. Risk Assessment for Methylmercury. Washington DC.

US EPA. 2000a. Exposure and Human Health Reassessment of tetrachlorodibenso-p-dioxin (TCDD) and Related Compounds. Draft Final Report. Washington DC.

US EPA. 1999a. Polychlorinated Biphenyls (PCBs) Update: Impact on Fish Advisories. EPA-823-F-99-019.

US EPA. 1999b. Polychlorinated Dibenzo-p-dioxins and Related Compounds Update: Impact on Fish Advisories. EPA-823-F-99-015.

USFDA. 1984. Bacteriological analytical methods, 6th edition. pp. 28.03. Association of Official Analytical Chemists. Arlington, VA.

Van Egmond, H.P., M.E. Van Apeldoorn and G.J.A. Speijers. 2004. Food andAgriculture Organization of the united nations. Marine Biotoxins.

Varnam, A.H. and M.G. Evans 1991. Foodborne Pathogens. Wolfe Publishing Ltd.

Venkitanarayanan, K.S. and P. Doyle. 2002. Food-borne infections and infestations. *In:* C. Berdanier (ed.). Handbook of Food and Nutrition. CRC Press, Boca Raton.

Venugopal, V. 2009. Safety hazards with Marine Products and their control. pp. 467–494. *In:* Marine products for healthcare: Functional and Bioactive Nutraceutical Compounds from the Ocean. CRC Press, Taylor & Francis Group, LLC.

Virtanen, J.K., T.H. Rissanen, S. Voutilainen et al. 2007. Mercury as a risk factor for cardiovascular diseases. The Journal of Nutritional Biochemistry. 18: 75–85.

Viviani, R. 1992. Eutrophication, Marine Biotoxins, Human Health. Sci. Total Environ. 631–662.

Wang, D.Z. 2008. Neurotoxins from marine dinoflagellates: A brief review. Mar. Drugs. 6: 349.

Weihe, P., P. Grandjean, F. Debes et al. 1996. Health implications for Faroe islanders of heavy metals and PCBsfrom pilot whales. The Science of The Total Environment. 186: 141–148.

Whanger, P.D. 1985. Metabolic interactions of selenium with cadmium, mercury, and silver. Advances in Nutritional Research. 7: 221–250.

WHO Surveillance Programme. 1995. Sixth report of WHO surveillance programme for control of foodborne infections and intoxications in Europe. FAO/WHO Collaborating Centre for Research and Training in Food Hygiene and Zoonoses, Berlin.

Willett, W.C. 2006. Fish: Balancing health risks and benefits. American Journal of Preventive Medicine 29(4): 320–321.

Yasumoto, T. and M. Murata. 1993. Marine toxins. Chem. Rev. 93(5): 1897–909.

Yasumoto, T. and M. Murata. 1990. Polyether toxins involved in seafood poisoning. *In:* S. Hall and G. Stricharty (eds.). Marine Toxins. Origin, Structure and Molecular Pharmacology. American Chemical Society, Washington DC.

Yasumoto, T., I. Nakajima, R. Bagnis and R. Adachi. 1977. Finding of a dinoflagellate as a likely culprit of ciguatera. Jpn. Soc. Sci. Fish. 43: 1021–1026.

Yasumoto, T., M. Satake, M. Fukui, H. Nagai, M. Murata and A.M. Legrand. 1993. A turning point in ciguatera study. pp. 455–461. *In:* T.J. Smayda and Y. Shimizu (eds.). Toxic Phytoplankton Blooms in the Sea. Elsevier, New York.

Yoshikawa-Ebesu, J.S.M. 2001. Tetrodotoxin. pp. 253–286. *In:* Y.H. Hui, D. Kitts and P.S. Stanfield (eds.). Foodborne disease handbook. 2nd ed., Vol. IV: Seafood and environmental Toxins, Marcel Dekker, New York, Basel.

Yoshizumi, M., H. Houchi, Y. Ishimura, Y. Masuda, K. Morita and M. Oka. 1991. Mechanism of palytoxin induced Na+ influx into cultured bovine adrenal chromaffin cells: possible involvement of Na+/H+ exchange system. Neurosci. Left. 130: 103–l06.

Yoshizumi, M., M. Minocci, A. Beran and L. Ivena. 2007. First record of Ostreopsis cfr. Ovata on macroalgae in the northern Adriatic Sea. Mar. Pol. Bull. 54: 598–601.

Zwahlen, A., M.H. Blanc and M. Robert. 1977. Epidémie d'intoxication par les moules ("paralytic shellfish poisoning"). Schweiz. Med. Wschr. 107: 226–230.

Index

16S rRNA gene sequencing 156, 157

A

Agar 361–364, 366–370, 372
Agar powder 368, 370
Agar sheets 368, 370
Agarophyte 361
Alginate 361, 364–366, 370–372
Alginic acid 371, 372
Alginophyte 361
Alkali treated cottonii (ATC) 369, 370
ANN 242, 243, 248, 249
Anticancer 415, 419, 420
Antidiabetes 418
Antiinflamation 417
Antimicrobial 406, 414
Antimicrobial compounds 188
Antioxidant 406, 407, 416, 418, 419
Antioxidant activity 354–356
Antiviral 415
application 170–181
Aquaculture 1–3, 5–8, 203, 207–209, 234
Arsenic 276–351
Arsenobetaine 278, 280, 300, 312, 314, 315, 329
Arsenocholine 278, 280
Arsenosugars 279, 280, 291, 295–302, 304, 306, 309–313, 320, 322, 327, 329, 330, 342, 343
Astaxanthin 264, 265, 404, 407–409, 416–420

B

β-carotene 264–266
Bacteria 458–482
Bacterial classification 157
Bacterial fermentation 352–358
Bacterial identification 134–139, 141, 142, 145, 146, 148, 153, 157, 160, 161
Bacteriocins 186, 189–193, 195, 196

Bioaccessibility 317–323, 327
Bioactive compounds 261–275
Bioavailability 276–351
Biological activity 397–432
Biomarkers 137
Biopreservation 183, 188, 192–195, 471
Biotechnology 214
BP 249
Breaking force 120, 123, 124, 126
Brochothrix 464
Brown algae 352–358
Brown algae (*Phaephyceae*) 361
By-products 376–396

C

Callinectes sapidus 8
Cantaxanthin 265
Carnobacteriocins 193
Carnobacterium 461, 462, 465, 467, 468, 471, 472
Carotenoids 263–266
Carrageenan 361, 362, 366, 369–372
Carrageenan powder 370
Carragenophyte 361, 369, 370
Cephalopods 6
Chamois 80–89
Chamois leather 80–89
Chemical contaminants 483, 484, 525, 551
Chilled 248
Chilling 235, 236, 246
Chitin 269–272, 402–406, 409, 414–417, 419, 421, 422
Chitosan 211–231, 399, 403–406, 414–421
Chlorella vulgaris 264, 267
Chlorophyta 9
Clam 51, 52, 56
Clarias sp. 176, 177
Clostridium 460, 468, 470, 474
Cod liver oil 80, 81, 84
Cold Chain Management 249

Color 237, 239, 240, 243, 245, 251
Cooking 285, 316, 320, 327
Cooking method 47–79
co-planner PCBs (Co-PCBs) 202–207, 209
Crab 53, 57, 59, 68, 74
Crustacean 53
Crustaceans 1–3, 5, 7–9, 11

D

Debaryomyces hansenii 19, 31
Deformation 120, 123, 124, 126
Delivery 213–215, 218, 224, 226, 228
Demersal 4
DHA 267, 268
Diadromous 2, 4
Dinoflagellates toxins 491
Dioxin-reduced treatment 204
Dioxins 202, 203, 205, 207–209
DMA 279, 280, 289–306, 309, 311, 313, 316,
        318, 322, 326–333, 336–341, 343
Docosachexenoic acid (DHA) 4

E

Echinoderms 1, 10
Eicosapentaenoic acid (EPA) 4, 267, 268
Enterocins 191, 194
Enzyme inhibitors 452
Ethanol fermentation 15, 19, 39, 41, 42
Euthynnus spp. 175, 176, 178, 179
Evaluation 232–260

F

Fatty acids 81, 82, 84, 85
Fermentation 14–46, 182, 183, 186–188, 482
Fiber 48, 58
Finfish shellfish 549
Fish 54, 59, 81, 458–465, 467–474
Fish muscle structural proteins 95, 112
Fish oil 81, 204
Fish protein 117–131
Food industry 377–379, 381–383, 385–389,
        391
Foodborne pathogens 133, 147–149
Freezing 233, 236–238, 240, 245, 246, 248,
        254, 255
Freezing rate 236, 237, 248
fresh water fish 171, 176
Frozen 233, 235, 237, 239, 240, 245, 248–250,
        252, 254, 256
Frozen stability 126, 127
Fucoidan 269, 271
Functional foods 385

G

GANN 249
Gastropods 6
Gel strength 367–370
gelling properties 171, 172, 176, 177
Green algae (*Chlorophyceae*) 361
Growth performance 205

H

Halophilic lactic acid bacteria 24, 25
Health benefits 47–79
Histamine 467–469, 472
hydrolysate cassava starch 172
hydrolysate of surimi liquid waste 172, 174

I

Image analysis e-nose 242
Immunoenhancement 262, 263, 265, 267,
        269, 271–272
Immunostimulants 263
In Vitro 318–328, 342
In Vivo 277, 318, 319, 328
Inorganic arsenic 277, 279, 280, 282, 288,
        301, 307, 316, 320–322, 328–331, 339
Iodine values 85, 86
Ionic strength 122, 123, 127
Isoelectric precipitation 117
Isopropyl alcohol (IPA) 370

K

Kelps 48

L

Lactic acid bacteria 182–201, 461
Lactic acid fermentation 15, 19, 22, 24, 32,
        33, 36, 39
Lactobacillus 182–185, 187–189
Lactobacillus brevis 19, 23
Laminaran 270, 271
liquid waste tofu 172, 173
Listeria 468, 469, 472
low economical marine fish (by catch) 171
Lysinoalanine 124, 126

M

Macroalgae 276, 277
MALDI-TOF MS 137–142, 145–150, 152–161
Marine 47–51, 58, 71
Marine algae 263, 271
Marine oils 81, 82, 85, 86

Marine silage 36
Mass spectral database 166
Mass spectral fingerprinting 160, 161
Maternal 433, 434, 438, 439, 441, 443–447, 449, 452
Mechanisms 437, 439
Mercury 434, 435, 438, 439, 448, 452
Mercury toxicity 439
Metacarcinus magister 8
Methylmercury 433, 438, 440, 452
Microalgae 276, 277
Microbial ecology 183
Microbiological 238, 239, 241
Microbiological risks 534
Microbiota 459–462, 465, 474
minced fish 170, 171, 174–179
MMA 280, 290–293, 296–299, 301, 306, 309, 311, 316, 322, 327–329, 331–333, 336–340
Molluscs 1–3, 5–7, 11, 51, 53
MTGase (microbial transglutaminase) 170, 172–179
Myofibrillar protein 121–123

N

Nanocomposite 211–231
Nanoencapsulation 214
Nisin 189, 190, 193, 194
Nutraceuticals 377, 385, 386
Nutrients 47, 48, 52, 75
Nutrition 382

O

Ocean fish 433–436, 440–446, 449
Oil tannage 80–89
Optimization 232, 243, 244, 248
Oreochronis niloticus 176, 177
Organic pollutants 525, 528
Oyster 51, 55, 59, 61, 62

P

Packaging film 218
Pagrus major 203
Pangasius sp. 177, 178
Parasites in seafood 541
Pelagic 4, 11
Perishability 234, 244, 245
Ph shifting process 117–119, 124–126
Phaeophyta 9
Photobacterium 460, 462, 464, 465, 474
Phylogenetics 136, 147, 148, 152, 157, 159, 160

Phyloproteomics 147
Pilot whales 441, 442, 445
Poisoning syndromes 484, 492
polychlorinated biphenyls (PCBs) 202–207, 209
Polychlorinated dibenzodioxins (PSDDs) 202
polychlorinated dibenzofurans (PCDFs) 202, 204–207
Polyunsaturated fatty acids (PUFA) 4, 7
Porphyra yezoensis 17, 33, 37
Prawn 50
Predictive modeling 244
Preservation 229
Priacanthus macracanthus 174, 175
Probiotics 194
Production 170–181
Protein 48, 53, 54, 58–61, 63, 66, 67, 69, 73, 74
Protein dispersion 94, 98, 101, 107, 108, 110, 111
Protein films and coatings 93
Provitamin A 265
Public health 500, 521, 527, 533, 534, 546, 548, 550–553

Q

Quality 211–213, 232–260
Quality Index 238

R

Rancidity control 91, 111
Rate of Thawing 248
Red algae (*Rhodophyceae*) 361
Red sea bream 202–210
Refined Carrageenan (RC) 369, 370
RFID 250, 252, 253
Rhodophyta 9
Risks of natural toxins 484, 490
Risks reducing measures 546

S

S. ladakanum NRL 3191 172–174
Saccharides 269
Safety 233–235, 238, 239, 251, 468, 471
Safety hazards 490
Sarcoplasmic protein 119–123
Scallop 53, 57, 59, 69–71
Sea cucumber 50, 51, 55, 59–61, 71, 72
Seafood 47–79, 211–231, 376–396, 433–482, 483–570
Seafood allergens 544

Seafood consumers 528, 549
Seafood products   182–201
Seafood quality 165
Seafood safety 135, 148, 149, 157
Seafood spoilage 132, 133, 148
Seagrass seed 39–41
Seaweed 14–46, 49, 50, 53, 276–351
Seaweed culture 359, 360, 365, 372
Seaweed sauce 33
Security, supply chain 234, 250, 251
Selenium 433–457
Selenium physiology 435, 437
Selenocysteine 435, 436
Selenoenzymes 436–439, 447
Semirefined carrageenan (SRC) 369, 370
Shelf life enhancement 105
Shewanella 460–465, 474
Shrimp 50–53, 55, 56, 59, 63–66, 73, 232–260,
    397–432
Shrimp alkaline phosphatase 411
Single cell detritus 27, 30
Sodium alginate 270–272
Softness 86
Speciation 276–351
SpectraBank 142–144, 148, 153, 154, 158,
    160, 161
Spoilage 183, 185, 186, 188, 189, 191, 459,
    462–468, 471–474
Spoilage bacteria 133, 142, 148, 150, 161
Super Chilling 236
Supply Chain 232–236, 241, 244–246, 249–
    252, 254, 256, 257
surimi liquid waste 172–174
SVM 249

T

Tanning 80–82, 84–86, 88
Tetragenococcus halophilus 24

Texture 235, 237, 239–242
Thawing 237, 240, 245, 248
Thermostable proteins 102, 103
Time Temperature monitoring (TTI) 249
Total volatile basic nitrogen (TVB-N) 235,
    236, 238, 239, 243, 464
Traceability 232–260
Trymethylamine (TMA) 236, 238, 316, 329,
    336, 337, 462–464, 467, 473, 474

U

Ulva 19, 20, 32, 33, 37–40
Undaria pinnatifida 20, 40
Urine 279, 318, 328–337, 340–342

V

Valorization 399
Vibrio 460–462, 465, 467, 468, 470, 474
Viral risks 439
Viscosity value 370, 371
Vitamin 60, 62, 64, 67, 70, 72–75

W

Waste 212, 213, 397–432

X

Xanthophyll 263

Z

Zeaxanthin 264, 265
Zostera marina 20, 40

About the Editor

Professor Se-Kwon Kim, Ph.D., currently serves as a distinguished Professor in the Department of Marine-bio Convergence Science in Specialized Graduate School of Convergence Science and Technology and also acts as a director of the Marine Bioprocess Research Center (MBPRC) at Pukyong National University in the Republic of Korea. He received his BSc, MSc, and PhD from the Pukyong National University and joined as a faculty member. He has previously served as a scientist in the University of Illinois, Urbana-Champaign, Illinois (1988–1989), and was a visiting scientist at the Memorial University of Newfoundland, Canada (1999–2000).

Professor Se-Kwon Kim was the first president of the Korean Society of Chitin and Chitosan (1986–1990) and the Korean Society of Marine Biotechnology (2006–2007). He was also the chairman for the *7th Asia-Pacific Chitin and Chitosan Symposium,* which was held in South Korea in 2006. He is one of the board members of the International Society of Marine Biotechnology and the International Society for Nutraceuticals and Functional Foods. Moreover, he was the editor in chief of the *Korean Journal of Life Sciences* (1995–1997), the *Korean Journal of Fisheries Science and Technology* (2006–2007), and the *Korean Journal of Marine Bioscience and Biotechnology* (2006–present). His research has been credited with the best paper award from the American Oil Chemist's Society (AOCS) and the Korean Society of Fisheries Science and Technology in 2002.

Professor Se-Kwon Kim's major research interests are investigation and development of bioactive substances derived from marine organisms and their application in oriental medicine, nutraceuticals, and cosmeceuticals via marine bioprocessing and mass-production technologies. He has also conducted research on the development of bioactive materials from marine organisms for applications in oriental medicine, cosmeceuticals, and nutraceuticals. To date, he has authored over 650 research papers and holds 140 patents. In addition, he has written or edited more than 50 books.

Color Plate Section

Chapter 1

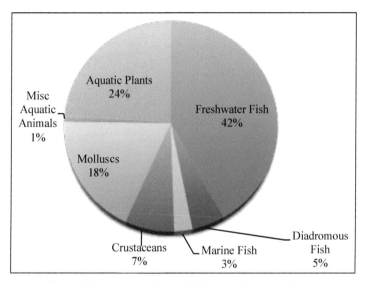

Figure 1. Proportion of total global aquaculture production.

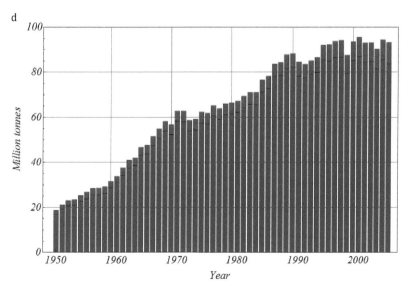

Figure 3. Marine (blue) and inland water wild fish catches 1950–2005.

Chapter 2

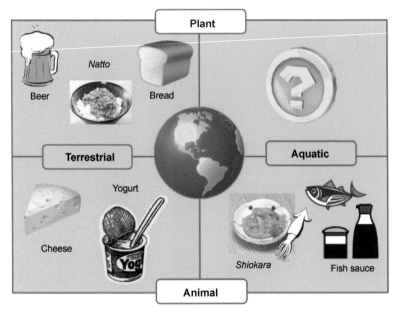

Figure 1. Categories of fermented foods based on raw materials (Uchida and Miyoshi 2012). Fermented foods prepared from aquatic plants (algae) are yet to be developed.

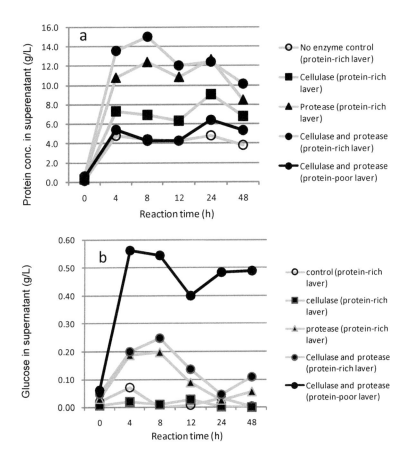

Figure 2. Production of soluble protein and glucose from laver after enzyme treatment (unpublished). Three grams of laver were mixed with or without enzymes, 0.1 g cellulase and/or 0.14 g protease, and total weight was adjusted to 50 g with distilled water. Protein-rich (23.2% protein) and -poor (14.9% protein) laver was used. The mixtures were reacted for 48 h at 50°C with moderate agitation. Protein (a) and glucose contents (b) in the supernatant were measured using commercial kits (Pierce BCA Protein Assay Kit and F kit, respectively). Data is shown as a mean of triplicate experiments.

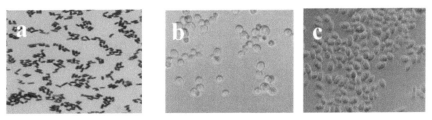

Figure 3. Microorganisms initially isolated from an algal fermented culture as a starter culture (Uchida and Murata 2004). A bacterium (a: *Lactobacillus brevis*) and two kinds of yeast (b: *Debaryomyces hansenii*, and c: *Candida zeylanoides*-related yeast).

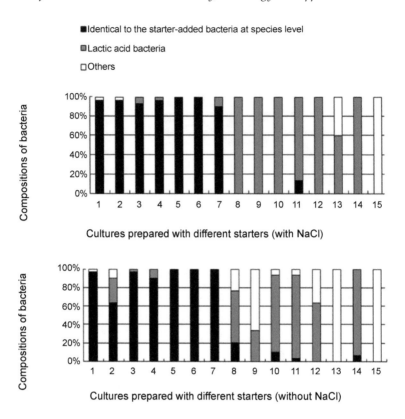

■Identical to the starter-added bacteria at species level

▨Lactic acid bacteria

☐Others

Cultures prepared with different starters (with NaCl)

Cultures prepared with different starters (without NaCl)

Figure 4. Results of the test for examining suitable starter culture of lactic acid bacteria for seaweed fermentation (Uchida et al. 2007). To prepare cultures with NaCl, 2.0 g of the commercial product of *Undaria* powder (Hamamidori, Riken Shokuhin) was mixed with 40 ml of autoclaved 3.5 % (wt/vol) NaCl solution, 40 mg of cellulase (12S, Yakult Honsha Co., Ltd.) and 0.4 ml of bacterial cell suspension. The bacterial cell suspension was prepared for the 14 LAB strains: No. 1; *Lactobacillus brevis* FRA 000033, 2; *Lact. brevis* IAM 12005, 3; *Lact. plantarum* ATCC 14917T, 4; *Lact. plantarum* IAM 12477T, 5; *Lact. casei* IFO 15883T, 6; *Lact. casei* FRA 000035, 7; *Lact. rhamnosus* IAM 1118T, 8; *Lact. zeae* IAM 12473T, 9; *Lact. acidophilus* IFO 13951T, 10; *Lact. kefir* NRIC 1693T, 11; *Lact. fermentum* ATCC 14931T, 12; *Lact. delbrueckii* subsp. *bulgaricus* ATCC 11842T, 13; *Streptococcus thermophilus* NCFB 2392, and 14; *Leuconostoc mesenteroides* IAM 13004T. These strains were pre-cultured with MRS medium (Merck Co.), collected by centrifuge (8,000 g x 20 min.), washed with autoclaved 0.85% NaCl solution, re-suspended to make a concentration of O.D.660 nm = 1.0 (containing 7.3 x 10^7–1.1 x 10^9 CFU/ml), and then used. To prepare culture without NaCl, autoclaved distilled water was used instead of 3.5% NaCl solution. Cultures without the inoculation of lactic acid bacteria were prepared as being without starter culture controls (No. 15). After incubating for 11 days at 20°C, the microbial composition was investigated. Ten colonies, each formed on the SMA plates prepared for viable counting, were chosen at random from the triplicated trials (Total n = 30), and then transferred to the BCP plates with the % proportion of yellow-colored colonies shown as average ± SE of lactic acid bacteria. *Lact. brevis*, (Trial Nos. 1, 2), *Lact. plantarum* (2, 4), *Lact. casei*, (5, 6), and *Lact. rhamnosus* (7) showed marked ability to be dominant in the *Undaria* cultures.

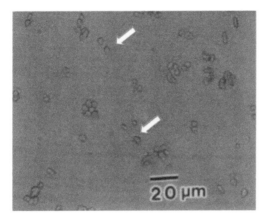

Figure 5. Microscopic observation of single cell detritus (SCD) prepared from *Undaria pinnatifida* (arrows, photo after 8 days of incubation) (Uchida et al. 2004a).

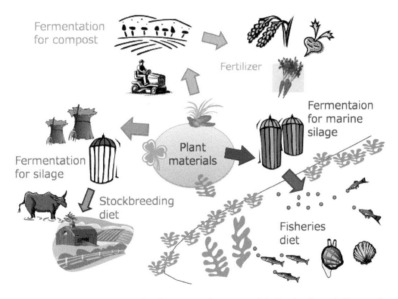

Figure 8. Use of fermentation technologies on plant materials for foods and diet production (Uchida and Miyoshi 2010). Fermentation is rarely or never used in the fisheries industry, and Uchida proposed a conceptual diet of marine silage.

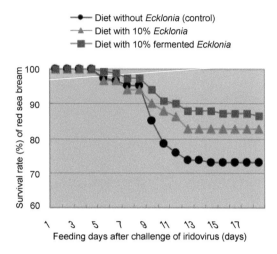

Figure 9. Results of the feeding test for red sea bream (Technical research association for new food creation 2005). The fish were challenged by iridovirus and then fed a control diet, a diet containing *Ecklonia* sp. at 10% wt/wt, and a diet containing fermented *Ecklonia* sp. at 10% wt/wt.

Chapter 3

<u>Sea cucumber</u>

Total Weight of Sea Cucumber: 100 g

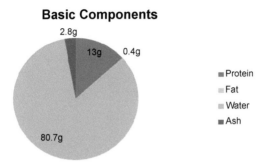

Figure 1. Shows the basic components in 100 g of Sea cucumber.

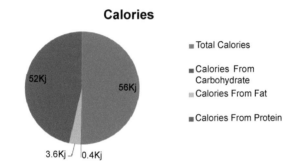

Figure 2. Shows the amount of calories in 100 g of sea cucumber.

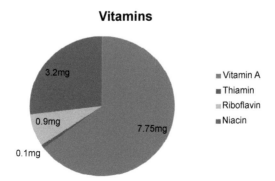

Figure 3. Shows the contents of vitamins in 100 g of sea cucumber.

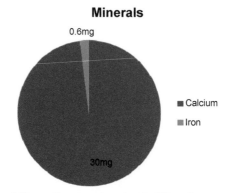

Figure 4. Shows the mineral content in 100 g of sea cucumber.

Oyster
Amount of Oyster (raw): 1 cup
Total Weight of Oyster (raw): 248 g

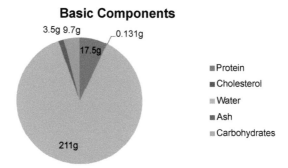

Figure 5. Shows the basic components in 248 g of oyster.

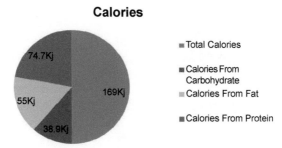

Figure 6. Shows the amount of calories in 248 g of oyster.

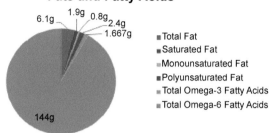

Figure 7. Shows the contents of fats and fatty acids in 248 g of oyster.

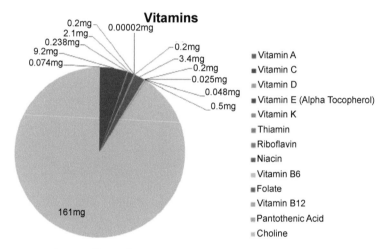

Figure 8. Shows the contents of vitamins in 248 g of oyster.

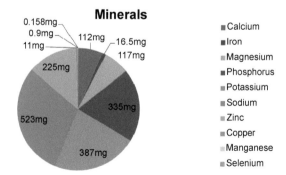

Figure 9. Shows the contents of minerals in 248 g of oyster.

Shrimp
Amount of Shrimp: 4.00 oz.
Total Weight of Shrimp: 113.40 g

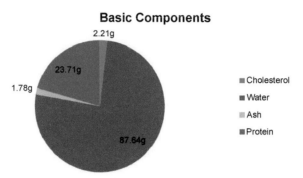

Figure 10. Shows the contents of basic components in 113.40 g of shrimp.

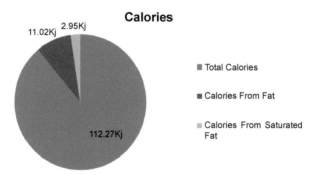

Figure 11. Shows the contents of calories in 113.40 g of shrimp.

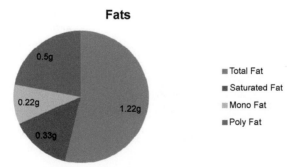

Figure 12. Shows the contents of fats in 113.40 g of shrimp.

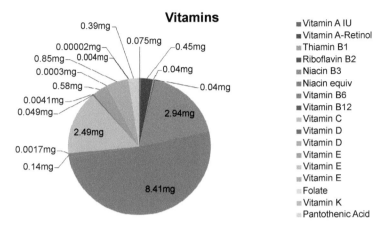

Figure 13. Shows the contents of vitamins in 113.40 g of shrimp.

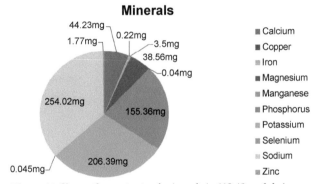

Figure 14. Shows the contents of minerals in 113.40 g of shrimp.

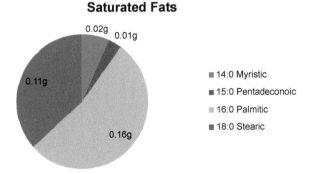

Figure 15. Shows the contents of saturated fats in 113.40 g of shrimp.

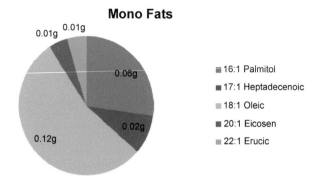

Figure 16. Shows the contents of mono fats in 113.40 g of shrimp.

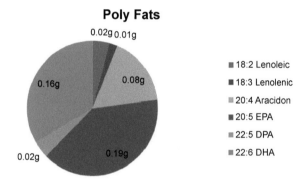

Figure 17. Shows the contents of poly fats in 113.40 g of shrimp.

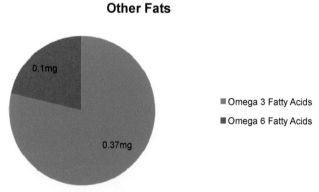

Figure 18. Shows the contents of other fats in 113.40 g of shrimp.

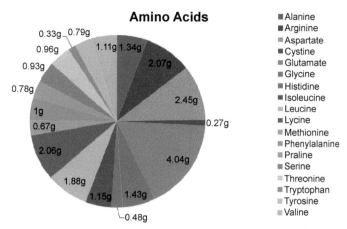

Figure 19. Shows the contents of amino acids in 113.40 g of shrimp.

Clams
Amount of Clams: 227 g (1 cup)

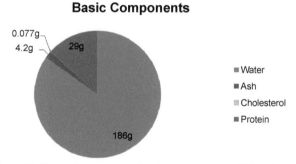

Figure 20. Shows the contents of basic components in 227 g of clams.

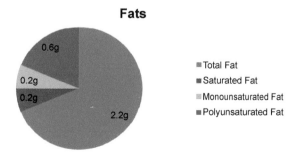

Figure 21. Shows the contents of fats in 227 g of clams.

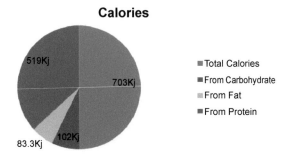

Figure 22. Shows the contents of calories in 227 g of clams.

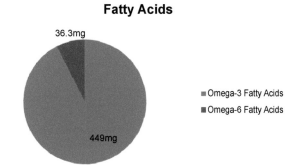

Figure 23. Shows the contents of fatty acids in 227 g of clams.

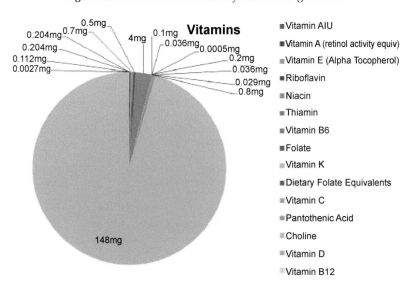

Figure 24. Shows the contents of vitamins in 227 g of clams.

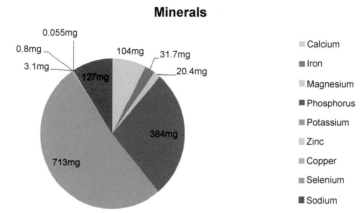

Figure 25. Shows the contents of minerals in 227 g of clams.

Crab
Amount of Crab: 85 g

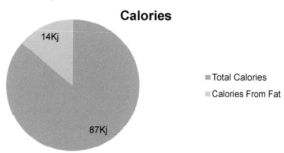

Figure 26. Shows the contents of calories in 85 g of clams.

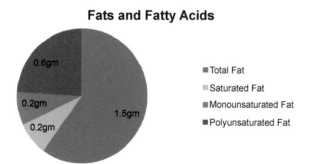

Figure 27. Shows the contents of fats and fatty acids in 85 g of clams.

Protein and Minerals

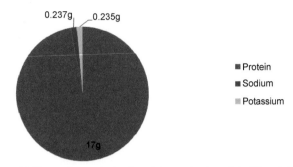

Figure 28. Shows the contents of proteins and minerals in 85 g of clams.

Scallop
Amount of Scallop: 4.00 oz-wt
Total Weight of Scallop: 113.40g

Basic Components

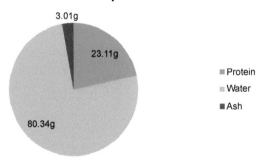

Figure 29. Shows the contents of basic components in 113.40 g of scallop.

Calories

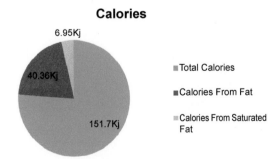

Figure 30. Shows the contents of calories in 113.40 g of scallop.

Fats

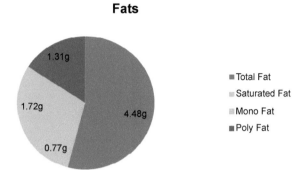

Figure 31. Shows the contents of fats in 113.40 g of scallop.

Vitamins

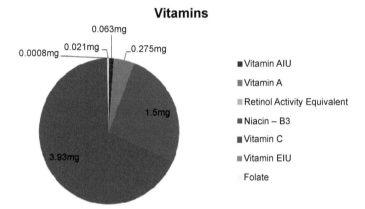

Figure 32. Shows the contents of vitamins in 113.40 g of scallop.

Fats

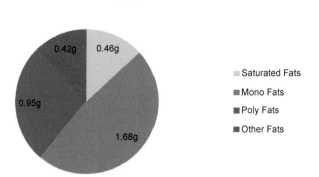

Figure 33. Shows the contents of fats in 113.40 g of scallop.

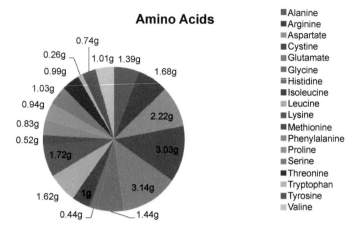

Figure 34. Shows the contents of amino acids in 113.40 g of scallop.

Chapter 7

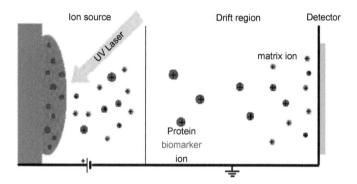

Figure 1. Principle of MALDI-TOF mass spectrometry.

Chapter 11

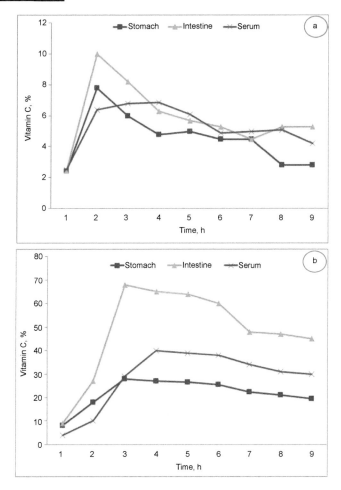

Figure 4. Controlled release of vitamin C in rainbow trout. (a) rainbow trout fed with feed supplemented with vitamin C, and (b) fish fed with feed supplemented with chitosan nanoparticles loaded with vitamin C. Adapted from Alishahi et al. (2011a,b).

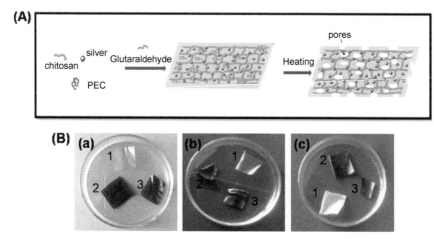

Figure 7. (A) Schematic illustration of preparation of porous chitosan–silver nanocomposite (PCSSNC) films. (B) (1) chitosan, (2) chitosan–silver nanocomposite, and (3) porous chitosan–silver nanocomposite films, (a–c) different composition of chitosan:PEG, 1:1, 2:1, and 5:1, respectively. Adapted from Vimala et al. (2010) with permission.

Chapter 12

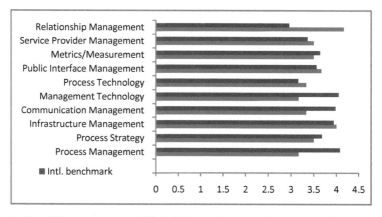

Figure 2. Capability comparison of Thai shrimp industry and international benchmark of food supply chain security. Ahmad and Komolavanij (2010).

Chapter 18

Figure 1. Volume and value of Indonesian shrimp export from 2002 to 2011 (Ministry of Marine Affairs and Fisheries 2012).